Stability and Bifurcation of Structures

Angelo Luongo • Manuel Ferretti • Simona Di Nino

Stability and Bifurcation of Structures

Statical and Dynamical Systems

 Springer

Angelo Luongo
DICEAA, University of L'Aquila
L'Aquila, Italy

Manuel Ferretti
DICEAA, University of L'Aquila
L'Aquila, Italy

Simona Di Nino
DICEAA, University of L'Aquila
L'Aquila, Italy

ISBN 978-3-031-27571-5   ISBN 978-3-031-27572-2   (eBook)
https://doi.org/10.1007/978-3-031-27572-2

This Springer imprint is published by the registered company Springer Nature Switzerland AG
The registered company address is: Gewerbestrasse 11, 6330 Cham, Switzerland

To my wife Fiorella (A.L.)
In memory of my beloved
father Raffaele (M.F.)
To my family (S.D.N.)

Preface

Structures are mechanical systems designed to carry loads. The stability of their equilibrium positions is an essential requisite for safety and functioning. Depending on the intensity and nature (static or dynamic) of loads, stability can be lost, with possible dramatic consequences. Aimed to optimize the mechanical behavior of the structure, it is convenient to examine not a specific system, but a family of systems, characterized by parameters (called bifurcation parameters) consisting of forces, geometric dimensions, and material properties. By browsing all members of the family in a given domain of the parameter space, singular points, named critical, can be encountered, in correspondence of which the stability of the equilibrium is modified with respect to that possessed by the system at close points. Moreover, the change of stability is accompanied by the birth of new equilibria or periodic motions, which can be, in turn, stable or unstable. The qualitative change of dynamics of the system, as the parameters are varied, is known as a bifurcation. The study of the behavior of the system in the parameter space is the subject of the Bifurcation Theory. This is a branch of Mathematical Physics, in which use is made of mathematical models and analytical and/or numerical tools, to classify the behavior of the family members near the singular points.

Stability and bifurcation (which are different concepts, often confused in the common language) are pillars of Solid Mechanics. Pioneering studies in this field were carried out by Leonardo da Vinci (1452–1519), but the first rigorous solutions were obtained by Euler (1707–1783), together with Bernoulli (1700–1782). In spite of his ancient roots, the discipline is conceived today as a "special topic," prerogative of few specialized technicians or scientists. Its importance is universally recognized, especially nowadays, in which slender structures are employed (at the macro-scale) and new meta-material are designed (stimulating a modern re-discovery of the problem at the micro-scale). However, the dissemination of the knowledge, from the few scholars of the discipline to the learners, suffers the following drawbacks, which the book aims to overcome.

1. In civil and mechanical engineering, stability of equilibrium is often conceived as synonymous of "buckling," i.e., of a static bifurcation. Indeed, when the attention

is focused on conservative systems, and according to the Lagrange-Dirichlet Theorem, dynamics does not play any role on stability, and the problem can (correctly) be attacked in static regime. There exist several books, mainly devoted to "Buckling and Postbuckling," in which the static point of view is the only one illustrated. Such an approach had in Koiter (1914–1997) its forerunner, followed by many other important scientists (Budiansky, Hutchinson, Thompson,...). The reader of these books ignores that buckling is just an aspect of the stability of mechanical systems. There exist many scientists working in buckling which never wrote a paper on dynamics (and vice-versa); the two worlds seem to be not communicating.

2. On the other hand, the general theory of stability was developed only after Poincaré (1854–1912) and Lyapunov (1857–1918), in an environment of physicists and mathematicians. General theorems (mainly the Center Manifold Method, CMM) have been proved only in recent decades. These tools are rarely used by engineers, except for researchers working in the System Theory and Control fields. However, they usually ignore buckling. In contrast, mathematicians almost exclusively use the CMM, since this is based on a theorem, and disdain perturbation methods (proposed by Koiter in statics, but also by Poincaré in dynamics), since they are funded on heuristic bases.

3. Parallel to stability, Nonlinear Dynamics has had an explosive growth in the last decades (mainly thanks to Nayfeh (1933–2017)). In this environment, Stability and Bifurcation play an important role, but they are not the main focus. The approach followed is consistent with that of the stability theory, but, once again, is difficult to find an expert of buckling in the nonlinear dynamic community.

This book is aimed to overcome the separation existing in literature between the static and the dynamic worlds. The contact point between the two environments is identified in a mathematical tool, the perturbation method, which allows solving static and dynamic problems virtually in the same way. Since the approach is heuristic, it is easily understandable by engineers and practitioners, and immediately applicable to model real problems. However, in order to keep the techniques at a low level of difficulty, and to make the book usable in graduation courses, more attention is devoted to linear problems, although the nonlinear counterpart is often treated for some emblematic cases.

The book is mainly devoted to graduate-level students, acquainted with fundamentals of Solid Mechanics and Structural Mechanics. However, it is hoped it will be of interest for PhD students and researchers working in the broad field of structures and metamaterials. Moreover, it could attract the attention of some professional engineers, operating in Civil and Mechanical engineering.

L'Aquila, Italy   Angelo Luongo
May 2023   Manuel Ferretti
                                                                                   Simona Di Nino

Contents

Chapter 1
Introduction

1.1 Basic Concepts

The concepts of stability and bifurcation are introduced, and the areas of application of the discipline are illustrated, both in statics and dynamics.

Stability
Stability is a primitive notion, widely used even in the common, nontechnical, language. It indicates that the steady state of a system, either an equilibrium point or a periodic motion, is "robust" to perturbations, in the sense that, if disturbed, it persists over time, possibly with small deviations. Stability is an essential quality of equilibrium. An example is offered by a thermally non-insulated environment, heated by a heater machine, in which an equilibrium temperature is reached. At the equilibrium, the steady flow of heat emitted by the machine equals that dispersed through the walls. If the equilibrium were not stable, a small perturbation, such as an opening and closing of a window, would significantly deviate the temperature, which would evolve from the unstable equilibrium value toward other stable values. In contrast, the small decrease/increase in temperature, due to the disturbance, reduces/increases the dispersion toward the outside, so that, with the heat engine which continues to produce the same heat in the unit of time, the temperature tends to return to that of stable equilibrium. Likewise, the arm of a robot performing cyclical operations on an assembly line moves on a closed orbit, which, if unstable, would be irremediably abandoned, as a result of a small accidental bump. The stability of the orbit, on the other hand, ensures that the motion remains confined to a neighborhood of the same orbit and, due to dissipation, tends to return to that.

The common notion of stability, however, does not include the concept of *"smallness" of the disturbance*, which must instead be appropriately defined in the scientific field. *An equilibrium position is stable if it is possible to confine the deviations from the state as much as desired*, by acting appropriately on the smallness of the initial disturbance. This means that, if a certain "tolerated"

A. Luongo et al., *Stability and Bifurcation of Structures*,
https://doi.org/10.1007/978-3-031-27572-2_1

maximum deviation is fixed, the initial perturbation must be smaller than a limit perturbation (to be determined). In other words, the equilibrium is stable if the deviation from it tends to zero with the amplitude of the perturbation. The obvious consequence is that, for sufficiently large perturbations, the motion can diverge even from a stable equilibrium position. Stability is therefore a *local* property of equilibrium. To illustrate this, the graph of a function of a single variable is considered, which, at a stationary point, is concave upward. The function is locally increasing, but this does not imply that, in a wider interval, it can instead decrease.

Bifurcation

Bifurcation is a concept completely different from stability. In fact, it refers not to a specific system but to *families of systems*, depending on one or more parameters. If it happens that, as the parameters are quasi-statically varied, the dynamics of the system suddenly changes, then a *bifurcation* is said to have occurred. Bifurcation, therefore, implies that either an equilibrium position changes stability or some equilibrium positions and/or periodic motions appear/disappear.

Indeed, the idea of the bifurcation is, from a merely lexical point of view, intimately linked to *branching*, which is a purely geometric concept. Accordingly, as a parameter changes, the system visits a regular equilibrium path, along which a single tangent is defined at each point. However, at a critical value of the parameter, this curve branches out, in the sense that the tangent is no longer unique at a criticality. Note, however, that branching is a particular case of bifurcation, as introduced before, because the number of equilibrium points and/or stability changes at the singular point, entailing a change in the dynamics of the system.

Static and Dynamic Views of Bifurcation

The question just discussed exemplifies the different views existing in the literature, regarding bifurcation. If we look at the problem exclusively from a static point of view, that is, if we ignore the dynamics, and we focus only on the admitted equilibrium positions, bifurcation is synonymous of branching. The singularity can only manifest itself with the birth of new equilibrium positions, other than those initially visited, as the parameter varies. This is the philosophy on which a very wide range of studies have been founded, concerning *buckling* and *postbuckling* of elastic structures, subjected to gravitational loads: straight beams, planar or spatial frames, arches, thin-walled beams with open profiles, plates, and shells. The gallery is so large to induce someone to erroneously to think that bifurcation concerns exclusively this class of problems.

However, there is a further form of static bifurcation, in which two equilibrium points approach each other, as a parameter changes, collide and then disappear (as it happens for the graph of a function having a horizontal tangent at a point). This is called a *fold bifurcation*, or *limit point*. Although it is a static bifurcation, it cannot be included in the branching class, and in fact it occupies a special chapter in the buckling narrative, although it fully falls within the general definition of bifurcation.

The dynamic bifurcation of equilibrium, on the other hand, opens up new scenarios. The singularity, in fact, can manifest itself through the birth of new periodic motions, which did not exist before the bifurcation. This is the case of the *Hopf bifurcation*.[1] According to it, a family of periodic orbits, also known as *limit cycles*, stable or unstable, branches off at the bifurcation from an equilibrium path. Since the tangent to the path remains unique, the simple geometric inspection of the curve does not explain the presence of a singularity, which in fact can only be detected when the problem is addressed in a dynamic context. The dynamic bifurcation is of much more recent study than the static one, and the related analysis has mainly developed in the mathematical field; therefore, it is much less known in the engineering community.

Limits of the Static Theory

A static problem (i.e., governed by an algebraic or differential system in space) is certainly easier to study than a dynamic one (in which the same becomes, respectively, an ordinary differential system, or a partial differential system in space and time). However, it should be understood *when* it is possible to reduce the problem to static and *whether* this operation furnishes correct results.

To answer these questions, the problem of stability is examined first. The Lagrange-Dirichlet theorem (valid for discrete systems and extended by Koiter to continuous ones) states that "an equilibrium position of a *conservative system* is stable if its Total Potential Energy therein assumes a local minimum (i.e., it is there defined positive); if the energy is indefinite or definite negative, the equilibrium is unstable." It emerges that the theorem makes no mention of kinetic energy, which therefore plays no role on the stability of this class of systems. It is concluded that the stability of the equilibrium of a conservative system (typically elastic, subject to dead loads) can be studied in the static field, regardless of any kinetic considerations. If, however, the system is *nonconservative*, due to the presence of forces that depend on position and/or velocity, the Lagrange-Dirichlet theorem can no longer be invoked. The only possible investigation, to ascertain the stability of the equilibrium of nonconservative systems, consists therefore in studying the motion caused by a perturbation. A separate discussion is reserved to nonconservative systems by nature of the material (e.g., elasto-plastic beams subjected to dead loads, which will be briefly dealt with in this book), for which the problem is usually analyzed in the static field only.

The question is now examined from the point of view of bifurcation. The study is aimed at determining the motions admitted by a family of systems around a singular point. These can be (a) periodic motions, which develop on closed orbits or (b) aperiodic motions (mainly transient), which, for example, bring the system from an unstable equilibrium point to a stable point. Now, it is necessary to distinguish the case of static bifurcation from that of dynamic bifurcation. In the case of static bifurcation of conservative systems, if transient motions are ignored, attention can be limited to detect the equilibrium equations, which permits to identify the equilibrium paths. Subsequently, stability is stated via the Lagrange-Dirichlet

---

[1] Eberhard Hopf (1902–1983), mathematician and astronomer. The dynamic bifurcation is also called Poincaré-Andronov-Hopf.

theorem. On the other hand, in case of dynamic bifurcation of nonconservative systems, the analysis necessarily requires the construction of the family of limit cycles. This calls for using nonlinear equations of motion and for searching periodic solutions.

Overall, (a) if the system is conservative, the analysis of its stability and bifurcation can be conducted in the static context, to determine the (correct) bifurcation points, the branched equilibrium paths, and their stable/unstable nature, but it is not possible to analyze the evolution of the system from one state to another; (b) if the system is nonconservative, the analysis of its stability and bifurcation necessarily requires the study of the equations of motion, which allow to determine the equilibrium points, transient motions, and limit cycles as well as ascertain the stability of the motions.

Stability Criteria

It is common, in the literature, to come across the following classification of the *stability equilibrium criteria*: (a) dynamic criterion, (b) energy criterion, and (c) static (or adjacent equilibrium) criterion. Here the meaning of these locutions is illustrated, and the existing interrelations are explained.

By *dynamic criterion* it is meant the application of the definition of equilibrium stability (due to Lyapunov), based on the observation of the evolution of the trajectory following an initial disturbance.[2] It can be applied to systems of all types, conservative or nonconservative.

By *energy criterion* it is meant the application of the Lagrange-Dirichlet theorem, based on the examination of the sign definiteness of the total potential energy. It can be applied to conservative systems only (since nonconservative systems do not admit a potential). For these systems, the energy criterion provides the same results as the dynamic criterion, since it is easy to verify that the kinetic energy, which is always positive definite, does not play any role on stability.

By *static criterion* it is meant a mechanical interpretation of the phenomenon, based on balance of forces, according to which the position is stable if the internal forces, produced by the perturbation, tend to bring the system back to its original position, that is, they are "opposite" to the disturbance that generates them. In order for stability to hold, the Jacobian matrix of the equilibrium equations, expressed in terms of displacement, i.e., the tangent stiffness matrix $\mathbf{K}$, must be positive definite.[3] When the system is conservative, and only in that case, the tangent stiffness matrix is also the Hessian of the total potential energy, so that the static criterion coincides with the energy one. The static criterion is, however, broader than the energy one, as it is applicable (in statics) to any systems, conservative or not, and for this reason, it

---

[2] An alternative approach, called the "direct method" of Lyapunov, will not be addressed in this book. It is based on the search for a suitable scalar function, called "Lyapunov function."

[3] Indeed, for infinitesimal displacements $\delta\mathbf{q}$, the internal forces $-\mathbf{K}\delta\mathbf{q}$ must form an obtuse angle with $\delta\mathbf{q}$, i.e., it must be $\delta\mathbf{q}^T\mathbf{K}\delta\mathbf{q} > 0 \quad \forall\delta\mathbf{q}$.

is widely used in the analysis of inelastic systems.[4] However, it suffers from a strong limitation, since a statically stable system may not be dynamically stable. This is often the case of elastic structures subjected to nonconservative forces, which, indeed, lose stability because of kinetic effects.

According to the static criterion, as a load parameter μ increases, the tangent stiffness matrix $\mathbf{K}(\mu)$ passes, in correspondence with a *critical value* $\mu = \mu_c$, from a condition of positive definiteness (which characterizes a stable state), to one of semi-definiteness (which characterizes the incipient loss of stability). In this state (called *neutral equilibrium*), the linearized equilibrium equations, $\mathbf{K}(\mu_c)\,\delta\mathbf{q} = \delta\mathbf{f}$, which link the incremental forces $\delta\mathbf{f}$ to the incremental displacements $\delta\mathbf{q}$, both infinitely small, generally admit no solution, because the operator is singular (i.e., its determinant is null). In the critical state, therefore, the structure behaves as if it were kinematically undetermined (i.e., labile). Consequently, even if the incremental forces are absent, i.e., $\delta\mathbf{f} = \mathbf{0}$, there exist infinite equilibrium configurations $\delta\mathbf{q} \neq \mathbf{0}$ under the same value of the load μ_c. The vectors $\delta\mathbf{q}$ lie in the kernel of the operator, i.e., in the space (here generally assumed to be one dimensional) spanned by the eigenvectors associated with the null eigenvalue, that is, the uniqueness of the equilibrium, which generally exists when $\mu < \mu_c$, is disrupted at $\mu = \mu_c$. Since the criterion refers to the equilibrium of an infinitely close configuration, it is called the *adjacent equilibrium criterion*. It has been presented here as deduced from stability conditions, but it can also be seen as the result of a bifurcation analysis, consistently with the implicit function theorem, which ensures the existence of at least one other path of equilibrium passing for the same point or denotes a *fold* of the existing path (i.e., a double equilibrium point). The interpretation of the adjacent equilibrium is considered very suggestive in the engineering world, because it explains the occurrence of a dangerous situation (the loss of stability and the associated bifurcation) through the labilization of the linearized system. This vision achieved a great success in the technical literature, often interested only in the critical value of the parameter, to make the criterion assume a prevalent role in the discipline, capable of obscuring other significant aspects.

On the Eigenvalue Problems Supplied by the Dynamic, Energy, and Static Criteria

One aspect, only apparently secondary, to be discussed here concerns a semantic question that sometimes generates confusion in the engineering community and that can make the communication difficult between scholars with different cultural backgrounds.

As it will be seen later in the book, the dynamic stability criterion requires the evaluation of the (generally complex) λ eigenvalues of the linearized dynamic system. Eigenvalues with a positive real part indicate modes diverging over time

---

[4] The tangent stiffness matrix $\mathbf{K}$ depends on the position $\mathbf{q}$ but, in the inelastic case, also on the direction of the incremental displacement $\delta\mathbf{q}$, i.e., $\mathbf{K} = \mathbf{K}(\mathbf{q}, \delta\mathbf{q})$. Furthermore, the symmetry of $\mathbf{K}$ is not guaranteed for some constitutive laws.

(and therefore instability); eigenvalues with a negative real part indicate motions decaying over time (and therefore stability, if all of this type).

On the other hand, when, for a conservative system, the problem is faced with the energy criterion, it is necessary to check the definiteness of the stiffness matrix at the bifurcation. The simplest way (for finite-dimensional systems) to perform the task consists in examining the real eigenvalues Λ of the (symmetric) stiffness matrix. If all the eigenvalues Λ are positive, the matrix is positive definite, and therefore the equilibrium is stable; if even only one of them is negative, the matrix is indefinite, and therefore the equilibrium is unstable. It is superfluous to remark that the eigenvalues λ and Λ, although correlated in the conservative case, are different quantities, which must in no way be confused.

When the problem is faced in the spirit of adjacent equilibrium, it is necessary, as mentioned above, to determine which is the smallest value of the bifurcation parameter μ for which the equation $\mathbf{K}(\mu)\delta\mathbf{q} = \mathbf{0}$ admits a non-trivial solution. Since, under generally accepted simplifying hypotheses,[5] the stiffness matrix is of the type $\mathbf{K}(\mu) := \mathbf{A} + \mu\mathbf{B}$, with $\mathbf{A}$ the elastic part and $\mathbf{B}$ the geometric part, both independent of μ, the equilibrium equation in the current configuration appears in the form $(\mathbf{A} + \mu\mathbf{B})\delta\mathbf{q} = \mathbf{0}$. This is also a (nonstandard) eigenvalue problem, in which, however, the eigenvalue μ has a meaning that is further different from those previously mentioned.

Although the nature of the three problems discussed should appear quite distinct, it is not uncommon to come across a certain confusion that reigns around the word "eigenvalue." This confusion is made even greater by the fact that, in almost all the texts dealing with buckling, the load multiplier μ, assumed as a bifurcation parameter, is instead indicated by λ, i.e., with a symbol almost universally accepted as an eigenvalue of the dynamic problem.

1.2 Overview of the Book

To make clear the flow of presentation, a general overview of the book is provided. The subject is organized into 14 chapters (including this Introduction), ideally divided into three segments, plus a chapter devoted to solved problems, and four appendices. In the first part, concepts and methods are illustrated; in the second, some static bifurcation problems, relevant to continuous structures, are analyzed; in the third part, some mechanical systems manifesting dynamic bifurcations are studied.

Concepts and Methods
Five chapters are dedicated to the enunciation of basic notions and to the illustration of analytical and numerical methods.

---

[5] Namely, the precritical deformations are ignored, so that the (unknown) configuration taken by the system at the bifurcation is identified with the (known) natural configuration.

In Chap. 2, after introducing the definitions of equilibrium stability and bifurcation, the mechanical behaviors of some sample structures are described on a phenomenological ground: the Euler beam, the trussed beam, the beam with elastic bracing, the flat arch, the Beck beam, the aeroelastic oscillator, and the Bolotin beam. The examples allow to illustrate important concepts that will be developed in the following: branching, snap-through, effects of imperfections, global and local instability, interaction between multiple modes, limit cycles, and parametric resonance.

Chapter 3 provides tools for stability and bifurcation analysis. With reference to a generic mechanical system, the nonlinear equations of motion, their linearization, and the local stability analysis are described. Conservative and *circulatory* systems are discussed, and the effect of damping on their stability is investigated. The bifurcation mechanisms of conservative, circulatory, and viscously damped systems are illustrated.

Chapter 4 is devoted to the *static bifurcations* of discrete and continuous conservative systems. With reference to the former, the equilibrium points are classified, and a sketch on numerical continuation methods for computing equilibrium paths is given. The *perturbation* (or asymptotic) *method* is introduced for the construction of paths that branch off from the trivial one and for the analysis of imperfections. Systems with precritical deformations are studied, for which (a) the nontrivial path is built up asymptotically and (b) the branching from this path is investigated.

In Chap. 5, some sample systems, called *paradigmatic*, are reviewed, namely, discrete nonlinear systems, with one or two degrees of freedom, which exhibit branching from trivial path; interaction between simultaneous modes; *snap-through* and branching from non-trivial path; and the nonlinear Euler beam, as a prototype of a continuous system. For all discrete systems, both exact and perturbation solutions are determined; for the continuous system, only linear exact solutions, and nonlinear perturbation solutions are provided.

In Chap. 6, the *linearized theory* of the buckling of (discrete and continuous) elastic structures is formulated, based on the hypothesis, widely invoked in applications, of negligibility of the precritical deformations. The theory allows to determine, with a good approximation, the critical load of axially high-stiff structures. Similarly, it permits to evaluate, with lower accuracy, the response of imperfect systems to incremental forces, via the use of an *amplification factor* of the linear response. In addition to the variational formulation, alternative approaches are discussed: (a) the virtual work principle, useful for the study of nonconservative systems, as well as (b) the direct equilibrium, useful for giving mechanical meaning to the equations derived in a variational way.

Buckling of Beams and Plates

The next four chapters deal with the elastic buckling of solid beams, open thin-walled beams and plates, where attention is limited to the critical load. Moreover, a short chapter is dedicated to the elastoplastic buckling of beams. The elastic models are always deduced from the stationarity of the *total potential energy* (in this regard, Appendix A briefly recalls the basic notions of the *calculus of variations*). The

Eulerian equations of the problem, however, are always reinterpreted on the basis of the balance of forces, reached by the structure in the adjacent configuration, in order to provide a physical meaning to the reader. The topics discussed in this part usually form the core of classical buckling texts. Alongside these, however, some less-known examples are dealt with, and the analytical treatment is often accompanied by a numerical finite element analysis, here developed *ad hoc* and not taken up by commercial software. The analysis is sided by examples, the results of which are always commented in a mechanical key.

Chapter 7 deals with buckling of elastic beams and planar frames. The critical load of single beams, constrained at the ends, is determined. The effect of transverse forces acting on beams, compressed by forces smaller than the critical ones (the so-called *column-beams*), is analyzed. Beams of step-wise variable cross-sections, or not uniformly compressed, are addressed, by pursuing exact solutions (when they exist) or approximate solutions, via the *Ritz method* (summarized, in its general aspects, in Appendix B). Uniformly compressed beams are then studied, equipped with elastic, concentrated or diffused, constraints (*Winkler-like elastic soil*), showing how the critical mode depends on the spring-to-beam stiffness ratio. The buckling of *prestressed reinforced concrete beams*, with external or internal cables, is analyzed, clarifying in which cases the prestress affects (or not) the critical load. The global and local forms of buckling, exhibited by a compressed trussed beam, are discussed. Finally, two different *finite elements*, polynomial and exact, are developed for buckling analysis of planar frames.

In Chap. 8, a short overview is given on the *elasto-plastic buckling* of a single beam, through the illustration of the classic *tangent* and *reduced elastic modulus* theories. The *push-over analysis* of beam systems is also addressed, in the presence of unstabilizing effects induced by compression. Some applications are shown, where the pushover curves, accounting or neglecting the geometric effects, are compared.

Preliminary to the study of Chap. 9, is the Appendix C, dedicated to the *Vlasov theory of nonuniform torsion* of open thin-walled beams. This topic, although not specific to the bifurcation, has been introduced here, since it is generally lacking in the engineer's cultural background. Even more significantly, the finite element treated here can be rarely found in textbooks. In recalling Vlasov's theory, particular attention is paid to the formulation of a one-dimensional model, which derives from the three-dimensional one, in which the notions of *bimoment* and *complementary torsional moment* find their natural definition. Space is also dedicated to the formulation of two finite elements, one exact and the other polynomial. Numerical results obtained from these techniques are hoped to increase the reader's confidence on the role played by constraints, either allowing or forbidding warping.

Chapter 9 is dedicated to the buckling of open thin-walled beams (TWB). Preliminary, and taking advantage of the notions provided in Appendix C, a model is formulated of an elastic (not prestressed) beam, having *two axes*: that of the *twist centers* and that of the centroids. The differential operator of geometric stiffness is successively derived, in case of constant axial force and variable bending moments. Referring to particular constraint conditions, the buckling of (i) uniformly compressed beams, with and without symmetry properties, (ii) beams undergoing uniform bending, or (iii) extension plus bending is analyzed. TWBs subjected to transverse forces are then examined, for which it is necessary to resort to approximate analytical solutions, via *Frobenius* or *Ritz methods*. Finally, a *polynomial finite element* is developed, which makes it possible to determine the critical load of a TWB, subject to uniform normal stress and piecewise linear bending moments. Numerical applications of the method are shown and commented on.

In Chap. 10, the buckling of *rectangular plates* and *plate assemblies*, prestressed in their own planes, is analyzed. The linear theory of the *Kirchhoff plate* is resumed, by following the variational procedure. Taking into account the prestress, the geometric operator is obtained. The buckling of rectangular plates is studied under various load and constraint conditions, even in the presence of a longitudinal stiffening. For the plate compressed in the longitudinal direction and simply supported on two opposite sides, after separation of variables, an *exact finite element* is formulated, to describe the *transverse elastic line*. The problem of the plate subjected to membrane shear is posed, for which a sketch of the exact solution, valid only for the infinitely long plate, is given. Then, plates of finite dimensions are considered, whose relevant problem is solved via the Ritz method. In the second part of the chapter, the buckling of prismatic shells is studied, manifesting itself in three different patterns of instability: *global*, *distortional* and *local*. The *finite strip method* for the analysis of such shells is described. By limiting the attention to the local critical load, an exact one-dimensional finite element is introduced (sectional model). Finally, the mechanical behavior to local buckling of different plate assemblies is discussed.

Dynamic Bifurcation

The last three chapters are devoted to the dynamic bifurcation of *autonomous systems* (i.e., systems whose characteristics are time-invariant), or *non-autonomous systems* (i.e., explicitly dependent on time). The topics covered concern (a) non-conservative forces depending on the position (in particular, *follower*), but not on velocity; (b) nonconservative forces depending both on velocity and position (as due to wind flow); (c) nonconservative forces explicitly dependent on time, this latter entering, as a geometric effect, in the mechanical characteristics of the system.

In Chap. 11 follower forces applied to elastic beams are considered, and their nonconservative character is discussed. As the conservation of energy fails, the equations of motion are taken from the *extended Hamilton principle* (illustrated in the Appendix D, together with the *Lagrange equations* of motion), which extends the principle of virtual works to the dynamic case. Two sample structures are studied: (a) one discrete (the *Ziegler column*, or double pendulum) and (b) one continuous (the *Beck beam*), both solicited by a compression follower force. The analysis is mainly (but not exclusively) aimed at determining the critical value of the force. The bifurcation occurs through the *flutter* mechanism (or *circulatory*, or reversible, *Hopf bifurcation*), caused by the collision of two purely imaginary pairs of eigenvalues. The effect of an added damping is discussed, which changes the circulatory into a *generic Hopf* bifurcation, involving only one pair of eigenvalues. Damping, although dissipative, is generally detrimental (unstabilizing) and induces discontinuity between the critical value of the undamped system and that of the damped system, when damping tends to zero (*Ziegler paradox*). Finally, for the Ziegler column, a nonlinear analysis is developed, aimed at determining the limit cycle arising in the postcritical regime, evolving as a function of the bifurcation parameter.

In Chap. 12 the phenomena of *aeroelastic bifurcation*, induced by an incident flow of stationary wind, are analyzed. The aerodynamic forces exerted on a long, rigid, and fixed cylinder, and their dependence on the angle (said *of attack*) that the flow forms with a material direction of the cross-section, are described. For elastically constrained rigid cylinders, the validity conditions of the *quasi-steady theory* are discussed, according to which the motion of the structure is assumed to be slow with respect to that of the fluid. Single degree of freedom cylinders, moving cross-wind, are studied, and the *galloping* phenomenon is explained. The critical and postcritical behavior of such systems, in which a limit cycle is reached, is analyzed. These results are then applied, with some approximations, to strings and beams, reduced to single degree of freedom systems. Planar aeroelastic systems are successively considered, represented by rigid cylinders with three degrees of freedom, which can translate transversely and rotate around their axis. The aeroelastic forces are determined according to two theories, one of which ignores, and the other approximately takes into account, the effect of the twist rate of the cylinder. Investigation of unidirectional motions reveals the existence of two new bifurcation mechanisms: *rotational divergence* (static bifurcation) and *rotational galloping* (dynamic bifurcation). Bifurcations involving two degrees of freedom are then addressed, to detect (a) the *translational galloping*, which couples the two translations, and (b) the *roto-translational galloping*, which couples the cross-wind translation and rotation. The chapter closes with a brief note to the *non-steady theory*. This accounts, in a semi-experimental way, for the influence that the motion of the cylinder (supposed harmonic) exerts on the motion of the fluid and, therefore, on the aeroelastic forces.

In Chap. 13, the *parametric excitation* is studied. It occurs in mechanical systems whose parameters (mass, damping, stiffness) are varied over time with an assigned periodic law. *Floquet's theorem*, for a generic continuous-time and finite-dimensional system, is stated, and the resulting stability conditions are discussed. These results are reinterpreted in the light of a discrete-time view of the system, obtained through the *Poincaré map*. The study of single degree of freedom systems, in the absence or presence of damping, is conducted in detail. It highlights two different bifurcation mechanisms, called *divergence* and *flip*. The celebrated *Mathieu equation*, which is a paradigm of these systems, is studied. As an example, the *Bolotin beam* (i.e., a beam subjected to pulsating axial load), reduced to a single degree of freedom system, is analyzed. The effects of nonlinearities are discussed for single degrees of freedom systems, using a perturbation method. Finally, a two degrees of freedom system is studied, which reveals a further bifurcation mechanism, called *Neimark-Sacker*; it involves two modes of the system, in *combination resonance* with the excitation frequency.

Solved Problems

In Chap. 14, solved problems are proposed. They concern static bifurcations exhibited by systems of beams, thin-walled beams, plates, and plate assemblies; dynamic bifurcations induced by follower or aerodynamic forces and parametrically excited systems. The problem is given and then solved step-by-step, via exact or Ritz approaches; then the results are illustrated and commented. Problems are intended to stimulate the reader to check its degree of knowledge, acquired by studying the book, and therefore are referenced at proper places in the main text.

1.3 Book Style

The style of the book is commented, and some information concerning the symbology adopted is given.

Style

The book is written, as far as possible, in a colloquial form, trying to make concepts simple, reducing mathematics to the essential. On the other hand, the rigor is not overlooked. The attempt is to find a balance between simplicity and precision, without indulgence toward formalism.

The book guides the reader to learn the theory, stimulating the development of its own logical process. To this end, numerous *remarks* point out the discussion, which, although they may be omitted in the first reading, are strongly recommended for the purpose of a deeper understanding of the subject. In the same spirit, the book contains a limited number of *supplements* which analyze particular aspects in greater detail, or which provide further information, not strictly necessary for the continuation of the reading. Likewise, some *examples*, exemplify principles and methods, referring to cases of applicative interest.

The methodological aspects are privileged over "tricks" or "shortcuts," which could simplify the solution of the specific problem. Consistently, the examples are illustrated not with the spirit of building a gallery of results, as in a handbook, but with the purpose of exemplifying general methods, which may be useful for solving similar problems.

The book is mainly aimed at readers who are approaching the theme of stability and bifurcation for the first time (except for the universal knowledge of the emblematic case of the Euler beam). However, it is the hope of the authors that even the professional engineer, who has developed his own knowledge in the static and dynamic fields of structural mechanics, will benefit from the unified vision of the problem that inspires this book.

Symbology and Conventions

The mathematical symbology is the standard one, according to which the scalars are indicated with italic, Latin, or Greek letters; matrices are indicated with capital letters, straight bold; and vectors and column matrices (as clarified by the context) with lowercase straight bold, with rare exceptions noted in the text.

All equations are numbered, with numbering reinitialized in each chapter. For example, Eq. 1.10 is the equation 10 of Chap. 1. In the case of a reference to a package of equations, the reference to Eq. 1.10 should be understood as the whole package; the reference to Eq. 1.10a or Eqs. 1.10a,b indicates the first or the first two equations of the package. The same convention applies to the figures, which are labeled in the format Fig. 1.10 and possibly divided into sub-figures Fig. 1.10a, 1.10b, ...

The bibliography refers almost exclusively to books, given the educational nature of the book. Very few articles are cited for their specificity. The bibliography is placed at the end of each chapter, listed in alphabetical order, and cited in the text at the proper place. It refers to the specific topics discussed in the chapter and is possibly repeated in other chapters.

Chapter 2
Phenomenological Aspects of Bifurcation of Structures

2.1 Introduction

Stability is a *dynamic concept* characterizing the "quality" of equilibrium (or motion). According to Lyapunov, it states that the equilibrium (or motion) is "robust" under small perturbations of its initial state, in the sense that the time-dependent deviations from the unperturbed state can be confined, simply acting on the initial conditions. *Bifurcation* concerns families of systems, governed by one or more parameters, whose dynamics abruptly change when the parameters assume critical values, while quasi-steady varied. Bifurcations can be *static*, if new equilibria are generated from an equilibrium path, or *dynamic*, if new periodic motions arise from the path. Static bifurcations are typically exhibited by elastic systems under compression dead loads, as the *Euler beam*; they are *sensitive* to imperfections, except for fold (i.e., *limit point*) bifurcations. Dynamic bifurcations, instead, exclusively occur (in the autonomous case) in nonconservative systems, as a consequence, e.g., of follower forces (as it happens for the *Beck beam*), or aeroelastic forces, which lead to *galloping* or *flutter* instability phenomena; they are *insensitive* to imperfections. Non-autonomous elastic systems, instead, can suffer dynamic bifurcations caused by *parametric excitation*, e.g., produced by axially pulsating dead loads (as it occurs in the *Bolotin beam*).

All these concepts are introduced in this chapter, by discussing phenomenological aspects and anticipating results to be proved in the remainder of the book.

2.2 Stability and Bifurcation

The concepts of stability and bifurcation are introduced and illustrated via some elementary examples. The discussion is merely qualitative; all the results illustrated here will be justified in successive chapters.

© The Author(s), under exclusive license to Springer Nature Switzerland AG 2023   13
A. Luongo et al., *Stability and Bifurcation of Structures*,
https://doi.org/10.1007/978-3-031-27572-2_2

2.2.1　Equilibrium Points

A generic, finite-dimensional (discrete) system is considered. The physical quantities $\mathbf{x}(t) \in \mathbb{R}^N$, describing the system evolution over the time t, are called *state variables*. In mechanics, they consist of *positions* $\mathbf{q}(t)$ (Lagrangian parameters) and *velocities* $\dot{\mathbf{q}}(t) := \frac{d\mathbf{q}}{dt}$, i.e., $\mathbf{x} = (\mathbf{q}, \dot{\mathbf{q}})$. The state of the system at time t is represented by a point in the $\mathbb{R}^N$ space, whose coordinates are $\mathbf{x}(t)$. As the state changes, the point describes a curve, called a *trajectory* or *orbit*. Particular interest is taken by the trajectories degenerating into a point, called *points of equilibrium*, or simply *equilibria*. If the system is brought to one of these points and left with zero speed, the state does not evolve but remains stationary. The "quality" of equilibrium is expressed by its *stability*.

2.2.2　Stability of Equilibrium

The notion of stability of equilibrium, according to Lyapunov [10], is here introduced informally (Fig. 2.1).

In the state-space, an equilibrium point E is considered and a neighborhood of it, of *arbitrarily small radius* ϵ, is chosen. If there exists a second neighborhood of radius $\delta(\epsilon) < \epsilon$, such that the trajectories originating from the smaller neighborhood remain confined to the larger neighborhood, then equilibrium is stable (S). The idea of stability is therefore that, for initial conditions $\mathbf{x}_0$ "close enough" to the equilibrium point, the trajectories emanating from $\mathbf{x}_0$ remain "sufficiently close" to E. If it is not possible to find a neighborhood of radius δ, as one or more trajectories originating from $\mathbf{x}_0$ move out of the domain of radius ϵ, then the equilibrium is *unstable* (U). If the trajectories, in addition to remaining confined, also tend to return (in an infinite time) to the equilibrium point, then the equilibrium is said to be *asymptotically stable* (AS). The asymptotic character is a feature that reinforces stability.

Fig. 2.1 Stability of the equilibrium point E: S stable, U unstable, AS asymptotically stable

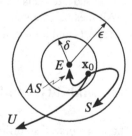

Lagrange-Dirichlet Theorem

If the mechanical system is conservative, the following theorem, said of Lagrange-Dirichlet, holds [5]:[1]

> *If the system is conservative, an equilibrium point is stable if the total potential energy (TPE) is minimal at that point.*

Example 2.1 (Point Mass on a Plane Curve) A point mass, subject to gravity, and constrained to belong to a plane curve of equation $y = f(x)$, is considered. If $f'(0) = 0$ (where a prime indicates differentiation), the origin is an equilibrium point. The equilibrium is stable if the point is a minimum ($f''(0) > 0$) and unstable if it is a maximum ($f''(0) < 0$). It is unstable also when the equilibrium occurs at a flex point ($f''(0) = 0$, $f'''(0) \neq 0$), as the trajectories that originate to the right or to the left of the equilibrium point are not confined.

In the example in question, and in agreement with the Lagrange-Dirichlet theorem, the stability of the equilibrium depends only on the *gravitational energy* of the system, not on its kinetic energy. □

2.2.3 Bifurcation

So far, attention has been focused on a specific system (e.g., a pendulum of given length and mass). In mechanics, however, one is often interested in considering *families of systems*, i.e., systems which depend on one or more parameters (e.g., infinite pendulums of the same mass but of different lengths). These parameters are usually free design quantities (e.g., geometric dimensions, mass, damping, elastic modulus, etc.) that the designer chooses in order to optimize the behavior.

A system depending on a single parameter μ is considered, and attention is focused to the whole, or to a part, of the state-space. As μ varies, the equilibrium points generally move with continuity and their stability properties can change. At particular values $\mu = \mu_c$, their number, or the quality of equilibrium, can modify, or particular trajectories (e.g., periodic motions) can appear or disappear. When this happens, a *bifurcation* is said to have occurred. The μ_c values are called *critical*, or bifurcation values.

Summarizing, the bifurcation is a *qualitative change of the dynamics* of the system, which occurs when the parameter passes through a critical value. The parameters responsible for bifurcations are called *bifurcation parameters*; the remaining ones, not affecting bifurcation, are called auxiliary parameters.

---

[1] The theorem was originally formulated for discrete systems and then generalized by Koiter [7] to continuous systems.

Bifurcation of Equilibrium

The previous definition refers to *global* bifurcations, in which the modification of the dynamics occurs on a large scale (e.g., in the whole or in a part of state-space). When the interest is instead confined to a neighborhood of a trajectory (or equilibrium point), the bifurcation is called *local*. In this book, only *local bifurcations of the equilibrium* will be considered.

The position of E in the state-space generally depends on μ; however, in many cases, it is independent of this parameter (e.g., the equilibrium positions of a pendulum do not depend on the magnitude of the acceleration of gravity). If it happens that, when $\mu = \mu_c$, the trajectories originating from the neighborhood of E (depending or not on μ) undergo a qualitative change (e.g., the stability of E is modified, and/or new equilibrium points or trajectories originate from E), then a (local) bifurcation of equilibrium is said to have occurred.

Remark 2.1 It should be noticed that a mere movement of the equilibria with μ is not a bifurcation, if the change is not accompanied by a qualitative modification of the dynamics. The new dynamics, indeed, are equivalent to the old ones, in the sense that there exists a deformation of the state-space which brings the former to overlap to the latter.

Static and Dynamic Bifurcations

The bifurcations are effectively represented in *bifurcation diagrams*, which are graphs that describe the evolution of one (or more) state variables versus the bifurcation parameter. If q represents the position of an equilibrium point, the graph of $\mu = \mu(s), q = q(s)$, with s a parameter, is also called an *equilibrium path*.

The bifurcations of equilibrium are distinguished in *static bifurcations* and *dynamic bifurcation*, according to the following properties [8, 13]:

- A bifurcation is static when, as the bifurcation parameter μ changes, two (or more) equilibrium points collapse in the same point when $\mu = \mu_c$. Two cases can occur: (a) two or more equilibrium paths intersect at one point (Fig. 2.2a) and (b) the path has a tangent orthogonal to the μ axis (Fig. 2.2b). In the first case, called *branch point*, the two points of equilibrium approach, collapse, and then separate; in the second case, called *fold bifurcation* or *limit point*, the two points approach, collide, and then disappear.
- A bifurcation is dynamic when one (or more) periodic orbits originate from an equilibrium point. In a $(\mu, q, \dot{q})$ space, there is an equilibrium branch lying in the plane $\dot{q} = 0$, which, at $(\mu_c, q_c, 0)$, intersects a surface, whose sections $\mu = \text{cost}$ are closed orbits, traveled by the system with a periodic law $q = q(t; \mu), \dot{q} = \dot{q}(t; \mu)$ (Fig. 2.2c).

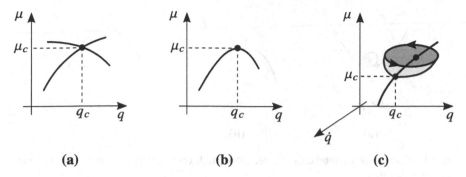

Fig. 2.2 Bifurcations: (**a**) branching (static), (**b**) fold (static), (**c**) dynamic

Conservative systems can only exhibit static bifurcations; nonconservative systems, both static and dynamic bifurcations. In this text, attention will be focused on (a) static bifurcations of elastic systems subject to conservative forces, and (b) dynamic bifurcations of elastic systems subject to nonconservative forces.[2]

The analysis of conservative systems, as suggested by the theorem of Lagrange-Dirichlet, can be conducted in a purely static context, by examining all the possible equilibria of the system and investigating stability through the positive definiteness of the TPE. This type of analysis is also known as *elastic buckling*. The analysis of nonconservative systems is much more complex, because it requires the study of dynamics in the nonlinear field. If, however, the interest is limited to the incipient bifurcation (i.e., to the evaluation of the critical value of the bifurcation parameter), it is sufficient to compute the eigenvalues of the linearized system, which govern the small motions around the equilibrium.

Remark 2.2 The definitions given above conform to the modern *bifurcation theory* (which is a branch of the *dynamical system theory*). In civil and mechanical engineering, where the theory has mainly been developed with reference to static bifurcations, the idea of bifurcation is instead linked to the appearance of *branch points* along equilibrium paths. However, the first definition is more general and includes the second one as a particular case.

Remark 2.3 The difference between stability and bifurcation of equilibrium is the following: stability refers to the quality of equilibrium of a specific system, while bifurcation refers to the change of dynamics of a family of systems. The two concepts are often confused, because, at the bifurcation, the qualitative change of dynamics often concerns the stability of an equilibrium point. Indeed, as the parameter changes, if the equilibrium point passes from stable to unstable, the dynamics qualitatively change; and *then a change of stability entails a bifurcation*. Conversely, the bifurcation does not necessarily imply a change in stability

---

[2] An exception concerns elastoplastic systems, which will be examined in the static field only.

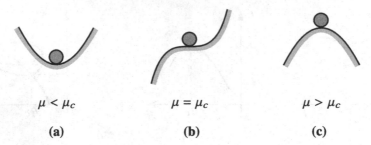

$$\mu < \mu_c \qquad\qquad \mu = \mu_c \qquad\qquad \mu > \mu_c$$

$$\textbf{(a)} \qquad\qquad\qquad \textbf{(b)} \qquad\qquad\qquad \textbf{(c)}$$

Fig. 2.3 Example of equilibria: (**a**) stable, (**b**) neutral, (**c**) unstable; μ_c critical value of the bifurcation parameter

(because, in principle, a different dynamic could occur in which all equilibria maintain their stability[3]). However, in all cases that will be studied here, the bifurcation will correspond to a stability change.

Example 2.2 (Point Mass on a Family of Plane Curves) Let us consider the example of the heavy point mass, now constrained to move not on a specific $y = f(x)$ curve but on a curve belonging to the one-parameter family $y = f(x; \mu)$ (Fig. 2.3). If it is $f'(0; \mu) = 0 \ \forall \mu$, the origin is an equilibrium point for each curve of the family. As μ increases, if the origin goes from a relative minimum to a maximum, then the value $\mu = \mu_c$ for which the curve is "locally flat" (i.e., for which $f''(0, \mu_c) = 0$) identifies a bifurcation of the equilibrium. $\square$

Example 2.3 (Point Mass on a Family of Surfaces) A similar example is offered by the heavy mass constrained to belong to the family of surfaces $f = \frac{1}{2}(x^2 + \mu y^2)$ (Fig. 2.4). Since $\left.\frac{\partial f}{\partial x}\right|_O = 0$, $\left.\frac{\partial f}{\partial y}\right|_O = 0$, $\forall \mu$, the origin $O := (0, 0)$ is an equilibrium point for each μ. Stability is governed by the Hessian matrix evaluated at the origin, $\mathbf{H}_O = \text{diag}[1, \mu]$. If $\mu > 0$, the Hessian is positive definite (the surface is an elliptical paraboloid, Fig. 2.4a) and the origin is stable; if $\mu < 0$ the Hessian is undefined (hyperbolic or saddle paraboloid, Fig. 2.4c) and the origin is unstable. Hence, $\mu_c = 0$, for which the Hessian is semi-definite positive, is the critical value of the bifurcation parameter (parabolic cylinder, Fig. 2.4b). $\square$

---

[3] As an example, the single degree of freedom system $\ddot{x} + c(\mu)\dot{x} + x = 0$, with $c(\mu) = 0$ when $\mu \leq 0$ and $c(\mu) = 1$ when $\mu > 0$, admits in $(x, \dot{x}) = (0, 0)$: (i) a center point, when $\mu \leq 0$, and (ii) a stable focus, when $\mu > 0$. Therefore $\mu = 0$ is a bifurcation point at which stability is preserved.

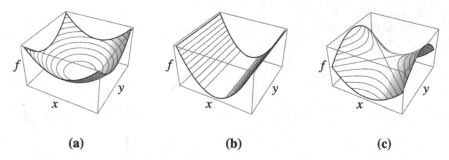

Fig. 2.4   Hessian matrix: (**a**) positive definite; (**b**) positive semi-definite; (**c**) indefinite

2.3   An Example of Static Bifurcation: The Euler Beam

A more complex, but universally known, example concerns the *Euler beam*. It is an internally constrained beam, undergoing bending only (i.e., not admitting extensional nor shear strains), supported at the ends, uniformly compressed by a load μP, whose multiplier μ is taken as the bifurcation parameter. As the load increases from zero, when a critical value μ_c is reached, the beam buckles in a shape of an arch of sag q (transverse displacement at the midspan), depending on μ. Figure 2.5 (which will be justified later) illustrates the pairs (q, μ), i.e., the equilibrium paths; it is therefore a *bifurcation diagram*.

There are two different equilibrium configurations: (a) trivial, $q = 0 \; \forall \mu$, in which the beam remains straight; (b) non-trivial, $\mu = \mu(q)$, in which the beam buckles. In the *precritical phase* ($\mu < \mu_c$), only one equilibrium position exists; in the *postcritical phase* ($\mu > \mu_c$), there are three equilibrium points. It can be proved, that:

- the equilibrium point $q = 0$ is stable when $\mu \leq \mu_c$;
- the equilibrium point $q = 0$ is unstable when $\mu > \mu_c$;
- the equilibrium points $q \neq 0$ are stable.

Since the dynamics of the beam changes qualitatively when μ crosses μ_c (as the stability of the trivial equilibrium changes, as well as the existing equilibria pass from one to three), a bifurcation occurs at μ_c. This definition includes the original idea of the bifurcation, associated with the occurrence of a "branch point" at $\mu = \mu_c$, $q = 0$.

Remark 2.4 The *path branching from the trivial path is not simply flat* ($\mu = \mu_c$, $\forall q$), as described in the elementary texts of mechanics of structures, where a linearized analysis is performed. The determination of the bifurcated path calls for using a *nonlinear model*. Although often the interest of the engineer is limited to loads smaller than the critical one, the determination of which only requires the simpler linearized model, the pattern of the branched path has great importance in

Fig. 2.5 Bifurcation
diagram for the Euler beam;
S stable, U unstable

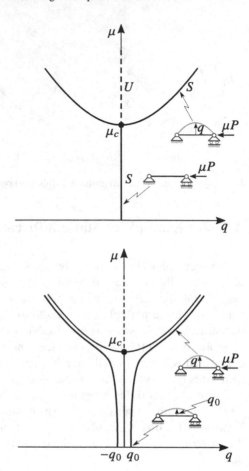

Fig. 2.6 Bifurcation
diagram for the imperfect
Euler beam

establishing the sensitivity of the system to imperfections, as will be made clear
soon.

Effect of Imperfections

If the beam is "imperfect," for example, it is not straight, but a small initial curvature
is present (sag $q_0 \neq 0$ when the load is zero), the bifurcation diagram changes as
shown in Fig. 2.6. Now a single equilibrium branch exists for each value q_0 (positive
or negative), which indefinitely approaches the branches of the perfect system. The
imperfections are always present in a system, and they *destroy the branch point*.
However, in the specific case, they do not entail dangerous states, because the
behavior of the imperfect system remains qualitatively similar to that of the perfect
system ($q_0 = 0$). For these reasons, systems behaving as the Euler beam are said to
be *insensitive to imperfections*.

Remark 2.5 The branch points *are not robust*, that is, a small perturbation is enough
to change qualitatively the bifurcation diagram. It is said that the geometric structure

of the diagram is unstable (*structural instability*, not to be confused with equilibrium instability).

2.4   Static Bifurcations of Elastic Structures

The different types of static bifurcations are classified [1, 6, 9, 11, 12]. The bifurcations are described in (q, μ) bifurcation diagrams, where μ usually is a load parameter and q is a displacement of a selected material point of the structure.

2.4.1   Fork and Transcritical Bifurcations

The Euler beam exhibits a kind of bifurcation which is very recurrent in *symmetric* elastic structures, i.e., systems whose mechanical behavior is insensible to the sign of displacements. The bifurcation is called a *stable* or *super-critical fork bifurcation* (Fig. 2.7a). The denomination alludes to the shape of the bifurcation diagram, reminiscent of a three-pronged gallows, and to the quality of the equilibrium on the bifurcated path. The adjective super-critical refers to the fact that the bifurcated branch exists only above the bifurcation value, that is, for $\mu \geq \mu_c$. The so-called fundamental path, i.e., that one passing through the origin, is stable below the branch point and unstable above. It should be noticed that there is at least one stable equilibrium position for all values of the bifurcation parameter.

Symmetric structures, however, do not always posses a stable postcritical behavior, but they may manifest an *unstable* or *sub-critical fork bifurcation* (Fig. 2.7b). This implies the birth of a symmetric equilibrium path with downward concavity, which exists only below the bifurcation value, i.e., for $\mu \leq \mu_c$. The equilibrium along the bifurcated path is unstable, as well as along the fundamental path, above the bifurcation. It follows that there are not stable equilibria when $\mu \geq \mu_c$. Furthermore, even when $\mu < \mu_c$, for which trivial equilibrium is stable, perturbations of large amplitude, which carry the system beyond the unstable equilibrium points, generate diverging motions. In other words, the sub-critical bifurcation limits the extension of the *basin of attraction* of the origin.[4] For this reason, the sub-critical fork bifurcations are extremely dangerous.

When the structure is non-symmetric, its behavior changes with the sign of the displacement. The fork-shaped bifurcations, therefore, being symmetric, cannot occur. Instead, a new geometric pattern appears, called *non-symmetric* or *transcritical* bifurcation (Fig. 2.7c). It implies the birth of a branched path (approximated in

---

[4] The basin of attraction of an asymptotically stable equilibrium point is the locus of all the initial conditions, in the state-space, which bring the system to come back to the equilibrium point. For asymptotic stability, damping must be added to an elastic structure.

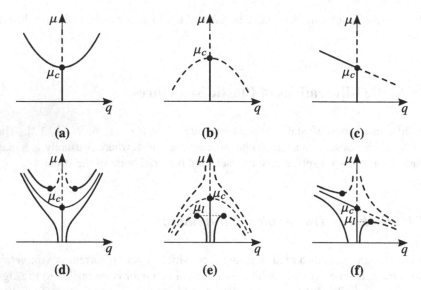

Fig. 2.7 Static bifurcations: (**a**) stable fork; (**b**) unstable fork; (**c**) transcritical; (**d**) imperfect stable fork; (**e**) imperfect unstable fork; (**f**) imperfect transcritical

figure by a straight line) that develops both below and above the critical point (from which the name). The bifurcated path consists of a stable branch, for $\mu > \mu_c$, and an unstable branch, for $\mu \leq \mu_c$, so that there always exist a stable position for any μ.

Effect of Imperfections

When the structure is imperfect, the bifurcation diagrams change as schematized in Fig. 2.7d–f. In all the cases, the paths of the imperfect structure approach those of the associated perfect structure, provided the imperfections are so small to be considered as perturbations. Imperfections, however, *destroy the branch points*, which, therefore, cannot never be observed in experiments. This circumstance is due to the ideal character of systems suffering from branching point, which is not physically achievable.

Specifically, in case of super-critical fork bifurcation (Fig. 2.7d), as the parameter μ increases, the structure follows a monotonically increasing path, which depends on the sign of the imperfection. Actually, there are other equilibrium paths, which, not originating from $\mu = 0$, cannot be visited in this type of experience, and therefore are called non-natural. The structure can rest at such non-natural state only if it is "brought" to it, with appropriate constraints to be successively removed. All imperfect natural paths are stable. For this reason, the structures suffering super-critical fork bifurcations are called *insensitive to imperfections*.

When, however, the fork bifurcation is sub-critical, a new phenomenon is observed, namely, the imperfect structure exhibits a *non-monotonic path* (Fig. 2.7e). The representative point reaches a maximum corresponding to a *limit value* $\mu_l < \mu_c$, which depends on the magnitude of the imperfection, and then descends.

The limit point is itself a bifurcation point, called *fold bifurcation* (which will be commented on later), as in it the number of equilibrium points change. This phenomenon is very dangerous, because it reduces the value of the load that makes the structure unstable, from μ_c to μ_l, with μ_l the smaller, the greater the amplitude of imperfection. The phenomenon is also called *erosion of the critical load*. The structures manifesting sub-critical bifurcations are called *sensitive to imperfections*.

When the bifurcation of the perfect structure is transcritical, the imperfections modify the relative bifurcation diagram such as in Fig. 2.7f. Here, asymmetry plays a fundamental role. Depending on the sign of the imperfection, the structure can follow a monotonically growing path, always stable, or a path which exhibits a limit point, at which stability is lost. Since the sign of imperfections is unpredictable, the structure always collapses when the limit load is reached. For this reason, all the non-symmetric structures are *very sensitive to imperfections*, to a greater extent than those manifesting unstable symmetric bifurcations, as the slope of the perfect bifurcated path enhances the erosion.

2.4.2 Snap-Through Phenomenon

The bifurcations illustrated above do not exhaust the gallery of static bifurcation suffered by the elastic structures. A further important phenomenon is the so-called *snap-through*, which is exhibited by very low arches or shells, subject to gravitational forces inducing compression. The structure, which at rest or for small loads is concave downward, when the multiplier of loads reaches and slightly exceeds a critical value, suddenly changes its shape, becoming concave upward. The change of configuration phenomenon is obviously dynamic[5] and generates oscillations which, due to friction, fade over time. However, if one is not interested in the transient motion, but only in equilibria, he can perform a simpler static analysis, which only provides the starting (at zero time) and the final position (reached after an infinite time).

The bifurcation diagram, schematically represented in Fig. 2.8, consists of a nonlinear *non-monotonic* curve. When the load μ increases from zero, the structure undergoes a q displacement that grows on the branch I in the figure. When the load reaches a limit value μ_{l1}, at which the tangent to the curve is horizontal (point L_1 in the figure), the structure leaves the point of equilibrium and instantly "jumps" to the new stable equilibrium S_1, existing under the same load. As the load continues to increase, the system travels along the stable branch II. If, starting from some point above S_1, the load is decreased, the structure descends branch II, until it undergoes a new snap-through at S_2, i.e., when $\mu = \mu_{l2}$. This latter jump brings the system back to branch I. It is understood how *hysteresis loops* can be run; the energy is dissipated during jumps and oscillations that follow them. If the

---

[5] The jump is usually accompanied by a "snap," as occurring for "click-clack" of metal capsules.

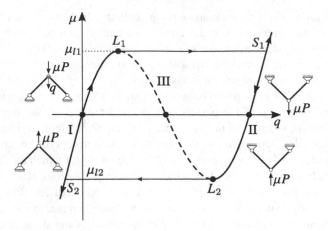

Fig. 2.8   Snap-through phenomenon exhibited by a low arch

structure is imperfect, for example, the initial curvature is affected by perturbations, the bifurcation diagram remains qualitatively the same, although the limit loads are modified by small quantities. Unlike branch points, therefore, *imperfections do not qualitatively modify the fold bifurcation*, which is therefore robust, or *structurally stable*.

2.4.3   Interaction Between Simultaneous Modes

Mechanical systems, except for pathological cases, have as many critical loads as their degrees of freedom (DOF), finite or infinite. Thus, a system with a single DOF exhibits a bifurcation point at $\mu = \mu_c$; systems with n DOF suffer from n bifurcations, which occur for μ equal to $\mu_{c1} \leq \mu_{c2} \leq \cdots \leq \mu_{cn}$; and continuous systems manifest infinite bifurcations, at $\mu_{c1} \leq \mu_{c2} \leq \cdots \leq \infty$. The critical values are called, respectively, "first critical load," "second critical load," and so on. The most interesting of all these bifurcations is the one corresponding to the *lowest value* of the parameter μ, as it is encountered "naturally" when the load is increased from zero. Accordingly, the first critical load is simply called the "critical load," $\mu_c := \mu_{c1}$, and the other ones are called the "higher critical loads" or "successive."

Each of these $n \leq \infty$ bifurcations is associated with a specific deflection shape, or *bifurcation mode*, which represents the *incipient deformation* assumed by the structure when it leaves the fundamental path, to move to a branched path. Traveling on this path, and moving away from the branch point, the structure gradually changes its shape while remaining "close" to the bifurcation mode. This latter is therefore an *act of motion*, similarly to what happens in dynamics for a natural mode of vibration. For example, the Euler beam buckles in one half-wave of sine, upon reaching the first critical load, two half-waves at the second critical load, and n half-waves at the nth critical load.

Generally, the successive critical loads are distinct, that is, $\mu_{c1} < \mu_{c2} < \cdots$; however, it is not rare the occurrence in which the first two (or more) lowest critical loads are coincident, i.e., $\mu_{c1} = \mu_{c2} < \mu_{c3} < \cdots$. In these cases, the associated critical modes are called (albeit with abuse of language) *simultaneous modes*, to allude to the fact that they manifest themselves "together," when the common values $\mu_c = \mu_{c1} = \mu_{c2}$ of the critical load are reached. This bifurcation naturally occurs in the presence of symmetries (e.g., an Euler beam with square cross-section, which can buckle in both principal inertia planes under the same load) or in families of more complex structures, in which it is possible to vary *two* bifurcation parameters. For this reason, it is also called *codimension-2 bifurcation*, because it occurs at an isolated point of a two-parameter family,[6] as it will be immediately made clear.

An Example of a Two-Parameter Family: The Compressed Truss

A plane truss beam is considered, whose currents are made of all equal beams of bending stiffness EI, except for the kth, of stiffness νEI. The beam is uniformly compressed and supported at the ends (Fig. 2.9a). For a fixed structure (ν assigned), when the load parameter μ is increased, two different forms of bifurcation can occur:

- of *global type*, at $\mu = \mu_{gl}(\nu)$, in which the whole beam buckles like an Euler beam;
- of *local type*, at $\mu = \mu_{loc}(\nu)$, in which only the kth member buckles, while the others remain straight.

Between the two critical modes, global and local, the one associated with the lowest critical load prevails, that is, $\mu_c = \min\left(\mu_{gl}(\nu), \mu_{loc}(\nu)\right)$. If, however, a two-parameter (μ, ν) family of trusses is considered, in which not only the load μ but also the stiffness ν of the kth beam is varied, the scenario depicted in Fig. 2.9b occurs; this is called a *linear stability diagram*.

- When ν is small (soft kth beam), the local mode manifests itself before the global one, that is, $\mu_c = \mu_{loc}(\nu)$;
- when ν is large (stiff kth beam), the global mode manifests itself before the local one, that is, $\mu_c = \mu_{gl}(\nu)$;
- when ν takes a special value ν_c, local and global modes manifest simultaneously, i.e., $\mu_c = \mu_{loc}(\nu_c) = \mu_{gl}(\nu_c)$.

This type of behavior is characteristic of many thin structures, e.g., open or closed metallic profiles, compressed or bent, which can behave either (i) as beams with undeformable cross-sections (undergoing global mode, in which the cross-section

---

[6] Or on a geometric manifold of dimension $m - 2$ belonging to a m-parameter space.

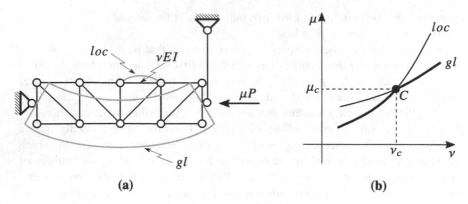

(a)   **(b)**

Fig. 2.9 Compressed truss: (**a**) local and global modes; (**b**) linear stability diagram (C bifurcation point of codimension-2)

translates and rotates, with negligible deformation), or (ii) as plate assemblies (undergoing local mode, in which the cross-section significantly modifies its own shape, with small translations and rotations).

Structural Optimization in the Linear Optics

Although the bifurcation of two simultaneous modes is, in principle, a rare occurrence, it is not difficult to find it in applications, as a result of a *structural optimization*. The following example is useful to clarify the problem.

An Euler beam is considered, subjected to a compression force μP, and equipped at midspan with an elastic bracing of stiffness vk (Fig. 2.10a). The behavior of the two-parameter family (μ, v) is investigated. When $v = 0$ (no bracing), the critical load is the Eulerian load (associated with a single half-wave), $\mu = \mu_{c1}$. As v is grown up, however, the critical load monotonically increases according to a $\mu = \mu_s(v)$ law, and it is associated with a symmetric bifurcation mode, with non-zero displacement at midspan (Fig. 2.10b). When $v \to \infty$, the displacement at midspan tends to zero, and, since the mode is symmetric, the beam buckles like a clamped-supported beam of half-length. However, this upper limit is not reachable, since, in addition to the symmetric mode, a two half-wave antisymmetric mode also exists, whose associated critical load is independent of the stiffness of the bracing, not involved in the deformation. This critical load equates the second Eulerian critical load, $\mu_a(v) = \mu_{c2}$. There is, therefore, a pair of values, (μ_c, v_c), to which two simultaneous modes correspond, one symmetric and the other antisymmetric.

When $v < v_c$, the symmetric mode prevails, whose critical load increases with v by the law $\mu_s(v)$; when $v > v_c$, the antisymmetric mode prevails, whose critical load μ_{c2} is independent of v.

The bracing stiffness v_c is an *optimal* value, since it maximizes the critical load with the least use of material. From a linear perspective, therefore, it seems

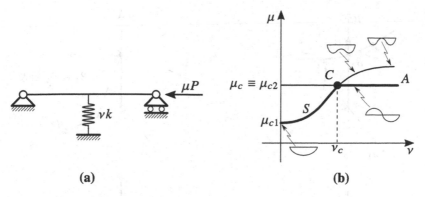

Fig. 2.10 Elastically braced beam: (**a**) bifurcation parameters μ, ν; (**b**) linear stability diagram (S symmetric mode, A antisymmetric mode, C codimension-2 bifurcation point)

inappropriate to increase the stiffness of the bracing beyond ν_c, as the critical load does not increase beyond μ_{c2}.

Nonlinear Interaction Between Simultaneous Modes

The previous considerations, however, are no longer valid when the problem is approached from a nonlinear perspective. Indeed, it happens that the simultaneous modes often interact in the nonlinear field, by coupling themselves and modifying the postcritical behavior from stable (as it would be in the case of separate modes) to unstable.

Figure 2.11 schematizes the phenomenon for an ideal structure. When the critical loads are distinct ($B_1 \neq B_2$, Fig. 2.11a), the structure buckles in one of two different patterns. The lowest mode is associated with a stable postcritical behavior (of fork-type). It, however, suffers from a secondary bifurcation at point S, at which an unstable coupled mode arises. This phenomenon is not dangerous, as long as the secondary bifurcation S is far away from the primary one B_1. However, if the two critical loads are made coincident ($B_1 \equiv B_2$, Fig. 2.11b), the primary and secondary bifurcations also coincide and give rise to four paths, all unstable, which branch off from the fundamental path. Optimization, in this case, makes the otherwise stable postcritical behavior unstable.

Although the above considerations *do not* have general character, there exist in literature many cases falling into this phenomenology. It is therefore concluded that structural optimization, based on the exam of the linear behavior, *could* instead be detrimental and that this effect can only be ascertained in a nonlinear context.

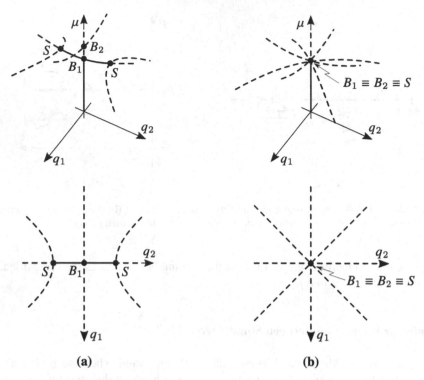

Fig. 2.11 Interaction between simultaneous modes; bifurcation diagram for (**a**) separate modes ($\nu \neq \nu_c$); (**b**) simultaneous modes ($\nu = \nu_c$); 3D paths and their projection onto the displacement plane

2.5   Dynamic Bifurcations of Elastic Structures Subject to Nonconservative Forces

The examples so far illustrated relate to conservative systems, since the structures are elastic and subject to gravitational forces. These systems, as mentioned, exhibit static bifurcations only. However, when the conservativeness of the system is destroyed, e.g., by external forces depending on position, or velocity, or time (and, possibly, by internal dissipative forces), other forms of loss of stability may arise, via *dynamic bifurcations* [4, 11, 13, 14]. These latter generate new periodic motions (i.e., new closed trajectories in the state-space) close to the equilibrium points and not merely new equilibria. The analysis of nonconservative systems is much more complex, as it necessarily requires the use of the equation of motion, which allows one "to observe" the trajectories. In other words, in these systems, the kinetic energy plays an essential role on stability. The examples commented ahead describe the phenomenology.

2.5.1 Flutter Induced by Follower Forces

The forces usually considered in static systems are of gravitational type; they keep their magnitude and direction over time and therefore admit a potential. However, there exist problems in which forces need to be to modeled as *follower*, i.e., as vectors of constant modulus that vary in direction, by "following" the structure on which they act. Examples are offered: (a) by the jet thrust of a rocket; (b) by the force exerted at the extremity of a pipe conveying a fluid; (c) by the friction forces acting on blades/skates/rollers traveling across a rough surface (e.g., during the "derapage," in which the frictional force follows the body); and (d) by frictional forces on disc brakes, through a mechanism similar to the previous one.

When the magnitude of the follower force reaches a critical value, periodic oscillations of exponentially increasing amplitude occur, eventually stabilizing on a limit cycle. The phenomenon is called *flutter* (literally, flapping of wings), in the technical literature, or "circulatory Hopf bifurcation", in the mathematical environment. Two degrees of freedom, at least, are required for the occurrence of flutter. The origin of the technical name derives from studies conducted in the field of aeroelastic instability of aircrafts and bridges, where the cross-section oscillates in the vertical plane and twists (the two abovementioned DOF), recalling the beating of the wings of a butterfly.

The paradigmatic model is the *Beck beam*, that is, a cantilever subjected at the free end to a compression force μF, which, in the current configuration, remains tangent to the beam at the tip (Fig. 2.12a) [3]. The stability of the trivial equilibrium position is observed by looking at the eigenvalues λ of the linearized problem. Purely imaginary eigenvalues denote a harmonic motion; purely real eigenvalues an exponential motion; complex eigenvalues an exponentially modulated harmonic motion. For small values of the force, the eigenvalues of the system belong to

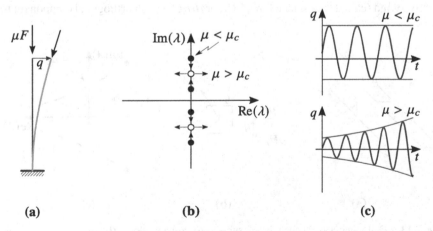

Fig. 2.12 Flutter: (a) undamped Beck beam; (b) locus of the eigenvalues; (c) time histories for $\mu < \mu_c$ and $\mu > \mu_c$

the imaginary axis of the Argand-Gauss plane (Fig. 2.12b), as it happens for conservative systems, so that the small motions are harmonic (Fig. 2.12c). In this case, the trivial equilibrium is called *marginally stable*, because just at the limit of stability, represented by the imaginary axis itself. However, as the force increases, two pairs of eigenvalues approach each other, collide, and then separate into complex eigenvalues, with real parts of opposite sign. The pair of eigenvalues with negative real part represents a damped harmonic motion and the pair with positive real part a harmonic motion with divergent amplitude (Fig. 2.12c); the equilibrium is therefore unstable. If the nonlinearities are opportune, the motion stabilizes on a limit cycle. Hence, the bifurcation gives rise to periodic orbits, which do not exist for sub-critical values of the force.

2.5.2 Galloping Induced by Aerodynamic Flow

A phenomenon similar to the previous one, but which can manifest itself also in systems with a single DOF, is the so-called *galloping*, according to the technical literature, or (generic) Hopf bifurcation, in the mathematical literature. This occurs in the presence of nonconservative forces which depend on the speed of the structure.

To explain the phenomenon, a beam is considered, internally damped, constrained to oscillate in the vertical plane, having non-circular section and subject to horizontal wind flow of velocity μU, normally incident on the beam axis (Fig. 2.13a) [2]. As the beam oscillates (e.g., due to an initial disturbance), its midspan possesses a vertical velocity $\dot{q}$; hence, the wind hits the structure with a *relative velocity* μU_{rel} inclined on the horizontal direction of an angle β, called *angle of attack*, which depends on the structural velocity $\dot{q}$ (Fig. 2.13b). Now, the aerodynamic force acting on the beam consists of a component aligned to the relative wind (called "resistance force" F_d, or *drag*) and an orthogonal component to

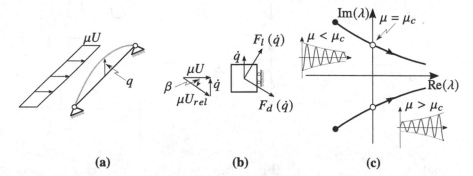

Fig. 2.13   Galloping: (a) damped beam subject to a wind flow; (b) angle of attack and aerodynamic forces; (c) locus of the eigenvalues and time histories

the wind (said "lift force" F_l, or *lift*), whose projections onto the vertical direction of motion load the structure. If one linearizes kinematics for small motions, the vertical aerodynamic force turns out to be proportional to the speed of the structure, so that it acts as an *aerodynamic damping force*, the intensity of which depends on the velocity μU. Unlike the structural damping, which is always dissipative, the aerodynamic damping can be exciting, according to the shape of the cross-section. If this is the case, the cross-sections is said to be *aerodynamically unstable*, since, at sufficiently high wind speed, the aerodynamic excitation prevails over the structural damping, causing dynamic instability.

The stability of the equilibrium position is governed by the eigenvalues of the system (Fig. 2.13c). When the wind speed μU is zero, the eigenvalues are all complex with negative real part (entailing damped harmonic motion, also plotted in the figure). However, if the beam cross-section is aerodynamically unstable, as the wind speed increases, a pair of eigenvalues approaches the imaginary axis and, at a critical value, crosses it. Upon the bifurcation, harmonic oscillations of exponentially increasing amplitude occur, which eventually stabilize on a limit cycle, as in the case of flutter. During the phenomenon, the beam seems to "gallop," from which the technical name.

2.5.3 Parametric Excitation Induced by Pulsating Loads

A system is said parametrically excited, or subject to *parametric excitation*, when its intrinsic characteristics vary over time with a known law. An example is offered by a pendulum, whose length is made to vary over time. Parametric excitation is distinguished from external excitation: in the latter, time appears in the right-hand member of the equation, to model the interaction of the system with the environment; in the former, time appears in the left side of the equation, namely, in its coefficients, to model the *non-invariance* of the system.

A case of parametric excitation of great interest occurs in the *Bolotin beam* [3] (Fig. 2.14a). This is a hinged-supported beam subject to a compression force (which, as far as known from the Euler beam, enters the first member of the equation), however not constant over time, but pulsating with the law $P(\Omega t) = P_s + \mu P_d \cos(\Omega t)$, where P_s is the static component and μP_d the dynamic component. When $\mu = 0$, the static bifurcation occurs at the Eulerian critical load P_E. However, if μ takes a critical value $\mu_c(\Omega)$, which depends on the excitation frequency Ω, a *dynamic bifurcation* manifests itself at a peak value of the force $P_c(\Omega) = P_s + \mu_c(\Omega) P_d$ which is *smaller* than the Eulerian one, i.e., $P_c < P_E$.

To study parametric excitation, it needs to use *two bifurcation parameters*, the excitation amplitude μ and the excitation frequency Ω. The pairs (μ, Ω), which induce dynamic bifurcation, delimit an unstable region in the two-parameter plane, which takes the name of *domain of stability* (Fig. 2.14b). The greater the μ load, the larger the instability interval on the Ω frequency axis. The phenomenon is caused by a form of *resonance* which occurs between the excitation frequency Ω and a

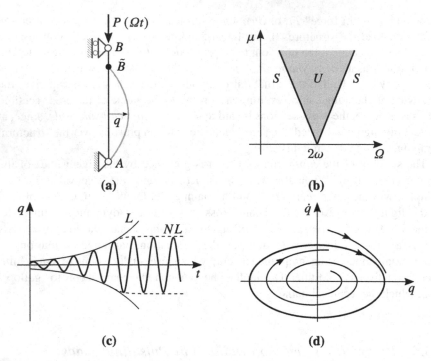

Fig. 2.14 Parametric excitation: (**a**) Bolotin beam; (**b**) stability domain (*S* stable, *U* unstable); (**c**) time history in the unstable region (*L* linear, *NL* nonlinear); (**d**) limit cycle in the phase portrait

natural frequency ω of the system. Indeed, while the beam oscillates with frequency ω, its longitudinally moving end, *in nonlinear kinematics*, describes a motion $B\tilde{B}$ of *double frequency* (i.e., it performs two cycles while the beam completes a single cycle). If therefore the parametric frequency of the force is double the natural one, that is, if $\Omega = 2\omega$, then the axial force resonates with the structure. Outside the unstable region, the beam oscillates with limited amplitudes; in the unstable region, it oscillates with exponentially divergent amplitude (Fig. 2.14c). These oscillations can eventually stabilize on a *cycle limit*, that is, on a periodic orbit of large but finite amplitude, if and only if the nonlinearities are taken into account in the model (Fig. 2.14d).

References

1. Bigoni, D.: Nonlinear solid mechanics: bifurcation theory and material instability. Cambridge University Press, Cambridge (2012)
2. Blevins, R.D.: Flow-induced vibration. Van Nostrand Reinhold Co., New York (1977)
3. Bolotin, V.V.: The dynamic stability of elastic systems. Holden Day, San Francisco (1964)

4. Cesari, L.: Asymptotic behavior and stability problems in ordinary differential equations. Springer, Berlin (1971)
5. Gantmakher, F.R.: Lectures in analytical mechanics. Mir, Moscow (1970)
6. Hoff, N. J.: Buckling and Stability. J. Roy. Aeron. Soc. **58** (1954)
7. Koiter, W. T.: On the stability of elastic equilibrium. National Aeronautics and Space Administration, Washington D.C. (1967)
8. Lacarbonara, W.: Nonlinear structural mechanics: theory, dynamical phenomena and modeling. Springer, New York (2013)
9. Leipholz, H.H.: Stability of elastic systems. Sijthoff & Noordhoff, Alphen aan den Rijn (1980)
10. Lyapunov, A.M.: Probleme General de La Stabilite du Mouvement (in French). Ann. Fac. Sci. Toulouse **9**, 203–474 (1907)
11. Panovko, Ya.G., Gubanova, I. I.: Stability and oscillation of elastic systems: modern concepts, paradoxes and errors. National Aeronautics and Space Administration, Washington D.C. (1973)
12. Thompson, J.M.T.: Instabilities and catastrophes in science and engineering. Wiley, New York (1982)
13. Troger, H., Steindl, A.: Nonlinear stability and bifurcation theory: an introduction for engineers and applied scientists. Springer, Wien (1991)
14. Ziegler, H.: Principles of structural stability. Blaisdell Publishing Co, Waltham (1968)

Chapter 3
Stability and Bifurcation Linear Analysis

3.1 Introduction

The first step, in carrying out the stability analysis of an equilibrium point, consists in linearizing the motion around it, and observing the character (real or complex) of the eigenvalues of the Jacobian matrix at that point. Except for critical cases, such a *linear stability analysis* is sufficient to establish the nature of the equilibrium, independently of the nonlinear terms. A study of the eigenvalues of finite-dimensional mechanical systems reveals that: (a) *conservative systems* (i.e., possessing a symmetric stiffness matrix) can lose stability only when one of their eigenvalues crosses the zero from the left, and this occurrence is not affected by the mass matrix, so that conservative systems can be studied in the static field (which is a quite common approach in the literature devoted to *buckling*); (b) *circulatory systems* (i.e., having a non-symmetric stiffness matrix) generally lose stability when a pair of eigenvalues crosses from the left the imaginary axis of the complex plane; (c) *damping* cannot restabilize an unstable conservative system, while it can either *restabilize* or *destabilize* an undamped circulatory system, which is respectively unstable or stable.

When the system depends on parameters, the eigenvalues of the Jacobian matrix also depend on them, so that, by quasi-steady varying the parameters, one or more of the eigenvalues may cross the imaginary axis. Based on the previous study, the mechanisms leading to bifurcations of conservative, circulatory, and damped systems is predicted. This analysis is called *linear bifurcation analysis*, which is propedeutic to the nonlinear analysis to be carried out ahead in the book.

In this chapter, all these concepts are discussed for general systems and illustrated by examples. The treatment starts with a brief introduction to dynamical systems, soon specialized to mechanics. The notion of *geometric stiffness matrix*, which plays an important role in bifurcation, is given; it is the tangent operator of the position-dependent external forces. Similarly, the *geometric damping matrix* is the tangent operator of the velocity-dependent external forces.

A. Luongo et al., *Stability and Bifurcation of Structures*,
https://doi.org/10.1007/978-3-031-27572-2_3

3.2   Dynamical Systems

A (discrete) *dynamical system* (of chemical, biological, economic, mechanical, nature, and so on) is a system whose state is described by time-dependent *state variables* (or phases) $\mathbf{x}(t) \in \mathbb{R}^N$, whose evolution is ruled by differential equations. The space spanned by $\mathbf{x}$ is the *state-space* (or the phase-space) [2, 9]; at each point $\mathbf{x}$, a specific state assumed by the system is associated. The dynamics of the system is governed by a vector-valued ordinary differential equation, either

$$\dot{\mathbf{x}} = \mathbf{f}(\mathbf{x}, t), \tag{3.1}$$

or

$$\dot{\mathbf{x}} = \mathbf{f}(\mathbf{x}), \tag{3.2}$$

where a dot denotes differentiation with respect to time t. The differential equation is accompanied by initial conditions specifying the state at $t = 0$:

$$\mathbf{x}(0) = \mathbf{x}_0. \tag{3.3}$$

The system of Eq. 3.1 is called *non-autonomous* (or time-variant); the system of Eq. 3.2 is called *autonomous* (or time-invariant). The column matrix $\mathbf{f}(\mathbf{x}, t)$, or $\mathbf{f}(\mathbf{x})$, is the *vector field*, because it assigns a (possibly time-dependent) velocity $\dot{\mathbf{x}}$ to each point $\mathbf{x}$ (Fig. 3.1a). In this book, reference is mainly made to autonomous systems.

The solutions $\mathbf{x}^*(t; \mathbf{x}_0)$ of Eqs. 3.2 or 3.3 are called the *trajectories* or *orbit* of the dynamical system. They are geometrically represented by curves belonging to the state-space, describing the evolution of the system consequent to the given initial condition (Fig. 3.1b). Closed orbits indicate *periodic motions* and are called *cycles* (Fig. 3.1b). The set of all the trajectories is said the *phase diagram* (or phase portrait, in the $N = 2$ case).

Construction of the Phase Diagram
In the autonomous case, the phase diagram is constructed as follows. Chosen a regular $\mathbf{x}_0$ point, for which it is $\mathbf{f}(\mathbf{x}_0) \neq \mathbf{0}$, the velocity $\mathbf{f}(\mathbf{x}_0)$ is uniquely defined at that point. After an infinitesimal time interval dt, the system moves to $\mathbf{x}_1 := \mathbf{x}_0 + \mathbf{f}(\mathbf{x}_0)dt$. At this new point, a new velocity $\mathbf{f}(\mathbf{x}_1)$ is defined, so that the system, after another dt, goes to $\mathbf{x}_2 := \mathbf{x}_1 + \mathbf{f}(\mathbf{x}_1)dt$, and so on. In this way, the (unique) trajectory passing through $\mathbf{x}_0$, is constructed.[1] Repeating the operation for different $\mathbf{x}_0$'s, the phase diagram is built up.

---

[1] As a consequence, the trajectories cannot intersect each other, except at points at which $\mathbf{f}(\mathbf{x}_0) = \mathbf{0}$, where the tangent is undetermined.

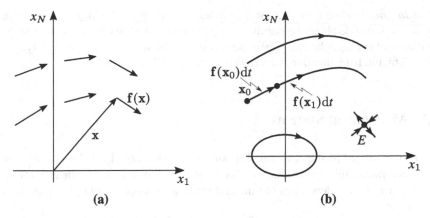

Fig. 3.1 Phase diagram: **(a)** vector field, **(b)** trajectories and equilibrium points

Equilibrium Points
A point $\mathbf{x}_E$, at which the vector field vanishes, i.e., $\mathbf{f}(\mathbf{x}_E) = \mathbf{0}$, is an *equilibrium point* (or a singular point) (Fig. 3.1b). If the system is placed at $\mathbf{x}_E$, since the velocity is zero there, the state does not evolve, and therefore the trajectory degenerates into the point itself. At an equilibrium point, the tangent to the trajectory is not defined (because the zero velocity vector is arbitrarily directed), and therefore, more trajectories or no trajectory can originate from it.[2] The nature of the trajectories around the equilibrium point determines the stability of equilibrium.

Linearization of the Vector Field
For stability purposes, it is of interest to study *small motions* in the neighborhood of equilibrium. Therefore, the vector field can be linearized around the equilibrium point $\mathbf{x}_E$. By truncating the Taylor series to the first order, one gets:

$$\mathbf{f}(\mathbf{x}) = \mathbf{f}(\mathbf{x}_E) + \left.\frac{\partial \mathbf{f}}{\partial \mathbf{x}}\right|_{\mathbf{x}_E} (\mathbf{x} - \mathbf{x}_E). \tag{3.4}$$

Since, for equilibrium, $\mathbf{f}(\mathbf{x}_E) = \mathbf{0}$, the linearized equation of motion read:

$$\delta\dot{\mathbf{x}} = \mathbf{J}\delta\mathbf{x}, \tag{3.5}$$

in which $\delta\mathbf{x} := \mathbf{x} - \mathbf{x}_E$ is the deviation of $\mathbf{x}$ from $\mathbf{x}_E$ and

$$\mathbf{J} := \left.\frac{\partial \mathbf{f}}{\partial \mathbf{x}}\right|_{\mathbf{x}_E}, \quad \text{or} \quad [J_{ij}] := \left[\left.\frac{\partial f_i}{\partial x_j}\right|_{\mathbf{x}_E}\right], \tag{3.6}$$

---

[2] For example, a *center point* is not crossed by any trajectory; a *saddle point* in the plane is crossed by two trajectories (separatrices).

is the *Jacobian matrix* of the vector field at $\mathbf{x}_E$. The properties of $\mathbf{J}$, in particular its eigenvalues, decide about the stability of $\mathbf{x}_E$, as they determine the evolution of the state around the equilibrium. Albeit the discussion of stability of a general system is not difficult, here attention is narrowed soon to mechanical systems.

3.3   Mechanical Systems

The form assumed by the equation of motion of a n degrees of freedom (DOF) viscoelastic system, subjected to forces of general nature, is discussed. The constraints are assumed to be independent of time and such as to suppress all rigid motions of the system.

Nonlinear Equation of Motion
The Lagrangian[3] vector equation of motion reads:[4]

$$\mathbf{M}\ddot{\mathbf{q}} = \mathbf{f}^{int}\left(\mathbf{q}, \dot{\mathbf{q}}\right) + \mathbf{f}^{ext}\left(\mathbf{q}, \dot{\mathbf{q}}; t\right), \tag{3.7}$$

where:

- $\mathbf{q} \in \mathbb{R}^n$ is the vector of Lagrangian parameters, taken as displacements and rotations of particular material points, measured from a *reference configuration* $\mathbf{q} = \mathbf{0}$;
- $\mathbf{M} \equiv \mathbf{M}^T$ is the *mass* (or inertia) *matrix*, symmetric and positive definite, because the kinetic energy $T := \frac{1}{2}\dot{\mathbf{q}}^T \mathbf{M}\dot{\mathbf{q}}$ is positive for any $\dot{\mathbf{q}}$;
- $\mathbf{f}^{int}\left(\mathbf{q}, \dot{\mathbf{q}}\right)$ are forces internal to the system, given by the sum of elastic forces $\mathbf{f}^{el}\left(\mathbf{q}\right)$, depending on the position only, and viscous forces $\mathbf{f}^v\left(\mathbf{q}, \dot{\mathbf{q}}\right)$, mainly depending on velocity, but, also on the position, when in finite kinematics;
- $\mathbf{f}^{ext}\left(\mathbf{q}, \dot{\mathbf{q}}; t\right)$ are external forces, generally nonconservative, which depend on the position (e.g., a follower force), velocity (e.g., an aerodynamic force), and on time.

Example 3.1 (Elementary Examples of Internal Forces) A linear spring of stiffness k, which connects a point of a body to the ground, exerts a force whose modulus is proportional to the elongation $\Delta\left(\mathbf{q}\right)$ and directed as the current direction $\mathbf{a}\left(\mathbf{q}\right)$ of the spring, i.e., $\mathbf{f}^{el} = -k\Delta\left(\mathbf{q}\right)\mathbf{a}\left(\mathbf{q}\right)$; therefore, $\mathbf{f}^{el} = \mathbf{f}^{el}\left(\mathbf{q}\right)$ is nonlinear in $\mathbf{q}$, although the spring is linear. Similarly, a dashpot of linear impedance c, applied at a point of the body, exerts a force proportional to the elongation rate $\dot{\Delta} = \left[\frac{\partial\Delta}{\partial\mathbf{q}}\left(\mathbf{q}\right)\right]^T \dot{\mathbf{q}} =:$ $\dot{\Delta}\left(\mathbf{q}, \dot{\mathbf{q}}\right)$, and directed as the current direction $\mathbf{a}\left(\mathbf{q}\right)$ of the dashpot, that is, $\mathbf{f}^v =$

---

[3] The adjective Lagrangian refers to the fact that the balance equation does not contain any reactive force.

[4] Here, a large class of systems not containing *inertial nonlinearities* is considered. An exception to this case will be considered in Sect. 11.4.

$-c\dot{\Delta}\,(\mathbf{q},\dot{\mathbf{q}})\,\mathbf{a}\,(\mathbf{q})$; hence, $\mathbf{f}^v = \mathbf{f}^v\,(\mathbf{q},\dot{\mathbf{q}})$ nonlinearly depends on the position and on the velocity, although the dashpot is linear. In the linear theory, in contrast, in which displacements are infinitesimal and the body configuration is assumed to be frozen at $\mathbf{q} = \mathbf{0}$, it is $\Delta\,(\mathbf{q}) = \boldsymbol{\alpha}^T\mathbf{q}$, $\dot{\Delta} = \boldsymbol{\alpha}^T\dot{\mathbf{q}}$, with $\boldsymbol{\alpha}$ a vector of coefficients independent of $\mathbf{q}$, and $\mathbf{a}\,(\mathbf{q}) = \mathbf{a}\,(\mathbf{0}) =: \mathbf{a}_0$; therefore, $\mathbf{f}^{el} = -k\boldsymbol{\alpha}^T\mathbf{q}\,\mathbf{a}_0$ and $\mathbf{f}^v = -c\boldsymbol{\alpha}^T\dot{\mathbf{q}}\,\mathbf{a}_0$, so that the forces are proportional to $\mathbf{q}$ and $\dot{\mathbf{q}}$, respectively. □

Motion Around the Origin

It is assumed that, in the state-space $(\mathbf{q},\dot{\mathbf{q}})$, the point $O := (\mathbf{0},\mathbf{0})$ is an equilibrium point, i.e., $\mathbf{f}^{int}\,(\mathbf{0},\mathbf{0}) = \mathbf{f}^{ext}\,(\mathbf{0},\mathbf{0};t) = \mathbf{0}\ \forall t$. This circumstance occurs when O is the natural state, at which all the internal forces vanish, while the external forces, in this state, do not contribute to the Lagrangian equation of motion. An example is offered by the Euler (or Bolotin) beam, whose axial force *does not* induces bending of the beam in the straight configuration, unlike what happens in the curved configuration.

Aimed to analyzing the motion in the small but finite neighborhood of O, it is convenient to expand the internal and external forces in Maclaurin series, as:

$$\mathbf{f}^{int}\,(\mathbf{q},\dot{\mathbf{q}}) = -\mathbf{K}_e\mathbf{q} - \mathbf{C}_s\dot{\mathbf{q}} + \mathbf{n}^{int}\,(\mathbf{q},\dot{\mathbf{q}}), \tag{3.8a}$$

$$\mathbf{f}^{ext}\,(\mathbf{q},\dot{\mathbf{q}};t) = -\mathbf{K}_g\,(t)\,\mathbf{q} - \mathbf{C}_g\,(t)\,\dot{\mathbf{q}} + \mathbf{n}^{ext}\,(\mathbf{q},\dot{\mathbf{q}};t). \tag{3.8b}$$

In the previous expressions, the following matrices and vectors appear:

- $\mathbf{K}_e := -\left.\frac{\partial \mathbf{f}^{int}}{\partial \mathbf{q}}\right|_O$ is the *elastic stiffness matrix*, evaluated at O, symmetric and positive definite, since, under the hypothesis of monotonic elastic law, the elastic energy $U := \frac{1}{2}\mathbf{q}^T\mathbf{K}_e\mathbf{q}$ is positive for any $\mathbf{q}$.

- $\mathbf{C}_s := -\left.\frac{\partial \mathbf{f}^{int}}{\partial \dot{\mathbf{q}}}\right|_O$ is the *structural damping matrix*, symmetric and positive definite, since the dissipated energy $\mathcal{D} := \frac{1}{2}\dot{\mathbf{q}}^T\mathbf{C}_s\dot{\mathbf{q}}$ is positive for any $\dot{\mathbf{q}}$.

- $\mathbf{K}_g := -\left.\frac{\partial \mathbf{f}^{ext}}{\partial \mathbf{q}}\right|_O$, is the *geometric stiffness matrix*, which describes the linear part of the positional external forces (whose effect manifests itself when a change of geometry occurs). If the external forces do not possess special properties, $\mathbf{K}_g$ is non-symmetric and indefinite. As a special case, the gravity forces (e.g., the compression force acting on the Euler beam), being potential, lead to a symmetric and positive definite matrix $\mathbf{K}_g$.[5] If $\mathbf{f}^{ext}$ explicitly depends on time, then so does $\mathbf{K}_g$.

- $\mathbf{C}_g := -\left.\frac{\partial \mathbf{f}^{ext}}{\partial \dot{\mathbf{q}}}\right|_O$, is the *geometric damping matrix*,[6] which describes the linear part of the external forces depending on velocity (which manifest themselves

---

[5] If the external force depends on a potential V, it is $\mathbf{f}^{ext} = -\frac{\partial V}{\partial \mathbf{q}}$, so that $\mathbf{K}_g := \left[k_{ij}^g\right] = \left.\frac{\partial^2 V}{\partial \mathbf{q}^2}\right|_O$; therefore $k_{ij}^g = \left.\frac{\partial^2 V}{\partial q_i \partial q_j}\right|_O = k_{ji}^g$.

[6] When the external force is aerodynamic, $\mathbf{C}_g$ is also called the *aerodynamic matrix* and denoted by $\mathbf{C}_a$.

through the acquisition of an act of motion $\dot{\mathbf{q}}$ of the system). $\mathbf{C}_g$ is generally non-symmetric and indefinite; purely dissipative forces (e.g., the air resistance) are an exception, leading to a symmetric and positive definite matrix. If $\mathbf{f}^{ext}$ is explicitly time-dependent, then so does $\mathbf{C}_g$.

- $\mathbf{n}^\alpha\,(\cdot)$ are *vectors of nonlinearities*, sum of quadratic, cubic, and other forms, each homogeneous in its own arguments.[7]

Taking into account the expansions in Eq. 3.8, the equation of motion, Eq. 3.7, reads:

$$\mathbf{M}\ddot{\mathbf{q}} + \left(\mathbf{C}_s + \mathbf{C}_g\right)\dot{\mathbf{q}} + \left(\mathbf{K}_e + \mathbf{K}_g\right)\mathbf{q} = \mathbf{n}^{int}\,(\mathbf{q},\dot{\mathbf{q}}) + \mathbf{n}^{ext}\,(\mathbf{q},\dot{\mathbf{q}};t)\,. \tag{3.9}$$

The problem is completed by the initial conditions:

$$\mathbf{q}\,(0) = \mathbf{q}_0, \qquad \dot{\mathbf{q}}\,(0) = \dot{\mathbf{q}}_0. \tag{3.10}$$

The sum of the homonymous structural and geometric matrices,

$$\mathbf{K} := \mathbf{K}_e + \mathbf{K}_g, \qquad \mathbf{C} := \mathbf{C}_s + \mathbf{C}_g, \tag{3.11}$$

will be referred in this book as the *total stiffness matrix* and *the total damping matrix*, respectively. It will usually be assumed, ahead, that external forces *do not* explicitly depend on time, i.e., the system is *autonomous*.[8]

Motion Around a Generic Equilibrium Point

Since the equation of motion, Eq. 3.7, is nonlinear, it admits one or more non-trivial solutions $E := (\mathbf{q}_E, \mathbf{0})$. Accordingly (an example is given in Fig. 3.2, where only the stiffness is represented):

Fig. 3.2 Equilibrium points O and E and relevant elastic and geometric stiffnesses

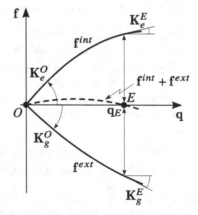

---

[7] For example, the quadratic part of $\mathbf{n}^{int}\,(\mathbf{q},\dot{\mathbf{q}})$ contains product of displacements $\sum_{i,j} a_{ij}q_iq_j$, of velocities $\sum_{i,j} b_{ij}\dot{q}_i\dot{q}_j$, and mixed $\sum_{i,j} c_{ij}q_i\dot{q}_j$.

[8] As an exception, time-periodic systems will be addressed in the Chap. 13.

$$\mathbf{f}^{int}\,(\mathbf{q}_E, \mathbf{0}) + \mathbf{f}^{ext}\,(\mathbf{q}_E, \mathbf{0}; t) = \mathbf{0}, \qquad \forall t. \tag{3.12}$$

When the motion in the neighborhood of E must be analyzed, Taylor series expansions of initial point $\mathbf{q}_E$ must be performed, leading to:

$$\mathbf{f}^{int}\,(\mathbf{q}, \dot{\mathbf{q}}) = \mathbf{f}^{int}\,(\mathbf{q}_E, \mathbf{0}) - \mathbf{K}_e^E\,(\mathbf{q} - \mathbf{q}_E) - \mathbf{C}_s^E\dot{\mathbf{q}} + \mathbf{n}^{int}\,(\mathbf{q} - \mathbf{q}_E, \dot{\mathbf{q}}), \tag{3.13a}$$

$$\mathbf{f}^{ext}\,(\mathbf{q}, \dot{\mathbf{q}}; t) = \mathbf{f}^{ext}\,(\mathbf{q}_E, \mathbf{0}; t) - \mathbf{K}_g^E\,(t)\,(\mathbf{q} - \mathbf{q}_E) - \mathbf{C}_g^E\,(t)\,\dot{\mathbf{q}} \tag{3.13b}$$
$$+ \mathbf{n}^{ext}\,(\mathbf{q} - \mathbf{q}_E, \dot{\mathbf{q}}; t).$$

Here, the following positions have been introduced:

$$\mathbf{K}_e^E := -\left.\frac{\partial \mathbf{f}^{int}}{\partial \mathbf{q}}\right|_E, \quad \mathbf{C}_s^E := -\left.\frac{\partial \mathbf{f}^{int}}{\partial \dot{\mathbf{q}}}\right|_E,$$

$$\mathbf{K}_g^E := -\left.\frac{\partial \mathbf{f}^{ext}}{\partial \mathbf{q}}\right|_E, \quad \mathbf{C}_g^E := -\left.\frac{\partial \mathbf{f}^{ext}}{\partial \dot{\mathbf{q}}}\right|_E, \tag{3.14}$$

denoting the *stiffness and damping matrices at* E; they have the same meaning and properties of the homologous $\mathbf{K}_e$, $\mathbf{C}_s$, $\mathbf{K}_g$, $\mathbf{C}_g$ (in which the "privileged" apex O was omitted).

By replacing the series in Eq. 3.13 in the equation of motion, Eq. 3.7, and taking into account the equilibrium condition in Eq. 3.12, it follows:[9]

$$\mathbf{M}\ddot{\mathbf{q}} + \left(\mathbf{C}_s^E + \mathbf{C}_g^E\right)\dot{\mathbf{q}} + \left(\mathbf{K}_e^E + \mathbf{K}_g^E\right)(\mathbf{q} - \mathbf{q}_E) = \mathbf{n}^{int}\,((\mathbf{q} - \mathbf{q}_E), \dot{\mathbf{q}})$$
$$+ \mathbf{n}^{ext}\,((\mathbf{q} - \mathbf{q}_E), \dot{\mathbf{q}}; t). \tag{3.15}$$

The matrices sum of structural and geometric effects:

$$\mathbf{K}^E := \mathbf{K}_e^E + \mathbf{K}_g^E, \quad \mathbf{C}^E := \mathbf{C}_s^E + \mathbf{C}_g^E, \tag{3.16}$$

are the total stiffness and damping matrices at E.

---

[9] In case of equilibrium at the origin, internal and external forces both vanish; in case of equilibrium at a point different from the origin, the two forces are equal and opposite in sign.

3.4 Linear Stability Analysis

Equations 3.9 and 3.15 govern the motion in the small, but finite, neighborhood of O or E. If, however, one is interested just in the stability of these points, it is sufficient to analyze only the linear part of the equation, i.e.:

$$\mathbf{M\ddot{q}} + \mathbf{C\dot{q}} + \mathbf{Kq} = \mathbf{0}, \tag{3.17}$$

where $\mathbf{K}$, $\mathbf{C}$ take the expressions in Eqs. 3.11 or 3.16 (with the superscript E omitted), respectively, and $\mathbf{q}$ measures the deviation from the equilibrium point. It happens, indeed, that when the motion is of small amplitude, the linear terms prevail over the quadratic ones, and these, in turn, over cubic terms, so that the character of the equilibrium (except in pathological cases), is governed by the linear part of the equation, only[10] [3–9].

Supplement 3.1 (State Form of the Equation of Motion) Equation 3.17 can also be recast in the first-order form, Eq. 3.2 (also said state form), in which the unknowns, displacements, and velocities are collected in a vector $\mathbf{x} \in \mathbb{R}^{2n}$. After a simple rearrangement, they read:

$$\dot{\mathbf{x}} = \mathbf{Jx}, \tag{3.18}$$

where the state vector $\mathbf{x}$ and the Jacobian matrix $\mathbf{J}$ are defined as follows:

$$\mathbf{x} := \begin{pmatrix} \mathbf{q} \\ \dot{\mathbf{q}} \end{pmatrix}, \quad \mathbf{J} := \begin{bmatrix} \mathbf{0} & \mathbf{I} \\ -\mathbf{M}^{-1}\mathbf{K} & -\mathbf{M}^{-1}\mathbf{C} \end{bmatrix}. \tag{3.19}$$

The second-order form, Eq. 3.17, will be usually used in this book.[11]   □

Stability Conditions
The equation of motion, Eq. 3.17, admits the solution:

$$\mathbf{q} = \mathbf{u} \exp\left(\lambda t\right), \tag{3.20}$$

with $\mathbf{u} \in \mathbb{C}^n$ and $\lambda \in \mathbb{C}$. Since the equation is real, if $(\lambda, \mathbf{u})$ is a solution, also its complex conjugate $(\bar{\lambda}, \bar{\mathbf{u}})$ is a solution.[12] Substituting Eq. 3.20 into the equation of

---

[10] The circumstance is identical to that encountered in the study of a scalar function $y = f(x)$ around the origin. If $f'(0) \neq 0$, its sign is sufficient to establish whether the function is locally increasing or decreasing, independently from the value of the successive derivatives. Only if $f'(0) = 0$ (pathological or critical case) it is necessary to examine the higher derivatives: if $f''(0) \neq 0$, this sets the local character of the function, regardless of $f'''(0)$ and higher derivatives.

[11] Except for the Chap. 13, where, in dealing with parametrically excited systems, the first order form is found to be more convenient.

[12] So that $\mathbf{u} \exp(\lambda t) + \bar{\mathbf{u}} \exp(\bar{\lambda} t)$ is real.

motion, Eq. 3.17, an *eigenvalue problem* of degree $2n$, in nonstandard form, follows, i.e.:[13]

$$\left(\lambda^2 \mathbf{M} + \lambda \mathbf{C} + \mathbf{K}\right) \mathbf{u} = \mathbf{0}. \tag{3.21}$$

Here $\mathbf{u}$ is an eigenvector of the linear system and λ the associated eigenvalue. Distinguishing the complex eigenvalues λ_k ($k = 1, 2, \cdots, n_c$) from the real ones λ_h ($h = 1, 2, \cdots, n_r$), with $2n_c + n_r = 2n$, and assuming that all the eigenvalues are distinct,[14] the general solution of the equation of motion is written as:

$$\mathbf{q} = \sum_{k=1}^{n_c} \left[A_k \mathbf{u}_k \, \exp\left(\lambda_k t\right) + \bar{A}_k \bar{\mathbf{u}}_k \, \exp\left(\bar{\lambda}_k t\right) \right] + \sum_{h=1}^{n_r} a_h \mathbf{u}_h \exp\left(\lambda_h t\right), \tag{3.22}$$

where $A_k \in \mathbb{C}$ and $a_h \in \mathbb{R}$ are arbitrary constants, to be determined by the initial conditions in Eq. 3.10. By letting $A_k := \frac{1}{2} a_k \exp\left(i\varphi_k\right)$, $\lambda_k := \delta_k + i\omega_k$, $\mathbf{u}_k := \mathbf{v}_k + i\mathbf{w}_k$, with δ_k, ω_k, $\mathbf{v}_k$, $\mathbf{w}_k$ real quantities and i the imaginary unit, the former equation is recast in real form, as:

$$\mathbf{q} = \sum_{k=1}^{n_c} a_k \exp\left(\delta_k t\right) \left[\mathbf{v}_k \cos\left(\omega_k t + \varphi_k\right) - \mathbf{w}_k \sin\left(\omega_k t + \varphi_k\right)\right] + \sum_{h=1}^{n_r} a_h \mathbf{v}_h \exp\left(\delta_h t\right). \tag{3.23}$$

Each of the n addends of Eq. 3.23 is said to be a *mode* of oscillation.[15] Complex modes are harmonic oscillations of decreasing amplitude, if $\delta_k < 0$, or increasing amplitude, if $\delta_k > 0$. Real modes are monotonously decaying ($\delta_h < 0$) or divergent ($\delta_h > 0$) motions. Harmonic modes (in which λ_k is purely imaginary) consist of harmonic oscillations.

From the law of motion, Eq. 3.22, the following conclusions are drawn:

1. If all the eigenvalues have a negative real part, then the motion amplitude decreases exponentially over time, i.e., $\|\mathbf{q}\| \to 0$ when $t \to \infty$. In this case, the equilibrium point is *asymptotically stable*.
2. If even just one of the eigenvalues, for example, the jth, has a positive real part, while all the other ones have negative real parts, then the jth mode diverges, while the others decay, so that $\|\mathbf{q}\| \to \infty$ when $t \to \infty$. In this case, the equilibrium point is *unstable*.
3. If one (or more) of the eigenvalues, for example, the jth, has zero real part, while the others have real negative parts, then the *critical case of equilibrium* occurs,

---

[13] If, in contrast, the state form in Eq. 3.18 is used, the standard eigenvalue problem $\mathbf{A}\mathbf{u} = \lambda\mathbf{u}$ is found, still of degree $2n$.

[14] Or coincident, provided that a complete system of eigenvectors $\mathbf{u}_k$ exists.

[15] This locution is in disagreement with what used in engineering jargon, where mode means the *spatial shape* $\mathbf{u}_k$.

for which the linear stability analysis is not sufficient to determine the quality of equilibrium.[16]

Remark 3.1 It is important noticing that *stability is exclusively decided by the real part δ of the eigenvalues*, which establishes the growth or degrowth rate of the motion, not from the imaginary part ω, which determines just the frequency of oscillation.

Remark 3.2 The divergence to infinity of the motion is a false information, consequent to linearization. Indeed, when the amplitude is large, linearization loses its validity, and the evolution is governed from nonlinearities, ignored here. Thus, divergence should be understood as an indication of the fact that the motion cannot be confined by acting on the initial conditions, so that, according to Lyapunov's definition, the equilibrium is unstable.

Remark 3.3 The linear stability analysis has been developed here according to the so-called dynamic criterion of stability. Two different criteria will be discussed ahead, the *energy* and *static* criteria, which, however, apply with limitations, as discussed in detail in Chap. 1.

3.4.1   Conservative Systems

Conservative systems are undamped, neither structurally nor geometrically. The related eigenvalue problem, Eq. 3.21, reduces to:

$$(\mathbf{K} - \Lambda \mathbf{M})\,\mathbf{u} = \mathbf{0}, \tag{3.24}$$

where $\mathbf{K} := \mathbf{K}_e + \mathbf{K}_g$ and the following quantity has been introduced:

$$\Lambda := -\lambda^2. \tag{3.25}$$

Since the external forces, by hypothesis, admit a potential, the geometric stiffness matrix, which descends from this, is symmetric, i.e., $\mathbf{K}_g = \mathbf{K}_g^T$. Due to $\mathbf{K}_e = \mathbf{K}_e^T$, the total stiffness matrix $\mathbf{K}$ is symmetric, too. Since the eigenvalues of a symmetric matrix are real together with their eigenvectors, it follows that $\Lambda \in \mathbb{R}$ and $\mathbf{u} \in \mathbb{R}^n$.[17] The eigenvalues $\lambda_k = \pm\sqrt{-\Lambda_k}$ are double in number with respect to the

---

[16] Some authors (e.g., [8]) state that the *linear* system is stable at criticality when all the eigenvalues with zero real part are simple or semi-simple (non-defective system). This circumstance, indeed, excludes that the linearized motion diverges with polynomial-harmonic law. However, stability of the linear system in the critical state does not entail stability of the nonlinear system. A special case occurs when the system is conservative, as it will be discussed in Sect. 3.4.1.

[17] It is known from algebra that $\mathbf{M}$ can be decomposed as $\mathbf{M} = \mathbf{Q}^T\mathbf{Q}$, where $\mathbf{Q}$ is real and non-singular. By letting $\mathbf{w} := \mathbf{Q}\mathbf{u}$, Eq. 3.24 is rewritten in the standard form as $(\mathbf{L} - \Lambda\mathbf{I})\,\mathbf{w} = \mathbf{0}$, with

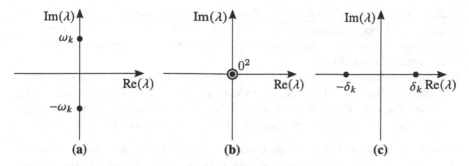

Fig. 3.3 Pairs of eigenvalues of conservative systems: (a) $\Lambda_k > 0$; (b) $\Lambda_k = 0$; (c) $\Lambda_k < 0$

eigenvalues Λ_k; they can be (a) purely imaginary and conjugate, if $\Lambda_k > 0$; (b) double zero, if $\Lambda_k = 0$; and (c) reals and opposite in sign, if $\Lambda_k < 0$ (Fig. 3.3). Both cases, (a) and (b), appear to be critical, according to the general definitions given above. However, a distinction will be made soon, based on the special character of the conservative systems, namely, the equilibrium is *marginally stable* in case (a) and (properly) critical in case (b) and unstable in case (c).

In order to relate stability and properties of matrices, a discussion is conducted. If (Λ_k, $\mathbf{u}_k$) is an eigensolution of Eq. 3.24, then the following identity holds:[18]

$$\Lambda_k = \frac{\mathbf{u}_k^T \mathbf{K} \mathbf{u}_k}{\mathbf{u}_k^T \mathbf{M} \mathbf{u}_k}, \qquad (3.26)$$

known as *Rayleigh's ratio*, in which all the quantities are real. Since the mass matrix is positive definite, the sign of Λ_k solely depends on the numerator.

Three cases can occur:

1. **K** is *positive definite*, as it happens in the absence of geometric stiffness, or when the geometric coefficients are small compared to the elastic ones. Then $\Lambda_k > 0 \; \forall k$, that is, $\lambda_k = \pm i\omega_k$. The modes in Eq. 3.23 are all harmonic,[19] so it is possible to confine the motion by taking the initial conditions sufficiently small. The equilibrium is said *marginally stable*, indicating that the eigenvalues are all on the margin of stability (i.e., on the imaginary axis). However, *this situation is robust*, in the sense that, if the conservativeness of the system is not altered, marginal stability persists.

$\mathbf{L} := \mathbf{Q}^{-T} \mathbf{K} \mathbf{Q}^{-1}$ symmetric. Λ is therefore an eigenvalue of a symmetric matrix. Moreover, since $\mathbf{w}^T \mathbf{L} \mathbf{w} = \mathbf{u}^T \mathbf{K} \mathbf{u}$, if **K** is positive definite, **L** is also defined positive.

[18] Written in Eq. 3.24 for the kth eigensolution, premultiplying it by $\mathbf{u}_k^T$ and finally solving for Λ_k, Eq. 3.26 follows.

[19] This circumstance is identical to that one usually analyzed in books devoted to mechanical vibrations, in which the geometric effects are generally ignored.

2. **K** is *positive semi-definite*, i.e., there is an eigenvector $\mathbf{u}_k$ for which the quadratic form, otherwise positive, is equal to zero. Since $\Lambda_k = 0$, there are *two* eigenvalues coalescing at the origin. The linear analysis is insufficient to determine the stability.[20]

3. **K** is *undefined*, in the sense that the quadratic form at numerator assumes negative values for at least one eigenvector $\mathbf{u}_k$. Then $\Lambda_k < 0$, that is, $\lambda_k = \pm \delta_k$ is real. There are two real modes in Eq. 3.23, one of which decays exponentially, while the other diverges exponentially. The equilibrium is therefore unstable.

Remark 3.4 The stability of an equilibrium point of a conservative system depends exclusively on the total stiffness matrix (elastic plus geometric), not on the mass matrix. This is in agreement with the Lagrange-Dirichlet theorem, which attributes the stability property of these systems to the total potential energy only, independently of the kinetic energy.

3.4.2 Circulatory Systems

An undamped elastic system, subjected to nonconservative positional forces (such as the Beck beam), is said to be *circulatory*[21] [1, 10]. The related eigenvalue problem is written again as in Eq. 3.24, but, due to nonconservativeness of the forces, it is $\mathbf{K}_g \neq \mathbf{K}_g^T$, so that the total stiffness matrix is non-symmetric. This implies that the eigenvalues Λ_k are generally complex, together with their eigenvectors $\mathbf{u}_k$. By letting $\Lambda_k = \rho_k \exp(i\theta_k)$, where $\rho_k > 0$ is the modulus and $\theta_k \in [0, \pi]$ the phase of Λ_k, and by taking into account that $-\Lambda_k = \rho_k \exp(i(\theta_k + \pi))$, from Eq. 3.25 a pair of eigenvalues follows (Fig. 3.4a):

$$\lambda_{k_{1,2}} = \sqrt{\rho_k} \exp\left(i\frac{\theta_k}{2} \pm i\frac{\pi}{2}\right). \tag{3.27}$$

These latter, together with their complex conjugates $\bar{\lambda}_{k_{1,2}}$, constitute a *quadruplet* of eigenvalues (Fig. 3.4b):

$$\left(\lambda_{k_{1,2}}, \bar{\lambda}_{k_{1,2}}\right) = \sqrt{\rho_k}\left[\pm \sin\left(\frac{\theta_k}{2}\right) \pm i \cos\left(\frac{\theta_k}{2}\right)\right]. \tag{3.28}$$

---

[20] The linearized system, indeed, is defective. Because of the presence of polynomial terms correcting Eq. 3.22, it could be concluded that the equilibrium is unstable. However, this may, or may not, be true, depending on the higher-order terms of the potential elastic energy U. For example, the equilibrium position $x = 0$ of the cubic oscillator, of equation $\ddot{x} + \kappa x^3 = 0$, is stable if $\kappa > 0$ and unstable if $\kappa < 0$, as it follows from $U = \frac{\kappa}{4}x^4$ and from the Lagrange-Dirichlet theorem.

[21] The minimum dimension of the circulatory system is 2, for which the stiffness matrix can be non-symmetric.

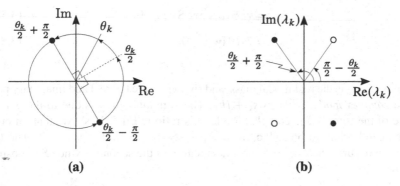

Fig. 3.4 Eigenvalues of circulatory systems: (**a**) complex square roots of $\exp(i(\theta_k + \pi))$; (**b**) quadruplet of eigenvalues

As a special case, when $\theta_k = 0, \pi$, $\Lambda_k = \pm \rho_k$ is real (together with the associated eigenvector). In this case, there exist just a pair (not a quadruplet) of eigenvalues $\lambda_{k_{1,2}}$, having the same properties of the conservative systems.

The previous Eq. 3.28 shows that the *equilibrium is stable if and only if* $\theta_k = 0\ \forall k$; *it is unstable in any other case*. Instability manifests itself through oscillations of increasing amplitude.

Supplement 3.2 (Rayleigh's Ratio in Complex Form) To relate stability and matrix properties, the Rayleigh's ratio in Eq. 3.26 is rewritten in complex form:[22]

$$\Lambda_k = \frac{\bar{\mathbf{u}}_k^T \mathbf{K} \mathbf{u}_k}{\bar{\mathbf{u}}_k^T \mathbf{M} \mathbf{u}_k}. \tag{3.29}$$

By letting $\mathbf{u}_k := \mathbf{v}_k + i\mathbf{w}_k$, and denoting by:

$$\mathbf{S} := \text{sym}[\mathbf{K}] = \frac{1}{2}\left(\mathbf{K} + \mathbf{K}^T\right), \quad \mathbf{H} := \text{skw}[\mathbf{K}] = \frac{1}{2}\left(\mathbf{K} - \mathbf{K}^T\right), \tag{3.30}$$

the symmetric and antisymmetric part of $\mathbf{K}$, respectively, the ratio assumes the form:

$$\Lambda_k = \frac{s_k + i h_k}{m_k}, \tag{3.31}$$

in which:

$$m_k := \mathbf{v}_k^T \mathbf{M} \mathbf{v}_k + \mathbf{w}_k^T \mathbf{M} \mathbf{w}_k > 0, \tag{3.32a}$$

---

[22] The complex ratio in Eq. 3.29 is obtained as the real ratio in Eq. 3.26, simply by premultiplying Eq. 3.24 by $\bar{\mathbf{u}}_k^T$ instead of $\mathbf{u}_k^T$.

$$s_k := \mathbf{v}_k^T \mathbf{S} \mathbf{v}_k + \mathbf{w}_k^T \mathbf{S} \mathbf{w}_k \gtrless 0, \qquad (3.32b)$$

$$h_k := 2\mathbf{v}_k^T \mathbf{H} \mathbf{w}_k \gtrless 0. \qquad (3.32c)$$

Here, (i) m_k is the real modal mass, and (ii) s_k, h_k are the real and imaginary parts of the *complex modal stiffness* $s_k + i h_k$. The sign ambiguity is due to the generic nature of the forces. The complex Rayleigh's ratio in Eq. 3.31 shows that, in order for the equilibrium to be stable, $h_k = 0$, $s_k > 0$ must hold for all k's. The first condition occurs when $\mathbf{w}_k = \mathbf{0}$, i.e., $\mathbf{u}_k$ is real, and the second is when $\mathbf{S}$ is positive definite.   □

3.4.3 Influence of Damping

It is interesting to check if structural damping, which is purely dissipative, can, or cannot, stabilize an equilibrium position which is unstable for the undamped system. The cases of conservative and circulatory systems will be dealt with separately.

Damped Conservative Systems

The eigenvalue problem is governed by Eq. 3.21, with $\mathbf{K}$ and $\mathbf{C} \equiv \mathbf{C}_s$ both symmetric. After having written the problem for the complex eigensolution $(\lambda_k, \mathbf{u}_k)$, and premultiplied it by $\bar{\mathbf{u}}_k^T$, it follows:

$$m_k \lambda_k^2 + c_k \lambda_k + k_k = 0, \qquad (3.33)$$

where, by taking $\mathbf{u}_k := \mathbf{v}_k \pm i\mathbf{w}_k$:

$$m_k := \mathbf{v}_k^T \mathbf{M} \mathbf{v}_k + \mathbf{w}_k^T \mathbf{M} \mathbf{w}_k > 0, \qquad (3.34a)$$

$$c_k := \mathbf{v}_k^T \mathbf{C} \mathbf{v}_k + \mathbf{w}_k^T \mathbf{C} \mathbf{w}_k > 0, \qquad (3.34b)$$

$$k_k := \mathbf{v}_k^T \mathbf{K} \mathbf{v}_k + \mathbf{w}_k^T \mathbf{K} \mathbf{w}_k \gtrless 0. \qquad (3.34c)$$

They are, respectively, mass, damping, and *real modal stiffness*; the first two of them are positive, and the third one has an indefinite sign.

The solution of the second-degree equation reads:

$$\lambda_{k_{1,2}} = \frac{1}{2m_k} \left(-c_k \pm \sqrt{c_k^2 - 4m_k k_k} \right). \qquad (3.35)$$

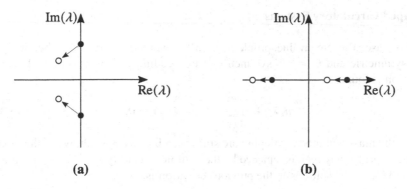

Fig. 3.5 Effect of damping on the eigenvalues of a conservative system: (**a**) $k_k > 0$, (**b**) $k_k < 0$;
● undamped, ○ damped

To discuss stability, two cases must be examined (Fig. 3.5):

- If the equilibrium position is stable for the undamped system, it is $k_k > 0$ for any
  k index. Hence, the two roots λ_k both have a real negative part. It is concluded that
  *the addition of damping to a conservative system in marginally stable equilibrium
  is beneficial, because damping changes marginal to asymptotic stability.*
- If the equilibrium position is unstable for the undamped system, it is $k_k < 0$ at
  least for one k index. Then, while one of the two roots λ_k has a negative real
  part, the other one has a real positive part. It is concluded that *adding damping
  to a conservative system in unstable equilibrium does not produce any beneficial
  effect, because the equilibrium position remains unstable.* The same conclusions
  are drawn in the critical case, in which $k_k = 0$.

By summarizing, adding damping to a conservative system shifts the eigenvalues
to the left; in the unstable case, however, such translation is insufficient to regain
stability, and this happens *regardless of the magnitude of the damping.*

Remark 3.5 Viscous forces, depending on velocity, cannot change the static nature
of the loss of stability.

Example 3.2 (Point Mass Immersed in a Fluid) An elementary example is offered
by a point ball moving on a rigid profile, immersed in a viscous fluid. If the ball is on
a crest, the motion consequent to a perturbation, although slowed down, diverges,
whatever the viscosity of the fluid. If, instead, the ball is located in a valley, the
motion slowly subsides. □

Damped Circulatory Systems

In this case, the eigenvalue problem is still governed by Eq. 3.21, but with $\mathbf{K}$ non-symmetric and $\mathbf{C} \equiv \mathbf{C}_s$ symmetric.[23] Proceeding as before, a second-degree equation is found:

$$m_k \lambda_k^2 + c_k \lambda_k + (s_k + i h_k) = 0, \tag{3.36}$$

where the mass and modal damping are still given by Eqs. 3.34a,b, while the modal stiffness, previously real, is replaced by the complex quantity $s_k + i h_k$, according to Eqs. 3.32b, c. The solution to the previous equation is:

$$\lambda_k = \frac{c_k}{2m_k} \left(-1 \pm \sqrt{1 - 4 m_k \frac{s_k + i h_k}{c_k^2}} \right). \tag{3.37}$$

By confining to dynamic instability, it must be assumed that $s_k > 0$. Two cases are considered:

- The equilibrium position of the undamped system is unstable, i.e., there exists at least one k for which $h_k \neq 0$. It can be checked that *a sufficiently large damping c_k can restabilize the equilibrium*. Indeed, by expanding the root for $\frac{s_k + i h_k}{c_k^2} \to 0$, it turns out that:

$$\lambda_k \simeq -\frac{s_k + i h_k}{c_k}, \quad \text{or} \quad \lambda_k \simeq -\frac{c_k}{m_k} + \frac{s_k + i h_k}{c_k}, \tag{3.38}$$

  whose real part is negative.
- The equilibrium position of the undamped system is stable, i.e., $h_k = 0 \; \forall k$. This case is more difficult to study, since, due to the fact the damping changes the eigenvectors of the system, any of the h_k can become non-zero. Thus, *a small damping can have an unstable effect*, even if its nature is dissipative. The phenomenon is known as the *Ziegler paradox* [10], which will be studied later (Chap. 11). When, however, damping becomes large, one falls back into the previous case, that is, damping has a beneficial effect.

Remark 3.6 The viscous forces, depending on the velocity, influence the phenomenon of dynamic instability. They can make an unstable position stable or *vice versa*. No conclusions can be drawn in general.

---

[23] The study of this case requires reading the Supplement 3.2; otherwise, the reader can directly go to the conclusions, summarized in Remark 3.6.

3.5   An Illustrative Example: The Planar Mathematical Pendulum

The concepts introduced above are now exemplified with reference to a single DOF nonlinear system: the planar mathematical pendulum. This is characterized by the length ℓ of the thread and the point-mass m attached to it. The current configuration is described by the angle of deviation of the thread from the vertical direction, θ, which is the only Lagrangian parameter (Fig. 3.6).

3.5.1   Equation of Motion and the Phase Portrait

The motion of the pendulum is governed by the first cardinal equation of the dynamic, $\mathbf{F} = m\mathbf{a}$, or, consistently with d'Alembert principle, $\mathbf{F} - m\mathbf{a} = \mathbf{0}$. Here, $\mathbf{F}$ is the resulting external force, including (i) the downward weight mg (with g the gravity acceleration), and (ii) the reaction R of the thread, directed along the thread itself, in its current configuration; moreover, $\mathbf{a}$ is the acceleration of the mass. Projecting the equation along the tangent to the circular trajectory (i.e., onto the normal to the thread), and taking into account that the modulus of the tangent acceleration is $a_t = m\ddot{\theta}\ell$,[24] it follows, after division by $m\ell$:

$$\ddot{\theta} + \frac{g}{\ell} \sin \theta = 0. \tag{3.39}$$

The equation of motion is of the type of Eq. 3.7, i.e., it appears in the form of a second-order differential equation; it can also be written as a pair of differential equations of the first order:

Fig. 3.6 Planar
mathematical pendulum

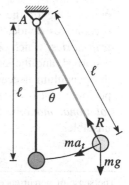

---

[24] The magnitude of the tangent velocity is $v_t = \dot{\theta}\ell$, whose time derivative is a_t.

$$\begin{pmatrix} \dot{\theta} \\ \dot{\psi} \end{pmatrix} = \begin{pmatrix} \psi \\ -\frac{g}{\ell}\sin\theta \end{pmatrix}. \tag{3.40}$$

These latter are of the type of Eq. 3.2, where $\mathbf{x} := (\theta, \psi)^T$ is the vector of state variables, consisting of the position θ and of the angular velocity $\psi = \dot{\theta}$; moreover, $\mathbf{f}(\mathbf{x}) := (\psi, -\frac{g}{\ell}\sin\theta)^T$ is the vector field. The problem is completed by the initial conditions, $\theta(0) = \theta_0$, $\dot{\theta}(0) = \dot{\theta}_0$, or $\mathbf{x}(0) = \mathbf{x}_0$.

Equilibrium Points

The equilibrium occurs at the points where the vector field vanishes, that is, when $\psi = 0$ and $\theta = \pm k\pi$, with $k = 0, 1, 2, \cdots$, i.e., when the pendulum occupies the lower or the upper vertical position with zero velocity. The values of $k > 1$ correspond to the same physical positions, θ being a cyclic variable.

Phase Portrait

The phase portrait of the pendulum lies on the (θ, ψ) plane. Proceeding as described in Sect. 3.2, the diagram in Fig. 3.7 is obtained, from which the following results are drawn.

- The equilibrium point $\theta = 0$ (lower position) is a *center*; it is surrounded by *closed trajectories*, on which the motion is periodic.[25] No trajectory passes through the center. If the equilibrium is disturbed by a small deviation and/or a small initial velocity, the motion develops on a closed trajectory, remaining confined. The equilibrium is therefore stable (but not asymptotically).
- The equilibrium points $\theta = \pm\pi$ (upper position), are called *saddles*; they are crossed by two trajectories, called *separatrices*,[26] which connect more saddle points. One of these is entering the point, the other one is outgoing. The separatrices divide the region of the closed trajectories from that of the *open trajectories*, which occupies the remaining part of the plan. If the equilibrium is disturbed, and the system is carried into the region internal to the separatrices, the pendulum executes large-amplitude oscillatory motions around the lower position; if it is carried into the region external to the separatrices, it performs *rotational motions*. Since the motion cannot be confined, the equilibrium is unstable.

---

[25] The sense of percurrence is clockwise, since, for conservation of the energy, as the rotation increases in magnitude, the velocity decreases in magnitude.

[26] The separatrices are trajectories joining equilibrium points. However, they describe motions that cannot be reproduced in laboratory, because the time needed to travel them is infinite, since the velocity goes to zero close to the equilibrium.

Fig. 3.7 Phase portrait of
the planar mathematical
pendulum

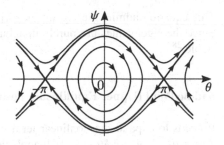

In any cases, due to the absence of damping, the motion (both oscillatory and rotatory) persists over time.

3.5.2   Local Stability Analysis

The phase portrait provides complete information about the stability of the equilibrium points. However, such a diagram can be easily built up only if the system has a single DOF. In case of multiple DOF, the increased size of the state-space makes this type of analysis not viable. It is therefore preferable to execute a *local analysis* of stability, based on the examination of the dynamics around the equilibrium point. It is important to stress that, by following this approach, any information about the *global* behavior of the system, as that given by the phase portrait, is lost.

To perform a local analysis, the equation of motion must be linearized around the equilibrium point of interest; then, the relevant eigenvalue problem must be solved. The two equilibrium points of the pendulum are now investigated.

Center Point (Lower Equilibrium Position)

Expanding the nonlinear term around $\theta = 0$, one has $\sin\theta = \theta - \frac{\theta^3}{6} + \cdots$; by retaining only the linear term, the equation of motion, Eq. 3.39, is written as:

$$\ddot{\theta} + \omega^2\theta = 0, \tag{3.41}$$

where $\omega^2 := \frac{g}{\ell}$ is the *circular frequency*. To solve the differential equation, $\theta = \Theta e^{\lambda t}$ is set, where Θ is a constant, from which the algebraic eigenvalue problem follows:

$$(\lambda^2 + \omega^2)\,\Theta = 0. \tag{3.42}$$

This equation admits the eigenvalues $\lambda_{1,2} = \pm i\omega$ and the eigenvector $\Theta = 1$.[27] Since the eigenvalues are purely imaginary, the equilibrium position is marginally stable.[28]

Saddle Point (Upper Equilibrium Position)

It needs to expand the nonlinear term in Taylor series around $\theta = \pi$. By letting $\theta = \pi + \delta\theta$, where $\delta\theta$ is a small angle denoting the deviation from the equilibrium position, one gets $\sin\theta = \sin\pi + (\theta - \pi)\cos\pi + \cdots = -\delta\theta$. The linearized equation of motion, Eq. 3.39, consequently, reads:

$$\delta\ddot{\theta} - \omega^2\delta\theta = 0. \tag{3.43}$$

After having expressed the solution as $\delta\theta = \Theta e^{\lambda t}$, an eigenvalue problem is drawn:

$$(\lambda^2 - \omega^2)\Theta = 0, \tag{3.44}$$

which admits the eigenvalues $\lambda_{1,2} = \pm\omega$ and the eigenvector $\Theta = 1$.[29] Since an eigenvalue is real and positive, the equilibrium point is unstable.[30]

3.5.3 Energy Criterion of Stability

The dynamic analysis developed above (according to the *dynamic criterion* of stability) represents the most general investigation tool. However, when the system is conservative, it appears exuberant, according to the theorem of Lagrange-Dirichlet, and can be replaced by a simpler energy analysis (*energy criterion* of stability). This requires (a) writing the total potential energy of the system Π, (b) determining the equilibria as stationary points of Π, and (c) checking that, for stability, Π attains a minimum at the equilibrium.

With reference to the pendulum, since the active forces are exclusively gravitational, it is $\Pi = \Pi_0 + mgh$. Here, Π_0 is an inessential constant, representing the energy at $\theta = 0$ (which therefore can be taken equal to zero), and $h = \ell(1 - \cos\theta)$

---

[27] In the state-space, the eigenvector has two components, the position and the velocity, hence $\mathbf{u} = (1, \pm i\omega)^T$.

[28] The small motions around the lower equilibrium point are described by $\theta = c_1\cos(\omega t) + c_2\sin(\omega t)$, with c_1, c_2 arbitrary constants, determined by the initial disturbance.

[29] In the state-space, it is $\mathbf{u} = (1, \pm\omega)^T$. These vectors are *tangent to the separatrices* at the equilibrium point. In the direction $(1, \omega)^T$ the motion is divergent, in the direction $(1, -\omega)^T$ it is convergent, consistently with Fig. 3.7.

[30] The small motions around the upper equilibrium point are described by $\delta\theta = c_1 e^{\omega t} + c_2 e^{-\omega t}$, with c_1, c_2 arbitrary constants.

Fig. 3.8 Diagram of the
total potential energy of the
pendulum

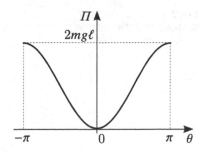

is the current height of the mass, measured from the lowest point. Therefore:

$$\Pi = mg\ell(1 - \cos\theta), \tag{3.45}$$

whose diagram is shown in Fig. 3.8.

The equilibrium points are determined by the condition:[31]

$$\frac{\mathrm{d}\Pi}{\mathrm{d}\theta} = mg\ell\sin\theta = 0, \tag{3.46}$$

which is satisfied at $\theta = 0, \pm\pi$. Stability is determined from the sign of the second derivative of the energy, $\frac{\mathrm{d}^2\Pi}{\mathrm{d}\theta^2} = mg\ell\cos\theta$, evaluated at the equilibrium. Accordingly:

- at $\theta = 0$, it is $\left.\frac{\mathrm{d}^2\Pi}{\mathrm{d}\theta^2}\right|_0 = mg\ell > 0$; the point is of local minimum, so the equilibrium is stable;
- at $\theta = \pm\pi$, it is $\left.\frac{\mathrm{d}^2\Pi}{\mathrm{d}\theta^2}\right|_\pi = -mg\ell < 0$; the point is of local maximum, so the equilibrium is unstable.

The results provided by the dynamic criterion are thus recovered.

Supplement 3.3 (Phase Portrait Derived from the Total Potential Energy) For conservative systems with a single (and no more) DOF, the total potential energy allows to build up the phase portrait, too, as it will be illustrated soon for the pendulum.

By the law of conservation of the energy, the sum of the kinetic energy $T = \frac{1}{2}m(\dot{\theta}\ell)^2$ and of the potential energy $\Pi(\theta) = mg\ell(1 - \cos\theta)$ is constant over time, i.e.:

$$T\left(\dot{\theta}\right) + \Pi(\theta) = E = \text{const}, \tag{3.47}$$

---

[31] It should be noticed that the energy approach replaces the direct writing of equilibrium.

Fig. 3.9 Phase portrait of the pendulum, derived from the total potential energy

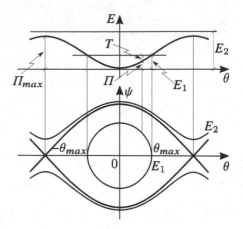

where the constant E is the total energy, established by the initial conditions, i.e., $E = T\left(\dot{\theta}_0\right) + \Pi\left(\theta_0\right)$. From the previous law, $\dot{\theta}$ can be evaluated as:

$$\dot{\theta} = \pm\frac{1}{\ell}\sqrt{2\left(E - \Pi\left(\theta\right)\right)}. \qquad (3.48)$$

This equation expresses the trajectories in the form $\dot{\theta} = \dot{\theta}\left(\theta; E\right)$; therefore, there exists a distinct trajectory for any energy level E. Motion is real only if the radicand is positive.

Figure 3.9 illustrates a graphical construction, from which it follows:

- If $E < \Pi_{max} := 2mg\ell$ ($E = E_1$ in the figure), the motion develops in the range $(-\theta_{max}, +\theta_{max})$, where θ_{max} is the angle for which the radicand of Eq. 3.48 vanishes, i.e., $\Pi(\theta_{max}) = E$. At $\theta = \theta_{max}$ the energy is entirely gravitational, so θ_{max} is the inversion point of the (oscillatory) motion.
- If, instead, $E > \Pi_{max}$ ($E = E_2$ in the figure), the radicand of the Eq. 3.48 is positive for any θ, so the motion is not confined (rotatory motion).

$\square$

3.5.4 Effect of Damping

In the event that the pendulum interacts with a fluid at rest (e.g., air), the dynamics are also influenced by dissipative forces. However, since the instability of pendulum is of the static type (being the system conservative), damping cannot stabilize the equilibrium (as stated in the general theory, Sect. 3.4.3). Here, the property is checked for the system under study.

Fig. 3.10 Damped
pendulum

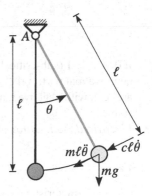

Equation of Motion

Compared to the conservative case, an additional force of modulus $f^v = c\dot{\theta}\ell$ and
direction opposite to velocity acts on the pendulum. It models the resistance of
medium, proportional to the velocity of the mass through the impedance $c > 0$
(Fig. 3.10). The equation of motion of the damped pendulum, according to the
principle of d'Alembert, is:

$$m\ell\ddot{\theta} + c\ell\dot{\theta} + mg\sin\theta = 0, \tag{3.49}$$

or:

$$\ddot{\theta} + 2\xi\omega\dot{\theta} + \omega^2\sin\theta = 0, \tag{3.50}$$

having divided by $m\ell$ and set $2\xi\omega := \dfrac{c}{m}$, with ξ the damping ratio and $\omega^2 := \dfrac{g}{\ell}$.

Local Stability Analysis

Since the system is nonconservative, stability cannot be analyzed through the energy
criterion, but the dynamic criterion must be applied. To this end, the equation of
motion is linearized around the equilibrium point, and the relevant eigenvalues are
computed. The two equilibrium points are considered separately.

- *Linearized motion around $\theta = 0$.* The governing equation is:

$$\ddot{\theta} + 2\xi\omega\dot{\theta} + \omega^2\theta = 0. \tag{3.51}$$

From the associated characteristic equation, $\lambda^2 + 2\xi\omega\lambda + \omega^2 = 0$, the eigenvalues
are obtained as:

$$\lambda_{1,2} = \left(-\xi \pm \sqrt{\xi^2 - 1}\right)\omega. \tag{3.52}$$

If $0 < \xi < 1$ (sub-critical damping), the eigenvalues are complex conjugate, with negative real part;[32] if $\xi \geq 1$ (critical or super-critical damping), the eigenvalues are real, with negative real part.[33] The equilibrium is, in any case, asymptotically stable.

- *Linearized motion around $\theta = \pi$.* The governing equation is:

$$\delta\ddot{\theta} + 2\xi\omega\delta\dot{\theta} - \omega^2\delta\theta = 0, \tag{3.53}$$

whose characteristic equation $\lambda^2 + 2\xi\omega\lambda - \omega^2 = 0$ provides the roots:

$$\lambda_{1,2} = \left(-\xi \pm \sqrt{\xi^2 + 1}\right)\omega. \tag{3.54}$$

Whatever $\xi > 0$ (sub-critical or super-critical), the roots are both real, one negative, the other positive.[34] The equilibrium is unstable.

According to the general theory, it is concluded that the mathematical pendulum being a conservative system: (a) damping modifies marginal stability in asymptotic stability; (b) damping cannot stabilize the unstable equilibrium position.

3.6 Bifurcations of Autonomous Systems

Mechanical systems depend on parameters (e.g., geometry, mass, stiffness, damping, and forces), which must be chosen in the design phase. Hence, instead of studying a specific system of fixed parameters, it is more convenient to browse a *family* of systems, in which one or more parameters are left free. This makes possible to investigate the dynamic properties of all the members of the family and to choose the optimal system. In particular, it is of interest analyzing how (a) the equilibrium points, admitted by the family, change with the parameters and (b) how their stability depends on parameters.

For the sake of simplicity, attention is here limited to *autonomous mechanical systems, dependent on a single parameter μ,* having the meaning of a *load*

---

[32] The motion is described by $\theta = e^{-\xi\omega t}\left(c_1 e^{i\omega\sqrt{1-\xi^2}t} + c_2 e^{-i\omega\sqrt{1-\xi^2}t}\right)$, so the motion decays exponentially while oscillating.

[33] The motion is described by $\theta = c_1 e^{\left(-\xi+\sqrt{\xi^2-1}\right)\omega t} + c_2 e^{\left(-\xi-\sqrt{\xi^2-1}\right)\omega t}$, so the motion decays exponentially, without oscillations.

[34] The motion is described by $\delta\theta = c_1 e^{\left(-\xi+\sqrt{\xi^2+1}\right)\omega t} + c_2 e^{\left(-\xi-\sqrt{\xi^2+1}\right)\omega t}$, so the perturbation (if generic) grows exponentially, without oscillations.

multiplier. The nonlinear equation of motion, Eq. 3.7 is therefore modified as follows:

$$\mathbf{M\ddot{q}} = \mathbf{f}^{int}(\mathbf{q}, \dot{\mathbf{q}}) + \mu\mathbf{f}^{ext}(\mathbf{q}, \dot{\mathbf{q}}). \tag{3.55}$$

The μ parameter is called the *bifurcation parameter*.

3.6.1 Equilibrium Paths

The equilibrium positions of the system $(\mathbf{q}_E, \mu)$ are solutions to the algebraic equation:

$$\mathbf{f}^{int}(\mathbf{q}_E, \mathbf{0}) + \mu\,\mathbf{f}^{ext}(\mathbf{q}_E, \mathbf{0}) = \mathbf{0}, \tag{3.56}$$

obtained by setting $\mathbf{q} = \mathbf{q}_E = \text{cost}$ (and therefore $\dot{\mathbf{q}} = \ddot{\mathbf{q}} = \mathbf{0}$) in the equation of motion (Eq. 3.55). The solutions $\mathbf{q}_E = \mathbf{q}_E(\mu)$,[35] of Eq. 3.56 describe the *equilibrium paths*, represented by curves in the $(\mathbf{q}_E, \mu)$ space. The diagram of these curves is called the *bifurcation diagram*. Among all the curves, particular importance is assumed by that one passing through the origin $O := (0, 0)$, where the system is stable in its natural state, and subject to no forces. This curve takes the name of *fundamental path* $\mathbf{q}_E = \mathbf{q}_E^f(\mu)$, or, for simplicity of notation, $\mathbf{q}_E = \mathbf{q}^f(\mu)$.

Since the computation of all the equilibrium paths of the nonlinear system is generally difficult, or even impossible, the analysis is often limited (a) to determine the fundamental path and (b) to investigate possible bifurcations occurring along it.

It is convenient, for clarity of exposition, to first consider a particular case and then to come back to the general case, according to the following classification:

1. *Structures with trivial fundamental path* (Fig. 3.11a). This is a very frequent case occurring in applications, in which the system remains in equilibrium in its natural configuration, whatever the value of the load multiplier is, namely, $\mathbf{q}^f = \mathbf{0}\ \forall\mu$. In order for this case to occur, $\mathbf{f}^{int}(\mathbf{0}, \mathbf{0}) = \mathbf{f}^{ext}(\mathbf{0}, \mathbf{0}) = \mathbf{0}$ must hold. An example of such a circumstance is offered by the *inextensible* Euler beam, which, under the action of the compression force, is in equilibrium in the straight configuration, whatever the value of the force is (obviously, regardless of stability).

2. *Structures with non-trivial fundamental path* (Fig. 3.11b). This is the general case, in which the system deviates from its natural configuration while the loads are increased, i.e., $\mathbf{q}_E = \mathbf{q}^f(\mu)$. This circumstance occurs when $\mathbf{f}^{int}(\mathbf{0}, \mathbf{0}) = \mathbf{0}$ (i.e., when the reference configuration is also natural), but $\mathbf{f}^{ext}(\mathbf{0}, \mathbf{0}) \neq \mathbf{0}$. An

---

[35] Or, more generally expressed in the parametric form $\mathbf{q}_E = \mathbf{q}_E(s)$, $\mu = \mu(s)$.

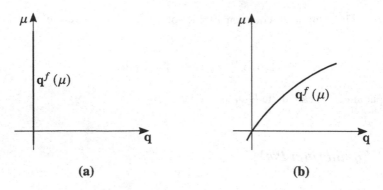

Fig. 3.11   Fundamental path: (**a**) trivial, (**b**) non-trivial

example is offered by the low arch, subject to a vertical load, or still by the Euler beam, but *extensible*, which is shortened under the action of the compression force. Generally, determination of $\mathbf{q}^f(\mu)$ is not simple and requires the use of numerical or perturbation methods. For the moment, this difficulty is ignored, and it is assumed that the non-trivial fundamental path has already been determined.

3.6.2   Bifurcations from a Trivial Path

A bifurcation from the trivial equilibrium path occurs when, as the parameter μ increases from zero, the equilibrium point $\mathbf{q} = \mathbf{0}$ loses its initial stability. To analyze bifurcation, the equation of motion, Eq. 3.55, is linearized around $\mathbf{q} = \mathbf{0}$ and μ arbitrary. Remembering Eq. 3.9, it follows:[36]

$$\mathbf{M}\ddot{\mathbf{q}} + \left(\mathbf{C}_s + \mu\mathbf{C}_g\right)\dot{\mathbf{q}} + \left(\mathbf{K}_e + \mu\mathbf{K}_g\right)\mathbf{q} = \mathbf{0}. \qquad (3.57)$$

By letting the solution as $\mathbf{q} = \mathbf{u}\,\exp(\lambda t)$, an algebraic eigenvalue problem is derived:

$$\left(\lambda^2\mathbf{M} + \lambda\mathbf{C}(\mu) + \mathbf{K}(\mu)\right)\mathbf{u} = \mathbf{0}, \qquad (3.58)$$

similar to that in Eq. 3.21; here, however, the matrices:

$$\mathbf{C}(\mu) := \mathbf{C}_s + \mu\mathbf{C}_g, \qquad \mathbf{K}(\mu) := \mathbf{K}_s + \mu\mathbf{K}_g, \qquad (3.59)$$

---

[36] The nonlinear terms in Eq. 3.9 were ignored and the external forces multiplied by μ.

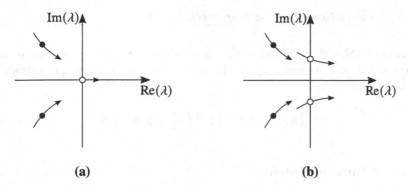

Fig. 3.12 Eigenvalue loci: (**a**) static bifurcation, (**b**) dynamic bifurcation

depend on the bifurcation parameter μ, via their geometric component. Consequently, also the eigenvalues of system depend on μ, i.e., $\lambda = \lambda\,(\mu)$.

The analytic determination of $\lambda\,(\mu)$ is usually impossible, except for systems with few DOF. It is therefore necessary to solve the problem numerically, by choosing a sequence of values of μ, in order to follow the evolution of the eigenvalues. Graphically, as μ increases from zero, the eigenvalues move on the Argand-Gauss plane, describing curves, called *eigenvalue loci*. If one of these eigenvalues crosses from the left the imaginary axis,[37] the equilibrium path loses stability, and a bifurcation occurs (Fig. 3.12). The value $\mu = \mu_c$ for which this occurs is called the *critical value of the bifurcation parameter*, or the *critical load*.

Two types of bifurcation can occur:

- *static bifurcation*, or *divergence*, if $\lambda\,(\mu_c) = 0$, i.e., when an eigenvalue crosses the imaginary axis at the origin of the complex plane (Fig. 3.12a);
- *dynamic bifurcation*, or *flutter* (or Hopf bifurcation), if $\lambda\,(\mu_c) = \pm i\omega$, i.e., when *two* complex conjugate eigenvalues cross the imaginary axis with non-zero imaginary part (Fig. 3.12b).[38]

---

[37] It should be remembered that, when $\mu = 0$, by hypothesis the eigenvalues all lie in the stable half-plane, or, at most, on the imaginary axis.

[38] It will be seen, later on in this book, that (a) the static bifurcation implies the branching of a new equilibrium path originating from $(\mathbf{0}, \mu_c)$, as it happens for the Euler beam; (b) the dynamic bifurcation implies the birth of a periodic motion, emanating from the same point, as it happens for the Beck beam. Such results, however, call for a *nonlinear analysis*, to be performed ahead (Chaps. 4–5, 11–13).

3.6.3 Bifurcations from a Non-trivial Path

The case in which the equilibrium path is non-trivial, i.e., $\mathbf{q}_E = \mathbf{q}^f(\mu)$, is now addressed. In this occurrence, Eq. 3.56 is satisfied for any μ belonging to an interval of interest, i.e.:

$$\mathbf{f}^{int}\left(\mathbf{q}_E^f(\mu), \mathbf{0}\right) + \mu \mathbf{f}^{ext}\left(\mathbf{q}_E^f(\mu), \mathbf{0}\right) = \mathbf{0}. \tag{3.60}$$

Linearized Equation of Motion

To analyze the stability of a non-trivial equilibrium path, $\mathbf{q}^f(\mu)$, it needs *to linearize the motion around a generic equilibrium point $E = E(\mu)$, which evolves with the bifurcation parameter*. Therefore, differently from the case of trivial path, the configuration assumed by the system at the bifurcation (in which stability is lost) it is not a priori known. To solve the problem, a generic μ is fixed, and the current configuration $\mathbf{q}(t; \mu)$ assumed by the system at the time t under the load μ, is expressed as the known equilibrium position $\mathbf{q}^f(\mu)$, plus an unknown *small deviation* $\delta\mathbf{q}(t; \mu)$, i.e.:

$$\mathbf{q}(t; \mu) = \mathbf{q}^f(\mu) + \delta\mathbf{q}(t; \mu). \tag{3.61}$$

Substituting this expression in the nonlinear equation of motion (Eq. 3.55), by accounting for $\dot{\mathbf{q}} = \delta\dot{\mathbf{q}}$ and linearizing in $\delta\mathbf{q}$, it follows:[39]

$$\mathbf{M}\delta\ddot{\mathbf{q}} + \left(\mathbf{C}_s^E(\mu) + \mu\mathbf{C}_g^E(\mu)\right)\delta\dot{\mathbf{q}} + \left(\mathbf{K}_e^E(\mu) + \mu\mathbf{K}_g^E(\mu)\right)\delta\mathbf{q} = \mathbf{0}, \tag{3.62}$$

where the matrices are thus defined:

$$
\begin{aligned}
\mathbf{K}_e^E(\mu) &:= -\left.\frac{\partial \mathbf{f}^{int}}{\partial \mathbf{q}}\right|_{(\mathbf{q}^f(\mu),0)}, \quad
\mathbf{C}_s^E(\mu) := -\left.\frac{\partial \mathbf{f}^{int}}{\partial \dot{\mathbf{q}}}\right|_{(\mathbf{q}^f(\mu),0)}, \\
\mathbf{K}_g^E(\mu) &:= -\left.\frac{\partial \mathbf{f}^{ext}}{\partial \mathbf{q}}\right|_{(\mathbf{q}^f(\mu),0)}, \quad
\mathbf{C}_g^E(\mu) := -\left.\frac{\partial \mathbf{f}^{ext}}{\partial \dot{\mathbf{q}}}\right|_{(\mathbf{q}^f(\mu),0)}.
\end{aligned} \tag{3.63}
$$

Equation 3.62 is also called the *variational equation of motion* around the equilibrium point $\mathbf{q}^f(\mu)$, as it can be formally obtained by operating the variation of Eq. 3.55, according to the rules of variational calculus. It should be observed that, since the matrices are evaluated at a point E that depends on μ, they also depend on μ, not only in their geometric part but also in their structural parts (differently from the case of trivial path).

---

[39] Equation 3.62 should be compared with Eq. 3.15.

Bifurcation Analysis

The stability of the non-trivial fundamental path is governed by the variational equation (Eq. 3.62). Proceeding as in the trivial case, i.e., by letting $\delta \mathbf{q} = \mathbf{u} \exp(\lambda t)$, the eigenvalue problem in Eq. 3.58 is recovered, but with the matrices now defined as:

$$\mathbf{C}(\mu) := \mathbf{C}_s^E(\mu) + \mu \mathbf{C}_g^E(\mu), \quad \mathbf{K}(\mu) := \mathbf{K}_e^E(\mu) + \mu \mathbf{K}_g^E(\mu). \qquad (3.64)$$

All the previous considerations about the eigenvalue loci still hold, in particular the type of bifurcation, static or dynamic.

3.6.4 Bifurcation Mechanisms for Conservative and Circulatory Systems, without or with Damping

The analysis of the stability of the equilibrium, developed in Sect. 3.4, allows discussing the bifurcation mechanisms which occur in important classes of mechanical systems, namely, (a) conservative, (b) circulatory, (c) damped conservative, and (d) damped circulatory.

Conservative Systems

If the system is conservative, the eigenvalue problem in Eq. 3.58 which governs the motion around the equilibrium path reduces to:

$$\left(\lambda^2 \mathbf{M} + \mathbf{K}(\mu) \right) \mathbf{u} = \mathbf{0}, \qquad (3.65)$$

with $\mathbf{K}(\mu)$ symmetric, equal to $\mathbf{K}(\mu) := \mathbf{K}_e + \mu \mathbf{K}_g$ (if the fundamental path is trivial), or $\mathbf{K}(\mu) := \mathbf{K}_e^E(\mu) + \mu \mathbf{K}_g^E(\mu)$ (if it is non-trivial). As seen in the Sect. 3.4.1, the eigenvalues of a conservative system are (a) purely imaginary or (b) real and opposite in sign (Fig. 3.3). When $\mu = 0$, since $\mathbf{K}(0) \equiv \mathbf{K}_e$ is positive definite, the equilibrium is marginally stable, i.e., all the eigenvalues lie on the imaginary axis. As μ increases, the eigenvalues move along the imaginary axis until, for a critical value μ_c of the load, two of them collide at the origin and successively separate, traveling on the real axis in opposite senses (Fig. 3.13a). Upon collision (which corresponds to the positive semi-definiteness of the stiffness matrix), a *static bifurcation* occurs, at which the fundamental path loses stability. This type of bifurcation is called *Hamiltonian divergence*.

Example 3.3 (Hamiltonian System with Single DOF) An elementary example of Hamiltonian static bifurcation is offered by a single DOF system, whose characteristic equation reads:

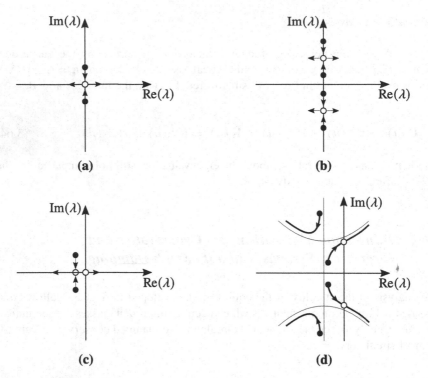

Fig. 3.13 Bifurcation mechanisms of systems: (**a**) conservative, (**b**) circulatory, (**c**) damped conservatives, (**d**) damped circulatory (thin line, singular case; thick line, generic case)

$$\lambda^2 + k_e - \mu k_g = 0, \tag{3.66}$$

where $k_e > 0$, $k_g > 0$ (destabilizing geometric stiffness, since it reduces the total stiffness) and where the (irrelevant) value of the mass has been set equal to 1. From the previous equation, it follows that

$$\lambda_{1,2} = \pm\sqrt{-k_e + \mu k_g}. \tag{3.67}$$

When $\mu < \frac{k_e}{k_g}$ (i.e., the discriminant $\Delta := -k_e + \mu k_g < 0$), the eigenvalues are imaginary; when $\mu = \mu_c := \frac{k_e}{k_g}$ (i.e., $\Delta = 0$), it is $\lambda_{1,2} = 0$; when $\mu > \frac{k_e}{k_g}$ (i.e., $\Delta > 0$), the eigenvalues are real and opposite.   □

Circulatory Systems

If the system is circulatory, the eigenvalue problem is still written as in Eq. 3.65 but with a *non-symmetric* stiffness matrix. As discussed in Sect. 3.4.2, the eigenvalues of a circulatory system are (a) pairs of purely imaginary numbers or (b) quadruplets of complex conjugate numbers, with opposite real parts. When $\mu = 0$, the equilibrium is marginally stable, i.e., all the eigenvalues lie on the imaginary axis. As the load increases, the eigenvalues move, remaining on this axis. If it happens that, when $\mu = \mu_c$, two eigenvalues with positive imaginary parts collide (together with their conjugates ones, having negative imaginary parts), a further increase in the load separates the eigenvalues in the two half-planes, stable and unstable (Fig. 3.13b). At the collision, a *dynamic bifurcation* occurs, at which the fundamental path loses stability. This type of bifurcation is called *flutter bifurcation* (or *circulatory Hopf bifurcation*).

Example 3.4 (Circulatory System with Two DOF) As an elementary example of dynamic circulatory bifurcation, a two DOF system is considered, with $\mathbf{M} = \mathbf{I}$, whose characteristic equation is biquadratic, of the type:

$$\lambda^4 + 2a_2 (\mu) \lambda^2 + a_4 (\mu) = 0, \qquad (3.68)$$

with $2a_2 (\mu) := \text{tr} (\mathbf{K} (\mu))$, $a_4 (\mu) := \det (\mathbf{K} (\mu))$. From this, it follows that:

$$\lambda_{1,2}^2 = -a_2 (\mu) \pm \sqrt{a_2^2 (\mu) - a_4 (\mu)}. \qquad (3.69)$$

For small values of μ, it is $a_2 (\mu) > 0$, $a_4 (\mu) > 0;$[40] moreover, it is $\Delta (\mu) := a_2^2 (\mu) - a_4 (\mu) > 0$, so that both roots $\lambda_{1,2}^2$ are real and negative. Therefore, the four roots $\lambda_{1,2,3,4}$ are imaginary. However, when the load increases, $\Delta (\mu_c) = 0$, so the four roots coincide in pairs. For a further increase of load, $\Delta (\mu) < 0$, so λ_1^2 and λ_2^2 both become complex, and their square roots separate into a quadruplet: two on the right and two on the left of the imaginary axis. □

Damped Conservative Systems

If damping is added to a conservative system, the eigenvalue problem in Eq. 3.58 is complete, but the stiffness matrix is symmetric. As seen in the Sect. 3.4.3, damping transforms marginally stable equilibria in asymptotically stable, by shifting the eigenvalues to the left of the imaginary axis. As the solicitation increases, however, the imaginary part of these complex eigenvalues tends to zero, so a collision occurs on the real axis (Fig. 3.13c). A further load increment separates the real eigenvalues, until one of these ones, at $\mu = \mu_c$, crosses the origin of the complex plane, giving rise to a static bifurcation. The critical load value is identical to that of the undamped system, since, as already observed, damping cannot stabilize an equilibrium position which is statically unstable. At the bifurcation, as in the undamped case, the

---

[40] The trace of a matrix is equal to the sum of its eigenvalues and the determinant is the product.

fundamental path loses stability. The mechanism of bifurcation that entails a single real eigenvalue which crosses zero is called *divergence bifurcation* (of generic type).

Remark 3.7 Damping, when added to a conservative system, shifts the eigenvalue locus to the left of the imaginary axis, without changing its shape. A collision always occurs on the real axis, which *precedes the bifurcation*.

Remark 3.8 Adding damping to a conservative system makes the equilibrium path asymptotically stable, which however bifurcates at the same value of the critical load of the undamped system.

Example 3.5 (Damped Hamiltonian System with Single DOF) A generic static bifurcation occurs, e.g., in a single DOF damped system, whose characteristic equation is:

$$\lambda^2 + c\lambda + k_e - \mu k_g = 0, \tag{3.70}$$

with $k_e > 0$, $k_g > 0$, $c > 0$ and unitary mass; consequently

$$\lambda_{1,2} = -c \pm \sqrt{c^2 - k_e + \mu k_g}. \tag{3.71}$$

By limiting the discussion to sub-critical damping, it is $c^2 < k_e$. Thus, when $\mu = 0$, the eigenvalues are complex conjugate with negative real parts. For small μ, it is $\Delta := c^2 - k_e + \mu k_g < 0$, so the eigenvalues are still complex conjugate, but with imaginary part decreasing in modulus. When $\mu = \frac{k_e - c^2}{k_g}$, it is $\Delta = 0$, so the two eigenvalues collide at the point $(-c, \ 0)$ of the complex plan. For a further increase of μ, it is $\Delta > 0$, so the two eigenvalues separate on the real axis. The largest of the two, successively, crosses zero when $\mu_c := \frac{k_e}{k_g}$, which is independent of c. The same result is obtained if c is super-critical. □

Damped Circulatory Systems

When the system is circulatory and damped, the eigenvalue problem is represented by Eq. 3.58, with non-symmetric stiffness matrix. As seen in the Sect. 3.4.3, the eigenvalues are generally complex, and damping can be beneficial or detrimental. The effect of damping at $\mu = 0$ shifts the eigenvalues of the circulatory system to the left. When the load is increased, two couples of eigenvalues reduce their mutual distances, as in the undamped system; however, *collision is avoided by damping*, which causes the eigenvalues to turn, one pair to the left, the other on the right (Fig. 3.13d). As μ further increases, the pair closest to the imaginary axis crosses it at $\mu = \mu_c$ (critical value which can be smaller or greater than the critical value of the circulatory system). Thus, a *dynamic bifurcation* occurs, at which the fundamental path loses stability. This bifurcation mechanism, in which only one pair of complex eigenvalues crosses the imaginary axis, is called (generic) *Hopf* bifurcation.

Example 3.6 (Damped Circulatory System with Two DOF) To exemplify a generic dynamic bifurcation, a two DOF damped system is considered, with $\mathbf{M} = \mathbf{I}$. The characteristic equation is a complete fourth degree equation, not easy to discuss. Therefore, here a particular case is considered, which admits a simple analytical solution, namely:

$$\det \begin{bmatrix} \lambda^2 + \lambda c + k_e & \frac{1}{2}\mu \\ -\frac{1}{2}\mu & \lambda^2 + \lambda c + 2k_e \end{bmatrix} = 0. \tag{3.72}$$

It follows from considering a diagonal damping matrix, a diagonal structural stiffness matrix, and an antisymmetric geometric stiffness matrix. Here $c > 0$ is a (sub-critical) damping coefficient, $k_e > 0$ an elastic stiffness coefficient, and μ the amplitude of the nonconservative forces. The equation admits the roots:

$$\lambda_{1,2,3,4} = -\frac{c}{2} \pm \frac{1}{2}\sqrt{c^2 - 6k_e \pm 2\sqrt{k_e^2 - \mu^2}}, \tag{3.73}$$

where the positive and negative signs are taken in four different combinations. When $\mu = k_e$, and for any c, the collision occurs between two pairs of eigenvalues. If $c = 0$, the collision occurs on the imaginary axis; hence, $\mu_c = k_e$ is the critical load of the undamped system. If, however, $c > 0$, the two pairs of eigenvalues, at the collision, are to the left of the imaginary axis, so the load can still increase, until one of the roots crosses the axis (thin lines in Fig. 3.13d). In this example, therefore, damping has a beneficial effect, because it increases the critical load of the circulatory system. The collision among the eigenvalues of the damped system, however, is a rare case. A small perturbation of the damping matrix taken here, indeed, produces the loci represented with thick lines in Fig. 3.13d. □

References

1. Bolotin, V.V.: Nonconservative problems of the theory of elastic stability. Macmillan, London (1963)
2. Dyn, C.L.: Stability theory and its applications to structural mechanics. Noordhoff, Leyden (1974)
3. Hahn, W.: Stability of motion. Springer, Berlin (1967)
4. Iooss, G., Joseph, D.D.: Elementary stability and bifurcation theory. Springer, New York (1980)
5. Lacarbonara, W.: Nonlinear structural mechanics: theory, dynamical phenomena and modeling. Springer, New York (2013)
6. Leipholz, H.H.: Stability of elastic systems. Sijthoff & Noordhoff, Alphen aan den Rijn (1980)
7. Pignataro, M., Rizzi, N., Luongo, A.: Stability, bifurcation and postcritical behaviour of elastic structures. Elsevier, Amsterdam (1990)
8. Seyranian, A.P., Mailybaev, A.A.: Multiparameter stability theory with mechanical applications. World Scientific, Singapore (2003)
9. Troger, H., Steindl, A.: Nonlinear stability and bifurcation theory: an introduction for engineers and applied scientists. Springer, Wien (1991)
10. Ziegler, H.: Principles of structural stability. Blaisdell Publishing Co, Waltham (1968)

Chapter 4
Buckling and Postbuckling of Conservative Systems

4.1 Introduction

Conservative mechanical systems, i.e., elastic structures subject to potential loads (mainly dead loads), usually occupy a large space in treatises on bifurcation. Many books are entirely devoted to them. The relevant scientific framework, called *Buckling (and Post-Buckling) Theory*, is developed in an exclusively static field, since dynamics, as it was noticed, does not play any role on their stability. In dealing with static systems, one first needs to evaluate the *fundamental equilibrium path*, i.e., the sequence of the equilibrium states visited by the system when a parameter is increased, starting from zero. This task can be accomplished either numerically, through the so-called *continuation methods*, or analytically (or semi-analytically) by *perturbation* (or asymptotic) *methods*. While the former methods are practically exact, they require to re-perform the analysis for any set of parameters (e.g., for any imperfections); the latter methods, instead, are approximate, but provide closed form formulas, which permit to perform parametric analyses. The asymptotic approach will be systematically followed in this book, whatever a nonlinear system has to be studied.

Once the path has been determined, bifurcations occurring along it are examined. The analysis calls for solving: (a) an eigenvalue problem in the bifurcation parameter, and (b) a sequence of linear equations, all governed by the same singular operator. Bifurcations occur at *branch points*, denoting the existence of more paths crossing there, or at *fold points*, denoting the existence of limit loads. The perturbation method allows: (a) to build up asymptotic expressions for the branched paths, (b) to evaluate the role of imperfections, causing a lowering of the maximum load reachable, and (c) to construct non-trivial fundamental paths, by extrapolating them from a known equilibrium point, thus estimating the limit load.

In this chapter, a classification of the equilibria is performed, and a sketch of the continuation methods is given. Perturbation methods for discrete systems

A. Luongo et al., *Stability and Bifurcation of Structures*,
https://doi.org/10.1007/978-3-031-27572-2_4

are explained in detail. The (more difficult) case of systems suffering precritical deformation is also briefly treated.

4.2 Static Analysis of Conservative Systems

The nonlinear equilibrium equation of an undamped elastic system, possessing n degrees of freedom (DOF), subject to positional *conservative forces* affected by a single multiplier μ, can be deduced from the stationary condition of the *total potential energy* (TPE):

$$\Pi\,(\mathbf{q}) := U\,(\mathbf{q}) + \mu V\,(\mathbf{q})\,. \tag{4.1}$$

Here, $U\,(\mathbf{q})$ is the (positive definite) elastic potential energy of system, $V\,(\mathbf{q})$ is the potential energy of loads, and $\mathbf{q}$ is the vector of the Lagrangian parameters.[1] The stationary condition of Π is written as:

$$\delta\Pi\,(\mathbf{q}) = \left(\left[\frac{\partial U}{\partial \mathbf{q}}\right]^{T} + \mu\left[\frac{\partial V}{\partial \mathbf{q}}\right]^{T}\right)\delta\mathbf{q} = 0, \qquad \forall\delta\mathbf{q}, \tag{4.2}$$

which implies:

$$\frac{\partial U}{\partial \mathbf{q}} + \mu\frac{\partial V}{\partial \mathbf{q}} = \mathbf{0}. \tag{4.3}$$

This is the *nonlinear equilibrium equation of the system*. Given that $\mathbf{f}^{el}\,(\mathbf{q}) := -\frac{\partial U}{\partial \mathbf{q}}$ is the internal elastic force and $\mathbf{f}^{ext}\,(\mathbf{q}) = -\frac{\partial V}{\partial \mathbf{q}}$ is the positional external force, Eq. 4.3 is equivalent to:[2]

$$\mathbf{f}^{el}\,(\mathbf{q}) + \mu\mathbf{f}^{ext}\,(\mathbf{q}) = \mathbf{0}. \tag{4.4}$$

---

[1] The notation $\Pi = U - W$ will also be occasionally adopted in the book. In it, and to within an inessential constant, $W := -V$ represents the work spent by the force $\mathbf{f}^{ext}$ to bring the system from $\mathbf{q} = \mathbf{0}$ to the current configuration. Indeed, by taking into account that $\mathbf{f}^{ext} = -\frac{\partial V}{\partial \mathbf{q}}$, it follows:

$$W\,(\mathbf{q}) = \int_{0}^{\mathbf{q}} \mathbf{f}^{ext}\,(\tilde{\mathbf{q}})\,d\tilde{\mathbf{q}} = -\int_{0}^{\mathbf{q}} \frac{\partial V}{\partial \tilde{\mathbf{q}}}\,(\tilde{\mathbf{q}})\,d\tilde{\mathbf{q}} = -\,(V\,(\mathbf{q}) - V\,(\mathbf{0}))\,.$$

[2] Note that Eq. 4.4 has been changed in sign, like Eqs. 4.57, 4.61.

Alternatively, this can be written via a *direct equilibrium* approach, i.e., particularizing the cardinal equation of dynamics, Eq. 3.55, to statics.

Equilibrium Paths and Bifurcation
The (generally multiple) solutions of the nonlinear Eq. 4.4 describe the *equilibrium paths* of the system (e.g., in the parametric form $\mathbf{q} = \mathbf{q}(s)$, $\mu = \mu(s)$, where s is a parameter). The object of the static bifurcation analysis consists in the description of the neighborhoods of the *points of singularity* occurring on these paths. To this end, it needs (a) to characterize the points of singularity and (b), in case of branching points, to construct the paths emanating from these latter. Section 4.3 discusses the first aspect. Regarding the construction of the paths, this can be performed either numerically (to obtain an exact path, Sect. 4.4) or analytically (to build up an approximated path, Sect. 4.5). Here only a hint is given about the numerical approach, while the analytical procedure is described in detail.

4.3 Classification of the Equilibrium Points

An equilibrium point is classified as (a) *regular* point, (b) *limit point*, or (c) *branching point*, according to the local character of the equilibrium path. Here, the classification is introduced for systems with a single DOF; then, it is (shortly) generalized to multi-DOF systems [6].

Single DOF Systems
The equilibrium Eq. 4.4, in case of single DOF system, reads:

$$f^{el}(q) + \mu f^{ext}(q) = 0. \tag{4.5}$$

Here, the dependence on the load multiplier is linear; however, for the sake of writing conciseness, reference will be made to a more general equation:

$$f(q, \mu) = 0, \tag{4.6}$$

where $f(\cdot)$ is generally nonlinear in its arguments.

A *known* equilibrium point $E := (q_E, \mu_E)$ is considered, which is assumed to be *not isolated*. Hence, there exists at least an (unknown) path, of parametric equations $q = q(s)$, $\mu = \mu(s)$, which passes through E. The following identity holds:

$$f(q(s), \mu(s)) = 0, \qquad \forall s. \tag{4.7}$$

By repeatedly differentiating the identity with respect to the parameter, the following perturbation equations are derived:

$$f_{,q}\,\dot{q} + f_{,\mu}\,\dot{\mu} = 0, \tag{4.8a}$$

$$f_{,q}\,\ddot{q} + f_{,\mu}\,\ddot{\mu} + 2f_{,q\mu}\,\dot{q}\,\dot{\mu} + f_{,qq}\,\dot{q}^2 + f_{,\mu\mu}\,\dot{\mu}^2 = 0, \tag{4.8b}$$

$$\cdots, \tag{4.8c}$$

where the dot indicates s-differentiation. The previous equations, evaluated at the equilibrium point E, become:

$$f_{,q}^E\,\dot{q}_E + f_{,\mu}^E\,\dot{\mu}_E = 0, \tag{4.9a}$$

$$f_{,q}^E\,\ddot{q}_E + f_{,\mu}^E\,\ddot{\mu}_E + 2f_{,q\mu}^E\,\dot{q}_E\,\dot{\mu}_E + f_{,qq}^E\,\dot{q}_E^2 + f_{,\mu\mu}^E\,\dot{\mu}_E^2 = 0. \tag{4.9b}$$

From the first perturbation equation, information about the tangent to the unknown path at the point E is obtained. By assuming the path can be expressed in the Cartesian form $q = q\,(\mu)$, the tangent is written as:

$$\left.\frac{dq}{d\mu}\right|_E = \frac{\left.\frac{dq}{ds}\right|_E}{\left.\frac{d\mu}{ds}\right|_E} = \frac{\dot{q}_E}{\dot{\mu}_E} = -\frac{f_{,\mu}^E}{f_{,q}^E}. \tag{4.10}$$

Similarly, if the inverse Cartesian form $\mu = \mu\,(q)$ exists, the tangent is written as:

$$\left.\frac{d\mu}{dq}\right|_E = -\frac{f_{,q}^E}{f_{,\mu}^E}. \tag{4.11}$$

Below is the classification of the equilibrium point (Fig. 4.1).

1. The equilibrium E is a *regular* point if $f_{,q}^E \neq 0$ and $f_{,\mu}^E \neq 0$ (Fig. 4.1a);[3] since the tangent is there unique, a single path passes through E.
2. The equilibrium E is a *limit* point (with respect to the load[4]) if $f_{,q}^E = 0$ and $f_{,\mu}^E \neq 0$, i.e., if $\dfrac{d\mu}{dq} = 0$ (Fig. 4.1b); a single path passes through this point, having a tangent orthogonal to the load axis μ; the path is locally expressed as $\mu = \mu\,(q)$, i.e., *q is the control parameter.*
3. The equilibrium E is *limit point* (with respect to the displacement[5]) if $f_{,q}^E \neq 0$ e $f_{,\mu}^E = 0$, i.e., if $\dfrac{dq}{d\mu} = 0$ (Fig. 4.1c); a single path passes through this point, having a tangent orthogonal to the axis of the displacement q; the path is expressed locally as $q = q\,(\mu)$, i.e., *μ is the control parameter.*

---

[3] It is worth noticing, for what follows, that the rank of the matrix $\left[\,f_{,q}^E \,\middle|\, f_{,\mu}^E\,\right]$ is maximum.

[4] These points, in the technical literature, are sometimes improperly said of *snap-through.*

[5] These points, in the technical literature, are also called *snap-back*, as the evolution of the response is reversed as the load increases.

Fig. 4.1   Classification of
equilibria: (**a**) regular point,
(**b**) limit point with respect to
load, (**c**) limit point with
respect to displacement, (**d**)
branch point

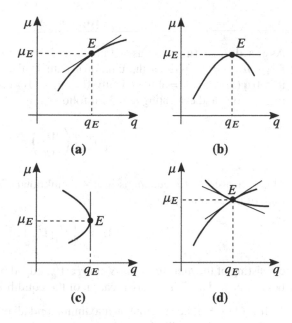

4. The equilibrium E is called a *branch point* if $f_{,q}^E = 0$ and $f_{,\mu}^E = 0$ (Fig. 4.1d).[6]
   In this case of greater degeneracy, *the tangent to the path is indeterminate to this order*, since it has the form $\frac{0}{0}$. Since the linear approximation is no more sufficient to determine it, it needs to resort to the perturbation equation involving the second derivatives, Eq. 4.9b. By accounting for the vanishing terms, this reduces to:

$$f_{,q}^E \ddot{q}_E + f_{,\mu}^E \ddot{\mu}_E + 2f_{,q\mu}^E \dot{q}_E \dot{\mu}_E + f_{,qq}^E \dot{q}_E^2 + f_{,\mu\mu}^E \dot{\mu}_E^2 = 0, \qquad (4.12)$$

i.e., to a quadratic equation in the ratio $\left.\frac{dq}{d\mu}\right|_E = \frac{\dot{q}_E}{\dot{\mu}_E}$ (or its inverse). Hence, *two paths pass through the point E*.[7] In the further degenerate case, where the two roots of the quadratic equations are coincident, the branch point is a *cusp*, that is, the two paths branch off from the same tangent line. In the maximally degenerate case, in which all the coefficients of the quadratic equation vanish, the tangents are determined by the cubic approximation. which returns *three* tangents to as many paths.

Multi-DOF Systems

The previous results are now generalized to a n DOF system, of equations (more details can be found in [6]):

---

[6] The rank of matrix $\left[\, f_{,q}^E \,\middle|\, f_{,\mu}^E \,\right]$ is no longer maximum.

[7] If $f(\cdot)$ is linear in μ, as in Eq. 4.5, since $f_{,\mu\mu}^E = 0$, one of the two paths has tangent orthogonal to the q axis.

$$\mathbf{f}(\mathbf{q}, \mu) = \mathbf{0}. \tag{4.13}$$

Assumed $E := (\mathbf{q}_E, \mu_E)$ as a known and not isolated equilibrium point, it is $\mathbf{f}(\mathbf{q}_E, \mu_E) = \mathbf{0}$. Written the unknown path in the form $\mathbf{q} = \mathbf{q}(s)$, $\mu = \mu(s)$, it is $\mathbf{f}(\mathbf{q}(s), \mu(s)) = \mathbf{0}$, identically for any s. By differentiating the identity with respect to s, and evaluating it at E, it follows:

$$\left[\mathbf{f}_{,\mathbf{q}}^{E} \,\middle|\, \mathbf{f}_{,\mu}^{E} \right] \begin{pmatrix} \dot{\mathbf{q}}_E \\ \dot{\mu}_E \end{pmatrix} = \mathbf{0}. \tag{4.14}$$

This is a system of n equations in $n + 1$ unknowns.[8] The solutions depend on the rank of the Jacobian matrix:

$$\mathbf{J}_E := \left[\mathbf{f}_{,\mathbf{q}}^{E} \,\middle|\, \mathbf{f}_{,\mu}^{E} \right], \tag{4.15}$$

consisting of the *tangent stiffness matrix*, $\mathbf{f}_{,\mathbf{q}}^{E}$, edged by the *velocity load vector* $\mathbf{f}_{,\mu}^{E}$, both evaluated at E. The classification of the equilibrium point follows.

1. If $\mathrm{rk}(\mathbf{J}_E) = n$, i.e., the rank is maximum, and all minors of $\mathbf{J}_E$ are non-zero, then E is a regular equilibrium point;[9]
2. If $\mathrm{rk}(\mathbf{J}_E) = n$, i.e., the rank is maximum, and a minor of order n is zero, then E is a limit equilibrium point with respect to the variable canceled to extract the null minor;[10]
3. If $\mathrm{rk}(\mathbf{J}_E) < n$, i.e., the rank is not maximum, i.e., *all* the minors of order n are null, then E is a branch point.[11]

Remark 4.1 The singularity of a limit point depends on a bad choice of the parameter used to describe the curve in Cartesian form; changing the parameter, the singularity disappears. Differently, a branch point is a singular point, regardless of the choice of the parameter.

---

[8] The redundancy of the unknowns depends on the arbitrariness of s, which could be removed with a normalization condition. However this it is inessential, because the tangent to the path is determined by the vector $\frac{\dot{\mathbf{q}}_E}{\dot{\mu}_E}$.

[9] Indeed, the components of $\frac{\dot{\mathbf{q}}_E}{\dot{\mu}_E}$ are determined and all non-zero.

[10] Indeed, the solution requires that one of the unknowns is null. If the zero minors are more than one, then E is a limit point with respect to more variables.

[11] Indeed, the ratios $\frac{\dot{\mathbf{q}}_E}{\dot{\mu}_E}$ are indeterminate.

4.4 Numerical Continuation Methods

To determine one or more equilibrium paths for the nonlinear vector Eq. 4.13, numerical algorithms are often used. Here a sketch of the methods is given, by first referring to a system with a single DOF and then by extending them to systems with several DOF [4].

4.4.1 Newton-Raphson Method

The equilibrium equation, Eq. 4.6, is considered, for which a position q_E is sought (if it exists, being not necessarily unique), in equilibrium with an assigned value $\bar{\mu}$ of the load. The equation to solve is $f(q) := f(q; \bar{\mu}) = 0$, in which the known load will be understood. This equation, as known from the numerical analysis, can be solved by the *Newton's method* (also called of Newton-Raphson, or *method of tangents*).

From a geometric point of view, solving the equation $f(q) = 0$ means to find an intersection q_E of the graph of the curve with the q axis (Fig. 4.2a). The method is based on the *predictor and corrector* approach, according to which an initially assumed "tentative" solution (predictor phase) is iteratively modified (corrector phase) until convergence is reached. The method develops as follows. Let q_k be a tentative solution; generally, it is $f(q_k) \neq 0$, called the *residual* of the equation (representing the imbalance of forces). If the curve $f(q)$ is approximated by its tangent at the point $(q_k, f(q_k))$, an improved q_{k+1} solution is found at the intersection of the tangent and the q axis. The ordinate $f(q_{k+1}) \neq 0$ constitutes the new residual of the equation, and the operation can be restarted. At the new point $(q_{k+1}, f(q_{k+1}))$, the curve is approximated by a new tangent, and a new approximation q_{k+2} is found, and so on. The iteration is truncated upon one or

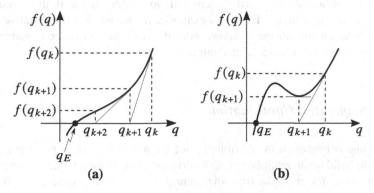

Fig. 4.2 Newton-Raphson method: (**a**) convergent, (**b**) divergent; iterations $q_k, q_{k+1}, \cdots$; solution q_E

both the convergence indicators are satisfied, namely: (a) the residual is smaller than a prefixed tolerance, i.e., $|f(q_k)| \leq tol_1$, and/or, (b) when two successive approximations are sufficiently close each other, i.e., $|q_{k+1} - q_k| \leq tol_2$.

To implement the algorithm, it needs to linearize the function around the initial point q_k; by using the Taylor series, it follows:

$$f(q_{k+1}) = f(q_k) + \left.\frac{\partial f}{\partial q}\right|_{q_k} (q_{k+1} - q_k). \qquad (4.16)$$

By requiring that $f(q_{k+1}) = 0$, the next approximation reads:

$$q_{k+1} = q_k - \left[\left.\frac{\partial f}{\partial q}\right|_{q_k}\right]^{-1} f(q_k), \qquad k = 0, 1, \cdots, \qquad (4.17)$$

with q_0 the initial choice ("initial guess" or "starting point"). This formula constitutes the "rule" for the successive iterations.

Generalizing to systems with several DOF (for which the previous geometric interpretation is no longer possible), the iteration is written:

$$\mathbf{q}_{k+1} = \mathbf{q}_k - \mathbf{J}^{-1}(\mathbf{q}_k)\mathbf{f}(\mathbf{q}_k), \qquad k = 0, 1, \cdots, \qquad (4.18)$$

where:

$$\mathbf{J}(\mathbf{q}_k) := \left.\frac{\partial \mathbf{f}}{\partial \mathbf{q}}\right|_{\mathbf{q}_k} \equiv \left[\left.\frac{\partial f_i}{\partial q_j}\right|_{\mathbf{q}_k}\right] \qquad (4.19)$$

is the Jacobian matrix evaluated at $\mathbf{q}_k$.

Remark 4.2 When the Jacobian matrix is singular (or quasi-singular), it cannot be inverted, and the iteration of Eq. 4.18 fails. With reference to Fig. 4.2b, the tangent becomes horizontal at q_{k+1}, and therefore it no longer intersects the q axis (or intersects it near to infinity). From this example, it emerges that the tentative value must be close enough to the unknown equilibrium point, in order to reduce the possibility that this drawback is encountered.

4.4.2 Sequential Continuation

Usually, one is interested in determining not a single equilibrium point at a given load but to build up an equilibrium path corresponding to load values belonging to a certain range, for example, from the natural state to a final state, $\mu \in [0, \mu_F]$. To this end, a *continuation method* is used, so-called since it "extends" a solution found for a specific $\bar{\mu}$ value to close values of the load. The simplest, among these methods, is that of the *sequential continuation*.

According to it, the interval of interest of the load is subdivided in sufficiently small sub-intervals, called *load steps*, of endpoints μ^i, with $i = 1, 2, \cdots$. Initially, one can choose to divide the interval into equal steps, except to use more refined methods in which the step is adaptively modified, according to the difficulty encountered. The basic idea is the following. By means of the Newton-Raphson method, an equilibrium point is obtained for a certain load value μ^i. The load is then increased, by passing from μ^i to μ^{i+1}. In this way, the equilibrium point obtained in the previous step is not anymore a solution to the problem; however, if the load step is small enough, it is to be expected that the equilibrium point changes slightly, so that the solution obtained at previous step can be considered a good starting point for the current step. The iteration is written as:

$$q_{k+1}^{i+1} = q_k^{i+1} - J^{-1}\left(q_k^{i+1}, \mu^{i+1}\right) f\left(q_k^{i+1}, \mu^{i+1}\right), \qquad k = 0, 1, \cdots, \qquad (4.20)$$

where the apex, indicating the load step, is fixed at $i + 1$, while the subscript, indicating the iteration in the same step, is variable. The starting point is chosen as $q_0^{i+1} = q^i$, with q^i the equilibrium point at which the iteration corresponding to μ^i has been truncated.

Sequential Continuation Failure

The method, very simple, suffers from the drawback inherent in the method of the tangents, that is, it fails in proximity to the limit points. Thus, if one tries to plot a $q(\mu)$ curve, and this curve possesses a limit point μ_l, he cannot find any close solutions when $\mu > \mu_l$. Under this circumstance, one could reduce the load step, until a good approximation of the limit point is obtained, but the *limit point cannot be crossed*. This is the result of a bad representation of the curve, since $q=q(\mu)$ *is not a single-valued* (or monodrome) *function*, and therefore μ is not an appropriate control parameter. In single DOF systems, since the inverse function $\mu = \mu(q)$, close to the limit point, is a a single-valued function, it is sufficient to exchange q and μ, controlling the process in terms of displacement, rather than of load. In n DOF systems, a displacement can be taken as a control parameter and the remaining $n - 1$ displacements and the load described as a function of the chosen displacement.[12]

Figure 4.3 shows an example of a two DOF system, which, in the range of interest, exhibits a limit point with respect to the load and one with respect to q_2, while it has no limit points with respect to q_1. The latter can therefore be chosen as a control parameter and the curve described as $q_2 = q_2(q_1)$, $\mu = \mu(q_1)$.

---

[12] The change of the control parameter can, of course, be programmed whenever the Jacobian matrix becomes quasi-singular.

Fig. 4.3 An example of equilibrium path (and its projections on the coordinated planes) having a limit point with respect to load ○ and one with respect to a displacement ●

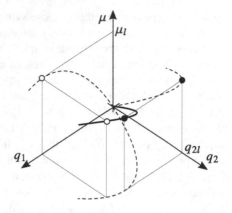

Remark 4.3 A circumstance similar to that just discussed is encountered in the experimental construction of the mechanical response curve of a material exhibiting a softening branch. For example, in the tensile test of steel, or compression test of concrete, upon reaching of the rupture load. The softening branch can only be visited if the experiment is displacement- and not load-controlled.

4.4.3 Arclength Method

The failure of the sequential continuation at limit points suggests the adoption of a control parameter which indefinitely grows during the loading process. However, this parameter cannot be chosen a priori among the n displacements, nor in the load. It is possible, otherwise, to introduce a $(n + 2)$th parameter s, having the meaning of *curvilinear abscissa*, or assimilated to it, which, by definition, is monotonically increasing along the curve, regardless of its shape.[13] The unknown path is therefore described, in parametric form, as $\mathbf{q} = \mathbf{q}(s)$, $\mu = \mu(s)$, with $s \in [0, s_F]$.

The introduction of the additional parameter requires the introduction of a further equation, giving a geometric meaning to the parameter itself. The system of equations to be solved is therefore written as:

$$\mathbf{f}(\mathbf{q}(s), \mu(s)) = 0, \tag{4.21a}$$

$$g(\mathbf{q}(s), \mu(s); \Delta s) = 0, \tag{4.21b}$$

where $g(\cdot)$ is a "constraint" scalar function and Δs is the measure of the "step" with which the points on the unknown curve are searched. After having determined an equilibrium point $E(s) := (\mathbf{q}_E(s), \mu_E(s))$, a new equilibrium point

---

[13] An extreme example is given by a curve several times knotted on itself, like a ball of yarn.

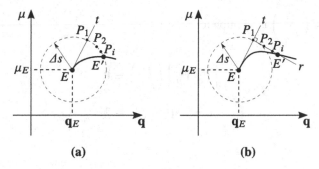

Fig. 4.4 Continuation methods: (a) arclength, (b) pseudo-arclength

$E\,(s + \Delta s) := (\mathbf{q}_E\,(s + \Delta s)\,, \mu_E\,(s + \Delta s))$ is sought, such as to fulfill equilibrium and the constraint. The research is carried out by performing iterations $(\mathbf{q}_k\,(s + \Delta s)\,, \mu_k\,(s + \Delta s))$ within the same step Δs, as in the sequential continuation. The iterations are written as (argument $s + \Delta s$ omitted):

$$\begin{pmatrix} \mathbf{q}_{k+1} \\ \mu_{k+1} \end{pmatrix} = \begin{pmatrix} \mathbf{q}_k \\ \mu_k \end{pmatrix} - \begin{bmatrix} \mathbf{f}_{,\mathbf{q}} & \mathbf{f}_{,\mu} \\ g_{,\mathbf{q}} & g_{,\mu} \end{bmatrix}^{-1}_{(\mathbf{q}_k, \mu_k)} \begin{pmatrix} \mathbf{f}(\mathbf{q}_k, \mu_k) \\ g(\mathbf{q}_k, \mu_k) \end{pmatrix}, \qquad k = 0, 1, \cdots . \quad (4.22)$$

The choice of the $g\,(\cdot)$ function can be done in different ways, two of which are mentioned here (Fig. 4.4).

1. *Arclength method.* The iteration is performed in such a way that the points $(\mathbf{q}_k\,(s + \Delta s)\,, \mu_k\,(s + \Delta s))$ all lie, in the same step Δs, on the circle of center $E\,(s)$ and radius Δs (Fig. 4.4a); hence:

$$g\,(\mathbf{q}\,(s)\,, \mu\,(s)\,; \Delta s) := (\mathbf{q} - \mathbf{q}_E)^2 + (\mu - \mu_E)^2 - \Delta s^2. \quad (4.23)$$

2. *Method of the pseudo-arclength.* The iteration is performed in such a way that the points $(\mathbf{q}_k\,(s + \Delta s)\,, \mu_k\,(s + \Delta s))$ all lie, in the same step Δs, on the line r, tangent to the circle of center $E\,(s)$ and radius Δs, at the point in which it is intersected by the line t, tangent to the curve at $E\,(s)$ (Fig. 4.4b). The equation of the line r is obtained by imposing that the scalar product between the vector $\left(\mathbf{q} - \mathbf{q}_E \quad \mu - \mu_E\right)$ and the tangent vector $\mathbf{t}$ is equal to Δs, i.e.:[14]

$$g\,(\mathbf{q}\,(s)\,, \mu\,(s)\,; \Delta s) := \left(\mathbf{q} - \mathbf{q}_E \quad \mu - \mu_E\right) \begin{pmatrix} \dot{\mathbf{q}}_E \\ \dot{\mu}_E \end{pmatrix} - \Delta s, \quad (4.24)$$

---

[14] With reference to Fig. 4.4b, the projections on $\overrightarrow{t}$ of the $\overrightarrow{E\,P_i}$ vectors are all equal to Δs.

where $\mathbf{t} := \begin{pmatrix} \dot{\mathbf{q}}_E \\ \dot{\mu}_E \end{pmatrix}$ is solution to:

$$\left[\mathbf{f}^E_{,\mathbf{q}} \,\middle|\, \mathbf{f}^E_{,\mu} \right] \begin{pmatrix} \dot{\mathbf{q}}_E \\ \dot{\mu}_E \end{pmatrix} = \mathbf{0}, \tag{4.25a}$$

$$\dot{\mathbf{q}}^2_E + \dot{\mu}^2_E = 1. \tag{4.25b}$$

4.5 Asymptotic Analysis of Bifurcation from Trivial Path

An analytical procedure for the construction of the paths, alternative to the numerical approach, is offered by the asymptotic (or perturbation) analysis [1–3, 5–9]. This consists of a *polynomial extrapolation of a path* from a known equilibrium point. Since the solution is approximate, the method gives technically valid results in a sufficiently small neighborhood of the extrapolation point. On the other hand, being the method analytic, it represents the solution in parametric form, which does not require the re-execution of the numerical calculation for each choice of the parameter set.

The more recurrent case in applications concerns *bifurcation from a trivial path*, which is first addressed here. This calls for extrapolation from a branch point whose coordinates are known in terms of displacement ($\mathbf{q} = \mathbf{0}$) and unknown only in terms of load ($\mu = \mu_c$). Later on (Sect. 4.8), more general cases will be analyzed, i.e., systems suffering from precritical deformations, for which either limit points or branching points exist, whose coordinates are all unknown.

It is therefore assumed that the equilibrium equation, Eq. 4.4, admits the trivial fundamental solution $\mathbf{q}^f = \mathbf{0}$, $\forall \mu$, i.e., it is $\mathbf{f}^{el}(\mathbf{0}) = \mathbf{f}^{ext}(\mathbf{0}) = \mathbf{0}$. Expanding in MacLaurin series, forces are written as:

$$\mathbf{f}^\alpha = \mathbf{f}^\alpha(\mathbf{0}) + \left.\frac{\partial \mathbf{f}^\alpha}{\partial \mathbf{q}}\right|_0 \mathbf{q} + \mathbf{n}^\alpha(\mathbf{q}), \qquad \alpha = el, ext. \tag{4.26}$$

By introducing the (symmetric) elastic and geometric stiffness matrices (evaluated at $\mathbf{q} = \mathbf{0}$):

$$\mathbf{K}_e := -\left.\frac{\partial \mathbf{f}^{el}}{\partial \mathbf{q}}\right|_0 = \left.\frac{\partial^2 U}{\partial \mathbf{q}^2}\right|_0, \quad \mathbf{K}_g := -\left.\frac{\partial \mathbf{f}^{ext}}{\partial \mathbf{q}}\right|_0 = \left.\frac{\partial^2 V}{\partial \mathbf{q}^2}\right|_0, \tag{4.27}$$

as well as the column vectors of polynomial nonlinearities (sum of quadratic, cubic, ..., homogeneous forms):

$$\mathbf{n}^\alpha(\mathbf{q}) = \mathbf{n}^\alpha_2(\mathbf{q}, \mathbf{q}) + \mathbf{n}^\alpha_3(\mathbf{q}, \mathbf{q}, \mathbf{q}) + \cdots, \qquad \alpha = el, ext, \tag{4.28}$$

the equilibrium equation assumes the form:

$$\left(\mathbf{K}_e + \mu \mathbf{K}_g\right) \mathbf{q} = \mathbf{n}^{el}\left(\mathbf{q}\right) + \mu \mathbf{n}^{ext}\left(\mathbf{q}\right). \tag{4.29}$$

Remark 4.4 The elastic and geometric stiffness matrices are, respectively, the *Hessian* matrices of the elastic and load potential energies. The total stiffness matrix $\mathbf{K} := \mathbf{K}_e + \mu \mathbf{K}_g$ is the *Hessian matrix of total potential energy*, $\mathbf{K} := \left.\frac{\partial^2 \Pi}{\partial \mathbf{q}^2}\right|_0$. By virtue of the Lagrange-Dirichlet theorem, $\mathbf{K}$ decides the stability of the trivial equilibrium path.

4.5.1 Linear Stability Analysis

The trivial fundamental path, according to the *energy criterion*, is stable for all values of μ for which $\mathbf{K}\left(\mu\right) := \mathbf{K}_e + \mu\,\mathbf{K}_g$ is positive definite. It is incipiently unstable for the critical value of the load, μ_c, for which $\mathbf{K}\left(\mu_c\right)$ is undefined. It is unstable for $\mu > \mu_c$. For $\mu = \mu_c$, therefore, a bifurcation of equilibrium occurs.

To determine μ_c it is necessary to require that $\mathbf{K}\left(\mu_c\right)$ has a zero eigenvalue,[15] i.e., that there exists a nontrivial solution to the equation:

$$\left(\mathbf{K}_e + \mu\,\mathbf{K}_g\right) \mathbf{u} = \mathbf{0}. \tag{4.30}$$

The eigenvalue problem, Eq. 4.30, admits n real roots $\mu_1 < \mu_2 < \cdots < \mu_n$ (assumed positive[16] and distinct); the smallest one is the critical load μ_c. The associated eigenvector, $\mathbf{u}_c$, is the *critical mode*, whose mechanical meaning will be made immediately clear.

Remark 4.5 Equation 4.30 appears in the form of a (static) eigenvalue problem in the unknown μ, not to be confused, however, with the (dynamic) eigenvalue problem in the unknown λ (characteristic exponent), which was discussed in Chap. 3 in the context of the dynamic stability criterion.

Remark 4.6 The energy criterion of stability gives the same result than the dynamic criterion. If, indeed, $\lambda = 0$ is set in Eq. 3.21, this latter coincides with Eq. 4.30, after having taken into account that $\mathbf{K} = \mathbf{K}_e + \mu\,\mathbf{K}_g$.

---

[15] The case of multiple bifurcations is excluded. It will be examined further on, in Sect. 5.3.2, with reference to an example.

[16] For example, if the Euler beam is taut, the roots are all negative.

Adjacent Equilibrium Criterion

Equation 4.30 is susceptible to a remarkable mechanical interpretation. Since it is the linear part of Eq. 4.4, it expresses the balance of forces in a configuration which is infinitely close to the trivial one, called *adjacent configuration*. According to this interpretation, the critical load is the smallest value of the load for which, next to the trivial configuration, there is an additional equilibrium position, adjacent to it. It can also be said that at $\mu = \mu_c$, the uniqueness of the solution of the elastic problem is lost. This is called the *adjacent equilibrium criterion*, and usually it is the argument with which, in technical textbooks, the notion of bifurcation of equilibrium is introduced.

Interpretation is useful in giving a physical meaning to the critical mode: it represents the (infinitesimal) displacement field undergone by the system upon reaching the critical load, in the act of motion with which it leaves the trivial configuration to assume a new (bifurcated) configuration. It is important to note that $\mathbf{u}_c$ represents only the *shape* of the incipient deformation, not the actual deformation. Indeed, although the result is valid only in the infinitesimal neighborhood of the trivial configuration, the displacement remains *defined to within an arbitrary constant* (as if $\mathbf{u}_c$ is the solution of the homogeneous problem, also $\alpha\mathbf{u}_c$ is a solution, with arbitrary α).[17]

4.5.2 Nonlinear Bifurcation Analysis

The determination of the equilibrium paths requires the solution of the nonlinear Eq. 4.4. The problem is here addressed by the *perturbation method*.

Asymptotic Expression of the Bifurcated Path

A point $C := (\mathbf{0}, \mu_c)$ of the space $(\mathbf{q}, \mu)$ is considered, belonging to the known fundamental path, from which an unknown bifurcated path originates (Fig. 4.5a).

This latter has parametric equations $\mathbf{q} = \mathbf{q}(\epsilon)$, $\mu = \mu(\epsilon)$ where ϵ is a parameter, chosen in such a way $\epsilon = 0$ locates the point C, i.e., $\mathbf{q}(0) = \mathbf{q}_c = \mathbf{0}$, $\mu(0) = \mu_c$. Since the interest consists in building *not* the whole bifurcated path, but only its portion in the neighborhood of $\epsilon = 0$ (i.e., to perform a *local analysis* of the bifurcation), it will be assumed that $0 < \epsilon \ll 1$ (hence the name of *perturbation parameter*, since small). Expanding in MacLaurin series, the parametric equations of the path are written as:

---

[17] The problem is analogous to the free dynamics of a linear system, where natural modes are shapes defined to within the amplitude.

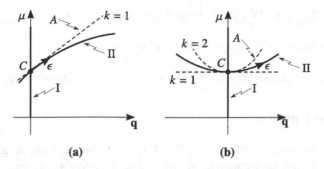

Fig. 4.5 Asymptotic construction of the bifurcation diagram for (**a**) non-symmetric systems and (**b**) symmetric systems; trivial fundamental path (I), bifurcated path (II), and its asymptotic approximation (A) of order k

$$\mathbf{q}(\epsilon) = \mathbf{q}_c + \epsilon \mathbf{q}_1 + \epsilon^2 \mathbf{q}_2 + \cdots, \tag{4.31a}$$

$$\mu(\epsilon) = \mu_c + \epsilon \mu_1 + \epsilon^2 \mu_2 + \cdots, \tag{4.31b}$$

where $\mathbf{q}_k := \frac{1}{k!} \frac{d^k \mathbf{q}}{d\epsilon^k}\big|_0$, $\mu_k := \frac{1}{k!} \frac{d^k \mu}{d\epsilon^k}\big|_0$, $k = 1, 2, \cdots$, are quantities proportional to the kth derivatives of displacement and load, evaluated at C. The asymptotic construction (meaning for $\epsilon \to 0$) of the path requires the computation of the unknowns $\mathbf{q}_k$, μ_k. If the series in Eq. 4.31 are truncated to $k = 1$, the bifurcated path is approximated by a straight line; if to $k = 2$, by a parabola, and so on.

Normalization

The ϵ parameter, which appears in the parametric equations of the bifurcated curve, does not necessarily have to assume a geometric meaning. If, however, it is wished to assign one, it needs to introduce an additional condition, called *normalization*. Among the many possible choices, one is taken here, in which the parameter is identified with the hth Lagrangian coordinate (provided it is significant[18]), i.e., $\epsilon \equiv q_h$, or $\epsilon \equiv \mathbf{e}_h^T \mathbf{q}$, with $\mathbf{e}_h := (0, 0, \cdots, 0, 1, 0, \cdots 0)^T$ a column matrix of zeros which has 1 in hth position. From the normalization condition, taking into account Eq. 4.31a, it follows:

$$\epsilon = \mathbf{e}_h^T \left(\epsilon \mathbf{q}_1 + \epsilon^2 \mathbf{q}_2 + \cdots \right); \tag{4.32}$$

---

[18] Since it is not possible to take a displacement which is zero (or nearly zero) at the leading order

by requiring this to hold for any ϵ, several conditions are drawn:

$$\mathbf{e}_h^T \mathbf{q}_1 = 1, \qquad \mathbf{e}_h^T \mathbf{q}_2 = 0, \qquad \mathbf{e}_h^T \mathbf{q}_3 = 0, \qquad \cdots , \tag{4.33}$$

called *normalization conditions at the order* ϵ, ϵ^2, ϵ^3, $\cdots$, respectively.

Perturbation Equations

To determine the series coefficients in Eqs. 4.31, the following procedure is applied. First, Eqs. 4.31 are substituted in the expanded nonlinear equilibrium equation, Eq. 4.29; then, by requiring equilibrium is satisfied for any ϵ (in order to obtain a path, and not a single point), the terms affected by the same powers of ϵ must be vanished separately. Thus, a sequence of equations, called *perturbation equations*, is obtained, as:

$$\epsilon^1 : \quad \left(\mathbf{K}_e + \mu_c \mathbf{K}_g \right) \mathbf{q}_1 = \mathbf{0}, \tag{4.34a}$$

$$\epsilon^2 : \quad \left(\mathbf{K}_e + \mu_c \mathbf{K}_g \right) \mathbf{q}_2 = - \mu_1 \mathbf{K}_g \mathbf{q}_1 + \mathbf{n}_2^{el} \left(\mathbf{q}_1, \mathbf{q}_1 \right) + \mu_c \mathbf{n}_2^{ext} \left(\mathbf{q}_1, \mathbf{q}_1 \right), \tag{4.34b}$$

$$\epsilon^3 : \quad \left(\mathbf{K}_e + \mu_c \mathbf{K}_g \right) \mathbf{q}_3 = - \mu_1 \mathbf{K}_g \mathbf{q}_2 - \mu_2 \mathbf{K}_g \mathbf{q}_1 + 2 \mathbf{n}_2^{el} \left(\mathbf{q}_1, \mathbf{q}_2 \right) \tag{4.34c}$$

$$+ 2 \mu_c \mathbf{n}_2^{ext} \left(\mathbf{q}_1, \mathbf{q}_2 \right) + \mu_1 \mathbf{n}_2^{ext} \left(\mathbf{q}_1, \mathbf{q}_1 \right)$$

$$+ \mathbf{n}_3^{el} \left(\mathbf{q}_1, \mathbf{q}_1, \mathbf{q}_1 \right) + \mu_c \mathbf{n}_3^{ext} \left(\mathbf{q}_1, \mathbf{q}_1, \mathbf{q}_1 \right).$$

Here, it was taken advantage of the fact that $\mathbf{n}_j^\alpha$ are homogeneous polynomials.[19] The perturbation equations, Eqs. 4.34, have the following properties:

- they are *linear equations*, all governed by the same matrix $\mathbf{K} \left(\mu_c \right) = \mathbf{K}_e + \mu_c \mathbf{K}_g$;
- the first equation (called *generatrix*) is homogeneous, while the successive ones are non-homogeneous;
- any equation always contains *just a new unknown displacement* with respect to the previous ones; hence, the first equation determines $\mathbf{q}_1$, the second $\mathbf{q}_2$, the third $\mathbf{q}_3$, and so on, except for the μ_k parameters.
- The μ_k parameters allow the *solvability* (or compatibility) conditions to be enforced, in order to solve the singular equations, as it will be explained soon.

Solution to the Perturbation Equations

Equation 4.34a is an eigenvalue problem, which coincides with the linearized bifurcation problem, Eq. 4.30. Therefore, the first step of the nonlinear bifurcation

---

[19] Thus, for example, $\mathbf{n}_2^\alpha \left(\epsilon \, \mathbf{q}_1 + \epsilon^2 \mathbf{q}_2, \epsilon \, \mathbf{q}_1 + \epsilon^2 \mathbf{q}_2 \right) = \epsilon^2 \mathbf{n}_2^\alpha \left(\mathbf{q}_1, \mathbf{q}_1 \right) + 2 \epsilon^3 \mathbf{n}_2^\alpha \left(\mathbf{q}_1, \mathbf{q}_2 \right) + \epsilon^4 \mathbf{n}_2^\alpha \left(\mathbf{q}_2, \mathbf{q}_2 \right)$.

analysis coincides with the linear stability analysis, denoting that at the loss of stability, a bifurcated path branches off. The equation admits the solution:

$$\mathbf{q}_1 = \mathbf{u}_c, \tag{4.35}$$

where $\mathbf{u}_c$ is the critical mode, associated with the critical load μ_c. Since the eigenvector is defined to within a constant, the first of the normalization conditions in Eqs. 4.33 is used to remove indeterminacy, by requiring the hth component of $\mathbf{u}_c$ is equal to 1.

Passing to solve Eq. 4.34b, it is observed that this is a *non-homogeneous equation with a singular matrix*, since $\det(\mathbf{K}(\mu_c)) = 0$ has been enforced at the previous step). The problem is generally impossible, unless the known term is *compatible* with the matrix. According to the Rouchè-Capelli Theorem, compatibility is expressed as follows:[20]

> *Necessary and sufficient condition for the algebraic problem* $\mathbf{Ax} = \mathbf{b}$ *admits solutions, is that the known term* $\mathbf{b}$ *is orthogonal to all the solutions* $\mathbf{y}$ *of the transpose homogeneous problem* $\mathbf{A}^T\mathbf{y} = \mathbf{0}$, *i.e.,* $\mathbf{y}^T\mathbf{b} = 0, \forall\, \mathbf{y}\,|\, \mathbf{A}^T\mathbf{y} = \mathbf{0}$.

Since, in the problem at hand, the singular matrix $\mathbf{K}(\mu_c)$ is symmetric, the transpose homogeneous problem coincides with Eq. 4.34a itself, and therefore solvability requires that the known term is orthogonal to the critical eigenvector. The condition can be interpreted as follows: the "forces" to the right-hand member of the equation must spend zero virtual work in the act of motion allowed by the "labile" system. This is written as:

$$\mathbf{u}_c^T \left(-\mu_1 \mathbf{K}_g \mathbf{u}_c + \mathbf{n}_2^{el}(\mathbf{u}_c, \mathbf{u}_c) + \mu_c \mathbf{n}_2^{ext}(\mathbf{u}_c, \mathbf{u}_c) \right) = \mathbf{0}, \tag{4.36}$$

from which the unknown is drawn:

$$\mu_1 = \frac{\mathbf{u}_c^T \left(\mathbf{n}_2^{el}(\mathbf{u}_c, \mathbf{u}_c) + \mu_c \mathbf{n}_2^{ext}(\mathbf{u}_c, \mathbf{u}_c) \right)}{\mathbf{u}_c^T \mathbf{K}_g \mathbf{u}_c}. \tag{4.37}$$

This information is sufficient to determine the *tangent to the bifurcated path* (Fig. 4.5a); by truncating Eqs. 4.31 at the ϵ order, the path reads:

$$\mathbf{q} = \epsilon\, \mathbf{u}_c, \tag{4.38a}$$

$$\mu = \mu_c + \epsilon\, \mu_1, \tag{4.38b}$$

---

[20] This is not the unique form for compatibility, but it is the most appropriate one for the purposes to be discussed here.

or, in the Cartesian form obtained by eliminating ϵ:

$$\mathbf{q} = \frac{\mu - \mu_c}{\mu_1} \mathbf{u}_c. \tag{4.39}$$

The asymptotic solution can be improved by (a) solving the perturbation Eq. 4.34b, (b) using the second of the normalization conditions in Eqs. 4.33 to remove arbitrariness,[21] and, finally, (c) enforcing solvability to Eq. 4.34c. From this latter, μ_2 is computed. However, since the procedure is tedious, analysis is here truncated to the linear approximation.

Case of Symmetric Systems

It is common, in applications, to encounter systems behaving symmetrically, for which $\mathbf{f}^\alpha (-\mathbf{q}) = -\mathbf{f}^\alpha (\mathbf{q})$, $\alpha = el, ext$ (i.e., the forces, internal and external, acting in opposite configurations, are themselves opposite, as it happens, e.g., in the Euler beam). The bifurcation diagram for such systems is also symmetric (Fig. 4.5b), since if the equilibrium Eq. 4.4 is satisfied by the pair $(\mathbf{q}, \mu)$, it is also satisfied from the pair $(-\mathbf{q}, \mu)$, for which all the forces change of sign. When forces are expanded in series, the even nonlinearities (in particular, the quadratic ones) identically vanish, i.e., $\mathbf{n}_2^\alpha (\mathbf{q}, \mathbf{q}) = 0$ $\forall \mathbf{q}$. Consequently, the previous analysis supplies $\mu_1 = 0$, i.e., the bifurcated path has a horizontal tangent at the bifurcation point. It is therefore necessary to accomplish an additional step in the asymptotic analysis, in order to determine μ_2 (i.e., the curvature of the bifurcated path at C).

The perturbation equations, Eqs. 4.34, in the symmetric case reduce to:

$$\epsilon^1 : \quad \left(\mathbf{K}_e + \mu_c \mathbf{K}_g\right) \mathbf{q}_1 = \mathbf{0}, \tag{4.40a}$$

$$\epsilon^2 : \quad \left(\mathbf{K}_e + \mu_c \mathbf{K}_g\right) \mathbf{q}_2 = - \mu_1 \mathbf{K}_g \mathbf{q}_1, \tag{4.40b}$$

$$\epsilon^3 : \quad \left(\mathbf{K}_e + \mu_c \mathbf{K}_g\right) \mathbf{q}_3 = - \mu_1 \mathbf{K}_g \mathbf{q}_2 - \mu_2 \mathbf{K}_g \mathbf{q}_1 + \mathbf{n}_3^{el} (\mathbf{q}_1, \mathbf{q}_1, \mathbf{q}_1) \tag{4.40c}$$

$$+ \mu_c \mathbf{n}_3^{ext} (\mathbf{q}_1, \mathbf{q}_1, \mathbf{q}_1) .$$

Solvability of Eq. 4.40b gives $\mu_1 = 0$; solving the equation, with the associate normalization condition (the second of Eqs. 4.33b), provides $\mathbf{q}_2 = \mathbf{0}$. The successive perturbation equation is then written as:

$$\left(\mathbf{K}_e + \mu_c \mathbf{K}_g\right) \mathbf{q}_3 = -\mu_2 \mathbf{K}_g \mathbf{q}_1 + \mathbf{n}_3^{el} (\mathbf{q}_1, \mathbf{q}_1, \mathbf{q}_1) + \mu_c \mathbf{n}_3^{ext} (\mathbf{q}_1, \mathbf{q}_1, \mathbf{q}_1) , \tag{4.41}$$

---

[21] Indeed, since the problem is of the type $\mathbf{Ax} = \mathbf{b}$, with $\mathbf{A}$ singular, after having satisfied solvability, the solutions are infinite in number, of the type $\mathbf{x} = \hat{\mathbf{x}} + \beta \mathbf{u}$, $\forall \beta$, where $\hat{\mathbf{x}}$ is a particular solution and $\mathbf{u}$ is a solution of the homogeneous problem $\mathbf{Au} = \mathbf{0}$. The normalization condition removes the arbitrariness of β.

whose solvability allows evaluation of the unknown:

$$\mu_2 = \frac{\mathbf{u}_c^T \left(\mathbf{n}_3^{el} \left(\mathbf{u}_c, \mathbf{u}_c, \mathbf{u}_c \right) + \mu_c \mathbf{n}_3^{ext} \left(\mathbf{u}_c, \mathbf{u}_c, \mathbf{u}_c \right) \right)}{\mathbf{u}_c^T \mathbf{K}_g \mathbf{u}_c}. \qquad (4.42)$$

The bifurcated path is thus approximated by the parabola:

$$\mathbf{q} = \epsilon \, \mathbf{u}_c, \qquad (4.43a)$$

$$\mu = \mu_c + \epsilon^2 \mu_2, \qquad (4.43b)$$

or, in Cartesian form:

$$\mathbf{q} = \sqrt{\frac{\mu - \mu_c}{\mu_2}} \, \mathbf{u}_c. \qquad (4.44)$$

4.6 Effect of Imperfections

The mechanical systems so far analyzed have been considered as "perfect" or "ideal." In particular, the geometry of the structure, in the reference configuration, was assumed to be free of "defects," such as a small curvature of a (supposed) straight beam or an unwanted misalignment between bodies. Likewise, external forces were considered ideal, too, for example, free from small eccentricities of their points of application or angular deviations of their lines of action. All these factors, geometric or of load, are called *imperfections*.[22]

Defects, however, are always present in real systems and in any laboratory experience. It is therefore legitimate to ask how much the imperfections can change the results of the ideal model and if, in particular, their influence is only of quantitative type or rather they qualitatively alter the structural response.

Remark 4.7 As it will be seen soon, the imperfections add known terms (forces) to the equation of motion of autonomous mechanical systems. These forces, being constant over time, have *no influence on dynamic bifurcations*, which are therefore said *robust to imperfections*. Instead, they remarkably affect static bifurcations, which depend only on balance of internal and external forces, and are therefore *sensitive to imperfections*. For this reason, the effect of the imperfections is studied only in the context of static bifurcations.

---

[22] Sometimes also called "initial imperfections," alluding to the fact that are already present in the system in the reference state.

4.6.1   Equilibrium Equations

The equilibrium equations of a mechanical system are established, in presence of *small* imperfections, both of geometric and load type.

Geometric Imperfections

A system is considered, which, in its natural state, assumes a nontrivial configuration $\mathbf{q} = \mathbf{q}_0$, due to small geometric imperfections. Examples are a slightly curved beam or a straight rod with a small angular deviation. The elastic forces are function of the difference $\mathbf{q} - \mathbf{q}_0$, i.e., of the displacement experienced by the body to move from the natural to current configuration, i.e., $\mathbf{f}^{el} = \mathbf{f}^{el}(\mathbf{q} - \mathbf{q}_0)$.[23] Since the imperfections are small by hypothesis, the elastic forces can be Taylor-expanded around $\mathbf{q}_0 = \mathbf{0}$, thus obtaining:

$$\mathbf{f}^{el}(\mathbf{q} - \mathbf{q}_0) = \mathring{\mathbf{f}}^{el}(\mathbf{q}) - \frac{\partial \mathring{\mathbf{f}}^{el}(\mathbf{q})}{\partial \mathbf{q}} \mathbf{q}_0 + \cdots, \tag{4.45}$$

where the "math ring" indicates evaluation at $\mathbf{q}_0 = \mathbf{0}$, i.e., in absence of imperfections. By retaining only terms linear in $\mathbf{q}_0$, and ignoring any product between $\mathbf{q}_0$ and $\mathbf{q}$, it must be taken $\frac{\partial \mathring{\mathbf{f}}^{el}(\mathbf{q})}{\partial \mathbf{q}} \simeq \frac{\partial \mathring{\mathbf{f}}^{el}(\mathbf{0})}{\partial \mathbf{q}} = -\mathbf{K}_e$; consequently:[24]

$$\mathbf{f}^{el}(\mathbf{q} - \mathbf{q}_0) = \mathring{\mathbf{f}}^{el}(\mathbf{q}) + \alpha \, \mathbf{K}_e \mathbf{q}_0, \tag{4.46}$$

where a *geometric imperfection multiplier* $\alpha \ll 1$ has been introduced (by contextually rescaling $\|\mathbf{q}_0\| = O(1)$).

Remark 4.8 Equation 4.46 shows how the geometric imperfections modify the internal forces $\mathring{\mathbf{f}}^{el}(\mathbf{q})$ of a perfect system: they add to these an "equivalent force" $\mathbf{K}_e \mathbf{q}_0$, equal and opposite to that necessary to move back the (linearized) system to the configuration which is exempt from imperfections.

---

[23] For example, the Euler beam may be slightly curved in the natural state, described by the function $v_0(x)$. If the constitutive law is linear, the bending moment $M(x)$ is proportional to the change in curvature with respect to the initial one, i.e., $M(x) = EI\left(v''(x) - v_0''(x)\right)$, with EI the bending stiffness.

[24] For example, for a cubic spring, it is $f^{el} = -k_1(q - q_0) - k_3(q - q_0)^3$. Expanding for small q_0, this reads: $f^{el} = -k_1 q - k_3 q^3 + (k_1 + 3k_3 q^2)q_0$. By ignoring the product $q^2 q_0$ one has: $f^{el} = -k_1 q - k_3 q^3 + k_1 q_0$.

Load Imperfections

The load can also differ from perfect condition, due to small eccentricities or angular deviations, described by a scalar parameter $\beta \ll 1$, so that $\mathbf{f}^{ext} = \mathbf{f}^{ext}(\mathbf{q}; \beta)$. Expanding in series around $\beta = 0$, it is:

$$\mathbf{f}^{ext}(\mathbf{q}; \beta) = \mathbf{f}^{ext}(\mathbf{q}; 0) + \beta \frac{d\mathbf{f}^{ext}}{d\beta}(\mathbf{q}, 0) + \cdots . \tag{4.47}$$

Here, $\mathbf{f}^{ext}(\mathbf{q}; 0) =: \overset{\circ}{\mathbf{f}}{}^{ext}(\mathbf{q})$ is the perfect force; moreover, $\frac{d\mathbf{f}^{ext}}{d\beta}(\mathbf{q}, 0) =: \mathbf{g}(\mathbf{q})$ is a disturbing force, which, when expanded for small $\mathbf{q}$, takes the form $\mathbf{g}(\mathbf{q}) = \mathbf{g}(\mathbf{0}) + \frac{\partial \mathbf{g}(\mathbf{0})}{\partial \mathbf{q}}\mathbf{q} + \cdots$. By ignoring the powers of β greater than 1, and the products between β and the powers of $\mathbf{q}$, ultimately it results:

$$\mathbf{f}^{ext}(\mathbf{q}; \beta) = \overset{\circ}{\mathbf{f}}{}^{ext}(\mathbf{q}) + \beta \mathbf{g}_0, \tag{4.48}$$

where $\mathbf{g}_0 := \mathbf{g}(\mathbf{0})$ is a force independent of the configuration.

Equilibrium Equation for Imperfect System

Taking into account both types of imperfections, described by Eqs. 4.46 and 4.48, the equilibrium Eq. 4.29 is modified by two additional terms proportional to α, β, becoming:

$$\left(\mathbf{K}_e + \mu \mathbf{K}_g\right)\mathbf{q} = \alpha \mathbf{K}_e \mathbf{q}_0 + \mu \beta \mathbf{g}_0 + \overset{\circ}{\mathbf{n}}{}^{el}(\mathbf{q}) + \mu \overset{\circ}{\mathbf{n}}{}^{ext}(\mathbf{q}). \tag{4.49}$$

4.6.2 Asymptotic Construction of the Imperfect Equilibrium Paths

The solution to Eq. 4.49, for assigned α, β, can be pursued through the asymptotic method previously used for the perfect system. The two cases of non-symmetric and symmetric systems are tackled separately.

Non-symmetric Systems

The imperfections are assumed to be small of order ϵ^2, by letting $(\alpha, \beta) = \epsilon^2\left(\hat{\alpha}, \hat{\beta}\right)$, with $\hat{\alpha}, \hat{\beta}$ quantities of order 1.[25] By using again the series in Eq. 4.31,

---

[25] Rescaling is done in order the imperfections appear in the first significant perturbation equation, from which the slope of the perfect bifurcated path is determined.

but taking into account the imperfections, the perturbation Eqs. 4.34a,b change as follows:

$$\epsilon^1: \quad \left(\mathbf{K}_e + \mu_c \mathbf{K}_g\right) \mathbf{q}_1 = 0, \tag{4.50a}$$

$$\epsilon^2: \quad \left(\mathbf{K}_e + \mu_c \mathbf{K}_g\right) \mathbf{q}_2 = -\mu_1 \mathbf{K}_g \mathbf{q}_1 + \mathbf{n}_2^{el}\left(\mathbf{q}_1, \mathbf{q}_1\right) + \mu_c \mathbf{n}_2^{ext}\left(\mathbf{q}_1, \mathbf{q}_1\right) \tag{4.50b}$$

$$+ \hat{\alpha}\, \mathbf{K}_e \mathbf{q}_0 + \mu_c\, \hat{\beta}\, \mathbf{g}_0.$$

The generating solution is still given by Eq. 4.35. The solvability condition of Eq. 4.50b then gives:

$$\mu_1 = \mathring{\mu}_1 + C_\alpha\, \hat{\alpha} + \mu_c\, C_\beta\, \hat{\beta}, \tag{4.51}$$

where $\mathring{\mu}_1$ refers to the perfect system (Eq. 4.37), and moreover:

$$C_\alpha := \frac{\mathbf{u}_c^T \mathbf{K}_e \mathbf{q}_0}{\mathbf{u}_c^T \mathbf{K}_g \mathbf{u}_c}, \qquad C_\beta := \frac{\mathbf{u}_c^T \mathbf{g}_0}{\mathbf{u}_c^T \mathbf{K}_g \mathbf{u}_c}. \tag{4.52}$$

The parametric equations of the path, in terms of the originals parameters α, β, are therefore written as:

$$\mathbf{q} = \epsilon\, \mathbf{u}_c, \tag{4.53a}$$

$$\mu = \mu_c + \epsilon\, \mathring{\mu}_1 + \frac{1}{\epsilon}\left(C_\alpha\, \alpha + \mu_c\, C_\beta\, \beta\right). \tag{4.53b}$$

Remark 4.9 The asymptotic expression in Eq. 4.53b is not valid when $\epsilon \to 0$ and α, β *are fixed*, since these latter have been linked to ϵ. If, instead, for a fixed ϵ, the magnitudes α, β of the imperfection are made to tend to zero, the imperfect path tends to the perfect one.

Symmetric Systems

The imperfections are assumed to be small of order ϵ^3, by taking $(\alpha, \beta) = \epsilon^3\left(\hat{\alpha}, \hat{\beta}\right)$, with $\hat{\alpha}, \hat{\beta}$ quantities of order 1.[26] The perturbation Eqs. 4.40 change as follows:

$$\epsilon^1: \quad \left(\mathbf{K}_e + \mu_c \mathbf{K}_g\right) \mathbf{q}_1 = 0, \tag{4.54a}$$

$$\epsilon^2: \quad \left(\mathbf{K}_e + \mu_c \mathbf{K}_g\right) \mathbf{q}_2 = -\mu_1 \mathbf{K}_g \mathbf{q}_1, \tag{4.54b}$$

---

[26] Rescaling is done in order the imperfections appear in the first significant perturbation equation, from which the curvature of the perfect path is determined.

$$\epsilon^3 : \quad \left(\mathbf{K}_e + \mu_c \mathbf{K}_g\right) \mathbf{q}_3 = -\mu_1 \mathbf{K}_g \mathbf{q}_2 - \mu_2 \mathbf{K}_g \mathbf{q}_1 + \mathbf{n}_3^{el} (\mathbf{q}_1, \mathbf{q}_1, \mathbf{q}_1) \qquad (4.54c)$$

$$+ \mu_c \mathbf{n}_3^{ext} (\mathbf{q}_1, \mathbf{q}_1, \mathbf{q}_1) + \hat{\alpha} \, \mathbf{K}_e \mathbf{q}_0 + \mu_c \, \hat{\beta} \, \mathbf{g}_0.$$

The generating solution is still given by Eq. 4.35, while Eq. 4.54b gives $\mu_1 = 0$ and $\mathbf{q}_2 = \mathbf{0}$. From the solvability of Eq. 4.54c, it follows:

$$\mu_2 = \mathring{\mu}_2 + C_\alpha \, \hat{\alpha} + \mu_c \, C_\beta \, \hat{\beta}, \qquad (4.55)$$

where $\mathring{\mu}_2$ refers to the perfect system (Eq. 4.42) and C_α, C_β are still given by Eq. 4.52. The branch is therefore described by:

$$\mathbf{q} = \epsilon \, \mathbf{u}_c, \qquad (4.56a)$$

$$\mu = \mu_c + \epsilon^2 \mathring{\mu}_2 + \frac{1}{\epsilon} \left(C_\alpha \, \alpha + \mu_c \, C_\beta \, \beta\right). \qquad (4.56b)$$

Remark 4.10 Equations 4.53b and 4.56b show that the imperfections destroy the branch points, as discussed, on qualitative ground, in Fig. 2.7.

4.7 Stability of the Equilibrium Paths

Once an equilibrium path $\mathbf{q} = \mathbf{q}(\epsilon)$, $\mu = \mu(\epsilon)$ has been built up, both for the perfect and imperfect system, it is of interest to analyze its stability. According to the energy criterion, the stability of the equilibrium point identified by the ϵ parameter is governed by the local character that the total potential energy $\Pi(\epsilon) := \Pi(\mathbf{q}(\epsilon); \mu(\epsilon))$ assumes at that point. Equilibrium is stable if the energy has a local minimum (i.e., it is there positive definite); it is unstable if only stationary (indefinite) or possesses a local maximum (negative definite).

To verify that $\Pi(\epsilon)$ assumes a local minimum at ϵ, it is sufficient to analyze the Hessian matrix $\mathbf{H}(\epsilon) := \left[\frac{\partial^2 \Pi(\epsilon)}{\partial q_i \partial q_j}\right]$ (representing the *stiffness matrix* evaluated at that point). If all eigenvalues $\Lambda_k = \Lambda_k(\epsilon)$ of $\mathbf{H}(\epsilon)$ (which are roots of the characteristic equation $\det(\mathbf{H}(\epsilon) - \Lambda \mathbf{I}) = \mathbf{0}$) are positive, then $\mathbf{H}(\epsilon)$ is defined positive, and the equilibrium point is stable. If at least one of them is negative, the matrix is undefined, and the point is unstable.

If $\Lambda_k(\epsilon) > 0$ for each ϵ belonging to a certain interval, then the equilibrium branch is stable in that range. If at least one of the eigenvalues changes sign at $\epsilon = \epsilon^*$, then the branch loses stability at ϵ^*.

4.8   Systems with Precritical Deformations

So far, perfect structures were analyzed, which, in the phase that precedes the bifurcation, admit the trivial solution $\mathbf{q} = \mathbf{0}$, $\forall \mu$.[27] The hypothesis holds when an infinite stiffness has been introduced in the model (e.g., an infinite axial stiffness in the Euler beam), so that no precritical deformations are allowed. The instability of these structures manifests itself through the occurrence of *branch points* from the trivial path. On the other hand, it was seen that imperfections destroy the branch points, and trigger the occurrence of *fold bifurcations* at which the structure loses stability. It is therefore of interest to investigate which are the consequences of removing the strong simplifying hypothesis of infinite stiffness, by looking for an answer to the following questions: (a) does a structure suffering from precritical deformations (like an inverted elastic pendulum or an axially deformable beam) still manifest branch points? (b) Are there any structures which, even in the *absence of imperfections*, admit fold bifurcations instead of branch points?

The answers, as will be seen, are both affirmative. To understand this, the following examples are considered:

1. An axially deformable Euler beam shortens in the precritical phase *remaining straight*. Since at the bifurcation its geometry is qualitatively the same of the original one, the extensible beam becomes unstable at a branch point, distinguishing itself from the inextensible beam only for the changed length.[28]
2. A three-hinged low arch, consisting of two elastic rods and subject to a force inducing compression, undergoes shortening of the rods but, at the same time, *a change of their angle of inclination*, thus modifying its initial geometry. If this shortening is such that the three hinges can align on the horizontal before the rods individually buckle, the system becomes kinematically degenerate (labile), and therefore instability manifests upon reaching a limit load.

To analyze the two types of bifurcation and to distinguish branch and limit point, the fundamental path $\mathbf{q} = \mathbf{q}^f (\mu)$ must first be determined, in exact or asymptotic form. On this path, it is then possible to classify the equilibrium points, as discussed in Sect. 4.3. Finally, in case of a branch point, it is necessary to build up the bifurcated path.

The analysis, rather onerous from the computational point of view, is rarely performed in engineering. Here, just some hints on the problem are given; in the next chapter, elementary examples will illustrate the methodology.

---

[27] Reference is made to a discrete system, although the concepts hold also for continuous systems.

[28] The beam *extremely axially soft* is an exception to this type of behavior, as it will be clarified in the next chapter (Sect. 5.6) by an example.

4.8.1 Asymptotic Construction of the Non-trivial Fundamental Path

A single DOF system is considered,[29] whose equilibrium equation, Eq. 4.4, is rewritten as:

$$f(q) + \mu g(q) = 0, \tag{4.57}$$

where, for notation simplicity, $f := f^{el}$, $g := f^{ext}$ were introduced. A non-trivial solution to this equation is sought in the Cartesian form $q = q^f(\mu)$ or in the parametric form $q = q^f(\epsilon)$, $\mu = \mu(\epsilon)$. Since the exact solution of the problem is generally impossible, asymptotic methods are used, which are based on an extrapolation from a known equilibrium point. Due to $f(0) = 0$, the natural state $(q, \mu) = (0, 0)$ is a solution to Eq. 4.57. The simplest (but sometimes grossly approximate) choice consists in taking this point as the starting point for the expansion.[30] Consistently, the parametric equations are expressed in the form:

$$q = \cancel{q_0} + \epsilon q_1 + \epsilon^2 q_2 + \epsilon^3 q_3 + \cdots, \tag{4.58a}$$

$$\mu = \cancel{\mu_0} + \epsilon \mu_1 + \epsilon^2 \mu_2 + \epsilon^3 \mu_3 + \cdots, \tag{4.58b}$$

where $\epsilon = 0$ was taken in the natural state. By replacing these series in the equilibrium equations, expanding for small ϵ, and separately zeroing the terms with the same power of ϵ, the perturbation equations are obtained as:

$$\epsilon^1: \quad -k_e q_1 + \mu_1 g^0 = 0, \tag{4.59a}$$

$$\epsilon^2: \quad -k_e q_2 + \mu_2 g^0 = -\frac{1}{2} f^0_{,qq} q_1^2 - \mu_1 g^0_{,q} q_1, \tag{4.59b}$$

$$\epsilon^3: \quad -k_e q_3 + \mu_3 g^0 = -f^0_{,qq} q_1 q_2 - \frac{1}{6} f^0_{,qqq} q_1^3 - \mu_2 g^0_{,q} q_1 - \mu_1 g^0_{,q} q_2 \tag{4.59c}$$

$$-\frac{1}{2} \mu_1 g^0_{,qq} q_1^2.$$

Here, $k_e := -f^0_{,q}$ is the elastic stiffness, and an apex 0 indicates evaluation at the natural state.[31]

---

[29] The generalization of the procedure to multi-DOF systems is immediate.

[30] In the next chapter (Sect. 5.5), with reference to an example, it will be shown how to perform the extrapolation from an equilibrium point different from the natural state.

[31] Equations 4.59 are of opposite sign to those descending from the variational procedure, as a consequence of the change of sign already observed in Eq. 4.4.

The previous equations must be accompanied by the normalization condition. For example, by choosing $q = \epsilon$, from this follows $q_1 = 1, q_2 = q_3 = \cdots = 0$. By solving in sequence the perturbation equations, one gets:

$$\mu_1 = \frac{g^0}{k_e}, \tag{4.60a}$$

$$\mu_2 = -\frac{1}{g^0}\left(\frac{1}{2}f^0_{,qq} + \frac{g^0}{k_e}g^0_{,q}\right), \tag{4.60b}$$

$$\mu_3 = -\frac{1}{g^0}\left[\frac{1}{6}f^0_{,qqq} - \frac{1}{g^0}\left(\frac{1}{2}f^0_{,qq} + \frac{g^0}{k_e}g^0_{,q}\right)g^0_{,q} + \frac{1}{2}\frac{g^0}{k_e}g^0_{,qq}\right], \tag{4.60c}$$

which, substituted in Eq. 4.58b, provide the equilibrium path in the form $\mu = \mu(q)$.

If it happens that $\left.\frac{d\mu}{dq}\right|_{q_l} = 0$ at some q_l close to the origin (where the asymptotic solutions holds), then $q = q_l$ is a limit point, whose corresponding load $\mu_l := \mu(q_l)$ is the limit load. An example of a single DOF system which exhibits a limit load will be illustrated in the next chapter (Sect. 5.5).

Remark 4.11 It is worth noticing that Eqs. 4.59 are *non-singular*, and therefore *do not require solvability*, as instead was the case of branching analysis. This is a consequence of the fact that, in the latter case, the extrapolation is carried out from a singular point, while in the current case, it is performed from a regular point.

4.8.2 Bifurcation from Non-trivial Path

A general n DOF system is considered, whose equilibrium equation is:

$$\mathbf{f}(\mathbf{q}) + \mu\,\mathbf{g}(\mathbf{q}) = \mathbf{0}, \tag{4.61}$$

where $\mathbf{f} := \mathbf{f}^{el}$, $\mathbf{g} := \mathbf{f}^{ext}$. By hypothesis, the system admits a *known* non-trivial path I, of equation $\mathbf{q} = \mathbf{q}^f(\mu)$. Consequently:

$$\mathbf{f}(\mathbf{q}^f(\mu)) + \mu\,\mathbf{g}(\mathbf{q}^f(\mu)) = \mathbf{0}. \tag{4.62}$$

Incremental Variable

Along path I, a point C (if any) is sought, from which a bifurcated path II branches off. To this end, the following change of variable is introduced:[32]

---

[32] Here $\mathbf{y}$ has a meaning similar to $\delta\mathbf{q}$, as introduced in Eq. 3.61 of Chap. 3, with the difference that now $\mathbf{y}$ is a finite quantity, while $\delta\mathbf{q}$ is infinitesimal.

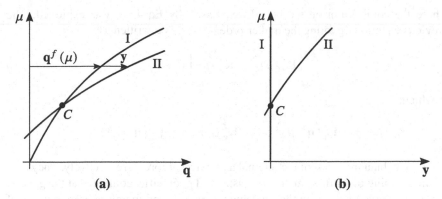

Fig. 4.6 Incremental variables **y** and their geometric meanings: (**a**) paths in the space $(\mathbf{q}, \mu)$, (**b**) paths in the space $(\mathbf{y}, \mu)$

$$\mathbf{q} = \mathbf{q}^f(\mu) + \mathbf{y}, \tag{4.63}$$

where **y** represents the offset between the positions of equilibrium II and I, which the system can assume under a given load μ (Fig. 4.6a).

The new variable **y** is called *incremental* (or *sliding*, to indicate that its origin "slips" on the fundamental path).

The equilibrium equation, after the transformation in Eq. 4.63, becomes:

$$\mathbf{f}\left(\mathbf{q}^f(\mu) + \mathbf{y}\right) + \mu\, \mathbf{g}\left(\mathbf{q}^f(\mu) + \mathbf{y}\right) = \mathbf{0}. \tag{4.64}$$

It is observed that, in this form, the equation:

1. *depends nonlinearly on* μ, albeit the original form was linear in μ;
2. *admits the trivial solution* $\mathbf{y} = \mathbf{0}$, $\forall \mu$.

Thus, the change of variables in Eq. 4.63 *transforms the non-trivial in a trivial path* (Fig. 4.6b). Therefore, the asymptotic methods already illustrated in Sect. 4.5.2 for bifurcation from a trivial path also applies to Eq. 4.64, with the unique (slight) complication of the nonlinear dependence on the μ parameter. Here, the detailed illustration of the method is omitted, and the analysis is limited to the determination of the critical load.

Critical Load

To determine the branch point C, a solution on branch II is sought for $\mathbf{y} \to \mathbf{0}$; thus, Eq. 4.64 can be linearized in **y** as:

$$\cancel{\mathbf{f}\left(\mathbf{q}^f(\mu)\right)} + \mathbf{f}_{,\mathbf{q}}\left(\mathbf{q}^f(\mu)\right)\mathbf{y} + \mu\left[\cancel{\mathbf{g}\left(\mathbf{q}^f(\mu)\right)} + \mathbf{g}_{,\mathbf{q}}\left(\mathbf{q}^f(\mu)\right)\mathbf{y}\right] + \cdots = \mathbf{0}, \tag{4.65}$$

where the equilibrium on the path I, expressed by Eq. 4.62, was exploited. The previous equation, ignoring the higher-order terms, is rewritten as:

$$\left[\mathbf{K}_e^E(\mu) + \mu \mathbf{K}_g^E(\mu) \right] \mathbf{y} = \mathbf{0}, \tag{4.66}$$

in which:

$$\mathbf{K}_e^E(\mu) := -\mathbf{f}_{,\mathbf{q}}\left(\mathbf{q}^f(\mu) \right), \qquad \mathbf{K}_g^E(\mu) := -\mathbf{g}_{,\mathbf{q}}\left(\mathbf{q}^f(\mu) \right), \tag{4.67}$$

are the Jacobian matrices, of the internal and external forces, respectively. They take on the meaning of stiffness matrices, elastic and geometric, computed at the generic point $E = E(\mu)$ belonging the fundamental path I, and therefore are functions of the parameter μ. Equation 4.66 has the form of a (nonlinear) eigenvalue problem in μ. The smallest root determines the sought critical load.

Remark 4.12 When the precritical deformations can be ignored (because small), the unknown geometry of the structure at the bifurcation is confused with the known geometry at the reference configuration (i.e., at $\mu = 0$). Therefore $\mathbf{K}_e^E(\mu) \simeq \mathbf{K}_e^E(0) =: \mathbf{K}_e$, and $\mathbf{K}_g^E(\mu) \simeq \mathbf{K}_g^E(0) =: \mathbf{K}_g$, which are known quantities. Consequently, the eigenvalue problem in Eq. 4.66 becomes linear in μ.

References

1. Budiansky, B.: Theory of buckling and post-buckling behavior of elastic structures. Adv. Appl. Mech. **14**, 1–65 (1974)
2. Godoy, L.A.: Theory of elastic stability: analysis and sensitivity. Taylor and Francis, Philadelphia (2000)
3. Koiter, W. T.: On the stability of elastic equilibrium. National Aeronautics and Space Administration, Washington D.C. (1967)
4. Lacarbonara, W.: Nonlinear structural mechanics: theory, dynamical phenomena and modeling. Springer, New York (2013)
5. Nayfeh, A.H.: Perturbation methods. Wiley, New York (1973)
6. Pignataro, M., Rizzi, N., Luongo, A.: Stability, bifurcation and postcritical behaviour of elastic structures. Elsevier, Amsterdam (1990)
7. Thompson, J.M.T., Hunt, G.W.: A general theory of elastic stability. Wiley, London (1973)
8. Thompson, J.M.T., Hunt, G.W.: Elastic instability phenomena. Wiley, Chichester (1984)
9. van Der Heijden, A.M.A.: W.T. Koiter's elastic stability of solids and structures. Cambridge University Press, Cambridge (2009)

Chapter 5
Paradigmatic Systems of Buckling and Postbuckling

5.1 Introduction

Discrete, as well as continuous, conservative systems manifest a limited number of bifurcation scenarios, named: (a) *stable* fork, (b) *unstable fork*, (c) *transcritical*, and (d) *fold bifurcation*. The first three of them denote *branching* from the fundamental path, of symmetric (a,b) or generic (c) systems; the last one concerns imperfect systems or perfect systems suffering precritical deformation (i.e., admitting a *non-trivial* fundamental path). More complex bifurcations, involving more than two paths, occur when the Jacobian matrix possesses a zero eigenvalue of multiplicity larger than 1. This case is referred, in the technical literature, as *interaction among simultaneous modes*. All these behaviors can be exemplified by single or two degrees of freedom mechanical systems, admitting a *trivial* fundamental path, namely a pendulum, or double pendulum, with different arrangements of the elastic constrains. These prototypes, therefore, assume the meaning of *paradigmatic systems*, since they encompass all the characteristic behaviors of bifurcations occurring in larger systems. They are browsed in this chapter, with the aim to investigate mechanical aspects, and to exemplify the perturbation algorithms illustrated in the previous chapter. Moreover, as an example of continuous system, the *inextensible Euler beam* is discussed here, for which the asymptotic construction of the branched path is illustrated. The analysis reveals that the perturbation method applies essentially in the same way to discrete and continuous systems, although the latter call for a proper algorithmic treatment of the boundary conditions.

When the system undergoes precritical deformations, the *non-trivial* fundamental path must first be evaluated, in exact form, when possible, or in asymptotic form. This path can exhibit a limit point, as it happens in the *snap-through* phenomenon of a single degree of freedom low arch, or a branch point, as it occurs for an extensible elastic pendulum. The last form of bifurcation reveals some unexpected and non-standard phenomena, which are worthy of being studied.

All these paradigmatic systems are analyzed in this chapter, most of them also affected by imperfections, and exact (when viable) and asymptotic solutions pursued.

5.2   Single Degree of Freedom Systems with Trivial Fundamental Path

Paradigmatic systems with a single degree of freedom (DOF) are studied, possessing a trivial fundamental path. *Stable fork, unstable fork* and *transcritical bifurcations* are illustrated [2, 4, 10]. Fold (or *limit point*) bifurcations are analyzed in the presence of *imperfections* [3, 11]. For each system, both the perfect and imperfect models are studied, by performing exact and asymptotic analyses.

5.2.1   Inverted Elastic Pendulum

The inverted pendulum in Fig. 5.1a is considered, consisting of a rigid rod of length ℓ, hinged at the ground at the lower end, elastically constrained there by a torsional spring of stiffness k, subject to the free end to a dead load of intensity P. The rotation θ of the pendulum is taken as the unique Lagrangian parameter.

Exact Analysis of the Perfect System

The elastic potential energy of the system is stored by the torsional spring and is equal to $U = \frac{1}{2}k\theta^2$. The potential energy of the external force is $V = -P\Delta$, equal to the opposite of the work expended by the force P in the lowering $\Delta :=$ $\ell(1 - \cos\theta)$ of the free end of the rod. The total potential energy (TPE) of the system is therefore:

Fig. 5.1   Rigid rod, elastically constrained by a torsional spring, subject to a dead load: (**a**) perfect system, (**b**) imperfect system

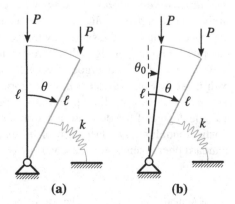
(a)   (b)

$$\Pi = U + V = \frac{1}{2}k\,\theta^2 - P\ell\,(1 - \cos\theta).$$ (5.1)

The equilibrium equation is drawn by the stationary condition of the energy, $\frac{d\Pi}{d\theta} = 0$, which provides:

$$k\theta - P\ell\sin\theta = 0.$$ (5.2)

Alternatively, direct balance of the torques with respect the grounded end can be imposed, when the body occupies the current position.

All the equilibrium positions of the pendulum, which are solutions of Eq. 5.2, are expressed by:

$$\theta = 0, \qquad \forall P, \qquad\qquad\qquad (I),$$ (5.3a)

$$P = \frac{k}{\ell}\frac{\theta}{\sin\theta}, \qquad\qquad\qquad (II),$$ (5.3b)

in which (I) is the trivial fundamental path and (II) the bifurcated one (Fig. 5.2a). The two paths intersect each other at the critical load:

$$P_c := \frac{k}{\ell}.$$ (5.4)

The stability of the equilibrium is determined by the local nature, of minimum or maximum, of the TPE. To evaluate it, the graph of the TPE is drawn in Fig. 5.2b–d, in the sub-critical, critical, and super-critical cases, respectively. It follows from the plots:

- in the sub-critical case, there is a single stationary point at $\theta = 0$; since the function has a local minimum, the equilibrium is stable;
- the previous situation persists even in the critical case, in which the trivial equilibrium remains stable;
- in the super-critical case, there are three stationary points, corresponding to an (unstable) maximum at the origin and two (stable) minima at the non-trivial positions.

Alternatively, the construction of the TPE graph can be avoided and replaced by the evaluation of the sign of its second derivative, computed on the two equilibrium paths. Since:

$$\frac{d^2\Pi}{d\theta^2} = k - P\ell\cos\theta,$$ (5.5)

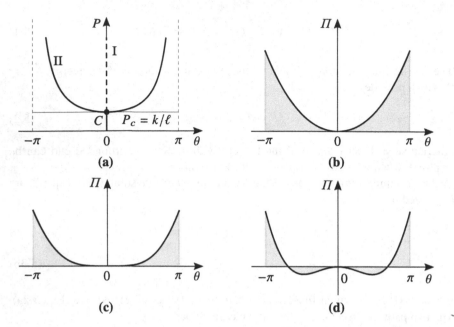

Fig. 5.2 Exact analysis of the inverted pendulum: (**a**) bifurcation diagram; TPE diagram, when (**b**) $P < P_c$, (**c**) $P = P_c$, (**d**) $P > P_c$

making system between this expression and Eqs. 5.3, it follows:

$$\frac{\mathrm{d}^2 \Pi}{\mathrm{d}\theta^2} = \begin{cases} k - P\ell \lessgtr 0, & \text{if } P \gtrless P_c, \quad \text{(I)}, \\ k\left(1 - \frac{\theta}{\tan\theta}\right) > 0, & \forall \theta. \qquad \text{(II)}. \end{cases} \tag{5.6}$$

The trivial equilibrium path, therefore, is stable when $P < P_c$ and unstable when $P > P_c$; the bifurcated path is stable for any P. Since $\frac{\mathrm{d}^2 \Pi}{\mathrm{d}\theta^2} = 0$ at the critical state, its stability is determined by the successive derivatives. Since the third derivative vanishes and the fourth is positive, the bifurcation point is stable. This type of bifurcation is called a *super-critical fork*, or stable fork.

Supplement 5.1 (Interpretation of the geometric nonlinearity) The following interpretation can be given about the behavior of the inverted pendulum, useful to illustrate the effects of the geometric nonlinearities. The rotation θ of the rod, around the hinge A, can be described, either (a) in infinitesimal or (b) in finite kinematics. In the first case, the free end B moves horizontally to B'; in the second case, B describes an arc of circumference of center A and moves to a point B'' belonging to the same straight line AB', but at a height lower than B'. Consequently, the arm of the external force with respect to the hinge, in finite kinematics, is smaller than that estimated in linear kinematics (since $|\sin\theta| < |\theta|$), i.e., *linearization overestimates the mechanical action of the external force*. Nonlinearity, therefore, has a hardening (stabilizing) effect. The restoring couple $k\theta$, instead, is the same in both kinematics. In particular, when the force assumes the critical value, the total stiffness vanishes, but the system is still stable, because of the geometric nonlinearities.   □

Exact Analysis of the Imperfect System

It is assumed that the rod, in its natural configuration, is inclined on the vertical of a small angle θ_0, representing a geometric imperfection (Fig. 5.1b). Consequently, the free end of the imperfect rod is lower than that of the perfect rod, of the amount $\Delta_0 := \ell (1 - \cos \theta_0)$. By measuring the angular deviation θ still from the vertical, the dead load expends work in the $\Delta - \Delta_0$ excursion; the total potential energy is therefore modified as follows:

$$\Pi = \frac{1}{2}k \, (\theta - \theta_0)^2 - P\ell \, (1 - \cos \theta) + P\ell \, (1 - \cos \theta_0) \, . \tag{5.7}$$

The equilibrium is expressed by $\frac{d\Pi}{d\theta} = 0$, or:

$$k \, (\theta - \theta_0) - P\ell \sin \theta = 0, \tag{5.8}$$

which represents the balance of torques with respect to the grounded hinge. This equation admits, for an assigned θ_0, only one path, having equation:

$$P = \frac{k}{\ell} \frac{\theta - \theta_0}{\sin \theta}. \tag{5.9}$$

The family of curves $P = P (\theta; \theta_0)$, obtained for several values of the imperfection parameter θ_0, is plotted in Fig. 5.3a. Each curve consists of a *natural branch* (passing through the point $(\theta, P) = (\theta_0, 0)$) and a *non-natural branch*, external to this point. Only the non-natural branch admits a *lower limit point* (θ_l, P_l). By enforcing $\frac{dP}{d\theta} = 0$, the equation $\theta_l - \theta_0 = \tan \theta_l$ follows, using which the limit load is found to be:

$$P_l = \frac{k}{\ell} \frac{1}{\cos \theta_l}. \tag{5.10}$$

Figure 5.3b–d show the TPE diagrams, from which information on the stability of the imperfect path is drawn. Alternatively, the second derivative of the energy is computed as:

$$\frac{d^2 \Pi}{d\theta^2} = k - P\ell \cos \theta. \tag{5.11}$$

By requiring that $\frac{d^2 \Pi}{d\theta^2} = 0$, the *critical locus*:

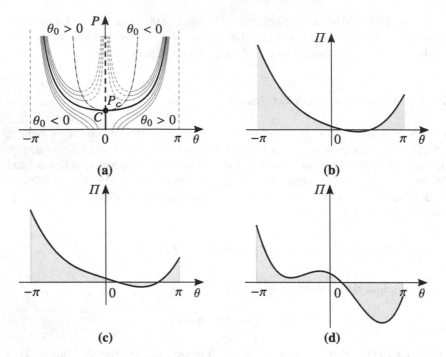

Fig. 5.3 Exact analysis of the imperfect inverted pendulum: **(a)** bifurcation diagram; TPE diagram, for $\theta_0 > 0$, when **(b)** $P < P_c$, **(c)** $P = P_c$, **(d)** $P > P_c$; critical locus (dash-point curve)

$$P = \frac{k}{\ell} \frac{1}{\cos \theta} \tag{5.12}$$

is obtained (Fig. 5.3a), which separates stable (lower region) from unstable (upper region) equilibria, on the (θ, P) plane. It is easy to check that such a curve coincides with the locus of the relative minima of the non-natural equilibrium paths.[1] Therefore, at the limit points, the stability of the path changes.

Remark 5.1 Structures with stable postcritical behavior are, for practical purposes, *insensitive to imperfections*, because they exhibit limit points only on non-natural branches.

Supplement 5.2 (Perturbation analysis of the perfect system) In the simple example dealt with here, it has been possible to perform an exact analysis. This, however, is not viable for more complex systems. It is therefore appropriate to repeat the previous study by the perturbation method, which can instead be applied to any

---

[1] Indeed, making system between the equilibrium path and the locus, it is found that these intersect each other at the abscissa θ^*, root of the equation $\theta^* - \theta_0 = \tan \theta^*$, from which $\theta^* = \theta_l$.

system. The equilibrium Eq. 5.2, expanded up to the third degree, is rewritten as:

$$(k - P\ell)\theta + \frac{1}{6}P\ell\theta^3 = 0. \tag{5.13}$$

Expressed the bifurcated path in Taylor series, as:

$$\theta = \epsilon\,\theta_1 + \epsilon^2\,\theta_2 + \epsilon^3\,\theta_3 + \cdots, \tag{5.14a}$$

$$P = P_c + \epsilon\,P_1 + \epsilon^2\,P_2 + \cdots, \tag{5.14b}$$

the following perturbation equations in the unknowns θ_k, P_k are obtained:

$$\epsilon^1 : (k - P_c\ell)\,\theta_1 = 0, \tag{5.15a}$$

$$\epsilon^2 : (k - P_c\ell)\,\theta_2 = P_1\ell\,\theta_1, \tag{5.15b}$$

$$\epsilon^3 : (k - P_c\ell)\,\theta_3 = P_2\ell\,\theta_1 + P_1\ell\,\theta_2 - \frac{1}{6}P_c\ell\,\theta_1^3. \tag{5.15c}$$

The normalization $\epsilon = \theta$ is chosen, which implies:

$$\theta_1 = 1, \quad \theta_2 = 0, \quad \theta_3 = 0. \tag{5.16}$$

By solving the perturbation equations in sequence, the following results are drawn:

$$\epsilon^1 : P_c = \frac{k}{\ell}, \tag{5.17a}$$

$$\epsilon^2 : P_1 = 0, \tag{5.17b}$$

$$\epsilon^3 : P_2 = \frac{1}{6}P_c\,\theta_1^2, \tag{5.17c}$$

by virtue of which the bifurcated path is written as:

$$P = \frac{k}{\ell}\left(1 + \frac{\theta^2}{6}\right). \tag{5.18}$$

This is the series expansion of the exact path in Eq. 5.3. The asymptotic path is a parabolic extrapolation from the bifurcation point; the two paths are compared in Fig. 5.4. □

Supplement 5.3 (Perturbation analysis of the imperfect system) The equilibrium Eq. 5.8, expanded up to the terms of the third degree, appears in the form:

$$k\,(\theta - \theta_0) - P\ell\left(\theta - \frac{1}{6}\theta^3\right) = 0. \tag{5.19}$$

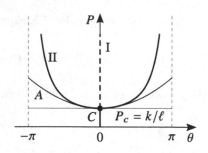

Fig. 5.4 Comparison between exact (II) and asymptotic (A) paths of the inverted pendulum

Since the system is symmetric, the imperfection is rescaled to the cubic order, by setting $\theta_0 = \epsilon^3 \hat{\theta}_0$. Introduced the series expansions (Eqs. 5.14) in the equilibrium equation, the perturbation equations follow:

$$\epsilon^1 : (k - P_c\ell)\,\theta_1 = 0, \tag{5.20a}$$

$$\epsilon^2 : (k - P_c\ell)\,\theta_2 = P_1\ell\,\theta_1, \tag{5.20b}$$

$$\epsilon^3 : (k - P_c\ell)\,\theta_3 = P_2\ell\,\theta_1 + P_1\ell\,\theta_2 - \frac{1}{6}P_c\ell\,\theta_1^3 + k\,\hat{\theta}_0. \tag{5.20c}$$

Using the same normalization as the system perfect, Eq. 5.16, and solving the equations in sequence, the unknowns are determined as:

$$\epsilon^1 : P_c = \frac{k}{\ell}, \tag{5.21a}$$

$$\epsilon^2 : P_1 = 0, \tag{5.21b}$$

$$\epsilon^3 : P_2 = \frac{1}{6}P_c\,\theta_1^2 - \frac{k\,\hat{\theta}_0}{\ell\,\theta_1}, \tag{5.21c}$$

from which the imperfect path takes the expression:

$$P = \frac{k}{\ell}\left(1 + \frac{\theta^2}{6} - \frac{\theta_0}{\theta}\right). \tag{5.22}$$

The family of asymptotic paths of the imperfect system is presented in Fig. 5.5a. To determine the limit point of the non-natural branch, $\frac{dP}{d\theta} = 0$ is imposed, entailing $\theta_l = -(3\theta_0)^{1/3}$, at which the load takes the limit value:

$$P_l = \frac{k}{\ell}\left(1 + \frac{1}{2}(3\theta_0)^{2/3}\right). \tag{5.23}$$

The relevant graph is presented in Fig. 5.5b, known as *imperfection sensitivity* plot.

□

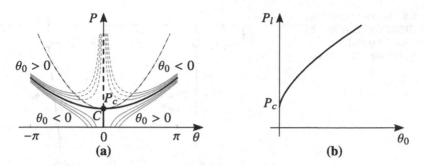

Fig. 5.5 Perturbation analysis of the imperfect inverted pendulum: (a) equilibrium paths, (b) sensitivity of the limit load with respect to imperfection; critical locus (dash-point curve)

5.2.2 *Inverted Pendulum with Sliding Spring*

An inverted pendulum is considered, consisting of a rigid rod of length ℓ, hinged at the bottom, elastically constrained at the top by an extensional spring of stiffness k, which is free to slide (without friction) at its grounded end, subject to a dead load P (Fig. 5.6a). The rotation θ is taken as the unique Lagrangian parameter.

Exact Analysis of the Perfect System

The elastic potential energy of the sliding spring is $U = \frac{1}{2}k\Delta a^2$, where $\Delta a := \ell \sin\theta$ is its extension. The TPE of the system reads:

$$\Pi = U + V = \frac{1}{2}k\,\ell^2 \sin^2\theta - P\ell\,(1 - \cos\theta). \qquad (5.24)$$

The equilibrium condition follows from the stationary condition $\frac{d\Pi}{d\theta} = 0$, namely:

$$\ell\sin\theta\,(k\ell\cos\theta - P) = 0, \qquad (5.25)$$

which represents the balance of torques with respect to the hinge. From this equation, the equilibrium paths are derived:

$$\theta = 0, \qquad \forall P, \qquad\qquad \text{(I)}, \qquad\qquad (5.26a)$$
$$P = k\ell\cos\theta, \qquad\qquad\qquad \text{(II)}, \qquad\qquad (5.26b)$$

presented in Fig. 5.7a. The intersection point between the two branches determines the critical load:

$$P_c = k\ell. \qquad (5.27)$$

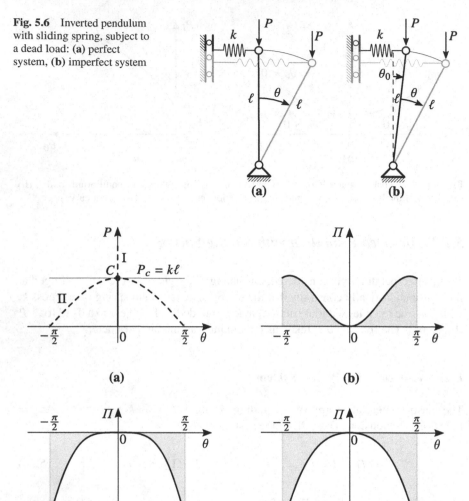

Fig. 5.6 Inverted pendulum with sliding spring, subject to a dead load: (a) perfect system, (b) imperfect system

Fig. 5.7 Exact analysis of the inverted pendulum with sliding spring: (a) bifurcation diagram; TPE diagram, when (b) $P < P_c$, (c) $P = P_c$, (d) $P > P_c$

The diagrams of the TPE are presented in Fig. 5.7b–d for $P \lesseqgtr P_c$, showing that the fundamental path is stable when $P < P_c$ and unstable when $P \geq P_c$; the bifurcated path is unstable. Alternatively, the second derivative of the TPE is:

$$\frac{\mathrm{d}^2 \Pi}{\mathrm{d}\theta^2} = k\ell^2 \left(2\cos^2\theta - 1\right) - P\ell\cos\theta, \qquad (5.28)$$

which, evaluated on the two equilibrium paths, provides:

$$\frac{d^2 \Pi}{d\theta^2} = \begin{cases} k\ell^2 - P\ell \lessgtr 0 & \text{if } P \gtrless P_c, \quad (\text{I})\,, \\ \frac{1}{k} \left(P^2 - k^2 \ell^2 \right) \leq 0 & \text{if } P \leq P_c \quad (\text{II})\,. \end{cases} \tag{5.29}$$

Since $\frac{d^2 \Pi}{d\theta^2} = 0$ at the critical state, its stability is determined by the successive derivatives. Since the third derivative vanishes and the fourth is negative, the bifurcation point is unstable. This type of bifurcation is called a *sub-critical fork*, or unstable fork.

Supplement 5.4 (Interpretation of geometric nonlinearity) The unstable behavior of the inverted pendulum with sliding spring can be explained as follows. Due to the translation of the spring, its arm with respect to the hinge is smaller than that estimated in linear theory, so the *linear theory overestimates the elastic reaction* (soft spring). Moving on to consider the external force, it is observed that its torque with respect the hinge is also smaller of the linear case (hardening effect); however, this geometric effect is less important than the previous one (indeed, the arm of the internal force, $\ell \cos \theta \simeq \ell \left(1 - \frac{\theta^2}{2} \right)$, deviates from its linear approximation more than the arm of the external force does, $\ell \sin \theta \simeq \ell \left(\theta - \frac{\theta^3}{6} \right)$). Therefore, between the two effects, the softening of the spring prevails. When the load reaches the critical value, the total stiffness is null, but the system is unstable, because the nonlinear elastic reaction is smaller than the linear one. $\square$

Exact Analysis of the Imperfect System

A θ_0 deviation of the rod is considered as a geometric imperfection (Fig. 5.6b). The TPE changes as follows:

$$\Pi = \frac{1}{2} k \left[\ell \sin \theta - \ell \sin \theta_0 \right]^2 - P\ell \left(1 - \cos \theta \right) + P\ell \left(1 - \cos \theta_0 \right), \tag{5.30}$$

from which, by letting $\frac{d\Pi}{d\theta} = 0$, the equilibrium condition is derived:

$$\ell \left[k\ell \cos \theta \left(\sin \theta - \sin \theta_0 \right) - P \sin \theta \right] = 0. \tag{5.31}$$

This admits the solution:

$$P = k\ell \frac{\sin \theta - \sin \theta_0}{\tan \theta}, \tag{5.32}$$

which describes the natural and non-natural paths plotted in Fig. 5.8a.

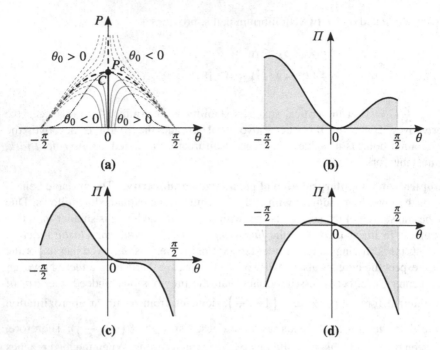

Fig. 5.8 Exact analysis of the imperfect inverted pendulum with sliding spring: (a) bifurcation diagram; TPE diagram, for $\theta_0 > 0$, when (b) $P < P_c$, (c) $P = P_c$, (d) $P > P_c$; critical locus (dash-point curve)

An *upper limit point* appears on the natural branches. The stationary condition $\frac{\mathrm{d}P}{\mathrm{d}\theta} = 0$ supplies $\sin\theta_l = \sin^{\frac{1}{3}}\theta_0$, which, introduced in the previous equation, furnishes the limit load value:

$$P_l = k\ell \left(1 - \sin^{\frac{2}{3}}\theta_0\right)^{\frac{2}{3}}. \tag{5.33}$$

To analyze the stability, the TPE is plotted in Fig. 5.8b–d; alternatively:

$$\frac{\mathrm{d}^2\Pi}{\mathrm{d}\theta^2} = k\ell^2 \left(\cos^2\theta - \sin^2\theta + \sin\theta\sin\theta_0\right) - P\ell\cos\theta, \tag{5.34}$$

is evaluated on the equilibrium paths of the imperfect structure, to furnish:

$$\frac{\mathrm{d}^2\Pi}{\mathrm{d}\theta^2} = k\ell^2 \left(\frac{\sin\theta_0}{\sin\theta} - \sin^2\theta\right). \tag{5.35}$$

The critical locus, which satisfies the condition $\frac{\mathrm{d}^2\Pi}{\mathrm{d}\theta^2} = 0$, is the curve of equation $\sin\theta_0 = \sin^3\theta$, coinciding with the locus of the maxima of the natural equilibrium paths (Fig. 5.8a). This curve separates stable states (below the curve) from unstable states (above the curve). At the limit points, stability changes.

Supplement 5.5 (Perturbation analysis of the perfect system) The equilibrium Eq. 5.25, expanded up to terms of the third degree, is:

$$k\ell^2 \left(\theta - \frac{2}{3}\theta^3 \right) - P\ell \left(\theta - \frac{1}{6}\theta^3 \right) = 0. \tag{5.36}$$

Substitution of the series expansions (Eqs. 5.14) leads to the perturbation equations:

$$\epsilon^1 : \left(k\ell^2 - P_c\ell \right)\theta_1 = 0, \tag{5.37a}$$

$$\epsilon^2 : \left(k\ell^2 - P_c\ell \right)\theta_2 = P_1\ell\,\theta_1, \tag{5.37b}$$

$$\epsilon^3 : \left(k\ell^2 - P_c\ell \right)\theta_3 = P_2\ell\,\theta_1 + P_1\ell\,\theta_2 + \left(\frac{2}{3}k\ell - \frac{1}{6}P_c \right)\ell\,\theta_1^3. \tag{5.37c}$$

Using again the normalization Eq. 5.16 and solving in sequence, the unknowns are evaluated as:

$$\epsilon^1 : P_c = k\ell, \tag{5.38a}$$

$$\epsilon^2 : P_1 = 0, \tag{5.38b}$$

$$\epsilon^3 : P_2 = \frac{1}{6}(P_c - 4k\ell)\theta_1^2, \tag{5.38c}$$

from which the parabolic approximation of the path is obtained:

$$P = k\ell \left(1 - \frac{\theta^2}{2} \right). \tag{5.39}$$

The comparison between exact and asymptotic path is shown in Fig. 5.9.   □

Fig. 5.9 Comparison between exact (II) and asymptotic (A) paths of the inverted pendulum with sliding spring

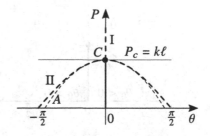

Supplement 5.6 (Perturbation analysis of the imperfect system) The equilibrium Eq. 5.31, expanded in series, becomes:

$$k\ell^2 \left(\theta - \frac{2}{3}\theta^3 - \theta_0 + \frac{1}{2}\theta^2\theta_0 + \frac{1}{6}\theta_0^3 \right) - P\ell \left(\theta - \frac{1}{6}\theta^3 \right) = 0. \qquad (5.40)$$

Making use of the series in Eq. 5.14, and rescaling the imperfection as $\theta_0 = \epsilon^3 \hat{\theta}_0$, with $\hat{\theta}_0 = O(1)$, the perturbation equations follow:

$$\epsilon^1 : \left(k\ell^2 - P_c\ell \right)\theta_1 = 0, \qquad (5.41a)$$

$$\epsilon^2 : \left(k\ell^2 - P_c\ell \right)\theta_2 = P_1\ell\,\theta_1, \qquad (5.41b)$$

$$\epsilon^3 : \left(k\ell^2 - P_c\ell \right)\theta_3 = P_2\ell\,\theta_1 + P_1\ell\,\theta_2 + \left(\frac{2}{3}k\ell - \frac{1}{6}P_c \right)\ell\,\theta_1^3 + k\ell^2\hat{\theta}_0. \qquad (5.41c)$$

Solution in sequence, with the help of the normalization Eq. 5.16, provides:

$$\epsilon^1 : P_c = k\ell, \qquad (5.42a)$$

$$\epsilon^2 : P_1 = 0, \qquad (5.42b)$$

$$\epsilon^3 : P_2 = \frac{1}{6}(P_c - 4k\ell)\theta_1^2 - k\ell\frac{\hat{\theta}_0}{\theta_1}, \qquad (5.42c)$$

from which the path is found in the form (Fig. 5.10a):

$$P = k\ell \left(1 - \frac{\theta^2}{2} - \frac{\theta_0}{\theta} \right). \qquad (5.43)$$

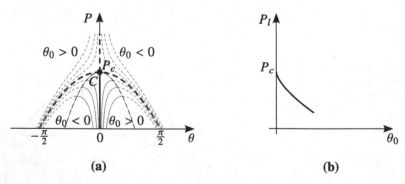

(a)   **(b)**

Fig. 5.10 Perturbation analysis of the inverted pendulum with sliding spring: (a) equilibrium paths, (b) sensitivity of the limit load with respect to imperfection; critical locus (dash-dot curve)

To determine the limit point, $\frac{\mathrm{d}P}{\mathrm{d}\theta} = 0$ is imposed, which implies $\theta_l = \theta_0^{1/3}$; substituting it in the previous equation:

$$P_l = k\ell \left(1 - \frac{3}{2}\theta_0^{2/3} \right) \tag{5.44}$$

is found, whose graph is plotted in Fig. 5.10b.   □

5.2.3 Cable-Stayed Inverted Pendulum

A single DOF system, prototype of non-symmetric systems, is now studied. It consists of an inverted pendulum of length ℓ, stayed by a spring of stiffness k, which connects the free end of the rod to the ground, inclined of an angle $\pi/4$ on the horizontal. The pendulum is subjected to a dead load P applied at the top (Fig. 5.11a). The θ rotation is assumed as a Lagrangian parameter.

Exact Analysis of the Perfect System

Denoting by $a^\star = \sqrt{2}\,\ell$ the initial length of the spring, and by $a = \sqrt{2}\,\ell\sqrt{1+\sin\theta}$ the current length, the elastic energy reads $U = \frac{1}{2}k\,(a-a^\star)^2$; consequently:

$$\Pi = U + V = k\,\ell^2 \left(\sqrt{1+\sin\theta} - 1 \right)^2 - P\ell\,(1-\cos\theta). \tag{5.45}$$

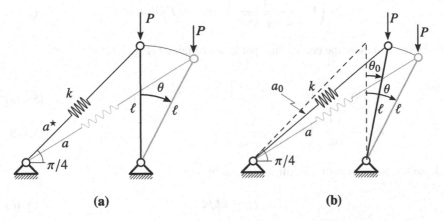

Fig. 5.11 Cable-stayed inverted pendulum under a dead load: (a) perfect system, (b) imperfect system

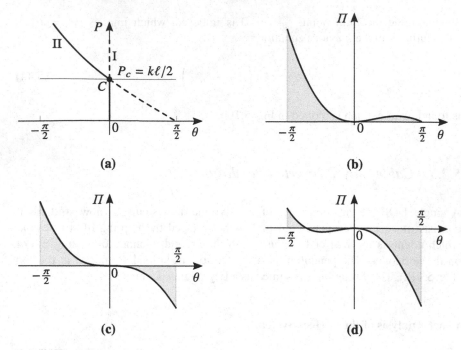

Fig. 5.12 Exact analysis of the cable-stayed inverted pendulum: (**a**) bifurcation diagram; TPE diagram, when (**b**) $P < P_c$, (**c**) $P = P_c$, (**d**) $P > P_c$

By requiring it to be stationary, i.e., $\frac{\mathrm{d}\Pi}{\mathrm{d}\theta} = 0$, the balance law follows:

$$k\ell \left(1 - \frac{1}{\sqrt{1 + \sin\theta}} \right) \cos\theta - P \sin\theta = 0. \tag{5.46}$$

From this equation, the equilibrium paths are obtained (Fig. 5.12a):

$$\theta = 0, \qquad \forall P, \qquad\qquad\qquad (\mathrm{I}), \tag{5.47a}$$

$$P = k\ell \frac{1}{\tan\theta} \left(1 - \frac{1}{\sqrt{1 + \sin\theta}} \right), \qquad (\mathrm{II}), \tag{5.47b}$$

whose intersection point identifies the critical load:

$$P_c = k\ell/2. \tag{5.48}$$

The TPE diagrams, in the sub-critical, critical, and super-critical cases (Fig. 5.12b–d), show that the fundamental path loses stability at the bifurcation and that the bifurcated path consists of a stable and an unstable branch. Stability can also be ascertained by the sign of:

$$\frac{d^2 \Pi}{d\theta^2} = k\ell^2 \left(\frac{1}{2}\sqrt{1 + \sin\theta} - \sin\theta \right) - P\ell\cos\theta, \qquad (5.49)$$

when evaluated on the two equilibrium paths; it turns out that:[2]

$$\frac{d^2 \Pi}{d\theta^2} = \begin{cases} \frac{k\ell^2}{2} - P\ell \lesseqgtr 0, \quad \text{if } P \gtreqless P_c, & \text{(I)}, \\[2mm] k\ell^2 \frac{2\sin\theta - 4\sqrt{1+\sin\theta} + \cos(2\theta) + 3}{4\sin\theta\,\sqrt{1+\sin\theta}} \gtreqless 0, \quad \text{if } \begin{cases} \theta \in (-\pi, 0), \\ \theta \in (0, \pi), \end{cases} & \text{(II)}. \end{cases}$$

$$(5.50)$$

At $P = P_c$, being the second derivative of Π equal to zero and the third derivative different from zero, the equilibrium is unstable. This type of bifurcation is called *transcritical*.

Remark 5.2 It is noticed that there is a stable equilibrium position for each value of P: trivial, if $P < P_c$, and non-trivial, if $P > P_c$.

Supplement 5.7 (Interpretation of geometric nonlinearity) To understand the reasons for the non-symmetric behavior of the cable-stayed pendulum, it needs to compare linear and nonlinear kinematics. By looking at the problem from a linear perspective, in order to evaluate the elongation of the spring, the horizontal displacement of the free end of the rod B is projected onto the original direction of the same spring. Hence, in infinitesimal kinematics, the elongation due to a clockwise rotation θ is equal in magnitude to the shortening produced by the same counterclockwise rotation. The system, therefore, behaves symmetrically. In nonlinear kinematics, however, due to finite rotations, the point B moves to a point B'' which is lower than B, for both clockwise and counterclockwise θ's. This effect increases contractions (hardening effect for counterclockwise θ) and reduces extensions (softening effect for clockwise θ), so the elastic behavior is non-symmetric. This explains why the path issued does not respect symmetry and why the equilibria are stable for counterclockwise θ, entailing hardening behavior.   □

Exact Analysis of the Imperfect System

A small angular deviation θ_0 of the inverted cable-stayed pendulum is introduced as a geometric imperfection (Fig. 5.11b). The initial stay length is $a_0 :=$

---

[2] Although the sign of $\frac{d^2 \Pi}{d\theta^2}$ on the bifurcated path is not of immediate determination, it is possible to prove what asserted.

$\sqrt{2}\,\ell\sqrt{1 + \sin\theta_0}$, and the elastic energy $U = \frac{1}{2}k\,(a - a_0)^2$; the TPE, therefore, reads:

$$\Pi = \ell^2 \left(\sqrt{1 + \sin\theta_0} - \sqrt{1 + \sin\theta_0}\right)^2 - P\ell\,(1 - \cos\theta) + P\ell\,(1 - \cos\theta_0). \tag{5.51}$$

From stationary $\frac{d\Pi}{d\theta} = 0$, the equilibrium equation is drawn:

$$k\ell \left(1 - \frac{\sqrt{1 + \sin\theta_0}}{\sqrt{1 + \sin\theta}}\right)\cos\theta - P\sin\theta = 0, \tag{5.52}$$

which is satisfied by:

$$P = k\ell\,\frac{1}{\tan\theta}\left(1 - \frac{\sqrt{1 + \sin\theta_0}}{\sqrt{1 + \sin\theta}}\right). \tag{5.53}$$

The family of curves of Eq. 5.53, taking θ_0 as the parameter, are plotted in Fig. 5.13a. From the TPE diagrams in the sub-critical, critical, and super-critical cases, information is obtained on stability, as shown in Fig. 5.13b–d. It is seen that:

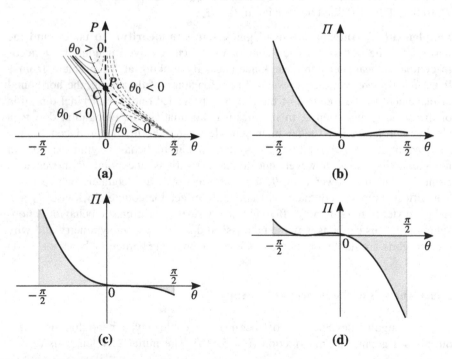

Fig. 5.13 Exact analysis of the imperfect cable-stayed pendulum: (a) bifurcation diagram; TPE diagram for $\theta_0 > 0$, when (b) $P < P_c$, (c) $P = P_c$, (d) $P > P_c$; critical locus (dash-point curve)

- if $\theta_0 < 0$, the natural branch is stable and the non-natural one unstable; these branches have no limit points;
- if $\theta_0 > 0$, both branches consist of a stable and an unstable piece, separated by a limit point (whose expression is not shown here).

The same result is obtained by studying the sign of $\frac{d^2 \Pi}{d\theta^2}$ on the path. The relevant discussion is omitted.

Supplement 5.8 (Perturbation analysis of the perfect system) The equilibrium Eq. 5.46, when expanded in series, reads:

$$k\ell^2 \left(\frac{1}{2}\theta - \frac{3}{8}\theta^2 - \frac{1}{48}\theta^3 \right) - P\ell \left(\theta - \frac{1}{6}\theta^3 \right) = 0. \tag{5.54}$$

By using the series expansions in Eqs. 5.14, the previous equation provides:

$$\epsilon^1 : \left(\frac{k\ell^2}{2} - P_c \ell \right) \theta_1 = 0, \tag{5.55a}$$

$$\epsilon^2 : \left(\frac{k\ell^2}{2} - P_c \ell \right) \theta_2 = P_1 \ell \theta_1 + \frac{3}{8} k\ell^2 \theta_1^2, \tag{5.55b}$$

$$\epsilon^3 : \left(\frac{k\ell^2}{2} - P_c \ell \right) \theta_3 = P_2 \ell \theta_1 + P_1 \ell \theta_2 + \left(\frac{1}{48} k\ell - \frac{1}{6} P_c \right) \ell \theta_1^3 + \frac{3}{4} k\ell^2 \theta_1 \theta_2. \tag{5.55c}$$

By solving in sequence, and exploiting the normalization Eq. 5.16, one finds:

$$\epsilon^1 : P_c = \frac{k\ell}{2}, \tag{5.56a}$$

$$\epsilon^2 : P_1 = -\frac{3}{8} k\ell \theta_1, \tag{5.56b}$$

$$\epsilon^3 : P_2 = \frac{1}{48} (8P_c - k\ell)\theta_1^2 - \frac{3}{8} k\ell \theta_2, \tag{5.56c}$$

hence, the parabolic approximation of the path is obtained (Fig. 5.14):

$$P = \frac{k\ell}{2} \left(1 - \frac{3}{4}\theta + \frac{1}{8}\theta^2 \right). \tag{5.57}$$

□

Supplement 5.9 (Perturbation analysis of the imperfect system) The equilibrium Eq. 5.52, when expanded up to the terms of the third degree, becomes:

Fig. 5.14 Comparison
between exact (II) and
asymptotic (A) paths of the
cable-stayed inverted
pendulum

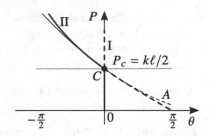

$$k\ell^2 \left[\frac{1}{2}(\theta - \theta_0) + \frac{1}{4}\theta\theta_0 + \frac{1}{8}\left(\theta_0^2 - 3\theta^2\right) + \frac{1}{16}\left(\theta^2\theta_0 - \theta\theta_0^2\right) + \frac{1}{48}\left(\theta_0^3 - \theta^3\right) \right]$$

$$- P\ell\left(\theta - \frac{1}{6}\theta^3\right) = 0.$$

$$\text{(5.58)}$$

Using again the series Eq. 5.14 and introducing the rescaling $\theta_0 = \epsilon^2\,\hat{\theta}_0$, $\hat{\theta}_0 = O(1)$, the perturbation equations are derived:

$$\epsilon^1 : \left(\frac{k\ell^2}{2} - P_c\ell\right)\theta_1 = 0,$$

$$\text{(5.59a)}$$

$$\epsilon^2 : \left(\frac{k\ell^2}{2} - P_c\ell\right)\theta_2 = P_1\ell\theta_1 + \frac{3}{8}k\ell^2\theta_1^2 + \frac{1}{2}k\ell^2\theta_0,$$

$$\text{(5.59b)}$$

where the analysis has been limited to the first significant terms. Solving in sequence with the normalization Eqs. 5.16, the unknowns are evaluated:

$$\epsilon^1 : \ P_c = \frac{k\ell}{2},$$

$$\text{(5.60a)}$$

$$\epsilon^2 : \ P_1 = -\frac{3}{8}k\ell\theta_1 - \frac{1}{2}k\ell\frac{\theta_0}{\theta_1},$$

$$\text{(5.60b)}$$

from which the imperfect path is drawn (Fig. 5.15a):

$$P = \frac{k\ell}{2}\left(1 - \frac{3}{4}\theta - \frac{\theta_0}{\theta}\right).$$

$$\text{(5.61)}$$

The limit point is easily assessed. By requiring $\frac{dP}{d\theta} = 0$, two roots $\theta_l^{\pm} = \pm\frac{2}{\sqrt{3}}\sqrt{\theta_0}$ are found, to which the following limit loads correspond, respectively:

$$P_l^{\pm} = \frac{k\ell}{2}\left(1 \pm \sqrt{3}\sqrt{\theta_0}\right),$$

$$\text{(5.62)}$$

plotted in Fig. 5.15b.

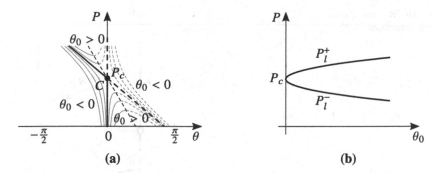

Fig. 5.15 Perturbation analysis of the imperfect cable-stayed inverted pendulum: (**a**) equilibrium paths, (**b**) sensitivity of the limit load with respect to imperfection; critical locus (dash-dot curve)

It is worth noticing that non-symmetric systems are more sensitive to imperfections than unstable symmetric systems. Indeed, the limit load decays with the power $\frac{1}{2}$ (Eq. 5.62) of the imperfection $\theta_0 \ll 1$, instead of the power $\frac{2}{3}$ (Eq. 5.44). □

5.3 Two Degrees of Freedom Systems with Trivial Fundamental Path

As a few examples of multi-DOF systems, structures with *two degrees of freedom* are considered here. These allow studying the cases of *distinct* or *quasi-coincident* critical loads. In the first example, the role of the critical eigenvectors is discussed, both on the incipient deformation and on the evolution of the deflection as the load increases. In the second example, the phenomenon of *interaction between quasi-simultaneous modes* is illustrated.

5.3.1 Reverse Elastic Double Pendulum

As a paradigmatic example of discrete systems with several degrees of freedom (and symmetric behavior), the elastic double pendulum is considered. The analysis is limited to the perfect structure and conducted through the perturbation method.

The system consists of two rigid rods of equal length ℓ, hinged to each other and to the ground, equipped with two elastic torsional springs of stiffness k, subject to a dead load of intensity P (Fig. 5.16). The system has two DOF, described by the Lagrangian parameters θ_1, θ_2, which represent the rotations of the lower and upper rod, respectively.

Fig. 5.16 Reverse elastic
double pendulum

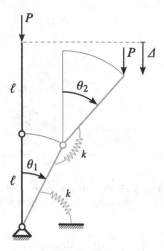

The Equilibrium Equations

The TPE reads:

$$\Pi = \frac{1}{2} k \, [\theta_1^2 + (\theta_2 - \theta_1)^2)] - P \, \Delta. \tag{5.63}$$

Here, the first two terms represent the elastic potential energy of the lower and upper spring, which undergo torsion θ_1 and $\theta_2 - \theta_1$, respectively; the third term is the gravitational potential energy, in which:

$$\Delta = \ell \, (2 - \cos \theta_1 - \cos \theta_2) \tag{5.64}$$

is the lowering of the free end of the double pendulum. In view of a perturbation solution, the TPE is expanded in series, up to the terms of the fourth degree, thus obtaining:

$$\Pi = \frac{1}{2} k \, [\theta_1^2 + (\theta_2 - \theta_1)^2] - P\ell \left(\frac{\theta_1^2}{2} + \frac{\theta_2^2}{2} - \frac{\theta_1^4}{24} - \frac{\theta_2^4}{24} \right). \tag{5.65}$$

Equilibrium requires TPE to be stationary, i.e., the partial derivatives with respect to the Lagrangian parameters vanish separately:

$$\frac{\partial \Pi}{\partial \theta_1} = k \, (2\theta_1 - \theta_2) - P\ell \left(\theta_1 - \frac{\theta_1^3}{6} \right) = 0, \tag{5.66a}$$

$$\frac{\partial \Pi}{\partial \theta_2} = k \, (\theta_2 - \theta_1) - P\ell \left(\theta_2 - \frac{\theta_2^3}{6} \right) = 0. \tag{5.66b}$$

In matrix form, the previous equations are written as:

$$\left(\begin{bmatrix} 2 & -1 \\ -1 & 1 \end{bmatrix} - \mu \begin{bmatrix} 1 & 0 \\ 0 & 1 \end{bmatrix} \right) \begin{pmatrix} \theta_1 \\ \theta_2 \end{pmatrix} + \frac{1}{6} \mu \begin{pmatrix} \theta_1^3 \\ \theta_2^3 \end{pmatrix} = \begin{pmatrix} 0 \\ 0 \end{pmatrix}, \tag{5.67}$$

in which, for convenience, a nondimensional load has been introduced:

$$\mu := \frac{P\ell}{k}. \tag{5.68}$$

In Eq. 5.67 one recognizes, in the exact order, the elastic stiffness matrix, the geometric stiffness matrix, and the nonlinearity vector (cubic, since the system is symmetric).

Bifurcation Analysis

The double pendulum admits the trivial path (I) of parametric equations $\theta_1 = 0$, $\theta_2 = 0 \; \forall \mu$, indicating that the rectilinear configuration is equilibrated whatever the intensity of the force is. A bifurcation analysis from the trivial path is then performed, which consists of the following steps: (a) first, a bifurcation point C is sought along the path, having coordinates $(\theta_1, \theta_2, \mu) = (0, 0, \mu_c)$, with μ_c the unknown critical load, and (b) a bifurcated path (II), branching off from C, is built up in the neighborhood of C.

The bifurcated path is described in the parametric form $\theta_1 = \theta_1(\epsilon)$, $\theta_2 = \theta_2(\epsilon)$, $\mu = \mu(\epsilon)$, where ϵ is a parameter vanishing at C; in series form and for ϵ small, it reads:

$$\theta_i = \epsilon \theta_{i1} + \epsilon^2 \theta_{i2} + \epsilon^3 \theta_{i3} + \cdots, \qquad i = 1, 2, \tag{5.69a}$$

$$\mu = \mu_c + \epsilon \mu_1 + \epsilon^2 \mu_2 + \cdots. \tag{5.69b}$$

The analysis is simplified if the symmetry properties of the system are exploited, which lead to the following conclusions: (a) in the θ_i (ϵ) series only the odd powers are non-zero; (b) in the μ (ϵ) series only the even powers are non-zero.[3] Hence:

$$\theta_i = \epsilon \theta_{i1} + \epsilon^3 \theta_{i3} + \cdots, \qquad i = 1, 2, \tag{5.70a}$$

$$\mu = \mu_c + \epsilon^2 \mu_2 + \cdots. \tag{5.70b}$$

Substituting Eq. 5.70 in the equilibrium Eq. 5.67 and separately zeroing the terms with the powers of ϵ, the perturbation equations are derived:

---

[3] Indeed, by changing the sign of ϵ, the displacement changes sign, while the load remains unchanged, so for any μ, two equilibria exist of opposite sign.

- Order ϵ:

$$\left(\begin{bmatrix} 2 & -1 \\ -1 & 1 \end{bmatrix} - \mu_c \begin{bmatrix} 1 & 0 \\ 0 & 1 \end{bmatrix} \right) \begin{pmatrix} \theta_{11} \\ \theta_{21} \end{pmatrix} = \begin{pmatrix} 0 \\ 0 \end{pmatrix}, \tag{5.71}$$

- Order ϵ^3:

$$\left(\begin{bmatrix} 2 & -1 \\ -1 & 1 \end{bmatrix} - \mu_c \begin{bmatrix} 1 & 0 \\ 0 & 1 \end{bmatrix} \right) \begin{pmatrix} \theta_{13} \\ \theta_{23} \end{pmatrix} = -\frac{1}{6} \mu_c \begin{pmatrix} \theta_{11}^3 \\ \theta_{21}^3 \end{pmatrix} + \mu_2 \begin{bmatrix} 1 & 0 \\ 0 & 1 \end{bmatrix} \begin{pmatrix} \theta_{11} \\ \theta_{21} \end{pmatrix}. \tag{5.72}$$

These equations should be accompanied by normalization conditions. Choosing to identify the ϵ parameter, for example, with θ_1, it turns out that:

$$\theta_{11} = 1, \qquad \theta_{13} = 0, \cdots . \tag{5.73}$$

Critical Loads and Modes

The perturbation equations of order ϵ, Eq. 5.71, constitutes a homogeneous problem (of the type of Eq. 4.30), whose eigenvalues are the critical loads μ_c. The characteristic equation follows from nullifying the determinant of the total stiffness matrix, i.e.:

$$\begin{vmatrix} 2 - \mu_c & -1 \\ -1 & 1 - \mu_c \end{vmatrix} = (2 - \mu_c)(1 - \mu_c) - 1 = 0. \tag{5.74}$$

By solving the equation, *two* critical loads are obtained:[4]

$$\mu_{c1} = \frac{3 - \sqrt{5}}{2} \simeq 0.382, \tag{5.75a}$$

$$\mu_{c2} = \frac{3 + \sqrt{5}}{2} \simeq 2.618. \tag{5.75b}$$

each associated with a different bifurcation point. The smallest of the two, $\mu_c := \mu_{c1}$ is called the (first) critical load.

By substituting the roots, one by one, in the eigenvalue problem, and solving the linear equations, it follows:

$$\begin{bmatrix} 2 - \mu_{c1} & -1 \\ -1 & 1 - \mu_{c1} \end{bmatrix} \begin{pmatrix} \theta_{11} \\ \theta_{21} \end{pmatrix} = \begin{pmatrix} 0 \\ 0 \end{pmatrix} \Rightarrow \begin{pmatrix} \theta_{11} \\ \theta_{21} \end{pmatrix} = \begin{pmatrix} 1 \\ 1.618 \end{pmatrix} =: \mathbf{u}_{c1}, \tag{5.76a}$$

---

[4] The symbols μ_{c1}, μ_{c2}, $\cdots$, denoting the critical loads, should not be confused with symbols μ_1, μ_2, denoting the coefficients of the series expansion of the load.

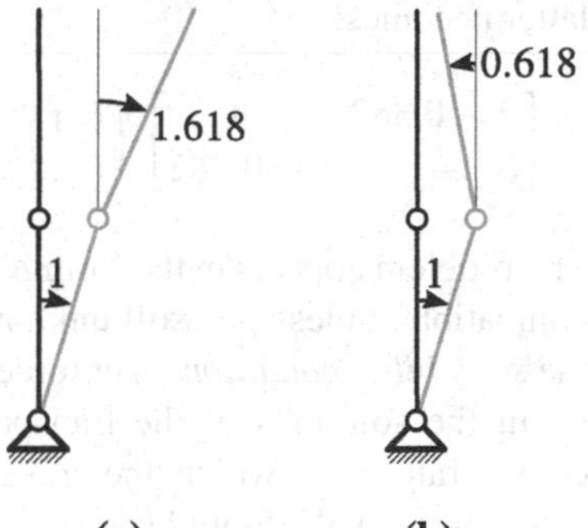

Fig. 5.17 First (**a**) and
second (**b**) critical mode of
the reverse elastic double
pendulum

Fig. 5.18 Tangents to the
bifurcated paths of the double
pendulum at the first and
second bifurcation point

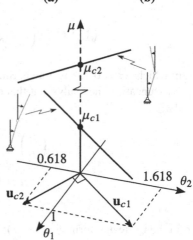

$$\begin{bmatrix} 2 - \mu_{c2} & -1 \\ -1 & 1 - \mu_{c2} \end{bmatrix} \begin{pmatrix} \theta_{11} \\ \theta_{21} \end{pmatrix} = \begin{pmatrix} 0 \\ 0 \end{pmatrix} \Rightarrow \begin{pmatrix} \theta_{11} \\ \theta_{21} \end{pmatrix} = \begin{pmatrix} 1 \\ -0.618 \end{pmatrix} =: \mathbf{u}_{c2}, \quad (5.76b)$$

where the first of the normalization Eq. 5.73 has been used for both the eigenvectors.
The two critical modes are presented in Fig. 5.17.

The critical modes, according to the series in Eq. 5.70, represent the projection
onto the displacement space of the straight lines tangent to the paths, branching off
from the trivial one (Fig. 5.18). In case of symmetric systems, for which $\mu_1 = 0$,
the tangents to the paths are horizontal and therefore completely identified by the
eigenvectors.

Bifurcated Path

The bifurcated path passing through the lowest branching point is now built up. The
procedure, of course, can be repeated for the upper path. By substituting $\mu_c = 0.382$
and $(\theta_{11}, \theta_{21}) = (1, 1.618)$ in the perturbation equations of order ϵ^3 (Eq. 5.72), this

latter becomes:

$$\begin{bmatrix} 2 - 0.382 & -1 \\ -1 & 1 - 0.382 \end{bmatrix} \begin{pmatrix} \theta_{13} \\ \theta_{23} \end{pmatrix} = -\frac{0.382}{6} \begin{pmatrix} 1^3 \\ 1.618^3 \end{pmatrix} + \mu_2 \begin{pmatrix} 1 \\ 1.618 \end{pmatrix}. \tag{5.77}$$

The problem appears in the form $\mathbf{A}\mathbf{x} = \mathbf{b}$, with $\mathbf{A}$ singular, so that it is generally not compatible, unless μ_2, still unknown, assumes a particular value. To determine it, the *solvability condition* is enforced, which requires the known term be orthogonal to all the solutions of the transposed homogeneous problem. In the case under consideration, in which the transposed homogeneous problem admits only the solution $\mathbf{u}_{c1}$, solvability leads to:

$$\left(1, 1.618 \right) \begin{pmatrix} -\frac{0.382}{6} + \mu_2 \\ -\frac{0.382}{6} 1.618^3 + 1.618 \mu_2 \end{pmatrix} = 0, \tag{5.78}$$

from which $\mu_2 = 0.138$ is drawn. Since $\mu_2 > 0$, the bifurcation is a *stable fork*.

By truncating the analysis at this order, the bifurcated path, in parametric form, reads:

$$\begin{pmatrix} \theta_1 \\ \theta_2 \end{pmatrix} = \epsilon \begin{pmatrix} 1 \\ 1.618 \end{pmatrix}, \tag{5.79a}$$

$$\mu = 0.382 + 0.138 \epsilon^2, \tag{5.79b}$$

or, in the Cartesian form $\theta_i = \theta_i(\mu)$:

$$\theta_1 = \pm \sqrt{\frac{\mu - 0.382}{0.138}}, \tag{5.80a}$$

$$\theta_2 = \pm 1.618 \sqrt{\frac{\mu - 0.382}{0.138}}, \tag{5.80b}$$

plotted in Fig. 5.19a.

Nonlinear Deformation

The approximation of Eq. 5.79 is generally sufficient to describe the behavior of the structure in the postcritical phase, as it expresses the critical load μ_c, the curvature of the bifurcated path μ_2, and the *incipient deformation*. However, it gives qualitatively incomplete information about this latter, asserting it is proportional to the eigenvector $\mathbf{u}_{c1} = (\theta_{11}, \theta_{21})^T$ via the ϵ parameter, i.e., to the square root of the increment of the load with respect the critical value. According to this expression,

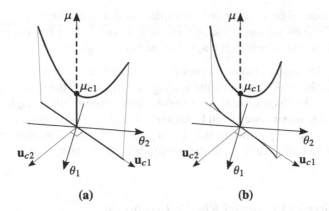

Fig. 5.19 Bifurcated paths of the elastic double pendulum, (**a**) without or (**b**) with modification of the deflection shape

only *the amplitude and not the shape* of the deformation changes with the load, so the displacement ratio $\frac{\theta_2}{\theta_1}$ remains unchanged as the load increases.

This result is a consequence of the truncation. Indeed, as the load grows, the shape of the deformation also changes, altering the ratio between the two rotations. To describe this phenomenon, it is necessary to solve the perturbation Eq. 5.77, after having imposed solvability. Taking into account the value of $\mu_2 = 0.138$, already determined, the same equation reads:

$$\begin{bmatrix} 2 - 0.382 & -1 \\ -1 & 1 - 0.382 \end{bmatrix} \begin{pmatrix} \theta_{13} \\ \theta_{23} \end{pmatrix} = -\frac{0.382}{6} \begin{pmatrix} 1^3 \\ 1.618^3 \end{pmatrix} + 0.138 \begin{pmatrix} 1 \\ 1.618 \end{pmatrix}, \quad (5.81)$$

whose solution is:

$$\begin{pmatrix} \theta_{13} \\ \theta_{23} \end{pmatrix} = -\begin{pmatrix} 0 \\ 0.075 \end{pmatrix} + \alpha \begin{pmatrix} 1 \\ 1.618 \end{pmatrix}, \quad (5.82)$$

where α, which multiplies the critical eigenvector, is an undetermined parameter. The ambiguity is removed by the second of the normalization conditions in Eq. 5.73, which supplies $\alpha = 0$. Consequently, the bifurcated path in Eq. 5.79, modifies as follows:

$$\begin{pmatrix} \theta_1 \\ \theta_2 \end{pmatrix} = \epsilon \begin{pmatrix} 1 \\ 1.618 \end{pmatrix} - \epsilon^3 \begin{pmatrix} 0 \\ 0.075 \end{pmatrix}, \quad (5.83a)$$

$$\mu = 0.382 + 0.138\,\epsilon^2. \quad (5.83b)$$

The additional term in Eq. 5.83a, of order ϵ^3, describes as the shape of the deflection changes with the load. The *enhanced* path is presented in Fig. 5.19b. It shows how the incipient deformation, which manifests along the first eigenvector, is, indeed, modified by a higher-order contribution along the direction of the second eigenvector. The projection of the path on the horizontal plane is therefore a third-degree algebraic curve, which is tangent to the critical eigenvector.

Remark 5.3 The result obtained shows that, if the deflection is decomposed in the modal space, the critical mode is the leading one. However, the second mode also contributes to the displacement as a higher-order quantity, the weight of which increases as one moves away from bifurcation. The first critical mode is also called *active* and the second *passive*, to indicate the fact that the contribution of the latter is a function (at least quadratic) of the amplitude of the former.

5.3.2 Spherical Inverted Elastic Pendulum

Usually, multi-DOF systems exhibit distinct critical loads $\mu_{c1} < \mu_{c2} < \cdots$. Under these circumstances, the first critical mode $\mathbf{u}_{c1}$ is active (which gives the largest contribution to the system response), and the successive ones, $\mathbf{u}_{c2}, \mathbf{u}_{c3}, \cdots$, are passive (which give small corrections). However, there exist situations in which, especially for particular symmetries, the first two (or more) critical loads coincide (or they are very close), i.e., *the critical load is a multiple root* of the characteristic equation of the eigenvalue problem [11, 12]. In such an occurrence, all the simultaneous modes contribute to the leading part of the response.

An example is offered by the Euler beam immersed in 3D space, when this has equal stiffness and equal constraints in the two principal inertia planes. It will be seen shortly that, if an eigenvalue is multiple, *several equilibrium paths* intersect the fundamental path at the bifurcation point. A brief mention of the problem is given here, by dealing with a two-DOF system, known in the literature as the *Augusti model* [2, 10].

Equilibrium Equations

A spherical inverted pendulum is considered, consisting of a rigid rod of length ℓ, immersed in the 3D space, initially oriented as the vertical axis z, and constrained to the ground by means of a spherical hinge and torsional springs of stiffnesses k_1, k_2, which, in the rest condition, lie in the (x, z) and (y, z) planes, respectively (Fig. 5.20a). The rod is subject at a dead load P.

In the current configuration, the rod forms with the axes x, y, z and the angles $\varphi_1, \varphi_2, \varphi_3$, respectively (Fig. 5.20b); however, only two of them are independent, as $\cos^2 \varphi_1 + \cos^2 \varphi_2 + \cos^2 \varphi_3 = 1$. The angles:

$$\theta_i := \frac{\pi}{2} - \varphi_i, \qquad i = 1, 2, \tag{5.84}$$

Fig. 5.20 Spherical inverted elastic pendulum: (**a**) reference configuration, (**b**) current configuration

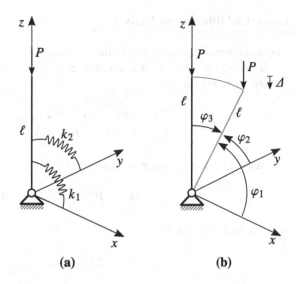

are chosen as Lagrangian parameters, measuring the variations of the initially right angles, formed by the rod, in its current configuration, with the x and y axes.

The total potential energy of the system is:

$$\Pi = \frac{1}{2}\left(k_1\theta_1^2 + k_2\theta_2^2\right) - P\Delta, \tag{5.85}$$

where:

$$\Delta = \ell\,(1 - \cos\varphi_3) = \ell\left(1 - \sqrt{1 - \sin^2\theta_1 - \sin^2\theta_2}\right), \tag{5.86}$$

is the lowering of the free end of the rod, having observed that $\cos\varphi_1 = \sin\theta_1$ and $\cos\varphi_2 = \sin\theta_2$. Expanding up to the fourth-degree terms, one finds:

$$\Pi = \frac{1}{2}\left(k_1\theta_1^2 + k_2\theta_2^2\right) - P\ell\left(\frac{1}{2}\theta_1^2 + \frac{1}{2}\theta_2^2 + \frac{1}{4}\theta_1^2\theta_2^2 - \frac{1}{24}\theta_1^4 - \frac{1}{24}\theta_2^4\right). \tag{5.87}$$

From stationary, the equilibrium equations follow:

$$\frac{\partial\Pi}{\partial\theta_1} = (k_1 - P\ell)\,\theta_1 + P\ell\left(\frac{1}{6}\theta_1^3 - \frac{1}{2}\theta_1\theta_2^2\right) = 0, \tag{5.88a}$$

$$\frac{\partial\Pi}{\partial\theta_2} = (k_2 - P\ell)\,\theta_2 + P\ell\left(\frac{1}{6}\theta_2^3 - \frac{1}{2}\theta_1^2\theta_2\right) = 0. \tag{5.88b}$$

Linearized Bifurcation Analysis

The equilibrium equations admit the trivial solution $\theta_1 = \theta_2 = 0 \forall \mu$, which describe the fundamental path. If only the critical load is of interest, the equations can be linearized, to provide:

$$\begin{bmatrix} k_1 - P\ell & 0 \\ 0 & k_2 - P\ell \end{bmatrix} \begin{pmatrix} \theta_1 \\ \theta_2 \end{pmatrix} = \begin{pmatrix} 0 \\ 0 \end{pmatrix}. \tag{5.89}$$

The relevant characteristic equation is:

$$(k_1 - P\ell)(k_2 - P\ell) = 0, \tag{5.90}$$

which admits two roots:

$$P_{c1} = \frac{k_1}{\ell}, \qquad P_{c2} = \frac{k_2}{\ell}, \tag{5.91}$$

respectively, associated with the critical modes:

$$\mathbf{u}_{c1} = \begin{pmatrix} 1 \\ 0 \end{pmatrix}, \qquad \mathbf{u}_{c2} = \begin{pmatrix} 0 \\ 1 \end{pmatrix}. \tag{5.92}$$

These latter describe the deflection of the pendulum, occurring in the plane (x, z) and (y, z), respectively.

If $k_1 \neq k_2$, the two critical loads are distinct: the (first) critical load is the smaller of the two, $P_c := \min (P_{c1}, P_{c2})$. Correspondingly, *the spherical pendulum buckles in the plane of minor stiffness*, that is, it behaves like a planar pendulum. If, however, $k_1 = k_2$, the two critical loads are coincident, i.e., P_c is a double root of the characteristic equation. In this case, *any linear combination of $\mathbf{u}_{c1}, \mathbf{u}_{c2}$ is still an eigenvector*:

$$\begin{pmatrix} \theta_1 \\ \theta_2 \end{pmatrix} = a_1 \mathbf{u}_{c1} + a_2 \mathbf{u}_{c2} = \begin{pmatrix} a_1 \\ a_2 \end{pmatrix}, \qquad \forall (a_1, a_2). \tag{5.93}$$

In this occurrence, the two critical modes are called *simultaneous*, because they manifest together, under the same load.

Remark 5.4 Differently from the case of simple root, in which the deformation of the structure is defined to within a constant, in the (degenerate) case of double root, *two* indeterminate constants appear. Consequently, not only the amplitude but even the direction along which buckling occurs is unknown at this order of analysis. As it will be seen immediately, the *indeterminacy is resolved by nonlinearities*, which decide the (no longer unique) direction of path.

Bifurcation Analysis in the Degenerate Case

Aimed to study the phenomenon of simultaneous modes, $k_1 = k_2 =: k$ will henceforth be assumed. The equilibrium Eq. 5.88 is rewritten as:

$$\begin{bmatrix} 1 - \mu & 0 \\ 0 & 1 - \mu \end{bmatrix} \begin{pmatrix} \theta_1 \\ \theta_2 \end{pmatrix} + \mu \begin{pmatrix} \frac{1}{6}\theta_1^3 - \frac{1}{2}\theta_1\theta_2^2 \\ \frac{1}{6}\theta_2^3 - \frac{1}{2}\theta_1^2\theta_2 \end{pmatrix} = \begin{pmatrix} 0 \\ 0 \end{pmatrix}, \tag{5.94}$$

where a load parameter has been defined:

$$\mu := \frac{P\ell}{k}. \tag{5.95}$$

The goals of the analysis are (a) to find the critical value of the load at which a multiple static bifurcation occurs and (b) to build up the bifurcated paths (whose number is a priori unknown) branching off from the bifurcation point. According to the perturbation method, the parametric equations of the bifurcated path are written as series expansions. By keeping in mind that the system has a symmetric behavior, these are taken in the form:

$$\begin{pmatrix} \theta_1 \\ \theta_2 \end{pmatrix} = \epsilon \begin{pmatrix} \theta_{11} \\ \theta_{21} \end{pmatrix} + \epsilon^3 \begin{pmatrix} \theta_{13} \\ \theta_{23} \end{pmatrix} + \cdots, \tag{5.96a}$$

$$\mu = \mu_c + \epsilon^2 \mu_2 + \cdots. \tag{5.96b}$$

When the series are substituted in the equilibrium Eq. 5.94, and terms with the same powers as ϵ are separately vanished, the perturbation equations follow:

- Order ϵ:

$$\begin{bmatrix} 1 - \mu_c & 0 \\ 0 & 1 - \mu_c \end{bmatrix} \begin{pmatrix} \theta_{11} \\ \theta_{21} \end{pmatrix} = \begin{pmatrix} 0 \\ 0 \end{pmatrix}, \tag{5.97}$$

- Order ϵ^3:

$$\begin{bmatrix} 1 - \mu_c & 0 \\ 0 & 1 - \mu_c \end{bmatrix} \begin{pmatrix} \theta_{13} \\ \theta_{23} \end{pmatrix} = \mu_2 \begin{pmatrix} \theta_{11} \\ \theta_{21} \end{pmatrix} + \mu_c \begin{pmatrix} -\frac{1}{6}\theta_{11}^3 + \frac{1}{2}\theta_{11}\theta_{21}^2 \\ -\frac{1}{6}\theta_{21}^3 + \frac{1}{2}\theta_{11}^2\theta_{21} \end{pmatrix}. \tag{5.98}$$

Remark 5.5 Unlike the cases discussed above, the imposition of the normalization condition is postponed. Indeed, as it will be seen ahead, there exist bifurcated paths in which one of the two rotations is zero, so it is not possible to identify the perturbation parameter always with the same of them.

Solution to the Perturbation Equations

Equation 5.97 admits the *double eigenvalue*:

$$\mu_c = 1 \tag{5.99}$$

and the (indeterminate) solution:

$$\begin{pmatrix} \theta_{11} \\ \theta_{21} \end{pmatrix} = \begin{pmatrix} a_1 \\ a_2 \end{pmatrix} \qquad \forall (a_1, a_2). \tag{5.100}$$

Equation 5.98, correspondingly, reads:

$$\begin{bmatrix} 0 & 0 \\ 0 & 0 \end{bmatrix} \begin{pmatrix} \theta_{13} \\ \theta_{23} \end{pmatrix} = \underbrace{\left(\begin{bmatrix} \mu_2 + \left(-\frac{1}{6}a_1^2 + \frac{1}{2}a_2^2 \right) \end{bmatrix} a_1 \\ \begin{bmatrix} \mu_2 + \left(-\frac{1}{6}a_2^2 + \frac{1}{2}a_1^2 \right) \end{bmatrix} a_2 \end{bmatrix} \right)}_{=:\mathbf{b}}. \tag{5.101}$$

This is a hyper-singular problem, as the matrix in the left member is twice singular (i.e., it has rank zero). Since the transposed homogeneous problem admits two independent solutions, $\mathbf{u}_{c1}$ and $\mathbf{u}_{c2}$, in order to solve Eq. 5.101, the known term $\mathbf{b}$ must be orthogonal to *both* these solutions, i.e., $\mathbf{u}_{c1}^T \mathbf{b} = 0$, $\mathbf{u}_{c2}^T \mathbf{b} = 0$ must hold. The condition leads to:[5]

$$\left[\mu_2 + \left(-\frac{1}{6}a_1^2 + \frac{1}{2}a_2^2 \right) \right] a_1 = 0, \tag{5.102a}$$

$$\left[\mu_2 + \left(-\frac{1}{6}a_2^2 + \frac{1}{2}a_1^2 \right) \right] a_2 = 0. \tag{5.102b}$$

Bifurcated Paths

The evaluation of the bifurcated paths calls for solving Eq. 5.102. Taken μ_2 as a parameter, each scalar equation describes an algebraic curve of degree 3 in the plane (a_1, a_2), whose intersections identify the solutions of problem. Their *maximum number*, by virtue of the Bézout theorem, is $3 \times 3 = 9$. Of the nine solutions, one is the trivial one; of the remaining eight, the essentially different solutions reduce to four, in that (a_1, a_2) and $(-a_1, -a_2)$ lie on the same path, given the

---

[5] In the simple example at hand, the solvability condition Eq. 5.102 can be trivially obtained by canceling the right members in Eq. 5.101. Here, however, the general method has been followed, to show how to proceed for more complex systems.

symmetry of the system. It is concluded that there exist *at most four essentially different bifurcated paths*.

Proceeding by inspection, the following solutions are found:

$$(T): \quad a_1 = a_2 = 0, \ \forall \mu_2, \tag{5.103a}$$

$$(MM_1): \quad a_2 = 0, \quad \mu = \frac{1}{6}a_1^2, \tag{5.103b}$$

$$(MM_2): \quad a_1 = 0, \quad \mu = \frac{1}{6}a_2^2, \tag{5.103c}$$

$$(BM_1): \quad a_1 = a_2, \quad \mu = -\frac{1}{3}a_2^2, \tag{5.103d}$$

$$(BM_2): \quad a_1 = -a_2, \quad \mu = -\frac{1}{3}a_2^2. \tag{5.103e}$$

The solution (T) is the trivial one. The solutions (MM) are called *mono-modal*, because only one critical mode participates in them; they correspond to a deflection of the pendulum in one of the two principal directions. The solutions (BM) are called *bimodal*, because both the critical modes contribute to the incipient deformation.

To build the parametric equations of the paths, one needs to replace the solutions listed in Eq. 5.103 in the series expansion Eqs. 5.96, together with the normalization condition. This can be chosen as $\epsilon = \theta_1$ on the paths in which $\theta_1 \neq 0$, and $\epsilon = \theta_2$ on the (unique) path in which $\theta_1 = 0$, i.e.:

$$a_1 = 1 \quad \text{on} \quad (MM_1), (BM_1), (BM_2), \tag{5.104a}$$

$$a_2 = 1 \quad \text{on} \quad (MM_2). \tag{5.104b}$$

The following paths, in parametric form, are thus obtained:

$$(T): \quad \theta_1 = \theta_2 = 0, \quad \forall \mu, \tag{5.105a}$$

$$(MM_1): \quad \theta_1 = \epsilon, \quad \theta_2 = 0, \quad \mu = \left(1 + \frac{1}{6}\epsilon^2\right), \tag{5.105b}$$

$$(MM_2): \quad \theta_1 = 0, \quad \theta_2 = \epsilon, \quad \mu = \left(1 + \frac{1}{6}\epsilon^2\right), \tag{5.105c}$$

$$(BM_1): \quad \theta_1 = \epsilon, \quad \theta_2 = \epsilon, \quad \mu = \left(1 - \frac{1}{3}\epsilon^2\right), \tag{5.105d}$$

$$(BM_2): \quad \theta_1 = \epsilon, \quad \theta_2 = -\epsilon, \quad \mu = \left(1 - \frac{1}{3}\epsilon^2\right), \tag{5.105e}$$

or, in Cartesian form:

$$(T): \quad \theta_1 = \theta_2 = 0, \quad \forall \mu, \tag{5.106a}$$

$$(MM_1): \quad \theta_2 = 0, \quad \mu = \left(1 + \frac{1}{6}\theta_1^2\right), \tag{5.106b}$$

$$(MM_2): \quad \theta_1 = 0, \quad \mu = \left(1 + \frac{1}{6}\theta_2^2\right), \tag{5.106c}$$

$$(BM_1): \quad \theta_1 = \theta_2, \quad \mu = \left(1 - \frac{1}{3}\theta_1^2\right), \tag{5.106d}$$

$$(BM_2): \quad \theta_1 = -\theta_2, \quad \mu = \left(1 - \frac{1}{3}\theta_1^2\right). \tag{5.106e}$$

The five paths are represented in the bifurcation diagram of Fig. 5.21. In addition to the straight trivial path, there are four parabolas, two upward concave, lying in the coordinated planes, and two downward concave, contained in the planes rotated of $\pm\frac{\pi}{4}$ radiants. To within higher-order effects, the pendulum buckles in one of the two principal planes, or in one of the two bisector planes.

Remark 5.6 The nonlinear analysis solves the ambiguity of the linear analysis, identifying four privileged directions, among the infinite ones predicted by linearized theory.

Stability of the Bifurcated Paths

To analyze the stability of the four bifurcated paths, it needs to study the definiteness of the TPE on each individual path, as discussed in the Sect. 4.7. The Hessian matrix, for the system at hand, reads:

$$\mathbf{H} := \begin{bmatrix} \frac{\partial^2 \Pi}{\partial \theta_1^2} & \frac{\partial^2 \Pi}{\partial \theta_1 \partial \theta_2} \\ \frac{\partial^2 \Pi}{\partial \theta_1 \partial \theta_2} & \frac{\partial^2 \Pi}{\partial \theta_2^2} \end{bmatrix} = k \begin{bmatrix} 1 - \mu + \frac{\mu}{2}\left(\theta_1^2 - \theta_2^2\right) & -\mu\theta_1\theta_2 \\ -\mu\theta_1\theta_2 & 1 - \mu + \frac{\mu}{2}\left(\theta_2^2 - \theta_1^2\right) \end{bmatrix}. \tag{5.107}$$

Fig. 5.21 Bifurcation diagram of the spherical inverted elastic pendulum

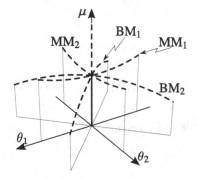

The sign of the eigenvalues Λ_k $(k = 1, 2)$ of $\mathbf{H}$, which are roots of the characteristic equation $\det (\mathbf{H} - \Lambda \mathbf{I}) = \mathbf{0}$, decides on the stability. In order for the path to be stable, $\Lambda_1 > 0$, $\Lambda_2 > 0$ must hold; if even one of the two eigenvalues is negative, the path is unstable. When $\mathbf{H}$ is evaluated on the different paths, it assumes the following expressions:

- On the path (T):

$$\mathbf{H}|_{(T)} = k \begin{bmatrix} 1 - \mu & 0 \\ 0 & 1 - \mu \end{bmatrix}, \tag{5.108}$$

hence, $\Lambda_1 = \Lambda_2 = k \left(1 - \mu \right)$. Therefore, the trivial path is stable if $\mu < \mu_c = 1$ and unstable if $\mu > \mu_c$.
- On path (MM$_1$):

$$\mathbf{H}|_{(MM_1)} = k \begin{bmatrix} \frac{1}{12}\theta_1^2 \left(4 + \theta_1^2 \right) & 0 \\ 0 & -\frac{1}{12}\theta_1^2 \left(8 + \theta_1^2 \right) \end{bmatrix}, \tag{5.109}$$

hence, $\Lambda_1 = \frac{1}{12}k\theta_1^2 \left(4 + \theta_1^2 \right)$, $\Lambda_2 = -\frac{1}{12}k\theta_1^2 \left(8 + \theta_1^2 \right)$; since $\Lambda_2 < 0$, $\forall \theta_1$, the path is unstable.
- On path (MM$_2$):

$$\mathbf{H}|_{(MM_2)} = \begin{bmatrix} -\frac{1}{12}k\theta_1^2 \left(8 + \theta_2^2 \right) & 0 \\ 0 & \frac{1}{12}k\theta_1^2 \left(4 + \theta_2^2 \right) \end{bmatrix}, \tag{5.110}$$

hence, $\Lambda_1 = -\frac{1}{12}k\theta_1^2 \left(8 + \theta_2^2 \right)$, $\Lambda_2 = \frac{1}{12}k\theta_1^2 \left(4 + \theta_2^2 \right)$. Since $\Lambda_1 < 0$, $\forall \theta_2$, the path is unstable.
- On path (BM$_1$):

$$\mathbf{H}|_{(BM_1)} = k \begin{bmatrix} -\frac{1}{6}\theta_2^2 & -\frac{1}{6} \left(6 + \theta_2^2 \right) \theta_2^2 \\ -\frac{1}{6} \left(6 + \theta_2^2 \right) \theta_2^2 & -\frac{1}{6}\theta_2^2 \end{bmatrix}, \tag{5.111}$$

hence, $\Lambda_1 = -\frac{1}{6}k\theta_2^2 \left(7 + \theta_2^2 \right)$, $\Lambda_2 = \frac{1}{6}k\theta_2^2 \left(5 + \theta_2^2 \right)$; since $\Lambda_1 < 0$, $\forall \theta_2$, the path is unstable.
- On path (BM$_2$):

$$\mathbf{H}|_{(BM_2)} = k \begin{bmatrix} -\frac{1}{6}\theta_2^2 & \frac{1}{6} \left(6 + \theta_2^2 \right) \theta_2^2 \\ \frac{1}{6} \left(6 + \theta_2^2 \right) \theta_2^2 & -\frac{1}{6}\theta_2^2 \end{bmatrix}, \tag{5.112}$$

hence, $\Lambda_1 = -\frac{1}{6}k\theta_2^2 \left(7 + \theta_2^2 \right)$, $\Lambda_2 = \frac{1}{6}k\theta_2^2 \left(5 + \theta_2^2 \right)$; since $\Lambda_1 < 0$, $\forall \theta_2$, the path is unstable.

By summarizing, *all four bifurcated paths are unstable*, even the super-critical ones (with upward concavity).

Remark 5.7 The interaction between simultaneous modes, which occurs when $k_1 = k_2$, makes unstable the super-critical equilibrium position, which would instead be stable (at least near the bifurcation) if it were $k_1 \neq k_2$. This type of behavior is common to many mechanical systems.

5.4 Euler Beam as a Paradigm of Continuous Systems

The analytical methods of bifurcation analysis were introduced in Chap. 4 with reference to discrete systems. There, for the sake of brevity, a general discussion on continuous systems was omitted. However, perturbation methods also work for continuous systems, essentially in the same way they do for discrete systems, with few complications related to the presence of boundary conditions, which, absent in discrete systems, now side the differential equations.

The problem illustrated here concerns a paradigmatic system: the inextensible *Euler beam* [1, 3, 5, 8, 10]. Preliminary, the nonlinear model of the beam is derived through a variational procedure; subsequently, by resorting to the perturbation analysis, the buckling and postbuckling behaviors are studied. Two beams, differently constrained, are considered, to highlight the role of the boundary conditions.

5.4.1 Inextensible and Shear-Undeformable Planar Beam Model

A straight prismatic beam is considered, *inextensible and shear undeformable*, constrained to move in a principal inertia plane, uniformly compressed by a potential force of intensity P and subject to transverse distributed loads of intensity $p(x)$ (Fig. 5.22a). The body is modeled as *one-dimensional polar continuum*[6] immersed in the 2D space [1, 9]. In the reference configuration, the beam lies on a segment of the axis x, of endpoints A, B. Its centerline is spanned by the material abscissa $x \in (0, \ell)$, where ℓ is the length of the beam.

Kinematics

The change of configuration (Fig. 5.22b) is described by the following displacement field:

$$\mathbf{u} = \begin{pmatrix} u(x) \\ v(x) \\ \varphi(x) \end{pmatrix}, \qquad (5.113)$$

---

[6] The continuum consists of material points endowed with orientation, which describe the attitude assumed by the cross-section of the underlying 3D model.

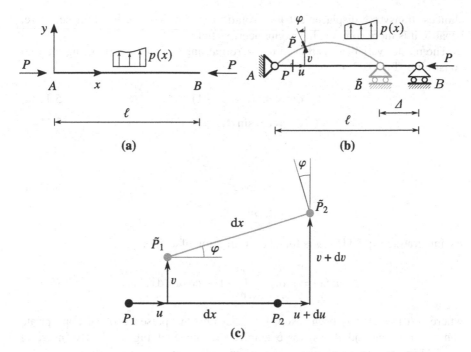

Fig. 5.22 Euler beam: **(a)** reference configuration, **(b)** current configuration, **(c)** displacement of an infinitesimal element of beam

where $u(x)$, $v(x)$ represent the longitudinal and transverse displacement components, respectively, and $\varphi(x)$ represents the rotation of the material point located at the abscissa x.

An infinitesimal beam element of length dx is considered, of endpoints P_1 and P_2 (Fig. 5.22c). Because of the geometric transformation, the element moves to a segment $\tilde{P}_1\tilde{P}_2$ of the plane, the length of which, due to inextensibility, remains unchanged, i.e., $\overline{P_1P_2} \equiv \overline{\tilde{P}_1\tilde{P}_2}$. Furthermore, due to the shear undeformability, the rotation of the cross-section coincides with the rotation of the centerline.

These two internal constraints make the three scalar configuration variables, u, v, and φ, *not independent*. It can be also said that the two constraint conditions reduce from three to one degree of freedom of each point. Among the three configuration variables, it is possible to choose one (privileged, also called "master"), with which to express the remaining two (also called "slave").[7] Between the two possible

---

[7] The choice of the master variable is not arbitrary. Indeed, the longitudinal displacement u cannot describe infinitesimal displacements, because it tends to zero as a second-order quantity (i.e., an infinitesimal deflection of the beam implies, at the first order, zero longitudinal displacement). Similarly, in the inverted pendulum of extremes A, B, it is possible to choose as a Lagrangian parameter either the rotation θ or the transverse displacement v_B, but not the longitudinal displacement u_B.

choices, transverse displacement v or rotation φ, the first one is performed here, because it is more common in the engineering field.[8]

The master variable v is related to the remaining two via the following internal constraints (Fig. 5.22c):

$$\mathrm{d}u = \mathrm{d}x\,(\cos\varphi - 1), \tag{5.114a}$$

$$\mathrm{d}v = \mathrm{d}x\,\sin\varphi, \tag{5.114b}$$

or:

$$u' = \cos\varphi - 1, \tag{5.115a}$$

$$v' = \sin\varphi. \tag{5.115b}$$

By integrating Eq. 5.115a, u is found as a function of φ, i.e.:

$$u\,(x) = u(0) - \int_0^x (1 - \cos\varphi)\,\mathrm{d}\hat{x}, \tag{5.116}$$

where $u\,(0) = 0$ has been taken, by assuming the presence of an appropriate constraint at the end A, as, for example, the hinge of Fig. 5.22b. By inverting Eq. 5.115b, the angle φ is expressed as a function of v:

$$\varphi = \arcsin v'. \tag{5.117}$$

Equations. 5.116 and 5.117 determine the variables u and φ as slave of the master variable v.

The only non-zero strain is the *flexural curvature* $\kappa := \varphi'$; it is evaluated by differentiating Eq. 5.117, thus obtaining:[9]

---

[8] Actually, choosing φ leads to simpler equations.

[9] It is possible to find in the literature the following (erroneous) expression for the flexural curvature:

$$\kappa = \frac{v''}{\left(1 + v'^2\right)^{3/2}}$$

This, however, has no physical sense, as it represents the *curvature of the graph* of the displacements v and not the curvature of the centerline of the beam. Indeed, a generic point belonging to the axis line experiences a transverse and a longitudinal displacement. However, the previous expression is exact if the derivatives are referred to the *spatial* coordinate $X = x + u$ (i.e., to the abscissa of the material point in the current configuration) and not to the *material* coordinate x (i.e., the abscissa measured in the reference configuration). As a matter of fact, from Fig. 5.22, it is found that $\tan\varphi = \frac{\mathrm{d}v}{\mathrm{d}X}$, $\cos\varphi = \frac{\mathrm{d}X}{\mathrm{d}x}$; moreover, by differentiating the first of them, it is $\frac{\mathrm{d}^2 v}{\mathrm{d}X^2} = \frac{1}{\cos^2\varphi}\frac{\mathrm{d}\varphi}{\mathrm{d}X}$. Since $\kappa = \frac{\mathrm{d}\varphi}{\mathrm{d}x} = \frac{\mathrm{d}\varphi}{\mathrm{d}X}\frac{\mathrm{d}X}{\mathrm{d}x}$, it follows:

$$\kappa = \frac{v''}{\sqrt{1 - v'^2}}. \tag{5.118}$$

It is useful, for future purposes, to calculate the "shortening"[10] of the beam (Fig. 5.22b), i.e., the longitudinal displacement of the end B (which *must* be allowed by the constraints). By using Eqs. 5.116 and 5.115b, it follows:

$$\Delta = -u_B = \int_0^\ell (1 - \cos\varphi)\, d\hat{x} = \int_0^\ell \left(1 - \sqrt{1 - v'^2}\right) d\hat{x}. \tag{5.119}$$

Total Potential Energy

The equilibrium equations are deduced from the stationary of the TPE. This is written as the sum of elastic energy of the beam and the potential energy of the loads, $\Pi = U + V$, where:

$$U = \int_0^\ell \frac{1}{2} EI\, \kappa^2\, dx, \tag{5.120a}$$

$$V = -P\,\Delta - \int_0^\ell p\, v\, dx. \tag{5.120b}$$

Here, EI is the bending stiffness of the beam and V is the work, changed in sign, expended by the axial force P and distributed forces p in the geometric transformation of the beam. Taking into account Eqs. 5.118 and 5.119, it is:

$$\Pi = \int_0^\ell \frac{1}{2} EI \left(\frac{v''}{\sqrt{1 - v'^2}}\right)^2 dx - P \int_0^\ell \left(1 - \sqrt{1 - v'^2}\right) d\hat{x} - \int_0^\ell p\, v\, dx. \tag{5.121}$$

By expanding in series up to the fourth-degree terms, the previous expression becomes:

$$\Pi = \int_0^\ell \frac{1}{2} EI \left[v''\left(1 + \frac{1}{2} v'^2\right)\right]^2 dx - P \int_0^\ell \left(\frac{1}{2} v'^2 + \frac{1}{8} v'^4\right) d\hat{x} - \int_0^\ell p\, v\, dx. \tag{5.122}$$

$$\kappa = \cos^3\varphi\, \frac{d^2 v}{dX^2} = \frac{\frac{d^2 v}{dX^2}}{\left(1 + \left(\frac{dv}{dX}\right)^2\right)^{\frac{3}{2}}}$$

i.e., the "wrong" expression, but with X replacing x.

[10] The word "shortening," also commonly used in the technical jargon, is incorrect: the beam, by virtue of inextensibility, cannot be shortened. It should be therefore understood as the shortening of the *projection* of the beam on the x axis.

Supplement 5.10 (Exact equilibrium equations) The exact equilibrium equations, and related boundary conditions, are obtained by imposing the exact functional in Eq. 5.121 to be stationary, i.e., $\delta\Pi = 0, \forall \delta v$. After integration by parts (Appendix A reports some reminders of calculus of variations), the field equilibrium equation is obtained:

$$
EI\frac{v''''}{1-v'^2} + P\frac{v''}{\sqrt{1-v'^2}} + v'^2\left[4EI\left(\frac{v''}{1-v'^2}\right)^3 + P\frac{v''}{\left(1-v'^2\right)^{3/2}}\right]
$$

$$
+ EI\frac{v''^3}{\left(1-v'^2\right)^2} + 4EI\frac{v'v''v'''}{\left(1-v'^2\right)^2} = p,
$$

(5.123)

with the boundary conditions:

$$
\left[\left(-EI\frac{v'''}{1-v'^2} - P\frac{v'}{\sqrt{1-v'^2}} - EI\frac{v'v''^2}{\left(1-v'^2\right)^2}\right)\delta v\right]_0^\ell = 0,
$$

(5.124a)

$$
\left[\left(EI\frac{v''}{1-v'^2}\right)\delta v'\right]_0^\ell = 0.
$$

(5.124b)

For different constraints at the ends $H = A, B$, these latter specialize into:

• Clamp (or slider, longitudinally scrolling):

$$
v_H = 0,
$$

(5.125a)

$$
v'_H = 0,
$$

(5.125b)

the latter following from $\varphi_H = \arcsin v'_H = 0$;

• Free end:

$$
-EI\frac{v'''_H}{1-v'^2_H} - P\frac{v'_H}{\sqrt{1-v'^2_H}} - EI\frac{v'_H v''^2_H}{\left(1-v'^2_H\right)^2} = 0,
$$

(5.126a)

$$
EI\frac{v''_H}{1-v'^2_H} = 0;
$$

(5.126b)

• Hinge:

$$
v_H = 0,
$$

(5.127a)

$$
EI\frac{v''_H}{1-v'^2_H} = 0;
$$

(5.127b)

- Slider, transversely scrolling:

$$- EI \frac{v_H'''}{1 - v_H'^2} - P \frac{v_H'}{\sqrt{1 - v_H'^2}} - EI \frac{v_H' v_H''^2}{\left(1 - v_H'^2\right)^2} = 0, \tag{5.128a}$$

$$v_H' = 0. \tag{5.128b}$$

□

Equilibrium Equation Expanded in Series

The equilibrium equations and boundary conditions, approximated at the third-degree, are obtained by imposing the functional in Eq. 5.122 to be stationary, i.e., $\delta\Pi = 0, \forall \delta v$ (or, alternatively, by expanding in series the exact equilibrium Eqs. 5.123, 5.124). The field equation is found to be:

$$EI \left[v'''' \left(1 + v'^2 \right) + 4 v' v'' v''' + v''^3 \right] + P v'' \left(1 + \frac{3}{2} v'^2 \right) = p, \tag{5.129}$$

or, in a more compact form:

$$EI \left\{ v'''' + \left[v' \left(v' v'' \right)' \right]' \right\} + P v'' \left(1 + \frac{3}{2} v'^2 \right) = p. \tag{5.130}$$

Regarding the boundary conditions, they are:

- Clamp (or slider, longitudinally scrolling):

$$v_H = 0, \tag{5.131a}$$

$$v_H' = 0; \tag{5.131b}$$

- Free end:

$$- EI \left[v_H''' \left(1 + v_H'^2 \right) + v_H' v_H''^2 \right] - P v_H' \left(1 + \frac{1}{2} v_H'^2 \right) = 0, \tag{5.132a}$$

$$EI v_H'' \left(1 + v_H'^2 \right) = 0; \tag{5.132b}$$

- Hinge:

$$v_H = 0, \tag{5.133a}$$

$$EI v_H'' \left(1 + v_H'^2 \right) = 0; \tag{5.133b}$$

- Slider, transversely scrolling:

$$- EI \left[v_H''' \left(1 + v_H'^2 \right) + v_H' v_H''^2 \right] - P v_H' \left(1 + \frac{1}{2} v_H'^2 \right) = 0, \tag{5.134a}$$

$$v_H' = 0. \tag{5.134b}$$

5.4.2 Linear Boundary Conditions: Simply Supported Beam

A bifurcation analysis is conducted for the hinged-supported Euler beam. The problem is governed by the expanded field equation, Eq. 5.130, and by the boundary conditions, as specified here:

$$EI \left\{ v'''' + \left[v' \left(v'v'' \right)' \right]' \right\} + P v'' \left(1 + \frac{3}{2} v'^2 \right) = \gamma p, \tag{5.135a}$$

$$v_H = 0, \qquad H = A, B, \tag{5.135b}$$

$$EI v_H'' = 0, \qquad H = A, B. \tag{5.135c}$$

In Eq. 5.135a, a load multiplier γ of the distributed load has been introduced, by letting $p \to \gamma p$. If γ is small and $p = O(1)$, the distributed load can be treated as an "imperfection."[11]

Perturbation Equations

The perturbation analysis, already illustrated in detail for discrete systems, is now applied to a continuous system. The method calls for expanding in series of a perturbation parameter ϵ both the dependent variable $v(x)$ and the load P. By separating the terms affected by the same powers of ϵ, linear perturbation equations are obtained, which consist of differential equations and related boundary conditions. The boundary value problems are solved in sequence, invoking solvability when they are singular, thus determining the series coefficients.

Since (due to the symmetry) only nonlinearities of the cubic type appear in the field equation and in the boundary conditions, the following odd/even series are introduced:

$$v(x) = \epsilon v_1(x) + \epsilon^3 v_3(x) + \cdots, \tag{5.136a}$$

---

[11] Here, to avoid a notation conflict, the symbol γ is used, instead of β, already introduced for discrete systems.

$$P = P_0 + \epsilon^2 P_2 + \cdots . \tag{5.136b}$$

Moreover, a rescaling of the distributed load multiplier is performed, as:

$$\gamma \to \epsilon^3 \hat{\gamma} . \tag{5.137}$$

Finally, a normalization condition is adopted, i.e., $v(\bar{x}) = \epsilon$, with $\bar{x}$ an appropriately chosen point. Consequently, $v_1(\bar{x}) = 1$, $v_3(\bar{x}) = 0$, $\cdots$. Here, $\bar{x}$ is selected as the abscissa at which the critical mode is maximum in modulus.

Substituting Eqs. 5.136 and 5.137 in Eq. 5.135, and collecting and zeroing the terms in the same power of ϵ, the perturbation equations are obtained:

- Order ϵ^1:

$$EI\, v_1'''' + P_0\, v_1'' = 0, \tag{5.138a}$$

$$v_{1H} = 0, \quad H = A, B, \tag{5.138b}$$

$$EI\, v_{1H}'' = 0, \quad H = A, B, \tag{5.138c}$$

- Order ϵ^3:

$$EI\, v_3'''' + P_0\, v_3'' = -P_2 v_1'' - EI\left[v_1'\left(v_1' v_1''\right)'\right]' - \frac{3}{2} P_0 v_1'^2 v_1'' + \hat{\gamma} p, \tag{5.139a}$$

$$v_{3H} = 0, \quad H = A, B, \tag{5.139b}$$

$$EI\, v_{3H}'' = 0, \quad H = A, B. \tag{5.139c}$$

Solution to the ϵ^1 Order Problem

A solution to the ϵ^1 order problem is searched in the form:[12]

$$v_1(x) = e^{\lambda x}. \tag{5.140}$$

Substituting the previous expression in the field Eq. 5.138a, the characteristic equation follows:

$$\lambda^2\left(\lambda^2 + \beta^2\right) = 0, \tag{5.141}$$

in which:

---

[12] Here, λ is the characteristic exponent and should not be confused with the eigenvalue of the dynamic problem, not addressed in this chapter.

$$\beta^2 := \frac{P_0}{EI} \tag{5.142}$$

is the unknown eigenvalue of the differential problem. The roots of Eq. 5.141 are:

$$\lambda_1 = i\,\beta, \quad \lambda_2 = -i\,\beta, \quad \lambda_3 = 0, \quad \lambda_4 = 0, \tag{5.143}$$

so that the general solution to the homogeneous problem is:

$$v_1(x) = c_1 \cos(\beta x) + c_2 \sin(\beta x) + c_3 x + c_4, \tag{5.144}$$

where c_i, $i = 1, \cdots, 4$ are arbitrary constants. The boundary conditions in Eq. 5.138b,c lead to the following linear algebraic system, homogeneous in the constants c_i:

$$\begin{pmatrix} 1 & 0 & 0 & 1 \\ -\beta^2 & 0 & 0 & 0 \\ \cos(\beta\ell) & \sin(\beta\ell) & \ell & 1 \\ -\cos(\beta\ell) & -\sin(\beta\ell) & 0 & 0 \end{pmatrix} \begin{pmatrix} c_1 \\ c_2 \\ c_3 \\ c_4 \end{pmatrix} = \begin{pmatrix} 0 \\ 0 \\ 0 \\ 0 \end{pmatrix}. \tag{5.145}$$

By zeroing the determinant of the coefficient matrix, a transcendental equation in the eigenvalue β follows:

$$\sin(\beta\ell) = 0, \tag{5.146}$$

which admits infinite solutions:[13]

$$\beta_k = \frac{k\pi}{\ell}, \qquad k = 1, 2, \cdots, \tag{5.147}$$

from which, by recalling the position in Eq. 5.142, the critical loads are determined:

$$P_{0k} = \left(\frac{k\pi}{\ell}\right)^2 EI, \qquad k = 1, 2, \cdots. \tag{5.148}$$

When $P_0 = P_{0k}$, the algebraic system admits the solution:

$$c_1 = 0, \quad \forall c_2, \quad c_3 = 0, \quad c_4 = 0; \tag{5.149}$$

hence, by taking into account the normalization, $v_{1k}(x) = \sin\left(\frac{k\pi}{\ell}x\right)$. The critical load, corresponding to the smallest root P_{0k}, is obtained by letting $k = 1$, i.e.:

---

[13] The root $\beta = 0$ is discarded, since it corresponds to the zero axial load.

$$P_c = \frac{\pi^2 EI}{\ell^2};$$

(5.150)

it is called the *Eulerian critical load*. The corresponding critical mode, $v_c = v_{11}$, is:

$$v_c(x) = \sin\left(\frac{\pi}{\ell}x\right) := \phi_c(x),$$

(5.151)

i.e., it is a sinusoidal wave of period 2ℓ.

Solution to the ϵ^3 Order Problem

By substituting the lower-order solution:

$$P_0 = P_c,$$

(5.152a)

$$v_1(x) = \sin\left(\frac{\pi}{\ell}x\right)$$

(5.152b)

in the ϵ^3 order equation, a non-homogeneous boundary value problem is found:

$$\epsilon^3: \quad EI\, v_3'''' + P_c\, v_3'' = P_2\left(\frac{\pi}{\ell}\right)^2 \sin\left(\frac{\pi x}{\ell}\right) - \frac{1}{8}P_c\left(\frac{\pi}{\ell}\right)^4 \sin\left(\frac{\pi}{\ell}x\right)$$

(5.153a)

$$- \frac{9}{8}P_c\left(\frac{\pi}{\ell}\right)^4 \sin\left(\frac{3\pi}{\ell}x\right) + \hat{\gamma}p,$$

$$v_{3H} = 0, \quad H = A, B,$$

(5.153b)

$$EI\, v_{3H}'' = 0, \quad H = A, B.$$

(5.153c)

The differential operator in Eq. 5.153a is singular, since P_c is an eigenvalue of the associated homogeneous problem. The beam, therefore, is kinematically degenerate (or "labile"), in the sense that it can experience not rigid displacement fields without expenditure of (quadratic) elastic energy.[14] Consequently, the non-homogeneous problem admits solution if and only if the known term is orthogonal to the field of displacement allowed by the constraints, i.e., to the critical mode. In other words, the problem admits solution if and only if the forces soliciting the beam expend zero virtual work on the critical mode $v_c(x) = \phi_c(x)$. By imposing this condition, it follows:

---

[14] These fields are also known as "floppy modes."

$$P_2 \left(\frac{\pi}{\ell}\right)^2 \int_0^\ell \phi_c\,(x) \sin\left(\frac{\pi x}{\ell}\right)\,\mathrm{d}x - \frac{1}{8} P_c \left(\frac{\pi}{\ell}\right)^4 \int_0^\ell \phi_c\,(x) \sin\left(\frac{\pi}{\ell}x\right)\,\mathrm{d}x$$

$$- \frac{9}{8} P_c \left(\frac{\pi}{\ell}\right)^4 \int_0^\ell \phi_c\,(x) \sin\left(\frac{3\pi}{\ell}x\right)\,\mathrm{d}x + \hat{\gamma} \int_0^\ell p\,(x)\,\phi_c\,(x)\,\mathrm{d}x = 0,$$

$$(5.154)$$

from which, the unknown is drawn:[15]

$$P_2 = \frac{1}{8} \left(\frac{\pi}{\ell}\right)^2 P_c - \hat{\gamma}\check{P}, \qquad\qquad (5.155)$$

where:

$$\check{P} := \frac{2\ell}{\pi^2} \int_0^\ell p(x)\phi_c\,(x)\,\mathrm{d}x \qquad\qquad (5.156)$$

represents the "projection" of $p\,(x)$ onto $\phi_c\,(x)$.[16]

Bifurcation Diagram

Reconstituting the solution through Eqs. 5.136 and 5.155, and coming back to the unrescaled transverse forces, one obtains:

$$P = P_c + \frac{1}{8}\epsilon^2 \left(\frac{\pi}{\ell}\right)^2 P_c - \frac{\gamma}{\epsilon}\check{P}, \qquad\qquad (5.157)$$

which relates the axial load P to the sag ϵ at midspan, for any given "imperfection" γ (i.e., it describes the bifurcation diagram). The result shows that the continuous nonlinear system under study is equivalent to a single DOF system with stable postcritical behavior (Sect. 5.2.1 should be remembered).[17] The Euler beam, in conclusion, exhibits a stable fork bifurcation.

Proceeding with the perturbation algorithm to the successive orders, it is possible to improve the solution and in particular to determine the shape corrections to the critical mode.

---

[15] It should observed that the harmonic of frequency $\frac{3\pi}{\ell}$ does not give contribution to the solvability condition, since it is orthogonal to ϕ_c. However, it contributes to the change of form $v_3(x)$, expressed by the solution of Eq. 5.153, not given here for brevity.

[16] In this particular case, the same result is obtained by expanding the transverse load in Fourier series and then canceling the coefficient of $\sin\left(\frac{\pi x}{\ell}\right)$ in Eq. 5.153a.

[17] This equivalence, however, should be meant in a broader sense, since higher-order modes *passively* contribute to the response.

5.4.3 Nonlinear Boundary Conditions: Cantilever Beam

Passing to consider a cantilever, it is necessary to side the field Eq. 5.130 with the following boundary conditions:

$$v_A = v'_A = 0, \qquad (5.158a)$$

$$-EI\left[v'''_B\left(1 + v'^2_B\right) + v'_B v''^2_B\right] - P\,v'_B\left(1 + \frac{1}{2}v'^2_B\right) = 0, \qquad (5.158b)$$

$$EI\,v''_B = 0. \qquad (5.158c)$$

It should be noticed that, differently from the case of supported beam, the *boundary conditions are nonlinear*.

Perturbation Equations

With the same series expansions of the previous section, the following perturbation equations are derived:

- Order ϵ^1:

$$EI\,v''''_1 + P_0\,v''_1 = 0, \qquad (5.159a)$$

$$v_{1A} = v'_{1A} = 0, \qquad (5.159b)$$

$$-EIv'''_{1B} - P_0\,v'_{1B} = 0, \qquad (5.159c)$$

$$EI\,v''_{1B} = 0, \qquad (5.159d)$$

- Order ϵ^3:

$$EI\,v''''_3 + P_0\,v''_3 = -P_2 v''_1 - EI\left[v'_1\left(v'_1 v''_1\right)'\right]' \qquad (5.160a)$$

$$-\frac{3}{2}P_0 v'^2_1 v''_1 + \hat{\gamma}\,p,$$

$$v_{3A} = v'_{3A} = 0, \qquad (5.160b)$$

$$-EI\,v'''_{3B} - P_0\,v'_{3B} = P_2\,v'_{1B} + EI\left[v'^2_{1B}v'''_{1B} + v'_{1B}v''^2_{1B}\right] \qquad (5.160c)$$

$$+\frac{1}{2}P_0\,v'^3_{1B},$$

$$EI\,v''_{3B} = 0. \qquad (5.160d)$$

Solution to the ϵ^1 Order Problem

The solution to the ϵ^1 order field equation is still expressed by Eq. 5.144. By imposing the boundary conditions and proceeding as in the previous section, the eigenvalues β for which there exist non-trivial solutions and their respective normalized eigenvectors are found. In particular, the first eigensolution is:

$$P_c = \frac{\pi^2 EI}{4\ell^2},$$
(5.161a)

$$v_c = 1 - \cos\left(\frac{\pi}{2\ell}x\right) =: \phi_c(x).$$
(5.161b)

Solution to the ϵ^3 Order Problem

Substituting the lower-order solution in the ϵ^3 order equations, a differential non-homogeneous problem follows:

$$\epsilon^3: \qquad EI\, v_3'''' + P_c\, v_3'' = -P_2\left(\frac{\pi}{2\ell}\right)^2 \cos\left(\frac{\pi}{2\ell}x\right)$$
(5.162a)

$$+ \frac{5}{4}P_c\left(\frac{\pi}{2\ell}\right)^4 \cos\left(\frac{\pi x}{2\ell}\right)$$

$$- \frac{9}{4}P_c\left(\frac{\pi}{2\ell}\right)^4 \cos\left(\frac{\pi x}{2\ell}\right)\cos\left(\frac{\pi x}{\ell}\right) + \hat{\gamma}p,$$

$$v_{3A} = v_{3A}' = 0,$$
(5.162b)

$$-EI\, v_{3B}''' - P_c\, v_{3B}' = P_2\frac{\pi}{2\ell} - \frac{1}{2}P_c\left(\frac{\pi}{2\ell}\right)^3,$$
(5.162c)

$$EI\, v_{3B}'' = 0.$$
(5.162d)

It is important noticing that the known term of the problem consists of two contributions, both generated by nonlinearities: *one in the domain and the other at the boundaries*, having respectively the meaning of distributed and concentrated forces at the ends. In order to make the problem solvable, a compatibility condition must be imposed, involving *both* known terms. The condition requires zeroing the work done *by all the external forces* (in the domain and at the boundaries) in the displacement field allowed by the singular operator (i.e., in the critical mode $\phi_c(x)$). Such condition reads:

$$\int_0^\ell \phi_c(x) \left(-P_2 \left(\frac{\pi}{2\ell} \right)^2 \cos \left(\frac{\pi}{2\ell} x \right) + \frac{5}{4} P_c \left(\frac{\pi}{2\ell} \right)^4 \cos \left(\frac{\pi x}{2\ell} \right) \right) dx$$

$$+ \int_0^\ell \phi_c(x) \left(-\frac{9}{4} P_c \left(\frac{\pi}{2\ell} \right)^4 \cos \left(\frac{\pi x}{2\ell} \right) \cos \left(\frac{\pi x}{\ell} \right) + \hat{\gamma} p(x) \right) dx \quad (5.163)$$

$$+ \phi_c(\ell) \left[P_2 \frac{\pi}{2\ell} - \frac{1}{2} P_c \left(\frac{\pi}{2\ell} \right)^3 \right] = 0,$$

By computing the integrals and solving with respect P_2, the following result is drawn:

$$P_2 = \frac{1}{8} \left(\frac{\pi}{2\ell} \right)^2 P_c - \hat{\gamma} \check{P}, \quad (5.164)$$

where:

$$\check{P} := \frac{8\ell}{\pi^2} \int_0^\ell \phi_c(x) p(x) \, dx. \quad (5.165)$$

Bifurcation Diagram

Reconstituting the solution through Eqs. 5.136 and 5.164, and returning to the unrescaled forces, the load-displacement law describing the bifurcation diagram is found:

$$P = P_c + \frac{1}{8} \epsilon^2 \left(\frac{\pi}{2\ell} \right)^2 P_c - \frac{\gamma}{\epsilon} \check{P}. \quad (5.166)$$

5.5 Systems with Non-trivial Path: The Snap-Through of the Three-Hinged Arch

A paradigmatic example of a class of structures suffering from limit load instability (*fold bifurcation*) is considered. It consists of a *low three-hinged arch*, symmetric and constrained to deform symmetrically (Fig. 5.23) [2].

The arch is made up of two identical rods, rigid to bending but axially elastic, of stiffness k, initially inclined on the horizontal of an angle α, subject to a vertical potential force of intensity P, positive downward. The angle θ is taken as Lagrangian parameter, positive counterclockwise, that the left rod forms with the horizontal in the current configuration.[18]

---

[18] The rotation of the left rod is therefore $\theta - \alpha$, positive counterclockwise.

Fig. 5.23 Three-hinged
symmetric low arch,
constrained to deform
symmetrically

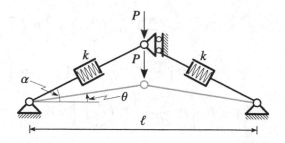

5.5.1　Exact Analysis

In this simple case, an exact analysis is viable. The total potential energy reads:

$$\Pi(\theta) = 2\frac{1}{2}k\,\Delta\ell^2 - P\Delta, \tag{5.167}$$

where $\Delta\ell$ is the elongation of both rods and Δ is the displacement of the apex of
arch, positive downward, respectively, given by:

$$\Delta\ell = \frac{\ell}{2}\left(\frac{1}{\cos\theta} - \frac{1}{\cos\alpha}\right), \tag{5.168a}$$

$$\Delta = \frac{\ell}{2}\left(\tan\alpha - \tan\theta\right). \tag{5.168b}$$

From the stationary condition $\frac{d\Pi}{d\theta} = 0$, an equilibrium equation is derived:

$$\frac{1}{2}\ell\frac{1}{\cos\theta}\left[k\ell\tan\theta\left(\frac{1}{\cos\theta} - \frac{1}{\cos\alpha}\right) + P\frac{1}{\cos\theta}\right] = 0, \tag{5.169}$$

whose solution is:

$$P = k\ell\sin\theta\left(\frac{1}{\cos\alpha} - \frac{1}{\cos\theta}\right). \tag{5.170}$$

It describes a non-monotonic smooth path (i.e., exempt of branching points),
presented in Fig. 5.24a for positive (downward) and negative (upward) force. There,
$\theta > 0$ denotes the upper position of the arch and $\theta < 0$ the lower position. To
determine the limit points, $\frac{dP}{d\theta} = 0$ is imposed, which implies:

$$\cos^3\theta = \cos\alpha. \tag{5.171}$$

This equation gives two equal and opposite roots, $\theta = \theta_l^{\pm}$, associated with the limit
loads:

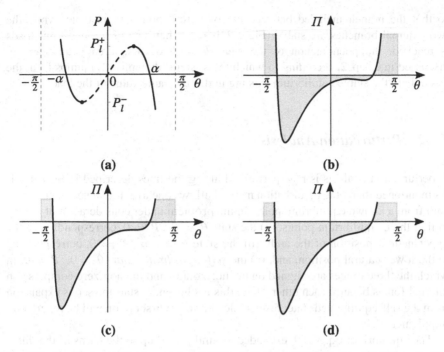

Fig. 5.24 Equilibrium of the low arch: (a) bifurcation diagram; TPE, when (b) $|P| < P_l$, (c) $|P| = P_l$, (d) $|P| > P_l$

$$P_l^{\pm} = \pm k\ell \tan^3 \theta_l. \tag{5.172}$$

To analyze the stability of the path, the second derivative of the potential energy is computed:

$$\frac{d^2 \Pi}{d\theta^2} = \frac{1}{4} \ell \frac{1}{\cos^2 \alpha} \left[k\ell \left(\frac{\cos (2\theta) - 3}{\cos \alpha \cos \theta} + 6\frac{1}{\cos^2 \alpha} - 4 \right) + 4P \, \tan \theta \right]. \tag{5.173}$$

To evaluate it along the path, $\cos \alpha$ is expressed via Eq. 5.170, thus obtaining:

$$\frac{d^2 \Pi}{d\theta^2} = \frac{k\ell^2}{\sin (2\theta)} \left(\tan^3 \theta - \frac{P}{k\ell} \right). \tag{5.174}$$

From this, it is concluded that:

$$|\theta| > \theta_l \Rightarrow \frac{d^2 \Pi}{d\theta^2} > 0, \tag{5.175a}$$

$$|\theta| < \theta_l \Rightarrow \frac{d^2 \Pi}{d\theta^2} < 0, \tag{5.175b}$$

so that the branch included between the two limit points is unstable, while the two external branches are stable (Fig. 5.24b–d). When one of the two limit loads is reached, the phenomenon of the *snap-through* occurs, already qualitatively discussed in Chap. 2, according to which the system (dynamically) jumps from the unstable to the stable equilibrium existing under the same value of the load.

5.5.2 Perturbation Analysis

A perturbation analysis is now performed along the lines described in Sect. 4.8.1. As mentioned there, the perturbation method allows one to extrapolate the unknown path from a known equilibrium point. In the problem under consideration, it is easy to find three equilibrium points: (a) the state $\theta = \alpha$, $P = 0$, corresponding to the upper natural position of the arch; (b) the state $\theta = -\alpha$, $P = 0$, corresponding to the lower natural position, and, (c) the *self-equilibrated state* $\theta = 0$, $P = 0$, in which the three hinges are aligned on the horizontal, and the non-zero compression internal forces balance each other. Here, this is chosen, to start the series expansion from the self-equilibrated state; results relevant to the first option will be reported in Supplement 5.11.

The equilibrium Eq. 5.169, expanded around $\theta = 0$ up to the terms of the third-degree, reads:

$$k\ell^2 \left[\frac{1}{2} \left(1 - \frac{1}{\cos\alpha} \right) \theta + \frac{1}{12} \left(8 - 5\frac{1}{\cos\alpha} \right) \theta^3 \right] + \frac{1}{2} P\ell \left(1 + \theta^2 \right) = 0. \quad (5.176)$$

Making use of the series:

$$\theta\,(\epsilon) = \epsilon\,\theta_1 + \epsilon^2\,\theta_2 + \epsilon^3\,\theta_3 + \cdots, \quad (5.177a)$$

$$P\,(\epsilon) = \epsilon\,P_1 + \epsilon^2\,P_2 + \epsilon^3\,P_3 + \cdots, \quad (5.177b)$$

the perturbation equations follow:

$$\epsilon^1 : \ k\ell^2 \left(1 - \frac{1}{\cos\alpha} \right) \theta_1 + P_1\ell = 0, \quad (5.178a)$$

$$\epsilon^2 : \ k\ell^2 \left(1 - \frac{1}{\cos\alpha} \right) \theta_2 + P_2\ell = 0, \quad (5.178b)$$

$$\epsilon^3 : \ k\ell^2 \left(1 - \frac{1}{\cos\alpha} \right) \theta_3 + P_3\ell = -P_1\ell\,\theta_1^2 + \frac{1}{6}k\ell^2 \left(\frac{5}{\cos\alpha} - 8 \right) \theta_1^3. \quad (5.178c)$$

These, resolved in sequence, provide:

$$\epsilon^1 : \quad P_1 = k\ell \left(\frac{1}{\cos \alpha} - 1 \right) \theta_1, \tag{5.179a}$$

$$\epsilon^2 : \quad P_2 = k\ell \left(\frac{1}{\cos \alpha} - 1 \right) \theta_2, \tag{5.179b}$$

$$\epsilon^3 : \quad P_3 = -P_1 \theta_1^2 + \frac{1}{6} k\ell \left(\frac{5}{\cos \alpha} - 8 \right) \theta_1^3 + k\ell\theta_3 \left(\frac{1}{\cos \alpha} - 1 \right). \tag{5.179c}$$

If the normalization $\theta = \epsilon$ is chosen, the following conditions must be enforced: $\theta_1 = 1$, $\theta_2 = \theta_3 = 0$. From these, by reconstituting, the asymptotic expression of the path is obtained:

$$P = k\ell \left[\left(\frac{1}{\cos \alpha} - 1 \right) \theta - \frac{1}{6} \left(\frac{1}{\cos \alpha} + 2 \right) \theta^3 \right]. \tag{5.180}$$

The graph of the path is presented in Fig. 5.25a and compared with the exact one. The evaluation of the limit point calls for solving $\frac{dP}{d\theta} = 0$, that is:

$$k\ell \left[\left(\frac{1}{\cos \alpha} - 1 \right) - \frac{1}{2} \left(\frac{1}{\cos \alpha} + 2 \right) \theta^2 \right] = 0, \tag{5.181}$$

that provides two roots:

$$\theta_l^{\pm} = \pm \sqrt{2} \sqrt{\frac{1 - \cos \alpha}{1 + 2 \cos \alpha}}. \tag{5.182}$$

The corresponding limit loads are:

$$P_l^{\pm} = \pm k\ell \frac{2\sqrt{2}}{3} \frac{\left(\frac{1}{\cos \alpha} - 1 \right)^{3/2}}{\sqrt{\frac{1}{\cos \alpha} + 2}}. \tag{5.183}$$

Supplement 5.11 (Extrapolation from the Upper Position) Alternatively to the procedure illustrated, extrapolation can be carried out from the point $\theta = \alpha$. The equilibrium equation, expanded up to the *fourth* power of the increment $\theta - \alpha$, is:

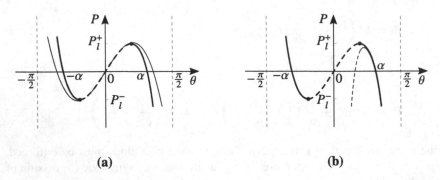

Fig. 5.25 Asymptotic equilibrium path of the low arch: (**a**) extrapolation from $\theta = 0$ (three hinges aligned), (**b**) extrapolation from $\theta = \alpha$ (arch in the natural upper position); exact solution (thick line), asymptotic solution (thin line)

$$\frac{1}{2}P\ell \sec^2 \alpha + \frac{\ell}{2}\tan \alpha \sec^2 \alpha \left(k\ell \tan \alpha + 2P\right)(\theta - \alpha)$$

$$-\frac{\ell}{16}\sec^5 \alpha \left(3k\ell \left(\sin(3\alpha) - 7\sin \alpha\right) - 12P\cos \alpha + 4P\cos(3\alpha)\right)(\theta - \alpha)^2$$

$$+\frac{\ell}{96}\sec^6 \alpha \left(7k\ell \cos(4\alpha) - 132k\ell \cos(2\alpha) + 149k\ell + 80P\sin(2\alpha)\right)$$

$$-8P\sin(4\alpha))(\theta - \alpha)^3 + \frac{\ell}{768}\sec^7 \alpha \left(3150k\ell \sin \alpha - 675k\ell \sin(3\alpha)\right)$$

$$+15k\ell \sin(5\alpha) + 640P\cos \alpha)(\theta - \alpha)^4 - \frac{\ell}{768}\sec^7 \alpha \left(400P\cos(3\alpha)\right)$$

$$-16P\cos(5\alpha))(\theta - \alpha)^4 = 0,$$

$$(5.184)$$

which, solved asymptotically (details omitted for brevity), provides:

$$P = -k\ell \left\{\tan^2 \alpha \left(\theta - \alpha\right) + \frac{1}{2}\tan \alpha \left(2\tan^2 \alpha + 3\right)(\theta - \alpha)^2 + \tan^4 \alpha \left(\theta - \alpha\right)^3 \right.$$

$$\left. +\frac{1}{6}\left(8\tan^2 \alpha + 3\right)(\theta - \alpha)^3 + \frac{1}{24}\tan \alpha \left(24\tan^4 \alpha + 40\tan^2 \alpha + 15\right)(\theta - \alpha)^4 \right\}.$$

$$(5.185)$$

The graph of this path is shown in Fig. 5.25b. Comparison with the asymptotic solution previously determined shows large errors, especially in the limit load. The current asymptotic solution is in good agreement with the exact solution only in a small neighborhood of the starting point $\theta = \alpha$, despite the fact that the perturbation scheme has been developed up to the fourth (rather than the third) degree.   □

5.6 Bifurcation from Non-trivial Path: The Extensible Pendulum

To show an example of bifurcation from a non-trivial path, an extensible inverted pendulum is considered (Fig. 5.26) [6, 7].

The system consists of a rigid telescopic rod, of extensional stiffness k_e, elastically hinged to the ground by a torsional spring of stiffness k_t, subject to a compression potential force P. Due to the extensibility of the rod, the system possesses two DOF; the rotation θ and the shortening d of the rod are chosen as Lagrangian parameters, i.e., $\mathbf{q} = (d, \theta)^T$.

Equilibrium Equations and Fundamental Path
The total potential energy of the system is:

$$\Pi = \frac{1}{2}k_t\theta^2 + \frac{1}{2}k_e d^2 - Pu_B, \tag{5.186}$$

in which:

$$u_B = \ell - (\ell - d)\cos\theta, \tag{5.187}$$

is the lowering of the free end of the rod. By requiring the TPE to be stationary, i.e., $\frac{\partial\Pi}{\partial d} = 0, \frac{\partial\Pi}{\partial\theta} = 0$, one gets:

$$k_e d - P\cos\theta = 0, \tag{5.188a}$$

$$k_t\theta - P(\ell - d)\sin\theta = 0. \tag{5.188b}$$

These represent, respectively, (a) the balance of the forces acting on the upper segment of the rod, in the tangent direction and (b) the balance of torques of all the forces with respect to the point A.

Fig. 5.26 Extensible inverted pendulum

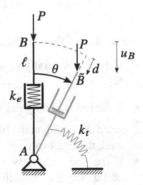

Equations 5.188 *do not* admit the trivial solution $d = 0$, $\theta = 0$, $\forall P$; however, it is easy (in this example) to determine the fundamental path by inspection, observing that, in the precritical phase, the rod undergoes an axial displacement without rotation, i.e.:

$$d^f = \frac{P}{k_e}, \qquad \theta^f = 0. \tag{5.189}$$

Linearized Analysis

To determine the critical loads, the incremental variables $\mathbf{y} = \left(\tilde{d}, \tilde{\theta}\right)^T$ are introduced, according to Eq. 4.63; hence, the Lagrangian coordinates are expressed as:

$$d = d^f + \tilde{d} = \frac{P}{k_e} + \tilde{d}, \tag{5.190a}$$

$$\theta = \theta^f + \tilde{\theta} = \tilde{\theta}. \tag{5.190b}$$

Substituting these latter in the equilibrium Eq. 5.188, and linearizing in the increments $\tilde{d}$, $\tilde{\theta}$, the linearized incremental equations are derived:

$$\begin{bmatrix} k_e & 0 \\ 0 & k_t - P\left(\ell - \dfrac{P}{k_e}\right) \end{bmatrix} \begin{pmatrix} \tilde{d} \\ \tilde{\theta} \end{pmatrix} = \begin{pmatrix} 0 \\ 0 \end{pmatrix}. \tag{5.191}$$

From Eq. 5.191a, it follows that $\tilde{d} = 0$, i.e., *the critical mode is inextensional*. The Eq. 5.191b then reads:

$$\left(P^2 - k_e\ell\, P + k_e k_t\right)\tilde{\theta} = 0, \tag{5.192}$$

thus appearing as a *nonlinear* eigenvalue problem in the load P. Its solutions are:

$$P_{c1,2} = \frac{1}{2}\left(k_e\ell \pm \sqrt{(k_e\ell)^2 - 4k_e k_t}\right), \tag{5.193}$$

the smaller of them is the critical load sought. The following considerations hold:

- If the rod is axially rigid ($k_e \to \infty$), Eq. 5.191b admits just one solution, $P_c^\infty := \frac{k_t}{\ell}$.
- If the rod is axially deformable, the flexibility reduces its length from ℓ to $\ell - \frac{P}{k_e}$; since the length at the bifurcation is not known, but it is a function of the unknown load, the eigenvalue problem is nonlinear.
- If the rod is axially very soft, and precisely if $k_e < \frac{4k_t}{\ell^2}$, the discriminant of Eq. 5.193 is negative, so there are no real solutions, that is, *the rod does*

not bifurcate. The phenomenon can be explained by observing that the large shortening reduces the slenderness of the rod, thus preventing the onset of instability. This circumstance, however, only manifests itself when the rod is *extremely soft* (e.g., when the system models a helical spring). As a matter of fact, if $k_e = \frac{4k_t}{\ell^2}$, the shortening is $d\,(P) = \frac{P}{k_e} = \frac{P\ell^2}{4k_t}$; thus, if the load is, e.g., taken equal to the double root $P_{c1} = P_{c2} = \frac{2k_t}{\ell}$, it results that $d\,(P_{c1,2}) = \frac{\ell}{2}$, i.e., the rod halves its initial length at the bifurcation.

The exact solution of the bifurcation problem can also be pursued, as illustrated in Supplement 5.12. This allows evaluation of the nonlinear equilibrium paths as the axial stiffness of the member is made to vary. The exact solution fully explains the disappearance of the bifurcation. The same problem can be tackled by an asymptotic method, as illustrated in Supplement 5.13.

Supplement 5.12 (Exact Solution to the Extensible Inverted Pendulum)   The total potential energy of the system, Eq. 5.186, is rewritten in the form:

$$\Pi = k_t \left\{ \frac{1}{2}\theta^2 + \frac{1}{2}\kappa\delta^2 - p\left[1 - (1-\delta)\cos\theta\right] \right\}, \tag{5.194}$$

in which the nondimensional quantities have been introduced:

$$\delta := \frac{d}{\ell}, \quad \kappa := \frac{k_e\ell^2}{k_t}, \quad p := \frac{P}{P_c^\infty} = \frac{P\ell}{k_t}. \tag{5.195}$$

Here, δ is the unit extension, κ is the spring stiffness ratio, $P_c^\infty := \frac{k_t}{\ell}$ is the critical load of the inextensible rod, and p is the normalized load (equal to 1 when $P = P_c^\infty$). By requiring the TPE to be stationary, i.e., $\dfrac{\partial\Pi}{\partial\delta} = 0$, $\dfrac{\partial\Pi}{\partial\theta} = 0$, one gets:

$$\kappa\,\delta - p\cos\theta = 0, \tag{5.196a}$$

$$\theta - p\,(1-\delta)\sin\theta = 0. \tag{5.196b}$$

The fundamental path (here named $\mathcal{P}_f$), is:

$$\delta^f = \frac{p}{\kappa}, \qquad \theta^f = 0. \tag{5.197}$$

Further paths bifurcating from $\mathcal{P}_f$ are sought. From Eq. 5.196a, it follows:

$$\delta = \frac{p}{\kappa}\cos\theta, \tag{5.198}$$

which, substituted in Eq. 5.196b, provides an equation in the rotation:

$$p^2\sin\theta\cos\theta - p\kappa\sin\theta + \kappa\,\theta = 0. \tag{5.199}$$

Only the solutions $\theta \neq 0$ are of interest (as $\theta = 0$ reproduces the fundamental path). Solving for p, two solutions are found:

$$p_{1,2} = \kappa \frac{1 \mp \sqrt{1 - \frac{4}{\kappa}\theta \cot \theta}}{2 \cos \theta}, \tag{5.200}$$

which, substituted in Eq. 5.198, give:

$$\delta_{1,2} = \frac{1}{2}\left(1 \mp \sqrt{1 - \frac{4}{\kappa}\theta \cot \theta}\right). \tag{5.201}$$

Equations 5.200 and 5.201 are the parametric equations of *two* paths (named $\mathcal{P}_{1,2}$), different from the fundamental one, which are expressed in the form $\delta = \delta(\theta)$, $p = p(\theta)$. In order for these paths to bifurcate from the fundamental path, their intersections with this latter must be real. By executing the limit $p_{c1,2} := \lim_{\theta \to 0} p_{1,2}$, the critical values are obtained:

$$p_{c1,2} = \frac{\kappa}{2}\left(1 \mp \sqrt{1 - \frac{4}{\kappa}}\right). \tag{5.202}$$

These are real if and only if $\kappa \geq 4$; if not, the paths $\mathcal{P}_{1,2}$ *do not* branch from the fundamental path, but they are external to it. If $\kappa \to \infty$, that is, if the rod is inextensible, $p_{c1} \to 1$ and $p_{c2} \to \infty$; therefore, just the unique critical load of the inextensible inverted pendulum is found, i.e., $P_{c1} \to P_c^\infty$.

A discussion is now carried out about the shape of the bifurcated paths, or on their projections onto the (p, θ) plane, as expressed by Eq. 5.200 (Fig. 5.27a,b). The concavity of the curves at the bifurcation point is decided by the sign of the second derivative:

$$p_{1,2}''(0) = \frac{\pm 8 + 3\kappa\left(\mp 1 + \sqrt{1 - \frac{4}{\kappa}}\right)}{6\sqrt{1 - \frac{4}{\kappa}}}. \tag{5.203}$$

The nondimensional critical loads, Eq. 5.202, and the concavity at the bifurcation, Eq. 5.203 are presented in Fig. 5.27c,d. It is seen that the two critical loads coincide at $p_{c1} = p_{c2} = 2$, when $\kappa = 4$; they vanish, when $\kappa < 4$; they separate from each other, when κ increases. The smaller of the two, p_{c1}, rapidly tends to 1, while the larger, $p_{c,2}$, tends to infinity. The extensibility of the rod, therefore, affects the critical load only in a small interval of κ, but it introduces a qualitatively important phenomenon, namely, it destroys the branching when $\kappa < 4$.

The analysis of the concavity of the bifurcated path is instead more complex: while the upper path always has positive concavity (upward), the lower one has negative concavity when $\kappa < \kappa^* :=\frac{16}{3}$ and positive concavity when $\kappa > \kappa^*$,

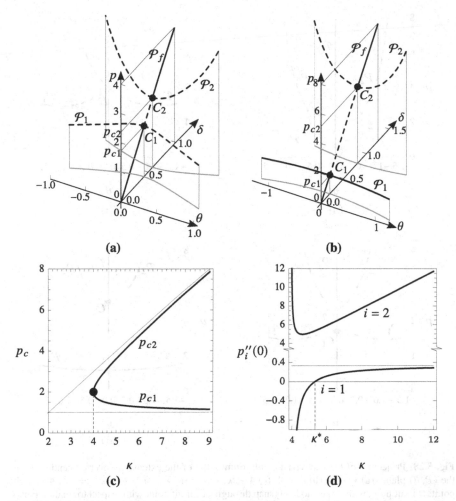

Fig. 5.27 Exact bifurcation analysis for the extensible inverted pendulum: 3D bifurcation diagrams, when (**a**) $4 < \kappa < \kappa^*$ and (**b**) $\kappa > \kappa^*$, with $\kappa^* := \frac{16}{3}$; (**c**) critical loads $p_{c1,2}$ *vs* $\kappa > 4$; (**d**) second derivative, at bifurcation, of the two bifurcated paths *vs* $\kappa > 4$

tending rapidly to the value $\frac{1}{3}$ (coinciding with that one found in the Sect. 5.2.1 for the inextensible pendulum). When $\kappa = \kappa^*$, the lowest path is locally flat. It is interesting to observe that for $\kappa \to 4^+$, the concavities of the two paths tend a $\mp\infty$, this denoting a singularity produced by the multiple bifurcation (discussed in Sect. 5.3.2 for the spherical pendulum).[19] The graph of the paths, for several values

---

[19] The two paths are described by two single-valued functions, $p_1(\theta) \leq 2$, $p_2(\theta) \geq 2$, respectively, below and above the bifurcation point, whose first derivative is discontinuous at $\theta = 0$.

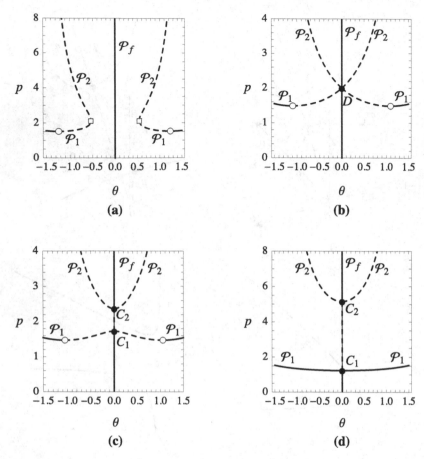

Fig. 5.28 Projection of the non-trivial equilibrium paths of the extensible inverted pendulum on the (P, θ) plane: (a) $\kappa < 4$; (b) $\kappa = 4$; (c) $4 < \kappa < \kappa^*$; (d) $\kappa > \kappa^*$, with $\kappa^* := \frac{16}{3}$; ● branch point, ○ limit point with respect to load (snap through), □ limit point with respect to rotation (snap back)

of κ, is presented in Fig. 5.28. It appears that, even when the path $\mathcal{P}_1$ has negative concavity at $\theta = 0$, this becomes positive for larger $|\theta|$.

To carry out the stability analysis of the paths, it is necessary to evaluate the properties of the Hessian matrix of the TPE. This is written (to within the inessential factor $k_t > 0$) as:

$$\mathbf{H} = \begin{bmatrix} \frac{\partial^2 \Pi}{\partial \delta^2} & \frac{\partial^2 \Pi}{\partial \delta \partial \theta} \\ \frac{\partial^2 \Pi}{\partial \theta \partial \delta} & \frac{\partial^2 \Pi}{\partial \theta^2} \end{bmatrix} = \begin{bmatrix} \kappa & p \sin \theta \\ p \sin \theta & 1 + p (\delta - 1) \cos \theta \end{bmatrix}. \tag{5.204}$$

According to Sylvester's criterion, the Hessian matrix is defined positive if the (1, 1) element of **H** is positive, together with its determinant $H := \det(\mathbf{H})$. The first of the two conditions is always satisfied, since $\kappa > 0$; the second one appears as:

$$H = \kappa \left[1 + p\,(\delta - 1)\cos\theta\right] - p^2 \sin^2\theta \quad > 0. \tag{5.205}$$

To evaluate H on the fundamental path, $\delta = \frac{p}{\kappa}$ must be substituted in the previous expression, thus obtaining:

$$H^f = p^2 - \kappa p + \kappa. \tag{5.206}$$

Since $H^f = 0$ in $p = p_{c1,2}$ (where the stiffness matrix becomes singular, according to the energy criterion of stability), it follows: (a) $H^f > 0$ when $p < p_{c1}$; (b) $H^f < 0$ when $p_{c1} < p < p_{c2}$; (c) $H^f > 0$ when $p < p_{c2}$. Therefore, the fundamental path is stable below the first bifurcation, unstable between the two bifurcations, and again stable above of the second bifurcation.

To evaluate H on the non-trivial $\mathcal{P}_{1,2}$ paths, it needs to substitute the parametric equations of the paths, $\delta = \delta(\theta)$, $p = p(\theta)$, as given by Eqs. 5.198 and 5.200, in Eq. 5.205, by getting $H_{1,2} = H_{1,2}(\theta; \kappa)$ (not shown here). For assigned κ values, the functions $H_{1,2}$ can be plotted and the range of θ found, in which $H_{1,2} > 0$ (stable portion of the paths) or $H_{1,2} < 0$ (unstable portions). The results are illustrated in Fig. 5.29, from which the following conclusions are drawn:

- the paths external to the fundamental one (existing only when $\kappa < 4$) are stable at low load values (entailing large rotations) and unstable at high load values (involving small rotations); the loss of stability occurs at a *point limit*, in which a fold bifurcation occurs;
- the lower branch path (which exists only when $\kappa \geq 4$) is unstable when $4 < \kappa < \kappa^*$ (in whose range it is downward concave) and it is stable when $\kappa > \kappa^*$ (i.e., when it is upward concave);
- the upper path (which exists only when $\kappa \geq 4$) is always unstable.

Overall, when the two critical loads are close together ($4 < \kappa < \kappa^*$), the two quasi-simultaneous critical modes interact in a way which is detrimental, making the postcritical behavior unstable. This is otherwise stable when the two critical loads are far apart enough (i.e., when $\kappa > \kappa^*$).

The example treated here, and as more fully discussed in [7], exhibits an "unconventional" mechanical behavior, which cannot occur in systems admitting a trivial fundamental path. In particular, the eigenvalues of the linearized problem (Fig. 5.27c) manifest a coalescing (or *merging*) point instead of the usual intersection point (or *crossing*), occurring in systems free of precritical deformation (an example, in this regard, is offered by the Augusti model). However, these unconventional behaviors exclusively manifest in the presence of large precritical deformations, rendering the problem of interest only for particular applications.   □

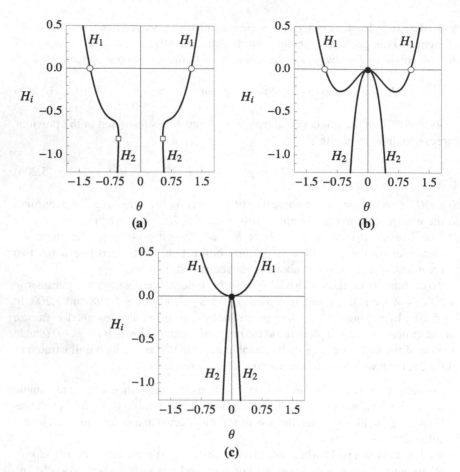

Fig. 5.29 Graph of the determinant $H = H_{1,2}$ of the Hessian matrix, evaluated at the non-trivial paths $\mathcal{P}_{1,2}$ of the extensible inverted pendulum: (a) $\kappa < 4$, (b) $4 < \kappa < \kappa^*$, (c) $\kappa > \kappa^*$, with $\kappa^* := \frac{16}{3}$; • branch point, ○ limit point with respect to load (snap through), □ limit point with respect to rotation (snap back)

Supplement 5.13 (Perturbation Solution to the Extensible Inverted Pendulum)
As discussed in Sect. 4.8.2, the following transformation is introduced:

$$\delta = \delta^f + \tilde{\delta}, \tag{5.207a}$$

$$\theta = \theta^f + \tilde{\theta}, \tag{5.207b}$$

where $\mathbf{y} = \left(\tilde{\delta}, \tilde{\theta}\right)^T$ are incremental variables. Substituting the latter equations in the equilibrium Eqs. 5.196 and taking into account Eq. 5.197, one gets the incremental equations:

$$p\left(1 - \cos\tilde{\theta}\right) + \kappa\tilde{\delta} = 0, \tag{5.208a}$$

$$\tilde{\theta} - p\sin\tilde{\theta}\left[1 - \left(\frac{p}{\kappa} + \tilde{\delta}\right)\right] = 0, \tag{5.208b}$$

admitting the trivial solution $\mathbf{y} = \mathbf{0}\ \forall p$.

Bifurcated solutions are sought in the parametric form $\mathbf{y} = \mathbf{y}(\epsilon)$, $p = p(\epsilon)$, with ϵ a perturbation parameter that zeroes at the bifurcation point. The parametric equations are then expanded in series. Here, the analysis is limited to the case of well-separated critical loads ($\kappa > 4$); a more detailed treatment is developed in [7]. To reduce the computational burden, symmetry considerations are introduced, suggested from the physics of the problem: if $\left(\tilde{\delta}, \tilde{\theta}, p\right)$ is a solution to Eqs. 5.208, then also $\left(\tilde{\delta}, -\tilde{\theta}, p\right)$ is a solution. It means that if, for a given load, the pendulum buckles on the left or on the right, it undergoes the same shortening. Accordingly, an odd power series of ϵ is taken for $\tilde{\theta}$ and an even power series for $\tilde{\delta}$, p, in the order that the two solutions are identified by ϵ and $-\epsilon$, respectively. Hence:

$$\tilde{\delta} = \epsilon^2\tilde{\delta}_2 + \cdots, \tag{5.209a}$$

$$\tilde{\theta} = \epsilon\tilde{\theta}_1 + \epsilon^3\tilde{\theta}_3 + \cdots, \tag{5.209b}$$

$$p = p_c + \epsilon^2 p_2 + \cdots. \tag{5.209c}$$

Substituting them in the incremental Eq. 5.208, expanding the circular functions, and separating the terms with the same power of ϵ, the following perturbation equations are derived:

$$\epsilon: \quad \left[1 - p_c\left(1 - \frac{p_c}{\kappa}\right)\right]\tilde{\theta}_1 = 0, \tag{5.210a}$$

$$\epsilon^2: \quad \kappa\,\tilde{\delta}_2 = -\frac{1}{2}p_c\tilde{\theta}_1^2, \tag{5.210b}$$

$$\epsilon^3: \quad \left[1 - p_c\left(1 - \frac{p_c}{\kappa}\right)\right]\tilde{\theta}_3 = -p_c\left[\frac{\tilde{\theta}_1^3}{6}\left(1 - \frac{p_c}{\kappa}\right) + \tilde{\theta}_1\tilde{\delta}_2\right] + p_2\tilde{\theta}_1\left(1 - 2\frac{p_c}{\kappa}\right). \tag{5.210c}$$

Equation 5.210a is a *nonlinear eigenvalue problem* in the critical load p_c. By solving it, the two critical loads p_{c1} and p_{c2} (Eq. 5.202) are found. The normalized critical mode is $\tilde{\theta}_1 = 1$ for both the eigenvalues. Passing to solve Eq. 5.210b, the shortening is evaluated as:

$$\tilde{\delta}_2^{(1,2)} = -\frac{p_{c1,2}}{2\kappa}, \tag{5.211}$$

along path $\mathcal{P}_1$ and $\mathcal{P}_2$. Substituting it in Eq. 5.210c, and zeroing the second member for solvability, the unknown p_2 on the two paths is drawn:

$$p_2^{(1,2)} = \frac{\pm 8 + 3\kappa \left(\mp 1 + \sqrt{1 - \frac{4}{\kappa}} \right)}{12\sqrt{1 - \frac{4}{\kappa}}}, \tag{5.212}$$

which is consistent with the exact result found in Eq. 5.203.[20]   □

References

1. Antman, S.S.: Bifurcation problems for nonlinearly elastic structures. In: Rabinowitz, P.H. (ed.) Symposium on Applications of Bifurcation Theory, pp. 73–125. Academic Press, New York (1977)
2. Bazant, Z.P., Cedolin, L.: Stability of Structures: Elastic, Inelastic, Fracture, and Damage Theories. World Scientific Publishing, Singapore (2010)
3. Britvec, S.J.: The Stability of Elastic Systems. Pergamon Press, New York (1973)
4. Croll, G.A., Walker, A.C.: Elements of Structural Stability. Macmillan, London (1972)
5. Dyn, C.L.: Stability Theory and Its Applications to Structural Mechanics. Noordhoff, Leyden (1974)
6. Feodosiev, V.I.: Advanced Stress and Stability Analysis: Worked Examples. Springer, Berlin (2005)
7. Ferretti, M., Di Nino, S., Luongo, A.: A paradigmatic system for non-classic interactive buckling. Int. J. Non Linear Mech. **134** (2021)
8. Leipholz, H.: Stability Theory. Academic Press, New York (1970)
9. Luongo, A., Zulli, D.: Mathematical Models of Beams and Cables. John Wiley & Sons, New York (2013)
10. Pignataro, M., Rizzi, N., Luongo, A.: Stability, Bifurcation and Postcritical Behaviour of Elastic Structures. Elsevier, Amsterdam (1990)
11. Thompson, J.M.T., Hunt, G.W.: A General Theory of Elastic Stability. John Wiley & Sons, London (1973)
12. Thompson, J.M.T., Hunt, G.W.: Elastic Instability Phenomena. John Wiley & Sons, Chichester (1984)

---

[20] Indeed $p_2^{(1,2)} = \frac{1}{2} \left. \frac{d^2 p_{1,2}}{d\theta^2} \right|_{\theta=0}$.

Chapter 6
Linearized Theory of Buckling

6.1 Introduction

The elastic structures, when loaded by conservative forces, suffer from static bifurcations. Among these, branching from the fundamental path is of particular importance. To study the phenomenon, a considerable simplification consists in modeling the system as rigid in one or more directions, in such a way to ensure the existence of a trivial fundamental path, whose bifurcation analysis is far simpler. Examples, already treated, are related to (a) axially rigid compressed rods, elastically constrained to the ground, or (b) the inextensible Euler beam. Another example concerns a beam with a thin rectangular cross-section, loaded in its plane of major inertia, assumed infinite, which buckles via a combination of out-of-plane bending and twist. A further example concerns a plate stressed in its own plan, assumed undeformable, which, at the bifurcation, buckles out-of-plan.

The infinite stiffness of a line or plane is, however, an unphysical hypothesis, even requiring a not always easy formulation of an internally constrained model. It is therefore interesting to investigate what changes in the buckling behavior of partially rigid structures when the unphysical hypothesis is removed and the body is modeled as *fully deformable* (i.e., free of internal constraints). In this regard, it was seen in the previous chapter that forces induce *precritical strains* in the body, which depend on the intensity of the parameter of bifurcation and that *change the geometry* of the body. To describe this transformation, it primarily needs to determine the non-trivial fundamental path, an operation that may be difficult, in case of large strains. At the critical state, in which the bifurcation manifests itself, the geometry of the body is unknown, since it depends on the load. This implies that the bifurcation analysis, already at the lowest level of mere determination of the critical load, requires the solution of a *nonlinear eigenvalue problem*, in which the stiffness matrix is a function of the eigenvalue, according to the pattern of the non-trivial path. Therefore, it is understood why, especially in the technical field (where simple, even if approximate, solutions are pursued), a procedure alternative

A. Luongo et al., *Stability and Bifurcation of Structures*, https://doi.org/10.1007/978-3-031-27572-2_6

to the exact analysis is desirable. This approach is actually viable, on the basis of the following considerations:

1. The precritical strains of structures, which are of interest in civil and mechanical engineering, are generally small, so that the *geometry of the system at the bifurcation can be confused with the initial geometry*. The following examples corroborate this idea: (a) the current length of the Euler beam, in the critical state, is approximately equal to its natural length, since the axial critical load induces a small precritical shortening; likewise, (b) the thin-walled beam, bent in the plane of largest stiffness, can be assimilated to a straight beam, since the transverse critical load induces a small precritical curvature. The idea, moreover, is comforted by the results of the previous chapter, relevant to the telescopic rod (Sect. 5.6). There, it was seen that, when the system possesses a large axial stiffness, the critical load provided by the more refined analysis differs little from that of the corresponding axially rigid structure.

2. The stresses generated by the precritical strains, however, are not negligible and are a function of the intensity of the external forces. Consistently with the hypothesis of small precritical displacements and strains, the stresses can be determined by a classic linear analysis (e.g., for a beam loaded at its ends or along its axis, by the de Saint-Venant theory and its technical extensions), so *they are proportional to the bifurcation parameter*.

With these hypotheses, the body does not deform in the precritical state (like the partially rigid body), but it is stressed (by active, rather than reactive, stresses). Such bodies are called *prestressed*; they are affected by a *known* state of stress, to within an unknown multiplication factor. The prestressed configuration is taken as a reference, and therefore strains and displacements are measured starting from it. For such bodies it is relatively easy to formulate a model that permits to evaluate the critical load and also the postcritical behavior. In this book, however, the analysis will be limited to compute the critical load, consistently with the *adjacent equilibrium criterion*. The influence of imperfections will also be taken into account, but always in the linearized context. The resulting framework is called *linearized theory*,[1] or, in a more technical jargon, the *second-order theory*.[2]

In this chapter, the linearized theory is illustrated by mainly following the Variational Formulation, both for discrete and continuous systems. However, the use of the Virtual Work Principle is also discussed, in view of analyzing nonconservative systems. Furthemore, the method of the direct equilibrium in the adjacent configuration is also mentioned, which will be useful ahead, when a physical interpretation of the equations ruling the buckling of beams and plates will be given.

---

[1] In this book, the terms "linearized theory" and "linear theory" are used with different meanings. The former accounts for prestress; the latter ignores it.

[2] The term should be understood as opposed to the *linear theory*, or of the *first order*, classic of the elementary textbooks of mechanics of solids and structures, in which the effects of prestress are ignored, by virtue of the "superposition principle."

6.2 Variational Formulation of the Equilibrium of Prestressed Bodies

The linearized equilibrium equations of a system, prestressed in the reference configuration, are derived via a variational method. Discrete systems are first addressed, and results are successively extended to continuous systems. This last step calls for introducing a nonlinear strain tensor.

6.2.1 Discrete Systems

A discrete elastic system is considered (Fig. 6.1a), whose current configuration is described by the Lagrangian parameters $\mathbf{q}$, measured from the natural configuration C_n, in which the stresses σ and the strains ε vanish. Due to the external forces $\mathbf{f}_0$, the system undergoes the $\mathbf{q}_0$ displacements and moves to a new configuration C_0. Here, $\sigma = \sigma_0$ is the *prestress state* and $\varepsilon = \varepsilon_0$ the corresponding strain state. *It is assumed that these strains and displacements are negligible, that is,* $\varepsilon_0 \simeq \mathbf{0}$, $\mathbf{q}_0 \simeq \mathbf{0}$, *so the prestressed configuration C_0 is confused with the natural configuration C_n*. Some *small* incremental forces $\tilde{\mathbf{f}}$ are then considered to act on the body, which, originating in C_0, bring the system to the current configuration $\tilde{C}$. Therefore, displacements, strains, and stresses undergo the increments $\tilde{\mathbf{q}}$, $\tilde{\varepsilon}$, $\tilde{\sigma}$, respectively, so the total quantities become:

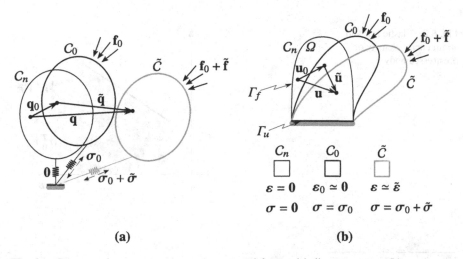

(a)   (b)

Fig. 6.1 Prestressed systems, subject to incremental forces: **(a)** discrete system, **(b)** continuous system

$$q = q_0 + \tilde{q}, \tag{6.1a}$$

$$\varepsilon = \varepsilon_0 + \tilde{\varepsilon}, \tag{6.1b}$$

$$\sigma = \sigma_0 + \tilde{\sigma}. \tag{6.1c}$$

Elastic Law and Total Potential Energy

By assuming that the incremental stresses and strains are proportional to each other, i.e., $\tilde{\sigma} = \mathbf{E}\tilde{\varepsilon}$, with $\mathbf{E} = \mathbf{E}^T$ the elastic matrix, and taking into account the prestress σ_0, the elastic law is written in a linear *non-homogeneous* form:

$$\sigma = \sigma_0 + \mathbf{E}\tilde{\varepsilon}. \tag{6.2}$$

It is then possible to write the elastic potential energy of the system as the sum of two contributions, namely, $U = U_0 + \tilde{U}$ (Fig. 6.2), where:

- $U_0 := \sigma_0^T \tilde{\varepsilon}$ is the *prestress elastic energy*, equal to the work expended by the prestresses in the incremental strains;
- $\tilde{U} := \frac{1}{2}\tilde{\sigma}^T \tilde{\varepsilon} = \frac{1}{2}\tilde{\varepsilon}^T \mathbf{E}\tilde{\varepsilon}$ is the *incremental elastic energy*, equal to the work done by the incremental stresses in the incremental strains.

The work of the external forces is also equal to the sum of two contributions, $W = W_0 + \tilde{W}$, where:[3]

- $W_0 := \mathbf{f}_0^T \left(q_0 + \tilde{q}\right)$ is the external work of the prestress forces, expended on the total displacement;

Fig. 6.2 Elastic law and strain energy density of a prestressed body

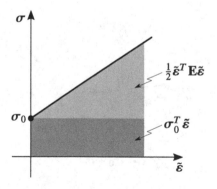

---

[3] Here, it is assumed that the work is linear in the Lagrangian parameters, as it is often possible to obtain with a suitable choice of them (an example is given in the Sect. 6.6); otherwise it is necessary to add some nonlinear terms (as it will be seen in the Sect. 9.7, dealing with forces applied to rigid cross-sections of thin-walled beams).

- $\tilde{W} := \tilde{\mathbf{f}}^T \tilde{\mathbf{q}}$ is the work of the incremental forces, expended on the incremental displacement only.

Given that $\mathbf{q} = \tilde{\mathbf{q}}$, $\boldsymbol{\varepsilon} = \tilde{\boldsymbol{\varepsilon}}$, the Total Potential Energy (TPE) of the system, $\Pi = U - W$, is written as:

$$\Pi = \frac{1}{2} \boldsymbol{\varepsilon}^T \mathbf{E} \boldsymbol{\varepsilon} + \boldsymbol{\sigma}_0^T \boldsymbol{\varepsilon} - \left(\mathbf{f}_0 + \tilde{\mathbf{f}} \right)^T \mathbf{q}. \tag{6.3}$$

Nonlinear Kinematics

To express the TPE in terms of displacement, it is needs to link the incremental strains to displacements, via the congruence equation $\boldsymbol{\varepsilon} = \boldsymbol{\varepsilon}(\mathbf{q})$.[4] In the regime of finite displacements, this relation is nonlinear. Expanding it in Taylor series, a polynomial expression is assumed:

$$\boldsymbol{\varepsilon} = \boldsymbol{\varepsilon}_1 + \boldsymbol{\varepsilon}_2 + \boldsymbol{\varepsilon}_3 + \cdots , \tag{6.4}$$

in which:

- $\boldsymbol{\varepsilon}_1 := \mathbf{D}\mathbf{q}$ is the *linear part* of the strain, with $\mathbf{D}$ the congruence matrix; it describes the infinitesimal strain resulting from infinitesimal displacements;
- $\boldsymbol{\varepsilon}_2 := \frac{1}{2}\mathbf{d}_2(\mathbf{q})$ is the *quadratic part* of the strain, where $\mathbf{d}_2(\cdot)$ is a homogeneous quadratic form in its arguments;
- $\boldsymbol{\varepsilon}_k := \frac{1}{k!}\mathbf{d}_k(\mathbf{q})$ is the *degree $k > 2$ part* of the strain, where $\mathbf{d}_k(\cdot)$ is a homogeneous form of degree k in its arguments.

TPE Truncated at the Second Degree

By limiting the series to the terms of the second degree, the following truncated form of TPE is obtained:

$$\Pi = \frac{1}{2} \boldsymbol{\varepsilon}_1^T \mathbf{E} \boldsymbol{\varepsilon}_1 + \boldsymbol{\sigma}_0^T (\boldsymbol{\varepsilon}_1 + \boldsymbol{\varepsilon}_2) - \left(\mathbf{f}_0 + \tilde{\mathbf{f}} \right)^T \mathbf{q}. \tag{6.5}$$

This expression can be further simplified if one considers that, by hypothesis, $\boldsymbol{\sigma}_0$, $\mathbf{f}_0$ are equilibrated in C_0 and that $\boldsymbol{\varepsilon}_1$, $\mathbf{q}$ describe an admissible act of motion applied to C_0. Therefore, for the *virtual work principle* (VWP), the work done by the stresses $\boldsymbol{\sigma}_0$ on the infinitesimal strains $\boldsymbol{\varepsilon}_1$ is equal to the work done by the forces $\mathbf{f}_0$ on the infinitesimal displacements $\mathbf{q}$, that is:

---

[4] In this book, the term "congruence equation" is used, instead of the more common (and pedestrian) "strain-displacement relationship." The locution "compatibility condition," also improperly used in literature, is instead reserved to the integrability conditions of the kinematic problem.

$$\sigma_0^T \varepsilon_1 = \mathbf{f}_0^T \mathbf{q}. \tag{6.6}$$

Consequently, the TPE simplifies into:

$$\Pi = \frac{1}{2}\varepsilon_1^T \mathbf{E}\varepsilon_1 + \mu\sigma_0^T \varepsilon_2 - \tilde{\mathbf{f}}^T \mathbf{q}, \tag{6.7}$$

in which the load multiplier μ (such that $\mathbf{f}_0 \to \mu\mathbf{f}_0$, and therefore $\sigma_0 \to \mu\sigma_0$) has been introduced. In terms of displacements, the previous expression reads:

$$\begin{aligned}\Pi &= \frac{1}{2}\mathbf{q}^T \mathbf{D}^T \mathbf{E}\mathbf{D}\mathbf{q} + \frac{1}{2}\mu\sigma_0^T \mathbf{d}_2(\mathbf{q}) - \tilde{\mathbf{f}}^T \mathbf{q} \\ &= \frac{1}{2}\mathbf{q}^T \mathbf{K}_e \mathbf{q} + \frac{1}{2}\mu\mathbf{q}^T \mathbf{K}_g \mathbf{q} - \tilde{\mathbf{f}}^T \mathbf{q}, \end{aligned} \tag{6.8}$$

where $\mathbf{K}_e := \mathbf{D}^T \mathbf{E}\mathbf{D} = \mathbf{K}_e^T$ is the *elastic stiffness matrix* and $\mathbf{K}_g = \mathbf{K}_g^T$ the *geometric stiffness matrix*, defined by the identity of the following two quadratic forms:[5]

$$\mathbf{q}^T \mathbf{K}_g \mathbf{q} = \sigma_0^T \mathbf{d}_2(\mathbf{q}). \tag{6.9}$$

Remark 6.1 The TPE of a prestressed body, truncated at the quadratic terms, differs from that of the linear theory *for the sole presence of the quadratic prestress energy*. It represents the work of the prestresses on the quadratic part of the incremental strains (hence, the term "second-order effects").

Equilibrium

By imposing the stationary of the TPE via $\dfrac{\partial \Pi}{\partial \mathbf{q}} = \mathbf{0}$, the following *linear* equations in the displacements are obtained:

$$(\mathbf{K}_e + \mu\mathbf{K}_g)\mathbf{q} = \tilde{\mathbf{f}}. \tag{6.10}$$

---

[5] The geometric stiffness matrix cannot be written by matrix symbolism, but indicial. By letting $\sigma_0 = (\sigma_k^0)$, $\mathbf{d}_2(\mathbf{q}) = \left(\sum_i \sum_j d_{kij}q_i q_j\right)$, with $d_{kij} = d_{kji}$, it follows:

$$\sigma_0^T \mathbf{d}_2(\mathbf{q}) = \sum_i \sum_j \left(\sum_k d_{kij}\sigma_k^0\right) q_i q_j$$

so the coefficient (i, j) of the matrix $\mathbf{K}_g$ is $K_{ij}^g = \left(\sum_k d_{kij}\sigma_k^0\right) = K_{ji}^g$.

Here, the incremental force $\tilde{\mathbf{f}}$, assumed to be small in module, plays the role of an *imperfection* (possibly depending on μ). If $\tilde{\mathbf{f}} = \mathbf{0}$, Eq. 6.10 reduces to a linear eigenvalue problem in μ (since the current configuration has been confused with the natural one).

Remark 6.2 The TPE, truncated at the second degree, gives equations *linear* in the displacements, which express the equilibrium in the configuration $\tilde{C}$, infinitely close (adjacent) to the prestressed configuration C_0.

Remark 6.3 Unlike what happens in axially rigid systems, *the geometric matrix is drawn from the elastic potential energy* via the prestress, rather than from the potential energy of the external forces.

6.2.2 Continuous Systems

The previous discussion can be straightforwardly extended to continuous systems [1–5]. An elastic body is considered (Fig. 6.1b), which, in the natural state C_n, occupies a domain Ω of the one-, two-, or three-dimensional space, of boundary Γ. The Γ_u portion of the boundary is constrained to the ground by bilateral constraints; the supplementary portion Γ_f is free. In the natural state, the strains $\boldsymbol{\varepsilon}(\mathbf{x})$ and the stresses $\boldsymbol{\sigma}(\mathbf{x})$ are functions of the point $\mathbf{x}$.[6] The configuration change is described by the displacement field $\mathbf{u}(\mathbf{x})$, measured by C_n, sufficiently regular, and such that $\mathbf{u}(\mathbf{x}) = \mathbf{0}$ on Γ_u. Due to the external forces $\mathbf{b}_0(\mathbf{x})$, applied in Ω, and $\mathbf{f}_0(\mathbf{x})$, applied on Γ_f, the body assumes a new configuration (called prestressed) C_0, reached by experiencing displacements $\mathbf{u}_0(\mathbf{x})$; in this configuration $\boldsymbol{\sigma} = \boldsymbol{\sigma}_0(\mathbf{x})$ and $\boldsymbol{\varepsilon} = \boldsymbol{\varepsilon}_0(\mathbf{x})$. By assuming that $\boldsymbol{\varepsilon}_0 \simeq \mathbf{0}$, $\mathbf{u}_0 \simeq \mathbf{0}$, *the prestressed configuration C_0 is confused with the natural one C_n.* Incremental forces $\tilde{\mathbf{b}}(\mathbf{x})$, $\tilde{\mathbf{f}}(\mathbf{x})$, respectively applied in Ω and Γ_f, are then considered to act on the body, producing incremental displacements, strains, and stresses, respectively, equal to $\tilde{\mathbf{u}}(\mathbf{x})$, $\tilde{\boldsymbol{\varepsilon}}(\mathbf{x})$, $\tilde{\boldsymbol{\sigma}}(\mathbf{x})$. In the current configuration $\tilde{C}$, it is therefore $\mathbf{u}(\mathbf{x}) \simeq \tilde{\mathbf{u}}(\mathbf{x})$, $\boldsymbol{\varepsilon}(\mathbf{x}) \simeq \tilde{\boldsymbol{\varepsilon}}(\mathbf{x})$ and $\boldsymbol{\sigma}(\mathbf{x}) = \boldsymbol{\sigma}_0(\mathbf{x}) + \tilde{\boldsymbol{\sigma}}(\mathbf{x})$.

Taken a linear elastic law:

$$\boldsymbol{\sigma}(\mathbf{x}) = \boldsymbol{\sigma}_0(\mathbf{x}) + \mathbf{E}\boldsymbol{\varepsilon}(\mathbf{x}), \qquad (6.11)$$

where $\mathbf{E} = \mathbf{E}^T$ is the elastic matrix; the TPE assumes the form:

---

[6] Here and in the following, the column matrices $\boldsymbol{\varepsilon}, \boldsymbol{\sigma}$ list the independent components of strain and stress. For example, for the Cauchy continuum, the six independent components of strain and stress tensors; for the Kirchhoff plate, the three curvatures (two flexural and one torsional) and the three moments (two bending and one torsional); and so on.

$$\Pi = \int_\Omega \left(\boldsymbol{\sigma}_0^T \boldsymbol{\varepsilon} + \frac{1}{2} \boldsymbol{\varepsilon}^T \mathbf{E} \boldsymbol{\varepsilon} \right) d\Omega - \int_\Omega \left(\mathbf{b}_0 + \tilde{\mathbf{b}} \right)^T \mathbf{u} d\Omega - \int_{\Gamma_f} \left(\mathbf{f}_0 + \tilde{\mathbf{f}} \right)^T \mathbf{u} d\Gamma.$$

(6.12)

The congruence equations, in finite kinematics, are written as $\boldsymbol{\varepsilon} = \mathcal{E}(\mathbf{u})$, with $\mathcal{E}$ a nonlinear differential operator. Expanding it in Taylor series up to the quadratic terms, $\boldsymbol{\varepsilon} = \boldsymbol{\varepsilon}_1 + \boldsymbol{\varepsilon}_2$ holds, with $\boldsymbol{\varepsilon}_1 = \mathcal{D}\mathbf{u}$ the linear and $\boldsymbol{\varepsilon}_2 = \frac{1}{2} d_2(\mathbf{u})$ the quadratic parts. With these expressions, the TPE, truncated at the second degree terms, reads:

$$\Pi = \int_\Omega \left(\frac{1}{2} \boldsymbol{\varepsilon}_1^T \mathbf{E} \boldsymbol{\varepsilon}_1 + \boldsymbol{\sigma}_0^T \boldsymbol{\varepsilon}_2 \right) d\Omega - \int_\Omega \tilde{\mathbf{b}}^T \mathbf{u} d\Omega - \int_{\Gamma_f} \tilde{\mathbf{f}}^T \mathbf{u} d\Gamma,$$

(6.13)

having taken into account that, for the virtual works principle, it is:

$$\int_\Omega \boldsymbol{\sigma}_0^T \boldsymbol{\varepsilon}_1 d\Omega - \int_\Omega \mathbf{b}_0^T \mathbf{u} d\Omega - \int_{\Gamma_f} \mathbf{f}_0^T \mathbf{u} d\Gamma = 0.$$

(6.14)

The TPE, in terms of displacement, is therefore written as:

$$\Pi = \int_\Omega \left(\frac{1}{2} (\mathcal{D}\mathbf{u})^T \mathbf{E} (\mathcal{D}\mathbf{u}) + \frac{1}{2} \mu \boldsymbol{\sigma}_0^T d_2(\mathbf{u}) \right) d\Omega - \int_\Omega \tilde{\mathbf{b}}^T \mathbf{u} d\Omega - \int_{\Gamma_f} \tilde{\mathbf{f}}^T \mathbf{u} d\Gamma,$$

(6.15)

where the load multiplier μ, affecting the prestress, has been introduced.

The stationary condition of the quadratic functional, which requires performing integration by parts (Appendix A), provides the differential (or field) equations and the related boundary conditions, all linear, of the type:

$$\left(\mathcal{K}_e + \mu \mathcal{K}_g \right) \mathbf{u} = \tilde{\mathbf{b}}, \qquad \text{in } \Omega,$$

(6.16a)

$$\left[\left(\mathcal{B}_e + \mu \mathcal{B}_g \right) \mathbf{u} - \tilde{\mathbf{f}} \right] \delta \mathbf{u} = \mathbf{0}, \qquad \text{on } \Gamma_f,$$

(6.16b)

$$\mathbf{u} = \mathbf{0}, \qquad \text{on } \Gamma_u.$$

(6.16c)

where $\mathcal{K}_e, \mathcal{K}_g, \mathcal{B}_e, \mathcal{B}_g$ are differential operators, in the domain and at the boundary, elastic and geometric. Equations 6.16 constitute a boundary value problem, which governs the equilibrium of the body in the configuration adjacent to the prestressed one.

The following supplements illustrate, with reference to the Cauchy continuum: (a) how to derive an exact expression for the strain, $\boldsymbol{\varepsilon} = \mathcal{E}(\mathbf{u})$ and (b) how to draw, from the variational procedure, the boundary value problem in Equations 6.16.

Supplement 6.1 (Green-Lagrange Strain Tensor) A material segment $d\mathbf{x}$ of the Cauchy continuum is considered. Due to the deformation, it changes into the segment $d\tilde{\mathbf{x}} = \mathbf{F} d\mathbf{x}$, where $\mathbf{F} := \frac{\partial \tilde{\mathbf{x}}}{\partial \mathbf{x}}$ is the *transport gradient tensor*. The square of the length of the segment, before the deformation, is $ds^2 = d\mathbf{x}^T d\mathbf{x}$; after that, it is $d\tilde{s}^2 = d\tilde{\mathbf{x}}^T d\tilde{\mathbf{x}} = d\mathbf{x}^T \mathbf{F}^T \mathbf{F} d\mathbf{x}$. By evaluating the difference between the two squares, one finds:

$$d\tilde{s}^2 - ds^2 = d\mathbf{x}^T \left(\mathbf{F}^T \mathbf{F} - \mathbf{I} \right) d\mathbf{x} =: 2 d\mathbf{x}^T \mathbf{H} d\mathbf{x}, \qquad (6.17)$$

in which:

$$\mathbf{H} := \frac{1}{2} \left(\mathbf{F}^T \mathbf{F} - \mathbf{I} \right), \qquad (6.18)$$

has been defined, called the *Green-Lagrange strain tensor.*

In terms of displacement $\mathbf{u}$, since $\tilde{\mathbf{x}} = \mathbf{x} + \mathbf{u}$, it is $\mathbf{F} = \mathbf{G} + \mathbf{I}$, with $\mathbf{G} := \frac{\partial \mathbf{u}}{\partial \mathbf{x}}$ the *displacement gradient tensor.* Hence:

$$
\begin{aligned}
\mathbf{H} :&= \frac{1}{2} \left[\left(\mathbf{G}^T + \mathbf{I} \right) (\mathbf{G} + \mathbf{I}) - \mathbf{I} \right] = \frac{1}{2} \left(\mathbf{G} + \mathbf{G}^T \right) + \frac{1}{2} \mathbf{G}^T \mathbf{G} \\
&= \frac{1}{2} \left[u_{i,j} + u_{j,i} \right] + \frac{1}{2} \left[u_{h,i} u_{h,j} \right],
\end{aligned}
\qquad (6.19)
$$

having used $\mathbf{G} = \left[u_{i,j} \right]$. Here, the comma indicates differentiation with respect to the following variable, and the summation with respect to the repeated indices is implied. The components of $\mathbf{H} = \left[\varepsilon_{ij} \right]$ consist of a linear and a quadratic part, namely:

$$\varepsilon_{ij} = \varepsilon_{ij}^{(1)} + \varepsilon_{ij}^{(2)} = \frac{1}{2} \left(u_{i,j} + u_{j,i} \right) + \frac{1}{2} u_{h,i} u_{h,j}. \qquad (6.20)$$

It is worth noticing that this definition of strain is exact; it does *not* descend from a series expansion.   □

Supplement 6.2 (Linearized Equilibrium Equations of the Prestressed Cauchy Continuum) As an example of derivation of the equilibrium equations of a prestressed continuous body, the Cauchy 3D continuum is considered. It occupies a volume $\mathcal{V}$ of the space, is bilaterally constrained to the ground on the surface S_u, and is free on the supplementary surface S_f. The body is subject to a state of prestress, described by the stress tensor $\mathbf{T}_0 = \left[\sigma_{ij}^0 \right]$, $i, j = 1, 2, 3$, in equilibrium with body forces $\mathbf{b}_0 = \left(b_i^0 \right)$ in $\mathcal{V}$ and surface forces $\mathbf{f}_0 = \left(f_i^0 \right)$ on S_f. Incremental forces, $\tilde{\mathbf{b}} = \left(\tilde{b}_i \right)$, $\tilde{\mathbf{f}} = \left(\tilde{f}_i \right)$, in $\mathcal{V}$ and on S_f, respectively, are then applied. In the transformation that brings the body from the prestressed to the current configuration, the body undergoes the displacement $\mathbf{u} = (u_i)$ (tilde omitted). These produce, in finite kinematics, strains described by the Green-Lagrange tensor $\mathbf{H} = \left[\varepsilon_{ij} \right]$ (tilde omitted), whose components are expressed by Eq. 6.20.

Assuming a linear elastic law:

$$\sigma_{ij} = \sigma_{ij}^0 + E_{ijhk}\varepsilon_{hk}, \qquad (6.21)$$

with E_{ijhk} fourth-order elastic tensor, the total potential energy, according to Eq. 6.13, reads:

$$\Pi = \int_V \left(\frac{1}{2} E_{ijhk}\varepsilon_{ij}^{(1)}\varepsilon_{hk}^{(1)} + \sigma_{ij}^0\varepsilon_{ij}^{(2)} \right) dV - \int_V \tilde{b}_i u_i dV - \int_{S_f} \tilde{f}_i u_i dS. \qquad (6.22)$$

Using Eq. 6.20, and taking into account the symmetries of E_{ijhk},[7] the TPE is expressed in terms of displacement:

$$\Pi = \frac{1}{2} \int_V \left(E_{ijhk} u_{i,j} u_{h,k} + \sigma_{ij}^0 u_{h,i} u_{h,j} \right) dV - \int_V \tilde{b}_i u_i dV - \int_{S_f} \tilde{f}_i u_i dS. \qquad (6.23)$$

Equilibrium requires Π to be stationary, i.e., $\delta\Pi = 0$, $\forall \delta\mathbf{u}$ kinematically admissible, entailing:[8]

$$\int_V \left(E_{ijhk} u_{h,k} + \sigma_{hj}^0 u_{i,h} \right) \delta u_{i,j} \, dV - \int_V \tilde{b}_i \delta u_i dV - \int_{S_f} \tilde{f}_i \delta u_i dS = 0, \qquad \forall \delta u_i.$$

$$\qquad (6.24)$$

After integration by parts of the first term,[9] taking into account that $\delta u_i = 0$ on S_u, the variational problem is rewritten as:

$$-\int_V \left[\left(E_{ijhk} u_{h,k} + \sigma_{hj}^0 u_{i,h} \right)_{,j} + \tilde{b}_i \right] \delta u_i \, dV$$

$$\qquad (6.25)$$

$$+ \int_{S_f} \left[\left(E_{ijhk} u_{h,k} + \sigma_{hj}^0 u_{i,h} \right) n_j - \tilde{f}_i \right] \delta u_i dS = 0, \qquad \forall \delta u_i,$$

---

[7] It should be remembered that the elastic tensor has *weak symmetries* $E_{ijhk} = E_{jihk} = E_{ijkh} = E_{jikh}$ (consequent to the symmetry of σ_{ij} and ε_{hk}), as well as *strong symmetries* $E_{ijhk} = E_{hkij}$, referred to the existence of an elastic energy.

[8] The symmetry of $\mathbf{T}$ has been exploited, and the indices i and h in the second term have been exchanged.

[9] This follows from Green's formula in space:

$$\int_V f \, g_{,j} dV = -\int_V f_{,j} \, g dV + \int_S f \, g \, n_j dS$$

where $\mathbf{n} = (n_i)$ is the outward normal.

where $\mathbf{n} = (n_i)$ is the normal to the surface, positive if outward to the body. From the previous equation, the indefinite equilibrium equation and boundary conditions are drawn:

$$E_{ijhk}u_{h,kj} + \left(\sigma_{hj}^0 u_{i,h}\right)_{,j} + \tilde{b}_i = 0, \qquad \text{in } \mathcal{V}, \qquad (6.26a)$$

$$\left(E_{ijhk}u_{h,k} + \sigma_{hj}^0 u_{i,h}\right) n_j = \tilde{f}_i, \qquad \text{on } S_f, \qquad (6.26b)$$

$$u_i = 0, \qquad \text{on } S_u, \qquad (6.26c)$$

which are of the form of Eq. 6.16. Equations 6.26 generalize the well-known *Navier equations* of an elastic body, by adding to them the geometric effect of the prestress.

□

6.3 Adjacent Equilibrium Through the Virtual Work Principle

The equilibrium equations, so far derived by the stationary condition of the TPE, can alternatively be drawn via the virtual work principle. If the use of this approach is merely a matter of taste when dealing with conservative systems, it becomes instead mandatory when studying nonconservative systems, for which a potential energy does not exist. Examples of this procedure will be shown later (Chap. 11), when using the VWP in a dynamic problem. To introduce the technique, static problems will be dealt with here, in the context of the linearized theory of prestressed bodies.

With the symbols defined in the previous sections, and with reference to a continuous body, the VWP states that:

$$\int_{\Omega} \boldsymbol{\sigma}^T \delta\boldsymbol{\varepsilon}\, d\Omega = \int_{\Omega} \mathbf{b}^T \delta\mathbf{u}\, d\Omega + \int_{\Gamma_f} \mathbf{f}^T \delta\mathbf{u}\, d\Gamma, \qquad \forall \delta\mathbf{u}, \qquad (6.27)$$

where forces and stresses are equilibrated and strains and displacements are congruent. Since $\boldsymbol{\sigma} = \boldsymbol{\sigma}_0 + \tilde{\boldsymbol{\sigma}}$ and $\delta\boldsymbol{\varepsilon} = \delta\boldsymbol{\varepsilon}_1 + \delta\boldsymbol{\varepsilon}_2$, by ignoring the product of higher-order terms, it is $\boldsymbol{\sigma}^T \delta\boldsymbol{\varepsilon} \simeq \boldsymbol{\sigma}_0^T (\delta\boldsymbol{\varepsilon}_1 + \delta\boldsymbol{\varepsilon}_2) + \tilde{\boldsymbol{\sigma}}^T \delta\boldsymbol{\varepsilon}_1$. Expressed also the forces in the incremental form, the VWP reads:

$$\int_{\Omega} \left(\boldsymbol{\sigma}_0^T (\delta\boldsymbol{\varepsilon}_1 + \delta\boldsymbol{\varepsilon}_2) + \tilde{\boldsymbol{\sigma}}^T \delta\boldsymbol{\varepsilon}_1\right) d\Omega = \int_{\Omega} \left(\mathbf{b}_0 + \tilde{\mathbf{b}}\right)^T \delta\mathbf{u}\, d\Omega - \int_{\Gamma_f} \left(\mathbf{f}_0 + \tilde{\mathbf{f}}\right)^T \delta\mathbf{u}\, d\Gamma.$$

$$(6.28)$$

Since the prestressed configuration is equilibrated, still invoking the VWP, it holds:

$$\int_\Omega \sigma_0^T \delta \varepsilon_1 \, d\Omega = \int_\Omega \mathbf{b}_0^T \delta \mathbf{u} \, d\Omega - \int_{\Gamma_f} \mathbf{f}_0^T \delta \mathbf{u} \, d\Gamma, \qquad (6.29)$$

so Eq. 6.28 reduces to:

$$\int_\Omega \left(\tilde{\sigma}^T \delta \varepsilon_1 + \sigma_0^T \delta \varepsilon_2 \right) d\Omega = \int_\Omega \tilde{\mathbf{b}}^T \delta \mathbf{u} \, d\Omega - \int_{\Gamma_f} \tilde{\mathbf{f}}^T \delta \mathbf{u} \, d\Gamma. \qquad (6.30)$$

It constitutes the *incremental form of the VWP*. Its terms represent, in the exact order, the virtual work that the incremental stresses expend on the infinitesimal virtual strains; the virtual work that the pre-existing stresses do on the second-order part of the virtual strains; and the virtual work that the incremental forces do on the virtual displacements.

If the system is conservative, Eq. 6.30 coincides with the condition $\delta \Pi = 0$ in Eq. 6.15, when using the constitutive law, $\tilde{\sigma} = \mathbf{E} \varepsilon_1$, the kinematic relationships $\varepsilon_1 = \mathcal{D} \mathbf{u}$, $\varepsilon_2 = \frac{1}{2} d_2(\mathbf{u})$, and the rescaling $\sigma_0 \to \mu \sigma_0$.

6.4 Direct Equilibrium of Prestressed Bodies

In Sect. 6.2, the equations ruling equilibrium in the adjacent configuration have been determined through a variational procedure. This approach will be systematically followed throughout the book. However, it is instructive, after the mathematical model has been formulated, to *interpret* the relevant equations by invoking the balance of forces. Albeit these arguments add nothing new to the formulation, they supply a key for understanding, on a physical ground, the mechanism underlying the birth of the geometric stiffness.

The exact nonlinear equation governing the indefinite equilibrium of a continuous body,[10] in its current configuration, is linear in the stresses σ but nonlinear in the displacements $\mathbf{u}$, measured from the natural state. It is of the type:

$$\mathcal{E}^* [\mathbf{u}] \, \sigma = \mathbf{b}, \qquad (6.31)$$

where $\mathcal{E}^* [\mathbf{u}]$ is the differential *equilibrium operator* and $\mathbf{b}$ are the volume forces. The equation, when expressed in scalar form, represents the balance of forces, internal and external, projected onto the reference basis. Under the action of prestress forces $\mathbf{b}_0$, the system undergoes a displacement field $\mathbf{u}_0$ and a prestress σ_0. Neglecting $\mathbf{u}_0$, as it is customary in the linearized theory, the balance equations in the prestressed state are written as:

---

[10] The discussion, for the sake of brevity, is limited to the field equation, by ignoring the boundary conditions. These latter, however, can be thought as incorporated in the operator.

$$\mathcal{E}_0^* \sigma_0 = \mathbf{b}_0, \tag{6.32}$$

where $\mathcal{E}_0^* := \mathcal{E}^*[0]$ is the equilibrium operator in the undeformed state. By introducing incremental stresses and forces, i.e., letting $\sigma = \sigma_0 + \tilde{\sigma}$ and $\mathbf{b} = \mathbf{b}_0 + \tilde{\mathbf{b}}$, and linearizing in $\mathbf{u}$, Eq. 6.31 becomes:

$$\left(\mathcal{E}_0^* + \mathcal{E}_{0,\mathbf{u}}^* \mathbf{u}\right)(\sigma_0 + \tilde{\sigma}) = \mathbf{b}_0 + \tilde{\mathbf{b}}. \tag{6.33}$$

Here, $\mathcal{E}_{0,\mathbf{u}}^* := \mathcal{E}_{,\mathbf{u}}^*[0]$ is the derivative of $\mathcal{E}^*$ with respect to $\mathbf{u}$, computed at zero;[11] it represents the *incremental equilibrium operator*. Taking into account the pre-existing equilibrium (Eq. 6.32), and ignoring the second-order quantities (described by the product of $\mathbf{u}$ and $\tilde{\sigma}$), the previous equation leads to:

$$\mathcal{E}_0^* \tilde{\sigma} + \left(\mathcal{E}_{0,\mathbf{u}}^* \mathbf{u}\right)\sigma_0 = \tilde{\mathbf{b}}. \tag{6.34}$$

This is the *incremental equilibrium equation*, which expresses the balance of forces in the adjacent configuration. It is possible to recognize in it (a) the contribution of the unknown incremental stress $\tilde{\sigma} = \mathbf{E}\tilde{\varepsilon}$, consistent with the linear theory, and (b) the contribution of the known prestress σ_0, which is proportional to the change of geometry described by $\mathbf{u}$.

From the previous considerations, it derives an operating procedure for direct writing the incremental equilibrium equations. These equations, in addition to the usual terms of the linear theory (referred to the initial geometry), are contributed by the *unbalanced part of prestress*, caused by the change of geometry. This contribution consists of (a) the components of the prestress in the directions orthogonal to the original ones (Example 6.1) and (b) the increment of their arm (Example 6.2), all consequent to the incremental displacements $\mathbf{u}$. In the following chapters (Supplements 9.4, 9.5, 9.7, 10.3), an extensive use of this procedure will be made while commenting the equations obtained by the variational method.

Remark 6.4 If Eq. 6.34 is compared with Eq. 6.16a, deduced from the variational procedure, it is concluded that $\mathcal{K}_e = \mathcal{E}_0^* \mathbf{E} \mathcal{D}$ (with $\mathcal{D}$ the congruence operator, already introduced) and that $\mathcal{K}_g \mathbf{u} = \left(\mathcal{E}_{0,\mathbf{u}}^* \mathbf{u}\right)\sigma_0$.

Example 6.1 (Direct Incremental Equations: Rotation of Prestress) The system in Fig. 6.3a is considered, consisting of two extensible rods hinged at the ends, solicited at the common joint by a prestress force F_0 and an incremental force $\tilde{F}$. The rotations θ_1, θ_2 of the rods are assumed as Lagrangian parameters, so $\mathbf{u} = (\theta_1, \theta_2)^T$. The stress state is described by $\sigma = (N_1, N_2)^T$, with N_i the internal axial force in the ith rod. The matrix form of the equilibrium equation in the current configuration (Fig. 6.3b) is:

---

[11] Since $\mathcal{E}^*$ is a two-dimensional matrix and $\mathbf{u}$ a column matrix, $\mathcal{E}_{0,\mathbf{u}}^*$ is a three-dimensional matrix.

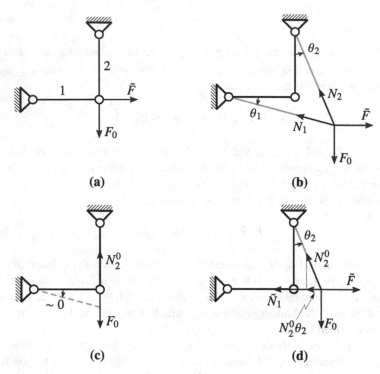

Fig. 6.3 Prestressed two DOF system: (**a**) structure, (**b**) equilibrium in the current configuration, (**c**) equilibrium in the prestressed configuration, (**d**) equilibrium in the adjacent configuration

$$
\begin{bmatrix} \cos\theta_1 & \sin\theta_2 \\ -\sin\theta_1 & -\cos\theta_2 \end{bmatrix} \begin{pmatrix} N_1 \\ N_2 \end{pmatrix} = \begin{pmatrix} \tilde{F} \\ -F_0 \end{pmatrix}. \tag{6.35}
$$

which are of the type of Eq. 6.31.

The force F_0 alone is first considered acting on the system (Fig. 6.3c). Freezing the configuration to the natural one $\mathbf{u} = \mathbf{0}$ (according to the hypothesis of negligibility of the prestrains), the prestress $\boldsymbol{\sigma}_0 = \left(N_1^0, N_2^0\right)^T = (0, F_0)$ is determined. When the incremental force is applied to the structure, the stress is incremented as $(N_1, N_2) = \left(0 + \tilde{N}_1, F_0 + \tilde{N}_2\right)$; by linearizing the equilibrium for small rotations, the previous equation reads:

$$
\begin{bmatrix} 1 & \theta_2 \\ -\theta_1 & -1 \end{bmatrix} \begin{pmatrix} \tilde{N}_1 \\ F_0 + \tilde{N}_2 \end{pmatrix} = \begin{pmatrix} \tilde{F} \\ -F_0 \end{pmatrix}. \tag{6.36}
$$

Ignoring the products between displacements and incremental forces, and accounting for the pre-existing equilibrium, the incremental equation is finally obtained (having the form of Eq. 6.34):

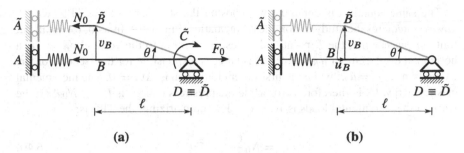

Fig. 6.4 Prestressed two DOF system: **(a)** equilibrium in the adjacent configuration, **(b)** nonlinear kinematics

$$\begin{bmatrix} 1 & 0 \\ 0 & -1 \end{bmatrix} \begin{pmatrix} \tilde{N}_1 \\ \tilde{N}_2 \end{pmatrix} + \begin{bmatrix} 0 & \theta_2 \\ -\theta_1 & 0 \end{bmatrix} \begin{pmatrix} 0 \\ F_0 \end{pmatrix} = \begin{pmatrix} \tilde{F} \\ 0 \end{pmatrix}, \tag{6.37}$$

or:

$$\tilde{N}_1 + F_0\theta_2 = \tilde{F}, \tag{6.38a}$$

$$-\tilde{N}_2 = 0. \tag{6.38b}$$

These equations express the equilibrium in the adjacent configuration. However, they can also be *directly written*, i.e., without passing for the linearization of the nonlinear equations (Fig. 6.3d). The procedure calls for considering the horizontal component $F_0\theta_2$ of the *rotated prestress* N_2^0, which is no more balanced by the external force F_0; this geometric force adds itself to the elastic force $\tilde{N}_1$, to balance the incremental external force $\tilde{F}$.   □

Example 6.2 (Direct Incremental Equations: Translation of Prestress) As a further example of direct writing of the equilibrium equations, the system represented in Fig. 6.4a is considered. It consists of a rigid horizontal rod BD, of length ℓ, constrained by a horizontal spring AB, whose end A is sliding. Because of a prestress force F_0, the spring is stressed by $N_0 = F_0$. The spring elongation is ignored and the prestressed configuration is confused with the natural one. On the prestressed system, an incremental couple $\tilde{C}$ is introduced. The rod then rotates by an angle θ around D, which remains fixed (indeed, a horizontal translation would involve an increment of stress in the spring, which could not be balanced). Limiting the kinematics to the infinitesimal field, due to the rotation, the point B undergoes the vertical translation $v_B = \theta\ell$, so the spring also translates vertically of the same amount. Balance in the adjacent configuration requires that the torque of the *translated prestress,* with respect to the point D, equilibrates the incremental couple, that is:

$$N_0\ell\theta = \tilde{C}. \tag{6.39}$$

The system therefore only possesses a geometric stiffness.

The same equation is obtained by imposing the stationary of the TPE, which, however, requires the study of nonlinear kinematics (Fig. 6.4b). Indeed, for small but finite displacements, the B point undergoes, in addition to the vertical $v_B = \theta\ell$, a horizontal displacement $u_B = \frac{\ell}{2}\theta^2$. This latter is responsible for the (second-order) elongation $\Delta\ell_2 := u_B$, while the first-order elongation is $\Delta\ell_1 = 0$. The incremental elastic energy, $\tilde{U}$, is therefore zero, while the prestress energy is $U_0 = N_0\Delta\ell_2$; the work of the incremental loads is $\tilde{W} = \tilde{C}\theta$. By summarizing, the TPE is:

$$\Pi = N_0\frac{\ell}{2}\theta^2 - \tilde{C}\theta. \tag{6.40}$$

whose stationary condition, $\frac{\partial\Pi}{\partial\theta} = 0$, returns Eq. 6.39.   □

6.5 Linearized Effects of Imperfections

The effects of geometric and load imperfections on the behavior of prestressed structures is examined here in the context of the linearized theory. For simplicity, reference will be made to discrete systems; later on in the book (Sect. 7.4), an example of a continuous structure will be shown, for which similar considerations hold.

It is observed, preliminary, that while the linearized theory provides an excellent approximation of the critical load of perfect structures (when sufficiently rigid, e.g., axially), the same does not happen in presence of imperfections, for which the theory leads to grossly approximated results. However, since this theory is of frequent use in technical applications, a mention of it is given here.

Linearized Analysis
The linearized equilibrium Eq. 6.10, in presence of geometric imperfections q_0, modifies as follows:

$$\left(\mathbf{K}_e + \mu\mathbf{K}_g\right)\mathbf{q} = \mathbf{K}_e\mathbf{q}_0 + \tilde{\mathbf{f}}, \tag{6.41}$$

where $\mathbf{K}_e\mathbf{q}_0$ is a force equivalent to the geometric imperfection.[12] It is a linear *non-homogeneous* algebraic system, whose matrix of coefficients depends on the parameter μ. Since it is generally impossible to analytically evaluate the inverse of the stiffness matrix, it is necessary to proceed numerically or to solve the problem in the modal space, as explained shortly.

Remark 6.5 When $\mu \to \mu_c$ in Eq. 6.41, the total stiffness matrix becomes singular, and therefore $\mathbf{q} \to \infty$. The result is clearly erroneous, as a consequence of

---

[12] Equation 6.41 should be compared with the linearized version of Eq. 4.49. Here, incremental forces have been considered as imperfections, and the multipliers α, β have been omitted.

linearization. Indeed, as soon as the response $\mathbf{q}$ becomes sufficiently large, the nonlinear terms are no longer negligible, and linearization is no longer valid. The problem is formally analogous to what happens in dynamics, when the frequency of the harmonic excitation tends to a natural frequency.

Single Degree of Freedom Systems
If the system has a single DOF, Eq. 6.41 can easily be solved. When the geometric effect is destabilizing, the equation appears in the form:

$$\left(k_e - \mu k_g\right) q = k_e q_0 + \tilde{f}, \tag{6.42}$$

with $k_e > 0$, $k_g > 0$. The associated homogeneous problem admits the critical eigenvalue $\mu_c := \frac{k_e}{k_g}$. Solving the non-homogeneous equation, one gets:

$$q = \frac{1}{1 - \frac{\mu}{\mu_c}} \left(q_0 + \frac{\tilde{f}}{k_e}\right). \tag{6.43}$$

Here, the term in parenthesis, $q_I := q_0 + \frac{\tilde{f}}{k_e}$, represents the *linear static response*, determined when the geometric stiffness is ignored. The solution of the linearized problem, q_{II}, is therefore equal to q_I multiplied by the *amplification* factor $\frac{1}{1-\frac{\mu}{\mu_c}}$.[13] The geometric effect, therefore (in the linearized context), is entirely expressed by this amplification factor.

Figure 6.5 illustrates the linearized response of the single DOF system (black curve), compared with the exact response (gray curve), in case of stable postcritical behavior. Both the responses consist of a sub-critical branch (natural curve) and a super-critical branch (non-natural curve). Linearized analysis provides the (erroneous) result $q \to \pm\infty$ when $\mu \to \mu_c$, because of the absence of nonlinearities. The natural curve of the linearized analysis is stable;[14] non-natural curves, on the other hand, are *entirely* unstable in the linearized analysis, in contrast with the exact analysis, that predicts an exchange of stability at the lower limit point (the unstable branch being the closest to the μ axis).[15] The drawbacks of linearized theory, albeit conceptually serious, are generally considered of little importance in the technical field, in which the interest is focused on the states $\mu \ll \mu_c$, where the natural curve provides a reasonable approximation of the exact solution. However, non-natural

---

[13] The analogy with the response of the harmonic oscillator to a harmonic excitation should be noticed again.

[14] Indeed, the second-order TPE, $\Pi = \frac{1}{2}\left(k_e - \mu k_g\right) q^2 - \left(k_e q_0 + \tilde{f}\right) q$, has second derivative $\frac{\partial^2 \Pi}{\partial q^2} = k_e - \mu k_g > 0$ when $\mu < \mu_c$, and second derivative $\frac{\partial^2 \Pi}{\partial q^2} < 0$ when $\mu > \mu_c$.

[15] This result is due to the fact that the stability limit curve on the bifurcation diagram, i.e., the locus on which the second derivative of the energy vanishes (e.g., the dash-dotted curve in Fig. 5.3), crushes on the line $\mu = \mu_c$, due to the truncation of the energy. Therefore, all the super-critical states are (erroneously) unstable in the linearized theory.

Fig. 6.5 Response of a
single DOF system suffering
from geometric
imperfections; linearized
analysis (black line),
nonlinear analysis (gray line)
in the presence of hardening
nonlinearity; stable
(continuous line), unstable
(dashed line)

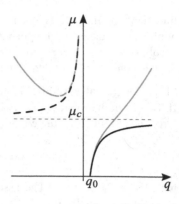

curves are relevant in the elastoplastic evolutive analysis, as it will be seen in the Chap. 8.

Response in the Modal Space

If the system has several DOF, it is convenient to conduct a (static) modal analysis, that is, to express the response *in the basis of its eigenvectors*. By denoting by $\mathbf{u}_k$ ($k = 1, 2, \cdots, n$) the real eigenvectors of the problem (i.e., the critical modes), and by μ_k the corresponding real eigenvalues (i.e., the critical loads), supposed to be distinct and ordered as $\mu_c := \mu_1 < \mu_2 < \cdots < \mu_n$, it holds:[16]

$$\left(\mathbf{K}_e + \mu_k \mathbf{K}_g\right) \mathbf{u}_k = \mathbf{0}. \tag{6.44}$$

The well-known orthogonality properties among the (normalized) modes apply:[17]

$$\mathbf{u}_h^T \mathbf{K}_e \mathbf{u}_k = -\mu_k \delta_{hk}, \qquad \mathbf{u}_h^T \mathbf{K}_g \mathbf{u}_k = \delta_{hk}, \tag{6.45}$$

---

[16] Here, the index c is suppressed on μ_{ck} and $\mathbf{u}_{ck}$, since, in the linearized context, no notational confusion can arise with the series expansions of load and displacements.

[17] Equations 6.45 are proved as follows. Written the eigenvalue problem for two distinct k and h, premultiplying for $\mathbf{u}_h$ and $\mathbf{u}_k$ respectively, one has:

$$\mathbf{u}_h^T \left(\mathbf{K}_e + \mu_k \mathbf{K}_g\right) \mathbf{u}_k = \mathbf{0}$$

$$\mathbf{u}_k^T \left(\mathbf{K}_e + \mu_h \mathbf{K}_g\right) \mathbf{u}_h = \mathbf{0}$$

By subtracting these equations member to member, and exploiting the symmetry of the two matrices, it follows:

$$(\mu_k - \mu_h) \mathbf{u}_h^T \mathbf{K}_g \mathbf{u}_k = \mathbf{0}$$

from which $\mathbf{u}_h^T \mathbf{K}_g \mathbf{u}_k = 0$ for $h \neq k$ (orthogonality property); hence, $\mathbf{u}_h^T \mathbf{K}_e \mathbf{u}_k = 0$ for $h \neq k$. On the other hand, if $h = k$, given the arbitrariness of the length of the eigenvector, it is possible to take $\mathbf{u}_k^T \mathbf{K}_g \mathbf{u}_k = 1$ (normalization condition), and consequently $\mathbf{u}_k^T \mathbf{K}_e \mathbf{u}_k = -\mu_k$.

where $\delta_{hk} = 1$ if $h = k$ and $\delta_{hk} = 0$ if $h \neq k$.

To express the solution of the non-homogeneous Eq. 6.41 in the modal basis, the solution is represented as a linear combination of the modes:

$$\mathbf{q} = \sum_{k=1}^{n} c_k \mathbf{u}_k, \tag{6.46}$$

where the unknown constants c_k are the projections of the response onto the unit vectors (modal amplitudes). To evaluate c_k, the modal solution, Eq. 6.46, is substituted in Eq. 6.41 and this latter is premultiplied by $\mathbf{u}_h^T$ $(h = 1, 2, \cdots, n)$:

$$\mathbf{u}_h^T \left[(\mathbf{K}_e + \mu \mathbf{K}_g) \sum_{k=1}^{n} c_k \mathbf{u}_k \right] = \mathbf{u}_h^T \left[\mathbf{K}_e \mathbf{q}_0 + \tilde{\mathbf{f}} \right]. \tag{6.47}$$

Taking into account the properties stated in Eqs. 6.45, it follows:

$$c_k = \frac{1}{1 - \frac{\mu}{\mu_k}} \frac{1}{\mu_k} \left(-\mathbf{u}_k^T \mathbf{K}_e \mathbf{q}_0 - \mathbf{u}_k^T \tilde{\mathbf{f}} \right), \tag{6.48}$$

from which the response, Eq. 6.46, is evaluated.

The ratio $\frac{1}{1 - \frac{\mu}{\mu_k}}$ is called the *kth amplification factor*. Since the range of interest of the load is $\mu \in [0, \mu_1]$, the amplification factors decrease as k increases. The contribution of the first mode is therefore larger than that of the second mode, this of the third, and so on.[18] It is then possible to approximate the response *by excess*, by replacing *all* the amplification factors by the largest of them, $\frac{1}{1 - \frac{\mu}{\mu_c}}$; thus, an *upper bound* $\|\mathbf{q}^+\| > \|\mathbf{q}\|$ is obtained, as:

$$\mathbf{q}^+ = \frac{1}{1 - \frac{\mu}{\mu_c}} \mathbf{q}_I, \tag{6.49}$$

where:

$$\mathbf{q}_I := \sum_{k=1}^{n} \left[-\frac{1}{\mu_k} \left(\mathbf{u}_k^T \mathbf{K}_e \mathbf{q}_0 + \mathbf{u}_k^T \tilde{\mathbf{f}} \right) \right] \mathbf{u}_k \tag{6.50}$$

is the *linear response of the system* (i.e., obtained by ignoring the geometric stiffness), $\mathbf{q}_I = \mathbf{q}_0 + \mathbf{K}_e^{-1} \tilde{\mathbf{f}}$, expressed in the modal basis.

---

[18] The imperfections are assumed to be of generic form. For particular forms, close to the higher modes, the contribution of these latter may be relevant. However, when $\mu \to \mu_1$ the first mode prevails, whatever the imperfection.

Equation 6.49 is a very simple (and roughly approximate) formula, which establishes an upper bound of the response of the imperfect system, equal to the response of the linear problem multiplied by the *amplification factor* $\frac{1}{1-\frac{\mu}{\mu_c}}$ of the *first mode only*. Thus, the law in Eq. 6.43, exactly holding for single DOF systems, is generalized (with approximation) to multi-DOF systems.

Remark 6.6 Based on Eq. 6.49, and consistently with linearization, it can be stated that any mechanical quantity g, *displacement, strain, or stress, obeys the amplification law*, that is, $g^+ = \frac{1}{1-\frac{\mu}{\mu_c}} g_I$.[19]

6.6 An Illustrative Example: The Extensible Inverted Pendulum

An example of linearized bifurcation analysis is now developed. The extensible inverted pendulum, already studied in the Sect. 5.6 to illustrate the bifurcation from non-trivial path, is considered again (Fig. 6.6).

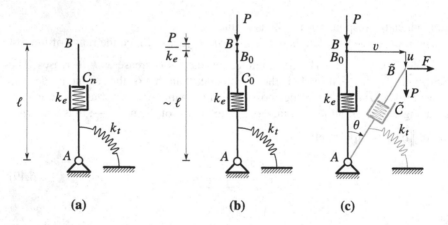

Fig. 6.6 Prestressed extensible pendulum: (**a**) natural, (**b**) prestressed, and (**c**) current configurations

---

[19] For example, when an Euler beam, compressed by a force P smaller than the critical load P_c, is subject to transverse loads, the relevant abscissa-dependent bending moment is $M^+(x) = \frac{1}{1-\frac{P}{P_c}} M_I(x)$, where $M_I(x)$ is the law established by the linear theory, in which the effects of the axial force are ignored. This example will be addressed in Sect. 7.4.

The pendulum is stressed by a (large) vertical force P applied at the top, which induces prestress, and is subject to a (small) horizontal incremental force F, applied at the same point. The prestress force shortens the pendulum by an amount P/k_e that, if the extensional stiffness k_e is large, can be considered negligible, compared to the initial length ℓ of the rod. The prestressed configuration is therefore confused with the natural one. However, the stress in the spring, $N_0 = -P$, is not negligible; it constitutes the prestress of the system.

Kinematics

When the pendulum moves to the current configuration, the rod rotates by an angle θ and its free end experiences a transverse displacement v and a longitudinal displacement u. The three kinematic parameters, however, are not independent. Chosen u, v as Lagrangian parameters, it follows:[20]

$$\tan\theta = \frac{v}{\ell - u}. \tag{6.51}$$

Correspondingly, the rod undergoes the elongation:

$$\Delta\ell = \sqrt{(\ell - u)^2 + v^2}. \tag{6.52}$$

For small displacements u, v, the previous relationships can be expanded in series, to give:[21]

$$\theta = \frac{v}{\ell} + \cdots, \tag{6.53a}$$

$$\Delta\ell = -u + \frac{1}{2}\frac{v^2}{\ell} + \cdots. \tag{6.53b}$$

Total Potential Energy

The TPE of the system, according to Eq. 6.3, reads:

$$\Pi = \frac{1}{2}\left(k_t\theta^2 + k_e\Delta\ell^2\right) + N_0\Delta\ell - (Fv + Pu). \tag{6.54}$$

whose terms represent, in the exact order, the elastic energy of the torsional spring; the incremental elastic potential energy of the extensional spring; the prestress elastic energy; and the work, changed of sign, of the external forces. As already

---

[20] Differently from what was done in the Sect. 5.6, here two Lagrangian parameters are adopted, which reproduce circumstances that will be systematically encountered in the analysis of continuous systems. In particular, these parameters allow to express the external work as *linear* in the displacements.

[21] The reasons for the different degree of truncation will be explained immediately.

noticed, the potential energy of the loads is linear in the Lagrangian parameters, by virtue of the choice made for these latter.

By substituting the kinematic relationship of Eqs. 6.53 in the TPE, and retaining terms up to the quadratic ones in the Lagrangian parameters, it follows:

$$\Pi = \frac{1}{2}k_t \frac{v^2}{\ell^2} + \frac{1}{2}k_e u^2 - \frac{1}{2}P\frac{v^2}{\ell} - Fv. \tag{6.55}$$

Here, due to the pre-existing equilibrium, and according to the virtual work principle, the linear terms have been eliminated (by exploiting $N_0 = -P$). Equation 6.55 expresses the TPE in the form of Eq. 6.7: in it, the incremental elastic energy involves only the linear part of the strain, while the prestress energy requires evaluation of the quadratic part of the elongation of the extensional spring.[22]

Equilibrium Equations and Solution

By imposing the stationary of the TPE, $\dfrac{\partial \Pi}{\partial u} = 0$, $\dfrac{\partial \Pi}{\partial v} = 0$, the equilibrium equations follow:

$$k_e u = 0, \tag{6.56a}$$

$$\left(\frac{k_t}{\ell} - P\right)v = F\ell. \tag{6.56b}$$

The first of them gives $u = 0$; hence, $\Delta\ell = 0$ to the first order, that is, the change of configuration is inextensional. This is a consequence of the fact that, at the first order, the axial force in the rod does not change in passing from the prestressed to the current configurations.

From the second equation, if $F = 0$, the critical load is obtained:

$$P_c := \frac{k_t}{\ell}, \tag{6.57}$$

which coincides with the limit value, for $k_e \to \infty$, of the exact solution (Sect. 5.6).

If $F \neq 0$, the linearized solution for the "imperfect" path is found:

$$v = \frac{1}{1 - \frac{P}{P_c}} \frac{F\ell^2}{k_t}, \tag{6.58}$$

consistent with the general form in Eq. 6.49.

---

[22] For this reason, in Eqs. 6.53, θ has been truncated at the first order and $\Delta\ell$ at the second order.

References

1. Bolotin, V.V.: Nonconservative problems of the theory of elastic stability. Macmillan, London (1963)
2. Como, M.: Theory of elastic stability (in Italian). Liguori, Napoli (1967)
3. Pignataro, M., Rizzi, N., Luongo, A.: Stability, bifurcation and postcritical behaviour of elastic structures. Elsevier, Amsterdam (1990)
4. Timoshenko, S.P., Gere, J.M.: Theory of elastic stability. McGraw-Hill, New York (1961)
5. van Der Heijden, A.M.A.: W.T. Koiter's elastic stability of solids and structures. Cambridge University Press, Cambridge (2009)

Chapter 7
Elastic Buckling of Planar Beam Systems

7.1 Introduction

The simplest continuous structure undergoing buckling is the uniformly compressed beam (i.e., the Euler beam), which plays an iconic role in the bifurcation theory. However, even a so simple system requires some care in modeling, since some variants are possible, that is: (a) shear deformable (Timoshenko) beam, or shear-undeformable (Euler-Bernoulli) beam; (b) inextensible or extensible beam. Usually, the influence of the shear flexibility is small in slender beams, the only of interest in dealing with elastic buckling problems, so that the shear-undeformable model is usually adopted. In contrast, the choice of an inextensible or extensible model is not so obvious. In Chap. 5, the inextensible model was used to analyze the post-buckling of a single beam, since it was found easier to tackle a single-displacement field (i.e., the transverse deflection), rather than a two-displacement field (including the longitudinal displacement). However, the inextensible beam is an internally constrained model, which entails some algorithmic difficulties (even in the linear problem), when an assembly of beams is considered, due to the fact the axial force has a reactive character. Therefore, in this chapter, reference is made to the *extensible Euler-Bernoulli* model, for which this drawback does not occur. The interest is confined to the linearized problem, in which only the buckling load and the associated mode are evaluated. To this end, the equations are developed in the context of the theory of prestressed bodies, introduced in Chap. 6. In this framework, when the beam is sub-critically compressed, it is easy to account for transverse loads, which trigger deflections and bending moments *magnified* with respect to those predicted by the linear theory, due to the destabilizing effect of the geometric stiffness.

When dealing with beams, a number of problems of technical interest are encountered, namely: stepped beams, non-uniformly compressed beams, beams with elastic constraints, beams on Winkler elastic soil, cable-prestressed beams, trussed beams, frames. Exact solutions are viable only in very simple cases;

© The Author(s), under exclusive license to Springer Nature Switzerland AG 2023
A. Luongo et al., *Stability and Bifurcation of Structures*,
https://doi.org/10.1007/978-3-031-27572-2_7

more often, semi-analytical analyses (according to the *Ritz method*), or numerical approaches (via the *Finite Element Method*), must be used. All these problems and algorithms are addressed in this chapter.

7.2 Extensible Beam Model

The linearized planar model of *extensible* and *shear undeformable* beam is formulated, referring to the theory of the prestressed elastic bodies, illustrated in the Chap. 6. This model is also known as the *extensible Euler beam*.

A rectilinear prismatic beam is considered, shear undeformable and immersed in the 2D space, uniformly compressed by an axial load of intensity P, applied at one end, and transversely loaded by distributed forces of linear density $p(x)$ (Fig. 7.1a). The 3D object is modeled as a *one-dimensional polar continuum*. In the prestressed configuration (confused with that natural one), the beam lies on the segment $x \in (0, \ell)$, of ends A, B (Fig. 7.1a).

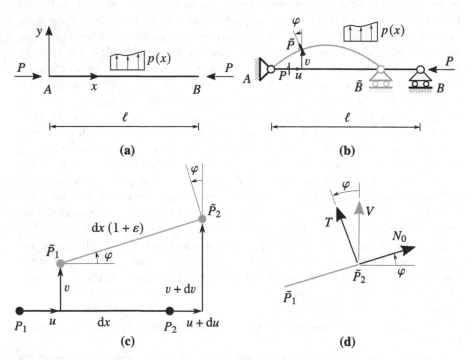

Fig. 7.1 Extensible beam: (**a**) prestressed reference configuration; (**b**) current configuration; (**c**) unit extension of an infinitesimal element of beam; (**d**) transverse component V of the internal force

Kinematics

According to the one-dimensional polar model, each material point P is endowed with orientation. The displacement of the point describes the translation of the centerline of the 3D beam, and the rotation of the point that of its cross-section. The generic change of configuration is described by the displacement field $\mathbf{u} := (u(x), v(x), \varphi(x))^T$, where u and v are the longitudinal and transverse components of the translation, respectively, and φ is the rotation (Fig. 7.1b). The three displacement components, however, are not independent, due to the internal constraint of shear undeformability (which ensures the preservation of orthogonality between cross-sections and centerline[1]).

To mathematically express the internal constraint, it is observed that (Fig. 7.1c):

$$dv = (dx + du) \tan \varphi, \tag{7.1}$$

from which[2]

$$\varphi = \arctan \left(\frac{v'}{1 + u'} \right), \tag{7.2}$$

where a dash denotes differentiation with respect to x. This relationship allows one to reduce to two the degrees of freedom of each point, e.g., u and v (commonly named the "master" variables, with φ the "slave" variable).

The strain magnitudes of the extensible beam are (a) the *unit extension* $\varepsilon := \frac{d\tilde{x} - dx}{dx}$ and (b) the *bending curvature* $\kappa := \varphi'$. By accounting for $d\tilde{x}^2 = dx^2 (1 + \varepsilon)^2 = (dx + du)^2 + dv^2$ (Fig. 7.1c), it follows that:

$$\varepsilon = \sqrt{(1 + u')^2 + v'^2} - 1. \tag{7.3}$$

On the other hand, by differentiating equation 7.2, one gets:

$$\kappa = \frac{(1 + u') v'' - u'' v'}{(1 + u')^2 + v'^2}. \tag{7.4}$$

Equations 7.3 and 7.4 are *kinematically exact* congruence equations. However, since the interest is here limited to a linearized analysis, it is sufficient to take strains

---

[1] In other words, the constraint implies the equality between the rotation of the axis line and the rotation of the cross-section.

[2] Equation 7.2 should be compared with Eq. 5.117, relevant to the inextensible model. Here, an element dx changes its length, i.e., $\overline{P_1 P_2} \neq \overline{\tilde{P}_1 \tilde{P}_2}$ (Fig. 7.1c).

expanded in series, truncated at the second order for extension and at first order for the curvature,[3] that is:

$$\varepsilon = u' + \frac{1}{2}v'^2 + \cdots , \tag{7.5a}$$

$$\kappa = v'' + \cdots . \tag{7.5b}$$

Therefore the linear strains are $\varepsilon_1 := u'$ and $\kappa_1 := v''$ and the quadratic extension is $\varepsilon_2 := \frac{1}{2}v'^2$.

Remark 7.1 A different measure of the extensional strain is offered by the $\varepsilon_{11} := \frac{1}{2}\left(d\tilde{s}^2 - ds^2\right)$ component of the Green-Lagrange tensor (Supplement 6.1), which is here rewritten:

$$\varepsilon_{11} = u_{1,1} + \frac{1}{2}\left(u_{1,1}^2 + u_{2,1}^2\right) \equiv u' + \frac{1}{2}\left(u'^2 + v'^2\right). \tag{7.6}$$

This strain is exact, as the unit extension in Eq. 7.3, but, unlike that, it is polynomial, and therefore does not need series expansion. It should be noticed that the two definitions coincide in their linear parts, but they differ by $\frac{1}{2}u'^2$ in their quadratic parts. However, this term is usually retained small, when compared to $\frac{1}{2}v'^2$, and generally neglected, so that the two constraints are believed to be coincident, for practical purposes.

Constitutive Law

The (active) generalized stresses of the extensible beam are (a) the axial force $N(x)$ and (b) the bending moment $M(x)$ (while the shear force $T(x)$ is a reactive internal force). By assuming an uncoupled linear elastic law, it follows:

$$\begin{pmatrix} N \\ M \end{pmatrix} = \begin{pmatrix} N_0 \\ 0 \end{pmatrix} + \begin{pmatrix} EA & 0 \\ 0 & EI \end{pmatrix}\begin{pmatrix} \varepsilon \\ \kappa \end{pmatrix}, \tag{7.7}$$

where $N_0 = -P$ is the axial prestress force and EA and EI are the axial and bending stiffnesses of the beam, respectively (with E the Young modulus, A the cross-section area, I the cross-section inertia moment).

Total Potential Energy

Recalling Eq. 6.12, the total potential energy (TPE) of the prestressed body, is

$$\Pi = \int_0^\ell \left[\frac{1}{2}EA\,\varepsilon_1^2 + \frac{1}{2}EI\,\kappa_1^2 + N_0\,(\varepsilon_1 + \varepsilon_2)\right]dx + P\,u\,(\ell) - \int_0^\ell p\,v\,dx. \tag{7.8}$$

---

[3] This approximation is consistent with the second order truncation of the total potential energy, Eq. 7.8, as it has already been observed in the Sect. 6.6, with reference to the extensible inverted pendulum.

Since the prestress $N_0(x)$ and the external force P are equilibrated in the reference configuration, the Virtual Power Principle states that:

$$\int_0^\ell N_0 \, \varepsilon_1 \mathrm{d}x = -P \, u \, (\ell), \tag{7.9}$$

so that the TPE in Eq. 7.8 becomes:

$$\Pi = \int_0^\ell \left(\frac{1}{2} EA \, \varepsilon_1^2 + \frac{1}{2} EI \, \kappa_1^2 + N_0 \, \varepsilon_2 \right) \mathrm{d}x - \int_0^\ell p \, v \, \mathrm{d}x, \tag{7.10}$$

which has the form of Eq. 6.13.

Equilibrium Equations and Boundary Conditions

By introducing the congruence Eqs. 7.5 into Eq. 7.10 and then imposing the stationary of the functional $\Pi [u, v]$ with respect to the independent variables, the following field equations and natural boundary conditions are derived:[4]

$$EA \, u'' = 0, \tag{7.11a}$$

$$\left[EA \, u' \, \delta u \right]_0^\ell = 0, \tag{7.11b}$$

and:

$$EI \, v'''' + P \, v'' = p, \tag{7.12a}$$

$$\left[(-EI \, v''' - P \, v') \, \delta v \right]_0^\ell = 0, \tag{7.12b}$$

$$\left[EI \, v'' \, \delta v' \right]_0^\ell = 0. \tag{7.12c}$$

Since Eqs. 7.11, in the u variable, are uncoupled from Eqs. 7.12, in the v variable, and, moreover, the first ones constitute a homogeneous and non-singular problem,

---

[4] Indeed, the first variation of the TPE in Eq. 7.10 is:

$$\delta\Pi = \int_0^\ell \left(EAu'\delta u' + EIv''\delta v'' + N_0 v'\delta v' - p\delta v \right) \mathrm{d}x$$

After integration by parts, and equating it to zero $\forall \, (\delta u, \delta v)$, one has:

$$\delta\Pi = \int_0^\ell \left[-EAu''\delta u + (EIv'''' - N_0 v'' - p) \, \delta v \right] \mathrm{d}x + \left[EAu'\delta u \right]_0^\ell + \left[EIv''\delta v' \right]_0^\ell$$

$$+ \left[(-EIv''' + N_0 v') \, \delta v \right]_0^\ell = 0, \quad \forall \, (\delta u, \delta v)$$

from which Eqs. 7.11 and 7.12 follow.

they only admit the trivial solution $u = 0$, $\forall x$, for which the buckling mode (at the first order) is inextensional. Hence, the critical buckling mode is *purely transversal*.[5]

The buckling problem of the extensible beam is therefore governed by Eqs. 7.12, in the v traverse displacement only. These equations coincide with the linear part of Eqs. 5.130–5.134, already obtained in the Chap. 5 for the inextensible beam. However, it is important noticing that the geometric stiffness contribution, either in the domain or at the boundary, now descends from the prestress energy, and not from the external force energy, as already mentioned in the Chap. 6.

Remark 7.2 The field Eq. 7.12a is susceptible to the following mechanical interpretation. In the adjacent configuration, close to the prestressed one, the external load p is balanced by the algebraic sum of two internal forces: (a) the *bending bearing capacity* $p_b := EIv''''$ and (b) the *funicular bearing capacity* $p_f := Pv''$. According to the latter, the bent beam supports an additional load, behaving like a compressed string.[6]

Remark 7.3 Equations 7.12 are of the type of Eqs. 6.16, already determined for a generic prestressed continuum. In particular, the geometric effect is present also at the boundary. Here $V(x) := -EIv''' - Pv'(x)$ has the meaning of *transverse component of the internal force* (which, at the boundary, expends virtual work in the displacement δv). This is different from shear force $T(x) = -EIv'''$, which is orthogonal to the bent centerline, being (Fig. 7.1 d): $V = T \cos \varphi + N_0 \sin \varphi \simeq T - Pv'$.

7.3   Critical Loads of Single-Span Beams

Single-span beams are considered, constrained at the ends $x = 0, \ell$ by clamps, hinges, sliders or rollers, as shown in Table 7.1. The critical load P_c is determined by solving the differential eigenvalue problem in Eq. 7.12. The general solution to Eq. 7.12a is (Eq. 5.144 should be remembered):

$$v = c_1 \cos(\beta x) + c_2 \sin(\beta x) + c_3 x + c_4, \tag{7.13}$$

where c_i $(i = 1, \ldots, 4)$ are arbitrary constants; moreover:

$$\beta := \sqrt{\frac{P_c}{EI}}, \tag{7.14}$$

is the *wave-number*, which is the eigenvalue of the boundary value problem.

---

[5] It is however possible to show that, at higher orders, neglected here, the beam also experiences longitudinal displacements.

[6] The equilibrium equation of the taut string is $N_0 v'' + p = 0$; for the compressed beam, it is $N_0 = -P$.

Table 7.1 Critical loads P_c of single-span beams and effective lengths ℓ_0

	Static scheme	Boundary conditions	P_c	ℓ_0
(a)		$v(0) = M(0) = 0$ $v(\ell) = M(\ell) = 0$	$\frac{\pi^2 EI}{\ell^2}$	ℓ
(b)		$v(0) = v'(0) = 0$ $M(\ell) = V(\ell) = 0$	$\frac{\pi^2 EI}{4\ell^2}$	2ℓ
(c)		$v(0) = v'(0) = 0$ $v(\ell) = v'(\ell) = 0$	$\frac{4\pi^2 EI}{\ell^2}$	$\ell/2$
(d)		$v(0) = v'(0) = 0$ $v(\ell) = M(\ell) = 0$	$2.05\,\frac{\pi^2 EI}{\ell^2}$	$0.7\,\ell$
(e)		$v'(0) = -T(0) = 0$ $v(\ell) = M(\ell) = 0$	$\frac{\pi^2 EI}{4\ell^2}$	2ℓ
(f)		$v'(0) = -T(0) = 0$ $v(\ell) = v'(\ell) = 0$	$\frac{\pi^2 EI}{\ell^2}$	ℓ

Boundary conditions are:

- Of *geometric* type, prescribing: (a) the displacement $v = 0$ at a hinge or roller and (b) the rotation $v' = 0$ at a clamp or slider both longitudinal (permitting the application of prestressing), or transverse.
- Of *mechanical* type, prescribing: (a) the bending moment $M(x) = EI\,v''(x) = 0$ at a hinge or roller, (b) the shear force $T(x) = -EI\,v'''(x) = 0$ at a transverse slider, and (c) the transverse component of the internal force, $V(x) = T(x) - P\,v'(x) = 0$, at a free end.

By enforcing four boundary conditions, a homogeneous system of algebraic equations in the unknown constants c_i is drawn. By vanishing the determinant of the coefficient matrix, a characteristic equation for the eigenvalue β follows, from which (infinite) solutions are derived, as many critical loads via Eq. 7.14. Among all these eigenvalues, the smallest one is of interest. Operating in this way, the critical loads P_c, shown in Table 7.1, are obtained.

The eigenvector admitted by the algebraic equations, collecting the constants c_i, $i = 1, \ldots, 4$, after arbitrary normalization and substitution into the general solution, Eq. 7.13, finally gives the associated critical mode, plotted in Fig. 7.2.

Effective Length

From the results displayed in Table 7.1, it is observed that the critical loads of single-span beams can always be written in the form:

$$P_c = \frac{\pi^2 EI}{\ell_0^2}, \tag{7.15}$$

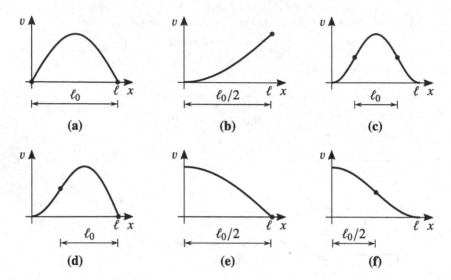

Fig. 7.2 Critical modes of the single-span beams in Table 7.1; inflection points and effective length ℓ_0

in which a suitable length ℓ_0 has been introduced, named *effective length*. Its geometric meaning emerges from the pattern of the modes plotted in Fig. 7.2, namely, ℓ_0, *is the distance between two inflection points of the critical deformation*. This result is justified by the fact that a single-span beam of length ℓ, which buckles with inflection points at distance ℓ_0, possesses the same critical load than the simply supported beam of length ℓ_0 (whose critical mode has, indeed, inflection points at the ends).

Critical Stress

It is useful to evaluate the *critical normal stress* $\sigma_c := \frac{P_c}{A}$, which occurs in the beam at the incipient bifurcation (more precisely, immediately before).[7] Taking into account Eq. 7.15, one has:

$$\sigma_c = \frac{\pi^2 E \rho^2}{\ell_0^2}, \tag{7.16}$$

where $\rho := \sqrt{\frac{I}{A}}$ is the inertia radius of the cross-section. The previous equation is also written as:

$$\sigma_c = \frac{\pi^2 E}{\lambda^2}, \tag{7.17}$$

---

[7] Immediately after buckling, the beam is bent, in addition to compressed.

Fig. 7.3 Eulerian critical
stress *vs* slenderness;
proportionality limit stress σ_p

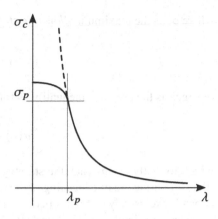

where a nondimensional quantity has been introduced:

$$\lambda := \frac{\ell_0}{\rho}, \tag{7.18}$$

called the *slenderness* of the beam. This latter summarizes the geometric characteristics of the structure, i.e., the cross-section properties A, I, and the beam length ℓ,
as well the constraint conditions.

In Fig. 7.3 the critical stress σ_c is represented *vs* the slenderness λ; this curve is
called the *Euler curve*. The value $\lambda = \lambda_p$, at which σ_c equates the *proportionality
limit stress* of the material, σ_p, divides the λ axis into two domains: (a) that of
squat beams, for which $\lambda < \lambda_p$, and (b) that of *slender beams*, for which $\lambda > \lambda_p$.
The Euler curve loses its validity in the domain of squat beams (and therefore it is
indicated by a dashed line), because the hypotheses of linear elasticity, on which the
model has been formulated, are violated. In this domain, an elasto-plastic analysis
must be carried out, which leads to determine the continuous curve, qualitatively
represented in the figure. The problem will be addressed in the next Chap. 8.

"Omega Method"

The compressed beam is in stable equilibrium until the load P is smaller than the
critical load P_c. The stability condition, $P < P_c$, can also be expressed in terms of
stress, as $\sigma < \sigma_c$. In the elastic field (i.e., when the beam is slender), this condition
reads:

$$\frac{P}{A} \leq \frac{\pi^2 E}{\lambda^2}, \tag{7.19}$$

or, equivalently:

$$\frac{P}{A} \leq \frac{\pi^2 E}{\lambda^2 \sigma_a} \sigma_a, \tag{7.20}$$

where σ_a is the maximum *allowable stress*, depending on material.[8] By defining:

$$\omega\,(\lambda) := \frac{\sigma_a\,\lambda^2}{\pi^2 E}, \qquad (7.21)$$

one recasts the previous inequality as follows:

$$\omega\,(\lambda)\,\frac{P}{A} \le \sigma_a, \qquad (7.22)$$

which formally brings back the stability verification to that of strength verification (thus obscuring the true meaning of the operation). The coefficient $\omega\,(\lambda) > 1$, which depends exclusively on the properties of the material and on the slenderness, can be regarded as a *load amplification factor* or as an *allowable stress reduction*. In the technical literature, this verification procedure is known as the *Omega method*.

7.4 Beams Transversely Loaded: Second Order Effects

Beams compressed by a sub-critical load $P < P_c$, subject to transverse loads (hence, also said *beam columns*), are considered. The response is evaluated taking into account the geometric effect of prestress. The results already obtained for discrete systems (Sect. 6.5) are here proved to hold also for a continuous system [2, 7, 11, 13–16, 25].

7.4.1 Simply Supported Beam Under Sinusoidal Transverse Load

The equilibrium of a beam, compressed by a load $P < P_c$, under sinusoidal transverse loads $p\,(x) = \hat{p}\,\sin\left(\frac{\pi x}{\ell}\right)$ (Fig. 7.4a), is governed by:

$$EI\,v'''' + P\,v'' = \hat{p}\,\sin\left(\frac{\pi x}{\ell}\right), \qquad x \in (0, \ell). \qquad (7.23)$$

If the beam is simply supported at the end, the boundary conditions are:

$$v\,(0) = 0, \qquad\qquad EIv''\,(0) = 0, \qquad (7.24a)$$

---

[8] Verification by allowable stresses, as well-known, is no longer accepted by many worldwide codes. However, it retains a profound mechanical meaning, which is the only one worth discussing here.

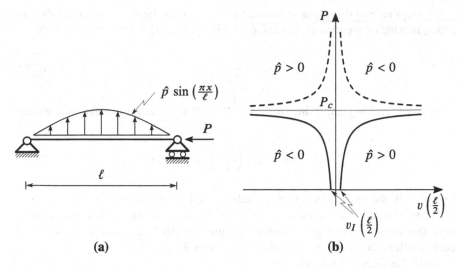

Fig. 7.4 Simply supported prestressed beam, under transverse sinusoidal load: (a) static scheme, (b) bifurcation diagram

$$v(\ell) = 0, \qquad\qquad EIv''(\ell) = 0. \qquad\qquad (7.24b)$$

In this simple case, the solution can be found "by inspection," as:

$$v = \hat{v} \sin\left(\frac{\pi x}{\ell}\right). \qquad\qquad (7.25)$$

Indeed, the guess solution satisfies the boundary conditions and reduces the differential equation to an algebraic equation in the unknown amplitude $\hat{v}$, i.e.:

$$\left(\frac{EI\,\pi^4}{\ell^4} - \frac{P\,\pi^2}{\ell^2}\right)\hat{v} = \hat{p}, \qquad\qquad (7.26)$$

whose solution is:

$$\hat{v} = \frac{1}{1 - \frac{P}{P_c}}\frac{\hat{p}}{\frac{\pi^4 EI}{\ell^4}} =: \alpha\,\hat{v}_I. \qquad\qquad (7.27)$$

Here, $P_c = \frac{\pi^2 EI}{\ell^2}$ is the critical load,

$$\hat{v}_I := \frac{\hat{p}}{\frac{\pi^4 EI}{\ell^4}} \qquad\qquad (7.28)$$

is the amplitude of the response according to the linear theory (also said of the first order, in which the geometric effects are neglected); moreover:

$$\alpha := \frac{1}{1 - \frac{P}{P_c}} \tag{7.29}$$

is an *amplification factor* (since $\alpha > 1$ if $P \le P_c$). Overall, the solution is written as:

$$v(x) = \alpha \, \hat{v}_I \sin\left(\frac{\pi x}{\ell}\right) =: \alpha \, v_I(x). \tag{7.30}$$

Figure 7.4b shows the deflection at midspan, $v\left(\frac{\ell}{2}\right)$, vs the load P. The diagram is extended to the super-critical values $P > P_c$, where unstable non-natural paths exist (the comments to Fig. 6.5 should be remembered). The role of the non-natural curves will be discussed in the next chapter (Sect. 8.3).

From Eq. (7.30), it follows that:

$$M(x) = EIv''(x) = \alpha \, M_I(x), \tag{7.31a}$$

$$T(x) = -EIv'''(x) = \alpha \, T_I(x), \tag{7.31b}$$

i.e., *all the first order internal forces are amplified by* α. The geometric effect, in the context of linearized theory, is sometimes referred to as the *second order* effect.

Remark 7.4 The constraint reactions, for equilibrium reasons and as the structure is isostatic, are *not* amplified by the geometric effect. For example, the reaction at the right support B is equal to the vertical component of the internal force:

$$V_B = T_B - P v'_B = -\alpha \hat{v}_I \frac{\pi}{\ell} \left[\frac{\pi^2 EI}{\ell^2} - P\right] = -\frac{\pi^3}{\ell^3} EI \hat{v}_I, \tag{7.32}$$

which coincides with the shear force of the linear theory. In other words, the shear force T_B is amplified, but the transverse component V_B of the internal force is not amplified.

7.4.2 Simply Supported Beam Under Generic Transverse Load

A simply supported beam is still considered but subject to a *generic* transverse load $p(x)$. The equilibrium equation is:

$$EI\,v'''' + P\,v'' = p(x), \qquad x \in (0, \ell), \tag{7.33}$$

which must be integrated with the boundary conditions in Eq. 7.24.

The load is expanded in Fourier series, as:

$$p(x) = \sum_{k=1}^{\infty} \hat{p}_k \sin\left(\frac{k\pi x}{\ell}\right),$$ (7.34)

where $\hat{p}_k$ are the coefficients of the series, expressed by:

$$\hat{p}_k = \frac{2}{\ell} \int_0^\ell p(x) \sin\left(\frac{k\pi x}{\ell}\right) dx.$$ (7.35)

In analogy to what was done in the case of a single harmonic, a solution is sought in the form:

$$v = \sum_{k=1}^{\infty} \hat{v}_k \sin\left(\frac{k\pi x}{\ell}\right).$$ (7.36)

By substituting it in the equilibrium equation, equating to zero the coefficients of the same harmonics, and solving for the unknowns $\hat{v}_k$, the following solution is found:

$$v(x) = \sum_{k=1}^{\infty} \alpha_k \hat{v}_{Ik} \sin\left(\frac{k\pi x}{\ell}\right),$$ (7.37)

where:

$$\alpha_k := \frac{1}{1 - \frac{P}{P_k}},$$ (7.38a)

$$\hat{v}_{Ik} := \frac{\hat{p}_k}{\frac{k^4 \pi^4 EI}{\ell^4}},$$ (7.38b)

and, moreover, $P_k = \frac{k^2 \pi^2 EI}{\ell^2}$ is the kth critical load. It is important noticing that, since $P_{k+1} > P_k$, $k = 1, \ldots$, *the amplification factors* α_k *decrease with* k. From this observation, it follows that:

$$v(x) < \alpha_1 \sum_{k=1}^{\infty} \hat{v}_{Ik} \sin\left(\frac{k\pi x}{\ell}\right) = \alpha_1 v_I(x);$$ (7.39)

therefore:

$$v^+(x) := \alpha_1 v_I(x)$$ (7.40)

is an *upper bound* for the response, that, from a precautionary point of view, can be adopted as *the* response $v(x) \simeq v^+(x)$. More generally:

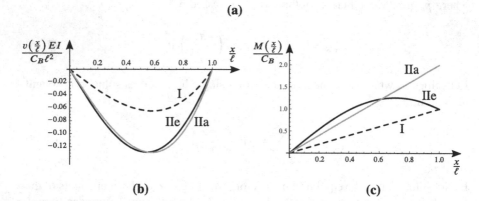

(a)

(b)   **(c)**

Fig. 7.5 Simply supported beam, under a compression force and a couple at one end: (**a**) static scheme, (**b**) displacements, (**c**) bending moments; first order (I), second order solutions, exact (IIe) and approximate (IIa) by the amplification rule; $P = \frac{1}{2} P_c$

$$S(x) \simeq \alpha_1 S_I(x), \tag{7.41}$$

where S_I represents a displacement or stress evaluated by the linear theory (e.g., a bending moment or a shear force). In this book, Eq. 7.41 will be referred to as the *amplification rule*.

Remark 7.5 The procedure followed here for a simply supported beam can be extended to beams with different constrains. Indeed, by expanding the unknown solution in the basis of the critical modes (no more sinusoidal) of the actual beam and then evaluating the upper bound, Eq 7.41 is found to still hold. Therefore, the amplification rule can be applied to any constraint conditions.

Example 7.1 (Simply Supported Beam Loaded by a Couple at One End) To test the accuracy of the amplification rule, the exact second order response is calculated for a simply supported beam, compressed by a force $P < P_c$, and subject to a counterclockwise couple C_B applied at the extreme B (Fig. 7.5a).

The exact second order problem is governed by the following equations:

$$EI\,v'''' + P\,v'' = 0, \qquad\qquad x \in (0, \ell), \tag{7.42a}$$

$$v(0) = 0, \tag{7.42b}$$

$$EI\,v''(0) = 0, \tag{7.42c}$$

$$v(\ell) = 0, \tag{7.42d}$$

$$EIv''(\ell) = C_B, \tag{7.42e}$$

admitting the solution:

$$v(x) = \frac{C_B}{EI\beta^2}\left(\frac{x}{\ell} - \frac{\sin(\beta x)}{\sin(\beta \ell)}\right), \tag{7.43a}$$

$$M(x) = EIv''(x) = C_B\frac{\sin(\beta x)}{\sin(\beta \ell)}. \tag{7.43b}$$

Displacements and bending moments are compared in Fig. 7.5b,c with the approximate solution coming from the amplification rule. It is seen that, while the exact displacement is well described by the rule, the exact bending moment is not well approximated near the point of application of the couple, where the solicitation, being prescribed, cannot be amplified. However the approximation is acceptable far from the couple.   □

7.5 Stepped Beams

Very often, the stiffness of a beam is not constant over the length, but it is stepwise variable, that is, $EI(x) = EI_i = \text{cost}$, $(i = 1, 2, \ldots, n)$, in sub-intervals $\mathcal{I}_i$ of the domain $(0, \ell)$. In this case, it is necessary to integrate the field Eq. 7.12a in each sub-interval $\mathcal{I}_i$ and to impose appropriate boundary conditions at the ends of any intervals [23].

As an alternative approach, the (variational) Ritz method can be applied, which supplies an approximate solution. This method is especially useful when the stiffness of the beam is variable with a generic law $EI(x)$, continuous or discontinuous.

7.5.1 Exact Analysis

As an example, the case of a compressed cantilever is developed here, having different stiffnesses EI_i, $i = 1, 2$, in two sub-intervals of length ℓ_i (Fig. 7.6a). Since the jump of stiffness constitutes a point of singularity, integration must be carried out into two sub-intervals. The relevant equilibrium equations are written as:

$$EI_1 v_1'''' + P v_1'' = 0, \qquad x_1 \in (0, \ell_1), \tag{7.44a}$$

$$EI_2 v_2'''' + P v_2'' = 0, \qquad x_2 \in (0, \ell_2). \tag{7.44b}$$

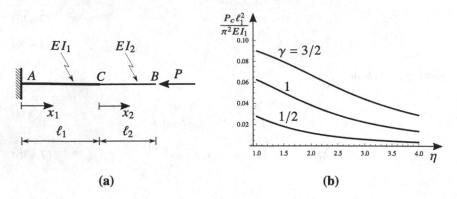

(a)   **(b)**

Fig. 7.6 Stepped cantilever: (a) static scheme, (b) critical load *vs* the stiffness ratio η, for different length ratios γ

The boundary conditions at the ends A and B are the usual ones for clamped and free end, respectively; the conditions at the singular point C, instead, must express (a) the continuity of translation and rotation and (b) the equilibrium of the infinitesimal element of beam containing the singular point. They are written as:

$$v_1(0) = 0, \qquad\qquad\qquad v_1'(0) = 0, \qquad\qquad (7.45a)$$

$$M_2(\ell_2) = 0, \qquad\qquad\qquad V_2(\ell_2) = 0, \qquad\qquad (7.45b)$$

$$v_1(\ell_1) = v_2(0), \qquad\qquad\qquad v_1'(\ell_1) = v_2'(0), \qquad\qquad (7.45c)$$

$$M_1(\ell_1) = M_2(0), \qquad\qquad\qquad V_1(\ell_1) = V_2(0), \qquad\qquad (7.45d)$$

where:

$$M_i(x_i) = E I_i\, v_i''(x_i), \qquad\qquad i = 1, 2, \qquad\qquad (7.46a)$$

$$T_i(x_i) = -E I_i\, v_i'''(x_i), \qquad\qquad (7.46b)$$

$$V_i(x_i) = T_i(x_i) - P\, v_i'(x_i). \qquad\qquad (7.46c)$$

The general solution to Eqs. 7.44 is:

$$v_1(x_1) = c_1 \cos(\beta_1 x_1) + c_2 \sin(\beta_1 x_1) + c_3 x_1 + c_4, \qquad\qquad (7.47a)$$

$$v_2(x_2) = c_5 \cos(\beta_2 x_2) + c_6 \sin(\beta_2 x_2) + c_7 x_2 + c_8, \qquad\qquad (7.47b)$$

where c_j, $j = 1, \ldots, 8$, are arbitrary constants and

$$\beta_i^2 = \frac{P}{E I_i}, \qquad i = 1, 2 \qquad\qquad (7.48)$$

are wave-numbers in each sub-interval, also piecewise variable, because of the different stiffnesses. By imposing the boundary conditions, one gets:

$$
\begin{pmatrix}
1 & 0 & 0 & 1 & 0 & 0 & 0 & 0 \\
0 & \beta_1 & 1 & 0 & 0 & 0 & 0 & 0 \\
0 & 0 & 0 & 0 & -\beta_2^2 C_2 & -\beta_2^2 S_2 & 0 & 0 \\
0 & 0 & 0 & 0 & 0 & 0 & -\beta_2^2 & 0 \\
C_1 & S_1 & \ell_1 & 1 & -1 & 0 & 0 & -1 \\
-\beta_1 S_1 & \beta_1 C_1 & 1 & 0 & 0 & -\beta_2 & -1 & 0 \\
-\beta_2^2 C_1 & -\beta_2^2 S_1 & 0 & 0 & \beta_2^2 & 0 & 0 & 0 \\
0 & 0 & -\beta_2^2 & 0 & 0 & 0 & \beta_2^2 & 0
\end{pmatrix}
\begin{pmatrix} c_1 \\ c_2 \\ c_3 \\ c_4 \\ c_5 \\ c_6 \\ c_7 \\ c_8 \end{pmatrix}
=
\begin{pmatrix} 0 \\ 0 \\ 0 \\ 0 \\ 0 \\ 0 \\ 0 \\ 0 \end{pmatrix},
$$

$$(7.49)$$

where $C_i := \cos(\beta_i \ell_i)$, $S_i := \sin(\beta_i \ell_i)$, $i = 1.2$. The relevant characteristic equation, which follows the vanishing of the determinant of the coefficient matrix, is found to be:

$$
\tan(\beta_1 \ell_1) \tan(\beta_2 \ell_2) = \frac{\beta_2}{\beta_1}. \tag{7.50}
$$

This can also be written as:

$$
\tan(\beta_1 \ell_1) \tan\left(\frac{\eta}{\gamma} \beta_1 \ell_1\right) = \eta, \tag{7.51}
$$

where the following nondimensional magnitudes have been introduced:

$$
\eta := \sqrt{\frac{E I_1}{E I_2}}, \quad \gamma := \frac{\ell_1}{\ell_2}. \tag{7.52}
$$

The parameters define the stiffness ratio η and the length ratio γ of the two beam segments. For a fixed pair of these parameters, the smallest root β_{1c} is computed, from which the critical load is evaluated as $P_c = \beta_{1c}^2 E I_1$. Figure 7.6b shows the nondimensional critical load $P_c \ell_1^2 / (\pi^2 E I_1)$, as a function of the stiffness ratio $\eta > 1$, for different values of the geometric ratio $\frac{1}{2} \le \gamma \le \frac{3}{2}$. It is seen that, keeping $E I_1$ fixed, the critical load reduces when $E I_2$ decreases and that, with the same stiffness and length ℓ_1, the critical load decreases as ℓ_2 increases, both as a consequence of the increased flexibility of the system.

7.5.2 Ritz Analysis

The previous example, due to its simplicity, was solved exactly. However, there are problems where the exact solution is very laborious (e.g., if the stiffness is variable in many intervals) or even impossible (if, e.g., the stiffness continuously varies). In these cases, it is appropriate to solve the problem in an approximate form, using the (variational) Ritz method (some short reminder of which are given in the Appendix B) [1, 2, 6, 21–23, 25]. Here the procedure is illustrated with reference to the same example of the Sect. 7.5.1, already solved exactly, in order to test the accuracy of the method.

Ritz Method
The TPE of the beam in Fig. 7.6a is written as:[9]

$$\Pi = \int_0^{\ell_1+\ell_2} \left(\frac{1}{2} EI(x)\, v''^2 - \frac{1}{2} P\, v'^2\right) dx, \quad x \in (0, \ell_1 + \ell_2), \tag{7.53}$$

where $I(x) = I_1$ in $(0, \ell_1)$ and $I(x) = I_2$ in $(\ell_1, \ell_1 + \ell_2)$. By letting:

$$v(x) = \sum_{i=1}^{n} \phi_i(x)\, q_i = \mathbf{\Phi q}, \tag{7.54}$$

where $\phi_i(x)$ are known trial functions and q_i are unknown amplitudes, and substituting in the TPE,

$$\Pi = \frac{1}{2} \mathbf{q}^T \left(\mathbf{K}_e + \mathbf{K}_g\right) \mathbf{q}, \tag{7.55}$$

is found, where:

$$\mathbf{K}_e := \int_0^{\ell_1+\ell_2} EI(x)\, \mathbf{\Phi}''^T \mathbf{\Phi}''\, dx, \tag{7.56a}$$

$$\mathbf{K}_g := -P \int_0^{\ell_1+\ell_2} \mathbf{\Phi}'^T \mathbf{\Phi}'\, dx. \tag{7.56b}$$

By imposing the stationary of the TPE, $\frac{\partial \Pi}{\partial \mathbf{q}} = \mathbf{0}$, algebraic equilibrium equations are derived:

$$\left(\mathbf{K}_e + \mathbf{K}_g\right) \mathbf{q} = \mathbf{0}. \tag{7.57}$$

An Example of Trial Functions
As an example, polynomial trial functions are chosen:

$$\phi_i(x) = x^{i+1}, \quad i = 1, 2, \ldots, n; \tag{7.58}$$

---

[9] Here, differently from what done before, a single abscissa x is defined.

they satisfy the geometric conditions at the clamp and the geometric continuity at the singular point (but do not satisfy any mechanical prescriptions). By performing integration, the following matrices are found:

$$
\mathbf{K}_e = E I_1 \ell_1 \begin{bmatrix} 4 & 6\ell_1 & \cdots \\ 6\ell_1 & 12\ell_1^2 & \cdots \\ \cdots & \cdots & \ddots \end{bmatrix} \tag{7.59a}
$$

$$
+ E I_2 \ell_2 \begin{bmatrix} 4 & 6(2\ell_1 + \ell_2) & \cdots \\ 6(2\ell_1 + \ell_2) & 12\left(3\ell_1^2 + 3\ell_2\ell_1 + \ell_2^2\right) & \cdots \\ \cdots & \cdots & \ddots \end{bmatrix},
$$

$$
\mathbf{K}_g = -P(\ell_1 + \ell_2)^3 \begin{bmatrix} \frac{4}{3} & \frac{3}{2}(\ell_1 + \ell_2) & \cdots \\ \frac{3}{2}(\ell_1 + \ell_2) & \frac{9}{5}(\ell_1 + \ell_2)^2 & \cdots \\ \cdots & \cdots & \ddots \end{bmatrix}. \tag{7.59b}
$$

If a single trial function is considered ($n = 1$) (i.e., a parabolic approximation is taken for the buckling mode), the following critical load is immediately obtained:

$$
P_c = \frac{3(E I_1 \ell_1 + E I_2 \ell_2)}{(\ell_1 + \ell_2)^3}. \tag{7.60}
$$

In terms of the nondimensional parameters defined in Eqs. 7.52, the previous result reads:

$$
\frac{P_c \ell_1^2}{\pi^2 E I_1} = \frac{3\gamma^2 \left(1 + \gamma\, \eta^2\right)}{\pi^2 \eta^2 (1 + \gamma)^3}. \tag{7.61}
$$

This is compared in Fig. 7.7a with the exact solution of Fig. 7.6b. It appears that, while the approximate solution describes quite well the exact solution in the interval $\eta \in (1, 2)$, the approximation becomes very rough for larger values of η. This is due to the fact that the larger the stiffness jump, the less accurate the approximation of parabolic deformation, which implies constant curvature of the beam.

If the analysis is repeated with $n = 2$, an eigenvalue algebraic problem of dimension 2 is drawn. The relative numerical solution is plotted in Fig. 7.7b. A slight improvement of the approximation is observed for small values of η. The accuracy of the solution can be further improved by using a larger number n of trial functions, as shown in the Problem 14.2.1. However, in the same Problem 14.2.1, it is found that, by taking different trial functions in each sub-interval, respectful of the continuity at the singular point, excellent results can be achieved with very few functions.

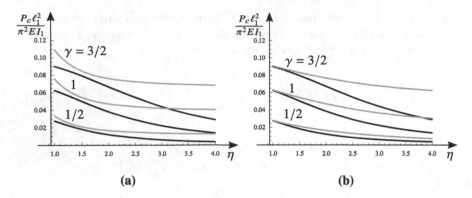

Fig. 7.7 Critical load of the stepped cantilever; comparison between the Ritz (in gray) and the exact solution (in black): (**a**) $n = 1$, (**b**) $n = 2$

Remark 7.6 The Ritz method, when used with a single trial function $\phi(x)$, provides a critical load appearing in a form called the *Rayleigh ratio*. Referring to a generic variation of the inertia moment $I(x)$, continuous or discontinuous, it is written as:

$$P_c = \frac{\int_0^\ell EI(x)\,\phi''^2(x)\,dx}{\int_0^\ell \phi'^2(x)\,dx}. \tag{7.62}$$

This is the ratio between the elastic energy and the prestress energy (due to a unit compression), possessed by the beam when subject to the $\phi(x)$ displacement field.

7.6   Beams Under Piecewise Variable Compression

So far, beams uniformly compressed along their entire length, for which $N_0(x) = -P = \text{cost}$ in the $(0, \ell)$ interval, have been considered. However, in applications, it is likely to encounter beams whose compression is piecewise constant, or even continuously variable. In this section, the first case is considered, postponing the second to the next section.

When the compression is piecewise constant, that is, $N_0(x) = N_{0i}$ $(i = 1, 2, \ldots, n)$ in sub-intervals I_i of the domain $(0, \ell)$, similarly to the case of piecewise variable stiffness, it needs to integrate the field equation in each I_i and to impose the boundary conditions at the relevant end points. Two examples will clarify the procedure.

7.6.1 Partially Compressed Beam

A simply supported beam is considered, of ends A and B, subject to a compressive load P applied at a generic interior point C, which divides the beam into two segments of length ℓ_1 and ℓ_2, and directed toward the hinged support A (Fig. 7.8a). The axial force law, in the prestressed configuration, is:

$$N_0(x) = \begin{cases} -P, & \text{in } x_1 \in (0, \ell_1), \\ 0, & \text{in } x_2 \in (0, \ell_2), \end{cases} \tag{7.63}$$

where x_1 and x_2 are abscissas of origin A and C, respectively. The equilibrium equation splits as follows:

$$EI\, v_1'''' + P\, v_1'' = 0, \qquad\qquad x_1 \in (0, \ell_1), \tag{7.64a}$$

$$EI\, v_2'''' = 0, \qquad\qquad x_2 \in (0, \ell_2), \tag{7.64b}$$

where $v_i(x_i)$ is the displacement field in the ith sub-interval. The boundary conditions at points $A, C,$ and B, in the order, read:

$$v_1(0) = 0, \qquad\qquad M_1(0) = 0, \tag{7.65a}$$

$$v_1(\ell_1) = v_2(0), \qquad\qquad v_1'(\ell_1) = v_2'(0), \tag{7.65b}$$

$$M_1(\ell_1) = M_2(0), \qquad\qquad V_1(\ell_1) = V_2(0), \tag{7.65c}$$

$$v_2(\ell_2) = 0, \qquad\qquad M_2(\ell_2) = 0, \tag{7.65d}$$

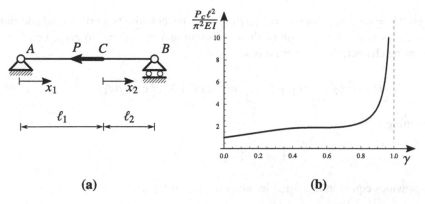

(a) **(b)**

Fig. 7.8 Simply supported beam partially compressed: (a) static scheme, (b) critical load *vs* the position γ of the force application point

in which the internal forces are expressed in terms of displacement through the following relations:[10]

$$M_i(x_i) = EI\, v_i''(x_i), \qquad\qquad\qquad i = 1, 2, \qquad\qquad (7.66a)$$

$$T_i(x_i) = -EI\, v_i'''(x_i), \qquad\qquad\qquad\qquad\qquad (7.66b)$$

$$V_i(x_i) = T_i(x_i) + N_{0i}\, v_i'(x_i). \qquad\qquad\qquad\qquad (7.66c)$$

The general solution of the homogeneous problem, by accounting for $\beta^2 := \frac{P}{EI}$, reads:

$$v_1(x_1) = c_1 \cos(\beta x_1) + c_2 \sin(\beta x_1) + c_3\, x_1 + c_4, \qquad (7.67a)$$

$$v_2(x_2) = c_5 + c_6\, x_2 + c_7\, x_2^2 + c_8\, x_2^3, \qquad\qquad\qquad (7.67b)$$

i.e., it is harmonic-polynomial in the compressed segment and simply polynomial in the uncompressed one. By imposing the boundary conditions, the following algebraic system in the eight unknown constants c_i is drawn:

$$\begin{pmatrix} 1 & 0 & 0 & 1 & 0 & 0 & 0 & 0 \\ -\beta^2 & 0 & 0 & 0 & 0 & 0 & 0 & 0 \\ C & S & \ell_1 & 1 & -1 & 0 & 0 & 0 \\ -\beta S & \beta C & 1 & 0 & 0 & -1 & 0 & 0 \\ -\beta^2 C & -\beta^2 S & 0 & 0 & 0 & 0 & -2 & 0 \\ 0 & 0 & -\beta^2 & 0 & 0 & 0 & 0 & 6 \\ 0 & 0 & 0 & 0 & 1 & \ell_2 & \ell_2^2 & \ell_2^3 \\ 0 & 0 & 0 & 0 & 0 & 0 & 2 & 6\ell_2 \end{pmatrix} \begin{pmatrix} c_1 \\ c_2 \\ c_3 \\ c_4 \\ c_5 \\ c_6 \\ c_7 \\ c_8 \end{pmatrix} = \begin{pmatrix} 0 \\ 0 \\ 0 \\ 0 \\ 0 \\ 0 \\ 0 \\ 0 \end{pmatrix}, \qquad (7.68)$$

where $C := \cos(\beta \ell_1)$ and $S := \sin(\beta \ell_1)$. The system admits a non-trivial solution if and only if the determinant of the coefficient matrix is equal to zero, i.e., if the following characteristic equation is satisfied:

$$\left[(\beta \ell_2)^2\, \ell_2 - 3\, (\ell_1 + 2\ell_2) \right] \sin(\beta \ell_1) = 3\beta \ell_2^2 \cos(\beta \ell_1). \qquad (7.69)$$

By letting:

$$\ell = \ell_1 + \ell_2, \quad \gamma = \ell_2/\ell, \qquad\qquad\qquad\qquad (7.70)$$

the previous equation is written in nondimensional form:

$$\left[(\beta \ell)^2\, \gamma^3 - 3\,(1 + \gamma) \right] \sin((1 - \gamma)\,\beta \ell) = 3\,(\beta \ell)\,\gamma^2 \cos((1 - \gamma)\,\beta \ell). \qquad (7.71)$$

---

[10] Since the beam is not prestressed in the interval $\mathcal{I}_2$, it is $V_2 \equiv T_2$.

For any given γ, the smallest root $\beta_c \ell$ is found numerically; for example, for $\gamma = \frac{1}{2}, \frac{1}{3}$, it is found $\beta_c \ell = 1.38\,\pi,\ 1.31\,\pi$, from which $P_c = 1.90\pi^2 \frac{EI}{\ell^2},\ 1.72\pi^2 \frac{EI}{\ell^2}$. By repeating the analysis for different values of γ, the (nondimensional) critical load $\frac{P_c \ell^2}{\pi^2 EI}$, represented in Fig. 7.8b, is obtained. It is observed that, when the point of application of the force C tends to the roller at B, it is $\gamma \to 0$, and consequently the critical load tends to that of the uniformly compressed beam, i.e., $\frac{P_c \ell^2}{\pi^2 EI} = 1$. When, instead, the point C tends to the fixed support, it is $\gamma \to 1$, so that the critical load rapidly tends to infinity, because the compressed segment of beam becomes evanescent.[11]

7.6.2 Beam Under Independent Compressive Forces: The Domain of Interaction

A simply supported AB beam is now considered, subject to a force P_1 applied at an interior point C, and a force P_2 applied at the rolled end B, both of compression (Fig. 7.9a).

Since the axial force, in the prestressed configuration, is:

$$N_0(x) = \begin{cases} -(P_1 + P_2), & \text{in } x_1 \in (0, \ell_1), \\ -P_2, & \text{in } x_2 \in (0, \ell_2), \end{cases} \tag{7.72}$$

the equilibrium equation, for each of the two segments, reads:

$$EI\, v_1'''' + (P_1 + P_2)\, v_1'' = 0, \qquad x_1 \in (0, \ell_1), \tag{7.73a}$$

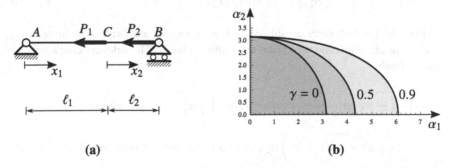

(a)   **(b)**

Fig. 7.9 Simply supported beam compressed by two independent forces: (a) static scheme, (b) stability domains for different length ratios γ

---

[11] The curve must, of course, be interrupted as soon as the critical stress reaches its proportionality limit.

$$EI\, v_2'''' + P_2\, v_2'' = 0, \qquad\qquad x_2 \in (0, \ell_2). \qquad (7.73b)$$

The boundary conditions are:

$$v_1(0) = 0, \qquad\qquad\qquad M_1(0) = 0, \qquad\qquad (7.74a)$$

$$v_1(\ell_1) = v_2(0), \qquad\qquad v_1'(\ell_1) = v_2'(0), \qquad\qquad (7.74b)$$

$$M_1(\ell_1) = M_2(0), \qquad\qquad V_1(\ell_1) = V_2(0), \qquad\qquad (7.74c)$$

$$v_2(\ell_2) = 0, \qquad\qquad\qquad M_2(\ell_2) = 0, \qquad\qquad (7.74d)$$

where:

$$M_i(x_i) = EI\, v_i''(x_i), \qquad\qquad i = 1, 2. \qquad (7.75a)$$

$$T_i(x_i) = -EI\, v_i'''(x_i), \qquad\qquad (7.75b)$$

$$V_i(x_i) = T_i(x_i) + N_{0i}\, v_i'(x_i), \qquad\qquad (7.75c)$$

By letting:

$$\beta_1^2 := \frac{P_1}{EI}, \qquad\qquad \beta_2^2 := \frac{P_2}{EI}, \qquad\qquad (7.76)$$

the general solution of the homogeneous problem is written as:

$$v_1(x_1) = c_1 \cos\left(\sqrt{\beta_1^2 + \beta_2^2}\, x_1\right) + c_2 \sin\left(\sqrt{\beta_1^2 + \beta_2^2}\, x_1\right) \qquad (7.77a)$$

$$+ c_3\, x_1 + c_4,$$

$$v_2(x_2) = c_5 \cos(\beta_2 x_2) + c_6 \sin(\beta_2 x_2) + c_7\, x_2 + c_8. \qquad (7.77b)$$

By imposing the boundary conditions, a homogeneous algebraic system of eight equations in eight unknowns is obtained (omitted here). The relevant characteristic equation reads:

$$\alpha_2^2 \sqrt{\alpha_1^2 + \alpha_2^2}\left(\alpha_1^2 \gamma + \alpha_2^2\right) \sin(\alpha_2\gamma) \cos\left(\sqrt{\alpha_1^2 + \alpha_2^2}(1 - \gamma)\right)$$

$$+ \sin\left(\sqrt{\alpha_1^2 + \alpha_2^2}(1 - \gamma)\right) \alpha_2 \left(\alpha_1^2 + \alpha_2^2\right)\left(\alpha_1^2 \gamma + \alpha_2^2\right) \cos(\alpha_2\gamma) \qquad (7.78)$$

$$- \sin\left(\sqrt{\alpha_1^2 + \alpha_2^2}(1 - \gamma)\right) \alpha_1^4 \sin(\alpha_2\gamma) = 0,$$

where:

$$\ell := \ell_1 + \ell_2, \qquad\qquad i = 1, 2. \qquad\qquad (7.79a)$$

$$\alpha_i := \beta_i \, \ell, \qquad\qquad\qquad (7.79b)$$

$$\gamma := \frac{\ell_2}{\ell}, \qquad\qquad\qquad (7.79c)$$

Equation 7.78 defines infinite combinations of forces P_1 and P_2 (i.e., of the nondimensional load parameters α_1 and α_2), which are critical for the beam (i.e., of incipient static bifurcation). For a fixed geometric parameter γ, the numerical solution of Eq. 7.78 supplies a value of α_1 for any chosen α_2. It is therefore possible to build up, point by point, a curve $\alpha_1 = \alpha_1 (\alpha_2; \gamma)$. For each γ, an *interaction domain* between the two forces is thus obtained, as shown in Fig. 7.9b. Points internal to the domain represent load conditions for which the beam is stable; points on the boundary represent load conditions of incipient instability. All the boundary curves, for each γ, emanate from $\alpha_1 = 0$, $\alpha_2 = \pi$, in which the beam is uniformly compressed by the force P_2, alone, applied at the end of the beam. In contrast, the curves intersect the α_1 axis in points depending on γ: these points represent states in which the beam is partially compressed by the force P_1, alone, applied inside the span (i.e., the case already studied in the Sect. 7.6.1), in which the critical load is the greater the closer to the fixed support is the point of application of the force (i.e., the greater is γ).

7.7 Beams Under Distributed Longitudinal Loads

If the beam is subject to distributed longitudinal loads, the axial force in the prestressed configuration is no longer constant, not even stepwise, but variable from point to point, with possible singularities. The problem is therefore much more complex, because it is governed by a differential equation with variable coefficients, possibly to be integrated piecewise. Here, only the case of continuously variable load is studied, and an application is illustrated [23].

Incremental Equilibrium Equation
A cantilever is considered, subject to distributed compressive loads $p(x)$, variables with continuity (Fig. 7.10). These loads induce an axial force which satisfies the equilibrium in the undeformed configuration, namely:

$$N_0(x) = -\int_x^\ell p(x)\,dx. \qquad\qquad (7.80)$$

Fig. 7.10 Beam compressed
by distributed forces

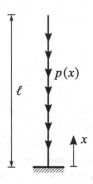

The total potential energy, consequently, reads:

$$\Pi = \int_0^\ell \left(\frac{1}{2} EI \, v''^2 + \frac{1}{2} N_0(x) \, v'^2 \right) dx. \tag{7.81}$$

By imposing the TPE to be stationary, the equilibrium equation follows:

$$EI \, v'''' - \left(N_0(x) \, v' \right)' = 0, \qquad x \in (0, \ell), \tag{7.82}$$

with the boundary conditions:

$$v(0) = 0, \qquad\qquad v'(0) = 0, \tag{7.83a}$$

$$EI \, v''(\ell) = 0, \qquad\qquad -EI \, v'''(\ell) = 0, \tag{7.83b}$$

where $N(\ell) = 0$ was taken into account.

In the particular case in which $p(x) = p = \text{cost}$, it is $N_0(x) = p(x - \ell)$, so that the equilibrium equation is written as:

$$EI \, v'''' + p \ell \left(1 - \frac{x}{\ell} \right) v'' - p \, v' = 0. \tag{7.84}$$

7.7.1 Power Series Solution

A solution to Eq. 7.84 is sought in the power series form:

$$v(x) = x^m \sum_{k=0}^{\infty} a_k \, x^k, \tag{7.85}$$

with m a positive integer and a_k constants, all unknown. By substituting the previous expression in the field equation, collecting the terms with the same power of x, and vanishing them separately, the following set of algebraic equations is derived:

$$x^{-4}: \quad m^4 - 6m^3 + 11m^2 - 6m = 0, \tag{7.86a}$$

$$x^{-3}: \quad a_1 = 0, \tag{7.86b}$$

$$x^{-2}: \quad a_0\beta^2 + a_2\left(m^2 + 3m + 2\right) = 0, \tag{7.86c}$$

$$\cdots, \tag{7.86d}$$

where $\beta := p\,\ell/EI$ has been used. Equation 7.86a admits four distinct roots:

$$m_1 = 0, \quad m_2 = 1, \quad m_3 = 2, \quad m_4 = 3. \tag{7.87}$$

Equations 7.86b,c, solved in sequence, provide the constants a_k (as a function of m):

$$a_1 = 0, \tag{7.88a}$$

$$a_2 = -\frac{a_0}{m^2 + 3m + 2}\beta^2, \tag{7.88b}$$

$$a_k = \frac{\frac{m+k-3}{\ell}a_{k-3} - (m + k - 2)\,a_{k-2}}{(m + k)\,(m + k - 1)\,(m + k - 2)}\beta^2, \quad k \geq 3, \tag{7.88c}$$

where a_0, being arbitrary, can be set equal to 1; therefore, there exist four sets of constants, one for each root m. Making use of Eq. 7.85, four independent solutions are built up, whose linear combination, with arbitrary constants c_i, reads:

$$v(x) = c_1\left(1 - \frac{1}{2}\beta^2 x^2 + \cdots\right) + c_2 x\left(1 - \frac{1}{6}\beta^2 x^2 + \cdots\right)$$
$$+ c_3 x^2\left(1 - \frac{1}{12}\beta^2 x^2 + \cdots\right) + c_4 x^3\left(1 - \frac{1}{20}\beta^2 x^2 + \cdots\right). \tag{7.89}$$

By imposing the geometric boundary conditions, $c_1 = c_2 = 0$ follow; by imposing the mechanical conditions, a homogeneous algebraic system is obtained, made of two equations in the unknowns c_3 and c_4. Such a system admits a solution different from the trivial one when the determinant of the coefficient matrix is equal to zero. By this condition, a characteristic equation in the unknown β is drawn, which, solved numerically, gives $\beta\ell^2 \simeq 7.83$; therefore the critical load is:

$$(p\,\ell)_c = \frac{7.83\,EI}{\ell^2}. \tag{7.90}$$

Remark 7.7 If the critical load in Eq. 7.90, relevant to distributed loads p, is compared with the critical load $P_c = \frac{\pi^2 EI}{4\ell^2}$ applied at the free end, it turns out that the former is 3.17 times larger than the latter. A qualitative reason of such a magnification is easily deduced from equilibrium considerations in the adjacent configuration.

7.7.2 Ritz Solution

The problem just examined, relevant to a cantilever under uniformly distributed longitudinal loads, is now tackled by the Ritz method (Appendix B). The transverse displacement field is expressed as:

$$v(x) = \sum_{i=1}^{n} \phi_i(x) q_i, \tag{7.91}$$

in which the following trial functions are chosen:

$$\phi_i(x) = x^{i+1}, \qquad i = 1, 2, \ldots, n. \tag{7.92}$$

The discrete representation, truncated to $n = 1$, or to $n = 2$, leads to the following results.

- If $n = 1$, the TPE in Eq. 7.81 becomes:

$$\Pi = 2EI\,\ell\,q_1^2 - \frac{1}{6}\ell^4\,p\,q_1^2. \tag{7.93}$$

By imposing stationary,

$$\left(4EI\,\ell - \frac{1}{3}\ell^4\,p\right)q_1 = 0 \tag{7.94}$$

follows, from which the critical load is evaluated as:

$$(p\ell)_c = \frac{12\,EI}{\ell^2}. \tag{7.95}$$

- If $n = 2$, the TPE in Eq. 7.81 takes the form:

$$\Pi = 2\,EI\,\ell\left(q_1^2 + 3\,\ell\,q_1\,q_2 + 3\,\ell^2\,q_2^2\right) - \frac{1}{60}\,p\,\ell^4\left(10\,q_1^2 + 18\,\ell\,q_1\,q_2 + 9\,\ell^2\,q_2^2\right). \tag{7.96}$$

From stationary, the equilibrium conditions follow:

$$\begin{pmatrix} 4\,EI\,\ell - \frac{p\,\ell^4}{3} & 6\,EI\,\ell^2 - \frac{3\,p\,\ell^5}{10} \\ 6\,EI\,\ell^2 - \frac{3\,p\,\ell^5}{10} & 12\,EI\,\ell^3 - \frac{3\,p\,\ell^6}{10} \end{pmatrix} \begin{pmatrix} q_1 \\ q_2 \end{pmatrix} = \begin{pmatrix} 0 \\ 0 \end{pmatrix}. \tag{7.97}$$

The system admits a non-trivial solution if the determinant of the coefficient matrix is equal to zero, from which:

$$(p\ell)_c = \frac{7.89\,EI}{\ell^2} \tag{7.98}$$

is drawn. This differs by less than 1% from that determined through the power series method.

A more complex problem, concerning a beam under combined concentrated and distributed loads, is treated in the Problem 14.2.2 by the Ritz method.

7.8   Elastically Constrained Beams

A technical strategy, aimed to increase the critical load, consists in equipping the beam with one or more *elastic bracings*, that is, flexible constraints of suitable stiffness, capable to countering the deflection at the bifurcation. Here, some examples are worked out, useful to illustrate the algorithmic aspects and to give indications about the design of bracings [1, 4, 5, 8, 12, 24, 25].

Contribution of the Elastic Springs to Equilibrium
To determine the equations of the problem, when elastic springs are present at the ends or at internal points of a beam, it needs to modify the TPE. This is the sum of the TPE of the beam, Π_b, and that of the springs, Π_s. With reference to a single spring of stiffness k, connected to the beam at the point H, the TPE of the device is $\Pi_s = \frac{1}{2}kv_H^2$. The stationary condition of the TPE (Appendix A) is therefore enriched with an additional term, $\delta\Pi_s = kv_H\delta v_H$, which enters the boundary conditions (and not the field equation, which remains unchanged).

As an alternative, the boundary conditions can be inferred by direct equilibrium considerations, without having to resort to the variational formulation. In particular:

- If $H = A, B$ is an end point, the boundary condition expresses the *equality* of the internal transverse force V_H *emerging* there, to the elastic force (external for the beam) $-kv_H$, that is, $-V_A = -kv_A$ or $V_B = -kv_B$.
- If $H = C$ is an interior point of the beam, the condition expresses the *equilibrium* between the internal transverse forces $V_C^{\pm}$, acting on the right and left of the point C, and the external elastic force exerted by the spring, that is, $V_C^+ - V_C^- - kv_C = 0$.

The following examples illustrate the procedure.

7.8.1   Beam Elastically Supported at One End

A cantilever is considered, compressed by a load P applied to the free end, braced by an elastic support of stiffness k placed at the same point (Fig. 7.11a). The adjacent equilibrium is governed by the field equation:

$$EI\,v'''' + P\,v'' = 0, \qquad x \in (0, \ell), \tag{7.99}$$

with the boundary conditions:

$$v(0) = 0, \qquad\qquad\qquad v'(0) = 0, \tag{7.100a}$$

$$EI v''(\ell) = 0, \qquad -EI v'''(\ell) - Pv'(\ell) + k\,v(\ell) = 0. \tag{7.100b}$$

The general solution of the homogeneous problem is:

$$v(x) = c_1 \cos(\beta x) + c_2 \sin(\beta x) + c_3\,x + c_4, \tag{7.101}$$

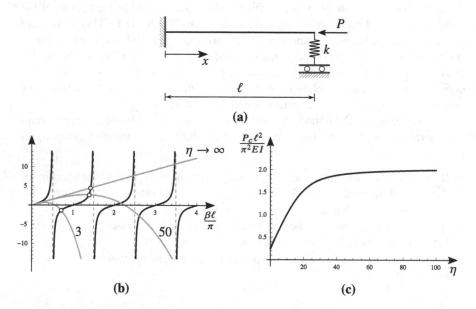

Fig. 7.11   Braced cantilever: (**a**) static scheme, (**b**) roots of the characteristic equation, (**c**) critical load *vs* the bracing-to-beam stiffness ratio

where $\beta := \sqrt{\frac{P}{EI}}$ is the wave-number and c_i, $i = 1, \ldots, 4$, are arbitrary constants. Imposition of the boundary conditions leads to:

$$
\begin{pmatrix}
1 & 0 & 0 & 1 \\
0 & \beta & 1 & 0 \\
-\beta^2 \cos(\beta\ell) & -\beta^2 \sin(\beta\ell) & \ell & 1 \\
\frac{k}{EI} \cos(\beta\ell) & \frac{k}{EI} \sin(\beta\ell) & \frac{k\ell}{EI} & \frac{k}{EI} - \beta^2 \frac{k}{EI}
\end{pmatrix}
\begin{pmatrix}
c_1 \\ c_2 \\ c_3 \\ c_4
\end{pmatrix}
=
\begin{pmatrix}
0 \\ 0 \\ 0 \\ 0
\end{pmatrix} .
\tag{7.102}
$$

Expanding the determinant of the coefficient matrix and equating it to zero supplies a transcendental characteristic equation for β:

$$
\tan(\beta\ell) = \beta\ell \left[1 - \frac{(\beta\ell)^2}{\eta} \right],
\tag{7.103}
$$

where:

$$
\eta := \frac{k\ell^3}{EI}
\tag{7.104}
$$

is a nondimensional parameter expressing the spring-to-beam stiffness ratio. Solved the transcendental equation in the unknown $\beta\ell$ and taken, among the infinite roots, the smallest one, $\beta_c\ell$, the critical load of the beam is computed as $P_c = (\beta_c\ell)^2 \frac{EI}{\ell^2}$.

To study the roots of Eq. 7.103, it is helpful to plot the graph of the functions to the left and to the right of the equality sign, as a function of the (nondimensional) independent variable $\beta\ell$, for different values of the assigned parameter η (Fig. 7.11b). The abscissa of the intersection points of the two graphs identify the roots. It appears that the unknown wave-number, β_c, depends on the stiffness ratio η. The limit cases are analyzed first:

- If $\eta \to 0$ (i.e., if the spring is very soft), then $\tan(\beta\ell) \to -\infty$, for which the smallest root is $\beta_c = \frac{\pi}{2\ell}$; the case of the unbraced cantilever is thus recovered.
- If $\eta \to \infty$ (i.e., if the spring is very stiff), then $\tan(\beta\ell) = \beta\ell$, for which the smallest root is $\beta_c = 1.43 \frac{\pi}{\ell}$; the case of the clamped-supported beam is thus recovered.

For values of $\eta \in (0, \infty)$, the wave-number β_c continuously increases in the interval $\left(\frac{\pi}{2\ell}, 1.43\frac{\pi}{\ell}\right)$; correspondingly, the critical load increases in the interval $\left(\frac{\pi^2 EI}{4\ell^2}, 2.05\frac{\pi^2 EI}{\ell^2}\right)$. Figure 7.11c illustrates the dependence of the critical load on the stiffness parameter. It is seen that, for values of η equal to about 30, the critical load reaches about 90% of its maximum value, so that any further increase of the bracing stiffness does not produce significant improvements in the structural behavior.

Example 7.2 (Beam with an Elastic Constraint of "Negative Stiffness") A further example, apparently different from that now studied, but which can be brought back

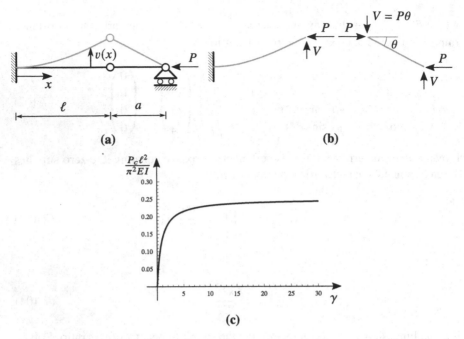

Fig. 7.12   Cantilever in series with a rigid rod: (**a**) critical mode; (**b**) forces; (**c**) critical load *vs* the length ratio

to it, is represented in Fig. 7.12a. It concerns a structure made of a horizontal elastic cantilever of length ℓ, arranged in series to a rigid rod of length a, compressed at one end by a load P.

In the current configuration, said $v(\ell)$ the displacement of the tip of the elastic beam, positive upward, the rigid rod rotates clockwise of the angle $\theta = \frac{v(\ell)}{a}$. From the current equilibrium of the rod (Fig. 7.12b), it follows that this latter exerts on the cantilever a horizontal compressive force P and a vertical force $V = P\theta = \frac{P}{a}v(\ell)$, positive upward, *of the same sign of the displacement* of the tip of beam. The rod, therefore, acts as a spring of "negative stiffness":

$$k := -\frac{P}{a}. \tag{7.105}$$

This circumstance allows one using as follows the results of the Sect. 7.8.1. Substituting Eq. 7.105 in Eq. 7.102 and equating to zero the determinant of the coefficient matrix, the following characteristic equation, transcendental in β, is found:

$$\tan(\beta\ell) = \beta\ell(1 + \gamma), \tag{7.106}$$

in which:

$$\gamma := \frac{a}{\ell}. \qquad (7.107)$$

Figure 7.12c describes the dependence of the critical load from the geometric parameter γ. It is observed that, when the length of the rigid rod is large (and therefore γ is also large), the critical load tends to that of the cantilever alone; this as the small rotation of the rod has little influence on the beam. On the other hand, when the rod length is small (and therefore $\gamma \to 0$), the critical load tends to zero, too. This surprising behavior is due to the fact the negative stiffness of the spring assumes very large values in modulus, inducing destabilizing effects that cannot be compensated by the finite stiffness of the beam, which therefore buckles even under an infinitely small load. □

7.8.2 Beam Elastically Supported in the Span

A beam of length 2ℓ, elastically braced by a spring of stiffness k, located at midspan, is now addressed (Fig. 7.13a). Since the spring introduces a singularity, it is necessary to integrate the differential problem in two sub-intervals, as follows.[12]

The equilibrium equations relevant to the two sub-intervals, are written as:

$$EI\, v_1'''' + P\, v_1'' = 0, \qquad x_1 \in (0, \ell), \qquad (7.108a)$$

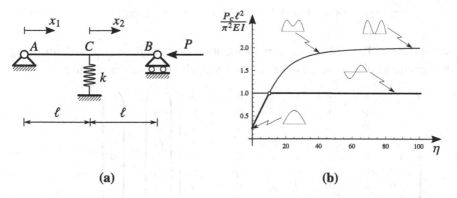

(a)   **(b)**

Fig. 7.13 Simply supported beam with elastic bracing at midspan: (a) static scheme, (b) critical load *vs* the spring-to-beam stiffness ratio

---

[12] In this particular example, it is also possible to consider only one half of the structure, equipped with suitable symmetry/antisymmetry constraints at midspan. However, here the general procedure is preferred, applicable even in cases where such symmetries do not exist.

$$EI\, v_2'''' + P\, v_2'' = 0, \qquad\qquad x_2 \in (0, \ell)\,, \qquad\qquad (7.108\text{b})$$

where $v_i\,(x_i)$ $i = 1, 2$ is the displacement field to the left and to the right of the singularity, respectively. The boundary conditions at the end points of the beam, A and B, are the usual ones; those at the midpoint C express (a) the continuity of displacement and rotation and (b) the translational and rotational equilibrium of the elastically constrained infinitesimal element of the beam, placed at midspan. They read:

$$v_1\,(0) = 0, \qquad\qquad M_1\,(0) = 0, \qquad\qquad (7.109\text{a})$$

$$v_2\,(\ell) = 0, \qquad\qquad M_2\,(\ell) = 0, \qquad\qquad (7.109\text{b})$$

$$v_1\,(\ell) = v_2\,(0)\,, \qquad\qquad v_1'\,(\ell) = v_2'\,(0)\,, \qquad\qquad (7.109\text{c})$$

$$M_1\,(\ell) = M_2\,(0)\,, \qquad V_1\,(\ell) + k\,v_1\,(\ell) = V_2\,(0)\,, \qquad (7.109\text{d})$$

where:

$$M_i\,(x_i) = EI\, v_i''\,(x_i)\,, \qquad\qquad i = 1, 2. \qquad\qquad (7.110\text{a})$$

$$T_i\,(x_i) = -EI\, v_i'''\,(x_i)\,, \qquad\qquad (7.110\text{b})$$

$$V_i\,(x_i) = T_i\,(x_i) - P\, v_i'\,(x_i)\,, \qquad\qquad (7.110\text{c})$$

The general solution of Eqs. 7.108 is written as:

$$v_1\,(x_1) = c_1\,\cos\,(\beta x_1) + c_2\,\sin\,(\beta x_1) + c_3\,x_1 + c_4, \qquad (7.111\text{a})$$

$$v_2\,(x_2) = c_5\,\cos\,(\beta x_2) + c_6\,\sin\,(\beta x_2) + c_7\,x_2 + c_8, \qquad (7.111\text{b})$$

where c_j, $j = 1, \dots, 8$ are arbitrary constants and where $\beta := \sqrt{\frac{P}{EI}}$. By imposing the boundary conditions, the following 8×8 algebraic system is obtained:

$$
\begin{pmatrix}
1 & 0 & 0 & 1 & 0 & 0 & 0 & 0 \\
-\beta^2 & 0 & 0 & 0 & 0 & 0 & 0 & 0 \\
0 & 0 & 0 & 0 & C & S & \ell & 1 \\
0 & 0 & 0 & 0 & -\beta^2 C & -\beta^2 S & 0 & 0 \\
C & S & \ell & 1 & -1 & 0 & 0 & -1 \\
-\beta S & \beta C & 1 & 0 & 0 & -\beta & -1 & 0 \\
-\beta^2 C & -\beta^2 S & 0 & 0 & \beta^2 & 0 & 0 & 0 \\
\frac{k}{EI} C & \frac{k}{EI} S & \frac{k\ell}{EI} - \beta^2 & \frac{k}{EI} & 0 & 0 & \beta^2 & 0
\end{pmatrix}
\begin{pmatrix}
c_1 \\ c_2 \\ c_3 \\ c_4 \\ c_5 \\ c_6 \\ c_7 \\ c_8
\end{pmatrix}
=
\begin{pmatrix}
0 \\ 0 \\ 0 \\ 0 \\ 0 \\ 0 \\ 0 \\ 0
\end{pmatrix},
\qquad (7.112)
$$

where $C := \cos\,(\beta\ell)$ and $S := \sin\,(\beta\ell)$. Expanding the determinant of the coefficient matrix, and equating it to zero, a transcendental equation in β is drawn:

$$\sin (\beta \ell) \left\{ \sin (\beta \ell) - (\beta \ell) \left[1 - \frac{(\beta \ell)^2}{\eta} \right] \cos (\beta \ell) \right\} = 0, \qquad (7.113)$$

in which a nondimensional stiffness parameter has been defined:

$$\eta := \frac{k \ell^3}{2 E I}. \qquad (7.114)$$

The characteristic Eq. 7.113 breaks in two equations, each of which admitting a set of different roots:

$$\sin (\beta \ell) = 0, \qquad (7.115\text{a})$$

$$\tan (\beta \ell) = (\beta \ell) \left[1 - \frac{(\beta \ell)^2}{\eta} \right]. \qquad (7.115\text{b})$$

Equation 7.115a is that of the simply supported beam of length ℓ, giving the roots $\beta_j = \frac{j\pi}{\ell}$, $j = 1, 2, \ldots$. The corresponding critical modes have a null point at the midspan C, so that *the associated critical load does not depend on the stiffness of the spring* (which, being unstretched, does not contribute to the elastic energy of the system). The lowest critical load is $P_c = \frac{\pi^2 E I}{\ell^2}$, equal to that of the single half-beam (and therefore four times greater than that of the unbraced beam). The critical mode is *antisymmetric* and consists of two half-waves.

Equation 7.115b gives roots, which, in contrast, do depend on the stiffness parameter η. It is formally identical to Eq. 7.103, already studied, from which it differs only in the different definition of η. Therefore, considerations similar to those already made with reference to Fig. 7.11b apply.

If the two sets of critical loads are plotted *vs* the stiffness ratio η, the graph in Fig. 7.13b is obtained. It is seen that for sufficiently small stiffnesses η, the critical load is associated with a symmetric mode (as it is smaller than the antisymmetric one). When, instead, the stiffness η is sufficiently large, the critical load corresponds to an antisymmetric mode (since smaller than the symmetric one). The intersection of the two curves defines the system whereby the two modes coexist under the same load, thus giving rise, in the nonlinear field, to interaction between simultaneous modes.

7.9 Beam on Winkler Soil

When a compressed beam rests on an elastic medium, the soil contributes to the stiffness of the system and therefore increases the critical load. Modeling the interaction between beam and medium, however, is a complex problem. The simpler model is offered by the so-called Winkler soil, i.e., an extreme schematization in which the ground reaction $r_f(x)$ to the abscissa x is assumed to be proportional, through a constant k_f (called Winkler constant[13]), to the displacement $v(x)$ at the same point (ignoring thus the nonlocal nature of the interaction between beam and ground); therefore, $r_f(x) = -k_f\, v(x)$. As a further hypothesis, the contact is assumed to be *bilateral*, that is, the soil is able to prevent detachment. This implicitly assumes the presence of transverse forces, which ensure the contact between beam and ground. Since these forces do not affect the critical load, but can be considered as they were imperfections, they will be ignored in the analysis.

In this section, the incremental equations of equilibrium are formulated and some applications developed in closed form [1, 3, 5, 8, 12, 21, 23]. An example of application of the Ritz method to a more complex problem of beam on partial foundation is shown in the Problem 14.2.3.

7.9.1   Model

The total potential energy of the extensible beam, resting on a Winkler soil, under uniform axial compression induced by an end load P and subject to transverse loads of intensity $p(x)$, is obtained by adding to the energy of the beam, Eq. 7.10, the energy stored by the soil in the deformation, that is:

$$\Pi = \int_0^\ell \left[\frac{1}{2} EA\, u'^2 - \frac{1}{2} P\, v'^2 + \frac{1}{2} EI\, v''^2 + \frac{1}{2} k_f\, v^2 \right] dx - \int_0^\ell p\, v\, dx. \qquad (7.116)$$

By imposing the stationary of the functional, the following equations of equilibrium are obtained:

$$EA\, u'' = 0, \qquad x \in (0, \ell), \qquad (7.117a)$$

$$EI\, v'''' + P\, v'' + k_f\, v = p, \qquad x \in (0, \ell), \qquad (7.117b)$$

with the boundary conditions:

$$\left[EA\, u'\, \delta u \right]_0^\ell = 0, \qquad (7.118a)$$

$$\left[\left(-EI\, v''' - P\, v' \right) \delta v \right]_0^\ell = 0, \qquad (7.118b)$$

---

[13] The Winkler constant has physical dimensions of a force per unit squared length $[\mathrm{ML^{-1}T^{-2}}]$; its values are available, for different types of soil, on specialized handbooks. There, it is generally expressed as a quantity $\tilde{k}_f$, having the physical dimensions $[\mathrm{ML^{-2}T^{-2}}]$, as it refers to a contact area. In a 1D model of beam, instead, where the contact occurs on a line, it must be taken $k_f = \tilde{k}_f b$, with b the transverse footprint dimension of the grounded beam.

$$\left[EI\,v''\,\delta v'\right]_0^\ell = 0. \tag{7.118c}$$

The effect of the soil on the equilibrium of the beam consists therefore in the introduction of the reaction $-k_f\,v$ in the right member of the field equation. As in the problem of the beam with no soil reactions, it is $u \equiv 0$.

7.9.2   Beam on Elastic Soil Simply Supported at the Ends

The bifurcation problem of the compressed beam on Winkler soil is governed by the differential equation 7.117b, made homogeneous:

$$EI\,v'''' + P\,v'' + k_f\,v = 0. \tag{7.119}$$

Before dealing with general constraint conditions at the ends (to be discussed in the following Sect. 7.9.3), a special case is premised, for the simplicity of interpretation of the relevant results. It concerns a beam, simply supported at the ends $x = 0, \ell$ (Fig. 7.14a), whose boundary conditions are:

$$v(0) = 0, \qquad\qquad v(\ell) = 0, \tag{7.120a}$$

$$EI\,v''(0) = 0, \qquad\qquad EI\,v''(\ell) = 0. \tag{7.120b}$$

The solution is sought in the form of a "tentative function":

$$v(x) = \hat{v}\,\sin\left(\frac{n\pi}{\ell}x\right), \tag{7.121}$$

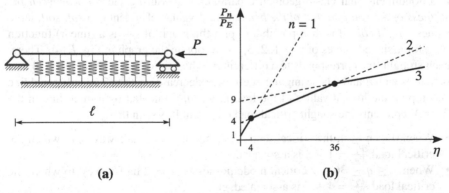

(a)   **(b)**

Fig. 7.14 Simply supported beam on Winkler soil: (**a**) static scheme, (**b**) critical load *vs* the soil-to-beam stiffness ratio

where n is an arbitrary integer (denoting the number of half-waves in which the beam buckles) and $\hat{v}$ is an inessential arbitrary constant. The boundary conditions are satisfied for any n, while the field equation admits solution different from the trivial one if and only if:

$$EI \left(\frac{n\pi}{\ell}\right)^4 - P \left(\frac{n\pi}{\ell}\right)^2 + k_f = 0. \tag{7.122}$$

This equation associates a critical value $P = P_n$ (i.e., an eigenvalue of the differential problem), to any integer n:

$$P_n = \left(\frac{n\pi}{\ell}\right)^2 EI + \left(\frac{\ell}{n\pi}\right)^2 k_f. \tag{7.123}$$

Among all the P_n's, it is of interest to determine the smallest one, $P_c := \min\{P_n\}$. It is observed that, being P_n a *non-monotone* function of n, P_c must be found by trial and error methods.[14] If $k_f = 0$ (i.e., if the beam is not constrained to the ground), the minimum P_n occurs for $n = 1$, i.e., it is equal to the Eulerian load $P_E := \frac{\pi^2 EI}{\ell^2} = P_1$. To study the case $k_f \neq 0$, it is convenient to nondimensionalize the unknown critical load with respect the Eulerian load, thus obtaining:

$$\frac{P_n}{P_E} = n^2 + \frac{\eta}{n^2}, \tag{7.124}$$

where:

$$\eta := \frac{k_f}{EI} \left(\frac{\ell}{\pi}\right)^4 \tag{7.125}$$

is a nondimensional elasto-geometric parameter measuring the *ratio between the stiffness of the soil and that of the beam*. Small values of η denote a *soft soil*, large values a *rigid soil*. It is then possible to plot the graph of $\frac{P_n}{P_E}$ as a (linear) function of η, for assigned values of $n = 1, 2, 3, \ldots$, to obtain the graph in Fig. 7.14b. There, each straight line corresponds to a different n, whose intercept with the vertical axis increases with n and whose angular coefficient decreases with n. The polygonal line enveloping the lowest values of the load at any η (indicated by a thick line in the figure) represents the sought critical load diagram. It is seen that:

- When $0 \leq \eta \leq 4$, the critical mode consists of $n = 1$ half-waves, to which the critical load $\frac{P_1}{P_E} = 1 + \eta$ is associated.
- When $4 \leq \eta \leq 36$, the critical mode possesses $n = 2$ half-waves, to which the critical load $\frac{P_2}{P_E} = 4 + \frac{\eta}{4}$ is associated, etc.

---

[14] Since n is not a continuous variable, it is not possible to differentiate P_n with respect to n.

The values η_n at which the eigenvalues occur in pair (i.e., *simultaneous critical modes* exist, with n and $n + 1$ half-waves), are:

$$\eta_n = [n(n+1)]^2. \tag{7.126}$$

Remark 7.8 The two contributions to the critical load, present in Eq. 7.123, are easily interpreted using the *Rayleigh ratio*:

$$P_c := \frac{\int_0^\ell \left[EI\, v''^2 + k_f\, v^2\right] dx}{\int_0^\ell v'^2 dx}. \tag{7.127}$$

This expresses the critical load as the ratio of the elastic energy of the system (beam plus soil) and the prestress energy of the beam. Taken v as in Eq. 7.121, the elastic bending energy increases with n^4, while the elastic energy of the soil is independent of n, and the prestress energy increases with n^2. So, to minimize P_c with respect to n, it needs to distinguish the following: (a) if the soil is soft (k_f small), the bending energy prevails at numerator of the ratio, so that n should be taken small, close to 1; and (b) if the ground is rigid (k_f large), the energy of the soil prevails at numerator, so that it is needs to take n large.

7.9.3 Beam on Elastic Soil Arbitrarily Constrained at the Ends

The general case of compressed beam on Winkler soil, arbitrarily constrained at the ends, is now considered. Since the tentative sinusoidal solution in Eq. 7.121 is not compatible with the constraints, it needs to find the general solution to Eq. 7.117b. This is here rewritten in the form:

$$v'''' + 2\mu\gamma^2 v'' + \gamma^4 v = 0, \tag{7.128}$$

where:

$$\gamma^4 := \frac{k_f}{EI} \tag{7.129}$$

is a stiffness parameter, and:

$$\mu := \frac{1}{2} \frac{P}{\sqrt{EI\, k_f}} \tag{7.130}$$

is the *nondimensional load*.

By looking for solutions $v(x) = e^{\lambda x}$, the differential equation gives the characteristic equation:

$$\lambda^4 + 2\mu\gamma^2 \lambda^2 + \gamma^4 = 0, \tag{7.131}$$

whose four roots are:

$$\lambda = \pm i \, \gamma \sqrt{\mu \mp \sqrt{\mu^2 - 1}}. \tag{7.132}$$

According to the values of μ, three cases occur, discussed ahead.

- **Case I:** $\mu > 1$. In this case, the following inequalities hold:

$$0 < \mu^2 - 1,$$

$$0 < \sqrt{\mu^2 - 1} < \mu, \tag{7.133}$$

$$0 < \mu \mp \sqrt{\mu^2 - 1},$$

which imply that the four roots in Eq. 7.132 are purely imaginary. By letting:

$$\omega_1 := \gamma \sqrt{\mu - \sqrt{\mu^2 - 1}}, \qquad \omega_2 := \gamma \sqrt{\mu + \sqrt{\mu^2 - 1}}, \tag{7.134}$$

they are written as:

$$\lambda_1 = i \, \omega_1, \quad \lambda_2 = -i \, \omega_1, \quad \lambda_3 = i \, \omega_2, \quad \lambda_4 = -i \, \omega_2. \tag{7.135}$$

The general solution of the homogeneous problem is therefore:

$$v(x) = c_1 \cos(\omega_1 x) + c_2 \sin(\omega_1 x) + c_3 \cos(\omega_2 x) + c_4 \sin(\omega_2 x), \tag{7.136}$$

with c_i arbitrary constants.
- **Case II:** $\mu = 1$. In this case, the roots in Eq. 7.132 have multiplicity two, and are equal to:

$$\lambda_1 = \lambda_3 = i \, \gamma, \qquad \lambda_2 = \lambda_4 = -i \, \gamma, \tag{7.137}$$

so that the general solution of the homogeneous problem is:

$$v(x) = c_1 \cos(\gamma x) + c_2 \sin(\gamma x) + c_3 x \cos(\gamma x) + c_4 x \sin(\gamma x), \tag{7.138}$$

with c_i arbitrary constants.
- **Case III:** $\mu < 1$. In this case, the roots in Eq. 7.132 are complex numbers, with real part different from zero, i.e.:

$$\lambda = \pm i \, \gamma \sqrt{\mu \mp i \sqrt{1 - \mu^2}}. \tag{7.139}$$

Through algebraic manipulations, the four roots can be rewritten as:

$$\lambda_1 = \alpha + i\,\omega, \qquad \lambda_2 = \alpha - i\,\omega, \qquad \lambda_3 = -\alpha + i\,\omega, \qquad \lambda_4 = -\alpha - i\,\omega,$$
$$(7.140)$$

where:

$$\alpha := \frac{\gamma}{\sqrt{2}}\sqrt{1-\mu}, \qquad \omega := \frac{\gamma}{\sqrt{2}}\sqrt{1+\mu}. \qquad (7.141)$$

The general solution of the homogeneous problem is:

$$v(x) = e^{\alpha x}\Big(c_1 \cos(\omega x) + c_2 \sin(\omega x)\Big) + e^{-\alpha x}\Big(c_3 \cos(\omega x) + c_4 \sin(\omega x)\Big),$$
$$(7.142)$$

with c_i arbitrary constants.

Since the critical load is unknown, μ is also unknown, and therefore it is not a priori known in which of the three cases discussed above the solution falls. It is therefore necessary to proceed by trial and error, adopting one of the three forms of the general solutions, imposing the boundary conditions, and looking for a real solution to the transcendental eigenvalue problem. The example below clarifies the procedure. A further example of exact solution, relevant to a free-free beam, is illustrated in the Problem 14.2.4.

Example 7.3 (Clamped-Sliding Beam on Elastic Soil) As an example of the general theory developed above, a clamped-clamped beam is considered, in which one of the two constraints permits longitudinal sliding (Fig. 7.15a). Given the symmetry of the problem, it is convenient to place the origin of the coordinates at midspan. The boundary conditions then read:[15]

$$v(-\ell/2) = 0, \qquad\qquad v'(-\ell/2) = 0, \qquad\qquad (7.143a)$$
$$v(\ell/2) = 0, \qquad\qquad v'(\ell/2) = 0. \qquad\qquad (7.143b)$$

To evaluate the critical load, the three cases discussed in the Sect. 7.9.3 must be analyzed. Assuming that the nondimensional critical load μ (Eq. 7.130) is larger than 1, case I occurs. Substituting the relevant general solution (Eq. 7.136) in the boundary conditions, four algebraic equations in the four unknown constants c_i, $i = 1, \dots, 4$ are drawn. By mutually adding and subtracting the equations, it is possible to obtain the following algebraic system, block diagonalized:

---

[15] Alternatively, a semi-interval $x \in (0, \frac{\ell}{2})$ can be taken, for example, on the right of midspan, for which (a) at $x = 0$, the symmetry/antisymmetry conditions must be enforced, implying $c_2 = c_4 = 0$, in the symmetric case or $c_1 = c_3 = 0$ in the antisymmetric case, and (b) at $x = \frac{\ell}{2}$, the clamp conditions $v(\ell/2) = 0$, $v'(\ell/2) = 0$ must be imposed. Thus, two uncoupled subproblems are found, identical to those contained in the following Eq. 7.144.

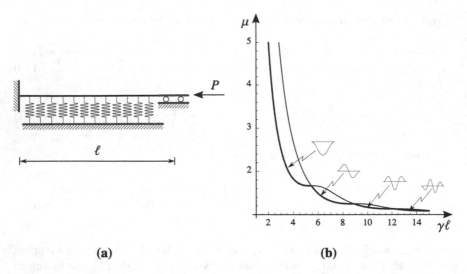

(a)   **(b)**

Fig. 7.15 Compressed clamped-clamped beam on Winkler soil: (**a**) static scheme, (**b**) critical load
vs the soil-to-beam stiffness ratio

$$
\begin{pmatrix}
\omega_1 S_1 & \omega_2 S_2 & 0 & 0 \\
C_1 & C_2 & 0 & 0 \\
0 & 0 & S_1 & S_2 \\
0 & 0 & \omega_1 C_1 & \omega_2 C_2
\end{pmatrix}
\begin{pmatrix}
c_1 \\ c_3 \\ c_2 \\ c_4
\end{pmatrix}
=
\begin{pmatrix}
0 \\ 0 \\ 0 \\ 0
\end{pmatrix},
\tag{7.144}
$$

where $C_i := \cos\left(\frac{\omega_i \ell}{2}\right)$, $S_i := \sin\left(\frac{\omega_i \ell}{2}\right)$, $i = 1, 2$. This system admits non-trivial
solutions if and only if the determinant of the coefficient matrix is zero. Due to the
structure of the matrix, its determinant is equal to the product of the determinants of
the individual blocks, so that the problem splits in two subproblems, each governed
by the following characteristic equations:

$$
\omega_1 \sin\left(\frac{\omega_1 \ell}{2}\right) \cos\left(\frac{\omega_2 \ell}{2}\right) - \omega_2 \cos\left(\frac{\omega_1 \ell}{2}\right) \sin\left(\frac{\omega_2 \ell}{2}\right) = 0,
\tag{7.145a}
$$

$$
\omega_2 \sin\left(\frac{\omega_1 \ell}{2}\right) \cos\left(\frac{\omega_2 \ell}{2}\right) - \omega_1 \cos\left(\frac{\omega_1 \ell}{2}\right) \sin\left(\frac{\omega_2 \ell}{2}\right) = 0.
\tag{7.145b}
$$

Equation 7.145a is associated with the symmetric buckling modes ($c_1 \neq 0$, $c_3 \neq 0$
and $c_2 = c_4 = 0$); Eq. 7.145b with antisymmetric modes ($c_1 = c_3 = 0$ and $c_2 \neq$
0, $c_4 \neq 0$). Substitution of Eqs. 7.134, for ω_1 and ω_2, in the previous equations
leads to transcendental equations in the two nondimensional parameters, μ and $\gamma \ell$.
Therefore, the nondimensional critical load μ can be determined as a function of
the stiffness ratio $\gamma \ell$. The relevant graph is shown in Fig. 7.15b, for symmetric and
antisymmetric modes, together with a sketch of the modal shapes. The envelope of

the lowest values is indicated with a thicker line. It is seen that, as in the case of the simply supported beam (Fig. 7.14b), the critical mode is alternately symmetric and antisymmetric where the two curves intersect, simultaneous modes occur.

If the analysis is repeated for the cases II and III discussed in the Sect. 7.9.3, different algebraic systems in the unknowns c_i are found, whose characteristic equations do not admit solution, in addition to the trivial one. The only significant solution, therefore, is that falling in case I.   □

7.10 Prestressed Reinforced Concrete Beams

It is well-known that the prestressing technique of reinforced concrete beams, either realized by posttensioned sliding cables or adherent pretensioned wires, permits to endow the concrete with an apparent tensile resistance, so that these beams behave as entirely reactive, when solicited in bending. Since prestressing, however, increases the compression of concrete, one wonders if, and how, it influences the buckling phenomenon, [9].

To discuss the problem, two simple but significant examples are illustrated here, relevant to posttensioning system: (a) the beam with *external prestressing cable* and (b) the beam with *internal prestressing cable*, which differ from each other in the way cable and beam mutually interact.

7.10.1 Externally Cable-Prestressed Beams

A column of length ℓ and bending stiffness EI is considered, clamped at the bottom and free at the top, prestressed by pretensioned *external cables*, whose resulting tensile force is centroidal, of module T_0 (Fig. 7.16a). The cables are anchored to the beam by a rigid head placed at the free end, through which they transmit to the beam the (internal) compressive force $-T_0$. The beam is also subject to an external compressive load P.

The equilibrium equations are determined by making stationary the TPE of the system, $\Pi = \Pi_b + \Pi_c$, equal to the sum of the TPE of the beam and that of the cable (Fig. 7.16 b). When the beam buckles (Fig. 7.16c), it lies on a *curve* $v(x)$; the resulting cable, instead, since it interacts with the beam only at the top end, lies on the *straight line* that joins the (fixed) foot and the (mobile) head of the beam. Beam and cable, therefore, undergo (second order) *different* unit extensions, respectively, equal to:

$$\varepsilon_b = \frac{1}{2}v'^2, \tag{7.146a}$$

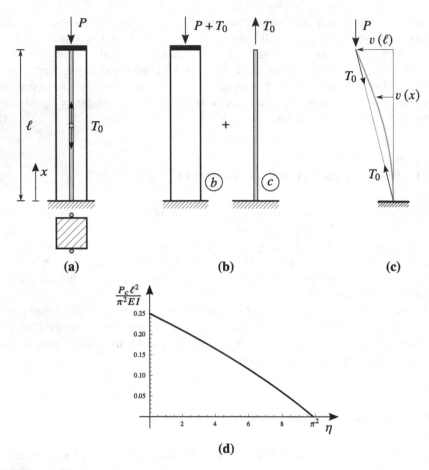

Fig. 7.16 Externally cable-prestressed beam: (**a**) prestressed configuration, (**b**) substructures, (**c**) adjacent configuration, (**d**) critical load *vs* the prestressing force

$$\varepsilon_c = \frac{1}{\ell}\left(\sqrt{\ell^2 + v\,(\ell)^2} - \ell\right) = \frac{1}{2}\frac{v\,(\ell)^2}{\ell^2} + \cdots,\tag{7.146b}$$

the first being deduced, as usual, from the Green-Lagrange tensor, the second from a MacLaurin series expansion. Since the beam, in the prestressed state, is compressed by a force $N_0 = -(P + T_0)$, his TPE is:

$$\Pi_b = \frac{1}{2}\int_0^\ell EIv''^2\mathrm{d}x - \frac{1}{2}\int_0^\ell (P + T_0)\,v'^2\mathrm{d}x.\tag{7.147}$$

The cable, on the other hand, being axially solicited by the force T_0, has TPE equal to:

$$\Pi_c = \int\limits_0^\ell T_0 \varepsilon_c \mathrm{d}x = \frac{1}{2} T_0 \frac{v(\ell)^2}{\ell}. \tag{7.148}$$

By requiring $\delta \Pi_b + \delta \Pi_c = 0$, $\forall \delta v$, the field equation follows:

$$EIv'''' + (P + T_0) v'' = 0, \tag{7.149}$$

with the boundary conditions:

$$v(0) = 0, \tag{7.150a}$$

$$v'(0) = 0, \tag{7.150b}$$

$$-EIv'''(\ell) - (P + T_0) v'(\ell) + T_0 \frac{v(\ell)}{\ell} = 0, \tag{7.150c}$$

$$EIv''(\ell) = 0. \tag{7.150d}$$

Remark 7.9 The field Eq. 7.149 can be easily interpreted as the equilibrium equation of a beam compressed not only by the external load P but also by the internal prestressing force T_0. The boundary condition in Eq. 7.150c has the meaning of an equilibrium condition at the free end of the beam, compressed by the force $P + T_0$, and *elastically constrained by a spring of* (positive) *stiffness*:

$$k := \frac{T_0}{\ell} \tag{7.151}$$

(Equation 7.100 should be remembered). This circumstance follows from the fact that, in the adjacent configuration (Fig. 7.16c), the force T_0 applied to the head of the beam is inclined of an angle $\frac{v(\ell)}{\ell}$; therefore, the cable exerts on the beam a transverse force equal to $T_0 \frac{v(\ell)}{\ell} = kv(\ell)$ and *opposite* to the displacement.

The previous observation permits to exploit the results of the Sect. 7.8.1, in particular, Eqs. 7.103 and 7.104. The characteristic equation of the problem at hand is therefore the same as the elastically constrained cantilever, i.e.:

$$\tan(\beta\ell) = (\beta\ell) \left[1 - \frac{(\beta\ell)^2}{\eta} \right], \tag{7.152}$$

where, however, the parameters are redefined as follows:

$$\beta := \sqrt{\frac{P + T_0}{EI}}, \quad \eta := \frac{T_0 \ell^2}{EI}. \tag{7.153}$$

For an assigned T_0 (and therefore η), the smallest root $\beta_c \ell$ of the characteristic equation is sought for, and finally the critical load computed as $P_c = (\beta_c \ell)^2 \frac{EI}{\ell^2} - T_0$; in nondimensional form:

$$\frac{P_c \ell^2}{\pi^2 EI} = \frac{(\beta_c \ell)^2 - \eta}{\pi^2}. \tag{7.154}$$

Figure 7.16d shows the nondimensional critical load as a function of η. It appears that, whatever T_0 (and therefore η), the effect of prestressing *reduces the critical load* of the unprestressed beam and that the reduction is much greater, the greater the prestressing force. Therefore, of the two antagonistic effects, one destabilizing, due to the increase in prestress, and the other stabilizing, due to the presence of an equivalent spring at the free end, the former prevails. When the prestressing is sufficiently large, the critical load P_c is zero, as the force T_0, alone, induces buckling. This occurs for $\eta = \eta_c := \pi^2$, i.e., for $T_0 = \frac{\pi^2 EI}{\ell^2}$, for which it is $\beta_c \ell = \pi$.

7.10.2 Internally Cable-Prestressed Beams

The same cantilever as in the previous case is considered but internally cable-prestressed. This cable interacts with the beam not only at the free end but at any abscissa x (Fig. 7.17). The two substructures, therefore, undergo the *same displacement field* $v(x)$, so that the unit extension is the same:

$$\varepsilon_b \equiv \varepsilon_c = \frac{1}{2} v'^2. \tag{7.155}$$

The relevant TPEs are:

$$\Pi_b = \frac{1}{2} \int_0^\ell EI v''^2 dx - \frac{1}{2} \int_0^\ell (P + T_0) v'^2 dx, \tag{7.156a}$$

$$\Pi_c = \int_0^\ell T_0 \frac{1}{2} v'^2 dx, \tag{7.156b}$$

from which it follows that $\Pi = \Pi_b + \Pi_c$ *does not depend on* T_0 and coincides with the TPE of the unprestressed beam. *The internal prestressing cable, therefore, has no influence on the critical load of the beam.*

Remark 7.10 This result is only apparently surprising. If, indeed, the interaction forces that cable and beam exchange in the domain and at boundary are considered, the following results are found (Fig. 7.17c). Since the cable assumes a curved configuration, described by the function $v(x)$, and it is solicited by a tension T_0,

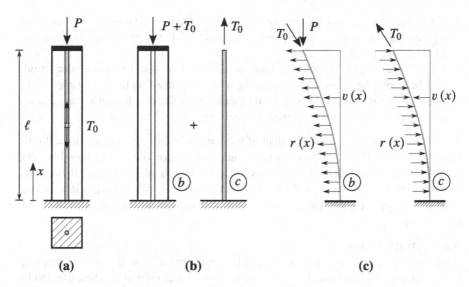

Fig. 7.17 Internally cable-prestressed beam: (**a**) prestressed configuration, (**b**) substructures, (**c**) adjacent configuration and interactive forces r

the funicular equilibrium states that distributed contact forces $r(x) = -T_0 v''(x)$ arise, together with a transverse concentrated force at the boundary $R = T_0 v'(\ell)$. These forces, changed in sign, act on the beam, whose field equation and boundary condition, consequently, read:

$$EIv'''' + (P + T_0) v'' = -r(x), \tag{7.157a}$$

$$-EIv'''(\ell) - (P + T_0) v'(\ell) = -T_0 v'(\ell). \tag{7.157b}$$

From which it follows that all forces proportional to T_0 cancel out each other, in the domain and at the boundary.

Remark 7.11 The previous observation better explains, by comparison, the functioning of the external cable. For this latter, it is $r(x) = 0$, so that the contribution of the prestressing survives in the equation field of the beam. Moreover, it is $R = T_0 \frac{v(\ell)}{\ell}$, which only partially balances $T_0 v'(\ell)$, since the slopes of beam and cable are *not* the same at the free end $x = \ell$.

7.11   Local and Global Instability of Compressed Truss Beams

A simple but significant example of buckling of beam systems, is offered by the truss, made of pinned-pinned members. This manifests two different forms of instability:

- *Local buckling*, in which only one (or few) members of the longerons instabilize, by bending themselves in just one half-wave; all the members participating in buckling are purely inflected.
- *Global buckling*, in which the whole structure is involved in the critical mode and behaves like a monolithic beam, bent with a wave-length of the order of the total length; in this modal shape, the members of the two longerons are mainly extended, lengthen and shortened, respectively.

The local critical load is independent of the length of the truss beam, since it only depends on the elastic and geometric characteristics of the single member. The global critical load, instead, decreases with the length of the beam. It happens, therefore, that, if the truss is short, the leading mode is of local type; if it is long is of global type. Hence, there exists a *critical length* at which the two modes are coincident.

An Illustrative Example

As an example, a simply supported truss is considered, of length $\ell = nh$, made up of n square modules of side h, each consisting of two triangular meshes, divided by the diagonal of the square (Fig. 7.18). The upper and lower longitudinal members are uniformly compressed by axial forces $P/2$. Named A_m the cross-section area of a single longitudinal member, the inertia moment of the truss, assimilated to a monolithic Euler-Bernoulli beam,[16] is $I_b :\simeq \frac{A_m h^2}{2}$. Given that the effective length is $\ell_0 := \ell$, the global critical load is:

$$P_{gl} := \frac{\pi^2 E I_b}{\ell^2} = \frac{E A_m}{2} \left(\frac{\pi}{n}\right)^2. \tag{7.158}$$

On the other hand, the inertia moment of the single member is $I_m := A_m \rho_m^2$, with ρ_m the radius of inertia of its cross-section. Since the effective length is $\ell_0 := h$ and

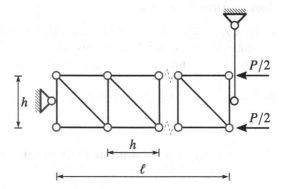

Fig. 7.18 Compressed truss beam

---

[16] A more refined model of equivalent beam, given the asymmetry of the square module with a single diagonal, should involve coupling between shear strain and axial extension [18].

the axial force is equal to $\frac{P}{2}$, the local critical load is:

$$P_{loc} = 2\frac{\pi^2 E I_m}{h^2} = 2EA_m \left(\frac{\pi \rho_m}{h}\right)^2, \tag{7.159}$$

and, therefore, it is independent of the number of modules n.

The relationship between the two critical loads is:

$$\frac{P_{loc}}{P_{gl}} = \left(\frac{2n\rho_m}{h}\right)^2. \tag{7.160}$$

The critical number of modules for which the two critical loads coincide (i.e., this being an integer, they are closer to each other) is $n_c := \frac{h}{2\rho_m}$. If $n < n_c$ then $P_{loc} < P_{gl}$, so that the instability is of local type; if $n > n_c$ then $P_{loc} > P_{gl}$, so that the instability is of global type.

Remark 7.12 If the members of the two longerons are all equal, local instability manifests itself with the *simultaneous buckling of all members*. The critical load therefore has a multiplicity $2n$, generally very high. However, since there are, inevitably, some nonuniformities in the state of stress (in addition to geometric differences), only one or a few members become unstable. For example, if the truss is subject to transverse forces, such as its own weight, the upper longeron is compressed, and the lower one is tensed; furthermore, the rods closest to midspan are more solicited than those near the supports. Hence, the local buckling affects only the most compressed rod. The closeness, however, of many other critical loads, can give rise to complex phenomena of nonlinear interaction [17, 20].

7.12   Finite Element Analysis of Buckling

The exact buckling analysis of a beam, albeit simple in principle, becomes very onerous when it needs to integrate the equilibrium equations in several sub-intervals, or even prohibitive in the case of frames, even just planar. It is appropriate, therefore, to address the problem numerically, through the *method of the finite elements*. This method follows the same steps as in linear elastic analysis (which is assumed to be known to the reader) but refers to axially prestressed beams, whose energy must also be taken into account. In this section two finite elements are briefly illustrated [2, 3, 5, 6, 10, 14, 19, 25]:

- The *polynomial finite element*, in which the displacement field is interpolated between the nodal values through polynomials of first and third degree.
- The *exact finite element*, in which the displacement field is described in each element by the transcendental solution of the differential equation governing the problem.

Obviously, the polynomial finite element is simpler, but, for the fact of being approximated, it requires a more dense discretization;[17] the exact element, on the other hand, can be used to describe, alone, the behavior of an entire member of a frame, provided that its elasto-geometric characteristics and prestress are uniform.

Remark 7.13 In a planar frame, there exist internal forces different from the axial one. However, these forces can generally be considered as "imperfections" with regard to the buckling phenomenon. Since they slightly modify the response of the system, but not the critical load, they will be ignored here.

7.12.1 Polynomial Finite Element

A single beam will be examined first, and then an assembly of beams will be considered, and an example illustrated.

Single Beam

An extensible and shear undeformable beam is studied, under uniform (positive if tensile) axial force N_0, subject to nodal forces and couples $f_{xi}, f_{yi}, m_i, i = 1, 2$. Its TPE, by Eqs. 7.10 and 7.5, reads:

$$\Pi = \int_0^\ell \left(\frac{1}{2} EA\, u'^2 + \frac{1}{2} EI\, v''^2 + \frac{1}{2} N_0\, v'^2 \right) dx$$
$$- \left(f_{x1}\, u\,(0) + f_{y1}\, v\,(0) + m_1\, v'\,(0) + f_{x2}\, u\,(\ell) + f_{y2}\, v\,(\ell) + m_2\, v'\,(\ell) \right). \tag{7.161}$$

In the spirit of the finite element method, the unknown displacement fields, longitudinal $u\,(x)$ and transverse $v\,(x)$, are expressed as linear combinations of known *interpolation functions* $\psi_i\,(x)$ and unknown nodal values (displacement u_i, v_i and rotations φ_i), these latter collected in the column matrix:

$$\mathbf{q} := \left(u_1\ v_1\ \varphi_1\ u_2\ v_2\ \varphi_2 \right)^T . \tag{7.162}$$

Accordingly:

$$u\,(x) = \psi_1\,(x)\, u_1 + \psi_2\,(x)\, u_2, \tag{7.163a}$$

$$v\,(x) = \psi_3\,(x)\, v_1 + \psi_4\,(x)\, \varphi_1 + \psi_5\,(x)\, v_2 + \psi_6\,(x)\, \varphi_2. \tag{7.163b}$$

Concerning the interpolation functions, polynomial of first and third degrees are taken here (whose graphs are plotted in Fig. 7.19):

---

[17] The critical load is approximated by excess.

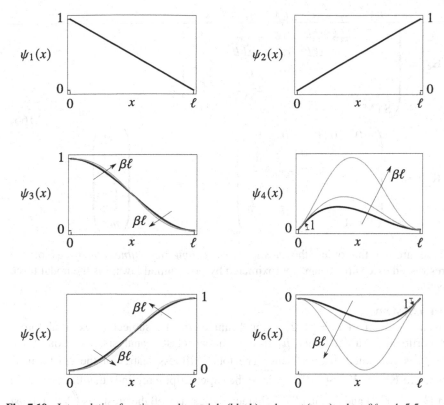

Fig. 7.19 Interpolation functions, polynomials (black) and exact (gray), when $\beta\ell = 4, 5.5$

$$\psi_1(x) = 1 - \frac{x}{\ell}, \qquad\qquad \psi_2(x) = \frac{x}{\ell}, \tag{7.164a}$$

$$\psi_3(x) = 1 - 3\left(\frac{x}{\ell}\right)^2 + 2\left(\frac{x}{\ell}\right)^3, \quad \psi_4(x) = \left[1 - 2\frac{x}{\ell} + \left(\frac{x}{\ell}\right)^2\right]x, \tag{7.164b}$$

$$\psi_5(x) = 3\left(\frac{x}{\ell}\right)^2 - 2\left(\frac{x}{\ell}\right)^3, \quad \psi_6(x) = \left[\left(\frac{x}{\ell}\right)^2 - \frac{x}{\ell}\right]x. \tag{7.164c}$$

Substituting Eqs. 7.163 and 7.164 in Eq. 7.161, performing integrations of the known functions, and requiring the TPE to be stationary with respect to the six free nodal displacements, the following equilibrium equation is drawn:

$$\left(\mathbf{K}_e + \mathbf{K}_g\right)\mathbf{q} = \mathbf{f}, \tag{7.165}$$

where:

$$
\mathbf{K}_e := \begin{pmatrix}
\frac{EA}{\ell} & 0 & 0 & -\frac{EA}{\ell} & 0 & 0 \\
 & \frac{12\,EI}{\ell^3} & \frac{6\,EI}{\ell^2} & 0 & -\frac{12\,EI}{\ell^3} & \frac{6\,EI}{\ell^2} \\
 & & \frac{4\,EI}{\ell} & 0 & -\frac{6\,EI}{\ell^2} & \frac{2\,EI}{\ell} \\
 & & & \frac{EA}{\ell} & 0 & 0 \\
 & & & & \frac{12\,EI}{\ell^3} & -\frac{6\,EI}{\ell^2} \\
\text{SYM} & & & & & \frac{4\,EI}{\ell}
\end{pmatrix},
$$

$$
\mathbf{K}_g := N_0 \begin{pmatrix}
0 & 0 & 0 & 0 & 0 & 0 \\
 & \frac{6}{5\ell} & \frac{1}{10} & 0 & -\frac{6}{5\ell} & \frac{1}{10} \\
 & & \frac{2\ell}{15} & 0 & -\frac{1}{10} & -\frac{\ell}{30} \\
 & & & 0 & 0 & 0 \\
 & & & & \frac{6}{5\ell} & -\frac{1}{10} \\
\text{SYM} & & & & & \frac{2\ell}{15}
\end{pmatrix}, \qquad
\mathbf{f} := \begin{pmatrix} f_{x1} \\ f_{y1} \\ m_1 \\ f_{x2} \\ f_{y2} \\ m_2 \end{pmatrix}.
$$

(7.166)

These are, in the order, the *elastic and geometric stiffness matrices* of the prestressed extensible beam, approximated by polynomial laws; $\mathbf{f}$ is the nodal force vector.

Beam System

Given a system of $e = 1, 2, \ldots, m$ beam elements, subject to external forces proportional to a multiplier μ, from a linear elastic analysis, the axial force $\mu N_0^{(e)}$ is computed in each beam. The total stiffness matrix of the eth beam is therefore $\mathbf{K}^{(e)} = \mathbf{K}_e^{(e)} + \mu \mathbf{K}_g^{(e)}$, where the superscript e reminds that the geometric $\left(A^{(e)}, I^{(e)}, \ell^{(e)}\right)$ and elastic $\left(E^{(e)}\right)$ characteristics, as well the prestress $\left(N_0^{(e)}\right)$, must actually be referred to the element e in question (although they appear without a superscript in Eq. 7.166).

These matrices link nodal forces and displacements, all expressed in the local basis. With the same technique used in the linear elastic analysis, it is necessary (a) to transform the matrices, to bring them back to the global basis, unique for all the elements; (b) to assembly these matrices according to the topology of the system, thus obtaining the *global stiffness matrix*; and (c) to introduce the external constraints, by canceling the rows and columns associated with the prevented displacements. Finally, an algebraic problem in the eigenvalue μ is obtained, whose smallest root is the critical load multiplier. The example that follows illustrates the procedure. Further examples are discussed in the Problems 14.2.5 and 14.2.6, where some considerations about convergence are made.

Remark 7.14 If the system includes *unprestressed springs* (of extensional and/or torsional type), these latter contribute to the elastic global stiffness matrix. This is obtained by adding the stiffness of the springs on the main diagonal of the beam system matrix, consistently with the ordering adopted for the configuration variables.

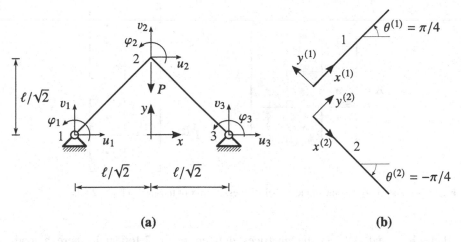

Fig. 7.20   Two-hinged arch: (a) static scheme, (b) bases and local coordinate systems

Example 7.4 (Critical Load of a Simple Frame)   The two-hinged arch in Fig. 7.20a is considered, consisting of two straight beams of equal length ℓ and equal elastic characteristics EA and EI, subject to a vertical concentrated load P applied at node 2.

This, according to the linear equilibrium analysis (i.e., by ignoring any deformations), induces compressive axial forces in the beams 1, 2, equal to:

$$N_0^{(1)} = N_0^{(2)} = -\frac{\sqrt{2}}{2} P. \tag{7.167}$$

In what follows, each of the two beams is modeled with a single finite element.

The (global) vector of configuration variables $\mathbf{q}$ consists of nine displacement components, three for each node, expressed in the global basis:

$$\mathbf{q} = (u_1, v_1, \varphi_1, u_2, v_2, \varphi_2, u_3, v_3, \varphi_3)^T. \tag{7.168}$$

Choosing a local basis for each beam (Fig. 7.20b), the total element stiffness matrices are:

$$\mathbf{K}^{(e)} = \mathbf{K}_e^{(e)} + \mu \, \mathbf{K}_g^{(e)}, \qquad e = 1, 2, \tag{7.169}$$

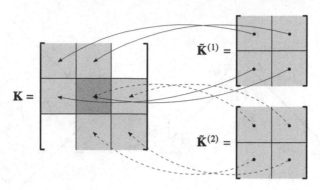

Fig. 7.21 Assembly scheme of the global stiffness matrix of the arch in Fig. 7.20

where $\mathbf{K}_e^{(e)}$ and $\mathbf{K}_g^{(e)}$ are the matrices defined in Eq. 7.166 and where a load multiplier μ (which only affects the geometric matrix) has been introduced. The stiffness matrices in the global basis are obtained as:

$$\tilde{\mathbf{K}}^{(e)} = \mathbf{R}^{(e)T} \mathbf{K}^{(e)} \mathbf{R}^{(e)}, \qquad e = 1, 2, \tag{7.170}$$

where:

$$\mathbf{R}^{(e)} := \begin{pmatrix} \cos\theta_e & \sin\theta_e & 0 & 0 & 0 & 0 \\ -\sin\theta_e & \cos\theta_e & 0 & 0 & 0 & 0 \\ 0 & 0 & 1 & 0 & 0 & 0 \\ 0 & 0 & 0 & \cos\theta_e & \sin\theta_e & 0 \\ 0 & 0 & 0 & -\sin\theta_e & \cos\theta_e & 0 \\ 0 & 0 & 0 & 0 & 0 & 1 \end{pmatrix}, \qquad e = 1, 2, \tag{7.171}$$

are element rotation matrices, and, moreover, $\theta_1 = \pi/4$ and $\theta_2 = -\pi/4$.

By assembling the two arrays $\tilde{\mathbf{K}}^{(e)}$, according to the scheme in Fig. 7.21, and imposing the constraints $u_1 = v_1 = u_3 = v_3 = 0$,[18] a system of algebraic equations is obtained, of the type:

$$\left(\bar{\mathbf{K}}_e + \mu\,\bar{\mathbf{K}}_g\right)\bar{\mathbf{q}} = \mathbf{0}. \tag{7.172}$$

By requiring that the determinant of the global stiffness matrix is equal to zero, $\mu = \frac{12\sqrt{2}\,EI}{P\ell^2}$ is determined, from which:

$$P_c = \frac{12\sqrt{2}\,EI}{\ell^2} \tag{7.173}$$

---

[18] That is, by deleting rows and columns corresponding to the suppressed displacements.

is evaluated. The corresponding critical mode is found to be antisymmetric.

The value in Eq. 7.173 of the critical load is grossly approximated, as the transverse deformation of the beams has been described as a cubic, rather than as a mixed linear-harmonic function. However, if the analysis is repeated by taking a total number of elements equal to six,

$$P_c = \frac{13.96 \, EI}{\ell^2},\tag{7.174}$$

is found, which is very close to the exact value $P_c = \frac{\sqrt{2}\pi^2 EI}{\ell^2} \cong \frac{13.94 \, EI}{\ell^2}$.[19]   □

7.12.2   Exact Finite Element

When polynomial interpolation functions, such as those in Eqs. 7.164, are adopted in the finite element technique, it follows that, while the differential equation 7.11a, which governs the longitudinal displacement, is punctually satisfied, the differential equation 7.12a, which governs the transverse displacement, is satisfied only *on average*, via the variational procedure. It is possible, however, to overcome this drawback by determining interpolation functions that *exactly* satisfy the equilibrium, as explained soon.

Exact Interpolation Functions
The transverse equilibrium of a beam, compressed by a load P, undergoing *prescribed displacements at the ends*, is governed by the field equation:

$$v'''' + \beta^2 v'' = 0, \qquad x \in (0, \ell),\tag{7.175}$$

where $\beta^2 := P/EI$, and by the non-homogeneous boundary conditions:

$$v(0) = v_1, \quad v'(0) = \varphi_1, \quad v(\ell) = v_2, \quad v'(\ell) = \varphi_2.\tag{7.176}$$

The solution of the boundary value problem is:

$$v(x) = \hat{\psi}_3(x)\, v_1 + \hat{\psi}_4(x)\, \varphi_1 + \hat{\psi}_5(x)\, v_2 + \hat{\psi}_6(x)\, \varphi_2,\tag{7.177}$$

---

[19] The exact critical load, due to the antisymmetry of the (inextensional) mode, which allows node 2 only to rotate, coincides with that of a simply supported beam of length ℓ, subject to an axial force $P/\sqrt{2}$.

where:

$$\hat{\psi}_3(x) = 1 - \hat{\psi}_5(x), \tag{7.178a}$$

$$\hat{\psi}_4(x) = x - \hat{\psi}_6(x) - \ell\,\hat{\psi}_5(x), \tag{7.178b}$$

$$\hat{\psi}_5(x) = \frac{\sin(\beta\ell)(\beta x - \sin(\beta x)) - (1 - \cos(\beta\ell))(1 - \cos(\beta x))}{\beta\ell\sin(\beta\ell) - 2(1 - \cos(\beta\ell))}, \tag{7.178c}$$

$$\hat{\psi}_6(x) = \frac{(\beta\ell - \sin(\beta\ell))(1 - \cos(\beta x)) - (1 - \cos(\beta\ell))(\beta x - \sin(\beta x))}{\beta^2\ell\sin(\beta\ell) - 2\beta(1 - \cos(\beta\ell))}, \tag{7.178d}$$

are the *exact interpolation functions* sought.

It is observed that functions in Eqs. 7.178 diverge to $\pm\infty$ when $\beta\ell$ takes singular values, the smallest of which is $(\beta\ell)^* = 2\pi$, that is, when P equals the critical load of the clamped-clamped beam of length ℓ, i.e., $P = P^* := \frac{4\pi^2 EI}{\ell^2}$. Since, in any other constraint condition, the critical load of a single element is smaller than P^*, the range of interest to explore for a member of the frame is $0 < \beta\ell < 2\pi$. Figure 7.19 reports the exact interpolation functions for some values of $\beta\ell$ in this range. It emerges that when $\beta\ell \to 0$, the exact functions tend to the polynomial ones, but when $\beta\ell$ is large enough, $\hat{\psi}_4(x)$ and $\hat{\psi}_6(x)$ remarkably differ from the homologous polynomials, while $\hat{\psi}_3(x)$ and $\hat{\psi}_5(x)$ remain almost unaltered.

Remark 7.15 The exact interpolation functions, unlike the polynomial ones, *depend on the axial force* through the parameter β, which appears as an argument of the circular functions.

Remark 7.16 Equations 7.178 also hold in case of tensile axial force, for which $\beta = i\,|\beta|$ is pure imaginary.[20] Since $\sin(i\,|\beta|\,\ell) = i\sinh(|\beta|\,\ell)$ and $\cos(i\,|\beta|\,\ell) = \cosh(|\beta|\,\ell)$, the expressions in Eqs. 7.178 can be checked to be real; they describe the response of the tensioned beam subject to displacements impressed at the nodes.

Equilibrium at Nodes

Given the displacement fields in Eqs. 7.177 and 7.163a, the following internal forces arise in the beam:

$$N(x) = EA\,u'(x), \tag{7.179a}$$

$$M(x) = EI\,v''(x), \tag{7.179b}$$

$$V(x) = -EI\,v'''(x) - P\,v'(x), \tag{7.179c}$$

---

[20] Here $i := \sqrt{-1}$ is the imaginary unit.

From them, it is possible to evaluate the external forces acting at nodes, dual of displacements:

$$f_{1x} = -N(0),$$ (7.180a)

$$f_{1y} = -V(0),$$ (7.180b)

$$m_1 = -M(0),$$ (7.180c)

$$f_{2x} = N(\ell),$$ (7.180d)

$$f_{2y} = V(\ell),$$ (7.180e)

$$m_2 = M(\ell).$$ (7.180f)

Such relations, in extended form, read:[21]

$$
\begin{pmatrix} f_{1x} \\ f_{1y} \\ m_1 \\ f_{2x} \\ f_{2y} \\ m_2 \end{pmatrix} =
\begin{pmatrix}
k_{11} & 0 & 0 & -k_{11} & 0 & 0 \\
 & k_{22} & k_{23} & 0 & -k_{22} & k_{23} \\
 & & k_{33} & 0 & -k_{23} & k_{36} \\
 & & & k_{11} & 0 & 0 \\
 & & & & k_{22} & -k_{23} \\
\text{SYM} & & & & & k_{33}
\end{pmatrix}
\begin{pmatrix} u_1 \\ v_1 \\ \varphi_1 \\ u_2 \\ v_2 \\ \varphi_2 \end{pmatrix}
=: \mathbf{Kq},
$$ (7.181)

where:

$$k_{11} := \frac{EA}{\ell}, \qquad k_{22} := \frac{EI\,\beta^3 \sin(\beta\ell)}{2 - 2\cos(\beta\ell) - \beta\ell\sin(\beta\ell)},$$

$$k_{23} := \frac{EI\,\beta^2 (1 - \cos(\beta\ell))}{2 - 2\cos(\beta\ell) - \beta\ell\sin(\beta\ell)}, \qquad k_{33} := \frac{EI\,\beta(\beta\ell\cos(\beta\ell) - \sin(\beta\ell))}{\beta\ell\sin(\beta\ell) + 2\cos(\beta\ell) - 2},$$

$$k_{36} := \frac{EI\,\beta(\beta\ell - \sin(\beta\ell))}{2 - 2\cos(\beta\ell) - \beta\ell\sin(\beta\ell)}.$$

(7.182)

The matrix $\mathbf{K}$ in Eq. 7.181 is the *exact stiffness matrix of the prestressed extensible beam.*

Beam System

By assembling the stiffness matrices and introducing the constraints, such as described above, an algebraic problem is drawn, which is linear in the displacements, but *transcendental in the eigenvalue* μ, as it appears as an argument of the matrix coefficients, i.e.:

---

[21] Alternatively, Eq. 7.181 can be inferred from the stationary of the TPE, as it was done for the polynomial finite element. Here, however, since the field equations are punctually satisfied, a more convenient procedure has been followed.

$$\bar{\mathbf{K}}(\mu)\,\bar{\mathbf{q}} = \mathbf{0}. \tag{7.183}$$

The characteristic equation $\det\left(\bar{\mathbf{K}}(\mu)\right) = 0$, also transcendental, is of the same type encountered in all examples exactly treated above.

References

1. Alfutov, N.A.: Stability of Elastic Structures. Springer, Berlin (2000)
2. Allen, H.G., Bulson, P.S.: Background to Buckling. McGraw-Hill, New York (1980)
3. Bazant, Z.P., Cedolin, L.: Stability of Structures: Elastic, Inelastic, Fracture, and Damage Theories. World Scientific Publishing, Singapore (2010)
4. Bleich, F.: Buckling Strength of Metal Structures. McGraw-Hill, New York (1952)
5. Brush, D.O., Almroth, B.O.: Buckling of Bars, Plates, and Shells. McGraw-Hill, New York (1975)
6. Chajes, A.: Principles of Structural Stability Theory. Prentice-Hall, New Jersey (1974)
7. Chen, W.F., Lui, E.M.: Structural Stability: Theory and Implementation. Elsevier, New York (1987)
8. Column Research Committee of Japan: Handbook of Structural Stability. Corona Publishing Company, Tokyo (1971)
9. Como, M.: Theory of Elastic Stability (in Italian). Liguori, Napoli (1967)
10. Corradi Dell'Acqua, L.: Meccanica delle strutture: La valutazione della capacità portante (in Italian), vol. 3. McGraw-Hill, Milano (1994)
11. Galambos, T.V., Surovek, A.E.: Structural Stability of Steel: Concepts and Applications for Structural Engineers. Wiley, Hoboken (2008)
12. Gerard, G.: Introduction to Structural Stability Theory. McGraw-Hill, New York (1962)
13. Hoff, N. J.: Buckling and stability. J. Roy. Aeron. Soc. **58** (1954)
14. Horne, M.R., Merchant, W.: The Stability of Frames. Pergamon Press, Oxford (1965)
15. Iyengar, N.G.R.: Elastic Stability of Structural Elements. Macmillan India, New Delhi (2007)
16. Kirby, P.A., Nethercot, D.A.: Design for Structural Stability. Granada Publishing, London (1979)
17. Luongo, A.: On the amplitude modulation and localization phenomena in interactive buckling problems. Int. J. Solids Struct. **27**(15), 1943–1954 (1991)
18. Luongo, A., Zulli, D.: Mathematical Models of Beams and Cables. Wiley, New York (2013)
19. Pignataro, M., Rizzi, N., Luongo, A.: Stability, Bifurcation and Postcritical Behaviour of Elastic Structures. Elsevier, Amsterdam (1990)
20. Potier-Ferry, M.: Foundations of elastic postbuckling theory. In: Buckling and Post-buckling, pp. 1–82. Springer, Berlin (1987)
21. Simitses, G.J.: An Introduction to the Elastic Stability of Structures. Prentice-Hall, New Jersey (1976)
22. Simitses, G.J., Hodges, D.H.: Fundamentals of Structural Stability. Elsevier, Burlington (2006)
23. Timoshenko, S.P., Gere, J.M.: Theory of Elastic Stability. McGraw-Hill, New York (1961)
24. Wang, C.M., Wang, C.Y., Reddy, J.N.: Exact Solutions for Buckling of Structural Members. CRC Press, Boca Raton (2005)
25. Yoo, C.H., Lee, S.: Stability of Structures: Principles and Applications. Elsevier, Amsterdam (2011)

Chapter 8
Elasto-Plastic Buckling of Planar Beam Systems

8.1 Introduction

When buckling occurs beyond the elastic limit, the relative treatment becomes extremely complex, for (a) algorithmic and (b) conceptual reasons. Regarding the former, the equations of equilibrium can no longer be derived from the stationary of the total potential energy (since this does not exist for a nonconservative system), but they must be determined by the direct method, or via the virtual work approach. Furthermore, the constitutive law implies a piecewise mechanical behavior, associated with successive plasticizations, which require evaluating the response in phases (or step-by-step), as it will become clear in the second part of the chapter.

Regarding the conceptual reasons, the static approach to the phenomenon of instability (also successfully applied to elastic systems) cannot be justified by the coincidence of the results offered by the energy criterion. An exception, albeit frequent, occurs when the stresses are monotonically increasing in any point of the body. In this case, indeed, since there are no elastic returns, the elasto-plastic system behaves as nonlinearly elastic, so that the tools already introduced in elasticity are still applicable. In the more general case, however, elastic unloadings occur, so that, aimed at limiting the (already complex) analysis to the static regime, it needs to invoke the *static criterion of stability* as an autonomous principle of stability, as discussed in the Chap. 1.[1] On the basis of this criterion, also called *of the adjacent equilibrium*, an elasto-plastic system loses stability when, as the load increases from

---

[1] A more rigorous approach to the stability of inelastic systems requires the use of thermodynamic principles, the discussion of which is beyond the limits of this book. Even in that case, however, the analysis can be traced back to the examination of the definiteness in sign of a tangent stiffness matrix (different for isothermal or adiabatic processes). This matrix, unlike the case of linear elasticity (i.e., not piecewise), depends on the "direction" of the incremental displacements, and, moreover, it can be non-symmetric (further details can be found, e.g., in [2]).

243
A. Luongo et al., *Stability and Bifurcation of Structures*,
https://doi.org/10.1007/978-3-031-27572-2_8

zero, the uniqueness of the solution to the fundamental mechanical problem (made of equilibrium, congruence, and constitutive law) is violated, i.e., when the tangent stiffness matrix, for the first time, becomes singular.

In this chapter, the elasto-plastic buckling of beams and frames is analyzed in the static field, consistently with the adjacent equilibrium criterion.

8.2   Elasto-Plastic Buckling of a Single Beam

The problem of loss of stability of a single compressed beam is dealt with here [1–3, 5, 6, 9, 10, 12, 14, 16], by referring to the classic treatments by Engesser [8], Von Kármán [15], and Shanley [13]. Two distinct theories are illustrated, and the relevant results are compared.

8.2.1   Tangent Elastic Modulus Theory

A beam made of hardening elasto-plastic material is considered, whose stress-strain law is represented in Fig. 8.1.

By denoting by σ_p the elastic limit stress (in which the proportionality limit and yield stresses are confused), it happens that if $|\sigma| \leq \sigma_p$, the material is in linear elastic phase, and if $|\sigma| > \sigma_p$, the material is in plastic phase. In the elastic phase, the Young modulus (equal to the slope of the straight line) is E; in the plastic phase, the *tangent elastic modulus* is defined as the slope of the tangent to the curve, i.e., $E_t(\sigma) := \frac{d\sigma}{d\varepsilon}$. The tangent elastic modulus is smaller than, or equal to, the Young modulus, i.e., $E_t(\sigma) \leq E$; moreover, unlike the latter, it is not constant, but it is a function of the stress σ.

When the beam, in its straight configuration, is compressed by an axial force P, a uniform stress $\sigma = \frac{P}{A}$ exists on the cross-section, of area A. If the stress has a modulus larger than σ_p, the beam still behaves as elastic, but with an apparent elastic modulus $E_t(\sigma)$, in place of the Young modulus E. Consequently, the Eulerian critical load changes to:

Fig. 8.1   Elasto-plastic hardening law, Young modulus E and tangent modulus E_t

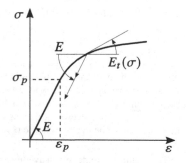

$$P_c = \pi^2 \frac{E_t\,(\sigma_c)\,I}{\ell_0^2}, \tag{8.1}$$

and the critical stress $\sigma_c = \frac{P_c}{A}$ becomes:

$$\sigma_c = \pi^2 \frac{E_t\,(\sigma_c)}{\lambda^2}, \tag{8.2}$$

where $\lambda := \frac{\ell_0}{\varrho}$ is the slenderness of the beam, and $\varrho := \sqrt{\frac{I}{A}}$ the inertia radius of the cross-section. The former is known as *Shanley critical stress*, though it was first introduced by Engesser.

It should be observed that Eq. 8.2 defines the critical stress in implicit way, as the unknown appears in both members of equation. Once the slenderness has been assigned, it needs to proceed by trial and error: (a) first, σ_c is estimated, from which the tangent modulus $E_t\,(\sigma_c)$ is computed through the constitutive law; (b) it is checked that, with this value of $E_t\,(\sigma_c)$, the right member of Eq. 8.2 provides the estimated critical stress; (c) if this does not happen, the value of the tangent modulus is updated with the new value of the critical stress, and the procedure is iterated, until convergence is reached. Alternatively, the values of the critical stress can be assigned and the corresponding slenderness determined (without iterations) from Eq. 8.2. In this way, the plot in Fig. 8.2 (curve E_t) is found, which corrects the Euler curve in the small slenderness range.

Remark 8.1 Since $E_t\,(\sigma_c)$ *does not* tend to zero as σ increases, the E_t curve in Fig. 8.2 tends to infinity when $\lambda \to 0$, as it happens in the elastic case (E curve). Of course, the curve must interrupted when the rupture limit stress of the material is reached.

Example 8.1 (Elasto-Perfectly-Plastic Law) As an example of application of the theory of the tangent elastic modulus, a beam is considered, made of elasto-perfectly-plastic material (piecewise linear curve in Fig. 8.3a):

Fig. 8.2 Critical stress *vs* slenderness; E, Euler curve; E_t, tangent elastic modulus theory; E_r, reduced elastic modulus theory

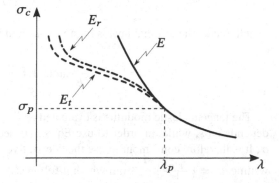

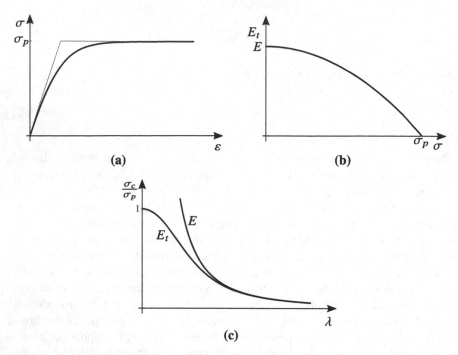

Fig. 8.3 Elasto-perfectly-plastic law: (**a**) regularization, (**b**) tangent modulus of elasticity as a function of stress, (**c**) critical stress *vs* slenderness (E_t curve), compared with the Euler law (E curve)

$$\sigma = \begin{cases} E\varepsilon, & \text{if} \quad |\varepsilon| \le \frac{\sigma_p}{E}, \\ \pm\sigma_p, & \text{if} \quad \varepsilon \gtrless \pm\frac{\sigma_p}{E}. \end{cases} \tag{8.3}$$

To make calculations easier, the law is regularized as follows (Fig. 8.3a):

$$\sigma = \sigma_p \tanh\left(\frac{E\varepsilon}{\sigma_p}\right), \tag{8.4}$$

which, unlike the original law, is monotonous and invertible, providing:

$$\varepsilon = \frac{\sigma_p}{E} \operatorname{arctanh}\left(\frac{\sigma}{\sigma_p}\right). \tag{8.5}$$

The tangent elastic modulus is evaluated as $E_t = \frac{d\sigma(\varepsilon)}{d\varepsilon}$; however, in this form, it depends on ε, while, in order to use Eq. 8.2, it needs to express it as a function of σ. It is therefore convenient to use the "derivative of the inverse function" theorem, writing $E_t = \left(\frac{d\varepsilon(\sigma)}{d\sigma}\right)^{-1}$, from which it follows:

$$E_t(\sigma) = E\left(1 - \left(\frac{\sigma}{\sigma_p}\right)^2\right). \tag{8.6}$$

Accordingly, the tangent elastic modulus parabolically decreases from E to 0, when σ grows from zero to σ_p (Fig. 8.3b). With this expression, Eq. 8.2 is written as:

$$\sigma_c = \pi^2 \frac{E}{\lambda^2}\left(1 - \left(\frac{\sigma_c}{\sigma_p}\right)^2\right). \tag{8.7}$$

In this simple case, the previous equation can be solved in closed-form (i.e., without iterations) with respect to σ_c, to give:

$$\frac{\sigma_c}{\sigma_p} = -\frac{\sigma_p}{E}\frac{\lambda^2}{2\pi^2} + \sqrt{1 + \left(\frac{\sigma_p}{E}\frac{\lambda^2}{2\pi^2}\right)^2}. \tag{8.8}$$

The graph of this function is plotted in Fig. 8.3c and compared with the Euler curve, evaluated with the elastic modulus E. Differently from hardening plasticity (Rmrk. 8.1), since $E_t(\sigma) \to 0$ as the stress increases, it turns out that $\sigma_c \to \sigma_p$ when $\lambda \to 0$. $\qquad\square$

8.2.2 Reduced Elastic Modulus Theory

The method of the tangent elastic modulus is susceptible of a well-founded objection. When the beam loses stability and buckles, it needs to distinguish the behavior of the compressed fibers (on the concave side of the beam) from that of the taut fibers (on the convex side of the beam). By taking into account that, along the fundamental path, all the fibers are equally compressed, the fibers at the concave side experience an *increment of compression*, described by the tangent elastic modulus E_t; the fibers on the convex side, in contrast, undergo a *reduction of compression*, i.e., an elastic unloading, described by the elastic modulus E (Fig. 8.1). Consequently, the relationship between the stress increment $\dot{\sigma}$ and the strain increment $\dot{\varepsilon}$ is:

$$\dot{\sigma} = \begin{cases} E_t\dot{\varepsilon}, & \text{if } \dot{\varepsilon} < 0, \\ E\dot{\varepsilon}, & \text{if } \dot{\varepsilon} > 0, \end{cases} \tag{8.9}$$

i.e., the material behaves like elastic, but *bi-modular*.

The generic cross-section of the beam is now considered, subject to a uniform compression and to an infinitesimal disturbance that induces bending. Said $\dot{\kappa}$ the incremental curvature and assuming that the cross-section remains planar, the elongation field is described by $\dot{\varepsilon} = \dot{\kappa}\,y_n$, where y_n is the oriented distance measured from the unknown neutral axis. Since the axial force is kept constant at $N_0 = -P$, the translational equilibrium of the cross-section requires that:

$$\dot{N} = \int_{\mathcal{A}} \dot{\sigma} \, dA = \int_{\mathcal{A}^+} E \, \dot{\kappa} \, y_n \, dA + \int_{\mathcal{A}^-} E_t \, \dot{\kappa} \, y_n \, dA = 0, \tag{8.10}$$

where $\mathcal{A}^+$ is the taut and $\mathcal{A}^-$ the compressed region and where use was made of Eq. 8.9. From the previous condition, the equation of the neutral axis follows:

$$E \, S_n^+ + E_t \, S_n^- = 0, \tag{8.11}$$

where $S_n^{\pm}$ are the static moments of the areas $\mathcal{A}^{\pm}$, evaluated with respect to the neutral axis.[2] Known the neutral axis, from the rotational equilibrium, one draws:

$$\dot{M} = \int_A \dot{\sigma} \, y_n \, dA = \int_{\mathcal{A}^+} E \, \dot{\kappa} \, y_n^2 \, dA + \int_{\mathcal{A}^-} E_t \, \dot{\kappa} \, y_n^2 \, dA, \tag{8.12}$$

from which the moment-curvature law reads:

$$\dot{M} = \left(E \, I_n^+ + E_t \, I_n^- \right) \dot{\kappa}, \tag{8.13}$$

where $I_n^{\pm}$ are the inertia moments of the areas $\mathcal{A}^{\pm}$ with respect to the neutral axis. This latter expression can more conveniently be rewritten as:

$$\dot{M} = E_r I_y \dot{\kappa}, \tag{8.14}$$

where:

$$E_r := \frac{E \, I_n^+ + E_t \, I_n^-}{I_y}, \tag{8.15}$$

is the *reduced elastic modulus* and I_y is the inertia moment of the entire section, *evaluated with respect to the centroidal axis*. It can be verified that $E_t < E_r < E$.

Since the law in Eq. 8.14 is formally analogous to that of a mono-modular material, it ensues that the critical stress of the bi-modular column is:

$$\sigma_c = \pi^2 \frac{E_r \, (\sigma_c)}{\lambda^2}, \tag{8.16}$$

that is, it is still given by Eq. 8.2 but with the reduced modulus in place of the tangent modulus. This formula is known as *Von Kármán critical stress*.[3]

---

[2] In other words, compared with the classic de Saint-Venant theory, here the $\mathcal{A}^{\pm}$ areas are "weighted" by their respective elastic moduli.

[3] This expression was later acknowledged by Engesser himself.

Fig. 8.4 Reduced elastic modulus of the ideal I-section

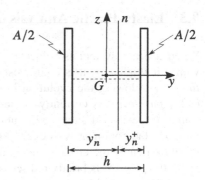

Proceeding as already described for the Shanley formula, the dotted line in Fig. 8.2 is obtained. This gives values only slightly higher than those previously found, so that, for practical purposes, the two formulas can be considered equivalent.[4]

Example 8.2 (Reduced Elastic Modulus of the Ideal I-Section) An ideal I-section is considered, consisting of two flanges, each of $A/2$ area, placed at a distance h, connected by a web of evanescent thickness (Fig. 8.4).

By denoting, respectively, by y_n^+ and $y_n^- = h - y_n^+$ the distances of the tensed and compressed flanges from the unknown neutral axis, Eq. 8.11 is written as:

$$\frac{A}{2}\left[Ey_n^+ - E_t\left(h - y_n^+\right)\right] = 0, \tag{8.17}$$

from which it follows:

$$y_n^+ = \frac{E_t}{E + E_t}h, \qquad y_n^- = \frac{E}{E + E_t}h. \tag{8.18}$$

The inertia moments of the two flanges, calculated with respect to the neutral axis, are $I_n^{\pm} = \frac{A}{2}y_n^{\pm 2}$, while the inertia moment of the whole cross-section with respect to the centroidal axis is $I_y = A\left(\frac{h}{2}\right)^2$. The reduced elastic modulus is derived from Eq. 8.15 as:

$$E_r = \frac{2E\,E_t}{E + E_t}. \tag{8.19}$$

□

---

[4] The Shanley formula, although obtained via not completely correct arguments, furnishes values closer to experimental results than the Von Kármán formula. This apparently strange circumstance is explained in literature with the fact that, in experimental tests, the load is *not* kept constant, but it is monotonously increasing. Therefore, small increments of P balance the decompression of the fibers induced by bending, so that the module E_t turns out to be sufficiently representative.

8.3   Elasto-Plastic Analysis of Beam Systems

When a beam system is subjected to forces, which (partly or totally) increase with a multiplier μ, it is of interest to determine its response *beyond the elastic limit*, by following the evolution in steps, up to the collapse. This analysis (also called *push-over*) is currently carried out to determine the nonlinear response of frames (especially if subjected to horizontal forces simulating the seismic action), in which the nonlinearity is exclusively of constitutive type, as consequent to the plasticization of the cross-sections, according to the idealization of plastic hinge. Here, the problem is briefly rediscussed in the context of the linearized (or second order) theory, in which the geometric effects are taken into account, in addition to the constitutive ones [4, 5]. The problem should therefore be framed as of interaction between *plasticity* and *instability* [7, 11].

8.3.1   Geometric Effects on the Elasto-Plastic Response of Planar Frames

A planar frame is considered, subject to external forces **p**, quasi-statically increasing with a multiplier μ (possibly keeping constant some of the forces, e.g., the vertical ones). A *push-over* curve $\mu = \mu(q)$, extended up to the collapse of the structure, is sought for, relating the load multiplier μ to a significant displacement component q (Fig. 8.5).

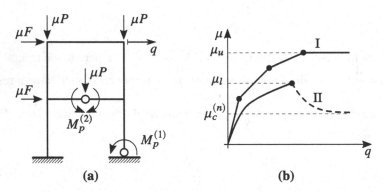

(a)   **(b)**

Fig. 8.5   Elasto-plastic response of a frame: (**a**) static scheme evolution; (**b**) response curve, (I) first order, (II) second order

As commonly accepted, it is assumed that (a) the beam is prevalently solicited in bending, and (b) the moment-curvature law $M = M(\kappa)$, at a generic abscissa of the beam, is elasto-perfectly-plastic:[5]

$$M = \begin{cases} EI\,\kappa, & \text{if} \quad |\kappa| \leq \frac{M_p}{EI}, \\ \pm M_p, & \text{if} \quad \kappa \gtrless \pm\frac{M_p}{EI}, \end{cases} \tag{8.20}$$

where M_p is the *plastic moment* of the cross-section. The mechanism leading the frame to collapse is related to the occurrence of successive plastic hinges, calling for a step-by-step analysis. In each of these steps, the static scheme of the structure evolves from that entirely elastic (here named *system-0*) to one modified from the birth of $k = 1, 2, \ldots$ plastic hinges (here named *system-k*). In this study, however, the geometric effects can be ignored (according to the first order analysis), as well included (i.e., in the second order analysis), as discussed immediately.

First Order Elasto-Plastic Analysis

By reducing the system to a finite number of degrees of freedom, for example, via a finite element discretization, the problem is initially governed by the elastic stiffness matrix $\mathbf{K}_e^{(0)}$ of the intact structure (system-0), loaded by the external forces $\mu\mathbf{p}$, i.e.:

$$\mathbf{K}_e^{(0)}\mathbf{q}^{(0)} = \mu\,\mathbf{p}, \tag{8.21}$$

where $\mathbf{q}^{(0)}$ are the initial Lagrangian parameters. As μ increases, a first plastic hinge forms at a $S^{(1)}$ cross-section, where the bending moment is "frozen" to its plastic (in module) value $M_p^{(1)}$. The static scheme, consequently, must be modified to account for the new hinge and the degrees of freedom increased in number by 1, now collected in the vector $\mathbf{q}^{(1)}$ (Fig. 8.5a). The new system-1 is governed by the updated stiffness matrix $\mathbf{K}_e^{(1)}$; moreover, it is loaded by additional nodal forces $\mathbf{b}^{(1)}$, represented by the couple(s) $M_p^{(1)}$(at the ground or internal), so that the equilibrium equations read:

$$\mathbf{K}_e^{(1)}\mathbf{q}^{(1)} = \mu\,\mathbf{p} - \mathbf{b}^{(1)}. \tag{8.22}$$

When, continuing the loading process, a second plastic hinge forms at a $S^{(2)}$ cross-section, the bending moment is there frozen to its plastic value $M_p^{(2)}$, and the static scheme must be further updated, to take into account an additional degree

---

[5] The phase of progressive plasticization of the cross-section, from the outermost to the innermost fibers, is ignored. Furthermore, reference is made to symmetric sections, for which the plastic moment does not depend from the sign of the solicitation.

of freedom and a new load, represented by $M_p^{(2)}$. After n plastic hinges have been formed (system-n), the equilibrium equations are:

$$\mathbf{K}_e^{(n)} \mathbf{q}^{(n)} = \mu \, \mathbf{p} - \sum_{k=1}^{n} \mathbf{b}^{(k)}, \tag{8.23}$$

The process stops when the system becomes kinematically undetermined (also said labile), due to the presence of an excessive number (or mispositioning) of the plastic hinges. The *ultimate* (or collapse) *value* μ_u of the multiplier is that for which $\det \left(\mathbf{K}_e^{(n)} \right) = 0$.

In the loading process, given the reduction in the number of constraints, the overall stiffness of the system decreases stepwise. Therefore, the load-displacement curve $\mu = \mu(q)$ is a piecewise linear curve, made of segments whose slope monotonously decreases to zero, when the ultimate load is reached (Fig. 8.5b).

Second Order Elasto-Plastic Analysis

The analysis previously carried out *does not* take into account that, as the static scheme evolves, the structure undergoes displacements. Consistently with the spirit of linear theory, the geometry of the structure is kept frozen to the initial one, and any geometric nonlinearity is ignored. In the framework of the linearized theory, instead, infinitesimal changes of geometry must be taken into account, under the assumption that equilibrium is achieved in a configuration adjacent to the reference one. These effects, called geometric, are relevant when the beams are strongly compressed, and may lead to results very different from those of the linear theory.

According to the linearized theory, the previously described equilibrium equations should be modified as follows:[6]

$$\left(\mathbf{K}_e^{(n)} + \mu \, \mathbf{K}_g^{(n)} \right) \mathbf{q}^{(n)} = \mu \, \mathbf{p} - \sum_{k=1}^{n} \mathbf{b}^{(k)}, \tag{8.24}$$

where $\mathbf{K}_g^{(n)}$ is the geometric stiffness matrix of the structure modified by the formation of n plastic hinges. Since the multiplier μ now also appears in the left member, *the response in each step is nonlinear*. Moreover, if the geometric effects are destabilizing, the response, under the same load, is amplified with respect that of the linear theory (Fig. 8.5b).

---

[6] Here, for simplicity, it is assumed that all forces are affected by a multiplier μ; later, in the Sect. 8.3.2, an example will be worked out in which only *some* of the forces proportionally increase with the multiplier. The Rmrk. 8.2 gives some hints in this regard.

The most interesting aspect of the problem, however, is the following. Equations 8.24 govern the response of the current system-n, subject to load "imperfections" $\mu\,\mathbf{p}$ and (formal) geometric imperfections $\mathbf{b}^{(k)}$, as discussed in the Sect. 6.5. If the imperfections are ignored, a critical load $\mu_c^{(n)}$ is determined, as the minimum eigenvalue of the associated homogeneous problem, i.e., for which $\det\left(\mathbf{K}_e^{(n)} + \mu_c^{(n)}\mathbf{K}_g^{(n)}\right) = 0$. The following cases may occur:

- If $\mu < \mu_c^{(n)}$, the response to the non-homogeneous problem follows the (approximate) amplification rule (Eq. 7.41), which reads:

$$\mathbf{q}^{(n)} = \frac{1}{1 - \dfrac{\mu}{\mu_c^{(n)}}}\left[\mathbf{K}_e^{(n)}\right]^{-1}\left(\mu\mathbf{p} - \sum_{k=1}^{n}\mathbf{b}^{(n)}\right). \tag{8.25}$$

It is, therefore, a portion of the imperfect response curve of the system-n, which approaches *from below* the horizontal asymptote $\mu_c^{(n)}$. The structure, however, does not generally reaches it, since a new plasticization occurs.

- If $\mu > \mu_c^{(n)}$, the previous expression still holds; it describes a curve that tends *from above* to the same asymptotic value $\mu_c^{(n)}$.[7] However, since this is an unstable branch, the system collapses at a limit point $\mu = \mu_l$, upon the formation of the $(n-1)$th plastic hinge.[8]

By summarizing, if the geometric effects are taken into account, the *structure collapses when the load multiplier exceeds for the first time the critical load of the current structure* (which progressively reduces, as plasticization progresses). If that doesn't happen for any of the current structures, the collapse occurs via labilization of the system, as in the first order theory, but generally at a smaller value of the multiplier.

Remark 8.2 It is also of interest the case in which a part of the external forces *does not* increase with the multiplier but remains constant. This circumstance occurs, for example, when a frame is subjected to constant vertical loads, while the horizontal forces quasi-statically increase. If the (small) variations of axial forces, induced by the horizontal forces, is ignored, the geometric stiffness matrix in Eq. 8.24, depending on the vertical forces only, is not affected by the μ factor. The push-over curve is therefore piecewise linear, as in the first order analysis, but the slopes of the segments are smaller, due to the destabilizing effects induced by the compression. Collapse is achieved, either: (a) by labilization, or (b) by overcoming the critical load of the current structure. The following example illustrates this circumstance.

---

[7] These curves were named *non-natural*, in discussing imperfect systems (Sect. 6.5). The comment on Fig. 7.4 should also be remembered.

[8] Here the wording "limit point" is kept, as the load reaches there a maximum, although the tangent to the path is not horizontal.

8.3.2   Column Subjected to a Constant Compression and Monotonically Increasing Transverse Forces

As an example of the theory developed in the previous section, a column AB is considered, of length ℓ and stiffness EI, clamped at both ends, subject to a vertical compression force P *held constant*, and to a horizontal force F, applied at a point C located at 2/3 of the height, quasi-statically increasing (Fig. 8.6a). The goal of the analysis is to build up the push-over curve $F = F(v_C)$,[9] with v_C the displacement of C, taken positive if concordant with the force.

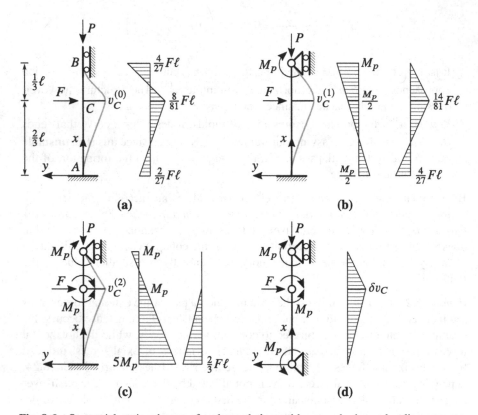

Fig. 8.6 Sequential static schemes of a clamped-clamped beam and relevant bending moment diagrams: (**a**) elastic column (system-0), (**b**) column plasticized at the head (system-1), (**c**) column plasticized at the head and in the span (system-2), (**d**) collapse kinematics

---

[9] Since the force is unique, it is not necessary to introduce a multiplier.

First Order Push-Over Response

The vertical force P is initially ignored, and the analysis is performed according to the linear theory. As the force F increases, the following sequential systems are distinguished, illustrated in Fig. 8.6:

- The column is entirely elastic, so that the static scheme is that of a clamped-sliding beam (system-0); the relative bending moment diagram is shown in Fig. 8.6a; the deflection at C is found to be:[10,11]

$$v_C^{(0)} = \frac{8}{2187} \frac{F\ell^3}{EI};$$ (8.26)

- The column plasticized at the top, so that the static scheme is that of a clamped-rolling beam (system-1); the bending moment diagram is the sum of that associated to the plastic moment M_p, acting at the head, and that equilibrated with the force F, shown in Fig. 8.6b; the deflection at C is equal to:

$$v_C^{(1)} = \frac{20}{2187} \frac{F\ell^3}{EI} - \frac{1}{27} \frac{M_p\ell^2}{EI};$$ (8.27)

- The column plasticized both at the head and at the point C, so that the static scheme is that of a cantilever beam in series with a simply supported beam (system-2); the bending moment diagram is the sum of (a) that induced by the three couples M_p, acting at the two plastic hinges (one, external, at the head, and one, internal, in the span), as well as (b) by the force F, both represented in Fig. 8.6c; the deflection at C is equal to:

$$v_C^{(2)} = \frac{8}{81} \frac{F\ell^3}{EI} - \frac{22}{27} \frac{M_p\ell^2}{EI};$$ (8.28)

- Successively, the structure collapses upon the formation of the third plastic hinge at the foot (Fig. 8.6d); the collapse kinematics is also indicated in this figure.

---

[10] The displacement $v_C^{(k)}$ of the point C, in the step k, is conveniently computed using the Virtual Works Principle, that provides:

$$v_C^{(k)} = \int_0^\ell M^{(k)}(x) M'(x) \frac{dx}{EI}$$

Here, $M^{(k)}(x)$ is the bending moment law in phase k, and $M'(x)$ *any* bending moment law which is equilibrated with a transverse force of unit intensity, $F_C' = 1$, applied at C and acting in the same sense of the unknown displacement. For example, one can take as $M'(x)$ the bending moment law of the simply supported AB beam, or that of the clamped-free AB beam.

[11] The displacements v_C, negative according to the axes in Fig. 8.6a, have instead been indicated in module in Eq. 8.26 and successive equations.

The step 0 ends when the first plastic hinge is formed, i.e., when the bending moment at the head, $-\frac{4}{27}F\ell$, equates the plastic moment $-M_p$; the corresponding force and displacement are:

$$F_1 := \frac{27}{4}\frac{M_p}{\ell} = 6.75\frac{M_p}{\ell}, \qquad v_{C1} := \frac{2}{81}\frac{M_p\ell^2}{EI} \simeq 0.025\frac{M_p\ell^2}{EI}, \qquad (8.29)$$

where v_{C1} denotes $v_C^{(0)}$ when $F = F_1$.

The step 1 ends when the second plastic hinge is formed, i.e., when the moment at C, $\frac{14}{81}F\ell - \frac{M_p}{2}$, equals the ultimate moment M_p; the force and displacement, correspondingly, are:

$$F_2 := \frac{243}{28}\frac{M_p}{\ell} \simeq 8.68\frac{M_p}{\ell}, \qquad v_{C2} := \frac{8}{189}\frac{M_p\ell^2}{EI} \simeq 0.042\frac{M_p\ell^2}{EI}, \qquad (8.30)$$

where v_{C2} denotes $v_C^{(1)}$ when $F = F_2$.

Collapse occurs when, at the end of step 2, the bending moment at foot, $-\frac{2}{3}F\ell + 5M_p$, equates the ultimate moment $-M_p$; the force and displacement, correspondingly, are:

$$F_u := F_3 := 9\frac{M_p}{\ell}, \qquad v_{C3} := \frac{2}{27}\frac{M_p\ell^2}{EI} \simeq 0.074\frac{M_p\ell^2}{EI}, \qquad (8.31)$$

where v_{C3} denotes $v_C^{(2)}$ when $F = F_3$.

Finally, reporting the points of coordinates (F_k, v_{Ck}) on the (F, v_C) plane and joining them with segments, the desired push-over curve is obtained (Fig. 8.7a). The last segment, horizontal, indicates labilization of the column.

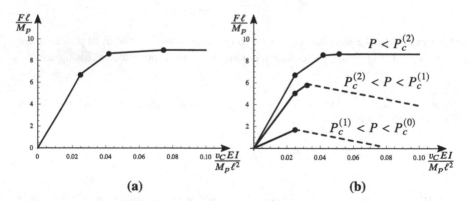

Fig. 8.7   Push-over curve for the clamped-clamped column: (**a**) first order analysis, (**b**) second order analysis, with $\frac{P\ell^2}{\pi^2 EI} = 0.05, 1, 3$

Remark 8.3 The value F_u of the ultimate force can also be obtained by applying the Virtual Work Principle to the collapse scheme represented in Fig. 8.6d. Said δv_C the virtual displacement of C, the lower member rotates of an angle (positive counterclockwise) $-\frac{3}{2}\frac{\delta v_C}{\ell}$, and the upper member of $+3\frac{\delta v_C}{\ell}$. By requiring that the sum of the virtual works, expended by the four couples M_p and by the force F_u, be equal to zero for any kinematically admissible displacement, it follows:

$$2M_p\left(-\frac{3}{2}\frac{\delta v_C}{\ell}\right) - 2M_p\left(3\frac{\delta v_C}{\ell}\right) + F_u\delta v_C = 0, \qquad \forall \delta v_C, \qquad (8.32)$$

from which the value of F_u, already determined, is drawn.

Second Order Push-Over Response

The geometric effects induced by the force P, hitherto neglected, are now taken into account. Preliminary, the critical loads $P_c^{(k)}$ $(k = 0, 1, 2)$ of the sequential systems-k, describing the elasto-plastic evolution of the column, are evaluated. They are:

$$P_c^{(0)} = 4\frac{\pi^2 EI}{\ell^2}, \qquad (8.33a)$$

$$P_c^{(1)} = 2.05\frac{\pi^2 EI}{\ell^2}, \qquad (8.33b)$$

$$P_c^{(2)} = 0.213\frac{\pi^2 EI}{\ell^2}. \qquad (8.33c)$$

As a matter of fact, $P_c^{(0)}$ is the critical load of the clamped-clamped beam (Fig. 8.6a, Table 7.1); $P_c^{(1)}$ is the critical load of the clamped-rolling beam (Fig. 8.6b, Table 7.1); $P_c^{(2)}$ is the critical load of the cantilever beam AB, connected in series with a rigid rod BC,[12] (Fig. 8.6c, using the results of the Example 7.2, and adapting the length to $\frac{2}{3}\ell$). Equations 8.33 show how the critical load decreases as plasticization progresses, that is, $P_c^{(0)} > P_c^{(1)} > P_c^{(2)}$.

When the axial force P acts on the column, the first order bending moments $M^{(k)}(x)$ and the first order deflections $v_C^{(k)}$ are (approximately) magnified by a factor

---

[12] The CB elastic beam can also become unstable, but at a value of the load, $P_c = \frac{9\pi^2 EI}{\ell^2}$, which is much higher than $P_c^{(2)}$; consequently, the CB beam behaves like a rigid rod.

$$\alpha^{(k)} := \frac{1}{1 - \frac{P}{P_c^{(k)}}}, \tag{8.34}$$

according to the amplification rule. Therefore:

$$\tilde{M}^{(k)}(x) = \alpha^{(k)} M^{(k)}(x), \qquad \tilde{v}_C^{(k)} = \alpha^{(k)} v_C^{(k)}, \tag{8.35}$$

where a tilde denotes a second order magnitude. By repeating the previous analysis, the updated values of the force $\tilde{F}_{k+1}$ and deflection $\tilde{v}_{Ck+1}$, relevant to the formation of the kth plastic hinge, are determined. These values are derived from the following conditions:

$$-\frac{4}{27}\alpha^{(0)}\tilde{F}_1\ell = -M_p, \qquad \tilde{v}_{C1} = \alpha^{(0)}\frac{8}{2187}\frac{\tilde{F}_1\ell^3}{EI}, \tag{8.36a}$$

$$\alpha^{(1)}\left(\frac{14}{81}\tilde{F}_2\ell - \frac{M_p}{2}\right) = M_p, \qquad \tilde{v}_{C2} = \alpha^{(1)}\left(\frac{20}{2187}\frac{\tilde{F}_2\ell^3}{EI} - \frac{1}{27}\frac{M_p\ell^2}{EI}\right), \tag{8.36b}$$

$$\alpha^{(2)}\left(-\frac{2}{3}\tilde{F}_3\ell + 5M_p\right) = -M_p, \qquad \tilde{v}_{C2} = \alpha^{(2)}\left(\frac{8}{81}\frac{\tilde{F}_3\ell^3}{EI} - \frac{22}{27}\frac{M_p\ell^2}{EI}\right). \tag{8.36c}$$

The push-over curves shown in Fig. 8.7b are thus obtained, each corresponding to a different value of the axial force (and, therefore, to a different set of factors $\alpha^{(k)}$). All the curves are piecewise linear; a negative slope indicates the overcoming of the critical load of the current system (for which $\alpha < 0$). The curves exemplify the cases discussed ahead, namely:

- $P < P_c^{(2)}$, i.e., the axial force is smaller than the smallest critical load. The phenomenon of instability does not manifest itself, so that the structure collapses for labilization, as in the first order theory. However, the geometric effects make the column more flexible, so that the push-over curve is lower.[13]
- $P_c^{(2)} < P < P_c^{(1)}$. The system-0 evolves into the system-1, as in the previous case, but with larger deformations, due to the increased geometric effect. When, however, system-1 changes into system-2, i.e., upon the formation of the second plastic hinge, since the axial force is larger than the current critical load, a collapse due to instability occurs (i.e., the push-over curve assumes a negative slope).

---

[13] In particular, it is $\tilde{F}_u := \tilde{F}_3 = \frac{3(1+5\alpha^{(2)})}{2\alpha^{(2)}}\frac{M_p}{\ell}$, which, for $\alpha^{(2)} > 1$, is smaller than $F_u = F_3 := 9\frac{M_p}{\ell}$.

- $P_c^{(1)} < P < P_c^{(0)}$. The collapse for instability occurs when the first plastic hinge appears.

The case $P_c^{(0)} < P$ is of no interest here, as instability occurs in the elastic regime.

8.3.3   Elastic Beam with Elasto-Plastic Bracing

In the previous section, a beam which undergoes plasticization under transverse loads was considered. Now, a different system is tackled, consisting of a braced beam, in which plasticization only affects the bracing, the beam remaining in the elastic regime during the whole loading process. The analysis is directly carried out at the second order, solving in *exact way* the relative equilibrium equations, thus renouncing to the simplified procedure (based on the amplification of the first order solution).

Elasto-Plastic Evolution of the Structure

A clamped-free column is considered, elastically constrained by an extensional spring (acting as a tie rod) applied at the free end (Fig. 8.8a). The column is subjected to a uniformly distributed transverse load $-\mu p$ and to a compression force of modulus μP, with μ a multiplier (which, differently from the problem treated in the Sect. 8.3.2, affects all forces, transverse and longitudinal).

The following elasto-perfectly-plastic law is assumed for the spring:

$$N = \begin{cases} k\, v\,(\ell)\,, & \text{if} \quad |v\,(\ell)| \leq \frac{N_p}{k}, \\ \pm N_p, & \text{if} \quad v\,(\ell) \lessgtr \pm \frac{N_p}{k}, \end{cases} \tag{8.37}$$

relating the axial force N to the elongation $-v\,(\ell)$; here, N_p is the plastic axial force.

For low values of the load multiplier, the structure behaves as fully elastic (system-0, Fig. 8.8a); however, when μ reaches a specific value μ_1, for which $N = N_p$, the spring plasticizes, and the structure evolves into a cantilever beam, loaded at its end by a constant transverse force N_p (system-1, Fig. 8.8b).

The critical loads of the two sequential systems are:

- *System-0*: the critical load has already been determined in the Sect. 7.8.1; by solving the relevant characteristic Eq. 7.103, i.e.:

$$\tan\,(\beta\ell) = (\beta\ell)\left[1 - \frac{(\beta\ell)^2}{\eta}\right], \tag{8.38}$$

Fig. 8.8 Compressed
column subjected to
transverse distributed loads,
constrained by an
elasto-plastic spring: (**a**)
elastic system (system-0); (**b**)
column with plasticized
spring (system-1)

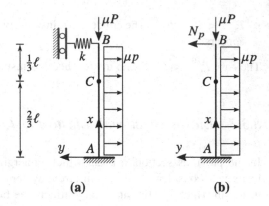

(**a**)   (**b**)

where $\eta := \frac{k\ell^3}{EI}$ is a nondimensional stiffness parameter and $\beta\ell$ the eigenvalue,
$\mu_c^{(0)} = (\beta\ell)^2 \frac{EI}{P\ell^2}$ is drawn. If, for example, $\eta = 10$, then $\beta\ell = 3.16$; hence, the
critical multiplier is $\mu_c^{(0)} = 1.01 \frac{\pi^2 EI}{P\ell^2}$.

- *System*-1: the critical multiplier of the cantilever beam is $\mu_c^{(1)} = \frac{1}{4} \frac{\pi^2 EI}{P\ell^2}$.

Response to Transverse Loads

The equilibrium equation of the beam, assumed to remain elastic in the whole range
of interest, is:

$$EI\, v'''' + \mu\, P\, v'' = -\mu\, p, \qquad x \in (0, \ell)\,. \tag{8.39}$$

This differential equation holds in the elastic ($k = 0$) as well as in the plastic ($k = 1$)
phase of the spring. Its general solution is:

$$v^{(k)}(x) = c_1^{(k)} \cos\left(\beta\,(\mu)\,x\right) + c_2^{(k)} \sin\left(\beta\,(\mu)\,x\right) + c_3^{(k)}\, x + c_4^{(k)} - \frac{p\, x^2}{2\, P}, \tag{8.40}$$

where $\beta\,(\mu) := \sqrt{\frac{\mu\, P}{EI}}$. The boundary conditions, however, differ in the two phases.
For system-0 they read (superscript omitted):

$$v\,(0) = 0, \qquad\qquad\qquad\qquad v'\,(0) = 0, \tag{8.41a}$$

$$EI\, v''\,(\ell) = 0, \qquad -EI\, v'''\,(\ell) - \mu P\, v'\,(\ell) + k\, v\,(\ell) = 0, \tag{8.41b}$$

and for system-1 they are (superscript omitted):

$$v\,(0) = 0, \qquad\qquad\qquad\qquad v'\,(0) = 0, \tag{8.42a}$$

$$EI\, v''\,(\ell) = 0, \qquad -EI\, v'''\,(\ell) - \mu P\, v'\,(\ell) - N_p = 0. \tag{8.42b}$$

Concerning system-0, by enforcing Eqs. 8.41, a non-homogeneous algebraic system in the four unknowns follows:

$$
\begin{pmatrix}
1 & 0 & 0 & 1 \\
0 & \beta & 1 & 0 \\
-\beta^2 C & -\beta^2 S & 0 & 0 \\
\frac{k}{EI}C & \frac{k}{EI}S & \frac{k\ell}{EI} - \beta^2 & \frac{k}{EI}
\end{pmatrix}
\begin{pmatrix}
c_1^{(0)} \\
c_2^{(0)} \\
c_3^{(0)} \\
c_4^{(0)}
\end{pmatrix}
=
\begin{pmatrix}
0 \\
0 \\
\frac{p}{P} \\
\frac{p\ell}{2P}\left(\frac{k\ell}{EI} - 2\beta^2\right)
\end{pmatrix},
\tag{8.43}
$$

where $C := \cos(\beta\ell)$ and $S := \sin(\beta\ell)$, which, solved, provides:

$$
c_1^{(0)} = -c_4^{(0)} = \frac{p}{2P\beta^2}\,\frac{2(\beta\ell)^3(\beta\ell S - 1) + 2\eta\beta\ell - \eta S\left[2 + (\beta\ell)^2\right]}{(\beta\ell)^3 C - \beta\ell C\eta + \eta S},
\tag{8.44a}
$$

$$
c_2^{(0)} = -\frac{1}{\beta}c_3^{(0)} = \frac{p}{2\beta^2 P}\,\frac{\eta\left[C\left(2 + (\beta\ell)^2\right) - 2\right] - 2C(\beta\ell)^4}{(\beta\ell)^3 C - \beta\ell C\eta + \eta S}.
\tag{8.44b}
$$

This solution is valid until $\mu \leq \mu_1$, where μ_1 is the first yielding load. To determine this limit, $v(\ell) = -\frac{N_p}{k}$ must be satisfied, i.e.:

$$
c_1^{(0)}\cos(\beta(\mu_1)\ell) + c_2^{(0)}\sin(\beta(\mu_1)\ell) + c_3^{(0)}\ell + c_4^{(0)} - \frac{p\,\ell^2}{2\,P} = -\frac{N_p}{k},
\tag{8.45}
$$

from which μ_1 is drawn.

Concerning system-1, by imposing Eqs. 8.42, the following system of equations is obtained:

$$
\begin{pmatrix}
1 & 0 & 0 & 1 \\
0 & \beta & 1 & 0 \\
-\beta^2 C & -\beta^2 S & 0 & 0 \\
0 & 0 & -\beta^2 & 0
\end{pmatrix}
\begin{pmatrix}
c_1^{(1)} \\
c_2^{(1)} \\
c_3^{(1)} \\
c_4^{(1)}
\end{pmatrix}
=
\begin{pmatrix}
0 \\
0 \\
\frac{p}{P} \\
\frac{N_p}{EI} - \frac{p\ell}{P}\beta^2
\end{pmatrix},
\tag{8.46}
$$

from which the constants are evaluated as:

$$
c_1^{(1)} = -c_4^{(1)} = \left(\frac{p\ell}{\beta P} - \frac{N_p}{EI\beta^3}\right)\tan(\beta\ell) - \frac{p}{\beta^2 P\cos(\beta\ell)},
\tag{8.47a}
$$

$$
c_2^{(1)} = -\frac{1}{\beta}c_3^{(1)} = \frac{N_p}{EI\beta^3} - \frac{p\ell}{\beta P}.
\tag{8.47b}
$$

This solution holds for $\mu > \mu_1$.

Second Order Push-Over Curve

The displacement $v_C := v\left(\frac{2\ell}{3}\right)$ is taken as descriptor of the structural response. Then, by making use of Eq. 8.40, the push-over curve $\mu = \mu\,(v_C)$ is derived, as represented in Fig. 8.9. Two cases can occur, discussed ahead.

1. *'Weak' spring*, for which $\mu_1 < \mu_c^{(1)}$ (Fig. 8.9a). The response develops initially along the imperfect path of system-0 (branch 0, having asymptote $\mu = \mu_c^{(0)}$), the imperfection being represented by the distributed load p.[14] When $\mu = \mu_1$, the spring plasticizes, and the structure evolves into a cantilever beam, disturbed from two forces: the load p and the force N_p (having effects of opposite signs on the displacement). The representative point, consequently, moves along the branch 1, which has the asymptote $\mu = \mu_c^{(1)}$. This branch originates from a positive value of v_C, caused by the force N_p alone, being $\mu = 0$; however, when μ is large enough, v_C changes sign, due to the effect of the distributed load. The structure collapses at $\mu = \mu_c^{(1)}$.

2. *'Strong' spring*, for which $\mu_1 \geq \mu_c^{(1)}$ (Fig. 8.9b). The response develops initially along the imperfect path of the system-0, already commented. When $\mu = \mu_1$, the spring plasticizes, and the structure evolves into the system-1. Here, since the compression load is higher than the critical load of the current system-1,

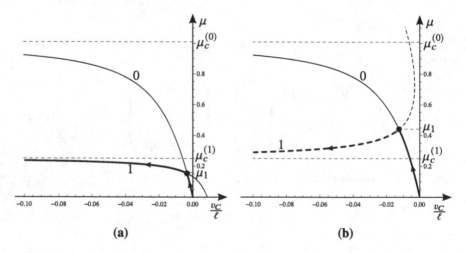

(a)   (b)

Fig. 8.9 Second order push-over curve of the cantilever beam with elasto-plastic spring: (a) "weak" spring, $\frac{N_p}{k\ell} = \frac{1}{200}$; (b) "strong" spring, $\frac{N_p}{k\ell} = \frac{1}{50}$; elastic phase (branch 0), elasto-plastic phase (branch 1), visited equilibrium states (thick curve); stable (continuous curve), unstable (dashed curve); $P = \frac{\pi^2 EI}{\ell^2}$, $p = \frac{\pi^2 EI}{10\ell^3}$, $\eta = 10$

---

[14] The curve emanates from the origin, because the imperfection disappears when $\mu = 0$.

the branch is unstable, so that the structure collapses at μ_1, i.e., upon the first plasticization.

Remark 8.4 A "weak" spring does not increase the critical load of the cantilever beam, but the collapse occurs after the occurrence of large displacements (*ductile* collapse). A "strong" spring, on the other hand, allows overcoming this critical load, but the collapse occurs instantaneously, at the first plasticization, when the displacements are still small (*fragile* collapse).

References

1. Allen, H.G., Bulson, P.S.: Background to Buckling. McGraw-Hill, New York (1980)
2. Bazant, Z.P., Cedolin, L.: Stability of Structures: Elastic, Inelastic, Fracture, and Damage Theories. World Scientific Publishing, Singapore (2010)
3. Bleich, F.: Buckling Strength of Metal Structures. McGraw-Hill, New York (1952)
4. Britvec, S.J.: The Stability of Elastic Systems. Pergamon Press, New York (1973)
5. Chajes, A.: Principles of Structural Stability Theory. Prentice-Hall, New Jersey (1974)
6. Chen, W.F., Lui, E.M.: Structural Stability: Theory and Implementation. Elsevier, New York (1987)
7. Corradi Dell'Acqua, L.: Meccanica delle strutture: La valutazione della capacità portante (in Italian), vol. 3. McGraw-Hill, Milano (1994)
8. Engesser, F.: Ueber die Knickfestigkeit gerader Stäbe. Z. Arch. Ing. Ver. Hannover **35**, 455–462 (1889)
9. Galambos, T.V., Surovek, A.E.: Structural Stability of Steel: Concepts and Applications for Structural Engineers. Wiley, Hoboken (2008)
10. Gerard, G.: Introduction to Structural Stability Theory. McGraw-Hill, New York (1962)
11. Horne, M.R., Merchant, W.: The Stability of Frames. Pergamon Press, Oxford (1965)
12. Iyengar, N.G.R.: Elastic Stability of Structural Elements. Macmillan India, New Delhi (2007)
13. Shanley, F.R.: Inelastic column theory. J. Aeronaut. Sci. **14**(5), 261–268 (1947)
14. Timoshenko, S.P., Gere, J.M.: Theory of Elastic Stability. McGraw-Hill, New York (1961)
15. von Karman, T.: Discussion of Shanley's paper. J. Aeronaut. Sci. **14**(5), 267–268 (1947)
16. Yoo, C.H., Lee, S.: Stability of Structures: Principles and Applications. Elsevier, Amsterdam (2011)

Chapter 9
Buckling of Open Thin-Walled Beams

9.1 Introduction

Thin-walled beams (TWB) with open cross-section exhibit forms of static instability quite peculiar, compared to beam with compact (or tubular) cross-section. When a compact cross-section beam, embedded in the 3D space, is compressed at a sufficiently high extent, it buckles in one of the two principal inertia planes (precisely, in that of lower stiffness), as the Euler planar beam does. This phenomenon is due to the high torsional stiffness of compact beams, which makes the two orthogonal bendings practically uncoupled. In contrast, when the beam has a thin open cross-section, due to the low torsional stiffness, it manifests a different form of buckling, called *flexural-torsional*, in which the beam bends into one or both the principal inertia planes, and, at the same time, twists.

The flexural-torsional coupling also occurs in beams solicited by forces transversal to the axis, contained in one of the two principal planes (generally the one with the larger stiffness). Upon reaching the critical load, the beam buckles into the plane orthogonal to that of solicitation and simultaneously twists. The same happens if the beam is simply bent, or eccentrically compressed, in a plane. This form of buckling is called *lateral instability*.

In order to study these phenomena, it is necessary to formulate a model of a 1D beam immersed in a 3D space, accurately describing the torsional behavior of the TWB, with respect to its elastic and geometric stiffness. The first requirement is fulfilled by the *Vlasov theory* of non-uniform torsion [7, 11, 12], which is assumed here to be known to the reader, but which is amply detailed in the Appendix C. According to this theory, the cross-section of the beam retains its natural shape in its own plane but undergoes *warping* out of this plane. Since warping is generally variable along the beam axis, as a consequence of the non-uniformity of the torsion, complementary normal and tangential stresses arise, which are not negligible compared to those predicted by the de Saint-Venant (DSV) model of uniform torsion.

265
A. Luongo et al., *Stability and Bifurcation of Structures*,
https://doi.org/10.1007/978-3-031-27572-2_9

The rotation of the cross-section takes place around the *center of torsion*, which is generally different from the centroid. The center of torsion is a geometric property of the cross-section, and it is found be coincident with the *shear center* of the Jourawsky theory of non-uniform bending. This circumstance suggests describing the displacement field of the TWB with reference not only to the centroid, as usually done for compact beams, but with reference (a) to the center of torsion, as regards the in-plane displacements, and (b) to the centroid, as regards the out-of-plane displacements. TWB, therefore, have *two different axes*, which is one of their distinctive characteristics.

In this chapter, the buckling problem is formulated in the framework of the linearized theory (Chap. 6), according to which the linear elastic and geometric stiffness operators are built-up. A number of sample problems are solved, to illustrate the mechanical behavior at bifurcation, i.e.: TWB uniformly compressed, uniformly bent, eccentrically compressed and non-uniformly bent by transverse loads. Exact solutions are provided in very simple cases; then the Ritz and the Finite Element methods are used in more complex problems.

9.2 Elastic Stiffness Operator

Reference is made to a 1D model of beam, with thin open cross-section. The analysis of the effects of the prestress is postponed, so that the elastic problem is formulated in the context of the linear theory, in order to identify the elastic stiffness operator.

9.2.1 Kinematics

A straight beam is considered, of centroidal axis z, whose open and thin cross-section has central inertia axes x and y, originating by the centroid G (Fig. 9.1a). A basis of orthonormal vectors $(\mathbf{a}_x, \mathbf{a}_y, \mathbf{a}_z)$, aligned with the homonym axes, is introduced. Along the *midline* Γ of the cross-section, a curvilinear abscissa s of origin O is taken. The (possibly variable) thickness of the cross-section is denoted by $b(s)$. It is assumed that all the quantities of interest are constant on the thickness, so that, e.g., the displacement is $\mathbf{u} = \mathbf{u}(s, z)$. The cross-section is assumed *undeformable in its own plan* π, but not out of this plane, orthogonally to which it undergoes a warping $w(s, z)$. The TWB is assumed to be *shear undeformable* on its middle surface S (i.e., on the surface of trace Γ on the cross-section). This implies that (a) in bending and extension, the beam behaves like an Euler-Bernoulli beam, (b) warping depends only on twist. The shear strains, however, are different from zero outside the middle surface, where they are described from the DSV torsion model.

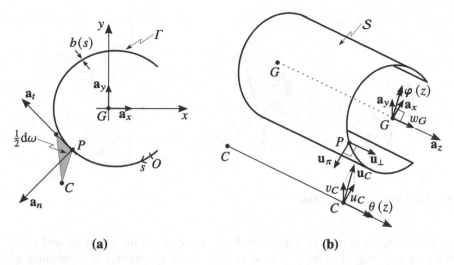

Fig. 9.1 Open thin-walled beam: (**a**) cross-section and sectorial area, (**b**) displacements

In-plane Displacements

As it is known from the theory of non-uniform torsion (Appendix C), the cross-section rotates in its own plane around a point C, said *center of torsion* (coincident with the shear center), whose coordinates x_C *and* y_C are a geometric property of the cross-section. Accordingly, it is convenient to describe the rigid plane displacement with reference to this point. By denoting by $\mathbf{u}_C(z) = u_C(z)\,\mathbf{a}_x + v_C(z)\,\mathbf{a}_y$ the in-plane translation of the center C and by $\theta(z)$ the angle of twist, the in-plane (infinitesimal) displacement $\mathbf{u}_\pi := u(s,z)\,\mathbf{a}_x + v(s,z)\,\mathbf{a}_y$ of a point $P(s)$ of coordinates $x(s)$, $y(s)$ reads (Fig. 9.1b):

$$\mathbf{u}_\pi = \mathbf{u}_C(z) + \theta(z)\,\mathbf{a}_z \times \mathbf{CP}(s). \tag{9.1}$$

Projecting this relationship onto the axes, one has:

$$u = u_C(z) - \theta(z)(y(s) - y_C), \tag{9.2a}$$

$$v = v_C(z) + \theta(z)(x(s) - x_C). \tag{9.2b}$$

Out-of-Plane Displacements

The displacement $\mathbf{u}_\perp := w(s,z)\,\mathbf{a}_z$, orthogonal to the cross-section plane, is due to the superposition of three effects: (a) a translation along z, (b) a rotation

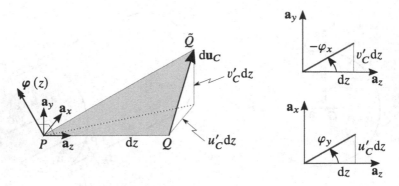

Fig. 9.2 Unshearability condition

$\boldsymbol{\varphi}(z) := \varphi_x(z)\,\mathbf{a}_x + \varphi_y(z)\,\mathbf{a}_y$ around an axis lying in the plane π, and (c) a warping proportional to the pure torsion, $\kappa_t = \theta'(z)$. Describing, for convenience, the *rotation around an axis passing through the centroid G* of the cross-section, and remembering the results of Vlasov theory (Eq. C.12 in the Appendix C), it follows (Fig. 9.1b):

$$w = w_G(z) + \boldsymbol{\varphi}(z) \times \mathbf{GP}(z) \cdot \mathbf{a}_z - \theta'(z)\,\omega(s), \tag{9.3}$$

where $w_G(z)$ is the axial displacement of the centroid and $\omega(s)$ is the *warping function* (equal to twice the sectorial area, Fig. 9.1a).

To evaluate the rotation $\boldsymbol{\varphi}(z)$, use is made of the undeformability condition in shear, when the beam experiences no torsion (Fig. 9.2). The internal constraint states that the cross-section (which remains planar in this type of motion) remains orthogonal to all the longitudinal fibers, which rotate by the same angle $\boldsymbol{\varphi}(z)$. Since the transverse displacement of each longitudinal fiber is $\mathbf{u}_C(z)$, the relative displacement between two points P and Q, dz away on the same fiber, is $d\mathbf{u}_C = \boldsymbol{\varphi}(z) \times dz\,\mathbf{a}_z$, hence $\mathbf{u}'_C(z) = \boldsymbol{\varphi}(z) \times \mathbf{a}_z = \varphi_y\mathbf{a}_x - \varphi_x\mathbf{a}_y$ (Fig. 9.2). By accounting for $\mathbf{u}'_C(z) = u'_C(z)\,\mathbf{a}_x + v'_C(z)\,\mathbf{a}_y$, it follows that $\varphi_x = -v'_C(z)$ and $\varphi_y = u'_C(z)$, so that:

$$w = w_G - v'_C(z)\,y(s) - u'_C(z)\,x(s) - \theta'(z)\,\omega(s). \tag{9.4}$$

Strains

Since the middle surface of the TWB is shear undeformable and the cross-section retains its shape in the π plane, the only non-zero strain on this surface is the axial unit extension $\varepsilon_z = \frac{\partial w}{\partial z}$; by using Eq. 9.4, it reads:

$$\varepsilon_z = w'_G(z) - u''_C(z)\,x(s) - v''_C(z)\,y(s) - \theta''(z)\,\omega(s). \tag{9.5}$$

In this expression, the following effects are recognized: an extension, two Euler-Bernoulli bendings, and a Vlasov warping. To the extension in Eq. 9.5, the shear strain occurring outside the middle surface S must be added, as provided by the DSV torsion theory.

Supplement 9.1 (Normal Stress in the 3D Model) The strains in Eq. 9.5 generate the normal stress:

$$\sigma_z = E\varepsilon_z = E(w'_G - u''_C x - v''_C y - \theta''\omega). \tag{9.6}$$

On the other hand, the following relationships hold for the Euler-Bernoulli and Vlasov beams (Eq. C.46 in the Appendix C):

$$w'_G = \frac{N}{EA}, \quad u''_C = \frac{M_y}{EI_y}, \quad v''_C = -\frac{M_x}{EI_x}, \quad \theta'' = \frac{B}{EI_\omega}, \tag{9.7}$$

where, in addition to the usual symbols, B is the bimoment and I_ω the sectorial inertia of the cross-section. Therefore, the normal stress in Eq. 9.6 also reads:

$$\sigma_z = \frac{N}{A} + \frac{M_x}{I_x}y - \frac{M_y}{I_y}x - \frac{B}{I_\omega}\omega. \tag{9.8}$$

This is a quadrinomy that generalizes the well-known trinomy Navier formula.   □

9.2.2 Equilibrium Equations

The equilibrium equations of the TWB are obtained by enforcing stationary of the total potential energy (TPE), $\Pi = U + V$. This is sum of the elastic potential energy, U,[1] and of the potential energy of the external forces, V. The two contributions are written ahead separately.

Elastic Potential Energy

The elastic potential energy of the beam, $U := U_\varepsilon + U_\gamma$, is the sum of two contributions, U_ε, due to the extensions, and U_γ, due to the shear strains. The first contribution reads:

---

[1] In this section, the tilde will be omitted on U, since the prestress is here ignored.

$$U_\varepsilon = \frac{1}{2} \int\limits_0^\ell dz \int\limits_\Gamma E\varepsilon_z^2 (s, z) \, b(s) \, ds = \frac{E}{2} \int\limits_0^\ell dz \int\limits_\Gamma \left(w_G' - u_C'' x - v_C'' y - \theta'' \omega \right)^2 b \, ds$$

$$= \frac{E}{2} \int\limits_0^\ell dz \int\limits_\Gamma \left(w_G'^2 + u_C''^2 x^2 + v_C''^2 y^2 + \theta''^2 \omega^2 \right) b \, ds$$

$$= \frac{1}{2} \int\limits_0^\ell \left(EAw_G'^2 + EI_y u_C''^2 + EI_x v_C''^2 + EI_\omega \theta''^2 \right) dz,$$

$$(9.9)$$

in which the *orthogonality properties* between the functions $1, x(s), y(s)$ and $\omega(s)$ have been exploited.[2] As usual, A is the cross-section area, I_x, I_y the central inertia moments, and I_ω the sectorial inertia moment.

The second contribution to the potential energy is brought by the DSV shear strains. It can be directly written in terms of the one-dimensional model as:

$$U_\gamma = \frac{1}{2} \int\limits_0^\ell GJ\theta'^2 dz, \qquad (9.10)$$

where GJ is the DSV torsional stiffness.

By adding the two contributions in Eqs. 9.9, 9.10, the strain energy is obtained, whose first variation reads:

$$\delta U = \int\limits_0^\ell \left(EAw_G' \delta w_G' + EI_y u_C'' \delta u_C'' + EI_x v_C'' \delta v_C'' + EI_\omega \theta'' \delta \theta'' + GJ\theta' \delta \theta' \right) dz.$$

$$(9.11)$$

Remark 9.1 The elastic potential energy is the sum of the energies of extension, bending, non-uniform torsion and DSV torsion, *without any coupling terms*. This occurrence is a consequence of the kinematic description adopted, referred to two axes.

Load Potential Energy

The following forces are consider acting on the TWB: (a) transverse loads of linear density $\mathbf{p} = p_x(z)\,\mathbf{a}_x + p_y(z)\,\mathbf{a}_y$ *applied to the center of torsion axis* and (b) twisting couples of linear density $c_z(z)$. By directly referring to the one-dimensional model, the load potential energy reads:

---

[2] The orthogonality entails (a) the vanishing of the centroidal static moments and of the mixed inertia central moment; (b) the properties of the sectorial area, given by Eqs. C.16 in the Appendix C.

$$V = - \int_0^\ell \left(p_x u_C + p_y v_C + c_z \theta \right) dz, \tag{9.12}$$

whose first variations is:

$$\delta V = - \int_0^\ell \left(p_x \delta u_C + p_y \delta v_C + c_z \delta \theta \right) dz. \tag{9.13}$$

The generalized forces, p_x, p_y, and c_z, can also be derived starting from a three-dimensional model (as described in the Supplement 9.2). The derivation reveals that, if distributed longitudinal forces are also considered, these latter lead not only to bending couples but also to a new generalized force (the *distributed bicouple*), dual of the twist gradient. These effects, although discussed in the cited supplement, will not be taken into account ahead in the book.[3]

Supplement 9.2 (Generalized Forces Deduced from the 3D Model) The load potential energy is evaluated referring to the 3D model, in which distributed forces per unit volume $\mathbf{f} = f_x(s, z)\, \mathbf{a}_x + f_y(s, z)\, \mathbf{a}_y + f_z(s, z)\, \mathbf{a}_z$ are applied. To make the discussion easier, in-plane and out-of-plane loads are considered separately. The potential energy of the in-plane loads is:

$$V_\pi = - \int_0^\ell dz \int_\Gamma \left(f_x u + f_y v \right) b\, ds$$

$$= - \int_0^\ell dz \int_\Gamma \left\{ f_x \left[u_C - \theta \left(y(s) - y_C \right) \right] + f_y \left[v_C + \theta \left(x(s) - x_C \right) \right] \right\} b\, ds, \tag{9.14}$$

where use has been made of Eq. 9.2. This can be rewritten in the form of Eq. 9.12, by defining:

$$p_x := \int_\Gamma f_x(s, z)\, b(s)\, ds, \tag{9.15a}$$

$$p_y := \int_\Gamma f_y(s, z)\, b(s)\, ds, \tag{9.15b}$$

---

[3] The Finite Element model is an exception, discussed in the Sect. 9.8, where *concentrated bicouples* will be used, which expend work on the torsion gradient, evaluated at the point to which they are applied.

$$c_z := \int_\Gamma \left[-f_x(s, z)(y(s) - y_C) + f_y(s, z)(x(s) - x_C)\right] b(s)\, ds. \qquad (9.15c)$$

The potential energy of the out-of-plane loads is:

$$V_\perp = -\int_0^\ell dz \int_\Gamma f_z w\, b\, ds = -\int_0^\ell dz \int_\Gamma f_z \left(w_G - v_C' y - u_C' x - \theta' \omega\right) b\, ds,$$

$$(9.16)$$

in which Eq. 9.4 has been exploited. This can be rewritten in the form:

$$V_\perp = -\int_0^\ell \left(p_z w_G + c_y u_C' - c_x v_C' + c_\omega \theta'\right) dz, \qquad (9.17)$$

after having introduced the following quantities:

$$p_z := \int_\Gamma f_z(s, z)\, b(s)\, ds, \qquad\qquad c_x := \int_\Gamma f_z(s, z)\, y(s)\, b(s)\, ds,$$

$$(9.18)$$

$$c_y := -\int_\Gamma f_z(s, z)\, x(s)\, b(s)\, ds, \qquad c_\omega := -\int_\Gamma f_z(s, z)\, \omega(s)\, b(s)\, ds.$$

Equations 9.18 define, respectively, the resultant centroidal force, the distributed couples of axes x and y, and the distributed bicouple c_ω. This latter is a new generalized force, absent in the classical model, that expends work on the torsion gradient (to represent, in the 1D model, the work done, in the 3D model, by the longitudinal forces on warping).

It is worth noticing that, as a consequence of the kinematic model, the couples c_z are evaluated with respect to the center of torsion, while the couples c_x and c_y with respect to the centroidal axes x and y.

By adding the two contributions in Eqs. 9.12, 9.17, performing the variation and then integrating by parts, it follows:

$$\delta V = -\int_0^\ell \left(p_x \delta u_C + p_y \delta v_C + p_z \delta w_G + c_z \delta\theta + c_y \delta u_C' - c_x \delta v_C' + c_\omega \delta\theta'\right) dz$$

$$= -\int_0^\ell \left[\left(p_x - c_y'\right) \delta u_C + \left(p_y + c_x'\right) \delta v_C + p_z \delta w_G + \left(c_z - c_\omega'\right) \delta\theta\right] dz$$

$$- \left[c_y \delta u_C - c_x \delta v_C + c_\omega \delta\theta\right]_0^\ell.$$

$$(9.19)$$

Equation 9.19 defines the distributed forces dual of the displacements, namely, $(p_x - c'_y,\ p_y + c'_x,\ p_z,\ c_z - c'_\omega)$, plus the boundary forces. It should be noted that, as a major difference with the theory of one-dimensional beams with compact cross-sections, the longitudinal forces f_z, via c_ω, bring a contribution to the couples c_z, and therefore, they induce torsion.   □

Equilibrium Equations

Taking into account Eqs. 9.11 and 9.13, the first variation of the TPE, equated to zero, reads:

$$\delta\Pi = \int_0^\ell \left(EAw'_G\delta w'_G + EI_y u''_C\delta u''_C + EI_x v''_C\delta v''_C + EI_\omega\theta''\delta\theta'' + GJ\theta'\delta\theta' \right) dz$$

$$- \int_0^l \left(p_x\delta u_C + p_y\delta v_C + c_z\delta\theta \right) dz = 0, \qquad \forall\,(\delta w_G, \delta u_C, \delta v_C, \delta\theta).$$

$$(9.20)$$

After integration by parts, the following field equations are obtained:

$$- EAw''_G = 0, \tag{9.21a}$$

$$EI_y u''''_C = p_x, \tag{9.21b}$$

$$EI_x v''''_C = p_y, \tag{9.21c}$$

$$EI_\omega\theta'''' - GJ\theta'' = c_z, \tag{9.21d}$$

with the relative boundary conditions, alternately geometric or mechanical:

$$\left[EAw'_G\ \delta w_G \right]_0^\ell = 0, \tag{9.22a}$$

$$\left[EI_y u''_C\ \delta u'_C \right]_0^\ell = 0, \tag{9.22b}$$

$$\left[-EI_y u'''_C\ \delta u_C \right]_0^\ell = 0, \tag{9.22c}$$

$$\left[EI_x v''_C\ \delta v'_C \right]_0^\ell = 0, \tag{9.22d}$$

$$\left[-EI_x v'''_C\ \delta v_C \right]_0^\ell = 0, \tag{9.22e}$$

$$\left[(GJ\theta' - EI_\omega\theta''')\ \delta\theta \right]_0^\ell = 0, \tag{9.22f}$$

$$\left[EI_\omega\theta''\ \delta\theta' \right]_0^\ell = 0. \tag{9.22g}$$

The field Eqs. 9.21, to be referred to as the *elastic line equations of the open TWB*, can be written in the operational form:

$$\mathcal{K}_e \mathsf{u}\,(z) = \mathsf{p}\,(z)\,, \tag{9.23}$$

where $\mathsf{u}\,(z) := (w_G\,(z)\,, u_C\,(z)\,, v_C\,(z)\,, \theta\,(z))^T$ collects the unknown displacement components, $\mathsf{p}\,(z) := \left(0,\, p_x\,(z)\,,\, p_y\,(z)\,,\, c\,(z)\right)^T$ the assigned loads, and

$$\mathcal{K}_e := \begin{bmatrix} -EA\partial_z^2 & 0 & 0 & 0 \\ 0 & EI_y\partial_z^4 & 0 & 0 \\ 0 & 0 & EI_x\partial_z^4 & 0 \\ 0 & 0 & 0 & EI_\omega\partial_z^4 - GJ\partial_z^2 \end{bmatrix} \tag{9.24}$$

is the diagonal *elastic stiffness operator*, in which $\partial_z := \frac{\mathrm{d}}{\mathrm{d}z}$.

For future developments, it is convenient to partition the displacements as $\mathsf{u}\,(z) =: (\mathsf{u}_\perp\,(z)\,,\ \mathsf{u}_\pi\,(z))$, where $\mathsf{u}_\perp\,(z) := (w_G\,(z))$ is the out-of-plane component, and $\mathsf{u}_\pi\,(z) := (u_C\,(z)\,, v_C\,(z)\,, \theta\,(z))^T$ the generalized displacements in the cross-section plane. Consequently:

$$\mathcal{K}_e := \begin{bmatrix} \mathcal{K}_e^\perp & \mathbf{0} \\ \mathbf{0} & \mathcal{K}_e^\pi \end{bmatrix}, \tag{9.25}$$

where:

$$\mathcal{K}_e^\perp := \begin{bmatrix} -EA\partial_z^2 \end{bmatrix}, \quad \mathcal{K}_e^\pi := \begin{bmatrix} EI_y\partial_z^4 & 0 & 0 \\ 0 & EI_x\partial_z^4 & 0 \\ 0 & 0 & EI_\omega\partial_z^4 - GJ\partial_z^2 \end{bmatrix}. \tag{9.26}$$

9.3 Geometric Stiffness Operator

Linearized stability analysis of a TWB requires building up the geometric stiffness operator. This arises from the variation of the quadratic part of the prestress energy [6, 9]:

$$U^0 = \int_0^\ell \mathrm{d}z \int_\Gamma \left(\sigma_z^0 \varepsilon_z^{(2)} + \tau_{zs}^0 \gamma_{zs}^{(2)}\right) b\mathrm{d}s, \tag{9.27}$$

which expresses the work of known normal prestresses $\sigma_z^0\,(s, z)$ and tangential stresses $\tau_{zs}^0\,(s, z)$, on the second order components of the unknown incremental strains $\varepsilon_z^{(2)}(s, z)$, $\gamma_{zs}^{(2)}(s, z)$, all evaluated on the middle surface S.[4] Since the strains depend on the displacements of the point $P(s)$ at the abscissa z and these in turn can be expressed in terms of the displacements of the two center lines of the TWB, which only depend on z, the model is transformed from three- to one-dimensional.

Prestresses
By limiting the attention to beams under extension and bending (but not torsion) and resorting to results of DSV and Jourawsky theories, the prestresses are evaluated as:

$$\sigma_z^0 = \frac{N^0(z)}{A} + \frac{M_x^0(z)}{I_x} y(s) - \frac{M_y^0(z)}{I_y} x(s), \tag{9.28a}$$

$$\tau_{zs}^0 = -\frac{T_x^0(z) S_y^*(s)}{I_y b(s)} - \frac{T_y^0(z) S_x^*(s)}{I_x b(s)}, \tag{9.28b}$$

where N^0, M_x^0, M_y^0, T_x^0, and T_y^0 are, respectively, the axial force, the bending moments, and the shear forces acting in the prestressed configuration. Here, the geometric characteristics of the cross-sections assume the usual meaning.

Quadratic Strains
The nonlinear part of the incremental strains is deduced from the quadratic component of the Green-Lagrange strain tensor:[5]

$$\varepsilon_z^{(2)} = \frac{1}{2}\left(u_{,z}^2 + v_{,z}^2 + \bcancel{w_{,z}^2}\right), \tag{9.29a}$$

$$\gamma_{zs}^{(2)} = u_{,z}u_{,s} + v_{,z}v_{,s} + \bcancel{w_{,z}w_{,s}}, \tag{9.29b}$$

where the barred terms are neglected, according to the hypothesis that out-of-plane displacements are smaller than in-plane displacements.[6] Equations 9.29 and their geometric meaning are discussed in the Supplement 9.3.

The evaluation of the geometric stiffness operator in the general case is somewhat complex and tedious. It is therefore preferred to derive the operator by successive steps, referring to simple solicitations. Then, taking into account the linearity of the problem, the general case will be built up by superposition.

Supplement 9.3 (Geometric Meaning of the Second Order Strains) Equation 9.29b is derived by observing that $\gamma_{zs}^{(2)}$ is the *resulting shear strain*, of components $\gamma_{zx}^{(2)}(x, y)$ and $\gamma_{zy}^{(2)}(x, y)$. It occurs in the direction of the tangent $\mathbf{a}_t$ to the midline

---

[4] On this surface, it is $\gamma_{zs}^{(1)} = 0$, as fundamental hypothesis by Vlasov.

[5] The Supplement 6.1 should be remembered.

[6] Moreover, $\sigma_z^0 w_{,z}^2$ is negligible when compared with $E w_{,z}^2$, appearing in the elastic part of the energy, being $\sigma_z^0 \ll E$.

Γ, having director cosines $(\cos{(\alpha{(s)})},\ \sin{(\alpha{(s)})}) = \left(\frac{\mathrm{d}x}{\mathrm{d}s}, \frac{\mathrm{d}y}{\mathrm{d}s}\right)$. Consequently:

$$
\begin{aligned}
\gamma_{zs}^{(2)} &= \gamma_{zx}^{(2)}\cos{(\alpha{(s)})} + \gamma_{zy}^{(2)}\sin{(\alpha{(s)})}\\
&= \left(u_{,z}u_{,x} + v_{,z}v_{,x} + w_{,z}w_{,x}\right)\frac{\mathrm{d}x}{\mathrm{d}s} + \left(u_{,z}u_{,y} + v_{,z}v_{,y} + w_{,z}w_{,y}\right)\frac{\mathrm{d}y}{\mathrm{d}s}\\
&= u_{,z}\left(u_{,x}\frac{\mathrm{d}x}{\mathrm{d}s} + u_{,y}\frac{\mathrm{d}y}{\mathrm{d}s}\right) + v_{,z}\left(v_{,x}\frac{\mathrm{d}x}{\mathrm{d}s} + v_{,y}\frac{\mathrm{d}y}{\mathrm{d}s}\right) + w_{,z}\left(w_{,x}\frac{\mathrm{d}x}{\mathrm{d}s} + w_{,y}\frac{\mathrm{d}y}{\mathrm{d}s}\right)\\
&= u_{,z}u_{,s} + v_{,z}v_{,s} + w_{,z}w_{,s}.
\end{aligned}
$$

(9.30)

The geometric meaning of $\varepsilon_z^{(2)}$ emerges from the calculation of the unit extension of a longitudinal segment, PQ, of the TWB, whose end Q undergoes the relative displacement $\mathrm{d}\mathbf{u}_Q = \left(u_{,z}\mathbf{a}_x + v_{,z}\mathbf{a}_y + w_{,z}\mathbf{a}_z\right)\mathrm{d}z$ with respect to P. Evaluating $\varepsilon_z := \frac{\overline{PQ}-\overline{PQ}}{\overline{PQ}} = \sqrt{\left(1+w_{,z}\right)^2 + u_{,z}^2 + v_{,z}^2} - 1$ and expanding in series for small displacements (as already done in the Remark 7.1 for the planar beam), it follows that, at the first order, $\varepsilon_z^{(1)} = w_{,z}$, and, at the second order, $\varepsilon_z^{(2)} = \frac{1}{2}\left(u_{,z}^2 + v_{,z}^2\right)$ (in which the neglected term in Eq. 9.29a does not appear). The latter magnitude, therefore, has the meaning of longitudinal unit extension produced from displacements transversal to the fiber.

To illustrate the meaning of $\gamma_{zs}^{(2)}$, a material element $\mathrm{d}s \times \mathrm{d}z$, around a point P, is considered, lying on the middle surface of the TWB. Let $(\mathbf{a}_t, \mathbf{a}_n, \mathbf{a}_z)$ be the basis intrinsic to the element, formed by the tangent, normal and binormal unit vectors to the midline. Taken a point Q at a distance $\mathrm{d}s$ from P, its displacement relative to P is $\mathrm{d}\mathbf{u}_Q = \mathbf{u}_{,s}\mathrm{d}s$. Consequently, the oriented material segment $\overrightarrow{PQ} = \mathbf{a}_t\mathrm{d}s$ changes into the segment $\overrightarrow{P\tilde{Q}} = \left(\mathbf{a}_t + \mathbf{u}_{,s}\right)\mathrm{d}s$. Similarly, taken a point R at a distance $\mathrm{d}z$ from P, it undergoes a relative displacement to P, equal to $\mathrm{d}\mathbf{u}_R = \mathbf{u}_{,z}\mathrm{d}z$, so that $\overrightarrow{PR} = \mathbf{a}_z\mathrm{d}z$ transforms into $\overrightarrow{P\tilde{R}} = \left(\mathbf{a}_z + \mathbf{u}_{,z}\right)\mathrm{d}z$. The initially right angle between $\overrightarrow{PQ}$ and $\overrightarrow{PR}$, after the transformation, measures $\frac{\pi}{2} - \gamma_{zs}$, with γ_{zs} the shear strain, which can be deduced from:

$$
\cos\left(\frac{\pi}{2} - \gamma_{zs}\right) = \frac{\overrightarrow{P\tilde{Q}} \cdot \overrightarrow{P\tilde{R}}}{\left|\overrightarrow{P\tilde{Q}}\right|\left|\overrightarrow{P\tilde{R}}\right|} = \frac{\left(\mathbf{a}_t + \mathbf{u}_{,s}\right) \cdot \left(\mathbf{a}_z + \mathbf{u}_{,z}\right)}{\left|\mathbf{a}_t + \mathbf{u}_{,s}\right|\left|\mathbf{a}_z + \mathbf{u}_{,z}\right|}.
$$

(9.31)

Expressed the displacement in the extrinsic basis, such as $\mathbf{u} = u\mathbf{a}_x + v\mathbf{a}_y + w\mathbf{a}_z$, neglecting the squares in the denominator, and observed that, for small γ_{zs}, is $\cos\left(\frac{\pi}{2} - \gamma_{zs}\right) = \sin\gamma_{zs} \simeq \gamma_{zs}$, it follows:

$$
\gamma_{zs} \simeq \frac{w_{,s} + u_{t,z} + u_{,z}u_{,s} + v_{,z}v_{,s} + w_{,z}w_{,s}}{\left(1+w_{,s}\right)\left(1+w_{,z}\right)},
$$

(9.32)

where $u_t := u \cos \alpha + v \sin \alpha$ is the tangent displacement. If it is assumed that $|w| \ll |u|$ and $|w| \ll |v|$, consistently with what has been done in Eqs. 9.29, the denominator of the previous expression is approximately equal to 1. Consequently, $\gamma_{zs}^{(1)} = w_{,s} + u_{t,z}$ and $\gamma_{zs}^{(2)} = u_{,z} u_{,s} + v_{,z} v_{,s}$. Within the limits of the approximations introduced, the physical measure of the shear strain coincides with that provided by the Green-Lagrange tensor. □

9.4 Uniformly Compressed Thin-Walled Beams

The bifurcation of uniformly compressed TWB is analyzed. This problem generalizes to a TWB immersed in the 3D space, the Euler problem relevant to a planar beam. First, the equilibrium equations are formulated; then, simple solutions are illustrated, useful to discuss important phenomenological aspects [4, 5, 8, 10, 12].

9.4.1 Formulation

A TWB, uniformly compressed by an axial force P, is considered . The state of prestress is described by:

$$\sigma_z^0 = -\frac{P}{A}. \tag{9.33}$$

The relevant prestress energy, by Eq. 9.27, is written as:

$$U^0 = \int_0^\ell dz \int_\Gamma \sigma_z^0 \varepsilon_z^{(2)} b\, ds = -\frac{1}{2} \frac{P}{A} \int_0^\ell dz \int_\Gamma \left(u_{,z}^2 + v_{,z}^2 \right) b\, ds, \tag{9.34}$$

where use has been made of Eq. 9.29a. By substituting Eqs. 9.2, $u\,(s, z)$, $v\,(s, z)$ are expressed in function of the generalized displacements $u_C\,(z)$, $v_C\,(z)$, $\theta\,(z)$ of the torsion center line; consequently:

$$
\begin{aligned}
U^0 [\mathbf{u}_\pi] &= -\frac{P}{2A} \int_0^\ell dz \int_\Gamma \left\{ \left[u_C' - \theta'\,(y - y_C) \right]^2 + \left[v_C' + \theta'\,(x - x_C) \right]^2 \right\} b\, ds \\
&= -\frac{P}{2} \int_0^\ell \left[u_C'^2 + v_C'^2 + r_C^2 \theta'^2 + 2 y_C u_C' \theta' - 2 x_C v_C' \theta' \right] dz.
\end{aligned}
\tag{9.35}
$$

Here, it has been taken into account, that the cross-section static moments are zero; moreover, the square of the polar inertia radius with respect at the center of torsion has been introduced, as

$$r_C^2 := \frac{1}{A} \int_\Gamma \left[(x - x_C)^2 + (y - y_C)^2 \right] b\,ds =: \frac{I_C}{A}. \tag{9.36}$$

It is observed that, because of the approximation introduced in Eq. 9.29a, the functional U^0 depends on the in-plane displacements $\mathsf{u}_\pi := (u_C, v_C, \theta)^T$ only, and not on the axial displacement w_G.

Geometric Stiffness Operator

By performing the first variation of the energy, and integrating by parts, one gets:

$$
\begin{aligned}
\delta U^0 &= -P \int_0^\ell \left[u_C' \delta u_C' + v_C' \delta v_C' + r_C^2 \theta' \delta\theta' + y_C \left(u_C' \delta\theta' + \theta' \delta u_C' \right) \right] dz \\
&\quad - P \int_0^\ell \left[-x_C \left(v_C' \delta\theta' + \theta' \delta v_C' \right) \right] dz \\
&= P \int_0^\ell \left[\left(u_C'' + y_C \theta'' \right) \delta u_C + \left(v_C'' - x_C \theta \right) \delta v_C \right] dz \\
&\quad + P \int_0^\ell \left[\left(r_C^2 \theta'' + y_C u_C'' - x_C v_C'' \right) \delta\theta \right] dz + \left[-P \left(u_C' + y_C \theta' \right) \delta u_C \right]_0^\ell \\
&\quad + \left[-P \left(v_C' - x_C \theta' \right) \delta v_C - P \left(r_C^2 \theta' + y_C u_C' - x_C v_C' \right) \delta\theta \right]_0^\ell.
\end{aligned}
\tag{9.37}
$$

By letting $\delta U^0 = \int_0^\ell \delta\mathsf{u}_\pi^T \boldsymbol{\mathcal{K}}_{g,c} \mathsf{u}_\pi\,dz + [\cdots]_0^\ell$, the *geometric stiffness operator of the compressed TWB* is recognized:

$$\boldsymbol{\mathcal{K}}_{g,c} := P \begin{bmatrix} 1 & 0 & y_C \\ 0 & 1 & -x_C \\ y_C & -x_C & r_C^2 \end{bmatrix} \partial_z^2. \tag{9.38}$$

It is *non-diagonal* and therefore induces coupling among the transverse displacement components, u_C, v_C and the twist θ.

Equilibrium Equations

By adding the elastic and geometric contributions, the indefinite equilibrium equations are obtained:

$$\mathcal{K}_e^\perp \mathsf{u}_\perp = 0, \tag{9.39a}$$

$$\left(\boldsymbol{\mathcal{K}}_e^\pi + \boldsymbol{\mathcal{K}}_{g,c} \right) \mathsf{u}_\pi = \mathbf{0}, \tag{9.39b}$$

or, in extended form:

$$- EAw_G'' = 0, \tag{9.40a}$$

$$EI_y u_C'''' + P\left(u_C'' + y_C\theta''\right) = 0, \tag{9.40b}$$

$$EI_x v_C'''' + P\left(v_C'' - x_C\theta''\right) = 0, \tag{9.40c}$$

$$EI_\omega\theta'''' - GJ\theta'' + P\left(r_C^2\theta'' + y_C u_C'' - x_C v_C''\right) = 0. \tag{9.40d}$$

Operating in the same way, the boundary conditions are found:

$$\left[EAw_G'\,\delta w_G\right]_0^\ell = 0, \tag{9.41a}$$

$$\left[EI_y u_C''\,\delta u_C'\right]_0^\ell = 0, \tag{9.41b}$$

$$\left[\left(-EI_y u_C''' - P\left(u_C' + y_C\theta'\right)\right)\delta u_C\right]_0^\ell = 0, \tag{9.41c}$$

$$\left[EI_x v_C''\,\delta v_C'\right]_0^\ell = 0, \tag{9.41d}$$

$$\left[\left(-EI_x v_C''' - P\left(v_C' - x_C\theta'\right)\right)\delta v_C\right]_0^\ell = 0, \tag{9.41e}$$

$$\left[\left(GJ\theta' - EI_\omega\theta''' - P\left(r_C^2\theta' + y_C u_C' - x_C v_C'\right)\right)\delta\theta\right]_0^\ell = 0, \tag{9.41f}$$

$$\left[EI_\omega\theta''\,\delta\theta'\right]_0^\ell = 0. \tag{9.41g}$$

The field equations, together with their boundary conditions, show that the axial displacement w_G is uncoupled from the other displacements. Since the $-EA\partial_z^2$ operator is non-singular and the boundary conditions are homogeneous, Eqs. 9.40a and 9.41a *only admit the trivial solution* $w_G = 0$. Hence, the buckling mode (at the first order) is inextensional, as already found for the Euler beam.

The remaining equations in u_C, v_C, θ constitute a twelfth order differential problem and therefore difficult to solve analytically. In the following, a particular case is illustrated, whose solution is very simple. Later on (Sect. 9.8), the general problem will be tackled by numerical methods.

Supplement 9.4 (Interpretation of the Adjacent Equilibrium) The Eqs. 9.40b–d express the equilibrium in the adjacent configuration, whose meaning will be discussed here. As illustrated in the Sect. 6.4, it needs to evaluate the unbalanced internal forces, proportional to the prestress, which are consequent to the infinitesimal change of configuration. For major clarity, the effects of translation and rotation will be discussed separately.

• When $\theta = 0$, the cross-section undergoes pure translations u_C, v_C, which entails the birth of distributed transverse forces (funicular bearing capacity) $p_x := Pu_C''$, $p_y := Pv_C''$, which are applied at the centroid and appear in the left member of

Fig. 9.3 Geometric torsional
bearing capacity of the
compressed beam

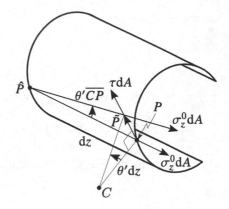

Eqs. 9.40b, c. When these forces are moved to the center of torsion, they give
rise to the transport couple $c_z := p_x y_C - p_y x_C = P \left(u''_C y_C - v''_C x_C \right)$, present in
Eq. 9.40d.

- When the cross-section undergoes a twist $\theta \neq 0$ with $u_C = 0$, $v_C = 0$, the
generic longitudinal fiber of the beam, of trace P and area dA, behaves as a
string presolicited by a tensile force $\sigma_z^0 dA$ (Fig. 9.3). By considering the relative
rotation $d\theta = \theta' dz$ between two cross-sections at distance dz, the fiber inclines
on the section of an angle $\theta' \overline{CP}$, where $\overline{CP}$ is its distance from the center of
torsion.

The inclination of the elementary presolicitation force, produces at P an
incremental tangential force, of modulus $\sigma_z^0 \theta' \overline{CP} dA$, directed along the normal
to $\overrightarrow{CP} = (x - x_C) \mathbf{a}_x + (y - y_C) \mathbf{a}_y$, and therefore equal to:

$$\tau dA = \sigma_z^0 \theta' \mathbf{a}_z \times \overrightarrow{CP} dA. \tag{9.42}$$

The field of these forces is statically equivalent to shear forces and an incremental
torque, given by:

$$T_x = \int_\Gamma \boldsymbol{\tau} \cdot \mathbf{a}_x b ds = \sigma_z^0 \theta' \int_\Gamma \mathbf{a}_x \times \mathbf{a}_z \cdot \overrightarrow{CP} b ds = \frac{P}{A} \theta' \int_\Gamma (y - y_C) b ds = -P\theta' y_C, \tag{9.43a}$$

$$T_y = \int_\Gamma \boldsymbol{\tau} \cdot \mathbf{a}_y b ds = \sigma_z^0 \theta' \int_\Gamma \mathbf{a}_y \times \mathbf{a}_z \cdot \overrightarrow{CP} b ds = -\frac{P}{A} \theta' \int_\Gamma (x - x_C) b ds = P\theta' x_C, \tag{9.43b}$$

$$M_z = \int_\Gamma \tau \, \overline{CP} \, b ds = \sigma_z^0 \theta' \int_\Gamma \overline{CP}^2 b ds = -\frac{P}{A} \theta' I_C = -P\theta' r_C^2. \tag{9.43c}$$

From the equilibrium equations of the one-dimensional beam:

$$T_x' + p_x = 0, \qquad\qquad M_y' + T_x = 0, \qquad\qquad (9.44a)$$

$$T_y' + p_y = 0, \qquad\qquad M_x' - T_y = 0, \qquad\qquad (9.44b)$$

$$M_z' + c_z = 0, \qquad\qquad (9.44c)$$

one gets: $p_x = -T_x' = P\theta''y_C$, $p_y = -T_y' = -P\theta''x_C$, $c_z = -M_z' = P\theta''r_C^2$, which appear in Eqs. 9.40 b–d, respectively. The first two terms represent the contribution of the torsion to the funicular bearing capacity; the last term is the *geometric torsional bearing* of the compressed TWB. The discussion clarifies the mechanism of the flexural-torsional coupling: torsion induces transverse displacements of the longitudinal fibers, which add themselves to the translations, thus participating to the funicular lift of each fiber.

□

9.4.2   Uniformly Compressed Beam, Simply Resting on Warping-Unrestrained Torsional Supports

Equations 9.40 and 9.41, as observed, are difficult to integrate in closed form for general boundary conditions. However, there exist a special case in which the solution is very easy, namely, when the TWB is simply supported in bending and torsionally clamped at the ends by devices not restraining warping, which, therefore, is free (Fig. 9.4). This simple case, far from representing an academic exercise, allows instead to draw important qualitatively information about the mechanical behavior of compressed TWB, which are valid even in more general constraint conditions.[7]

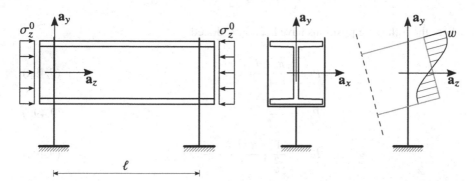

Fig. 9.4 Uniformly compressed beam, supported and torsionally clamped at the ends, with unrestrained warping

---

[7] A confirmation of this can be found in the Problem 14.3.1, concerning different boundary conditions, for which one has to resort to the Ritz method.

The boundary conditions in Eqs. 9.41, for the above described constraints become:

$$u_C = v_C = \theta = 0, \qquad\qquad \text{at } z = 0, \ell, \qquad\qquad (9.45a)$$

$$u_C'' = v_C'' = \theta'' = 0, \qquad\qquad \text{at } z = 0, \ell, \qquad\qquad (9.45b)$$

where Eqs. 9.45b express the zeroing of the bending moments and of the bimoment at the ends. These boundary conditions suggest a "guess" harmonic solution that satisfies them identically:[8]

$$\begin{pmatrix} u_C(z) \\ v_C(z) \\ \theta(z) \end{pmatrix} = \begin{pmatrix} \hat{u}_C \\ \hat{v}_C \\ \hat{\theta} \end{pmatrix} \sin\left(\frac{\pi z}{\ell}\right), \qquad\qquad (9.46)$$

where $\hat{u}_C$, $\hat{v}_C$, and $\hat{\theta}$ are unknown amplitudes. Since the field equations only involve even order derivatives of the displacements, Eq. 9.46 leads to the following *algebraic eigenvalue problem*:

$$\left\{ \begin{bmatrix} \frac{\pi^2 E I_y}{\ell^2} & 0 & 0 \\ 0 & \frac{\pi^2 E I_x}{\ell^2} & 0 \\ 0 & 0 & \frac{\pi^2 E I_\omega}{\ell^2} + GJ \end{bmatrix} - P \begin{bmatrix} 1 & 0 & y_C \\ 0 & 1 & -x_C \\ y_C & -x_C & r_C^2 \end{bmatrix} \right\} \begin{pmatrix} \hat{u}_C \\ \hat{v}_C \\ \hat{\theta} \end{pmatrix} = \begin{pmatrix} 0 \\ 0 \\ 0 \end{pmatrix}. \qquad (9.47)$$

These equations can be more conveniently rewritten as:

$$\begin{bmatrix} P_y - P & 0 & -P y_C \\ 0 & P_x - P & P x_C \\ -P y_C & P x_C & r_C^2 (P_\theta - P) \end{bmatrix} \begin{pmatrix} \hat{u}_C \\ \hat{v}_C \\ \hat{\theta} \end{pmatrix} = \begin{pmatrix} 0 \\ 0 \\ 0 \end{pmatrix}, \qquad\qquad (9.48)$$

where the following positions have been introduced:

$$P_x := \frac{\pi^2 E I_x}{\ell^2}, \qquad\qquad (9.49a)$$

$$P_y := \frac{\pi^2 E I_y}{\ell^2}, \qquad\qquad (9.49b)$$

$$P_\theta := \frac{1}{r_C^2} \left(\frac{\pi^2 E I_\omega}{\ell^2} + GJ \right). \qquad\qquad (9.49c)$$

Here:

---

[8] It is easy to check that harmonic solutions with a higher wave-number provide higher critical loads.

- P_x is the *Eulerian critical* load related to bending of axis x (in which the beam buckles in the y, z plane).
- P_y is the *Eulerian critical* load related to bending of axis y (in which the beam buckles in the x, z plane).
- P_θ is the *purely torsional critical load*, the meaning of which will emerge shortly from the analysis.

The algebraic problem in Eq. 9.48 admits non-trivial solution if and only if the determinant of coefficient matrix vanishes, i.e., if the following characteristic equation is satisfied:

$$f(P) := r_C^2 (P_\theta - P)(P_x - P)(P_y - P)$$
$$- P^2 \left[y_C^2 (P_x - P) + x_C^2 (P_y - P) \right] = 0. \qquad (9.50)$$

The roots of the polynomial $f(P)$ are the three critical loads of the compressed TWB, relative to the wave-number π/ℓ. The associated eigenvectors, computed solving Eq. 9.48, describe the critical modes.

Non-symmetric Cross-Section

Assuming, for example, that $P_x < P_y < P_\theta$ (Fig. 9.5), one wonders what relationship exists between the smallest solution of the characteristic equation, Eq. 9.50, and P_x, P_y, and P_θ. By observing that $f(0) = r_C^2 P_\theta P_x P_y > 0$, while $f(P_x) = -P_x^2 (P_y - P_x) x_C^2 < 0$, since the $f(P)$ function is continuous, the smallest root P_c falls between 0 and P_x (Fig. 9.5). By permuting the values of $P_x, P_y, and P_\theta$, similar results are obtained, so that $P_c < \min \left[P_x, P_y, P_\theta \right]$.

The critical mode generally contains all three displacement components:

$$\hat{\mathbf{u}} = \left(\hat{u}_C, \hat{v}_C, \hat{\theta} \right)^T, \qquad (9.51)$$

Fig. 9.5 Graph of the characteristic polynomial in Eq. 9.50, relevant to the eigenvalue problem in Eq. 9.48

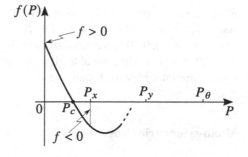

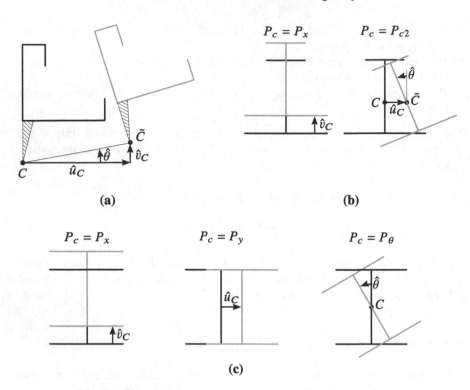

Fig. 9.6 Buckling modes of compressed thin-walled beams, with: (**a**) non-symmetric, (**b**) mono-symmetric, (**c**) bi-symmetric cross-sections

so that the TWB bends simultaneously in the two principal planes, and twists (Fig. 9.6a).

Remark 9.2 The flexural-torsional coupling *reduces the critical Eulerian load*. If coupling were ignored, and the smaller between the two Eulerian critical loads P_x and P_y were taken, one would overestimate the critical value, to detriment of safety.

Remark 9.3 If the TWB has high torsional stiffness, it is $P_\theta \gg P_x$, $P_\theta \gg P_y$. Assumed $P_x = O(1)$, $P_y = O(1)$, $P_\theta = O(\epsilon^{-1})$, with $\epsilon \ll 1$, in order for the first term in Eq. 9.50 is balanced by the other two, it must be $P_c = P_x + O(\epsilon)$, or $P_c = P_y + O(\epsilon)$. The critical load, therefore, is close to that of the Euler beam, and torsion does not play a relevant role.

Mono-symmetric Cross-Section

A mono-symmetric TWB, possessing an axis of symmetry, e.g, the y axis, is considered. Since the center of torsion lies on this axis, it is $x_C = 0$, $y_C \neq 0$, so that the characteristic Eq. 9.50 simplifies as follows:

$$(P_x - P)\left[r_C^2\,(P_\theta - P)\,(P_y - P) - P^2 y_C^2\right] = 0, \tag{9.52}$$

and factorizes into:

$$P_x - P = 0, \tag{9.53a}$$

$$r_C^2\,(P_\theta - P)\,(P_y - P) - P^2 y_C^2 = 0. \tag{9.53b}$$

From Eq. 9.53a, $P_{c1} = P_x$ is drawn, whose corresponding eigenvector is:

$$\hat{\mathbf{u}}_1 = \left(0,\ \hat{v}_C,\ 0\right)^T. \tag{9.54}$$

This describes an Eulerian mode, in which *the TWB buckles in the symmetry* plane (Fig. 9.6b). From Eq. 9.53b, two additional roots are found:

$$P_{c2,3} = \frac{P_y + P_\theta \mp \sqrt{\left(P_y + P_\theta\right)^2 - 4 P_y P_\theta \left(1 - \frac{y_C^2}{r_C^2}\right)}}{2\left(1 - \frac{y_C^2}{r_C^2}\right)}, \tag{9.55}$$

with $P_{c2} < P_{c3}$. The corresponding eigenvectors are both of flexural-torsional type:

$$\hat{\mathbf{u}}_{2,3} = \left(\hat{u}_C,\ 0,\ \hat{\theta}\right)^T, \tag{9.56}$$

in which the TWB *buckles orthogonally to the plane of symmetry and twists* (Fig. 9.6b). The critical load of the beam, ultimately, is:

$$P_c = \min\left[P_{c1},\ P_{c2}\right], \tag{9.57}$$

that is, is the smallest between (i) the Eulerian load in the plane of symmetry and (ii) the lowest flexural-torsional critical load.

Remark 9.4 The flexural-torsional coupling disappears in the plane of symmetry, while it persists in the plane of asymmetry.

Bi-symmetric Cross-Section

If the profile is bi-symmetric, the torsion center coincides with the centroid, that is, $x_C = y_C = 0$. Equation 9.50 further simplifies into:

$$r_C^2\,(P_x - P)\,(P_\theta - P)\,(P_y - P) = 0. \tag{9.58}$$

From this equation, it follows:

$$P_{c1} = P_x, \tag{9.59a}$$

$$P_{c2} = P_y, \tag{9.59b}$$

$$P_{c3} = P_\theta. \tag{9.59c}$$

The corresponding critical modes, respectively, are (Fig. 9.6c):

$$\hat{\mathbf{u}} = \left(0, \ \hat{v}_C, \ 0\right)^T, \tag{9.60a}$$

$$\hat{\mathbf{u}} = \left(\hat{u}_C, \ 0, \ 0\right)^T, \tag{9.60b}$$

$$\hat{\mathbf{u}} = \left(0, \ 0, \ \hat{\theta}\right)^T, \tag{9.60c}$$

which describe, in the order, two bendings in the principal inertia planes, and a *pure torsion*. The critical load is therefore:

$$P_c = \min\left[P_x, \ P_y, \ P_\theta\right]. \tag{9.61}$$

Remark 9.5 The flexural-torsional coupling disappears completely when the TWB is bi-symmetric, it being related to the eccentricity between the center of torsion and centroid.

Remark 9.6 The instability due to pure torsion is a peculiar characteristic of the bi-symmetric open TWB. In this motion, the elastic forces are balanced by the geometric torsional bearing previously discussed (Fig. 9.3). From Eq. 9.49c, if ℓ is large or I_ω is small, or even zero,[9] one has:

$$P_\theta = \frac{GJ}{r_C^2}. \tag{9.62}$$

In these cases, *the critical load is independent of the length of the beam*. This is a consequence of the fact that the DSV elastic forces and the geometric forces both depend on the second derivative of the twist angle.

Remark 9.7 According to the Vlasov theory, in the degenerate case $I_\omega = 0$, the critical mode has an indeterminate form. The model without warping is therefore inadequate to describe the phenomenon. Indeed, warping, in a thin rectangular section, although is zero on the midline, is non-zero and linearly variable on the chord, as predicted from the DSV theory.

---

[9] $I_\omega = 0$ occurs, e.g., for thin rectangular, T, or cruciform cross-sections, for which the warping function is identically zero, since the shear center is located at the intersection of the middle lines of the component walls.

9.5 Uniformly Bent Thin-Walled Beams

The critical value of the bending couple soliciting a TWB is determined [4, 10].

9.5.1 Formulation

A TWB is considered, which, in the prestressed state, is subject to bi-axial uniform bending, induced by couples $\pm \mathbf{C}$ applied at the ends, with $\mathbf{C} = C_x \mathbf{a}_x + C_y \mathbf{a}_y$. Since the bending moment is constant, equal to $M_x^0 (z) = C_x$, $M_y^0 (z) = C_y$, the stress is described by Navier binomial formula:

$$\sigma_z^0 = \frac{C_x}{I_x} y - \frac{C_y}{I_y} x. \tag{9.63}$$

The prestress energy is equal to the work expended by these stresses on the second order part of the unit extensions, $\varepsilon_z^{(2)}$. By using Eqs. 9.29a and 9.2, one obtains:

$$
\begin{aligned}
U^0 [\mathbf{u}_\pi] &= \frac{1}{2} \int_0^\ell dz \int_\Gamma \left(\frac{C_x}{I_x} y - \frac{C_y}{I_y} x \right) \Big\{ \left[u_C' - \theta' (y - y_C) \right]^2 \\
&\quad + \left[v_C' + \theta' (x - x_C) \right]^2 \Big\} b ds \\
&= \frac{1}{2} \int_0^\ell dz \int_\Gamma \left(\frac{C_x}{I_x} y - \frac{C_y}{I_y} x \right) \Big\{ u_C'^2 + v_C'^2 + \theta'^2 \left[(y - y_C)^2 + (x - x_C)^2 \right] \\
&\quad - 2 u_C' \theta' (y - y_C) + 2 v_C' \theta' (x - x_C) \Big\} b ds.
\end{aligned}
\tag{9.64}
$$

Taking into account that the static moments and the mixed inertia moment are zero, and by letting:

$$x_H := x_C - \frac{1}{2 I_y} \int_\Gamma x \left(x^2 + y^2 \right) b ds, \tag{9.65a}$$

$$y_H := y_C - \frac{1}{2 I_x} \int_\Gamma y \left(x^2 + y^2 \right) b ds, \tag{9.65b}$$

one gets:

$$U^0 [\mathbf{u}_\pi] = - \int_0^\ell \left[C_x \left(u_C' + y_H \theta' \right) \theta' + C_y \left(v_C' - x_H \theta' \right) \theta' \right] dz. \tag{9.66}$$

Remark 9.8 Equation 9.65 define the coordinates of a new point H (distinct from G and C), which is also a geometrical characteristic. If the cross-section has an axis of symmetry, H falls on this axis; if the cross-section has two axes of symmetry, H coincides with the centroid.

Geometric Stiffness Operator

Executing the first variation of the prestress energy, and integrating by parts, one finds:

$$
\begin{aligned}
\delta U^0 = &- \int_0^\ell \left\{ \left[C_x \left(u_C' + 2 y_H \theta' \right) + C_y \left(v_C' - 2 x_H \theta' \right) \right] \delta \theta' + C_x \theta' \delta u_C' \right. \\
&\left. + C_y \theta' \delta v_C' \right\} dz \\
= &\int_0^\ell \left\{ \left[C_x u_C'' + C_y v_C'' + 2 \left(y_H C_x - x_H C_y \right) \theta'' \right] \delta \theta + C_x \theta'' \delta u_C \right. \\
&\left. + C_y \theta'' \delta v_C \right\} dz \\
&+ \left[- \left(C_x u_C' + C_y v_C' + 2 \left(y_H C_x - x_H C_y \right) \theta' \right) \delta \theta - C_x \theta' \delta u_C \right. \\
&\left. - C_y \theta' \delta v_C \right]_0^\ell .
\end{aligned}
\tag{9.67}
$$

By letting $\delta U^0 = \int_0^\ell \delta \mathbf{u}_\pi^T \boldsymbol{\mathcal{K}}_{g,b} \mathbf{u}_\pi \, dz + [\cdots]_0^\ell$, with $\mathbf{u}_\pi := (u_C, v_C, \theta)^T$, the *geometric stiffness operator of the uniformly bent TWB* is recognized to be:

$$
\boldsymbol{\mathcal{K}}_{g,b} := \begin{bmatrix} 0 & 0 & C_x \\ 0 & 0 & C_y \\ C_x & C_y & 2 \left(y_H C_x - x_H C_y \right) \end{bmatrix} \partial_z^2 .
\tag{9.68}
$$

The geometric operator is generally coupled.

Equilibrium Equations

By adding the elastic and geometric contributions and taking into account that $w_G \equiv 0$, the indefinite equilibrium equations are obtained:

$$
\left(\boldsymbol{\mathcal{K}}_e^\pi + \boldsymbol{\mathcal{K}}_{g,b} \right) \mathbf{u}_\pi = \mathbf{0},
\tag{9.69}
$$

as well as the boundary conditions. In extended form, they read:

$$
E I_y u_C'''' + C_x \theta'' = 0,
\tag{9.70a}
$$

$$
E I_x v_C'''' + C_y \theta'' = 0,
\tag{9.70b}
$$

$$EI_\omega\theta'''' - GJ\theta'' + C_x u_C'' + C_y v_C'' + 2\left(y_H C_x - x_H C_y\right)\theta'' = 0, \qquad (9.70c)$$

together with:

$$\left[EI_y u_C'' \, \delta u_C'\right]_0^\ell = 0, \qquad (9.71a)$$

$$\left[\left(-EI_y u_C''' - C_x\theta'\right)\delta u_C\right]_0^\ell = 0, \qquad (9.71b)$$

$$\left[EI_x v_C'' \, \delta v_C'\right]_0^\ell = 0, \qquad (9.71c)$$

$$\left[\left(-EI_x v_C''' - C_y\theta'\right)\delta v_C\right]_0^\ell = 0, \qquad (9.71d)$$

$$\left[\left(GJ\theta' - EI_\omega\theta''' - C_x u_C' - C_y v_C' - 2\left(y_H C_x - x_H C_y\right)\theta'\right)\delta\theta\right]_0^\ell = 0, \qquad (9.71e)$$

$$\left[EI_\omega\theta'' \, \delta\theta'\right]_0^\ell = 0. \qquad (9.71f)$$

Supplement 9.5 (Interpretation of the Adjacent Equilibrium) The field Eqs. 9.70 express the equilibrium of the TWB in the deformed configuration, adjacent to the prestressed one. They can be interpreted as follows. By using the superposition of the effects, reference is made to a beam solicited by a mono-axial bending moment $M_x^0(z) = C_x$.

- When the cross-section rotates by an angle θ (Fig. 9.7a), the normal stresses σ_z^0, which are joined with the cross-section, also rotate, so that their resultant is a couple C_x rotated by an angle θ. This has a component in the $\mathbf{a}_x$ direction, equal to (about) C_x, as in the prestressed state, but also a $M_y := C_x\theta$ component in the $\mathbf{a}_y$ direction, which is an incremental moment of geometric type. According to the equilibrium Eqs. 9.44, $p_x := M_y'' = C_x\theta''$ is a distributed force in the $\mathbf{a}_x$ direction, which represents the funicular bearing capacity of the bent TWB, as it appears in Eq. 9.70a.
- When the beam buckles in the $\mathbf{a}_x$ direction, the generic cross-section at the abscissa z rotates by an angle $\varphi_y = u_C'$ around the $\mathbf{a}_y$ axis (Fig. 9.7b). The normal stresses, joined with the cross-section, give rise to an incremental bending moment of $\mathbf{a}_z$ axis, equal to $M_z := -C_x u_C'$. Its first derivative, $c_z := -M_z' = C_x u_C''$ is a distributed axial torque, which appears in Eq. 9.70c.
- In order to interpret the geometric torsional bearing capacity in Eq. 9.70c, it needs to repeat the reasoning developed in the Supplement 9.4 about the compressed TWB. Due to the twist, the longitudinal fibers incline, as a function of their distance from the center of torsion, so that the stresses σ_z^0, which follow the fibers, provide incremental forces tangent to the plane $(\mathbf{a}_x, \mathbf{a}_y)$, in modulus equal to $\tau\,dA = \sigma_z^0\theta'\overline{CP}dA$, directed normally to $\overline{CP}$, whose torque with respect to the center of torsion is $dM_z = \sigma_z^0\theta'\overline{CP}^2 dA$. Differently from the case examined above, however, σ_z^0 is not constant on the cross-section but variable with a linear law, according to Navier formula. Thus, integrating on the cross-section and taking into account the definition in Eq. 9.65b, one gets:

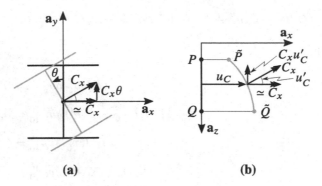

(a)   **(b)**

Fig. 9.7 Equilibrium in the adjacent configuration of the beam subject to constant bending moment of axis x: (**a**) incremental y-axis bending moment, induced by the twist angle; (**b**) incremental torsional moment, induced by the y-axis bending

$$M_z = \frac{C_x}{I_x}\theta' \int_\Gamma y\left[(x - x_C)^2 + (y - y_C)^2\right] b\,\mathrm{d}s$$

$$= \frac{C_x}{I_x}\theta'\left[\int_\Gamma y\left(x^2 + y^2\right) b\,\mathrm{d}s - 2I_x y_C\right] = -2y_H C_x^0\theta'. \tag{9.72}$$

By differentiating with respect to z, the distributed couple $c_z := -\frac{\mathrm{d}M_z}{\mathrm{d}z} = 2y_H C_x\theta''$ is found, which appears in the equilibrium equation. This, therefore, represents the *geometric torsional bearing of the bent TWB*. Finally, if the shear force T_x, equivalent to these tangential forces, is evaluated, similarly to what was done in Eq. 9.43a, one gets $T_x = -C_x\theta'$, hence $p_x = -T_x' = C_x\theta''$. This result, now acquired by integrating the stresses, is identical to that one previously found by operating on the solicitations, being based on the same kinematics.

□

9.5.2 Uniformly Bent Beam, Simply Resting on Warping-Unrestrained Torsional Supports

As an example of application of the model developed so far, a TWB is analyzed, simply supported and torsionally clamped at the ends by constraints which leave the warping free.[10] For this beam, the boundary conditions in Eqs. 9.71 specialize as follows:

---

[10] A uniformly bent beam, with different boundary conditions, is studied by the Ritz method in the Problem 14.3.2.

$$u_C = v_C = \theta = 0, \qquad\qquad \text{at } z = 0, \ell, \qquad\qquad (9.73a)$$

$$u_C'' = v_C'' = \theta'' = 0, \qquad\qquad \text{at } z = 0, \ell. \qquad\qquad (9.73b)$$

A harmonic solution is guessed, as in Eq. 9.46, that satisfies the boundary conditions and reduces the field equation to an algebraic eigenvalue problem, i.e.:

$$\begin{bmatrix} P_y & 0 & -C_x \\ 0 & P_x & -C_y \\ -C_x & -C_y & r_C^2 P_\theta - 2\left(y_H C_x - x_H C_y\right) \end{bmatrix} \begin{pmatrix} \hat{u}_C \\ \hat{v}_C \\ \hat{\theta} \end{pmatrix} = \begin{pmatrix} 0 \\ 0 \\ 0 \end{pmatrix}, \qquad (9.74)$$

where the definitions in Eqs. 9.49 have been used. Zeroing the determinant of the coefficient matrix, the characteristic equation is obtained:

$$P_x P_y \left[r_C^2 P_\theta - 2\left(y_H C_x - x_H C_y\right) \right] - C_x^2 P_x - C_y^2 P_y = 0. \qquad (9.75)$$

Equation 9.75 is the equation of an ellipse in the (C_x, C_y) plane, which contains the origin, of axes parallel to the coordinate ones (Fig. 9.8).[11] The ellipse represents the locus of the critical states, that is, of the combinations between the two components of the external couple that induce incipient instability of the TWB. The equilibrium of the bent beam is stable inside the ellipse and unstable out of it. If the solicitation is increased proportionally to a multiplier load μ, the representative point moves along a straight line outgoing from the origin, until it intersects the ellipse at a critical point.

Fig. 9.8 Stability domain of the thin-walled beam, simply resting on warping-unrestrained torsional supports; S stable region, U unstable region

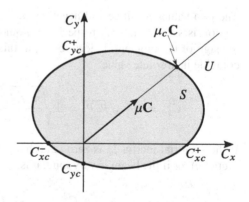

---

[11] Indeed, the Eq. 9.75 can also be written:

$$\frac{\left(C_x + y_H P_y\right)^2}{P_y \left(x_H^2 P_x + y_H^2 P_y + r_C^2 P_\theta\right)} + \frac{\left(C_y - x_H P_x\right)^2}{P_x \left(x_H^2 P_x + y_H^2 P_y + r_C^2 P_\theta\right)} = 1.$$

The buckling mode involves all three displacement components, whether the cross-section is non-symmetric or symmetric, so that the beam bends in both planes and twists.

Mono-axial Bending of a Generic Cross-Section

If bending is mono-axial, e.g., it is $C_x \neq 0$, $C_y = 0$, the critical values $C_{xc}^{\pm}$ are obtained from the intersection of the ellipse with the abscissa axis:

$$C_{xc}^{\pm} = -y_H P_y \pm \sqrt{\left(y_H P_y\right)^2 + r_C^2 P_\theta P_y}. \tag{9.76}$$

The above equation shows that, being $|C_{xc}^{+}| \neq |C_{xc}^{-}|$, the critical value is sensitive to the change in the sign of the couple, since it implies the exchange between the tensed and compressed regions of the cross-section.

At the bifurcation, since $\hat{\mathbf{u}} = \left(\hat{u}_C, 0, \hat{\theta}\right)^T$, the *beam buckles in the direction normal to the plane of inflection, while twisting*. For this reason the phenomenon is called of *lateral instability*. It should be noticed, that $\hat{v}_C = 0$ is not a consequence of symmetry, as it was in the case of compression, but it is rather a consequence the absence of C_y.

Mono-axial Bending of a Symmetric Cross-Section with Respect to the Moment Axis

The two values, positive and negative, of $C_{xc}^{\pm}$, are distinct. However, if the cross-section is symmetric with respect the moment axis x, the profile is insensitive to the sign of the moment. Indeed, being in this case $y_H = 0$, the two critical values coincide in absolute value:

$$C_{xc}^{\pm} = \pm r_C \sqrt{P_\theta P_y} = \pm \sqrt{\frac{\pi^2 E I_y}{\ell^2} \left(\frac{\pi^2 E I_\omega}{\ell^2} + G J\right)}. \tag{9.77}$$

If the sectorial inertia I_ω is zero (as it happens for rectangular or cruciform cross-sections), or if the beam is very long, it is:

$$C_{xc}^{\pm} = \pm \frac{\pi}{\ell} \sqrt{E I_y \, G J}. \tag{9.78}$$

This value is known as the *Prandtl critical load*.

Remark 9.9 The buckling mode of a bent TWB always involves twist, regardless of any symmetry of the cross-section. This occurrence is explained by the mechanical interpretation given in the Supplement 9.5, according to which a lateral displacement generates a geometric torsional couples, and a torsion generates lateral geometric forces.

9.6 Eccentrically Compressed Thin-Walled Beams

The critical value of the eccentric axial force soliciting a TWB is determined [3, 10, 12]. The possible occurrence of bifurcations induced by *tensile* forces is also investigated.

9.6.1 Formulation

When the TWB is subject to a compression force P, applied at a point $S = (x_S, \ y_S)$, called the *center of solicitation*, other than the centroid G, the normal stresses are given by the Navier trinomial formula:

$$\sigma_z^0 = -\frac{P}{A} + C_x \frac{y}{I_x} - C_y \frac{x}{I_y}$$
$$= -P \left(\frac{1}{A} + \frac{y_S y}{I_x} + \frac{x_S x}{I_y} \right),$$

(9.79)

being $C_x = -Py_S$, $C_y = Px_S$.

To determine the geometric operator $\mathcal{K}_{g,cb}$ relevant to this state of prestress, one can superimpose the effects of compression and bending, by exploiting the linearity of the problem. By remembering the expressions of $\mathcal{K}_{g,c}$ (compressed beam, Eq. 9.38) and of $\mathcal{K}_{g,b}$ (bent beam, Eq. 9.68), the *geometric stiffness operator of the eccentrically compressed TWB* is obtained as $\mathcal{K}_{g,cb} := \mathcal{K}_{g,c} + \mathcal{K}_{g,b}$, i.e.:

$$\mathcal{K}_{g,cb} := P \begin{bmatrix} 1 & 0 & (y_C - y_S) \\ 0 & 1 & -(x_C - x_S) \\ (y_C - y_S) & -(x_C - x_S) & \left[r_C^2 - 2(y_H y_S + x_H x_S) \right] \end{bmatrix} \partial_z^2. \quad (9.80)$$

To derive the differential equilibrium equations, the geometric operator must be added to the elastic operator (Eq. 9.26b), thus obtaining:

$$\left(\mathcal{K}_e^\pi + \mathcal{K}_{g,cb} \right) u_\pi = 0. \quad (9.81)$$

The same procedure must be followed for the boundary conditions.

9.6.2   Eccentrically Compressed Beam, Simply Resting on Warping-Unrestrained Torsional Supports

If the TWB is simply supported, torsionally clamped and free to warp at the ends, the relevant eigenvalue problem is derived by using the harmonic solution in Eq. 9.46, which leads to:[12]

$$
\begin{bmatrix}
P_y - P & 0 & -P(y_C - y_S) \\
0 & P_x - P & P(x_C - x_S) \\
-P(y_C - y_S) & P(x_C - x_S) & \left[r_C^2(P_\theta - P) + 2P(y_H y_S + x_H x_S) \right]
\end{bmatrix}
\begin{pmatrix} \hat{u}_C \\ \hat{v}_C \\ \hat{\theta} \end{pmatrix}
= \begin{pmatrix} 0 \\ 0 \\ 0 \end{pmatrix},
$$
(9.82)

where the definitions in Eqs. 9.49 have been used. This, of course, is a combination of the algebraic problems in Eqs. 9.48, 9.74.

The associated characteristic equation is written as:

$$
f(P) := P^3 \left[(x_C - x_S)^2 + (y_C - y_S)^2 + 2(x_H x_S + y_H y_S) - r_C^2 \right]
$$
$$
+ P^2 \left\{ (P_x + P_y) \left[r_C^2 - 2(x_H x_S + y_H y_S) \right] + r_C^2 P_\theta - P_y(x_C - x_S)^2 \right.
$$
$$
\left. - P_x(y_C - y_S)^2 \right\} - P \left[(P_x + P_y) r_C^2 P_\theta + P_x P_y \left(r_C^2 - 2x_H x_S - 2y_H y_S \right) \right]
$$
$$
+ r_C^2 P_\theta P_x P_y = 0,
$$
(9.83)

whose smallest root is the critical load P_c. With reasoning analogous to that done for the simply compressed beam, it is proved that $P_c < \min \left[P_x, P_y, P_\theta \right]$.

Instability Due to an Eccentric Tensile Force

It can be shown that the *critical load can be negative* (denoting extension and bending of the TWB), this happening when the *eccentricity of the solicitation center is large enough*. The condition is suggested by the fact that an eccentric tensile force generates normal stresses of both signs, which are stabilizing when of traction, and instabilizing when of compression. In order for the instabilizing effect to prevail, the compressed region must be sufficiently extended, and, consequently, the eccentricity sufficiently large.[13]

To determine the locus of the points S *limit of stability of tensile forces*, it needs to find solutions $P_c \to -\infty$ of the characteristic Eq. 9.83. In this limit, the equation requires to vanish the coefficient of the cubic term, i.e.:

---

[12] A compressed and uniformly bent beam, with different boundary conditions, is studied by the Ritz method in the Problem 14.3.3.

[13] Said in other words, the (always existing) instabilizing effect of bending should prevail on the stabilizing effect of the tensile force.

$$x_S^2 + y_S^2 - 2\left[x_S\left(x_C - x_H\right) + y_S\left(y_C - y_H\right)\right] = r_G^2, \tag{9.84}$$

where $r_G^2 := r_C^2 - \left(x_C^2 + y_C^2\right)$ is the squared inertia radius of the cross-section with respect to the centroid. This is the equation of a circumference with radius R_t and center T, given by:

$$R_t = \sqrt{r_G^2 + (x_C - x_H)^2 + (y_C - y_H)^2}, \tag{9.85a}$$

$$T = (x_C - x_H, y_C - y_H). \tag{9.85b}$$

Solicitation centers S inside the circumference *do not* induce instability by traction; S centers outside the circumference produce instability by traction, *provided the tensile force is sufficiently large* in modulus. Some remarkable particular cases are now examined.

Solicitation Center Coincident with the Torsion Center

If $S \equiv C$, the eigenvalue problem in Eq. 9.82 uncouples (even for non-symmetric cross-sections). The three eigenvalues are therefore equal to the coefficients on the main diagonal of the matrix, i.e.:

$$P_{c1} = P_x, \tag{9.86a}$$

$$P_{c2} = P_y, \tag{9.86b}$$

$$P_{c3} = \frac{r_C^2 P_\theta}{r_C^2 - 2\left(x_C x_H + y_C y_H\right)}. \tag{9.86c}$$

The relevant eigenvectors are also uncoupled. This case generalizes that of simple compression, where uncoupling occurs only for bi-symmetric cross-sections, the only ones for which the solicitation center (there coincident with the centroid) falls in the center of torsion.

Remark 9.10 The torsional critical load can assume both signs, in accordance with the discussion previously made. The phenomenon of torsional instability for traction is due to the onset of a *negative geometric torsional bearing*, which is caused by the mechanism already discussed for bending. The Example 9.1 shows an application.

Example 9.1 (Instability by Pure Torsion of an Open Annular TWB, Axially Solicited at the Torsion Center) A TWB is considered, consisting of a circular tube of radius R, cut along a generatrix, subject to a tensile/compression force P, applied at the center of torsion. Making use of the results illustrated in the Appendix C (Sect. C.4), it is $x_C = x_H = 0$, $y_C = -2R$. Using Eq. 9.65b, it follows that $y_H = y_C = -2R$, since the integral contained therein is zero. Being $r_G = R$, it turns out that $r_C =$

$\sqrt{r_G^2 + (x_C^2 + y_C^2)} = R\sqrt{5}$. The stability locus of the tensile forces (Eq. 9.85) is the circumference of center $T = (0, 0)$ and radius $R_t = R$; therefore it coincides with the midline of the annular section. Since the torsion center is outside this line, the profile can become unstable by traction. Equation 9.86c, indeed, gives $P_{c3} = -\frac{5}{3}P_\theta$.

$\square$

Solicitation Center Belonging to the Symmetry Axis of a Mono-symmetric Cross-Section

A mono-symmetric cross-section, with respect to the y axis, is considered. Being $x_C = x_H = x_S = 0$, the eigenvalue problem in Eq. 9.82 partially uncouples, leading to two different buckling patterns: (a) a bending in the symmetry plane, upon reaching $P_{c1} = P_x$; (b) a bending orthogonal to the symmetry plane, accompanied by a twist, upon reaching a load, $P_{c2,3}$, which is solution to:

$$P^2\left[r_C^2 - (y_C - y_S)^2 - 2y_H y_S\right] - P\left[(P_\theta + P_y)\,r_C^2 - 2P_y y_H y_S\right] + P_\theta P_y r_C^2 = 0.$$

$$(9.87)$$

The case generalizes that of simple compression of mono-symmetric cross-sections.

Solicitation Center Belonging to One of the Two Symmetry Axes of a Bi-symmetric Cross-Section

If the cross-section has two axes of symmetry, and the solicitation center falls on one of these axes, for example, y, it turns out that $x_C = y_C = x_H = y_H = x_S = 0$. The characteristic Eq. 9.83 then simplifies into:

$$(P_x - P)\left[\left(r_C^2 - y_S^2\right)P^2 - r_C^2\left(P_\theta + P_y\right)P + r_C^2 P_\theta P_y\right] = 0, \qquad (9.88)$$

and factorizes. One of the critical loads is of Eulerian type, $P_{c1} = P_x$; it corresponds to a buckling mode occurring in the plane of solicitation (y, z). The others two critical loads, $P_{c2,3}$, are the roots of a second degree equation; the relevant buckling modes are of flexural-torsional type, in which the beam bends in the (x, z) plane, orthogonal to that of solicitation.

Concerning the flexural-torsional critical loads, since the linear term in P is negative and the known term is positive, from Descartes rule of signs, one deduces that:

- If the eccentricity of the stress center is small, i.e., if $|y_S| < r_C$, it is $P_{c2} > 0$, $P_{c3} > 0$, i.e., the critical load is of compression only.
- If the eccentricity of the stress center is large, i.e., if $|y_S| > r_C$, it is $P_{c2} < 0$, $P_{c3} > 0$, i.e., the critical load can be either of traction and compression.

The same result is achieved by observing that, from Eq. 9.85, the limit circumference has radius $R_t = r_G \equiv r_C$ and center $T = (0, 0)$.

9.7 Non-uniformly Bent Thin-Walled Beams

The critical value of forces acting transversely to the TWB axis is determined. It is assumed that, in the precritical state, the loads induce *non-uniform bending without torsion* [1, 4, 5, 10, 12] .

9.7.1 Formulation

A TWB is considered, whose cross-section has midline Γ, with end points *M* and *N* and thickness $b(s)$. The beam is subjected to transverse forces[14] $\mathbf{p}(z) := p(z) \, \text{vers} \left(\overrightarrow{CQ} \right)$, applied to an axis parallel to z, of trace Q on the cross-section, directed along the straight line passing through the center of torsion C (Fig. 9.9a). These forces induce solicitations consisting of (a) bending moments $M_x^0(z)$ *and* $M_y^0(z)$, variable along the beam axis, and (b) shear forces, $T_x^0(z)$ *and* $T_y^0(z)$, generally also variables (Fig. 9.9b); the torsional moment, evaluated with respect to the center of torsion, is instead zero. These solicitations trigger normal prestresses, expressed by the Navier binomial formula:

$$\sigma_z^0 = \frac{M_x^0(z)}{I_x} y - \frac{M_y^0(z)}{I_y} x, \qquad (9.89)$$

and tangential prestresses τ_{zs}^0, given by the Jourawsky formula (Eq. 9.28b). The latter stresses, as known, are derived by integrating the differential equation:

$$\left(\tau_{zs}^0 b \right)_{,s} + \sigma_{z,z}^0 b = 0, \qquad (9.90)$$

which states the equilibrium along z of a segment of beam, of size $\mathrm{d}s \times \mathrm{d}z \times b(s)$ (Fig. 9.9c).[15]

---

[14] Also said "lateral forces" in the technical jargon.

[15] Equation 9.90 is the indefinite equilibrium equation, $\nabla \cdot \boldsymbol{\tau}^0 + \sigma_{z,z}^0 = 0$, averaged on thickness (membrane balance). Indeed, $\nabla = \mathbf{a}_t \frac{\partial}{\partial s} + \mathbf{a}_n \frac{\partial}{\partial n}$, with $\mathbf{a}_t, \mathbf{a}_n$ the tangent and normal unit vectors, respectively, and $\boldsymbol{\tau}^0 = \tau_{zs}^0 \mathbf{a}_t$, in the hypothesis of small thickness.

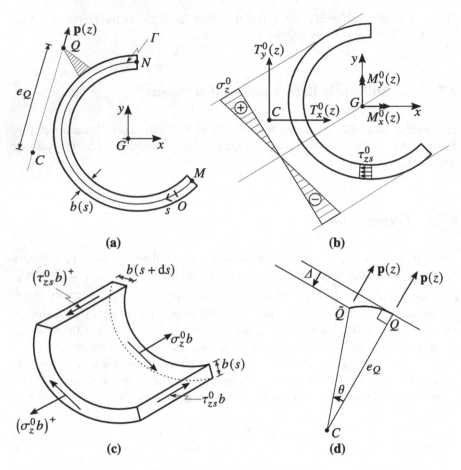

Fig. 9.9 Thin-walled beam subject to non-uniform bending: (**a**) transverse forces, (**b**) prestresses; (**c**) equilibrium of a beam segment (the apex + denotes an incremented quantity); (**d**) displacement of the point of application of the forces

Prestress and Load Energies

The prestress energy consists of two contributions:

$$U^0 := U^0_\sigma + U^0_\tau, \tag{9.91}$$

which express the work of the prestresses, respectively σ^0_z and τ^0_{zs}, in the second order components of the homologous strains (Eqs. 9.29), that is:

$$U_\sigma^0 := \int_0^\ell dz \int_\Gamma \sigma_z^0 \varepsilon_z^{(2)} \, bds = \frac{1}{2} \int_0^\ell dz \int_\Gamma \sigma_z^0 \left(u_{,z}^2 + v_{,z}^2 \right) bds, \qquad (9.92a)$$

$$U_\tau^0 := \int_0^\ell dz \int_\Gamma \tau_{zs}^0 \gamma_{zs}^{(2)} \, bds = \int_0^\ell dz \int_\Gamma \tau_{zs}^0 \left(u_{,z} u_{,s} + v_{,z} v_{,s} \right) bds. \qquad (9.92b)$$

A third contribution must be added to these two energies, namely, the work, changed in sign, expended by the loads $p(z)$ on the *second order components of the displacement of their point of application* Q (Fig. 9.9d).[16] Each contribution is separately developed ahead.

Normal Prestress Energy

The prestress energy associated with normal stresses has already been derived in the Sect. 9.5 with reference to constant bending moments (Eq. 9.66). However, by repeating those steps, it is easy to check that the expression obtained there is also valid for variable moments, so that, once the substitutions $C_x \to M_x^0(z)$, $C_y \to M_y^0(z)$ have been made, the cited equation gives:

$$U_\sigma^0 = -\int_0^\ell \left[M_x^0 \theta' \left(u_C' + y_H \theta' \right) + M_y^0 \theta' \left(v_C' - x_H \theta' \right) \right] dz, \qquad (9.93)$$

where x_H and y_H are given by Eqs. 9.65.

Tangential Stress Energy

The prestress energy associated with the tangential stresses requires, instead, a much more complex calculation. Since the planar motion of the cross-section is expressed by Eq. 9.2, repeated here:

$$u = u_C(z) - \theta(z)(y(s) - y_C), \qquad (9.94a)$$

$$v = v_C(z) + \theta(z)(x(s) - x_C), \qquad (9.94b)$$

it follows that:

---

[16] This energy, introduced here for the first time in the context of continuous systems, is of the same type frequently computed in Chap. 5, regarding gravitational loads applied to structures with finite number of degrees of freedom (e.g., the reverse pendulum).

$$u_{,s} = -\theta(z)\, y'(s), \qquad u_{,z} = u'_C(z) - \theta'(z)(y(s) - y_C), \qquad (9.95a)$$

$$v_{,s} = \theta(z)\, x'(s), \qquad v_{,z} = v'_C(z) + \theta'(z)(x(s) - x_C), \qquad (9.95b)$$

where the dash denotes differentiation with respect to the corresponding variable (s or z),[17] consequently:

$$U^0_\tau = \int_0^\ell dz \int_\Gamma \tau^0_{zs}\, \theta \left\{-y'\left[u'_C - \theta'(y - y_C)\right] + x'\left[v'_C + \theta'(x - x_C)\right]\right\} b\, ds$$

$$= \int_0^\ell \theta\left(v'_C - \theta' x_C\right) dz \int_\Gamma \tau^0_{zs}\, x'\, b\, ds - \int_0^\ell \theta\left(u'_C + \theta' y_C\right) dz \int_\Gamma \tau^0_{zs}\, y'\, b\, ds$$

$$+ \int_0^\ell \theta\, \theta'\, dz \int_\Gamma \tau^0_{zs}\, (x x' + y y')\, b\, ds.$$

$$(9.96)$$

By manipulating this expression, as described in the Supplement 9.6, one gets:

$$U^0_\tau = -\int_0^\ell \left[\left(M^0_x\right)'\theta\left(u'_C + y_H\,\theta'\right) + \left(M^0_y\right)'\theta\left(v'_C - x_H\,\theta'\right)\right] dz. \qquad (9.97)$$

Supplement 9.6 (Prestress Shear Energy) The integrals on Γ that appear in Eq. 9.96 can be executed by integrating by parts. Using the equilibrium Eq. 9.90 and successively the Navier formula in Eq. 9.89, it follows:

$$\int_\Gamma \tau^0_{zs} x'\, b\, ds = -\int_\Gamma \left(\tau^0_{zs} b\right)_{,s} x\, ds + [\cdots]^N_M = \int_\Gamma \sigma^0_{z,z}\, x\, b\, ds$$

$$= \int_\Gamma \left(\frac{\left(M^0_x\right)'}{I_x}\, y - \frac{\left(M^0_y\right)'}{I_y}\, x\right) x\, b\, ds = -\left(M^0_y\right)', \qquad (9.98)$$

$$\int_\Gamma \tau^0_{zs} y'\, b\, ds = -\int_\Gamma \left(\tau^0_{zs} b\right)_{,s} y\, ds + [\cdots]^N_M = \int_\Gamma \sigma^0_{z,z}\, y\, b\, ds$$

$$= \int_\Gamma \left(\frac{\left(M^0_x\right)'}{I_x}\, y - \frac{\left(M^0_y\right)'}{I_y}\, x\right) y\, b\, ds = \left(M^0_x\right)', \qquad (9.99)$$

---

[17] When there is ambiguity, as for the derivatives of stresses, the notation $(\cdot)_{,s}$ or $(\cdot)_{,z}$ will be used.

$$\int_\Gamma \tau_{zs}^0 \left(xx' + yy'\right) b \, ds = \frac{1}{2} \int_\Gamma \tau_{zs}^0 \left(x^2 + y^2\right)' b \, ds$$

$$= -\frac{1}{2} \int_\Gamma \left(\tau_{zs}^0 b\right)_{,s} \left(x^2 + y^2\right) ds + [\cdots]_M^N$$

$$= \frac{1}{2} \int_\Gamma \sigma_{z,z}^0 \left(x^2 + y^2\right) b \, ds \tag{9.100}$$

$$= \frac{1}{2} \int_\Gamma \left(\frac{\left(M_x^0\right)'}{I_x} y - \frac{\left(M_y^0\right)'}{I_y} x\right) \left(x^2 + y^2\right) b \, ds$$

$$= -\left(M_x^0\right)' (y_H - y_C) + \left(M_y^0\right)' (x_H - x_C),$$

where the boundary terms disappear, because $\tau_{zs}^0 = 0$ at M, N, and where the definitions in Eqs. 9.65 have been used. Therefore:

$$U_\tau^0 = \int_0^\ell \left[-\left(M_y^0\right)' \theta \left(v_C' - \theta' x_C\right) - \left(M_x^0\right)' \theta \left(u_C' + \theta' y_C\right)\right] dz$$

$$+ \int_0^\ell \theta \theta' \left[-\left(M_x^0\right)' (y_H - y_C) + \left(M_y^0\right)' (x_H - x_C)\right] dz, \tag{9.101}$$

from which Eq. 9.97 is derived. □

Total Prestress Energy

By adding the energy contributions by the normal stresses (Eq. 9.93) and tangential stresses (Eq. 9.97) and then by combining them, one obtains:

$$U^0 = -\int_0^\ell \left[\left(M_x^0 \theta\right)' \left(u_C' + y_H \, \theta'\right) + \left(M_y^0 \theta\right)' \left(v_C' - x_H \, \theta'\right)\right] dz. \tag{9.102}$$

Quadratic Load Energy

The quadratic part of the energy of the loads:

$$V = \int_0^\ell p(z) \, \Delta(z) \, dz, \tag{9.103}$$

is evaluated. Here, the forces $p(z)$ are *positive when outgoing* from Q,[18] moreover, the "shortening" $\Delta(z) > 0$ has been defined as (Fig. 9.9d):

$$\Delta = e_Q (1 - \cos\theta) = e_Q \left(\frac{1}{2}\theta^2 + \cdots \right), \tag{9.104}$$

in which a MacLaurin series expansion has been performed and the eccentricity $e_Q := \overline{CQ}$ introduced. Energy is therefore written as:

$$V := \frac{1}{2} e_Q \int_0^\ell p(z)\,\theta^2(z)\,\mathrm{d}z. \tag{9.105}$$

Remark 9.11 The sign of the work $W = -V$, done by the transverse forces $p(z)$ in the rotation $\theta(z)$, gives an indication of the stabilizing/instabilizing character of these forces. If $W < 0$, the loads are stabilizing, because they are opposite to the perturbation; if $W > 0$, they are instabilizing, because they are concordant with the perturbation. Therefore, if the transverse forces are outgoing from C (positive), they have a *stabilizing effect* (like the tensile force on the pendulum); if they are entering in C (negative), they have an *instabilizing effect* (such as the compression force on the pendulum).

Geometric Stiffness Operator

Executing the first variation of the prestress energy in Eq. 9.102, one finds:

$$\begin{aligned}
\delta U^0 = &-\int_0^\ell \left[\left(\delta u_C' + y_H\,\delta\theta'\right)\left(M_x^0\theta\right)' \right. \\
&\left. + \left(u_C' + y_H\,\theta'\right)\left(\left(M_x^0\right)'\delta\theta + M_x^0\delta\theta'\right) \right] \mathrm{d}z \\
&-\int_0^\ell \left[\left(\delta v_C' - x_H\,\delta\theta'\right)\left(M_y^0\theta\right)' \right. \\
&\left. + \left(v_C' - x_H\,\theta'\right)\left(\left(M_y^0\right)'\delta\theta + M_y^0\delta\theta'\right) \right] \mathrm{d}z.
\end{aligned} \tag{9.106}$$

Integrating by parts and simplifying:[19]

---

[18] The reason for adopting this convention is clarified by the next Remark 9.11.

[19] Indeed:

$$\delta U^0 = \int_0^\ell \left[\left(M_x^0 \theta \right)'' \delta u_C + \left(M_y^0 \theta \right)'' \delta v_C \right] dz + \int_0^\ell \left(M_x^0 u_C'' + M_y^0 v_C'' \right) \delta\theta \, dz$$

$$+ \int_0^\ell \left\{ \left[\left(y_H M_x^0 - x_H M_y^0 \right) \theta \right]'' + \left(y_H M_x^0 - x_H M_y^0 \right) \theta'' \right\} \delta\theta \, dz$$

$$+ \left[- \left(M_x^0 \theta \right)' \delta u_C \right]_0^\ell + \left[- \left(M_y^0 \theta \right)' \delta v_C \right]_0^\ell + \left[\left(- M_x^0 u_C' - M_y^0 v_C' \right. \right.$$

$$\left. \left. - \left(y_H M_x^0 \theta - x_H M_y^0 \theta \right)' - \left(y_H M_x^0 - x_H M_y^0 \right) \theta' \right) \delta\theta \right]_0^\ell .$$

$$(9.107)$$

Then, carrying out the variation of the load energy, one gets:

$$\delta V := e_Q \int_0^\ell p\theta \, \delta\theta \, dz . \tag{9.108}$$

By letting $\delta U^0 + \delta V = \int_0^\ell \delta \mathbf{u}_\pi^T \mathcal{K}_{g,l} \mathbf{u}_\pi \, dz + [\cdots]_0^\ell$, with $\mathbf{u}_\pi := (u_C, v_C, \theta)^T$, *the geometric stiffness operator of the TWB subjected to lateral forces* follows:[20]

$$\mathcal{K}_{g,l} := \begin{bmatrix} 0 & 0 & \partial_z^2 \left(M_x^0 \bullet \right) \\ 0 & 0 & \partial_z^2 \left(M_y^0 \bullet \right) \\ M_x^0 \partial_z^2 & M_y^0 \partial_z^2 & \partial_z^2 \left[\left(y_H M_x^0 - x_H M_y^0 \right) \bullet \right] + \left(y_H M_x^0 - x_H M_y^0 \right) \partial_z^2 + p e_Q \end{bmatrix} .$$

$$(9.109)$$

This expression, in a quite surprisingly simple way, generalizes to the case of variable bending moments the Eq. 9.68, holding for constant bending moments.

Remark 9.12 The $\mathcal{K}_{g,l}$ operator, although non-symmetric, is *self-adjoint* (as obvious, deriving it from a quadratic energy). Its non-symmetric form is due to the fact that it has *variable coefficients*. Indeed, by multiplying, e.g., the element $(1, 3)$ of the matrix by u and integrating by parts, one obtains:

$$\overline{\delta U^0 = \int_0^\ell \left[\left(M_x^0 \theta \right)'' \delta u_C + \left(M_y^0 \theta \right)'' \delta v_C \right] dz + \int_0^\ell \left[y_H \left(M_x^0 \theta \right)'' - \left(M_x^0 \right)' \left(u_C' + y_H \theta' \right) \right.}$$

$$\left. + \left(M_x^0 \right)' \left(u_C' + y_H \theta' \right) + M_x^0 \left(u_C'' + y_H \theta'' \right) \right] \delta\theta \, dz + \int_0^\ell \left[- x_H \left(M_y^0 \theta \right)'' \right.$$

$$\left. - \left(M_y^0 \right)' \left(v_C' - x_H \theta' \right) + \left(M_y^0 \right)' \left(v_C' - x_H \theta' \right) + M_y^0 \left(v_C'' - x_H \theta'' \right) \right] \delta\theta \, dz$$

$$+ \left[- \left(M_x^0 \theta \right)' \delta u_C \right]_0^\ell + \left[- \left(M_y^0 \theta \right)' \delta v_C \right]_0^\ell$$

$$+ \left[\left(- M_x^0 u_C' - M_y^0 v_C' + x_H \left(M_y^0 \right)' \theta - y_H \left(M_x^0 \right)' \theta + 2 \left(x_H M_y^0 - y_H M_x^0 \right) \theta' \right) \delta\theta \right]_0^\ell .$$

---

[20] The dot indicates the position of the operand; so, for example, $\partial_z^2 \left(M_x^0 \bullet \right) \theta = \partial_z^2 \left(M_x^0 \theta \right)$.

$$\int_0^\ell u \left(M_x^0 \theta\right)'' dz = \int_0^\ell u'' M_x^0 \theta \, dz + [\cdots]_0^\ell$$

that is, the element $(3, 1)$ scalarly multiplied by θ.

Equilibrium Equations

Adding the elastic (Eq. 9.26b) and geometric (Eq. 9.109) contributions, the indefinite equilibrium equations are obtained:

$$\left(\boldsymbol{\mathcal{K}}_e^\pi + \boldsymbol{\mathcal{K}}_{g,l}\right) \mathsf{u}_\pi = \mathbf{0}, \tag{9.110}$$

that is, in extended form:

$$EI_y u_C'''' + \left(M_x^0 \theta\right)'' = 0, \tag{9.111a}$$

$$EI_x v_C'''' + \left(M_y^0 \theta\right)'' = 0, \tag{9.111b}$$

$$EI_\omega \theta'''' - GJ\theta'' + M_x^0 u_C'' + M_y^0 v_C'' + \left(y_H M_x^0 \theta - x_H M_y^0 \theta\right)'' \tag{9.111c}$$

$$+ \left(y_H M_x^0 - x_H M_y^0\right)\theta'' + p_{eQ}\,\theta = 0.$$

Operating in the same way on the boundary conditions, one gets:

$$\left[EI_y u_C'' \, \delta u_C'\right]_0^\ell = 0, \tag{9.112a}$$

$$\left[\left(-EI_y u_C''' - \left(M_x^0 \theta\right)'\right) \delta u_C\right]_0^\ell = 0, \tag{9.112b}$$

$$\left[EI_x v_C'' \, \delta v_C'\right]_0^\ell = 0, \tag{9.112c}$$

$$\left[\left(-EI_x v_C''' - \left(M_y^0 \theta\right)'\right) \delta v_C\right]_0^\ell = 0, \tag{9.112d}$$

$$\left[\left(GJ\theta' - EI_\omega \theta''' - M_x^0 u_C' - M_y^0 v_C' - \left(y_H M_x^0 \theta - x_H M_y^0 \theta\right)'\right.\right. \tag{9.112e}$$

$$\left.\left. - \left(y_H M_x^0 - x_H M_y^0\right)\theta'\right)\delta\theta\right]_0^\ell = 0,$$

$$\left[EI_\omega \theta'' \, \delta\theta'\right]_0^\ell = 0. \tag{9.112f}$$

Supplement 9.7 (Interpretation of the Adjacent Equilibrium) The mechanical interpretation of the equilibrium equations of the non-uniformly bent TWB is based on reasoning similar to those developed for the uniformly bent beam. Thus, the geometric contributions in Eqs. 9.111a, b, that is, the funicular bearing capacity, are obvious extensions of the homologous terms in Eqs. 9.70a, b. Accordingly, since $M_y = M_x^0 \theta$ and taking into account that $M_x^0 = M_x^0(z)$ is no longer constant along the beam axis, it is $p_x := M_y'' = \left(M_x^0 \theta \right)''$. However, the shear forces and related shear stresses, which are absent when the bending is uniform, introduce some not trivial aspects that are now discussed. To this end, the effects of the translation and twist are separately investigated.

When the beam, for example, subject only to $M_x^0(z)$, buckles in the $\mathbf{a}_x$ direction, the generic cross-section at the abscissa z rotates by an angle $\varphi_y = u_C'$ around the $\mathbf{a}_y$ axis (Fig. 9.10a). The bending moment, joined to the cross-section, as it happens in the case in which it is constant, gives rise to an incremental torsional moment $M_z := -M_x^0 u_C'$, whose derivative $c_z^{(\sigma)} := -M_z' = \left(M_x^0 u_C' \right)'$ is a distributed torsional couple. To this latter, however, a further couple $c_z^{(\tau)}$ adds itself, due to misalignment of the shear force, which produces an increment of torsional moment $dM_z = T_y u_C' dz$, which, in turn, gives rise to $c_z^{(\tau)} = -\frac{dM_z}{dz} = -T_y u_C' = \left(M_x^0 \right)' u_C'$. The resulting couple is therefore $c_z := c_z^{(\sigma)} + c_z^{(\tau)} = -M_x^0 u_C''$, as it appears in Eq. 9.111c.

When the beam twists, the normal stresses σ_z^0, joined to the fibers, lean, as already discussed for uniform compression and bending. Recalling Eq. 9.72 of the Supplement 9.5, the tangential components of the inclined normal stresses are statically equivalent to a torsional moment $M_z = -2y_H M_x^0 \theta'$. Since, now, M_x^0 is a function of z, $c_z^{(\sigma)} := -\frac{dM_z}{dz} = 2y_H \left(M_x^0 \theta' \right)'$.

A further geometric effect must be added to that of the inclined normal stresses, due to the tangential stresses τ_{zs}^0. The twist θ, indeed, changes both the *position* and the *orientation* of the pre-existing tangential stresses, as indicated in Fig. 9.10b. Consequently, the tangential stresses (due to pure shear) give rise to a non-zero incremental torsional moment. To evaluate it, it is convenient to further separate the two kinematic effects, namely, (a) the tangential stresses move by remaining parallel to themselves, so that their contribution to the torque is proportional to normal translation; (b) the tangential stresses incline without moving, so that their contribution to the torque is proportional to their component in the normal direction. Since the analysis is linearized, the combined effect must be ignored, since of the second order. In this way, a distributed torsional couple $c_z^{(\tau)}$ is obtained (details of the calculation are omitted here for the sake of brevity) which, when added to $c_z^{(\sigma)}$, determined before, leads to the term $c_z = 2y_H \left(M_x^0 \theta' \right)' + y_H \theta \left(M_x^0 \right)''$ which is contained in Eq. 9.111c. □

Remark 9.13 Differently from the problems dealt with in the previous sections, the equilibrium equations of a TWB subject to transverse forces has *variable*

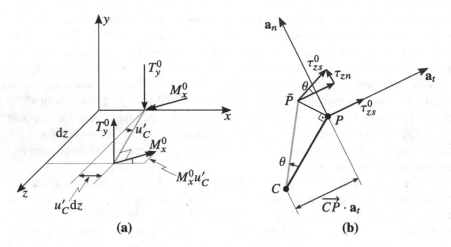

Fig. 9.10 Equilibrium in the adjacent configuration of the thin-walled beam subject to non-uniform bending: (a) lateral bending, (b) contribution of tangential stresses to torsional bearing capacity

coefficients, as $M_x^0 = M_x^0(z)$, $M_y^0 = M_y^0(z)$, $p = p(z)$. This makes not viable an exact integration in terms of elementary functions.

Bending in a Plane of Symmetry

A notable special case occurs when the beam cross-section has an axis of symmetry, for example, y, and the transverse loads $\mathbf{p} = p(z)\,\mathbf{a}_y$ are parallel to it. Then, it turns out that $M_y^0(z) \equiv 0$, $x_C = x_H = 0$, so that the equilibrium Eqs. 9.111 reduce to:

$$EI_y u_C'''' + \left(M_x^0 \theta\right)'' = 0, \qquad (9.113a)$$

$$EI_x v_C'''' = 0, \qquad (9.113b)$$

$$EI_\omega \theta'''' - GJ\theta'' + M_x^0 u_C'' + y_H \left(M_x^0 \theta\right)'' + y_H M_x^0 \theta'' + p\, e_Q\, \theta = 0, \qquad (9.113c)$$

with the related boundary conditions:

$$\left[EI_y u_C'' \, \delta u_C'\right]_0^\ell = 0, \qquad (9.114a)$$

$$\left[\left(-EI_y u_C''' - \left(M_x^0 \theta\right)'\right)\delta u_C\right]_0^\ell = 0, \qquad (9.114b)$$

$$\left[EI_x v_C'' \, \delta v_C'\right]_0^\ell = 0, \qquad (9.114c)$$

$$\left[\left(-EI_x v_C''' \right) \delta v_C \right]_0^\ell = 0, \qquad (9.114\text{d})$$

$$\left[\left(GJ\theta' - EI_\omega \theta''' - M_x^0 u_C' - y_H \left(M_x^0 \theta \right)' - y_H M_x^0 \theta' \right) \delta\theta \right]_0^\ell = 0, \qquad (9.114\text{e})$$

$$\left[EI_\omega \theta'' \, \delta\theta' \right]_0^\ell = 0. \qquad (9.114\text{f})$$

Since the indefinite equilibrium equations and boundary conditions in the variable v_C are homogeneous and independent of the load, they only admit the trivial solution $v_C(z) \equiv 0$, that is, the beam *buckles transversely to the plane of symmetry (containing the loads) and twists*. This type of phenomenon is said *lateral instability*, and it is similar to that already discussed for the uniform bending of TWB with symmetric cross-section.

9.7.2 Fixed-Free Beam with Thin Rectangular Cross-Section Subject to a Transverse Load Applied at the Free End

As an example, a comparatively simple case is studied, consisting of a fixed-free beam with thin rectangular cross-section, subject to a transverse load $\mathbf{F} = F\mathbf{a}_y$ (Fig. 9.11a) applied at the centroid $G \equiv C$ of the end cross-section. Due to $I_\omega = 0$, the equation is lowered in order, and the boundary conditions are reduced in number. The problem is governed by Eqs. 9.113, where $v_C \equiv 0$, $y_H = 0$, $p = 0$, i.e.:

$$EI_y u_C'''' + \left(M_x^0 \theta \right)'' = 0, \qquad (9.115\text{a})$$

$$-GJ\theta'' + M_x^0 u_C'' = 0, \qquad (9.115\text{b})$$

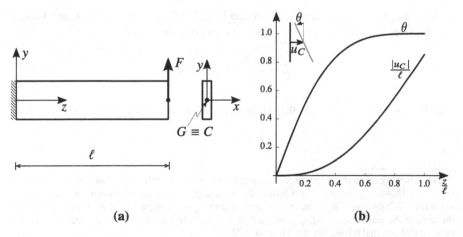

Fig. 9.11 Lateral instability of a fixed-free beam with thin rectangular cross-section, loaded by a force applied at the end: (**a**) system, (**b**) critical mode (with $\frac{GJ}{EI_y} = \frac{3}{2}$)

in which:

$$M_x^0 (z) = -F (\ell - z).$$

(9.116)

The boundary conditions in Eqs. 9.114 specialize as follows:

$$u_C (0) = 0, \qquad u_C' (0) = 0, \qquad \theta (0) = 0,$$

(9.117)

together with:

$$EI_y u_C'' (\ell) = 0,$$

(9.118a)

$$EI_y u_C''' (\ell) + F\theta (\ell) = 0,$$

(9.118b)

$$GJ\theta' (\ell) = 0,$$

(9.118c)

where $M_x^0 (\ell) = 0$ has been exploited.[21]

Reduction of the System to a Single Equation

Equation 9.115a, once integrated, provides:

$$EI_y u_C''' + \left(M_x^0 \theta \right)' = c_1,$$

(9.119)

with c_1 an arbitrary constant. Evaluated the previous equation at $z = \ell$ and enforced the condition in Eq. 9.118b,[22] one finds $c_1 = 0$. Integrating a second time, one gets:

$$EI_y u_C'' + M_x^0 \theta = c_2,$$

(9.120)

with c_2 another arbitrary constant. Evaluating also this latter at $z = \ell$ and taking into account Eq. 9.118a, $c_2 = 0$ follows. It turns out, then, that:

$$u_C'' = \frac{F\ell}{EI_y} \left(1 - \frac{z}{\ell} \right) \theta,$$

(9.121)

which allows one to write Eq. 9.115b in terms of the twist angle only, i.e.:

---

[21] The following mechanical interpretation can be done to Eq. 9.118b . When the end cross-section twists, the tangential stresses, which follow the cross-section, produce an incremental shear force $-F\theta (\ell)$ (i.e., opposite to $\mathbf{a}_x$). The internal force emerging to the right end of the beam is therefore the sum of the elastic force $-EI_y u_C''' (\ell)$ and of the geometric force $-F\theta (\ell)$; since there are no incremental external forces, Eq. 9.118b must hold.

[22] That is, taking into account Eq. 9.114b.

$$\theta'' + \frac{F^2 \ell^2}{E I_y \, G J} \left(1 - \frac{z}{\ell}\right)^2 \theta = 0. \tag{9.122}$$

This differential equation must be integrated with the boundary conditions in Eqs. 9.117c, 9.118c.

Once $\theta(z)$ has been determined, the lateral displacement is calculated by integrating twice Eq. 9.121, under the remaining boundary conditions $u_C(0) = u'_C(0) = 0$, thus obtaining:

$$u_C = \frac{F \ell}{E I_y} \int\limits_0^z d\hat{z} \int\limits_0^{\hat{z}} d\tilde{z} \left(1 - \frac{\tilde{z}}{\ell}\right) \theta(\tilde{z}) \, d\tilde{z}. \tag{9.123}$$

Solution by Power Series

To solve Eq. 9.122, it is convenient to introduce the nondimensional abscissa $\zeta := 1 - \frac{z}{\ell}$ and to rewrite the equations as:

$$\theta''(\zeta) + \mu \zeta^2 \, \theta(\zeta) = 0, \tag{9.124a}$$

$$\theta(1) = 0, \tag{9.124b}$$

$$\theta'(0) = 0, \tag{9.124c}$$

where differentiation with respect to ζ has still been denoted by a dash and a nondimensional load has been introduced:

$$\mu := \frac{F^2 \ell^4}{E I_y \, G J}. \tag{9.125}$$

Equations (9.124) constitute a differential eigenvalue problem with variable coefficients.[23] To solve it and according to the *Frobenius method*, a power series solution is sought in the form:

$$\theta = \sum_{n=0}^{\infty} A_n \zeta^n, \tag{9.126}$$

where A_n are unknown constants. Consequently:

---

[23] Equation 9.124a is a Bessel equation.

$$\theta'' = \sum_{n=2}^{\infty} (n-1)\, n\, A_n \zeta^{n-2} = \sum_{k=0}^{\infty} (k+1)\,(k+2)\, A_{k+2} \zeta^k, \qquad (9.127\text{a})$$

$$\zeta^2 \theta = \sum_{n=0}^{\infty} A_n \zeta^{n+2} = \sum_{k=2}^{\infty} A_{k-2} \zeta^k, \qquad (9.127\text{b})$$

where $n =: k+2$ has been used in the first equation and $n =: k-2$ in the second one. By substituting these expressions in the field equation, it follows:

$$2A_2 + 6A_3 \zeta + \sum_{k=2}^{\infty} [(k+1)\,(k+2)\, A_{k+2} + \mu A_{k-2}]\, \zeta^k = 0. \qquad (9.128)$$

Since this equation must be satisfied for any ζ, the coefficients of the different powers of ζ must vanish separately, i.e.:

$$A_2 = 0, \qquad (9.129\text{a})$$

$$A_3 = 0, \qquad (9.129\text{b})$$

$$A_{k+2} = -\mu \frac{A_{k-2}}{(k+1)\,(k+2)}, \qquad k = 2, 3, \cdots. \qquad (9.129\text{c})$$

From Eq. 9.129c, being $A_2 = 0$, it follows that $A_6 = A_{10} = \cdots = 0$; being $A_3 = 0$, it follows that $A_7 = A_{11} = \cdots = 0$. On the other hand, by the same Eq. 9.129c, each of the arbitrary A_0 and A_1 generates a chain of monomials, namely:

$$(A_0,\ A_4,\ A_8,\ A_{12},\ \cdots)$$

$$= A_0 \left(1,\ -\frac{\mu}{3\cdot 4},\ \frac{\mu}{3\cdot 4}\frac{\mu}{7\cdot 8},\ -\frac{\mu}{3\cdot 4}\frac{\mu}{7\cdot 8}\frac{\mu}{11\cdot 12},\ \cdots \right), \qquad (9.130\text{a})$$

$$(A_1,\ A_5,\ A_9,\ A_{13},\ \cdots)$$

$$= A_1 \left(1,\ -\frac{\mu}{4\cdot 5},\ \frac{\mu}{4\cdot 5}\frac{\mu}{8\cdot 9},\ -\frac{\mu}{4\cdot 5}\frac{\mu}{8\cdot 9}\frac{\mu}{12\cdot 13},\ \cdots \right). \qquad (9.130\text{b})$$

The general solution of the differential Eq. 9.124a, therefore, reads:

$$\theta = A_0 \left(1 - \frac{\mu}{12}\zeta^4 + \frac{\mu^2}{672}\zeta^8 - \frac{\mu^3}{88,704}\zeta^{12} + \frac{\mu^4}{21,288,960}\zeta^{16} + \cdots \right)$$

$$+ A_1 \left(\zeta - \frac{\mu}{20}\zeta^5 + \frac{\mu^2}{1440}\zeta^9 - \frac{\mu^3}{224,640}\zeta^{13} + \cdots \right). \qquad (9.131)$$

The two arbitrary constants are determined by the two boundary conditions in Eqs. 9.124b, c; an algebraic problem for the eigenvalue μ follows:

$$\left[\begin{array}{cc} 1 - \frac{\mu}{12} + \frac{\mu^2}{672} - \frac{\mu^3}{88,704} + \cdots & 1 - \frac{\mu}{20} + \frac{\mu^2}{1440} - \frac{\mu^3}{224,640} + \cdots \\ 0 & 1 \end{array}\right] \binom{A_0}{A_1} = \binom{0}{0},$$

(9.132)

which is governed by a triangular matrix. Its characteristic equation:

$$1 - \frac{\mu}{12} + \frac{\mu^2}{672} - \frac{\mu^3}{88,704} + \frac{\mu^4}{21,288,960} + \cdots = 0 \qquad (9.133)$$

provides a solution that depends on the number of terms considered in the truncated series. With two terms, the solution is $\mu_c = 12$; with three terms, $\mu_c = 17.417$; with four terms, $\mu_c = 16.033$. With a further term $\mu_c = 16.104$ is obtained, nearly equal to the value to which the series converges. Ultimately, from Eq. 9.125:

$$F_c = \pm \frac{4.013}{\ell^2} \sqrt{GJ\, EI_y}. \qquad (9.134)$$

The associated eigenvector is $(A_0, A_1) = (1, 0)$. The θ component of the critical mode, to within a constant, is given by Eq. 9.131. By coming back to the original variable z and then using Eq. 9.123 to evaluate u_C, the following result is finally drawn:[24]

$$u_C = \pm \ell \sqrt{\frac{GJ}{EI_y}} \left[1.869 \left(\frac{z}{\ell}\right)^3 - 0.926 \left(\frac{z}{\ell}\right)^4 + \cdots \right], \qquad (9.135a)$$

$$\theta = 1 - 1.342 \left(1 - \frac{z}{\ell}\right)^4 + 0.386 \left(1 - \frac{z}{\ell}\right)^8 - 0.047 \left(1 - \frac{z}{\ell}\right)^{12} \qquad (9.135b)$$

$$+ 0.003 \left(1 - \frac{z}{\ell}\right)^{16} + \cdots.$$

It is worth noticing that the constraint $\theta(0) = 0$ is not exactly satisfied but that the error tends to zero as the number of terms of the series increases. The critical mode is represented in Fig. 9.11b, where $\frac{GJ}{EI_y} = \frac{3}{2}$ (corresponding to a Poisson's ratio $\nu = \frac{1}{3}$) has been taken. It appears that the two displacement components are comparable.

9.7.3 Ritz Method

Pursuing a closed form solution to the differential Eqs. 9.111, which governs the bifurcation of a non-uniformly bent TWB, is an extremely laborious task, like the

---

[24] In Eq. 9.135a, the full result has not been reported, for the sake of brevity.

example of the Sect. 9.7.2 has shown. It is therefore convenient to tackle the problem
in an approximation perspective, as offered by the *Ritz method*, summarized in the
Appendix B.

To this end, the total potential energy of the TWB is written, as sum of the
elastic contributions in Eqs. 9.9, 9.10, of the prestress in Eq. 9.102 and of the loads
in Eq. 9.105, i.e.:

$$\Pi = \tilde{U} + U^0 + V. \tag{9.136}$$

Here $\tilde{U}$ is the (quadratic) elastic potential energy stored by the beam in moving
from the prestressed to the adjacent configuration; U^0 is the prestress energy, equal
to the (quadratic) work expended by the prestresses in the incremental strains; V
is the potential energy of the loads, which, in the absence of incremental loads,
is the (quadratic, and changed of sign) work that the pre-existing loads do in the
incremental displacements. In formulas:

$$
\begin{aligned}
\Pi = \frac{1}{2} \int_0^\ell & \left(EI_y u_C''^2 + EI_x v_C''^2 + EI_\omega \theta''^2 + GJ\theta'^2 \right) dz \\
& + \mu \int_0^\ell \left[-\left(M_x^0 \theta \right)' \left(u_C' + y_H \, \theta' \right) - \left(M_y^0 \theta \right)' \left(v_C' - x_H \, \theta' \right) \right] dz \\
& + \mu \frac{e_Q}{2} \int_0^\ell p(z) \theta^2(z) \, dz,
\end{aligned}
\tag{9.137}
$$

where a load multiplier μ has been introduced.

According to the Ritz method, the unknown displacement field is expressed as a
linear combination with unknown coefficients $\mathbf{q} := \left(\hat{\mathbf{u}}, \hat{\mathbf{v}}, \hat{\boldsymbol{\theta}} \right)$ (taking the meaning of
Lagrangian parameters) of known trial functions $\boldsymbol{\phi}_\alpha(z)$ $(\alpha = u, v, \theta)$, which fulfill
at least the geometric boundary conditions. Accordingly:[25]

$$u_C(z) = \boldsymbol{\phi}_u(z) \, \hat{\mathbf{u}}, \tag{9.138a}$$

$$v_C(z) = \boldsymbol{\phi}_v(z) \, \hat{\mathbf{v}}, \tag{9.138b}$$

$$\theta(z) = \boldsymbol{\phi}_\theta(z) \, \hat{\boldsymbol{\theta}}. \tag{9.138c}$$

Consequently, the total potential energy in Eq. 9.137 is written as:

---

[25] The $\boldsymbol{\phi}_\alpha$ are row matrices containing the trial functions related to the variable $\alpha = u, v, \theta$; instead,
$\hat{\mathbf{u}}, \hat{\mathbf{v}}, \hat{\boldsymbol{\theta}}$ are column matrices containing the unknown amplitudes.

$$\Pi = \frac{1}{2} \int_0^\ell \left(EI_y \hat{\mathbf{u}}^T \boldsymbol{\phi}_u''^T \boldsymbol{\phi}_u'' \hat{\mathbf{u}} + EI_x \hat{\mathbf{v}}^T \boldsymbol{\phi}_v''^T \boldsymbol{\phi}_v'' \hat{\mathbf{v}} \right.$$

$$\left. + EI_\omega \hat{\boldsymbol{\theta}}^T \boldsymbol{\phi}_\theta''^T \boldsymbol{\phi}_\theta'' \hat{\boldsymbol{\theta}} + GJ \hat{\boldsymbol{\theta}}^T \boldsymbol{\phi}_\theta'^T \boldsymbol{\phi}_\theta' \hat{\boldsymbol{\theta}} \right) \mathrm{d}z$$

$$+ \mu \int_0^\ell \left[-\hat{\boldsymbol{\theta}}^T \left(M_x^0 \boldsymbol{\phi}_\theta^T \right)' \left(\boldsymbol{\phi}_u' \hat{\mathbf{u}} + y_H \boldsymbol{\phi}_\theta' \hat{\boldsymbol{\theta}} \right) \right.$$

$$\left. - \hat{\boldsymbol{\theta}}^T \left(M_y^0 \boldsymbol{\phi}_\theta^T \right)' \left(\boldsymbol{\phi}_v' \hat{\mathbf{v}} - x_H \boldsymbol{\phi}_\theta' \hat{\boldsymbol{\theta}} \right) \right] \mathrm{d}z$$

$$+ \mu \frac{e_Q}{2} \int_0^\ell p(z) \hat{\boldsymbol{\theta}}^T \boldsymbol{\phi}_\theta^T \boldsymbol{\phi}_\theta \hat{\boldsymbol{\theta}} \, \mathrm{d}z. \tag{9.139}$$

Once the integrations have been executed, Π becomes an *algebraic* quadratic form, $\Pi = \frac{1}{2} \mathbf{q}^T \left(\mathbf{K}_e + \mu \mathbf{K}_g \right) \mathbf{q}$, in which:

$$\mathbf{K}_e := \begin{bmatrix} \mathbf{K}_{uu}^e & \mathbf{0} & \mathbf{0} \\ \mathbf{0} & \mathbf{K}_{vv}^e & \mathbf{0} \\ \mathbf{0} & \mathbf{0} & \mathbf{K}_{\theta\theta}^e \end{bmatrix}, \qquad \mathbf{K}_g := \begin{bmatrix} \mathbf{0} & \mathbf{0} & \mathbf{K}_{u\theta}^g \\ \mathbf{0} & \mathbf{0} & \mathbf{K}_{v\theta}^g \\ \mathbf{K}_{\theta u}^g & \mathbf{K}_{\theta v}^g & \mathbf{K}_{\theta\theta}^g \end{bmatrix}, \tag{9.140}$$

are the stiffness matrices, respectively elastic and geometric, whose elements are defined as follows:

$$\mathbf{K}_{uu}^e := \int_0^\ell EI_y \boldsymbol{\phi}_u''^T \boldsymbol{\phi}_u'' \, \mathrm{d}z, \qquad\qquad \mathbf{K}_{vv}^e := \int_0^\ell EI_x \boldsymbol{\phi}_v''^T \boldsymbol{\phi}_v'' \, \mathrm{d}z,$$

$$\mathbf{K}_{\theta\theta}^e := \int_0^\ell \left(EI_\omega \boldsymbol{\phi}_\theta''^T \boldsymbol{\phi}_\theta'' + GJ \boldsymbol{\phi}_\theta'^T \boldsymbol{\phi}_\theta' \right) \mathrm{d}z, \quad \mathbf{K}_{\theta u}^g := - \int_0^\ell \left(M_x^0 \boldsymbol{\phi}_\theta^T \right)' \boldsymbol{\phi}_u' \, \mathrm{d}z,$$

$$\mathbf{K}_{\theta v}^g := - \int_0^\ell \left(M_y^0 \boldsymbol{\phi}_\theta^T \right)' \boldsymbol{\phi}_v' \, \mathrm{d}z,$$

$$\mathbf{K}_{\theta\theta}^g := \int_0^\ell \left[e_Q \, p(z) \boldsymbol{\phi}_\theta^T \boldsymbol{\phi}_\theta - 2y_H \left(M_x^0 \boldsymbol{\phi}_\theta^T \right)' \boldsymbol{\phi}_\theta' + 2x_H \left(M_y^0 \boldsymbol{\phi}_\theta^T \right)' \boldsymbol{\phi}_\theta' \right] \mathrm{d}z,$$

$$\tag{9.141}$$

with $\mathbf{K}_{\alpha\beta}^g := \left(\mathbf{K}_{\beta\alpha}^g \right)^T$. The bifurcation condition requires that $\left(\mathbf{K}_e + \mu \mathbf{K}_g \right) \mathbf{q} = \mathbf{0}$ admits a non-trivial solution. The critical load multiplier is the smallest root of the characteristic equation $\det \left(\mathbf{K}_e + \mu \mathbf{K}_g \right) = 0$.

An example of application of the method follows. A further example, relevant to a distributed load, is illustrated in the Problem 14.3.4.

Example 9.2 (Lateral Instability of a Fixed-Free Beam with Thin Rectangular Cross-Section, Subject to Constant Shear) As an example of application of the Ritz method, the problem solved in the Sect. 9.7.2 is reconsidered. The relevant TPE, taking into account Eq. 9.116, reduces to:

$$\Pi = \frac{1}{2} \int_0^\ell \left[EI_y u_C''^2 + GJ\theta'^2 + 2F \left((\ell - z)\,\theta \right)' u_C' \right] dz. \tag{9.142}$$

A single trial function, for each of the two scalar fields, is adopted, i.e.:

$$u_C(z) = \hat{u}\,\phi_u(z), \tag{9.143a}$$

$$\theta(z) = \hat{\theta}\,\phi_\theta(z), \tag{9.143b}$$

so that the TPE reads:

$$\Pi = \frac{1}{2} \int_0^\ell \left[EI_y \hat{u}^2 \phi_u''^2 + GJ\hat{\theta}^2 \phi_\theta'^2 + 2F \left((\ell - z)\phi_\theta \right)' \phi_u' \,\hat{\theta}\hat{u} \right] dz. \tag{9.144}$$

Enforcing stationary, one gets:

$$\begin{bmatrix} k_{uu} & k_{u\theta} \\ k_{u\theta} & k_{\theta\theta} \end{bmatrix} \begin{pmatrix} \hat{u} \\ \hat{\theta} \end{pmatrix} = \begin{pmatrix} 0 \\ 0 \end{pmatrix}, \tag{9.145}$$

where:

$$k_{uu} := \int_0^\ell EI_y \phi_u''^2 dz, \quad k_{u\theta} := F \int_0^\ell \left((\ell - z)\phi_\theta \right)' \phi_u' dz, \quad k_{\theta\theta} := \int_0^\ell GJ\hat{\theta}^2 \phi_\theta'^2 dz.$$
$$\tag{9.146}$$

A possible choice for the trial functions is:

$$\phi_\theta = 2\frac{z}{\ell} - \left(\frac{z}{\ell}\right)^2, \tag{9.147a}$$

$$\phi_u = \ell \left[\frac{1}{3}\left(\frac{z}{\ell}\right)^3 - \frac{1}{4}\left(\frac{z}{\ell}\right)^4 + \frac{1}{20}\left(\frac{z}{\ell}\right)^5 \right]. \tag{9.147b}$$

The first of these is a parabola, which fulfills both the conditions on twist: $\phi_\theta(0) = 0$, $\phi_\theta'(\ell) = 0$; it is normalized by $\phi_\theta(\ell) = 1$. The second trial function is suggested by Eq. 9.123, which, to within an inessential factor, reads:

$$\phi_u = \frac{1}{\ell} \int\limits_0^z d\hat{z} \int\limits_0^{\hat{z}} d\tilde{z} \left(1 - \frac{\tilde{z}}{\ell}\right) \phi_\theta\left(\tilde{z}\right) d\tilde{z}. \tag{9.148}$$

With the trial functions chosen above, the equilibrium Eq. 9.145 specializes into:

$$\begin{bmatrix} \frac{8}{105} \frac{EI_y}{\ell} & -\frac{8}{105} F\ell \\ -\frac{8}{105} F\ell & \frac{4}{3} \frac{GJ}{\ell} \end{bmatrix} \begin{pmatrix} \hat{u} \\ \hat{\theta} \end{pmatrix} = \begin{pmatrix} 0 \\ 0 \end{pmatrix}. \tag{9.149}$$

By zeroing the determinant of the coefficients, the critical value of the force is found:

$$F_c := \pm 4.183 \frac{\sqrt{EI_y \, GJ}}{\ell^2}, \tag{9.150}$$

which is only about 4% higher than the exact value. Solving the linear system, the normalized critical eigenvector is computed:

$$\begin{pmatrix} \hat{u}_c \\ \hat{\theta}_c \end{pmatrix} = \begin{pmatrix} \pm 4.183 \sqrt{\frac{GJ}{EI_y}} \\ 1 \end{pmatrix}. \tag{9.151}$$

The critical mode drawn by the Ritz method is compared in Fig. 9.12 with that found in the previous section by the Frobenius method (both for $GJ = \frac{3}{2} EI_y$). The two solutions are normalized in such a way the twist at the free end is the same. It is seen that the approximate solution, although obtained with only one trial function per component, describes quite well both the displacements. □

Fig. 9.12 Lateral instability of a fixed-free beam with thin rectangular cross-section, loaded by a force applied at the end: critical mode (with $\frac{GJ}{EI_y} = \frac{3}{2}$); Ritz solution (in gray) and Frobenius solution (in black)

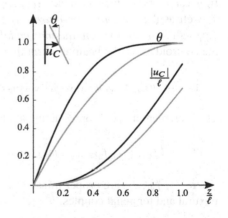

9.8   Finite Element Buckling Analysis of Thin-Walled Beams

The buckling analysis of an open TWB, when pursued in exact form, is often a very complex problem. This is due (i) to constraint conditions, which may or may not prevent warping, (ii) to the presence of variable coefficients in the equilibrium equations, when the beam is transversely loaded, and (iii) to the influence of the point of application of the same transverse forces. Here the problem is addressed in the context of the finite element theory, by limiting the treatment to a single beam of straight axis. Polynomial interpolation functions are only used, since the problem cannot be integrated in exact form, even in a subdomain [2]. The presoliciting bending moment, which is in principle variable with arbitrary law, is here approximated as piecewise linear, interpolated between the nodal values. The axial force is assumed piecewise constant.

9.8.1   Polynomial Finite Element

A single finite element of TWB is first considered; then, an assembly of elements arranged in series is addressed.

Total Potential Energy

An element e of an open TWB is given, of length ℓ, presoliced by a constant axial force N^0 and bending moments $M_x^0(z)$ and $M_y^0(z)$, variables along z. With reference to Fig. 9.13, the following Lagrangian parameters are taken: (a) the *transverse nodal displacements* u_{Ci}, v_{Ci}, θ_i $(i = 1, 2)$ of the torsion center axis, as well as the *torsional curvatures* $\kappa_{ti} := \theta_i'$ (according to the Sect. C.5.1 in the Appendix C); (b) the *longitudinal nodal displacements* $w_i, \varphi_{xi}, \varphi_{yi}$ $(i = 1, 2)$ of the centroid axis. The beam element therefore has 14 degrees of freedom:

$$\mathbf{q}^{(e)} := \left(u_{C1}, v_{C1}, w_{C1}, \varphi_{x1}, \varphi_{y1}, \theta_1, \kappa_{t1}; u_{C2}, v_{C2}, w_{C2}, \varphi_{x2}, \varphi_{y2}, \theta_2, \kappa_{t2}\right)^T .$$

$$(9.152)$$

The incremental nodal forces acting at the ends of the beam are:

$$\mathbf{f}^{(e)} := \left(f_{x1}, f_{y1}, f_{z1}, m_{x1}, m_{y1}, m_{z1}, b_1; f_{x2}, f_{y2}, f_{z2}, m_{x2}, m_{y2}, m_{z2}, b_2\right)^T ,$$

$$(9.153)$$

where f_{xi}, f_{yi}, f_{zi} are proper transverse and longitudinal forces; m_{x1}, m_{y1}, m_{z1} are flexural and torsional couples; b_i are *bicouples*,[26] dual of the torsional curvature.

---

[26] A bicouple is the external action which equilibrates the *bimoment* emerging from the TWB.

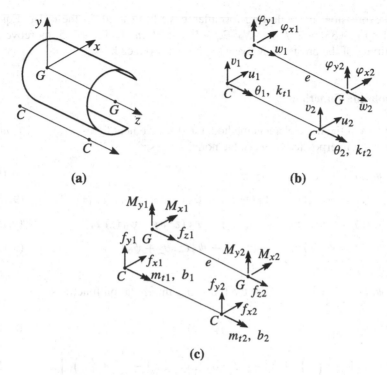

Fig. 9.13 Finite element of open TWB: (**a**) axes, (**b**) nodal displacements, (**c**) nodal forces

The total potential energy of the prestressed system[27] reads $\Pi = \tilde{U} + U^0 + V$, where $\tilde{U}$ is the incremental elastic energy, U^0 the prestress energy, and V the load potential energy (of prestressing and incremental loads), given by:

$$\tilde{U} = \frac{1}{2} \int_0^\ell \left(EAw_G'^2 + EI_y u_C''^2 + EI_x v_C''^2 + GJ\theta'^2 + EI_\omega \theta''^2 \right) dz, \tag{9.154a}$$

$$U^0 = \mu \frac{N_0}{2} \int_0^\ell \left[u_C'^2 + v_C'^2 + r_C^2 \theta'^2 + 2y_C u_C' \theta' - 2x_C v_C' \theta' \right] dz \tag{9.154b}$$

$$+ \mu \int_0^\ell \left[-\left(M_x^0 \theta \right)' \left(u_C' + y_H \theta' \right) - \left(M_y^0 \theta \right)' \left(v_C' - x_H \theta' \right) \right] dz,$$

$$V = \mu \frac{e_Q}{2} \int_0^\ell p(z) \theta^2(z) \, dz - \mathbf{f}^{(e)T} \mathbf{q}^{(e)}. \tag{9.154c}$$

---

[27] Already introduced in the Sect. 9.7.3 for a less general state of stress.

In this expression, the following formulas have been used, in the order: Eqs. 9.9 and 9.10; Eq. 9.35, with $P = -N_0$; Eq. 9.102; and finally, Eq. 9.105; moreover, the μ multiplier of the prestressing loads has been introduced.

Interpolation Functions

In the spirit of the finite element method, the displacement fields $u(z)$, $v(z)$, $w(z)$, and $\theta(z)$ are interpolated between the nodal values:[28]

$$w(z) = \psi_1(z)\, w_1 + \psi_2(z)\, w_2, \tag{9.155a}$$

$$u(z) = \psi_3(z)\, u_1 + \psi_4(z)\, \varphi_{y1} + \psi_5(z)\, u_2 + \psi_6(z)\, \varphi_{y2}, \tag{9.155b}$$

$$v(z) = \psi_3(z)\, v_1 - \psi_4(z)\, \varphi_{x1} + \psi_5(z)\, v_2 - \psi_6(z)\, \varphi_{x2}, \tag{9.155c}$$

$$\theta(z) = \psi_3(z)\, \theta_1 + \psi_4(z)\, \kappa_{t1} + \psi_5(z)\, \theta_2 + \psi_6(z)\, \kappa_{t2}, \tag{9.155d}$$

where $\psi_i(x)$, $i = 1, \ldots, 6$, are the following interpolation functions:

$$\psi_1(z) = 1 - \frac{z}{\ell}, \qquad\qquad \psi_2(z) = \frac{z}{\ell}, \tag{9.156a}$$

$$\psi_3(z) = 1 - 3\left(\frac{z}{\ell}\right)^2 + 2\left(\frac{z}{\ell}\right)^3, \quad \psi_4(z) = z\left[1 - \frac{2z}{\ell} + \left(\frac{z}{\ell}\right)^2\right], \tag{9.156b}$$

$$\psi_5(z) = 3\left(\frac{z}{\ell}\right)^2 - 2\left(\frac{z}{\ell}\right)^3, \qquad \psi_6(z) = z\left[\left(\frac{z}{\ell}\right)^2 - \frac{z}{\ell}\right], \tag{9.156c}$$

already used for constructing the polynomial finite element of an extensible and prestressed planar beam (Sect. 7.12).

The prestress bending moments are linearly interpolated between the nodal values M_{xi}^0, M_{yi}^0 $(i = 1, 2)$ as:

$$M_x^0(z) = \psi_1(z)\, M_{x1}^0 + \psi_2(z)\, M_{x2}^0, \tag{9.157a}$$

$$M_y^0(z) = \psi_1(z)\, M_{y1}^0 + \psi_2(z)\, M_{y2}^0. \tag{9.157b}$$

Stiffness Matrices

Substituting Eqs. 9.155, 9.156, and 9.157 in the energy, performing integrations on z, and imposing stationary with respect to the Lagrangian parameters, the following element equilibrium equations are obtained:

$$\left(\mathbf{K}_e^{(e)} + \mu \mathbf{K}_g^{(e)}\right) \mathbf{q}^{(e)} = \mathbf{f}^{(e)}, \tag{9.158}$$

---

[28] It should be remembered that $\varphi_x = -v_C'$, $\varphi_y = u_C'$, hence the minus signs in Eqs. 9.155.

where $\mathbf{K}_e^{(e)}$ is the elastic matrix and $\mathbf{K}_g^{(e)}$ the geometric matrix. The geometric matrix is the sum of three contributions: $\mathbf{K}_g^{(e)} =: \mathbf{K}_{g,N}^{(e)} + \mathbf{K}_{g,M}^{(e)} + \mathbf{K}_{g,p}^{(e)}$, respectively, due to axial prestress, bending prestresses and eccentricity of the transverse loads. The expressions of the coefficients of the matrices are given in the Supplement 9.8.

Supplement 9.8 (Coefficients of the Stiffness Matrices of the TWB Finite Element) The matrices $\mathbf{K}_e$, $\mathbf{K}_{g,N}$, $\mathbf{K}_{g,M}$, $\mathbf{K}_{g,p}$ (superscript e, of element, suppressed), all of dimension 14×14, are partitioned into four sub-matrices of dimension 7×7:

$$\mathbf{K}_\alpha := \begin{bmatrix} \mathbf{K}_\alpha^{(1,1)} & \mathbf{K}_\alpha^{(1,2)} \\ \mathbf{K}_\alpha^{(2,1)} & \mathbf{K}_\alpha^{(2,2)} \end{bmatrix}, \tag{9.159}$$

with $\mathbf{K}_\alpha^{(2,1)} = \left(\mathbf{K}_\alpha^{(1,2)} \right)^T$. The relevant structure (denoted by the dots, which indicate the non-null terms) and their explicit expressions are shown below. Some of the sub-matrices are symmetric, others non-symmetric.

- Elastic matrix:

$$\mathbf{K}_e^{(1,1)} := \begin{bmatrix} \bullet & \cdot & \cdot & \cdot & \cdot & \cdot & \cdot \\ & \bullet & \bullet & \cdot & \cdot & \cdot & \cdot \\ & & \bullet & \cdot & \cdot & \cdot & \cdot \\ & & & \bullet & \bullet & \cdot & \cdot \\ & & & & \bullet & \cdot & \cdot \\ & & & & & \bullet & \bullet \\ \text{SYM} & & & & & & \bullet \end{bmatrix}, \quad \mathbf{K}_e^{(2,1)} := \begin{bmatrix} \bullet & \cdot & \cdot & \cdot & \cdot & \cdot & \cdot \\ \cdot & \bullet & \bullet & \cdot & \cdot & \cdot & \cdot \\ \cdot & \bullet & \bullet & \cdot & \cdot & \cdot & \cdot \\ \cdot & \cdot & \cdot & \bullet & \bullet & \cdot & \cdot \\ \cdot & \cdot & \cdot & \bullet & \bullet & \cdot & \cdot \\ \cdot & \cdot & \cdot & \cdot & \cdot & \bullet & \bullet \\ \cdot & \cdot & \cdot & \cdot & \cdot & \bullet & \bullet \end{bmatrix}, \quad \mathbf{K}_e^{(2,2)} := \begin{bmatrix} \bullet & \cdot & \cdot & \cdot & \cdot & \cdot & \cdot \\ & \bullet & \bullet & \cdot & \cdot & \cdot & \cdot \\ & & \bullet & \cdot & \cdot & \cdot & \cdot \\ & & & \bullet & \bullet & \cdot & \cdot \\ & & & & \bullet & \cdot & \cdot \\ & & & & & \bullet & \bullet \\ \text{SYM} & & & & & & \bullet \end{bmatrix},$$

$$\tag{9.160}$$

where:

$$K_{e11}^{(1,1)} = \frac{EA}{\ell}, \qquad K_{e22}^{(1,1)} = \frac{12EI_y}{\ell^3}, \qquad K_{e23}^{(1,1)} = \frac{6EI_y}{\ell^2},$$

$$K_{e33}^{(1,1)} = \frac{4EI_y}{\ell}, \qquad K_{e44}^{(1,1)} = \frac{12EI_x}{\ell^3}, \qquad K_{e45}^{(1,1)} = -\frac{6EI_x}{\ell^2},$$

$$K_{e55}^{(1,1)} = \frac{4EI_x}{\ell}, \qquad K_{e66}^{(1,1)} = \frac{12EI_\omega}{\ell^3} + \frac{6GJ}{5\ell}, \qquad K_{e67}^{(1,1)} = \frac{6EI_\omega}{\ell^2} + \frac{GJ}{10},$$

$$K_{e77}^{(1,1)} = \frac{4EI_\omega}{\ell} + \frac{2GJ\ell}{15}, \tag{9.161}$$

$$K_{e11}^{(2,1)} = -\frac{EA}{\ell}, \qquad K_{e22}^{(2,1)} = -\frac{12EI_y}{\ell^3}, \qquad K_{e23}^{(2,1)} = -K_{e32}^{(2,1)} = -\frac{6EI_y}{\ell^2},$$

$$K_{e33}^{(2,1)} = \frac{2EI_y}{\ell}, \qquad K_{e44}^{(2,1)} = -\frac{12EI_x}{\ell^3}, \qquad K_{e45}^{(2,1)} = -K_{e54}^{(2,1)} = \frac{6EI_x}{\ell^2},$$

$$K_{e55}^{(2,1)} = \frac{2EI_x}{\ell}, \qquad K_{e66}^{(2,1)} = -\frac{12EI_\omega}{\ell^3} - \frac{6GJ}{5\ell}, \qquad K_{e67}^{(2,1)} = -K_{e76}^{(2,1)} = \frac{6EI_\omega}{\ell^2} - \frac{GJ}{10},$$

$$K_{e77}^{(2,1)} = \frac{2EI_\omega}{\ell} - \frac{GJ\ell}{30}, \tag{9.162}$$

$$K^{(2,2)}_{e11} = \frac{EA}{\ell}, \qquad K^{(2,2)}_{e22} = \frac{12EI_y}{\ell^3}, \qquad K^{(2,2)}_{e23} = -\frac{6EI_y}{\ell^2},$$

$$K^{(2,2)}_{e33} = \frac{4EI_y}{\ell}, \qquad K^{(2,2)}_{e44} = \frac{12EI_x}{\ell^3}, \qquad K^{(2,2)}_{e45} = \frac{6EI_x}{\ell^2},$$

$$K^{(2,2)}_{e55} = \frac{4EI_x}{\ell}, \qquad K^{(2,2)}_{e66} = \frac{12EI_\omega}{\ell^3} + \frac{6GJ}{5\ell}, \qquad K^{(2,2)}_{e67} = -\frac{6EI_\omega}{\ell^2} - \frac{GJ}{10},$$

$$K^{(2,2)}_{e77} = \frac{4EI_\omega}{\ell} + \frac{2GJ\ell}{15}. \tag{9.163}$$

- Geometric matrix, related to the axial force:

$$\mathbf{K}^{(1,1)}_{g,N} := N_0 \begin{bmatrix} \cdot & & & \cdot & & \cdot & \cdot \\ & \frac{6}{5\ell} & \frac{1}{10} & \cdot & \cdot & \frac{6y_C}{5\ell} & \frac{y_C}{10} \\ & & \frac{2\ell}{15} & \cdot & \cdot & \frac{y_C}{10} & \frac{2\ell y_C}{15} \\ & & & \frac{6}{5\ell} & -\frac{1}{10} & -\frac{6x_C}{5\ell} & -\frac{x_C}{10} \\ & & & & \frac{2\ell}{15} & \frac{x_C}{10} & \frac{2\ell x_C}{15} \\ & & & & & \frac{6r_C^2}{5\ell} & \frac{r_C^2}{10} \\ \text{SYM} & & & & & & \frac{2\ell r_C^2}{15} \end{bmatrix},$$

$$\mathbf{K}^{(2,1)}_{g,N} := N_0 \begin{bmatrix} \cdot & \cdot & \cdot & \cdot & \cdot & \cdot & \cdot \\ \cdot & -\frac{6}{5\ell} & -\frac{1}{10} & \cdot & \cdot & -\frac{6y_C}{5\ell} & -\frac{y_C}{10} \\ \cdot & \frac{1}{10} & -\frac{\ell}{30} & \cdot & \cdot & \frac{y_C}{10} & -\frac{\ell y_C}{30} \\ \cdot & \cdot & \cdot & -\frac{6}{5\ell} & \frac{1}{10} & \frac{6x_C}{5\ell} & \frac{x_C}{10} \\ \cdot & \cdot & \cdot & -\frac{1}{10} & -\frac{\ell}{30} & \frac{x_C}{10} & -\frac{\ell x_C}{30} \\ \cdot & -\frac{6y_C}{5\ell} & -\frac{y_C}{10} & \frac{6x_C}{5\ell} & -\frac{x_C}{10} & -\frac{6r_C^2}{5\ell} & -\frac{r_C^2}{10} \\ \cdot & \frac{y_C}{10} & -\frac{\ell y_C}{30} & -\frac{x_C}{10} & -\frac{\ell x_C}{30} & \frac{r_C^2}{10} & -\frac{\ell r_C^2}{30} \end{bmatrix}, \tag{9.164}$$

$$\mathbf{K}^{(2,2)}_{g,N} := N_0 \begin{bmatrix} \cdot & & & \cdot & & \cdot & \cdot \\ & \frac{6}{5\ell} & -\frac{1}{10} & \cdot & \cdot & \frac{6y_C}{5\ell} & -\frac{y_C}{10} \\ & & \frac{2\ell}{15} & \cdot & \cdot & -\frac{y_C}{10} & \frac{2\ell y_C}{15} \\ & & & \frac{6}{5\ell} & \frac{1}{10} & -\frac{6x_C}{5\ell} & \frac{x_C}{10} \\ & & & & \frac{2\ell}{15} & -\frac{x_C}{10} & \frac{2\ell x_C}{15} \\ & & & & & \frac{6r_C^2}{5\ell} & -\frac{r_C^2}{10} \\ \text{SYM} & & & & & & \frac{2\ell r_C^2}{15} \end{bmatrix}.$$

- Geometric matrix, related to the bending moments:

$$
\mathbf{K}_{g,M}^{(1,1)} :=
\begin{bmatrix}
\cdot & \cdots & \cdots & & \\
 & \cdots & \cdots & \bullet & \bullet \\
 & & \cdots & \bullet & \bullet \\
 & & \cdots & \bullet & \bullet \\
 & & & \bullet & \bullet \\
 & & & \bullet & \bullet \\
\text{SYM} & & & & \bullet
\end{bmatrix},
\quad
\mathbf{K}_{g,M}^{(2,1)} :=
\begin{bmatrix}
\cdot & \cdots & \cdots & & \\
 & \cdots & \cdots & \bullet & \bullet \\
 & & \cdots & \bullet & \bullet \\
\cdot & \cdots & \cdots & \bullet & \bullet \\
 & & \cdots & \bullet & \bullet \\
\cdot & \bullet\bullet\bullet\bullet\bullet\bullet & & \\
\cdot & \bullet\bullet\bullet\bullet\bullet\bullet & &
\end{bmatrix},
\quad
\mathbf{K}_{g,M}^{(2,2)} :=
\begin{bmatrix}
\cdot & \cdots & \cdots & & \\
 & \cdots & \cdots & \bullet & \bullet \\
 & & \cdots & \bullet & \bullet \\
 & & \cdots & \bullet & \bullet \\
 & & & \bullet & \bullet \\
 & & & \bullet & \bullet \\
\text{SYM} & & & & \bullet
\end{bmatrix},
$$

$$(9.165)$$

where is it:

$$
K_{g,M26}^{(1,1)} = -\frac{1}{10\ell}\left(11M_{x1}^0 + M_{x2}^0\right), \quad K_{g,M27}^{(1,1)} = -\frac{M_{x1}^0}{10},
$$

$$
K_{g,M36}^{(1,1)} = \frac{1}{10}\left(M_{x1}^0 - 2M_{x2}^0\right), \qquad K_{g,M37}^{(1,1)} = -\frac{\ell}{30}\left(3M_{x1}^0 + M_{x2}^0\right),
$$

$$
K_{g,M46}^{(1,1)} = -\frac{1}{10\ell}\left(11M_{y1}^0 + M_{y2}^0\right), \quad K_{g,M47}^{(1,1)} = -\frac{M_{y1}^0}{10},
$$

$$
K_{g,M56}^{(1,1)} = \frac{1}{10}\left(2M_{y2}^0 - M_{y1}^0\right), \qquad K_{g,M57}^{(1,1)} = \frac{\ell}{30}\left(3M_{y1}^0 + M_{y2}^0\right),
$$

$$
K_{g,M66}^{(1,1)} = \frac{1}{5\ell}\left[x_H\left(11M_{y1}^0 + M_{y2}^0\right) - y_H\left(11M_{x1}^0 + M_{x2}^0\right)\right],
$$

$$
K_{g,M67}^{(1,1)} = \frac{1}{5}\left(x_H M_{y2}^0 - y_H M_{x2}^0\right),
$$

$$
K_{g,M77}^{(1,1)} = \frac{\ell}{15}\left[x_H\left(3M_{y1}^0 + M_{y2}^0\right) - y_H\left(3M_{x1}^0 + M_{x2}^0\right)\right],
$$

$$(9.166)$$

$$
K_{g,M26}^{(2,1)} = \frac{1}{10\ell}\left(11M_{x1}^0 + M_{x2}^0\right), \quad K_{g,M27}^{(2,1)} = \frac{M_{x1}^0}{10},
$$

$$
K_{g,M36}^{(2,1)} = \frac{1}{10}\left(M_{x2}^0 - 2M_{x1}^0\right), \qquad K_{g,M37}^{(2,1)} = \frac{\ell M_{x2}^0}{30},
$$

$$
K_{g,M46}^{(2,1)} = \frac{1}{10\ell}\left(11M_{y1}^0 + M_{y2}^0\right), \quad K_{g,M47}^{(2,1)} = \frac{M_{y1}^0}{10},
$$

$$
K_{g,M56}^{(2,1)} = \frac{1}{10}\left(2M_{y1}^0 - M_{y2}^0\right), \qquad K_{g,M57}^{(2,1)} = -\frac{\ell M_{y2}^0}{30},
$$

$$
K_{g,M62}^{(2,1)} = \frac{1}{10\ell}\left(M_{x1}^0 + 11M_{x2}^0\right), \quad K_{g,M63}^{(2,1)} = \frac{1}{10}\left(2M_{x2}^0 - M_{x1}^0\right),
$$

$$
K_{g,M64}^{(2,1)} = \frac{1}{10\ell}\left(M_{y1}^0 + 11M_{y2}^0\right), \quad K_{g,M65}^{(2,1)} = \frac{1}{10}\left(M_{y1}^0 - 2M_{y2}^0\right),
$$

$$K^{(2,1)}_{g,M67} = \frac{1}{5}\left(y_H M^0_{x2} - x_H M^0_{y2}\right), \quad K^{(2,1)}_{g,M72} = -\frac{M^0_{x2}}{10},$$

$$K^{(2,1)}_{g,M73} = \frac{\ell M^0_{x1}}{30}, \qquad\qquad\qquad K^{(2,1)}_{g,M74} = -\frac{M^0_{y2}}{10},$$

$$K^{(2,1)}_{g,M75} = -\frac{1}{30}\left(\ell M^0_{y1}\right), \qquad\qquad K^{(2,1)}_{g,M76} = \frac{1}{5}\left(x_H M^0_{y1} - y_H M^0_{x1}\right),$$

$$K^{(2,1)}_{g,M66} = -\frac{6}{5\ell}\left[x_H\left(M^0_{y1} + M^0_{y2}\right) - 6y_H\left(M^0_{x1} + M^0_{x2}\right)\right],$$

$$K^{(2,1)}_{g,M77} = -\frac{1}{30}\ell\left[x_H\left(M^0_{y1} + M^0_{y2}\right) - y_H\left(M^0_{x1} + M^0_{x2}\right)\right],$$

$$(9.167)$$

$$K^{(2,2)}_{g,M26} = -\frac{1}{10\ell}\left(M^0_{x1} + 11M^0_{x2}\right), \quad K^{(2,2)}_{g,M27} = \frac{M^0_{x2}}{10},$$

$$K^{(2,2)}_{g,M36} = \frac{1}{10}\left(2M^0_{x1} - M^0_{x2}\right), \qquad K^{(2,2)}_{g,M37} = -\frac{\ell}{30}\left(M^0_{x1} + 3M^0_{x2}\right),$$

$$K^{(2,2)}_{g,M46} = -\frac{M^0_{y1} + 11M^0_{y2}}{10\ell}, \qquad K^{(2,2)}_{g,M47} = \frac{M^0_{y2}}{10},$$

$$K^{(2,2)}_{g,M56} = \frac{1}{10}\left(M^0_{y2} - 2M^0_{y1}\right), \qquad K^{(2,2)}_{g,M57} = \frac{\ell}{30}\left(M^0_{y1} + 3M^0_{y2}\right),$$

$$K^{(2,2)}_{g,M66} = \frac{1}{5\ell}\left[x_H\left(M^0_{y1} + 11M^0_{y2}\right) - y_H\left(M^0_{x1} + 11M^0_{x2}\right)\right],$$

$$K^{(2,2)}_{g,M67} = \frac{1}{5}\left(y_H M^0_{x1} - x_H M^0_{y1}\right),$$

$$K^{(2,2)}_{g,M77} = \frac{1}{15}\ell\left[x_H\left(M^0_{y1} + 3M^0_{y2}\right) - y_H\left(M^0_{x1} + 3M^0_{x2}\right)\right].$$

$$(9.168)$$

- Geometric matrix, related to the eccentricity, with respect to the center of torsion, of the application point of the transverse loads:

$$
\mathbf{K}_{g,p}^{(1,1)} =
\begin{bmatrix}
\cdot & \cdot\;\cdot\;\cdot\;\cdot & \cdot & \cdot \\
 & \cdot\;\cdot\;\cdot\;\cdot & \cdot & \cdot \\
 & \cdot\;\cdot\;\cdot & \cdot & \cdot \\
 & \cdot\;\cdot & \cdot & \cdot \\
 & & \frac{13}{35}e\varrho p\ell & \frac{11}{210}e\varrho p\ell^2 \\
\text{SYM} & & & \frac{1}{105}e\varrho p\ell^3
\end{bmatrix},
$$

$$
\mathbf{K}_{g,p}^{(2,1)} =
\begin{bmatrix}
\cdot\;\cdot\;\cdot\;\cdot\;\cdot & \cdot & \cdot \\
\cdot\;\cdot\;\cdot\;\cdot\;\cdot & \cdot & \cdot \\
\cdot\;\cdot\;\cdot\;\cdot\;\cdot & \cdot & \cdot \\
\cdot\;\cdot\;\cdot\;\cdot\;\cdot & \cdot & \cdot \\
\cdot\;\cdot\;\cdot\;\cdot\;\cdot & \frac{9}{70}e\varrho p\ell & \frac{13}{420}e\varrho p\ell^2 \\
\cdot\;\cdot\;\cdot\;\cdot\;\cdot & -\frac{13}{420}e\varrho p\ell^2 & -\frac{1}{140}e\varrho p\ell^3
\end{bmatrix},
\tag{9.169}
$$

$$
\mathbf{K}_{g,p}^{(2,2)} =
\begin{bmatrix}
\cdot & \cdot\;\cdot\;\cdot\;\cdot & \cdot & \cdot \\
 & \cdot\;\cdot\;\cdot\;\cdot & \cdot & \cdot \\
 & \cdot\;\cdot\;\cdot & \cdot & \cdot \\
 & \cdot\;\cdot & \cdot & \cdot \\
 & & \frac{13}{35}e\varrho p\ell & -\frac{11}{210}e\varrho p\ell^2 \\
\text{SYM} & & & \frac{1}{105}e\varrho p\ell^3
\end{bmatrix}.
$$

□

Matrix Assembly

A straight TWB is considered, divided into $e = 1, 2, \ldots, m$ finite elements. Once the prestress state at the nodes is known, by the previous expressions, the stiffness matrices of all the elements are evaluated; then, they are assembled and the constraints introduced, as in the usual linear elastic analysis. In particular, in the presence of torsional clamp at node i, it is $\theta_i = 0$; however, it is needs to distinguish whether the constraint prevents or leaves free the warping. In the former case, the torsional curvature must be zeroed, $\kappa_{ti} = 0$; in the latter case, κ_{ti} remains arbitrary (the examples shown in the Appendix C illustrate the topic).

The procedure leads to the eigenvalue problem:

$$
\left(\mathbf{K}_e + \mu \mathbf{K}_g \right) \mathbf{q} = \mathbf{0},
\tag{9.170}
$$

where μ is the multiplier of the prestresses. The smallest eigenvalue μ_c is the critical value sought for.

9.8.2 Numerical Examples

Some numerical examples are shown, in which, by applying the method of the finite elements, the critical load and the critical mode of open TWB are computed, in different constraint and load conditions. Unless otherwise specified, the beam is that analyzed in the Appendix C (Fig. C.8), i.e., tubular, of small thickness, cut along the generatrix intersected by the positive y semi-axis. Its dimensions are radius $R = 40\,\mathrm{mm}$, thickness $b = 4\,\mathrm{mm}$, and length $\ell = 4\,\mathrm{m}$ (except different specification). The material is steel, of moduli $E = 207,000\,\mathrm{N/mm^2}$, $G = 79,300\,\mathrm{N/mm^2}$.

Example 9.3 (Compressed Beam, Simply Supported by Warping-Unrestrained Torsional Constraints) The structure depicted in Fig. 9.14a, is studied, consisting of a TWB, presolicited by a compression load P, constrained at the ends by simple supports and torsional clamps that leave the warping free. The critical load of this

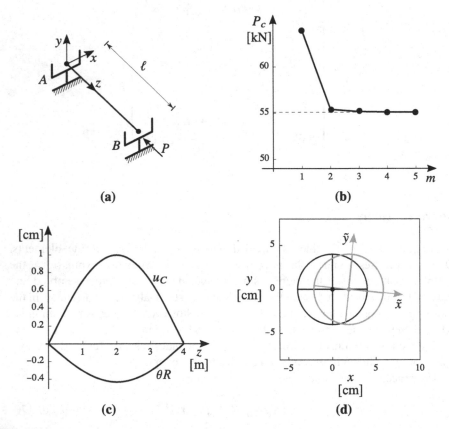

Fig. 9.14 Compressed TWB, simply supported, torsionally clamped with free warping: (**a**) structure, (**b**) convergence analysis, (**c**) critical mode, (**d**) midspan cross-section of the buckled beam

Fig. 9.15 Critical load P_c of
the beam in Fig. 9.14 vs the
length ℓ: FT
flexural-torsional, E Eulerian
critical load in one of the two
principal inertia planes
$(P_x \equiv P_y)$

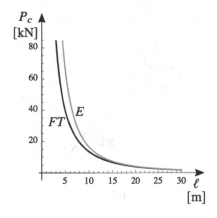

beam can be determined exactly, as discussed in the Sect. 9.4, and turns out to be
$P_c = 55.1$ kN. When the finite element method is used, the critical load depends on
the number m of the elements in which the beam is discretized. Figure 9.14b reports
the value of the critical load vs the number of elements (convergence analysis). It
emerges that the approximation is by excess and that the approximate value tends
fast and monotonously to the exact one. When $m = 4$, the result it is practically
correct. The following analyses, however, have been carried out by taking 50 finite
elements, to take into account more complex stress conditions.

The critical mode is found to be of the flexural-torsional type, in which the beam
buckles in the plane orthogonal to that of symmetry and twists. Figure 9.14c illustrates the two components of the critical mode, the transverse displacement $u_C(z)$,
and the twist $\theta(z) R$ (rendered dimensionally homogeneous to a displacement). The
pattern is sinusoidal, as predicted by the exact solution. Figure 9.14d describes the
attitude of the midspan cross-section in the deformed configuration. It appears that,
for the length considered ($\ell = 4$ m), the torsional contribution is of the same order
of magnitude as flexural.

In order to investigate the influence of the beam length, the analysis has been
repeated, by increasing ℓ to very high values, and the relevant critical load P_c,
together with the two Eulerian loads $P_x \equiv P_y$, has been determined and plotted
in Fig. 9.15. The first of these Eulerian loads is associated with a buckling occurring
in the plane of symmetry (y, z), the second with a buckling taking place in the
plane (x, z), orthogonal to that of symmetry, in which one erroneously ignores
the torsional contribution. It is seen that the (true) flexural-torsional critical load is
smaller than the two Eulerian loads, as already observed in the Sect. 9.4. However,
when the beam is very long, and according to the Remark 9.3, the lateral-torsional
coupling becomes less and less important, as $P_\theta / P_x \to \infty$. In this case, two nearly
simultaneous Eulerian modes take place.   □

Example 9.4 (Compressed Beam, Simply Supported by Warping-Prevented Tor-
sional Constraints) The structure illustrated in Fig. 9.16a is analyzed, which differs
from that considered in the previous example only for the addition of constraints

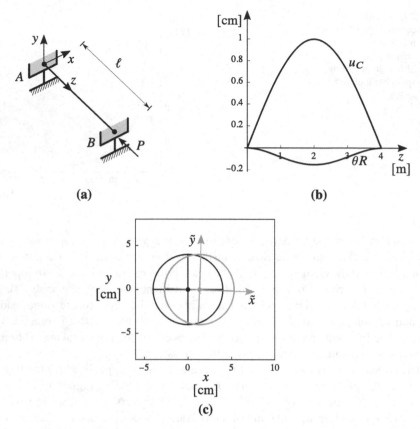

Fig. 9.16 Simply supported compressed twin walled beam, torsionally clamped with prevented warping: (**a**) structure, (**b**) critical mode, (**c**) midspan cross-section of the buckled beam

preventing warping. It is important to point out that no closed form solution is available for this problem.

From a finite element analysis, the critical load is determined to be $P = 81.04\,\text{kN}$, which *is approximately 1.5 times larger than that relevant to the free warping case*, showing the considerable stiffening effect exerted by the additional constraints. About the critical mode, which is still found to be flexural-torsional, the results shown in Fig. 9.16b, c are found. Comparing them with those of the previous example, it is observed that while the translation $u_C\,(z)$ maintains a sinusoidal pattern, the twist $\theta\,(z)$ changes significantly, mainly close to the supports, as the boundary conditions impose that $\kappa_t = \theta'$ zeroes there.   □

Example 9.5 (Uniformly Bent Beam, Simply Supported by Warping-Unrestrained Torsional Constraints) The structure illustrated in Fig. 9.17a is examined, consisting of a TWB of length $\ell = 4\,\text{m}$, having the same cut annular cross-section used

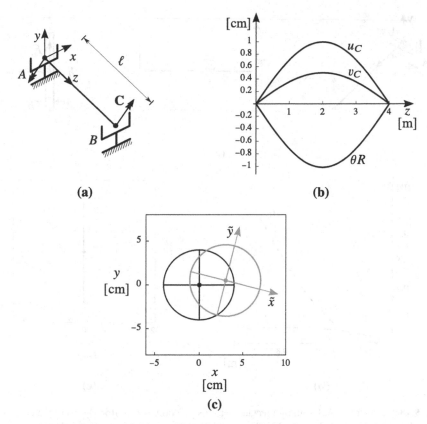

(a)   (b)

(c)

Fig. 9.17 Uniformly bent thin-walled beam, simply supported, torsionally clamped with pre-vented warping: (**a**) structure, (**b**) critical mode, (**c**) midspan cross-section of the buckled beam

in the previous examples, prestressed by bending couples $\mathbf{C} = \pm \left(C \, \mathbf{a}_x + \frac{1}{2} C \, \mathbf{a}_y \right)$ applied to the ends. Also in this case the exact solution is known (Sect. 9.5).

By discretizing the beam into 50 finite elements, $C_c = -4.06$ kN m is found, i.e., a value almost identical to that provided by the exact solution (Eq. 9.75, with $C_x = C$, $C_y = \frac{1}{2}C$). All three components of the critical mode, $u_C\,(z)$, $v_C\,(z)$, $and\theta\,(z)$, of sinusoidal type, are different from zero (Fig. 9.17b) and of the same order of magnitude and are in perfect agreement with the analytical solution. Figure 9.17c shows the displacement of the cross-section at the abscissa $z = \ell/2$.   □

Example 9.6 (Fixed-Free Beam Bent by a Transverse Force Applied at the End) The fixed-free beam of Fig. 9.18a is considered, prestressed by a force F applied at the free end, at the center of torsion. The cross-section is rectangular, of thickness b and height $h \gg b$. The geometric characteristics are $A = h\,b$, $I_x = \frac{1}{12}bh^3$, $I_y = \frac{1}{12}b^3h$, $J = \frac{1}{3}h\,b^3$, $andI_\omega = 0$. In the example, the following numerical values have been

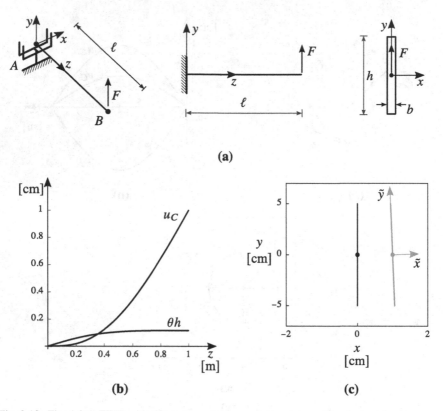

(a)

(b)   **(c)**

Fig. 9.18 Fixed-free TWB with thin rectangular cross-section, non-uniformly bent by a concentrated load: (**a**) structure, (**b**) critical mode, (**c**) midspan cross-section of the buckled beam

taken: $b = 2\,\text{mm}$, $h = 100\,\text{mm}$, $\ell = 1\,\text{m}$ and $E = 207{,}000\,\text{N/mm}^2$, and $G = 79{,}300\,\text{N/mm}^2$.

The analytic solution to this problem (obtained in the Sect. 9.7.2 by the Frobenius method) provides the critical load $F_c = 68.6\,\text{N}$. The finite element analysis yields $F_c = 68.9\,\text{N}$, with an error of 0.4%. Figure 9.18b, c shows the critical mode, which consists of a lateral displacement and a twist of the same order of magnitude. □

Example 9.7 (Simply Supported Beam with Thin Rectangular Cross-Section, Bent by a Uniformly Distributed Load) The simply supported beam in Fig. 9.19a is considered, having thin rectangular cross-section, as previously described, torsionally clamped. The beam is prestressed by a distributed load of intensity $\mathbf{p} = -p\mathbf{a}_y$, applied on a line parallel to the z axis. The case in which this line has zero eccentricity with respect to the torsion center axis is considered first (Fig. 9.20a). From the numerical analysis, the critical load is found to be equal to $p_c = 484\,\text{N/m}$. In Figs. 9.19b, c the critical mode is illustrated, for which the considerations already made for the concentrated load are still valid.

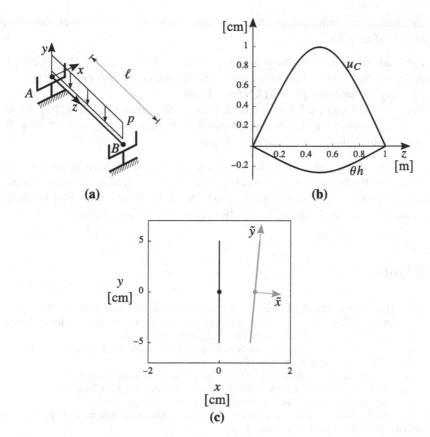

(a)

(b)

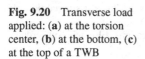

(c)

Fig. 9.19 Simply supported beam, with thin rectangular cross-section, non-uniformly bent by distributed loads: (**a**) structure, (**b**) critical mode, (**c**) midspan cross-section of the buckled beam

Fig. 9.20 Transverse load applied: (**a**) at the torsion center, (**b**) at the bottom, (**c**) at the top of a TWB

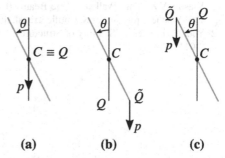

(a)   **(b)**   **(c)**

To analyze the influence of the load application line, two further cases are considered (Fig. 9.20b, c):

- The load p (positive, because outgoing from the torsion center, Fig. 9.20b) is applied at the bottom of the beam and therefore has eccentricity $e_Q = \overline{CQ} = \frac{h}{2}$; repeating the analysis, $p_c = 512 \, \text{N/m}$ is found.
- The $-p$ load (negative because it is entering in the center of torsion, Fig. 9.20c) is applied to the top of the beam and therefore has eccentricity $e_Q = \overline{CQ} = \frac{h}{2}$; from the analysis, $p_c = 457 \, \text{N/m}$ is obtained.

The position of the load, therefore, affects the critical load, increasing it (when the load is at the bottom) or by reducing it (when the load is at the top) of some percentage units, in accordance with the Remark 9.11. Figure 9.20 clearly shows the neutral, stabilizing or instabilizing effect of the position of the load.   □

References

1. Allen, H.G., Bulson, P.S.: Background to Buckling. McGraw-Hill, New York (1980)
2. Barsoum, R.S., Gallagher, R.H.: Finite element analysis of torsional and torsional–flexural stability problems. Int. J. Numer. Meth. Eng. **2**(3), 335–352 (1970)
3. Bazant, Z.P., Cedolin, L.: Stability of Structures: Elastic, Inelastic, Fracture, and Damage Theories. World Scientific Publishing, Singapore (2010)
4. Bleich, F.: Buckling Strength of Metal Structures. McGraw-Hill, New York (1952)
5. Chajes, A.: Principles of Structural Stability Theory. Prentice-Hall, New Jersey (1974)
6. Como, M.: Theory of Elastic Stability (in Italian). Liguori, Napoli (1967)
7. Corradi Dell'Acqua, L.: Meccanica delle strutture: La valutazione della capacità portante (in Italian), vol. 3. McGraw-Hill, Milano (1994)
8. Iyengar, N.G.R.: Elastic Stability of Structural Elements. Macmillan India, New Delhi (2007)
9. Pignataro, M., Rizzi, N., Luongo, A.: Stability, Bifurcation and Postcritical Behaviour of Elastic Structures. Elsevier, Amsterdam (1990)
10. Timoshenko, S.P., Gere, J.M.: Theory of Elastic Stability. McGraw-Hill, New York (1961)
11. Vlasov, V.Z.: Thin-Walled Elastic Beams (English translation of the 2nd Russian edition of 1959). Israel Program for Scientific Translation, Jerusalem (1961)
12. Yoo, C.H., Lee, S.: Stability of Structures: Principles and Applications. Elsevier, Amsterdam (2011)

Chapter 10
Buckling of Plates and Prismatic Shells

10.1 Introduction

It is well-known, from the linear elastic theory, that a plate subjected to transverse forces can be assimilated to a system of two orders of mutually orthogonal strips, of infinitesimal width and contiguous. Accordingly, the plate behaves like a grillage, made of an infinite number of infinitesimal beams, exchanging shear forces and bending and torsional moments at nodes. The same analogy existing in the linear field still holds in buckling. It is possible, indeed, to prove (e.g., via a Ritz or a finite element analysis), that a rectangular grillage, simply supported at the boundary, compressed by external forces acting in one direction, upon the achievement of a threshold value of the solicitation, buckles out-of-plane. The deflection manifests itself through *waves in the two directions*, whose length depends on the geometric characteristics of the beams and their mutual distances (similarly to what happens for a beam on Winkler soil). The phenomenon can be viewed as generated from buckling of some, or all, the constituent beams, which interact at the connecting nodes.

Buckling of a rectangular plate, compressed by normal forces at its edges, is a phenomenon similar to that described. The bifurcation also manifests when the stress is bi-axial (i.e., when both orders of strips are compressed, or one is compressed and the other is taut), or even when the plate is subject to in-plane shear stress (as happens, e.g., for the web of a I-cross-section beam, when non-uniformly bent).

When a thin-walled member (TWM), with open or closed profile, is analyzed, it is necessary to distinguish whether there are, or not, diaphragms or ribs, able of guarantee the undeformability of the cross-section.[1] In the presence of

---

[1] In this chapter, the locution "thin-walled member," instead of "thin-walled beam," is used, to stress the fact that the member can behave as a beam or as a shell.

A. Luongo et al., *Stability and Bifurcation of Structures*,
https://doi.org/10.1007/978-3-031-27572-2_10

reinforcements, the TWM behaves, with a good approximation, as an Euler or Vlasov beam; in contrast, in the absence of reinforcements, it acts as an assembly of plates, more precisely, a *prismatic shell*,[2] whose cross-section is free to deform in its own plane. When such a shell is axially or eccentrically compressed, the individual constituent plates can buckle out of their respective planes, although restrained to respect equilibrium and congruence along the interconnection nodal lines. The same lines, either remain almost straight and motionless (so that a *local buckling* is said to occur, in which the cross-section loses its shape), or they undergo significant displacements (so that buckling is called *distortional*, which implies loss of the transverse shape, or *global*, in which the loss of shape is negligible, and therefore peculiar of beams).

In order to study these phenomena, a model of a 2D plate, presolicited by membrane stresses, immersed in the 3D space, needs to be developed. As it is known from the linear theory, there exist in literature several models, including (a) the *Kirchhoff plate* (isotropic or orthotropic), also called "shear undeformable," in which it is assumed that the material segments, initially orthogonal to the middle surface, remains straight and orthogonal to this surface after the deformation; (b) the *Mindlin plate* (isotropic or orthotropic), also called "shear deformable," where it is assumed that such segments can rotate with respect to the middle surface; (c) the *fibered plate* (orthotropic, by definition), also called "infinitesimal mesh grillage," in which this is admitted, that the torsional moments acting in the two orthogonal directions can differ each other. For the introductory purposes of this book, the simpler isotropic model of Kirchhoff suffices, and therefore reference will be made to it. To deal with more complex problems, however, it is necessary to introduce at least the orthotropy (e.g., in case of a densely ribbed plate, in one or two directions, which may be modeled as homogeneous, but orthotropic).

In this chapter, the buckling problem of plates and assemblies of plates is formulated in the context of the linearized theory. The geometric stiffness operator is evaluated for plates presolicited in their own planes. The critical load of rectangular plates, in different load and constraint conditions, are computed, either exactly or by the Ritz method. Concerning prismatic shells, global, distortional and local forms of instability are discussed, and numerical procedures able to detect them, as the Finite Strip Method and the Finite Element Sectional Model, are described.

10.2   Kirchhoff Plate Model

The shear undeformable plate model, assumed to be known to the reader, is here shortly recalled in the context of the linear theory (i.e., by ignoring, for the moment, the state of prestress). Consistently with the general approach of the book, the fundamental equations are derived through a variational procedure, although an a posteriori interpretation, based on equilibrium, is provided.

---

[2] Also said "folded plate."

The plate is a cylindrical body, in which two dimensions (in the x, y directions) are much larger than the third one (the thickness, in the z direction). It is customary in the textbooks, to derive the model by three-dimensional elasticity, assuming that the material segments, which are initially parallel to the z axis, remain straight and orthogonal to the deformed middle plane (x, y). Here, instead, it is preferred a *direct formulation*, in which the plate is idealized ab initio as a two-dimensional *polar* continuum, i.e., made of material particles endowed with orientation, able to mutually exchange couples, as well as forces (thus generalizing concepts which are well-known from the 1D beam theory).

10.2.1 Kinematics

The two-dimensional plate, in its natural state, is made of material points P lying in the (x, y) plane. In the current configuration, it assumes a slightly deflected shape, in which the points P undergo displacements $w(x, y)$, orthogonal to the plane,[3] as well as rotations φ_x, φ_y, respectively around the x, y axes (Fig. 10.1a). If, however, the model is shear undeformable, rotations are not independent variables, but they are related to displacements by the relations (Fig. 10.1b):

$$\varphi_x = w_{,y}, \quad \varphi_y = -w_{,x}, \tag{10.1}$$

so that $w(x, y)$ is the only kinematic descriptor.

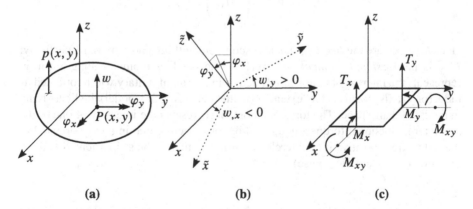

Fig. 10.1 Two-dimensional model of plate: (a) displacements, (b) shear undeformability, (c) stresses

---

[3] The in-plane displacements, triggering the plate membrane behavior, are here ignored, because they lead to equations uncoupled from those governing the out-of-plane motion. This circumstance can also be verified in buckling, where, as it happens for the Euler and Vlasov beams, the critical mode is inextensional.

The following (linear) generalized strains are defined:

$$\kappa := \begin{pmatrix} \kappa_x \\ \kappa_y \\ \kappa_{xy} \end{pmatrix} := \begin{pmatrix} w_{,xx} \\ w_{,yy} \\ 2w_{,xy} \end{pmatrix}, \tag{10.2}$$

where κ_x, κ_y are the *flexural curvatures* and κ_{xy} is the *torsional curvature*. The flexural curvatures are identical to those experienced by an Euler-Bernoulli beam oriented along the x or y axis and therefore positive when the curves have concavity toward the positive z semi-axis. The torsional curvature is instead the *sum* of the torsional curvatures of the two orthogonal strips (beams) crossing at P, equal in module, but taken one positive counterclockwise, the other positive clockwise.[4]

10.2.2 Internal Forces and Elastic Law

The internal forces (referred ahead as generalized stresses) are introduced and the elastic law stated.

Generalized Stresses
The strains in Eqs. 10.2 are associated with dual generalized stresses (having physical dimensions of a moment per unit of length, i.e., of a force):

$$\mathbf{m} := \begin{pmatrix} M_x \\ M_y \\ M_{xy} \end{pmatrix}. \tag{10.3}$$

Here M_x, M_y are the *bending moment* acting along the x and y axes, respectively; M_{xy} is the *torsional moment* (Fig. 10.1c). The bending moment along x (or y) represents the internal bending couple acting in a strip, of unitary width, oriented as the x (or y) axis, positive if it "extends" (in the underlying 3D model) the fibers lying in the $z < 0$ half-space. The torsional moment represents the intensity of one of the *two* internal torsional torques, equal and opposite, acting one in a strip oriented in the x direction (positive if counterclockwise), the other in the strip oriented in the y direction (positive if clockwise).[5]

---

[4] A beam parallel to x is (counterclockwise) twisted by $\kappa_{tx} = \varphi_{x,x} = \left(w_{,y}\right)_{,x}$; a beam parallel to y is clockwise twisted by $\kappa_{ty} = -\varphi_{y,y} = -\left(-w_{,x}\right)_{,y} = \kappa_{tx}$. Thus, it turns out, that $\kappa_{xy} = \kappa_{tx} + \kappa_{ty}$. The circumstance is similar to that occurring in a plane state of strain, in which $\gamma_{xy} = \varepsilon_{xy} + \varepsilon_{yx}$, where $\varepsilon_{xy} = \varepsilon_{yx}$ are the rotations, equal and *opposite*, of two initially orthogonal segments.

[5] The analogy with the plane state of stress holds, in which the stress τ acts on two orthogonal faces, positive in verses which induce *opposite* torques with respect to the center of the element.

In addition to the *active generalized stresses* **m**, *reactive stresses* $\mathbf{t} := \begin{pmatrix} T_x & T_y \end{pmatrix}^T$ also exist. They represent the generalized shear stresses in the two order of strips, positive if concordant with the z axis, when acting on the positive faces of the element (i.e., having outgoing normal equiverse with the x, y axes, Fig. 10.1c). These internal forces, however, expend zero work on the (inhibited) shear strains of the shear undeformable plate, so that they do not appear in the variational formulation, which provides equilibrium equations *pure* in displacements.[6] Their consideration, however, is important to interpret the boundary conditions, also provided by the variational process.

Isotropic Elastic Law

A linear isotropic hyperelastic constitutive law is assumed, between generalized stresses and strains, i.e.:

$$\mathbf{m} = \mathbf{E}\kappa, \tag{10.4}$$

where the *elastic matrix* is defined as follows:

$$\mathbf{E} := D \begin{bmatrix} 1 & v & 0 \\ v & 1 & 0 \\ 0 & 0 & \frac{1-v}{2} \end{bmatrix}. \tag{10.5}$$

In this,

$$D := \frac{Eh^3}{12\left(1 - v^2\right)} \tag{10.6}$$

is the *flexural stiffness of the plate*, E is the Young modulus, v the Poisson's ratio, and h the thickness of the plate. This law directly follows from that governing the plane stress elastic state, in which the coupling between the normal stresses is due to the Poisson effect.[7] The torsional moment is instead proportional to the torsional curvature.

Remark 10.1 When $v = 0$, the link between bending moments and curvatures reproduces that of a beam with rectangular $1 \times h$ cross-section, for which the flexural

---

[6] In other words, the internal shear forces play the same role of the shear force in the Euler-Bernoulli beam.

[7] Indeed:

$$\begin{pmatrix} \sigma_x \\ \sigma_y \\ \tau_{xy} \end{pmatrix} = \frac{E}{1 - v^2} \begin{bmatrix} 1 & v & 0 \\ v & 1 & 0 \\ 0 & 0 & \frac{1-v}{2} \end{bmatrix} \begin{pmatrix} \varepsilon_x \\ \varepsilon_y \\ \gamma_{xy} \end{pmatrix}$$

having taken into account that the tangential modulus of elasticity is $G = \frac{E}{2(1+v)}$.

stiffness is $EI = \frac{Eh^3}{12}$. The stiffening factor, $\frac{1}{1-v^2}$, is due to the *prevented transverse dilatation of the inflected strips*, caused by the presence of adjacent strips, exerting a confinement action.

Remark 10.2 The link between the torsional moment and curvature of the plate is quite more complex. By taking into account that the elastic tangential modulus is $G = \frac{E}{2(1+v)}$, it reads $M_{xy} = \frac{Gh^3}{12}\kappa_{xy}$; since, as mentioned, $\kappa_{xy} = 2\kappa_t$, the plate law states that $M_{xy} = \frac{Gh^3}{6}\kappa_t$. On the other hand, in a beam of rectangular $1 \times h$ cross-section, the same link reads $M_t = GJ\kappa_t = \frac{Gh^3}{3}\kappa_t$. Hence, with the same strain κ_t, the *torsional moment in the plate is half the torsional moment in the beam*. This occurrence is explained as follows. Whereas in the plate strip the *flow of tangential stresses is open*, i.e., the τ stresses run parallel to the long sides, in the beam cross-section the flow is closed, i.e., the τ's run (approximately) parallel to the entire perimeter. Now, it is well-known that the contribution brought by the τ's parallel to the short sides is equal to that of the τ's parallel to the long sides; this entails that the same strains generate in the two models the *same τ's, but the plate torsional moments is half* of that of beam. However, since the moments M_{xy} are two, acting in the two directions x and y, while the torsional moment M_t is just one, the two models are *energetically equivalent*. Indeed, the virtual works δW, done by the stresses of the two models in the same deformation, are equal, i.e., $\delta W^{plate} := M_{xy}\delta\kappa_{xy} = \left(\frac{M_t}{2}\right)(2\delta\kappa_t) = M_t\delta\kappa_t =: \delta W^{beam}$. It can also be said that, with the same deformation of the two models, the elastic potential energy density e is the same; indeed, $e^{plate} = \frac{1}{2}M_{xy}\kappa_{xy} = \frac{1}{2}\frac{Gh^3}{12}\kappa_{xy}^2$ and $e^{beam} = \frac{1}{2}M_t\kappa_t = \frac{1}{2}\frac{Gh^3}{3}\kappa_t^2$ are equal, given the relationship between the two curvatures.

10.2.3   Elastic Potential Energy and Equilibrium Equations

The equilibrium equations of the plate, subject to transverse forces $p(x, y)$, are derived here through a variational procedure. The density (per unit area) of the elastic potential energy, by taking into account Eq. 10.4, reads:

$$e = \frac{1}{2}\mathbf{m}^T\boldsymbol{\kappa} = \frac{1}{2}\boldsymbol{\kappa}^T\mathbf{E}\boldsymbol{\kappa} = \frac{D}{2}\left[\kappa_x^2 + \kappa_y^2 + 2v\kappa_x\kappa_y + \frac{1-v}{2}\kappa_{xy}^2\right], \tag{10.7}$$

or, by using the congruence relations in Eqs. 10.2:

$$e = \frac{D}{2}\left[w_{,xx}^2 + w_{,yy}^2 + 2vw_{,xx}w_{,yy} + 2(1-v)w_{,xy}^2\right]. \tag{10.8}$$

The total potential energy (TPE) is the sum of the elastic potential energy[8] $U = \iint_\Omega e \, dx \, dy$, where Ω is the plane domain occupied by the plate, and of the load potential energy; therefore:

$$\Pi = \frac{D}{2} \iint_\Omega \left[w_{,xx}^2 + w_{,yy}^2 + 2vw_{,xx}w_{,yy} + 2(1-v)w_{,xy}^2 \right] dx \, dy - \iint_\Omega p \, w \, dx \, dy.$$

$$(10.9)$$

Equating to zero the first variation of the TPE, one gets:

$$\delta\Pi = D \iint_\Omega \left[\left(w_{,xx} + vw_{,yy} \right) \delta w_{,xx} + \left(w_{,yy} + vw_{,xx} \right) \delta w_{,yy} \right.$$

$$\left. + 2(1-v) w_{,xy} \, \delta w_{,xy} \right] dx \, dy - \iint_\Omega p \, \delta w \, dx \, dy = 0, \qquad \forall \delta w.$$

$$(10.10)$$

To "free" the variations from the derivatives, it is necessary to integrate by parts, using the Green formula:[9]

$$\iint_\Omega f \, g_{,j} dx \, dy = - \iint_\Omega g \, f_{,j} dx \, dy + \oint_\Gamma f \, g \, n_j ds, \qquad (10.11)$$

where n_j are the director cosines of the normal outgoing from the domain and s is a curvilinear abscissa on Γ, oriented to leave the domain to the left (Fig. 10.2a). Equivalently, given that $n_x ds = dy$ and $n_y ds = -dx$, Eq. 10.11 can be written:

$$\iint_\Omega f \, g_{,x} dx \, dy = - \iint_\Omega g \, f_{,x} dx \, dy + \oint_\Gamma f \, g \, dy, \qquad (10.12a)$$

$$\iint_\Omega f \, g_{,y} dx \, dy = - \iint_\Omega g \, f_{,y} dx \, dy - \oint_\Gamma f \, g \, dx. \qquad (10.12b)$$

---

[8] Here, the tilde on U is omitted, because prestresses are absent.

[9] This follows from *Green lemma*:

$$\iint_\Omega F_{,j} dx \, dy = \oint_\Gamma F \, n_j \, ds,$$

where $F = f \, g$.

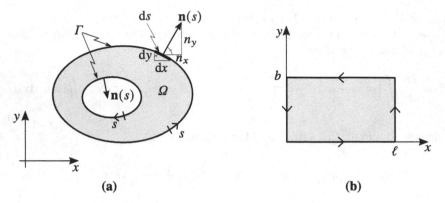

Fig. 10.2   Plate: (**a**) of general shape, (**b**) of rectangular shape; domain Ω, boundary Γ, outgoing normal **n**, abscissa s

By integrating Eq. 10.10 twice by parts, one obtains:

$$\delta\Pi = \iint_{\Omega} \left[D \left(w_{,xxxx} + 2w_{,xxyy} + w_{,yyyy} \right) - p \right] \delta w \, dx \, dy$$

$$+ \oint_{\Gamma} D \left[\cdots \right] ds = 0, \qquad \forall \delta w,$$

(10.13)

where the boundary terms will be derived later.

Field Equation

From Eq. 10.18, the indefinite equilibrium equation follows:

$$D \left(w_{,xxxx} + 2w_{,xxyy} + w_{,yyyy} \right) = p,$$

(10.14)

which, in operational form, reads:

$$\mathcal{K}_e w = p,$$

(10.15)

where

$$\mathcal{K}_e := \partial^4_{xxxx} + 2\partial^4_{xxyy} + \partial^4_{yyyy} \equiv \nabla^4$$

(10.16)

is the *elastic stiffness operator of the plate*. The mechanical meaning of this equation will be discussed in the Supplement 10.2.

Remark 10.3 The Poisson's ratio, although different from zero, does not appear in the indefinite equilibrium equation of the isotropic plate.

Boundary Conditions

From Eq. 10.10, the boundary conditions also follow. Limiting to consider rectangular plates, for which $\Omega := \{(x, y) \,|\, x \in [0, \ell], \; y \in [0, b]\}$ (Fig. 10.2b), after some transformations (described in detail in the Supplement 10.1), these are found to be:

$$D\left(w_{,xx} + \nu w_{,yy}\right) \delta w_{,x} = 0, \qquad \text{at} \quad x = 0, \ell, \tag{10.17a}$$

$$D\left(w_{,yy} + \nu w_{,xx}\right) \delta w_{,y} = 0, \qquad \text{at} \quad y = 0, b, \tag{10.17b}$$

$$-D\left(w_{,xxx} + (2 - \nu)\, w_{,xyy}\right) \delta w = 0, \qquad \text{at} \quad x = 0, \ell, \tag{10.17c}$$

$$-D\left(w_{,yyy} + (2 - \nu)\, w_{,xxy}\right) \delta w = 0, \qquad \text{at} \quad y = 0, b, \tag{10.17d}$$

$$2D\left(1 - \nu\right) w_{,xy}\delta w = 0, \qquad \text{at} \quad (0, 0),\, (\ell, 0), \tag{10.17e}$$

$$(\ell, b),\, (0, b).$$

In each of them, the (mechanical) term in parentheses must be vanished if the corresponding kinematic quantity is arbitrary. For example, if the plate is supported on the $x = \ell$ side, the virtual rotation $\delta\varphi_y = -\delta w_{,x}$ is non-zero there, so that $D\left(w_{,xx} + \nu w_{,yy}\right) = 0$ must hold, together with $w = 0$. If the plate is free on the same side, $\delta w_{,x}$ and δw are both non-zero, so $D\left(w_{,xx} + \nu w_{,yy}\right) = 0$ and $D\left(w_{,xxx} + (2 - \nu)\, w_{,xyy}\right) = 0$ must hold. The latest conditions, Eq. 10.17e, are punctual, and refer to the vertices of the plate: if the displacement δw is free at those points, then $2D\left(1 - \nu\right) w_{,xy} = 0$ must be satisfied at the same vertices. The mechanical meaning of the boundary conditions is discussed in the Supplement 10.2.

Supplement 10.1 (Boundary Conditions for the Rectangular Plate, Deduced by the Variational Procedure) To prove Eqs. 10.17, one has to proceed as follows. Equation 10.10, integrated by parts according to Eqs. 10.12, reads:

$$\delta\Pi = \iint_{\Omega} \left[D\left(w_{,xxxx} + 2w_{,xxyy} + w_{,yyyy}\right) - p\right] \delta w \, \mathrm{d}x \, \mathrm{d}y$$

$$+ \oint_{\Gamma} D\left(w_{,xx} + \nu w_{,yy}\right) \delta w_{,x}\mathrm{d}y - \oint_{\Gamma} D\left(w_{,xx} + \nu w_{,yy}\right)_{,x} \delta w \, \mathrm{d}y$$

$$- \oint_{\Gamma} D\left(w_{,yy} + \nu w_{,xx}\right) \delta w_{,y} \, \mathrm{d}x + \oint_{\Gamma} D\left(w_{,yy} + \nu w_{,xx}\right)_{,y} \delta w \, \mathrm{d}x$$

$$- \oint_{\Gamma} D\left(1 - \nu\right) w_{,xy} \, \delta w_{,x} \, \mathrm{d}x - \oint_{\Gamma} D\left(1 - \nu\right) w_{,xyy} \, \delta w \, \mathrm{d}y$$

$$+ \oint_{\Gamma} D\left(1 - \nu\right) w_{,xy} \, \delta w_{,y} \, \mathrm{d}y + \oint_{\Gamma} D\left(1 - \nu\right) w_{,xxy} \, \delta w \, \mathrm{d}x = 0, \qquad \forall \delta w. \tag{10.18}$$

By simplifying, the boundary integrals become:

$$\oint_\Gamma D\left(w_{,xx} + \nu w_{,yy}\right)\delta w_{,x}\mathrm{d}y - \oint_\Gamma D\left(w_{,yy} + \nu w_{,xx}\right)\delta w_{,y}\,\mathrm{d}x$$

$$- \oint_\Gamma D\left(w_{,xxx} + w_{,xyy}\right)\delta w\,\mathrm{d}y + \oint_\Gamma D\left(w_{,yyy} + w_{,xxy}\right)\delta w\mathrm{d}x$$

$$- \oint_\Gamma D\left(1 - \nu\right) w_{,xy}\,\delta w_{,x}\,\mathrm{d}x + \oint_\Gamma D\left(1 - \nu\right) w_{,xy}\,\delta w_{,y}\,\mathrm{d}y = 0, \qquad \forall \delta w.$$

$$(10.19)$$

Since $\mathrm{d}x = 0$ on sides $x = \text{cost}$ and $\mathrm{d}y = 0$ on sides $y = \text{cost}$, taking into account the counterclockwise traveling verse, one has:

$$\oint_\Gamma F\left(x, y\right)\mathrm{d}x = \int_0^\ell F\left(x, 0\right)\mathrm{d}x + \int_\ell^0 F\left(x, b\right)\mathrm{d}x = -\int_0^\ell [F\left(x, y\right)]_{y=0}^{y=b}\,\mathrm{d}x,$$

$$(10.20a)$$

$$\oint_\Gamma F\left(x, y\right)\mathrm{d}y = \int_0^b F\left(\ell, y\right)\mathrm{d}y + \int_b^0 F\left(0, y\right)\mathrm{d}y = \int_0^b [F\left(x, y\right)]_{x=0}^{x=\ell}\,\mathrm{d}y.$$

$$(10.20b)$$

Hence:

$$\oint_\Gamma \left(w_{,xx} + \nu w_{,yy}\right)\delta w_{,x}\mathrm{d}y = \int_0^b \left[\left(w_{,xx} + \nu w_{,yy}\right)\delta w_{,x}\right]_{x=0}^{x=\ell}\,\mathrm{d}y, \qquad (10.21a)$$

$$\oint_\Gamma \left(w_{,yy} + \nu w_{,xx}\right)\delta w_{,y}\,\mathrm{d}x = -\int_0^\ell \left[\left(w_{,yy} + \nu w_{,xx}\right)\delta w_{,y}\right]_{y=0}^{y=b}\,\mathrm{d}x, \qquad (10.21b)$$

$$\oint_\Gamma \left(w_{,xxx} + w_{,xyy}\right)\delta w\,\mathrm{d}y = \int_0^b \left[\left(w_{,xxx} + w_{,xyy}\right)\delta w\right]_{x=0}^{x=\ell}\,\mathrm{d}y, \qquad (10.21c)$$

$$\oint_\Gamma \left(w_{,yyy} + w_{,xxy}\right)\delta w\,\mathrm{d}x = -\int_0^\ell \left[\left(w_{,yyy} + w_{,xxy}\right)\delta w\right]_{y=0}^{y=b}\,\mathrm{d}x. \qquad (10.21d)$$

Moreover:

$$\oint_{\Gamma} w_{,xy}\, \delta w_{,x}\, \mathrm{d}x = -\int_0^{\ell} \left[w_{,xy}\delta w_{,x}\right]_{y=0}^{y=b}\, \mathrm{d}x = \int_0^{\ell} \left[w_{,xxy}\delta w\right]_{y=0}^{y=b}\, \mathrm{d}x - \left[w_{,xy}\delta w\right]_{y=0}^{y=b}\Big|_{x=0}^{x=\ell},$$

(10.22a)

$$\oint_{\Gamma} w_{,xy}\, \delta w_{,y}\, \mathrm{d}y = \int_0^{b} \left[w_{,xy}\delta w_{,y}\right]_{x=0}^{x=\ell}\, \mathrm{d}y = -\int_0^{b} \left[w_{,xyy}\delta w\right]_{x=0}^{x=\ell}\, \mathrm{d}y + \left[w_{,xy}\delta w\right]_{x=0}^{x=\ell}\Big|_{y=0}^{y=b},$$

(10.22b)

where a further integration by parts has been performed.[10] With these results, Eq. 10.19 becomes:

$$\int_0^{b} \left[D\left(w_{,xx} + \nu w_{,yy}\right)\delta w_{,x}\right]_{x=0}^{x=\ell}\, \mathrm{d}y + \int_0^{\ell} \left[D\left(w_{,yy} + \nu w_{,xx}\right)\delta w_{,y}\right]_{y=0}^{y=b}\, \mathrm{d}x$$

$$+ \int_0^{b} \left[-D\left(w_{,xxx} + (2-\nu)\, w_{,xyy}\right)\delta w\right]_{x=0}^{x=\ell}\, \mathrm{d}y + \int_0^{\ell} \left[-D\left(w_{,yyy} + (2-\nu)\, w_{,xxy}\right)\delta w\right]_{y=0}^{y=b}\, \mathrm{d}x$$

$$+ 2D\left(1-\nu\right)\left[w_{,xy}\delta w\right]_{x=0}^{x=\ell}\Big|_{y=0}^{y=b} = 0, \qquad \forall \delta w,$$

(10.23)

from which Eqs. 10.17 follow.   □

Supplement 10.2 (Interpretation of the Equilibrium in the Undeformed Configuration) The variational procedure followed, being limited to first order kinematics, leads to indefinite and boundary equilibrium equations which refer to the *undeformed configuration* of the plate. These, as usually done in the structural mechanic textbooks, may also be derived from *direct balance* conditions, taking into account, however, some complications, descending from the fact that the Kirchhoff plate is internally constrained.

---

[10] The integration by parts of integrals containing the *derivative* $\frac{\partial w}{\partial s}$, is a consequence of the fact that *the rotation around the normal, $\varphi_n := \frac{\partial w}{\partial s}$, is not an independent kinematic variable*, being it uniquely determined, once $w\,(s)$ is known. In contrast, the rotation around the tangent to the boundary, $\varphi_t := -\frac{\partial w}{\partial n}$ is an independent quantity. The kinematic parameters at the edge are therefore only two, w and $\frac{\partial w}{\partial n}$, this implying at most *two* mechanical conditions.

The direct equilibrium of the plate element, in which the reactive forces are condensed, read:[11]

$$M_{x,xx} + 2M_{xy,xy} + M_{y,yy} = p, \tag{10.24}$$

from which, making use of Eq. 10.4, Eq. 10.14 is recovered. Equation 10.24 establishes that the transverse load p is carried by the plate through three different mechanisms: (a) the *flexural bearing* $p_x := M_{x,xx}$ of the strips oriented along x, as well-known from the Euler-Bernoulli beam theory; (b) the analogous flexural bearing $p_y := M_{y,yy}$, relevant to the strips in the y direction; (c) the *torsional bearing* $p_{xy} := 2M_{xy,xy}$ of the two orders of strips. To explain the latter, which is less obvious than the other two contributions, reference is made to a strip of unitary width, oriented along y. Since its long sides are subject to torsional moments M_{xy} and M_{xy}^+ of opposite sign, these, added together, are equivalent to distributed pairs $c_x = M_{xy,x}$, whose derivative with respect to the axis of the strip, $\frac{1}{2}p_{xy} := c_{x,y} = M_{xy,xy}$, represents the desired torsional bearing of the strip;[12] the same value is found for the strip in the x direction.

When the constitutive law is used, the flexural bearings are written as $p_x = D\left(w_{,xx} + v w_{,yy}\right)_{,xx}$, $p_y = D\left(w_{,yy} + v w_{,xx}\right)_{,yy}$ and the torsional bearing as $p_{xy} = 2D\left(1 - v\right)w_{,xxyy}$; in the sum, the Poisson's ratio cancels out. Regardless of this circumstance, the three terms in Eq. 10.14 are often indicated as flexural bearings $p_x = Dw_{,xxxx}$, $p_y = Dw_{,yyyy}$ and torsional bearing $p_{xy} = Dw_{,xxyy}$, although this is only true when $v = 0$.

Regarding the mechanical boundary conditions in Eqs. 10.17, the first two of them represent the zeroing of the bending moments at the edge, $M_x = D\left(w_{,xx} + v w_{,yy}\right)$ and $M_y = \left(w_{,yy} + v w_{,xx}\right)$, which are the quantities dual of the rotations at the edge; the further two conditions express the zeroing of the so-called Kirchhoff shear force:

$$T_x^* := T_x - M_{xy,y} = -D\left(w_{,xxx} + (2 - v)w_{,xyy}\right), \tag{10.25a}$$

$$T_y^* := T_y - M_{xy,x} = -D\left(w_{,yyy} + (2 - v)w_{,xxy}\right), \tag{10.25b}$$

---

[11] From the equilibrium of the element one obtains (Fig. 10.1c):

$$T_{x,x} + T_{y,y} + p = 0$$

$$M_{y,y} + M_{xy,x} + T_y = 0$$

$$M_{x,x} + M_{xy,y} + T_x = 0$$

By condensing the shear forces, Eq. 10.24 follows.

[12] For a beam subject to distributed forces and couples, the equilibrium states that $M' + T + c = 0$ and $T' + p = 0$. Condensing the shear force, one has $M'' + c' = p$, where c' is a bearing capacity that adds itself to M''.

as quantities dual of the displacement at the edge. The Kirchhoff shear force, as it is known, arises from the fact that, since the plate is shear undeformable, the torsional moments at the edge are statically equivalent to (i) distributed shear forces, which add themselves to the properly called shear forces and (ii) forces concentrated at the vertices,[13] appearing in Eqs. 10.17e.   □

10.3 In-Plane Prestressed Plate

To perform the linearized stability analysis of the plate, it needs to build-up the geometric stiffness operator, which calls for evaluating the quadratic part of the prestress energy, U^0 [7, 13]. The procedure is illustrated ahead.

Prestress State

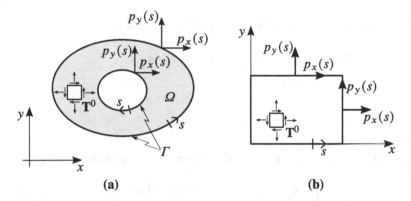

(a)   **(b)**

Fig. 10.3 Plate prestressed by in-plane edge forces: (**a**) of generic shape, (**b**) of rectangular shape

---

[13] The Kirchhoff shear forces can be more elegantly explained by using the virtual work equation. Indeed, the virtual work of the internal forces emerging, for example, at the edge $x = \ell$, is:

$$\delta W = \int_0^b \left[T_x \delta w + M_{xy} \delta \varphi_x\right]_{x=\ell} \mathrm{d}y = \int_0^b \left[T_x \delta w + M_{xy} \delta w_{,y}\right]_{x=\ell} \mathrm{d}y$$

$$= \int_0^b \left[(T_x - M_{xy,y})\, \delta w\right]_{x=\ell} \mathrm{d}y + \left[M_{xy} \delta w\right]_{y=0}^{y=b}\Big|_{x=\ell}$$

where, from Eq. 10.1a, use has been made of $\delta \varphi_x = \delta w_{,y}$ and integration by parts. The integral defines $T_x^* := T_x - M_{xy,y}$ as the mechanical quantity that expends work on the virtual displacement δw. The boundary term is the contribution that the torsional moments, acting at the side $x = \ell$, bring to the concentrated forces at the vertices $(\ell, 0)$ and (ℓ, b); to this one, the contribution coming from the orthogonal side must be added.

Here (Fig. 10.3), the attention is limited to plates prestressed by forces $p_x (s)$, $p_y (s)$, (i) contained in the plan and (ii) acting exclusively at the boundary; therefore, body forces are absent. The boundary forces induce a membrane state of stress, described by the stress tensor:

$$\mathbf{T}^0 = \begin{bmatrix} N_x^0 & N_{xy}^0 \\ N_{xy}^0 & N_y^0 \end{bmatrix}, \tag{10.26}$$

where $N_x^0 (x, y)$ and $N_y^0 (x, y)$ are the normal stresses and $N_{xy}^0 (x, y)$ the in-plane shear stresses, all having dimensions of forces per unit of length, generally variable in the domain. The stresses satisfy the (homogeneous) indefinite equilibrium equations:

$$N_{x,x}^0 + N_{xy,y}^0 = 0, \qquad\qquad \text{in} \quad \Omega, \tag{10.27a}$$

$$N_{xy,x}^0 + N_{y,y}^0 = 0, \tag{10.27b}$$

and the (non-homogeneous) boundary equilibrium conditions:

$$N_x^0 n_x + N_{xy}^0 n_y = p_x, \qquad\qquad \text{on} \quad \Gamma. \tag{10.28a}$$

$$N_{xy}^0 n_x + N_y^0 n_y = p_y, \tag{10.28b}$$

Prestress Energy

The prestress energy, associated with the stresses in Eq. 10.26, is:

$$U^0 = \iint\limits_{\Omega} \left[N_x^0 \varepsilon_x^{(2)} + N_y^0 \varepsilon_y^{(2)} + N_{xy}^0 \gamma_{xy}^{(2)} \right] dx\, dy, \tag{10.29}$$

where $\varepsilon_x^{(2)}$, $\varepsilon_y^{(2)}$, and $\gamma_{zs}^{(2)}$ are the second order components of the incremental strains. Using the Green-Lagrange strain tensor (Supplement 6.1), and taking into account that the only non-zero displacement is the transverse one, w, it follows:

$$\varepsilon_x^{(2)} = \frac{1}{2} w_{,x}^2, \tag{10.30a}$$

$$\varepsilon_y^{(2)} = \frac{1}{2} w_{,y}^2, \tag{10.30b}$$

$$\gamma_{xy}^{(2)} = w_{,x} w_{,y}. \tag{10.30c}$$

Consequently:

$$U^0 = \frac{1}{2} \iint\limits_{\Omega} \left[N_x^0 w_{,x}^2 + N_y^0 w_{,y}^2 + 2N_{xy}^0 w_{,x} w_{,y} \right] dx\, dy. \tag{10.31}$$

Geometric Stiffness Operator

The first variation of the energy is:

$$\delta U^0 = \iint_{\Omega} \left[\left(N_x^0 w_{,x} + N_{xy}^0 w_{,y} \right) \delta w_{,x} + \left(N_{xy}^0 w_{,x} + N_y^0 w_{,y} \right) \delta w_{,y} \right] \mathrm{d}x\,\mathrm{d}y,$$

$$(10.32)$$

which, after integration by parts, reads:

$$\delta U^0 = - \iint_{\Omega} \left[\left(N_x^0 w_{,x} + N_{xy}^0 w_{,y} \right)_{,x} + \left(N_{xy}^0 w_{,x} + N_y^0 w_{,y} \right)_{,y} \right] \delta w\,\mathrm{d}x\,\mathrm{d}y$$

$$+ \oint_{\Gamma} \left[\left(N_x^0 w_{,x} + N_{xy}^0 w_{,y} \right) n_x + \left(N_{xy}^0 w_{,x} + N_y^0 w_{,y} \right) n_y \right] \delta w\,\mathrm{d}s.$$

$$(10.33)$$

Taking into account the pre-existing equilibrium, expressed by Eqs. 10.27 and 10.28, the latter expression simplifies into:

$$\delta U^0 = - \iint_{\Omega} \left(N_x^0 w_{,xx} + 2 N_{xy}^0 w_{,xy} + N_y^0 w_{,yy} \right) \delta w\,\mathrm{d}x\,\mathrm{d}y$$

$$(10.34)$$

$$+ \oint_{\Gamma} \left(p_x w_{,x} + p_y w_{,y} \right) \delta w\,\mathrm{d}s.$$

By letting $\delta U^0 = \iint_{\Omega} \delta w\, \mathcal{K}_g w\, \mathrm{d}x\,\mathrm{d}y + \oint_{\Gamma} \delta w \mathcal{B}_g w\,\mathrm{d}s$, the *geometric stiffness operator*s of the Kirchhoff plate are obtained, in the domain:

$$\mathcal{K}_g := - \left(N_x^0 \partial_{xx}^2 + 2 N_{xy}^0 \partial_{xy}^2 + N_y^0 \partial_{yy}^2 \right),$$ $$(10.35)$$

and at the boundary:

$$\mathcal{B}_g := p_x \partial_x + p_y \partial_y.$$ $$(10.36)$$

Equilibrium Equations of the Prestressed Plate

The indefinite equilibrium of the prestressed plate is governed by:

$$\left(\mathcal{K}_e + \mathcal{K}_g \right) w = 0, \qquad \text{in } \Omega,$$ $$(10.37)$$

in which the elastic and geometric stiffnesses have been added up. In explicit form, remembering Eqs. 10.16 and 10.35, one gets:

$$D\left(w_{,xxxx} + 2w_{,xxyy} + w_{,yyyy}\right) - \left(N_x^0 w_{,xx} + 2N_{xy}^0 w_{,xy} + N_y^0 w_{,yy}\right) = 0.$$

$$(10.38)$$

Equilibrium at the boundary, for a rectangular plate (Eqs. 10.17, 10.36), requires that:

$$D\left(w_{,xx} + vw_{,yy}\right)\delta w_{,x} = 0, \quad \text{at } x = 0, \ell,$$

$$(10.39a)$$

$$D\left(w_{,yy} + vw_{,xx}\right)\delta w_{,y} = 0, \quad \text{at } y = 0, b,$$

$$(10.39b)$$

$$\left[-D\left(w_{,xxx} + (2-v)\,w_{,xyy}\right) + p_x w_{,x} + p_y w_{,y}\right]\delta w = 0, \quad \text{at } x = 0, \ell,$$

$$(10.39c)$$

$$\left[-D\left(w_{,yyy} + (2-v)\,w_{,xxy}\right) + p_x w_{,x} + p_y w_{,y}\right]\delta w = 0, \quad \text{at } y = 0, b,$$

$$(10.39d)$$

$$2D\,(1-v)\,w_{,xy}\delta w = 0, \quad \text{at } (0,0),\, (\ell,0),$$

$$(10.39e)$$

$$(\ell, b),\, (0, b).$$

The mechanical meaning of the geometric terms appearing in the equilibrium equations is discussed in the Supplement 10.3.

In the following sections, some examples of integration of Eqs. 10.38, 10.39 are illustrated, relative to rectangular plates in different constraint and loading conditions.

Supplement 10.3 (Interpretation of the Adjacent Equilibrium) Equation 10.38 expresses the equilibrium of a plate in a configuration adjacent to the reference one, described by the displacement field $w\,(x, y)$. In addition to the elastic bearing capacity already discussed in the Supplement 10.2, flexural and torsional geometric contributions are present, namely, (a) the *funicular bearings*, $-N_x^0 w_{,xx}$ and $-N_y^0 w_{,yy}$, and (b) the *membrane shear bearing*, $-2N_{xy}^0 w_{,xy}$, which are explained as follows. When the element is deformed, the prestresses, by following the body, incline. Their projection on the (x, y) plane, at the first order, remains unchanged, but that on the z axis is non-zero and represents transverse geometric shear forces, available to balance the incremental loads (Fig. 10.4). More specifically, by considering an element of membrane of dimensions 1×1, undergoing the rotation $w_{,y}$ around the x axis (Fig. 10.4a): (i) the normal prestress N_y^0 gives rise to the geometric shear forces $T_y := N_y^0 w_{,y}$; (ii) the in-plane shear force N_{xy}^0, acting on the faces of normal x, produces shear forces $T_x := N_{xy}^0 w_{,y}$. When a rotation $w_{,x}$ around the y axis is considered (Fig. 10.4b), with a similar mechanism, the geometrical forces $T_x := N_x^0 w_{,x}$ and $T_y := N_{xy}^0 w_{,x}$ arise. By superimposing the effects (Fig. 10.4c), it follows: $T_x := N_x^0 w_{,x} + N_{xy}^0 w_{,y}$, $T_y := N_y^0 w_{,y} + N_{xy}^0 w_{,x}$.

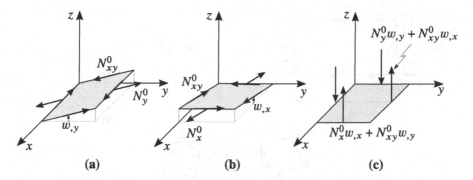

Fig. 10.4   Geometric bearing capacity of the plate: (**a**) rotation around x, (**b**) rotation around y, (**c**) geometric shear forces, sum of the two effects

The geometric bearing capacity is obtained as $p_g := -T_{x,x} - T_{y,y}$. Similar considerations apply to the boundary conditions.   □

10.4   Plate Simply Supported on Four Sides and Compressed in One Direction

A simple case is addressed, in which the plate is subjected to uniaxial compression, produced by forces $p_x = $ cost acting at the sides $x = 0, \ell$ (Fig. 10.5a) [1–5, 15, 17].

The state of prestress is described by $N_x^0 = -p_x$, $N_y^0 = N_{xy}^0 = 0$, uniformly in the body. The boundary value problem in Eqs. 10.38 and 10.39, therefore, reads:

$$D\left(w_{,xxxx} + 2w_{,xxyy} + w_{,yyyy}\right) + p_x w_{,xx} = 0, \tag{10.40a}$$

$$w = 0, \qquad w_{,xx} = 0, \qquad \text{at } x = 0, \ell, \tag{10.40b}$$

$$w = 0, \qquad w_{,yy} = 0, \qquad \text{at } y = 0, b, \tag{10.40c}$$

where the boundary conditions state the vanishing of the displacement and of the normal bending moment at the edge.[14]

It is observed that, since only even order derivatives in both the independent variables are involved in the problem, sine waves, of suitable length, satisfy the field equations as well as the boundary conditions.[15] As a "guess solution," a linear

---

[14] For example, on a side parallel to y, it is $M_x = D\left(w_{,xx} + vw_{,yy}\right) = 0$, where it has been taken into account, that w is there constant (zero).

[15] As the sine vanishes together with its second derivative.

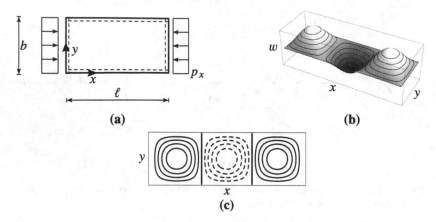

Fig. 10.5 Simply supported plate, compressed in one direction: (**a**) structure, (**b**) 3D view of the buckled configuration, (**c**) nodal lines (example, $n = 3$, $m = 1$)

combination of products of sinusoidal functions is taken (i.e., a double Fourier series of only sines):

$$w(x, y) = \sum_{n=1}^{\infty} \sum_{m=1}^{\infty} A_{nm} \sin\left(\frac{n\pi x}{\ell}\right) \sin\left(\frac{m\pi y}{b}\right), \qquad (10.41)$$

where A_{nm} are unknown amplitudes and n and m are integers. Each term of the series describes a deformed configuration, in which the plate bends in n half-waves (of length $\frac{\ell}{n}$) in the x direction and m half-waves (of length $\frac{b}{m}$) in the y direction (Fig. 10.5b,c).

Substituting the guess solution in Eqs. 10.40, it turns out that (i) the boundary conditions, as mentioned, are identically satisfied; (b) the field equation provides:

$$\sum_{n=1}^{\infty} \sum_{m=1}^{\infty} \left\{ D\left[\left(\frac{n\pi}{\ell}\right)^2 + \left(\frac{m\pi}{b}\right)^2\right]^2 - p_x \left(\frac{n\pi}{\ell}\right)^2 \right\}$$
$$\times A_{nm} \sin\left(\frac{n\pi x}{\ell}\right) \sin\left(\frac{m\pi y}{b}\right) = 0, \qquad \forall x, y. \qquad (10.42)$$

For it to be satisfied at any point of the domain, the coefficients of the double series must vanish separately, i.e.:

$$\left\{ D\left[\left(\frac{n\pi}{\ell}\right)^2 + \left(\frac{m\pi}{b}\right)^2\right]^2 - p_x \left(\frac{n\pi}{\ell}\right)^2 \right\} A_{nm} = 0 \qquad n, m = 1, 2, \cdots . $$
$$(10.43)$$

Hence, *the harmonics uncouple*, generating a system of ∞^2 linear algebraic equations, each in a single unknown amplitude.[16] In other words, each term of the series in Eq. 10.41 is an *eigenvector* of the differential problem; therefore, Eq. 10.41 is the *exact solution* to the boundary value problem, expressed in the basis of the eigenvectors. The eigenvalue $p_{xc}(n, m)$, associated with the pair (n, m) which characterizes the mode, zeroes the stiffness in such a mode; it is equal to:

$$p_{xc}(n, m) = D\frac{\pi^2}{b^2}\left(\frac{nb}{\ell} + m^2\frac{\ell}{nb}\right)^2. \tag{10.44}$$

The amplitude of the mode, as always happens in the linearized analysis, remains indeterminate.

By summarizing, there are ∞^2 bifurcation values, in correspondence of which the plate assumes a specific critical deflection.[17] Since the interest consists in finding the smallest of these loads, it needs to minimize $p_{xc}(n, m)$ with respect to the two variables.[18] It is immediately observed that, as $p_{xc}(n, m)$ increases monotonously with m, in order to minimize it, $m = 1$ must be taken, whereby *the plate inflects with only one half-wave in the direction transverse to the load*. The value of n that makes $p_{xc}(n, 1)$ minimum, conversely, cannot be deduced a priori, since it depends on the geometry of the plate. To highlight this dependence, it is convenient to put the critical load in the form:

$$p_{xc}(n, 1) := \mu_c\frac{\pi^2 D}{b^2}, \tag{10.45}$$

having introduced a *nondimensional critical load*:[19]

$$\mu_c := \left(\frac{n}{\alpha} + \frac{\alpha}{n}\right)^2, \tag{10.46}$$

which is a function of the number of longitudinal half-waves n and of the *plate aspect ratio* $\alpha := \frac{\ell}{b}$. For each value of $n = 1, 2, \cdots$, it is possible to plot the graph of μ_c as a function of α, obtaining the curves represented in Fig. 10.6. By taking the lowest value of μ_c for any α, the thick line in figure is drawn, which represents

---

[16] For this reason, the double sum in Eq. 10.41 is often omitted in the literature, and reference is made to the generic (n, m) term. This shortcut will be followed ahead, whenever possible.

[17] The result generalizes that relevant to the Euler beam, for which there are ∞^1 critical loads, each associated with a sinusoidal deflection of different wave-number.

[18] The minimization must be done by trial and error, since the variables n, m do not assume continuous, but discrete values.

[19] The nondimensional critical load in Eq. 10.45 is often denoted in the literature by the letter k. Here, however, this symbol is avoided, in order not to create confusion with others meanings attributed to it. On the other hand, μ has systematically been used in this book as a (nondimensional) multiplier of load and μ_c as its critical value.

Fig. 10.6 Nondimensional critical load $\mu_c = \frac{p_{xc}b^2}{\pi^2 D}$ of the simply supported plate, compressed in one direction, *vs* the aspect ratio $\alpha := \dfrac{\ell}{b}$

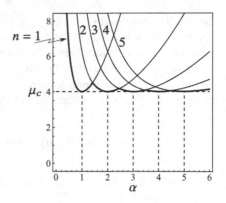

the dependence of the (lowest) critical load *vs* the aspect ratio of the plate.[20] It is observed that the smallest value of μ_c (on α) is $\mu_{c,min} = 4$ and that this occurs for an integer aspect ratio, $\alpha = 1, 2, \cdots$; in such cases, the plate buckles in *square fields*. If, on the other hand, the ratio is not an integer, the plate buckles in a number of longitudinal half-waves equal to the integer immediately below or above α. The associate critical load is greater than that of the square plate; however, the difference between μ_c and $\mu_{c,min}$ reduces as α increases and tends to zero when $\alpha \to \infty$. Therefore, if the plate is sufficiently long, one can take, with good approximation, $\mu_c = 4$.

Double eigenvalues exist at the intersection points of the curves relevant to n and $n + 1$, i.e., at the abscissas $\alpha = \sqrt{n(n+1)}$. In the nonlinear field, these eigenvalues cause *interaction between simultaneous modes*, in which the plate (to the first order) inflects in a combination of the two critical modes.

Remark 10.4 The factor 4 is easily explained for a square plate $\ell = b$, for which, given the mechanism of flexural bearing in the two directions and the torsional bearing of the two orders of strips, the critical load is four times larger than that of the Euler's beam of equal length. A rectangular plate, of integer aspect ratio, behaves like a square plate, similarly to what happens for a continuous beam with equal spans, which has the same critical load as the single-span.[21]

---

[20] This curve is sometimes called, in the technical language, 'the signature curve', thus alluding to an illegible signature.

[21] This is because, considering the elements as independent and in equilibrium in their deformed configurations, the conditions of continuity of the displacements at the internal supports (or nodal lines) are satisfied simply taking the arbitrary amplitudes all equal in modulus.

Fig. 10.7 Simply supported
plate subjected to a bi-axial
compression/traction

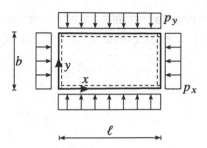

10.5 Plate Simply Supported on Four Sides and Subject to Bi-Axial Stress

A rectangular plate is considered, simply supported on the entire contour, pre-stressed by forces p_x and p_y uniformly distributed and normal to the edges, positive when they generate compression (Fig. 10.7) [15]. The state of prestress is $N_x^0 = -p_x$, $N_y^0 = -p_y$, $N_{xy}^0 = 0$. The equilibrium Eqs. 10.38 and 10.39 specialize as follows:

$$D\left(w_{,xxxx} + 2w_{,xxyy} + w_{,yyyy}\right) + p_x w_{,xx} + p_y w_{,yy} = 0, \tag{10.47a}$$

$$w = 0, \qquad w_{,xx} = 0, \qquad \text{at } x = 0, \ell, \tag{10.47b}$$

$$w = 0, \qquad w_{,yy} = 0, \qquad \text{at } y = 0, b. \tag{10.47c}$$

Still assuming the solution as in Eq. 10.41 (or making reference to the generic term of the series) and proceeding as in the Sect. 10.4, the bifurcation condition is obtained:

$$p_x\left(\frac{n\pi}{\ell}\right)^2 + p_y\left(\frac{m\pi}{b}\right)^2 = D\left[\left(\frac{n\pi}{\ell}\right)^2 + \left(\frac{m\pi}{b}\right)^2\right]^2. \tag{10.48}$$

By letting, in analogy with Eq. 10.45:

$$\alpha := \frac{\ell}{b}, \qquad \rho := \frac{p_y}{p_x}, \qquad p_{xc}(n, m) := \mu_c \frac{\pi^2 D}{b^2}, \tag{10.49}$$

a nondimensional load is defined as:

$$\mu_c := \frac{\left[\left(\frac{n}{\alpha}\right)^2 + m^2\right]^2}{\left(\frac{n}{\alpha}\right)^2 + \rho m^2}. \tag{10.50}$$

Table 10.1 Nondimensional critical load $\mu_c = \frac{p_{xc}b^2}{\pi^2 D}$ of the simply supported square plate ($\alpha = 1$), compressed/taut in two directions, for different values of the load ratio $\rho := \dfrac{p_y}{p_x}$

ρ	n	m	μ_c
0	1	1	4
1	1	1	2
-1	2	1	8.33

Here, α is the aspect ratio and ρ the ratio between the magnitudes of the two loads. Minimizing μ_c with respect to the two integers n and m is now a more complex problem, since μ_c depends not only on the aspect ratio but also by the relative magnitude between the loads. Moreover, Eq. 10.50 shows that one of the two loads (but not both, as it appears from Eq. 10.48), may assume negative values, denoting traction. In the following discussion, it will be assumed that $p_x > 0$ and $p_y \gtreqless 0$, so that $\rho \gtreqless 0$ and $\mu_c > 0$.

To make the study of Eq. 10.50 easier, it is convenient to fix the aspect ratio α and to seek, by trial and error, the values of n, m which minimize the nondimensional critical load μ_c, for any load ratio ρ. For example, for a square plate, $\alpha = 1$, the values in Table 10.1 are obtained. It is seen that, when $\rho = 0$ (plate compressed in one direction), it is $\mu_c = 4$, as already found. If $\rho = 1$, that is, if the plate is subjected to equal compression forces in the two directions, the nondimensional critical load is halved, $\mu_c = 2$; however, the critical mode remains that of the plate compressed in one direction, i.e., $(n, m) = (1, 1)$. If $\rho = -1$, that is, if the load in the y direction is changed of sign, so that the plate is compressed in one direction and tensed in the other, then the critical load more than doubles, becoming $\mu_c = 8.33$. Furthermore, the critical mode changes, as the plate buckles with two half-waves in the direction of compression, but still with a half-wave in the direction of traction. The increase of the critical load is explained by the stabilizing effect exerted by the tensed strips on the compressed ones; the same stiffening effect explains the increase of the number of the longitudinal half-waves, in analogy with the beam on Winkler soil (Sect. 7.9, Fig. 7.14).

Stability Domain

If the loads p_x and p_y are independent and their critical combinations are desired, it is more convenient to build up a geometric locus, namely, the *domain of stability*. Once the plate dimensions have been fixed, Eq. 10.48 establishes a linear relationship between the two loads and therefore defines a straight line in (p_x, p_y) plane. This, however depends on the two integers, m, n, so that there exist ∞^2 lines in the plane. The segments closest to the origin constitute a piecewise straight line that delimits the domain of stability. The internal points represent stable conditions, the points on the boundary incipient instability, and the external points instability.

As an example, by referring to a square plate, the critical values for uniaxial compression are equal to $\bar{p}_{xc} = \bar{p}_{yc} = 4\dfrac{\pi^2 D}{b^2}$. Equation 10.48 is therefore rewritten as:

Fig. 10.8 Stability domain of the simply supported square plate ($\alpha = 1$), subject to bi-axial compression/tractions. The stable region is grayed out

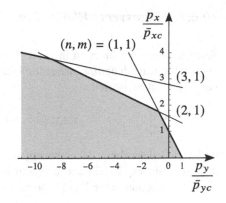

$$\left(\frac{p_x}{\bar{p}_{xc}}\right) n^2 + \left(\frac{p_y}{\bar{p}_{yc}}\right) m^2 = \frac{1}{4}\left(n^2 + m^2\right)^2, \tag{10.51}$$

which is represented in Fig. 10.8 for some (n, m) pairs. The domain of stability is bounded by (i) the polygonal line marked in figure and (ii) the $p_x = 0$ axis; moreover, it is unbounded on the left (i.e., for traction $|p_y|$ arbitrarily large, not accounting, of course, for limits of elasticity).

Examples of stability domains for plates of other aspect ratios, as well as on Winkler soil, are given in the Problems 14.4.1 and 14.4.3.

10.6 Separation of Variables and Exact Finite Element

When a rectangular plate is simply supported on two opposite sides, it is possible to pursue an analytical solution via *separation of variables* [2, 15, 16]. Here the procedure is illustrated with reference to a plate compressed in the direction orthogonal to the supported sides.[22] The method also suggests the formulation of an exact finite element, which will be introduced soon.

A rectangular plate is considered, of dimensions $\ell \times b$, supported along the edges parallel to the y axis, arbitrarily (but uniformly) constrained on the remaining two sides, compressed in the x direction by loads $p_x = $ const. The field equation and the boundary conditions at the sides $x = 0, \ell$ are still given by Eqs. 10.40, i.e.:

$$D\left(w_{,xxxx} + 2\,w_{,xxyy} + w_{,yyyy}\right) + p_x\,w_{,xx} = 0, \tag{10.52a}$$

$$w = 0, \quad \text{at} \quad x = 0, \ell, \tag{10.52b}$$

$$w_{,xx} = 0, \quad \text{at} \quad x = 0, \ell, \tag{10.52c}$$

while the boundary conditions at the sides $y = 0, b$ will be specified later.

---

[22] The procedure, however, can be extended to a compression parallel to the supported sides, or even to a bi-axial state.

10.6.1   Transverse Elastic Line Equation

Since a sinusoidal function in the x direction (from now on, called longitudinal), of suitable wave-length, satisfies the boundary conditions at $x = 0, \ell$, a separable variable solution is tried:

$$w\,(x, y) = \sin\left(\frac{n\pi}{\ell}x\right) Y\,(y), \tag{10.53}$$

with $Y\,(y)$ describing the unknown transverse deflection. Substituting Eq. 10.53 into the field Eq. 10.52a and by requiring this is satisfied for any x, an ordinary differential equation for $Y\,(y)$ follows, i.e.:

$$Y'''' - 2\left(\frac{n\pi}{\ell}\right)^2 Y'' + \left(\frac{n\pi}{\ell}\right)^2 \left[\left(\frac{n\pi}{\ell}\right)^2 - \frac{p_x}{D}\right] Y = 0, \tag{10.54}$$

where a dash denotes differentiation with respect to the independent variable y. This equation governs the transverse deflection of the plate; accordingly, it will be referred to as the *equation of the transverse elastic line* of the plate.

Remark 10.5 It should be noticed, that Eq. 10.54 is formally analogous to the equation of a beam, taut by an axial force N_0, resting on a Winkler soil of stiffness k_f, i.e.:

$$EIw'''' - N_0 w'' + k_f w = 0. \tag{10.55}$$

The analogy allows one to easily discuss Eq. 10.54, by comparing the coefficients of the two equations, i.e.: $N_0 := 2\left(\frac{n\pi}{\ell}\right)^2$, $k_f := \left(\frac{n\pi}{\ell}\right)^2 \left[\left(\frac{n\pi}{\ell}\right)^2 - \frac{p_x}{D}\right]$. Since $N_0 > 0$, the funicular bearing capacity is stabilizing. The effect of the ground, instead, is determined by the sign of k_f, which in turn depends on p_x. If the load is small, k_f is positive; thus the beam is in stable equilibrium; if p_x is large enough, k_f is negative, bringing a destabilizing contribution which contrasts the beneficial funicular effect. There is therefore a critical value of p_x for which the equilibrium is incipiently unstable.

General Solution

Equation 10.54 admits solutions of the type $Y\,(y) = e^{\lambda y}$, with λ a characteristic exponent. Substitution into the differential equation leads to:

$$\lambda^4 - 2\left(\frac{n\pi}{\ell}\right)^2 \lambda^2 + \left(\frac{n\pi}{\ell}\right)^2 \left[\left(\frac{n\pi}{\ell}\right)^2 - \frac{p_x}{D}\right] = 0, \tag{10.56}$$

which is a biquadratic algebraic equation for the unknown λ. To discuss its roots, it needs to establish the sign of the known term of the equation, which depends on p_x.

To this end, it is observed that when the plate is simply supported along the sides $x = 0, \ell$, and free on the sides $y = 0, b$, it behaves like a beam, whose critical Eulerian load is:[23]

$$p_{xE} := D \left(\frac{n\pi}{\ell}\right)^2. \tag{10.57}$$

Since the constraints existing on the sides $y = 0, b$, whatever they are, can only increase the critical load, it turns out that $p_{xc} \geq p_{xE}$, with the equality sign holding only when these sides are free. It is therefore of interest to investigate the solution to the problem in the open interval $p_x \geq D \left(\frac{n\pi}{\ell}\right)^2$, in which the known term of the biquadratic equation is negative, according to the mechanical interpretation given in the Remark 10.5. Solving the characteristic equation with respect to λ^2, one obtains:

$$\lambda_+^2 := \left(\frac{n\pi}{\ell}\right)^2 + \left(\frac{n\pi}{\ell}\right)\sqrt{\frac{p_x}{D}} > 0, \tag{10.58a}$$

$$\lambda_-^2 := \left(\frac{n\pi}{\ell}\right)^2 - \left(\frac{n\pi}{\ell}\right)\sqrt{\frac{p_x}{D}} < 0. \tag{10.58b}$$

There exist, therefore, two real roots $\lambda_{1,2} = \pm\sqrt{\lambda_+^2}$ and two imaginary roots $\lambda_{3,4} = \pm i\sqrt{-\lambda_-^2}$, i.e.:

$$\lambda_1 = \beta_1, \quad \lambda_2 = -\beta_1, \quad \lambda_3 = i\,\beta_2, \quad \lambda_4 = -i\,\beta_2, \tag{10.59}$$

in which:

$$\beta_1 := \sqrt{\left(\frac{n\pi}{\ell}\right)^2 + \left(\frac{n\pi}{\ell}\right)\sqrt{\frac{p_x}{D}}}, \tag{10.60a}$$

$$\beta_2 := \sqrt{\left(\frac{n\pi}{\ell}\right)\sqrt{\frac{p_x}{D}} - \left(\frac{n\pi}{\ell}\right)^2}. \tag{10.60b}$$

Hence, the general solution to the homogeneous problem in Eq. 10.54 has the following expression:

$$Y(y) = c_1 \sinh(\beta_1 y) + c_2 \cosh(\beta_1 y) + c_3 \sin(\beta_2 y) + c_4 \cos(\beta_2 y). \tag{10.61}$$

---

[23] Here D, in place of EI, accounts for the inhibited transverse expansion of the strips.

Boundary Conditions

The arbitrary constants c_i are determined by imposing the boundary conditions at the sides $y = 0, b$. For example, if the plate is clamped along $y = 0$, by taking into account Eq. 10.53, the conditions read:

$$w(x, 0) = \sin\left(\frac{n\pi}{\ell}x\right) Y(0) = 0, \qquad \forall x, \qquad (10.62a)$$

$$w_{,y}(x, 0) = \sin\left(\frac{n\pi}{\ell}x\right) Y'(0) = 0, \qquad \forall x, \qquad (10.62b)$$

which entail:

$$Y(0) = 0, \quad Y'(0) = 0. \qquad (10.63)$$

Similarly, if the plate is free at $y = b$, Eqs. 10.39b,d require:

$$w_{,yy}(x, b) + v w_{,xx}(x, b) = \left[Y''(b) - v\left(\frac{n\pi}{\ell}\right)^2 Y(b)\right] \qquad (10.64a)$$

$$\times \sin\left(\frac{n\pi}{\ell}x\right) = 0, \quad \forall x,$$

$$w_{,yyy}(x, b) + (2 - v) w_{,xxy}(x, b) = \left[Y'''(b) - (2 - v)\left(\frac{n\pi}{\ell}\right)^2 Y'(b)\right] \qquad (10.64b)$$

$$\times \sin\left(\frac{n\pi}{\ell}x\right) = 0, \quad \forall x,$$

which imply:

$$Y''(b) - v\left(\frac{n\pi}{\ell}\right)^2 Y(b) = 0, \qquad (10.65a)$$

$$Y'''(b) - (2 - v)\left(\frac{n\pi}{\ell}\right)^2 Y'(b) = 0. \qquad (10.65b)$$

10.6.2   Exact One-Dimensional Finite Element

The boundary conditions for the transverse elastic line Eq. 10.54 are not difficult to be enforced, in case of a single homogeneous plate. However, there exist several interesting problems, relevant to *plate assemblies*, in which such an approach is computationally burdensome. Among these are (i) plates constituted by inhomogeneous strips parallel to the x axis, (ii) homogeneous plates ribbed in the x direction, and (iii) prismatic shells. In all these cases, it is convenient to formulate an *exact one-dimensional finite element*, which automatizes the computation, as explained ahead.

Interpolation of Displacements

In view of determining the linear relationship existing between displacements and forces/couples at the edges, the former are assigned at the ends of the $(0, b)$ interval, i.e.:

$$Y(0) = Y_1, \quad Y'(0) = Y_1',$$
$$Y(b) = Y_2, \quad Y'(b) = Y_2', \tag{10.66}$$

where Y_j, Y_j', $(j = 1, 2)$ are generalized displacements. By enforcing these conditions to the general solution of the homogeneous problem, Eq. 10.61, one determines the arbitrary constants c_i, $i = 1, \cdots, 4$ and rewrites the solution in the form:

$$Y(y) = \hat{\psi}_1(y)\, Y_1 + \hat{\psi}_2(y)\, Y_1' + \hat{\psi}_3(y)\, Y_2 + \hat{\psi}_4(y)\, Y_2'. \tag{10.67}$$

Here, the $\hat{\psi}_i(z)$, $i = 1, \cdots, 4$ take the meaning of *interpolation functions* (although exact), and are equal to:

$$\hat{\psi}_i(z) = c_{i1} \sinh(\beta_1 y) + c_{i2} \cosh(\beta_1 y) + c_{i3} \sin(\beta_2 y) + c_{i4} \cos(\beta_2 y), \tag{10.68}$$

in which:

$$c_{11} := -\beta_2 C_5 (\beta_1 C_1 C_4 + \beta_2 C_2 C_3), \quad c_{12} := \beta_2 C_5 [\beta_1 (C_2 C_4 - 1) + \beta_2 C_1 C_3],$$
$$c_{13} := \beta_1 C_5 (\beta_1 C_1 C_4 + \beta_2 C_2 C_3), \quad c_{14} := \beta_1 C_5 [\beta_2 (C_2 C_4 - 1) - \beta_1 C_1 C_3],$$
$$c_{21} := C_5 [\beta_2 (C_2 C_4 - 1) - \beta_1 C_1 C_3], \quad c_{22} := C_5 (\beta_1 C_2 C_3 - \beta_2 C_1 C_4),$$
$$c_{23} := C_5 [\beta_1 (C_2 C_4 - 1) + \beta_2 C_1 C_3], \quad c_{24} := -c_{22},$$
$$c_{31} := \beta_2 C_5 (\beta_1 C_1 + \beta_2 C_3), \quad c_{32} := \beta_1 \beta_2 C_5 (C_4 - C_2),$$
$$c_{33} := -\beta_1 C_5 (\beta_1 C_1 + \beta_2 C_3), \quad c_{34} := -c_{32},$$
$$c_{41} := \beta_2 C_5 (C_4 - C_2), \quad c_{42} := C_5 (\beta_2 C_1 - \beta_1 C_3),$$
$$c_{43} := \beta_1 C_5 (C_2 - C_4), \quad c_{44} := -c_{42},$$

$$\tag{10.69}$$

and

$$C_1 := \sinh(\beta_1 b), \qquad\qquad C_2 := \cosh(\beta_1 b),$$
$$C_3 := \sin(\beta_2 b), \qquad\qquad C_4 := \cos(\beta_2 b),$$
$$C_5 := 1 / \left[C_1 C_3 \left(\beta_2^2 - \beta_1^2 \right) + 2\beta_1 \beta_2 (C_2 C_4 - 1) \right].$$

$$\tag{10.70}$$

Forces at the Edges

The displacement field in Eq. 10.53 induces generalized stress fields, namely, a transverse bending moment M_y and a Kirchhoff shear T_y^*,[24] given by:

$$M_y(x, y) = D\left(w_{,yy} + v w_{,xx}\right) \tag{10.71a}$$

$$= D\left[Y''(y) - v\left(\frac{n\pi}{\ell}\right)^2 Y(y)\right]\sin\left(\frac{n\pi x}{\ell}\right),$$

$$T_y^*(x, y) = -D\left[w_{,yyy} + (2 - v) w_{,xxy}\right] \tag{10.71b}$$

$$= D\left[-Y'''(y) + (2 - v)\left(\frac{n\pi}{\ell}\right)^2 Y'(y)\right]\sin\left(\frac{n\pi x}{\ell}\right),$$

By evaluating these stresses at the boundary, the external forces f_i and couples m_i ($i = 1, 2$), equilibrating them, are derived as:

$$f_1(x) = -T_y^*(x, 0), \tag{10.72a}$$

$$m_1(x) = -M_y(x, 0), \tag{10.72b}$$

$$f_2(x) = T_y^*(x, b), \tag{10.72c}$$

$$m_2(x) = M_y(x, b), \tag{10.72d}$$

that is:

$$f_1(x) = -D\left[-Y'''(0) + (2 - v)\left(\frac{n\pi}{\ell}\right)^2 Y'(0)\right]\sin\left(\frac{n\pi x}{\ell}\right) \tag{10.73a}$$

$$=: \hat{f}_1 \sin\left(\frac{n\pi x}{\ell}\right),$$

$$m_1(x) = -D\left[Y''(0) - v\left(\frac{n\pi}{\ell}\right)^2 Y(0)\right]\sin\left(\frac{n\pi x}{\ell}\right) =: \hat{m}_1 \sin\left(\frac{n\pi x}{\ell}\right),$$
$$\tag{10.73b}$$

$$f_2(x) = D\left[-Y'''(b) + (2 - v)\left(\frac{n\pi}{\ell}\right)^2 Y'(b)\right]\sin\left(\frac{n\pi x}{\ell}\right) \tag{10.73c}$$

$$=: \hat{f}_2 \sin\left(\frac{n\pi x}{\ell}\right),$$

$$m_2(x) = D\left[Y''(b) - v\left(\frac{n\pi}{\ell}\right)^2 Y(b)\right]\sin\left(\frac{n\pi x}{\ell}\right) =: \hat{m}_2 \sin\left(\frac{n\pi x}{\ell}\right).$$
$$\tag{10.73d}$$

---

[24] The Supplement 10.2 should be remembered.

It follows that the forces at the longitudinal edges (as the displacements) vary sinusoidally with x, $\hat{f}_i$, $\hat{m}_i$ being their amplitudes. Taking into account the interpolation law, Eq. 10.67, one obtains the link between the amplitudes of the forces/couples and the amplitudes of the displacements at the edges, i.e.:

$$\begin{pmatrix} \hat{f}_1 \\ \hat{m}_1 \\ \hat{f}_2 \\ \hat{m}_2 \end{pmatrix} = \begin{pmatrix} k_{11} & k_{12} & k_{13} & k_{14} \\ & k_{22} & -k_{14} & k_{24} \\ & & k_{11} & -k_{12} \\ \text{SYM} & & & k_{22} \end{pmatrix} \begin{pmatrix} Y_1 \\ Y_1' \\ Y_2 \\ Y_2' \end{pmatrix} =: \mathbf{K}^{(e)} \mathbf{q}^{(e)}. \tag{10.74}$$

The coefficients k_{ij} of the element stiffness matrix $\mathbf{K}^{(e)}$ take the following expressions:

$$k_{11} := - D\beta_1\beta_2 \left(\beta_1^2 + \beta_2^2\right) C_5 \left(\beta_1 C_1 C_4 + \beta_2 C_2 C_3\right),$$

$$k_{12} := - DC_5 \left\{ C_1 C_3 \left[\left(\frac{n\pi}{\ell}\right)^2 (v-2)\left(\beta_1^2 - \beta_2^2\right) + \left(\beta_1^4 + \beta_2^4\right) \right] \right.$$

$$\left. - \beta_1\beta_2 (C_2 C_4 - 1) \left(2\left(\frac{n\pi}{\ell}\right)^2 (v-2) + \beta_1^2 - \beta_2^2\right) \right\},$$

$$k_{13} := D\beta_1\beta_2 C_5 \left(\beta_1^2 + \beta_2^2\right) (\beta_1 C_1 + \beta_2 C_3), \tag{10.75}$$

$$k_{14} := - D\beta_1\beta_2 C_5 \left(\beta_1^2 + \beta_2^2\right) (C_2 - C_4),$$

$$k_{22} := D \left(\beta_1^2 + \beta_2^2\right) C_5 \left(\beta_2 C_1 C_4 - \beta_1 C_2 C_3\right),$$

$$k_{24} := D \left(\beta_1^2 + \beta_2^2\right) C_5 \left(\beta_1 C_3 - \beta_2 C_1\right).$$

It should be noticed that the coefficients of $\mathbf{K}^{(e)}$ depend on the load p_x in a transcendental way, through the wave-numbers β_i.

10.6.3 Critical Load of Single Plates, Simply Supported on Two Opposite Sides

The exact finite element developed above makes easy evaluating the critical loads of single plates, simply supported on two opposite sides, compressed on these sides, and in different constraint conditions on the remaining two [4, 8]. The solution requires performing the following steps :

1. The global stiffness matrix $\mathbf{K}$ of the structure is built up, by sequentially assembling one or more elements (with a scheme analogous to that of Fig. 7.21).

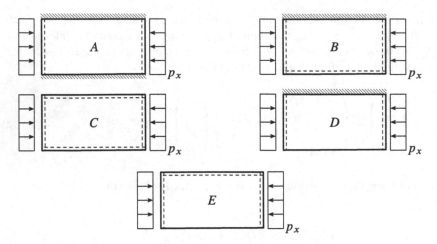

Fig. 10.9 Plate compressed in one direction, supported on the loaded sides and in different constraint conditions on the remaining sides: (A) clamped-clamped; (B) clamped-supported; (C) supported-supported; (D) clamped-free; (E) supported-free

2. The external constraints are introduced, by deleting the rows and the columns of $\mathbf{K}$ corresponding to the degrees of freedom suppressed at the ends $y = 0, b$; hence, a lower dimension stiffness matrix $\bar{\mathbf{K}}$ is obtained.
3. The matrix singularity condition, $\det \bar{\mathbf{K}} = 0$, is imposed, from which a critical load $p_{xc}(n)$ is obtained for any chosen n; the critical load of the plate is $p_{xc} := \min_{n}(p_{xc}(n))$.

The following Example 10.1 illustrates some numerical results.

Example 10.1 (Critical Load of Plates Under Different Constraint Conditions)
A number of plates are considered, in the constraint conditions illustrated in Fig. 10.9. The cases $B, C, D,$ and E, offer no difficulty; indeed, in them, at least one generalized displacement at the ends $y = 0, b$ is non-zero (one displacement for plate B, two for plates C, D, and three for plate E), so that it is sufficient using *a single finite element*. In these problems, the global stiffness matrix coincides with that of the single element (in Eq. 10.74), which, after the introduction of the constraints, becomes of dimension 1 (in case B), 2 (in case C, D) or 3 (in case E).

For the clamped-clamped plate A, however, since all the generalized displacements at the ends $y = 0, b$ are zero, if a single finite element were used, the matrix of the constrained system would have null dimension. The circumstance depends on the fact that it is not possible, for this plate, to describe the transverse deflection inside the domain in terms of the nodal displacements, since all of them are zero. It needs, therefore, to consider *at least two finite elements*, for example, each of length $b/2$, and to assembly their local stiffness matrices.

The results thus obtained are summarized in Fig. 10.10, which shows the value of the nondimensional load $\mu_c := \frac{p_x b^2}{\pi^2 D}$ vs the aspect ratio $\alpha := \ell/b$. The figure reveals a qualitative trend similar to that observed for the plate simply supported

Fig. 10.10 Critical load
$\mu_c = \frac{p_x b^2}{\pi^2 D}$ of plates
compressed in one direction,
supported on the loaded sides
and in different constraint
conditions on the remaining
sides, *vs* the aspect ratio
$\alpha = \ell/b; \nu = 0$

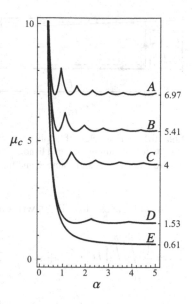

on four sides (case *C*); in particular, when α is sufficient large, the solution is practically independent of the number of half-waves. The limit solutions ($\alpha \to \infty$) also apply for constraint conditions on the loaded sides other than simple supports (not considered here), as the mechanical behavior of infinitely long plate is not affected by the constraints on the short sides. In contrast, the critical load depends on the constraints on the long sides.

The solution relevant to the plate *E* (having one side free and the other simply supported) differs from all the others, since the critical mode possesses a single longitudinal half-wave ($n = 1$), whatever the α aspect ratio is.   □

10.7 Plate Otherwise Solicited or Constrained

The exact solutions developed in the previous sections were obtained: (a) for uniform compression/traction states and (b) by separation of the variable, permitted by the presence of hinges on opposite sides. When these favorable circumstances do not occur, it is necessary to resort to approximate solutions, for example, supplied by the Ritz method [5, 10, 15, 17], discussed in detail in the Appendix B. Here, some applications are shown. A further case, concerning a plate with a punctual elastic constraint, is presented in the Problem 14.4.2.

In the following examples, the critical load is determined for (a) rectangular plates, simply supported at the boundary, subject to compression and in-plane bending in one direction (Fig. 10.11a), and (b) rectangular plates, uniformly compressed in one direction but under constraint conditions other than simple support (Fig. 10.11b).

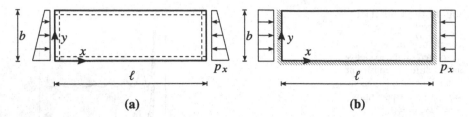

Fig. 10.11   Examples of application of the Ritz method: (**a**) simply supported plate under simultaneous compression and bending, (**b**) plate uniformly compressed in one direction, clamped on two loaded sides and clamped and free on the other two sides

In both cases, the TPE is written as:

$$\Pi = \frac{1}{2} \int_0^\ell \int_0^b \left\{ D \left[w_{,xx}^2 + w_{,yy}^2 + 2v\, w_{,xx} w_{,yy} + 2(1-v)\, w_{,xy}^2 \right] - p_x w_{,x}^2 \right\} \mathrm{d}x\,\mathrm{d}y,$$

(10.76)

where $p_x = p_x(y)$ is the prestressing load. The unknown deflection is expressed as:

$$w(x, y) = \sum_{i=1}^N \sum_{j=1}^N a_{ij} X_i(x)\, Y_j(y),$$

(10.77)

where a_{ij} are Lagrangian parameters and $X_i(x), Y_j(y)$ trial functions that satisfy the geometric boundary conditions. A possible choice for the trial functions consists in taking the buckling modes of a beam under constraint conditions equal to those of the plate (as discussed in the Appendix B, Example B.1). Once the integrations have been carried out, one arrives at an algebraic eigenvalue problem, $(\mathbf{K}_e + \mu \mathbf{K}_g)\, \mathbf{q} = \mathbf{0}$, with $\mathbf{q} = (a_{ij})$ and μ the load parameter, from which the critical load is drawn.

Example 10.2 (Plate Under Simultaneous Compression and In-Plane Bending, Simply Supported on Four Sides)   A rectangular plate is considered, simply supported on four sides, and subject to compression loads on the sides $x = 0, \ell$, variable with a linear law (Fig. 10.11a):

$$p_x(y) = p_0 \left(1 - \eta \frac{y}{b}\right).$$

(10.78)

The distribution of the loads reproduces the state of tension induced, for example, in the web of a I-beam, when it is compressed and bent (viz., $\eta = 0$ for uniform compression and $\eta = 2$ for bending). This case is relatively simple, since the field equation of the plate has coefficients varying with y, but not with x, so that it admits a separable variable solution. In agreement with the Ritz method, the solution can be taken as follows:[25]

$$w(x, y) = \sin\left(\frac{n\pi x}{\ell}\right) \sum_{m=1}^{N} q_m \sin\left(\frac{m\pi y}{b}\right).$$ (10.79)

Accordingly, the plate buckles in n half-waves in the longitudinal direction x, and in a generic way in the transverse direction y.

Due to the orthogonality of the sines, the elastic matrix $\mathbf{K}_e := \left[k_{ij}^e\right]$ turns out to be diagonal, of coefficients:

$$k_{ii}^e = \frac{\pi^4 D}{4} b\ell \left(\frac{i^2}{b^2} + \frac{n^2}{\ell^2}\right)^2.$$ (10.80)

The geometric matrix $\mathbf{K}_g := \left[k_{ij}^g\right]$, on the other hand, is full, of coefficients:

$$k_{ij}^g := \begin{cases} p_0 \frac{b}{\ell} \frac{(n\pi)^2}{8}(\eta - 2), & i = j, \\ -2\frac{b}{\ell} n^2 p_0 \frac{ij}{(i^2 - j^2)^2}\eta, & i \neq j, \quad i \pm j \text{ odd,} \\ 0, & i \neq j, \quad i \pm j \text{ even.} \end{cases}$$ (10.81)

After having introduced the nondimensional quantities $\mu := \frac{b^2 p_0}{\pi^2 D}$, $\alpha := \frac{\ell}{b}$ and $\hat{q}_m := q_m/b$ $(m = 1, 2, \cdots, N)$, the eigenvalue problem is written as:

$$\left(\begin{bmatrix} \hat{k}_{11}^e & \cdot & \cdots \\ & \hat{k}_{22}^e & \cdot \\ & & \hat{k}_{33}^e \\ \text{SYM} & & & \ddots \end{bmatrix} + \mu \begin{bmatrix} \hat{k}_{11}^g & \hat{k}_{12}^g & \cdot & \cdots \\ & \hat{k}_{22}^g & \hat{k}_{23}^g \\ & & \hat{k}_{33}^g \\ \text{SYM} & & & \ddots \end{bmatrix}\right) \begin{pmatrix} \hat{q}_1 \\ \hat{q}_2 \\ \hat{q}_3 \\ \vdots \end{pmatrix} = \begin{pmatrix} 0 \\ 0 \\ 0 \\ \vdots \end{pmatrix},$$ (10.82)

where:

$$\hat{k}_{ii}^e := \frac{\pi^2}{4\alpha}\left(\alpha i^2 + \frac{n^2}{\alpha}\right)^2,$$

$$\hat{k}_{ij}^g := \begin{cases} \frac{(n\pi)^2}{8\alpha}(\eta - 2), & i = j, \\ -\frac{2n^2}{\alpha} \frac{ij}{(i^2 - j^2)^2}\eta, & i \neq j, \quad i \pm j \text{ odd,} \\ 0, & i \neq j, \quad i \pm j \text{ even.} \end{cases}$$ (10.83)

---

[25] It can be verified that, by taking a double sines series, the stiffness matrices (in Eq. B.12) uncouple in blocks, each corresponding to a single harmonic in the x direction. These harmonics can therefore be studied separately, such as done here.

For a fixed aspect ratio α and for a given load shape η, the nondimensional critical load μ_c must be minimized with respect to the number of half-waves n. The result, of course, depends on the number N of terms in the series. For example, by fixing $\alpha = 3$, $\eta = 1$: with $N = 1$ terms, one finds $n = 3$, $\mu_c = 8$; with $N = 2$ terms, one gets $n = 3$, $\mu_c = 7.81225$; with $N = 3$ terms, one obtains $n = 3$, $\mu_c = 7.81205$. Repeating the analysis for $\alpha = 3$, $\eta = 2$, with $N = 5$ terms, the result is $n = 4$, $\mu_c = 24.1121$. Comparing these results with those (practically exact) provided by a finite element analysis, errors are found to be 0.36 % and 0.71 %, respectively.   □

Example 10.3 (Uniformly Compressed Plate, Clamped on Three Sides and Free on the Fourth) A plate is considered, uniformly compressed in the x direction, clamped on the sides $x = 0$, ℓ, clamped on $y = 0$ and free on $y = b$ (Fig. 10.11b). Limiting the analysis to few degrees of freedom, and taking into account the symmetry with respect to the axis $x = \frac{\ell}{2}$, one can take:

$$w(x, y) = a_{11}X_1(x)Y_1(y) + a_{12}X_1(x)Y_2(y) + a_{31}X_3(x)Y_1(y) + a_{32}X_3(x)Y_2(y),$$
(10.84)

or:

$$w(x, y) = a_{21}X_2(x)Y_1(y) + a_{22}X_2(x)Y_2(y) + a_{41}X_4(x)Y_1(y) + a_{42}X_4(x)Y_2(y),$$
(10.85)

where X_1, X_3 are the first and third (symmetric) modes and X_2, X_4 the second and fourth (antisymmetric) modes of the clamped-clamped beam (Eq. B.21); $Y_1, Y_2, \cdots$ are the first and second (non-symmetric) modes of the clamped-free beam (Eq. B.23). Executing the integrals numerically but leaving analytic the dependence on the parameter $\alpha := \frac{\ell}{b}$ and the Poisson's ratio v, one obtains, for the x-symmetric mode:

$$\left(\begin{bmatrix} k^e_{S11} & k^e_{S12} & k^e_{S13} & \cdot \\ & k^e_{S22} & \cdot & k^e_{S24} \\ & & k^e_{S33} & k^e_{S34} \\ \text{SYM} & & & k^e_{S44} \end{bmatrix} + \mu_S \begin{bmatrix} k^g_{S11} & k^g_{S12} & \cdot & \cdot \\ & k^g_{S22} & \cdot & \cdot \\ & & k^g_{S33} & k^g_{S34} \\ \text{SYM} & & & k^g_{S44} \end{bmatrix} \right) \begin{pmatrix} \hat{a}_{11} \\ \hat{a}_{12} \\ \hat{a}_{31} \\ \hat{a}_{32} \end{pmatrix} = \begin{pmatrix} 0 \\ 0 \\ 0 \\ 0 \end{pmatrix},$$
(10.86)

and for the x-anti-symmetric mode:

$$\left(\begin{bmatrix} k^e_{A11} & k^e_{A12} & k^e_{A13} & \cdot \\ & k^e_{A22} & \cdot & k^e_{A24} \\ & & k^e_{A33} & k^e_{A34} \\ \text{SYM} & & & k^e_{A44} \end{bmatrix} + \mu_A \begin{bmatrix} k^g_{A11} & k^g_{A12} & \cdot & \cdot \\ & k^g_{A22} & \cdot & \cdot \\ & & k^g_{A33} & k^g_{A34} \\ \text{SYM} & & & k^g_{A44} \end{bmatrix} \right) \begin{pmatrix} \hat{a}_{21} \\ \hat{a}_{22} \\ \hat{a}_{41} \\ \hat{a}_{42} \end{pmatrix} = \begin{pmatrix} 0 \\ 0 \\ 0 \\ 0 \end{pmatrix},$$
(10.87)

where $\mu_H := \frac{p_x b^2}{\pi^2 D}$ ($H = S, A$), $\hat{a}_{ij} := a_{ij}/b$ ($i = 1, 2, 3, 4$ and $j = 1, 2$). The stiffness coefficients are found to be:

$$k^e_{S11} := \frac{17.90}{\alpha^3} - \frac{6.28}{\alpha}\nu + 0.46\alpha + \frac{4.93}{\alpha}, \qquad k^e_{S12} := \frac{45.45}{\alpha^3} + \frac{6.28}{\alpha}\nu,$$

$$k^e_{S22} := \frac{151.95}{\alpha^3} + \frac{18.85}{\alpha}\nu + 37.47\alpha + \frac{44.41}{\alpha}, \qquad k^e_{S13} := 0.31\alpha,$$

$$k^e_{S33} := \frac{286.47}{\alpha^3} - \frac{25.13}{\alpha}\nu + 0.46\alpha + \frac{19.74}{\alpha}, \qquad k^e_{S24} := 24.98\alpha,$$

$$k^e_{S44} := \frac{2431.13}{\alpha^3} + \frac{75.40}{\alpha}\nu + 37.47\alpha + \frac{177.65}{\alpha}, \qquad k^e_{S34} := \frac{727.14}{\alpha^3} + \frac{25.13}{\alpha}\nu,$$

$$k^g_{S11} := -\frac{4.48}{\alpha}, \qquad k^g_{S12} := -\frac{11.36}{\alpha},$$

$$k^g_{S22} := -\frac{37.99}{\alpha}, \qquad k^g_{S33} := -\frac{17.90}{\alpha},$$

$$k^g_{S34} := -\frac{45.45}{\alpha}, \qquad k^g_{S44} := -\frac{151.95}{\alpha},$$

$$(10.88)$$

$$k^e_{A11} := \frac{74.93}{\alpha^3} - \frac{12.85}{\alpha}\nu + 0.26\alpha + \frac{10.10}{\alpha}, \qquad k^e_{A12} := \frac{190.20}{\alpha^3} + \frac{12.85}{\alpha}\nu,$$

$$k^e_{A22} := \frac{635.91}{\alpha^3} + \frac{38.56}{\alpha}\nu + 20.82\alpha + \frac{90.86}{\alpha}, \qquad k^e_{A13} := 0.10\alpha,$$

$$k^e_{A33} := \frac{654.65}{\alpha^3} - \frac{37.99}{\alpha}\nu + 0.26\alpha + \frac{29.84}{\alpha}, \qquad k^e_{A24} := 8.33\alpha,$$

$$k^e_{A44} := \frac{5555.70}{\alpha^3} + \frac{113.98}{\alpha}\nu + 20.82\alpha + \frac{268.56}{\alpha}, \qquad k^e_{A34} := \frac{1661.70}{\alpha^3} + \frac{37.99}{\alpha}\nu,$$

$$k^g_{A11} := -\frac{9.16}{\alpha}, \qquad k^g_{A12} := -\frac{23.24}{\alpha},$$

$$k^g_{A22} := -\frac{77.71}{\alpha}, \qquad k^g_{A33} := -\frac{27.07}{\alpha},$$

$$k^g_{A34} := -\frac{68.70}{\alpha}, \qquad k^g_{A44} := -\frac{229.70}{\alpha}.$$

$$(10.89)$$

For different values of α and for two different values of ν, one gets the critical loads in Table 10.2. It is noticed that, when ν is fixed and α increases, the critical mode changes from symmetric to antisymmetric (this occurs, with the adopted step, at $\alpha = 3$). Furthermore, when α is fixed and ν increases, the critical load reduces, since the shear elastic modulus G decreases. □

Table 10.2 Symmetric (μ_S) or antisymmetric (μ_A) nondimensional critical load of the plate in Fig. 10.11b, for different aspect ratios $\alpha := \frac{\ell}{b}$ and Poisson's ratio ν

	$\nu = 0$		$\nu = 0.3$	
α	μ_S	μ_A	μ_S	μ_A
1	**4.82**	8.86	**4.67**	8.76
1.5	**2.78**	4.39	**2.58**	4.26
2	**2.2**	2.88	**1.97**	2.72
2.5	**2.05**	2.24	**1.81**	2.05
3	2.02	**1.94**	1.79	**1.73**
3.5	1.94	**1.80**	1.74	**1.57**
4	1.82	**1.74**	1.62	**1.51**

10.8 Compressed Plate Stiffened by a Longitudinal Rib

The critical load of a homogeneous plate, whatever the constraint conditions, is directly proportional to the bending stiffness D. Fixing the constituent material and the dimensions of the plate in the (x, y) plane, the critical load can only be increased by augmenting the thickness h. However, this technical solution is not economic, so that it is more advantageous to stiffen the plate with suitable longitudinal and/or transverse ribs. The structure, in this case, consists of an assembly of plates and beams [3, 10, 15, 17].

Here, as an illustrative example, a simple case is dealt with. It concerns a plate simply supported on the two opposite sides, $x = 0, \ell$, at which compression loads p_x are applied, generically constrained on the sides $y = 0, b$, and ribbed by a single beam, having bending stiffness EI and torsional stiffness GJ, parallel to the load direction, placed at the abscissa $y = \bar{y}$ (Fig. 10.12a). To further simplify the problem, it is assumed that the centroidal axis of the beam is contained in the (x, y) plane (Fig. 10.12b), since a possible eccentricity would trigger the membrane behavior of the plate, now uncoupled from the flexural one.

By expressing the solution in a separable variable form (Eq. 10.53), the problem reduces to one dimension, spanned the coordinate y (Fig. 10.12c). In this transverse direction, as mentioned, the plate behaves as a beam, whose deflection $Y(x)$ is governed by Eq. 10.54. As it will be immediately proved, the rib is equivalent to a *uniform linear distributions of equivalent springs*. One of them has bending stiffness k_f;[26] the other has torsional stiffness k_t,[27] both determined as follows, via an energy equivalence.

Equivalent Spring Stiffnesses
Due to bending and torsion, the beam stores an amount of the elastic energy; moreover, since it is compressed, it also stores a prestress energy. By assuming that, in the precritical state, the compression stress on the rib, P/A (with P the unknown axial force and A its cross-section area), is equal to that of the plate p_x/h (with h the thickness of the plate), the axial force soliciting the rib is $P = p_x \frac{A}{h}$. The total

---

[26] k_f has physical dimensions of a force divided a length, per unit of length, i.e., $[\mathrm{ML^{-1}T^{-2}}]$.
[27] k_t has dimensions of a couple divided an angle, per unit of length, i.e., $[\mathrm{MLT^{-2}}]$.

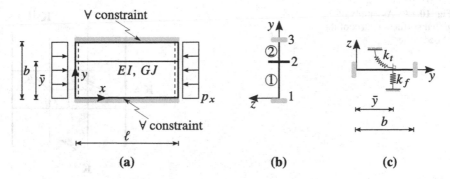

Fig. 10.12 Ribbed plate: (**a**) structure, (**b**) cross-section, nodes and finite elements, (**c**) transverse static scheme

potential energy of the beam, therefore, reads:

$$\Pi_b = \frac{1}{2} EI \int_0^\ell w_b''^2 (x) \, dx + \frac{1}{2} GJ \int_0^\ell \theta'^2 (x) \, dx - \frac{1}{2} P \int_0^\ell w_b'^2 (x) \, dx,$$
(10.90)

where $w_b (x)$ is the deflection and $\theta (x)$ the twist. By invoking the congruence between plate and beam and remembering Eq. 10.53, it follows:

$$w_b (x) = w (x, \bar{y}) = \sin \left(\frac{n\pi}{\ell} x \right) Y (\bar{y}),$$
(10.91a)

$$\theta (x) = w_{,y} (x, \bar{y}) = \sin \left(\frac{n\pi}{\ell} x \right) Y' (\bar{y}),$$
(10.91b)

so that, by performing the integrations, one obtains:

$$\Pi_b = \frac{1}{4} \ell \left\{ \left[EI \left(\frac{n\pi}{\ell} \right)^4 - P \left(\frac{n\pi}{\ell} \right)^2 \right] Y (\bar{y})^2 + GJ \left(\frac{n\pi}{\ell} \right)^2 Y' (\bar{y})^2 \right\}.$$
(10.92)

On the other hand, the energy stored by a linear and uniform distribution of springs, subject to the same displacement field experienced by the beam, is:

$$\Pi_s = \frac{1}{2} \int_0^\ell \left[k_f Y (\bar{y})^2 + k_t Y' (\bar{y})^2 \right] \sin^2 \left(\frac{n\pi}{\ell} x \right) \, dx$$

$$= \frac{1}{4} \ell \left[k_f Y (\bar{y})^2 + k_t Y' (\bar{y})^2 \right].$$
(10.93)

By requiring that $\Pi_b = \Pi_s$ for any generalized displacement, the equivalent stiffnesses for the spring are derived:

$$k_f = EI \left(\frac{n\pi}{\ell} \right)^4 - P \left(\frac{n\pi}{\ell} \right)^2,$$
(10.94a)

Fig. 10.13 Assembly of the
global stiffness matrix of the
ribbed plate

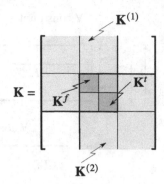

$$k_t = GJ \left(\frac{n\pi}{\ell}\right)^2 . \tag{10.94b}$$

Finite Element Solution

Referring to the static scheme in Fig. 10.12c, the critical load of the ribbed plate calls for integrating Eq. 10.54 in two intervals, with boundary conditions to be enforced at the edges $y = 0, b$ and at the singular point $y = \bar{y}$, by accounting for the presence of the springs. However, the calculation is easier if use is made of the exact finite element developed in Sect. 10.6.2. To this end, two elements are considered: one of length $\bar{y}$ and local stiffness matrix $\mathbf{K}^{(1)}$ and the other of length $b - \bar{y}$ and local stiffness matrix $\mathbf{K}^{(2)}$. The (global) vector of configuration variables $\mathbf{q}$ is made up of six displacement components, two for each node (Fig. 10.12b), i.e.:

$$\mathbf{q} = \left(Y_1, Y_1', Y_2, Y_2', Y_3, Y_3'\right)^T . \tag{10.95}$$

The flexural and torsional elastic springs have stiffness matrices of dimension 1×1, denoted by $\mathbf{K}^f = \left[k_f\right]$ and $\mathbf{K}^t = [k_t]$. The global stiffness matrix, of dimensions 6×6, is obtained by assembling the element stiffness matrices of plate and springs, according to the scheme in Fig. 10.13. By introducing the external constraints and then zeroing the determinant of the reduced matrix, the critical load is computed for any given number n of longitudinal half-waves.

Example 10.4 (Stiffened Plate Resting on Four Sides) An example is shown, concerning a plate simply supported on four sides, stiffened by a beam located at the abscissa $\bar{y} = b/2$, having negligible torsional stiffness (as it happens, e.g., for a thin rectangular cross-section). Proceeding as previously described and imposing the conditions at the edges, $Y_1 = Y_3 = 0$, a reduced stiffness matrix is obtained, of dimensions 4×4. By vanishing its determinant, a characteristic equation follows, which breaks into two equations, namely:

$$\sin\left(\frac{b\beta_2}{2}\right) = 0,$$
(10.96a)

$$\left[EI\left(\frac{n\pi}{\ell}\right)^4 - P\left(\frac{n\pi}{\ell}\right)^2\right]\left[\beta_1 \sin\left(\frac{b\beta_2}{2}\right)\cosh\left(\frac{b\beta_1}{2}\right)\right.$$
(10.96b)

$$\left.-\beta_2 \cos\left(\frac{b\beta_2}{2}\right)\sinh\left(\frac{b\beta_1}{2}\right)\right] + 2D\beta_1\beta_2\left(\beta_1^2 + \beta_2^2\right)\cos\left(\frac{b\beta_2}{2}\right)\cosh\left(\frac{b\beta_1}{2}\right) = 0.$$

The first of them supplies the critical loads associated with antisymmetric modes with respect the $y = b/2$ axis. These are independent of the geometric character-istics of the beam, since the rib remains undeformed. The second equation, on the other hand, provides the critical loads associated with the symmetric modes, which do depend on the characteristics of the stiffening beam, which bends together with the plate.

To analyze the solutions, use is made of the following nondimensional parame-ters:

$$\mu_c := \frac{P_{xc}\, b^2}{\pi^2 D}, \quad \alpha := \frac{\ell}{b}, \quad \gamma := \frac{EI}{bD}, \quad \delta := \frac{A}{bh}.$$
(10.97)

Figure 10.14 reports the values of the nondimensional critical load μ_c for several case studies. From the figure, the following considerations are drawn.

- If the bending stiffness of the beam is small ($\gamma = 0.15$, Fig. 10.14a), the buckling mode is transversely symmetric, with $n = 1$ longitudinal half-waves, when $\alpha < 1.5$, and $n = 2$, when $\alpha > 1.5$. Since the plate buckles in approximately square fields, the critical load μ_c turns out to be only slightly higher than 4, relevant to the unstiffned plate. In this interval, therefore, the beam has a little influence on the behavior of the plate. Ultimately, the rib does not bring a significant improvement to the mechanical performance of the plate.
- If the bending stiffness of the beam is moderately large ($\gamma = 5$, Fig. 10.14b), when $\alpha < 0.65$, the buckling mode is transversely antisymmetric, with $n = 1$ longitudinal half-waves; the minimum value of the associated critical load is $\mu = 16$, since the length of the transverse inflection is reduced to $b/2$. When $\alpha > 0.65$, the mode becomes symmetric, still remaining $n = 1$ the number of half-waves. For α larger, the critical load decreases but remains significantly above 4, as the beam contributes to the stiffening of the system.
- If the stiffness of the beam is very large ($\gamma = 20$, Fig. 10.14c), the behavior is more complex. When $\alpha < 1.65$, the mode is transversely antisymmetric, as in the previous case, but the number of the longitudinal half-waves passes, as α increases, from $n = 1$ to $n = 3$. When $\alpha > 1.65$, the buckling mode becomes symmetric, and the number of half-waves comes back to $n = 1$. In all cases, in

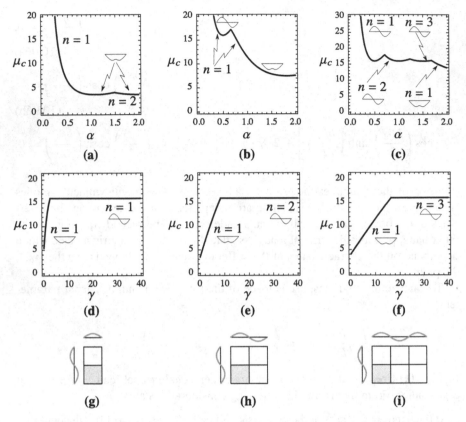

Fig. 10.14 Nondimensional critical load of the ribbed plate simply supported on four sides, when $\delta = 0.075$; (a)–(c) as a function of the α aspect ratio in case of (a) soft rib, $\gamma = 0.15$; (b) moderately stiff rib, $\gamma = 5$; (c) stiff rib, $\gamma = 20$; (d)–(f) as a function of the stiffness γ of the rib, for selected aspect ratios $\alpha = (0.5, 1, 1.5)$, respectively; (g)–(i) critical modes for $\alpha = (0.5, 1, 1.5)$ and γ sufficiently large; critical mode associated with each branch: stylized transverse deformation $Y(y)$, and number of half-waves n in the longitudinal direction

the range of α examined, the minimum value of μ_c is around 16, which proves the great effectiveness of the stiffening.

- If the aspect ratio is fixed (e.g., it is $\alpha = 0.5, 1, 1.5$ in Fig. 10.14d,e,f, respectively), when the stiffness of the rib increases, the transverse critical mode always changes from symmetric (for small γ) to antisymmetric (for large γ). The critical load increases on the first branch and remains constant, about equal to 16, on the second branch. The maximum value of μ_c, however, is attained at threshold values of γ which depend on α and increase with it. It is interesting to observe that, while the number of half-waves n keeps equal to 1 on the ascending branch, it is equal to 1,2,3 on the horizontal branch, depending on the value of α. This value of n is such that the plate always buckles in square fields of side $b/2$ (or closer to the square, if α is not a multiple of 0.5, Fig. 10.14g,h,i).   □

10.9 Plate Subject to Uniform Shear Force

A rectangular plate is considered, subject to tangential loads p_{xy} uniformly distributed at the edges (Fig. 10.15a), which induce the uniform prestress state $N_x^0 = 0$, $N_y^0 = 0$ and $N_{xy}^0 = p_{xy}$ [1, 2, 17]. The indefinite equilibrium Eq. 10.38 specializes as follows:

$$D \left(w_{,xxxx} + 2w_{,xxyy} + w_{,yyyy} \right) - 2p_{xy}w_{,xy} = 0. \tag{10.98}$$

If the plate is simply supported at the edges, the boundary conditions are:

$$w = 0, \qquad w_{,xx} = 0, \qquad \text{at} \quad x = 0, \ell, \tag{10.99a}$$

$$w = 0, \qquad w_{,yy} = 0, \qquad \text{at} \quad y = 0, b; \tag{10.99b}$$

if it is clamped, they are:

$$w = 0, \qquad w_{,x} = 0, \qquad \text{at} \quad x = 0, \ell, \tag{10.100a}$$

$$w = 0, \qquad w_{,y} = 0, \qquad \text{at} \quad y = 0, b. \tag{10.100b}$$

Differently from the cases examined so far (in which only normal stresses exist in the precritical state), the field Eq. 10.98 contains, in addition to the derivatives of even order, also the derivatives of odd order of the unknown function $w(x, y)$. This occurrence invalidates the "guess" harmonic solutions, based on the separation of the variables. It is therefore not possible to exactly solve the boundary value problem, even for simple supports.

An exception is represented by the infinitely long plate, for example, in the x direction (Fig. 10.15b). In this case, indeed, it is possible (a) to invoke the periodicity of the response in the x direction, i.e., to assume that $w(x + 2\ell_0, y) = w(x, y)$, with ℓ_0 the unknown half-wave-length, and (b) to ignore the boundary conditions

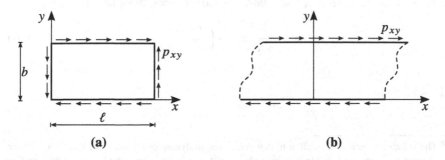

Fig. 10.15 Plate stressed by uniform shear: (**a**) of finite dimension, (**b**) infinitely long

on the "short" sides $x = \pm\infty$. For this problem it is possible to determine the exact value of the critical load, whether the plate is simply supported or clamped at the sides $y = 0, b$ [14] or in different boundary conditions, not dealt with here. These solutions, as it will be seen ahead, are technically valid for sufficiently long plates (e.g., $\alpha := \frac{\ell}{b} > 5$). If, however, the aspect ratio is closer to 1, it is necessary to resort to approximate solutions.

In the following, the case of the infinitely long plate will be addressed first; then some applications of the Ritz method will be discussed, for the evaluation of the critical load of plates of finite dimensions.

10.9.1 Infinitely Long Plate: Exact Solution

If the plate has an infinite dimension in the x direction (Fig. 10.15b), one can try a *separable variable solution in the complex field*, by letting:

$$w(x, y) = Y(y) \exp\left(i\frac{\gamma \pi x}{b}\right), \tag{10.101}$$

with $Y(y) \in \mathbb{C}$ a complex-valued function and $\gamma \in \mathbb{R}$ a real number, both unknown.[28] Since the field equation has real coefficients, if the previous complex quantity is a solution, its complex conjugate is also a solution; therefore, by adding the two solutions, one gets:

$$w(x, y) = f(y) \cos\left(\frac{\gamma \pi x}{b}\right) - g(y) \sin\left(\frac{\gamma \pi x}{b}\right), \tag{10.102}$$

where $f(y) := 2\mathrm{Re}[Y(y)]$, $g(y) := 2\mathrm{Im}[Y(y)]$. The deflection is therefore periodic in x, of half period $\ell_0 := \frac{b}{\gamma}$. The number $\gamma = \frac{b}{\ell_0}$ assumes the meaning of ratio between the plate width and the half-wave-length. The solution in Eq. 10.102 is a superposition of two different transverse shapes, $f(y), g(y)$, which "propagate" harmonically in the longitudinal direction, being out-of-phase by $\frac{\pi}{2}$.

Contour Lines
From Eq. 10.102, an important property follows, concerning the geometric locus $w(x, y) = 0$: *the deflection nodal lines are curved lines*, of equation:

$$x = \frac{b}{\gamma \pi} \arctan\left(\frac{f(y)}{g(y)}\right), \tag{10.103}$$

---

[28] The variable separation is similar to that performed in dynamics for gyroscopic systems, where the time t replaces the variable x, which also admit purely imaginary eigenvalues and complex eigenvectors $Y(y)$.

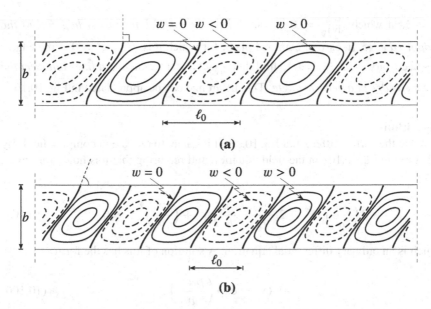

Fig. 10.16 Contour lines w = cost of the infinitely long plate, subject to uniform shear force: (**a**) supported plate, (**b**) clamped plate

and no longer straight lines parallel to the coordinate axes, as it happens for plates subject to normal stresses only.[29] Given the periodicity in x, the nodal lines divide the plate in non-rectangular fields.

The *iso-displacement lines* (i.e., the contour lines w = cost), since they cannot cross the nodal lines, nor the edges, are necessarily closed. They take the form shown in Fig. 10.16. The plate, therefore, buckles into adjacent fields, in which the deflection w assumes alternatively positive and negative sign.

Supplement 10.4 (Slope of the Nodal Lines at the Edges) The slope of the nodal lines at the edges $\bar{y} = 0, b$ is given by:

$$\left.\frac{\mathrm{d}x}{\mathrm{d}y}\right|_{\bar{y}} = \frac{b}{\gamma\pi}\left.\frac{gf' - fg'}{f^2 + g^2}\right|_{\bar{y}}. \tag{10.104}$$

Since $f = g = 0$ at the edges, for simply supported as well as for clamped plates, this ratio is indeterminated. To solve it, it needs to repeatedly apply de l'Hôpital Theorem, computing the ratio between the kth ($k = 1, 2, \cdots$) derivatives of the numerator and denominator, until a determined form is found. In case of supported plate, for which it is $f'' = g'' = 0$ at the edges, the indeterminacy is solved when

---

[29] Indeed, if the variables are separable in the real field, it is $w\,(x, y) = F\,(x)\,G\,(y)$, so that if, e.g., it is $F\,(x_0) = 0$, then $w\,(x_0, y) = 0\,\forall y$.

$k = 2$, at which $\left.\frac{dx}{dy}\right|_{\bar{y}} = 0$ is found, i.e., *the nodal lines are orthogonal to the supported edge.* In case of clamped plate, for which it is $f' = g' = 0$ at the edges, it is necessary to take $k = 4$, to find that $\left.\frac{dx}{dy}\right|_{\bar{y}} \neq 0$, i.e., *the nodal lines intersect the clamped edge at a generic angle.* The difference can be appreciated in Fig. 10.16.

□

Algorithm

To solve the partial differential Eq. 10.98, it is easier to operate in complex field. By substituting Eq. 10.101 in the field equation and requiring that this holds for any x, one gets:

$$Y'''' - 2\left(\frac{\gamma\pi}{b}\right)^2 Y'' - 2i\frac{P_{xy}}{D}\left(\frac{\gamma\pi}{b}\right) Y' + \left(\frac{\gamma\pi}{b}\right)^4 Y = 0, \qquad (10.105)$$

which is an ordinary differential equation. A solution of this has the form:

$$Y(y) = \exp\left(\frac{\beta\pi y}{b}\right), \qquad (10.106)$$

where $\beta \in \mathbb{C}$ is the unknown characteristic exponent. By substituting this latter into Eq. 10.105 and requiring this is satisfied for any y, an algebraic equation for β follows, i.e.:

$$\beta^4 - 2\gamma^2\beta^2 - 2i\mu\gamma\beta + \gamma^4 = 0, \qquad (10.107)$$

in which, as usual,

$$\mu := \frac{P_{xy}\,b^2}{\pi^2 D}. \qquad (10.108)$$

Equation 10.107 is a quartic equation, admitting (generally complex) roots $\beta_j = \beta_j(\gamma, \mu)$ $(j = 1, \cdots, 4)$, which can be found numerically for any given γ and μ. With these roots, the general solution is built up, as:

$$Y(y) = \sum_{j=1}^{4} C_j \exp\left(\frac{\beta_j(\gamma, \mu)\pi y}{b}\right), \qquad (10.109)$$

where $C_j \in \mathbb{C}$ are arbitrary complex constants. The constants must be chosen in such a way as to satisfy the boundary conditions at the points $y = 0, b$. If the plate is simply supported, from Eq. 10.99b, and by accounting for Eq.10.101, one draws:

$$Y = 0, \qquad Y'' = 0, \qquad \text{at} \quad y = 0, b; \qquad (10.110)$$

if the plate is clamped, from Eq. 10.100b, it follows that:

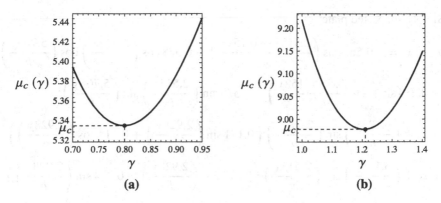

Fig. 10.17 Critical shear force $\mu_c(\gamma)$ vs the wave-length γ: (a) supported plate, (b) clamped plate

$$Y = 0, \qquad Y' = 0, \qquad \text{at} \quad y = 0, b. \tag{10.111}$$

The four boundary conditions lead to a system of four linear and homogeneous algebraic equations in the constants C_j. Zeroing the determinant of the matrix provides a transcendental equation, of the type $F(\gamma, \mu) = 0$. For each γ, the smallest root $\mu_c(\gamma)$ is of interest. The minimum value $\mu_c := \min_{\gamma}[\mu_c(\gamma)]$ is the nondimensional critical load sought for.

The somewhat laborious procedure requires the execution of the following steps:

1. A value of γ is fixed.
2. A "guess" value of μ is taken, and the quartic Eq. 10.107 is solved for β_j; then, it is checked that the characteristic equation $F(\gamma, \mu) = 0$ is satisfied to within a prefixed tolerance; if this is not the case, an iteration over μ is carried out, until the $\mu_c(\gamma)$ value is found; it determines the critical load which corresponds to the predetermined wave-length.
3. γ is changed, and the process is repeated, in order to build-up the $\mu_c(\gamma)$ curve in a certain interval, in which the minimum value μ_c is determined.

The application of the method described above leads to the following results [14]:

- Simply supported plate: $\mu_c = 5.34$ (corresponding to $\gamma = 0.8$; $\beta_1 = -1.949 + 0.932\,i$, $\beta_2 = -0.048\,i$, $\beta_3 = -1.816\,i$ and $\beta_4 = 1.949 + 0.932\,i$).
- Clamped plate: $\mu_c = 8.98$ (corresponding to $\gamma = 1.21$; $\beta_1 = -2.715 + 1.250\,i$, $\beta_2 = -0.1\,i$, $\beta_3 = -2.399\,i$ and $\beta_4 = 2.715 + 1.250\,i$).

The graph of the $\mu_c(\gamma)$ function is represented in Fig. 10.17 for the two constraint conditions considered.

Supplement 10.5 (Critical Modes of the Infinite Plate, Subject to Shear Forces)
Using Eqs. 10.109, together with Eqs. 10.110 or 10.111, and replacing the numerical values found above for the characteristic exponents, the following critical modes are obtained:

Simply supported plate

$$w\left(x, y\right) = -0.560 \cos\left(\frac{2.513x - 0.152y}{b}\right) + 0.388 \cos\left(\frac{2.513x}{b}\right) \cos\left(\frac{5.705y}{b}\right)$$

$$+ 0.214 \sin\left(\frac{2.513x - 0.152y}{b}\right) + 0.388 \sin\left(\frac{2.513x}{b}\right) \sin\left(\frac{5.705y}{b}\right)$$

$$- \cosh\left(\frac{6.123y}{b}\right) \sin\left(\frac{2.513x}{b}\right)\left(0.171 \sin\left(\frac{2.928y}{b}\right) + 0.214 \cos\left(\frac{2.928y}{b}\right)\right)$$

$$+ \cosh\left(\frac{6.123y}{b}\right) \cos\left(\frac{2.513x}{b}\right)\left(0.171 \cos\left(\frac{2.928y}{b}\right) - 0.214 \sin\left(\frac{2.928y}{b}\right)\right)$$

$$+ \sinh\left(\frac{6.123y}{b}\right) \sin\left(\frac{2.513x}{b}\right)\left(0.173 \sin\left(\frac{2.928y}{b}\right) + 0.214 \cos\left(\frac{2.928y}{b}\right)\right)$$

$$+ \sinh\left(\frac{6.123y}{b}\right) \cos\left(\frac{2.513x}{b}\right)\left(0.214 \sin\left(\frac{2.928y}{b}\right) - 0.173 \cos\left(\frac{2.928y}{b}\right)\right),$$

$$(10.112)$$

Clamped plate

$$w\left(x, y\right) = -0.535 \cos\left(\frac{3.801x - 0.314y}{b}\right) + 0.4 \cos\left(\frac{3.801x}{b}\right) \cos\left(\frac{7.537y}{b}\right)$$

$$- 0.272 \sin\left(\frac{3.801x - 0.314y}{b}\right) + 0.4 \sin\left(\frac{3.801x}{b}\right) \sin\left(\frac{7.537y}{b}\right)$$

$$+ \cosh\left(\frac{8.53y}{b}\right) \sin\left(\frac{3.801x}{b}\right)\left(0.272 \cos\left(\frac{3.926y}{b}\right) - 0.135 \sin\left(\frac{3.926y}{b}\right)\right)$$

$$+ \cosh\left(\frac{8.53y}{b}\right) \cos\left(\frac{3.801x}{b}\right)\left(0.272 \sin\left(\frac{3.926y}{b}\right) + 0.135 \cos\left(\frac{3.926y}{b}\right)\right)$$

$$+ \sinh\left(\frac{8.53y}{b}\right) \sin\left(\frac{3.801x}{b}\right)\left(0.135 \sin\left(\frac{3.926y}{b}\right) - 0.272 \cos\left(\frac{3.926y}{b}\right)\right)$$

$$- \sinh\left(\frac{8.53y}{b}\right) \cos\left(\frac{3.801x}{b}\right)\left(0.272 \sin\left(\frac{3.926y}{b}\right) + 0.135 \cos\left(\frac{3.926y}{b}\right)\right),$$

$$(10.113)$$

whose contour lines are represented in Figs. 10.16a,b.   □

10.9.2 Infinitely Long Plate: Ritz Approximate Solutions

Although exact solutions are available for infinitely long plates, many authors prefer to discuss easier approximate solutions, based on the Ritz method. Here, a mention of some of them is given, mainly aimed to introduce the use of *trial functions dependent on free parameters*, a general tool that could be useful in different circumstances. Reference will be made to simple supported plates.

First of all, the TPE Π must be written for the bent plate; in the case of uniform shear load, it is:

$$\Pi = \frac{1}{2} \int_0^\ell \int_0^b \left\{ D \left[w_{,xx}^2 + w_{,yy}^2 + 2v\, w_{,xx} w_{,yy} + 2\,(1-v)\, w_{,xy}^2 \right] + 2 p_{xy} w_{,x} w_{,y} \right\} dx\, dy.$$

(10.114)

Given the periodicity, it is sufficient to refer to a half period, replacing the length ℓ with the half-wave-length ℓ_0.

The Ritz method usually requires expressing the displacement field as a linear combination with unknown coefficients of known trial functions. Here, instead, a different strategy is adopted, namely, using a *single trial function*, which, however, depends on several free parameters, to be determined later, in the process of minimizing the critical load. Accordingly, the deflection is expressed as:

$$w\,(x, y) = q\, \phi\,(x, y; \ell_0, \eta)\,,$$

(10.115)

where q is the only Lagrangian parameter; ϕ is a trial function, which satisfies at least the geometric conditions at the edges $y = 0, b$; ℓ_0 and η are free parameters. Substituting it in the TPE, integrating and imposing the stationary condition $\frac{\partial \Pi}{\partial q} = 0$, the load $p_{xyc}\,(\ell_0, \eta)$ is got, as a function of the free parameters. Minimizing it with respect to the parameters, one obtains $p_{xyc} := \min_{\ell_0, \eta} \left[p_{xyc}\,(\ell_0, \eta) \right]$, which is an approximation (viz., an upper bound) for the exact critical load. The following examples illustrate the process.

Example 10.5 (Trial Function with Straight Nodal Lines) A first trial function is adopted as [13, 15, 16]:

$$\phi\,(x, y; \ell_0, \eta) := \sin\left(\pi \frac{x - \eta y}{\ell_0} \right) \sin\left(\frac{\pi y}{b} \right),$$

(10.116)

which satisfies the geometric conditions $w = 0$ on the supported sides, *but not the mechanical ones*, $w_{,yy} = 0$.[30] The nodal lines of the deflection are the geometric locus at which $\phi\,(x, y; \ell_0, \eta) = 0$; in addition to the edges $y = 0, b$, they are a family of parallel lines of equation:

$$x - \eta y = n \ell_0, \qquad n = 0, 1, 2, \cdots,$$

(10.117)

---

[30] It should be noticed, that this trial function assumes the form in Eq. 10.102, when the sine of the sum is expanded.

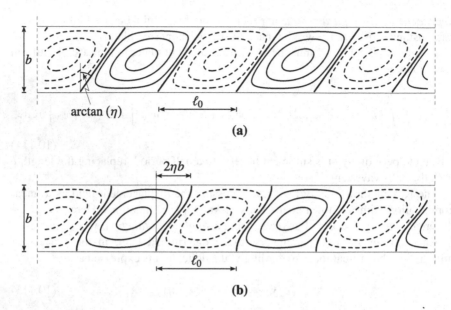

Fig. 10.18 Contour lines of the approximate buckling mode of the simply supported infinite plate, subject to shear: (a) straight nodal lines, (b) curved nodal lines, normal to the edges

whose slope with respect the y axis is arctan η and which detach segments of equal length ℓ_0 on the axis x (Fig. 10.18a). By replacing Eq. 10.116 into Eq. 10.114 and integrating, one gets:

$$\Pi = \frac{1}{2}\left\{\frac{\pi^4 D}{4b^3\ell_0^3}\left[\left(\eta^2+1\right)^2 b^4 + 2\left(3\eta^2+1\right)b^2\ell_0^2 + \ell_0^4\right] - \frac{\pi^2 b}{2\ell_0}\eta\, p_{xy}\right\}q^2,$$
(10.118)

which is the TPE of a single degree of freedom system. The critical load, obtained by zeroing the stiffness, is:

$$p_{xyc}\left(\ell_0,\eta\right) = \frac{1}{2\eta}\frac{\pi^2 D}{b^2}\left[\left(\eta^2+1\right)^2\left(\frac{b}{\ell_0}\right)^2 + 2\left(3\eta^2+1\right) + \left(\frac{\ell_0}{b}\right)^2\right].$$
(10.119)

Minimizing $p_{xyc}\left(\ell_0,\eta\right)$ with respect to ℓ_0, one finds $\ell_0 = b\sqrt{1+\eta^2}$; still minimizing it with respect to η, one gets $\eta = \frac{\sqrt{2}}{2}$.[31] With these values of the parameters, by using the definition in Eq. 10.108, $\mu_c = 4\sqrt{2} = 5.66$ is finally found. The error, compared to the exact solution, is about 6 %. With the same parameters, the contour lines of the critical mode, represented in Fig. 10.18a, are found.   □

---

[31] The slope of the straight nodal lines is about 35.3° on the y axis.

Example 10.6 (Trial Function with Curved Nodal Lines) To improve the approximation of the critical mode, it is necessary to search for a trial function that satisfies also the mechanical conditions at the edges. To this end, Eq. 10.116 is generalized as [13, 16]:

$$\phi\,(x,\,y;\,\ell_0) := \sin\left(\pi\,\frac{x f\,(y)}{\ell_0}\right)\sin\left(\frac{\pi y}{b}\right),\tag{10.120}$$

where $f\,(y)$ is an arbitrary regular function. The corresponding nodal lines are a family of curves of equation:

$$x - f\,(y) = n\ell_0,\qquad n = 0, 1, 2, \cdots.\tag{10.121}$$

Since $\phi_{,yy}$, evaluated at the edges, is proportional to $f'\,(0)$ or $f'\,(b)$, one needs to choose $f\,(y)$ in such a way its first derivative vanishes at the ends of the interval $(0,\,b)$. For example, the function:

$$f\,(y) := \eta\,b\left(1 - \cos\left(\frac{\pi y}{b}\right)\right),\tag{10.122}$$

satisfies this requirement, with η an arbitrary nondimensional parameter, assuming the meaning illustrated in Fig. 10.18b. With this choice, the nodal lines are orthogonal to the edges, as required from the discussion developed in the Supplement 10.4.

The corresponding TPE (Eq. 10.114) is found to be:

$$\Pi = \frac{1}{2}\left\{\frac{\pi^4 D}{32 b^3 \ell_0^3}\left[\left(5\pi^4\eta^4 + 12\pi^2\eta^2 + 8\right) b^4 + 2\left(15\pi^2\eta^2 + 8\right) b^2 \ell_0^2 + 8\ell_0^4\right]\right.$$
$$\left. -\frac{4\pi^2 b}{3\ell_0}\,\eta\,P_{xy}\right\} q^2,$$
$$\tag{10.123}$$

from which the critical load follows:

$$p_{xyc}\,(\ell_0,\,\eta) = \frac{3}{128\,\eta}\,\frac{\pi^2 D}{b^2}\left[\left(5\pi^4\eta^4 + 12\pi^2\eta^2 + 8\right)\left(\frac{b}{\ell_0}\right)^2\right.$$
$$\left. +2\left(15\pi^2\eta^2 + 8\right) + 8\left(\frac{\ell_0}{b}\right)^2\right].\tag{10.124}$$

Minimizing it with respect to the two free parameters, one finds $\ell_0 = \frac{b}{2^{3/4}}\sqrt[4]{5\pi^4\eta^4 + 12\pi^2\eta^2 + 8}$, $\eta = 0.276$. Consequently, the smallest critical load, in the notation of Eq. 10.108, is $\mu_c = 5.41$, with an error of 1 % with respect the exact solution. The contour lines of the buckling mode are represented in Fig. 10.18b. □

10.9.3 Plate of Finite Dimensions

The computation of the shear critical load of rectangular plates of finite dimensions $\ell \times b$ is now tackled (Fig. 10.15a). The problem is governed by the field Eq. 10.98, with the boundary conditions Eqs. 10.99 and 10.100. Since, as mentioned, there is no closed form solution, one has to resort to the Ritz method, in order to transform the problem from differential to algebraic. Using Eq.10.114 for the TPE and adopting trial functions with separable variables (such as in Eq. B.14 in the Appendix B), namely:

$$w(x, y) = \sum_{n=1}^{N} \sum_{m=1}^{N} a_{nm} X_n(x) Y_m(y), \tag{10.125}$$

one arrives at the following algebraic eigenvalue problem:

$$\left(\mathbf{K}_e + \mu \mathbf{K}_g\right) \mathbf{q} = \mathbf{0}, \tag{10.126}$$

where a load multiplier has been introduced via the substitution $p_{xy} \to \mu \, p_{xy}$ The smallest eigenvalue, μ_c, is the nondimensional critical load sought for.

Simply Supported Plate
If the plate is simply supported on four sides, one can choose the following trial functions:

$$X_n(x) = \sin\left(\frac{n\pi x}{\ell}\right), \qquad Y_m(y) = \sin\left(\frac{m\pi y}{b}\right), \tag{10.127}$$

for which the TPE assumes the form:

$$\Pi = \frac{D}{2} \frac{\pi^4 b \ell}{4} \sum_{n=1}^{N} \sum_{m=1}^{N} a_{nm}^2 \left(\frac{n^2}{\ell^2} + \frac{m^2}{b^2}\right)^2$$

$$+ 4 p_{xy} \sum_{n=1}^{N} \sum_{m=1}^{N} \sum_{h=1}^{N} \sum_{k=1}^{N} a_{nm} a_{hk} \frac{nmhk}{\left(n^2 - h^2\right)\left(k^2 - m^2\right)}, \tag{10.128}$$

where $n \pm h$ and $m \pm k$ are odd integers. By enforcing $\frac{\partial \Pi}{\partial a_{nm}} = 0$, the equilibrium equations are derived:

$$D \frac{\pi^4 b \ell}{4} \left(\frac{n^2}{\ell^2} + \frac{m^2}{b^2}\right)^2 a_{nm}$$

$$+ 8 p_{xy} \sum_{h=1}^{N} \sum_{k=1}^{N} a_{hk} \frac{nmhk}{\left(n^2 - h^2\right)\left(k^2 - m^2\right)} = 0, \qquad n, m = 1, \cdots, N. \tag{10.129}$$

It is noticed that the elastic stiffness matrix is diagonal, while the geometric stiffness matrix is full (although several coefficients are zero). This entails that all the trial functions (symmetric and antisymmetric) participate to the critical mode. Therefore, the antisymmetry of the mode with respect the longitudinal axis, which exists for the infinitely long plate (Fig. 10.18), is destroyed in the plate of finite dimensions.

As an example, when a the plate has aspect ratio $\alpha = \frac{\ell}{b} = 3$, the nondimensional critical load is found to be $\mu_c = 5.84$, with the associated critical mode illustrated in Fig. 10.19. The contour lines highlight the effect of the short sides, which break the periodicity of the deflection, which only holds for infinite plates.

Clamped Plate

If the plate is clamped on four sides, one can choose the following trial functions:

$$
X_n(x) = \begin{cases} 1 - \cos(\beta_{xn}x), & n = 1, 3, 5, \cdots, \\ 1 - \cos(\beta_{xn}x) - \frac{1-\cos(\beta_{xn}\ell)}{\sin(\beta_{xn}\ell)-\beta_{xn}\ell}\left(\sin(\beta_{xn}x) - \beta_{xn}x\right), & n = 2, 4, 6, \cdots, \end{cases}
$$
$$(10.130\text{a})$$

$$
Y_m(y) = \begin{cases} 1 - \cos\left(\beta_{ym}y\right), & m = 1, 3, 5, \cdots, \\ 1 - \cos\left(\beta_{ym}y\right) - \frac{1-\cos(\beta_{ym}b)}{\sin(\beta_{ym}b)-\beta_{ym}b}\left(\sin\left(\beta_{ym}y\right) - \beta_{ym}y\right), & m = 2, 4, 6, \cdots, \end{cases}
$$
$$(10.130\text{b})$$

where:

$$
\beta_{xn} = \begin{cases} \frac{2\pi}{\ell}, \frac{4\pi}{\ell}, \frac{6\pi}{\ell}, \cdots, & n = 1, 3, 5, \cdots, \\ \frac{2.86\pi}{\ell}, \frac{4.92\pi}{\ell}, \frac{6.94\pi}{\ell}, \cdots, & n = 2, 4, 6, \cdots, \end{cases}
$$
$$(10.131\text{a})$$

$$
\beta_{ym} = \begin{cases} \frac{2\pi}{b}, \frac{4\pi}{b}, \frac{6\pi}{b}, \cdots, & n = 1, 3, 5, \cdots, \\ \frac{2.86\pi}{b}, \frac{4.92\pi}{b}, \frac{6.94\pi}{b}, \cdots, & n = 2, 4, 6, \cdots. \end{cases}
$$
$$(10.131\text{b})$$

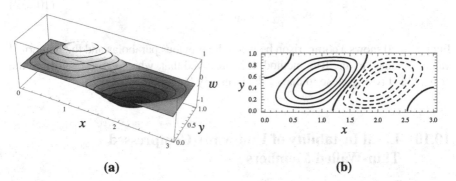

(a)   (b)

Fig. 10.19 Critical mode of a simply supported plate of aspect ratio $\alpha = \frac{\ell}{b} = 3$, subject to uniform shear ($N = 30$): (**a**) three-dimensional view, (**b**) contour lines

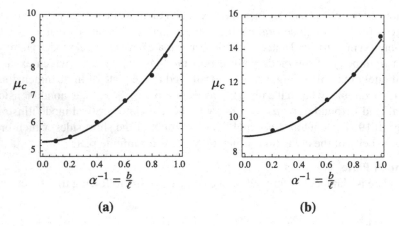

Fig. 10.20 Critical shear load of plates of finite dimensions *vs* the aspect ratio: (**a**) simply supported, (**b**) clamped; continuous curve: parabolic interpolation; dots: Ritz method, with $N = 30$

In this case, the TPE assumes a cumbersome expression, not reported here. Both the stiffness matrices, elastic and geometric, are full. Similar considerations hold for the destroyed symmetry of the buckling mode.

Approximate Interpolating Formulas

Repeating the Ritz analysis for several aspect ratio values $\alpha = \frac{\ell}{b}$, it is possible to plot the nondimensional critical load $\mu_c = \mu_c(\alpha)$ of simply supported and clamped plates [15]. To include very long plates in the graph, however, it is preferable to refer to the inverse ratio and to plot $\mu_c = \mu_c(\alpha^{-1})$, so that the infinite plate falls into the origin. It can be checked that the numerical results are well approximated by the parabolas:

$$\mu_c = \begin{cases} 5.34 + \frac{4}{\alpha^2}, & \text{simply supported,} \\ 8.98 + \frac{5.6}{\alpha^2}, & \text{clamped.} \end{cases} \tag{10.132}$$

Figure 10.20 shows a comparison between the previous parabolas and the numerical values obtained by the Ritz method. It is observed that, when $\alpha \simeq 5$ or greater, the critical load is very close to that of the infinitely long plate.

10.10 Local Instability of Uniformly Compressed Thin-Walled Members

Prismatic members of small thickness (TWM), with open or closed section, are assemblies of flat plates (ahead referred to as "walls"), whose major axis is parallel to the member axis.

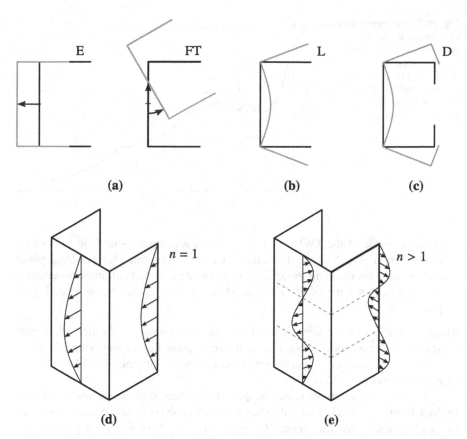

Fig. 10.21 Critical modes of a compressed TWM with open cross-section: (**a**) global Eulerian (E) and flexural-torsional (FT), (**b**) local (L), (**c**) distortional (D); wave-length of: (**d**) the Eulerian mode, (**e**) the local mode

When a TWM is longitudinally compressed, it can exhibit two types of instability: (a) *global* and (b) *local*, having the peculiar characteristics discussed below.

- The *global instability* manifests itself through bending, with or without torsion, of the TWM, *in absence of significant cross-section distortion* (Fig. 10.21a). The mode wave-length is of the same order of magnitude of the length of the beam (Fig. 10.21d). The single wall of the assembly are mainly extended and inflected in their own plane. The buckling mode and relative critical load are sufficiently well described by the beam theory (Euler-Bernoulli, if the cross-section is closed, or Vlasov, if it is open).
- The *local instability* manifests itself through a significant *loss of shape of the cross-section*, which almost leaves fixed the positions of the intermediate nodes (but not those of the end nodes, Fig. 10.21b). Since any wall, as seen in the study of the compressed plate, buckles in nearly square fields, the wave-length of the

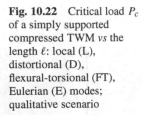

Fig. 10.22 Critical load P_c of a simply supported compressed TWM *vs* the length ℓ: local (L), distortional (D), flexural-torsional (FT), Eulerian (E) modes; qualitative scenario

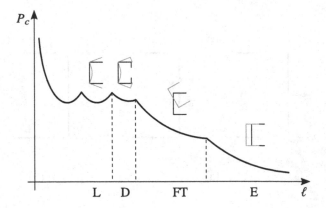

buckling mode of the TWM is of the order of a characteristic dimension of its cross-section (Fig. 10.21e). The individual wall, therefore, bends out-of-plane and twists. This buckling mode, and its associate critical load, must necessarily be described by a model of *prismatic shell*, i.e., constituted by an assembly of plates [9].

There exist also a hybrid form of buckling, called *distortional*, in which all, or part of, the cross-section nodes move, this implying in-plane flexure and extension of all, or some of, the walls, i.e., a mechanism which coexists with the plate-like behavior of the elements (Fig. 10.21c) [9].

Typically, local instability occurs in short or thin beams; global instability, in long or thick beams, or even in ribbed thin beams in which the stiffeners contrast the onset of local instability. As example (Fig. 10.22), if a family of simply supported open TWM is considered, in which only the length is increased, the following qualitative scenario is observed. For squat profiles, a plate behavior manifests, in which the local critical load weakly depends on the length and the critical mode consists of a number of longitudinal waves $n > 1$. Subsequently, when the TWM is long enough, and after a short transition phase in which a distortional mode is likely to manifest itself, instability occurs through a global mode, generally first of flexural-torsional and then of Eulerian type. In both global modes, the critical load strongly decreases with the length, while the number of the half-waves remains fixed at $n = 1$.

Remark 10.6 The local critical load, in addition to be practically independent of the length of the beam, is also independent of the boundary conditions at the bases of the TWM, which have a small influence on the deflection, except for a small neighborhood close to the constraints. The global critical load, on the other hand, is strongly dependent on the length as well as on the boundary conditions.

10.10.1   Finite Strip Method

To determine the critical load of a TWM, local or distortional, a prismatic shell model must be formulated. This, of course, also provides information on the global critical load, which, however, is more easily evaluated by a beam model.

With reference to a shell with axis x and directrix Γ spanned by the curvilinear abscissa s (Fig. 10.23a), the equilibrium for each wall is governed by (a) the equation of the prestressed plate, Eq. 10.38, in the unknown normal displacement $w(x, s)$, and (b) the membrane equations for the plate, in the unknown displacement $u(x, s)$, $v(x, s)$, respectively normal and tangent to the directrix. The boundary conditions along the interconnection lines of the plates prescribe continuity of displacements and rotations, as well as the equilibrium of the line. The boundary conditions at the longitudinal free edges (if any) are the usual plate conditions for bending moments and Kirchhoff shear forces, to which those involving the membrane stresses must be added. Although the two indefinite equations, describing the flexural and membrane behavior, are uncoupled, *the geometric conditions along the nodal lines couple the two types of behavior.*[32]

The differential problem, thus formulated, is extremely complex, and not soluble in practice. It is therefore necessary to resort to a discretization method, which transforms the problem from differential into algebraic. Among the existing ones, the *finite strip method* is specially oriented to the problem under consideration [6, 11]. Here, a hint of the method will be given, limiting the discussion to uniformly compressed profiles. It is also assumed that the TWM is simply supported at the ends, where it is free to warp, which is a nonrestrictive hypothesis with regard to the determination of the local critical load.

Displacement Field
The domain occupied by the shell is divided into rectangular fields (called *finite strips*), delimited by lines parallel to the x axis (called nodal lines) and by the end cross-sections (Fig. 10.23b). These straight lines include the interconnecting lines between the walls (also called "natural" nodal lines), but they segment more finely the walls themselves. In other words, one can think to fractionize the Γ directrix in adjacent segments, as it is customary in finite element discretization of a planar beam system, and to generate the finite strips by translating Γ in the x direction. Referring to the generic strip e, of dimensions $\ell \times b$ (Fig. 10.23c), the unknown displacement field is expressed as a sum of product of separable variable functions. By adopting harmonic functions in the longitudinal direction, which are compatible with the geometric constraints (i.e., $u \neq 0$ and $v = w = 0$ at $x = 0, \ell$), one has:

---

[32] The circumstance is similar to that occurring in a planar frame, where the continuity of the displacements at the node between two orthogonal beams prescribes that the normal displacement of a beam is equal to the tangent displacement of the other beam.

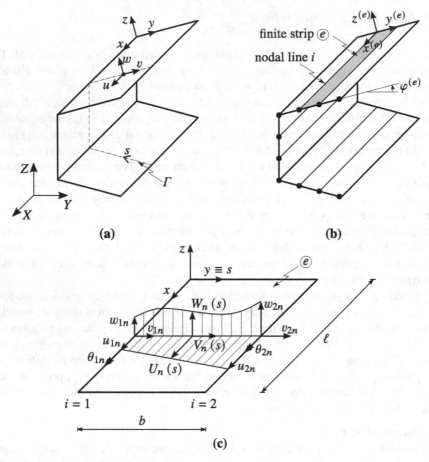

Fig. 10.23 Prismatic shell: (**a**) intrinsic displacement components $(u(x, s), v(x, s), w(x, s))$; (**b**) nodal lines i, finite strips e, local axes $\left(x^{(e)}, y^{(e)}, z^{(e)}\right)$ and orientation angle $\varphi^{(e)}$; (**c**) interpolation of the displacements on the directrix (superscript e suppressed)

$$u(x, s) = \sum_{n=1}^{N} U_n(s) \cos\left(\frac{n\pi x}{\ell}\right),$$ (10.133a)

$$v(x, s) = \sum_{n=1}^{N} V_n(s) \sin\left(\frac{n\pi x}{\ell}\right),$$ (10.133b)

$$w(x, s) = \sum_{n=1}^{N} W_n(s) \sin\left(\frac{n\pi x}{\ell}\right).$$ (10.133c)

The functions $U_n(s)$, $V_n(s)$, $W_n(s)$, which describe the field on the generatrix, must fulfill inter-elementary continuity, both in-plane and out-of-plane. For satisfying these conditions, it is convenient to adopt polynomial interpolation functions between the nodal values. By denoting by:

$$\mathbf{q}_n^{(e)} := (u_{1n}, v_{1n}, w_{1n}, \theta_{1n}, u_{2n}, v_{2n}, w_{2n}, \theta_{2n})^T \qquad (10.134)$$

the amplitudes of the translations and of twist of the nodal lines $i = 1, 2$ of the element (superscript e suppressed), one has:

$$U_n(s) := u_{1n}\psi_1(s) + u_{2n}\psi_2(s), \qquad (10.135a)$$

$$V_n(s) := v_{1n}\psi_1(s) + v_{2n}\psi_2(s), \qquad (10.135b)$$

$$W_n(s) := w_{1n}\psi_3(s) + \theta_{1n}\psi_4(s) + w_{2n}\psi_5(s) + \theta_{2n}\psi_6(s), \qquad (10.135c)$$

where:

$$\psi_1(s) := 1 - \frac{s}{b}, \qquad (10.136a)$$

$$\psi_2(s) := \frac{s}{b}, \qquad (10.136b)$$

$$\psi_3(s) := 1 - 3\left(\frac{s}{b}\right)^2 + 2\left(\frac{s}{b}\right)^3, \qquad (10.136c)$$

$$\psi_4(s) := \left[\frac{s}{b} - 2\left(\frac{s}{b}\right)^2 + \left(\frac{s}{b}\right)^3\right]b, \qquad (10.136d)$$

$$\psi_5(s) := 3\left(\frac{s}{b}\right)^2 - 2\left(\frac{s}{b}\right)^3, \qquad (10.136e)$$

$$\psi_6(s) := \left[-\left(\frac{s}{b}\right)^2 + \left(\frac{s}{b}\right)^3\right]b, \qquad (10.136f)$$

are the classic linear and cubic polynomials, usually adopted in finite element analysis of planar frames.

Element Stiffness Matrices

The elastic potential energy of the element, $\tilde{U}$, is sum of the flexural energy $\tilde{U}^f$ and of the membrane energy $\tilde{U}^m$,[33] given by:

---

[33] The latter is derived from the constitutive law in Note 7 and by the congruence equations $\varepsilon_x = u_{,x}$, $\varepsilon_s = v_{,s}$, $\gamma_{xs} = u_{,s} + v_{,x}$.

$$\tilde{U}^f = \frac{1}{2} \int_0^\ell \int_0^b D \left[w_{,xx}^2 + w_{,ss}^2 + 2\nu \, w_{,xx} w_{,ss} + 2 \left(1 - \nu\right) w_{,xs}^2 \right] dx \, ds,$$

(10.137a)

$$\tilde{U}^m = \frac{1}{2} \int_0^\ell \int_0^b C \left[u_{,x}^2 + v_{,s}^2 + 2\nu \, u_{,x} v_{,s} + \frac{1 - \nu}{2} \left(u_{,s} + v_{,x}\right)^2 \right] dx \, ds,$$

(10.137b)

where $C := \frac{Eh}{1-\nu^2}$ is the extensional stiffness of the plate.

A uniform axial compression stress $\sigma_x^0 = -\frac{P}{A}$ is considered, with A the TWM cross-section area. Hence, $N_x^0 = -\frac{Ph}{A}$ is the prestress in the plate, whose associate energy reads:[34]

$$U^0 = -\frac{1}{2} \frac{P h}{A} \int_0^\ell \int_0^b \left(u_{,x}^2 + v_{,x}^2 + w_{,x}^2 \right) dx \, ds.$$

(10.138)

Substituting the Eqs. 10.133, 10.135, and 10.136 in $\Pi = \tilde{U}^f + \tilde{U}^m + U^0$ and executing the integrations, the TPE of the element is obtained. Taking into account the orthogonality properties among harmonic functions, this reads:

$$\Pi = \frac{1}{2} \sum_{n=1}^N \mathbf{q}_n^{(e)T} \left(\mathbf{K}_{en}^{(e)} + \mu \mathbf{K}_{gn}^{(e)} \right) \mathbf{q}_n^{(e)},$$

(10.139)

where $\mathbf{K}_{en}^{(e)}$, $\mathbf{K}_{gn}^{(e)}$, of dimensions 8×8, are the *elastic and geometric matrices of the element*, respectively, *associated with the harmonic n* (i.e., relative to an inflection of the shell in n longitudinal half-waves); moreover, the load multiplier μ has been introduced. The explicit expressions of the stiffness matrices are reported in the Supplement 10.6. The favorable circumstance of uncoupling[35] allows one to refer to only one harmonic at a time and determining the critical load corresponding to it. The analysis must be repeated for several n, in order to minimize the critical value.

---

[34] In Eq. 10.138, the complete expression of the Green-Lagrange strain, including the in-plane displacements u, v, has been used. This differ from what done for the single plate, where the transverse displacements w was only considered. By dealing with shells, sometimes, $u_{,x}^2$ is neglected, with arguments similar to those used for cross-undeformable beams. Here, however, in view of a numerical solution, all terms have been retained in the analysis.

[35] Uncoupling fails in presence of shear prestresses, as already observed in Sect. 10.9 for the single plate.

Supplement 10.6 (Coefficients of the Element Stiffness Matrices) The element stiffness matrices, relative to the harmonic n, take the following form (superscript e suppressed on the coefficients):

$$
\mathbf{K}_{\alpha n}^{(e)} := \begin{bmatrix}
k_{\alpha 11} & k_{\alpha 12} & \cdot & \cdot & k_{\alpha 15} & k_{\alpha 16} & \cdot & \cdot \\
 & k_{\alpha 22} & \cdot & \cdot & k_{\alpha 25} & k_{\alpha 26} & \cdot & \cdot \\
 & & k_{\alpha 33} & k_{\alpha 34} & \cdot & \cdot & k_{\alpha 37} & k_{\alpha 38} \\
 & & & k_{\alpha 44} & \cdot & \cdot & k_{\alpha 47} & k_{\alpha 48} \\
 & & & & k_{\alpha 55} & k_{\alpha 56} & \cdot & \cdot \\
 & & & & & k_{\alpha 66} & \cdot & \cdot \\
 & & & & & & k_{\alpha 77} & k_{\alpha 76} \\
\text{SYM} & & & & & & & k_{\alpha 88}
\end{bmatrix}, \qquad \alpha = e, g,
$$

(10.140)

where:

$$
k_{e11} = k_{e55} := C \left[\frac{(n\pi)^2}{6} \frac{b}{\ell} + \frac{1-\nu}{4} \frac{\ell}{b} \right],
$$

$$
k_{e12} = -k_{e56} := C \frac{n\pi}{8} (3\nu - 1),
$$

$$
k_{e15} := C \left[\frac{(n\pi)^2}{12} \frac{b}{\ell} - \frac{1-\nu}{4} \frac{\ell}{b} \right],
$$

$$
k_{e16} = -k_{e25} := -C \frac{n\pi}{8} (1 + \nu),
$$

$$
k_{e22} = k_{e66} := C \left[\frac{(n\pi)^2}{12} (1 - \nu) \frac{b}{\ell} + \frac{1}{2} \frac{\ell}{b} \right],
$$

(10.141)

$$
k_{e26} := C \left[\frac{(n\pi)^2}{24} (1 - \nu) \frac{b}{\ell} - \frac{1}{2} \frac{\ell}{b} \right],
$$

$$
k_{e33} = k_{e77} := D \left[\frac{13}{70} (n\pi)^4 \frac{b}{\ell^3} + \frac{6}{5} (n\pi)^2 \frac{1}{b\ell} + 6 \frac{\ell}{b^3} \right],
$$

$$
k_{e34} = -k_{e78} := D \left[\frac{11}{420} (n\pi)^4 \frac{b^2}{\ell^3} + \frac{5\nu + 1}{10} (n\pi)^2 \frac{1}{\ell} + 3 \frac{\ell}{b^2} \right],
$$

$$
k_{e37} := D \left[\frac{9}{140} (n\pi)^4 \frac{b}{\ell^3} - \frac{6}{5} (n\pi)^2 \frac{1}{b\ell} - 6 \frac{\ell}{b^3} \right],
$$

$$
k_{e38} = -k_{e47} := D \left[-\frac{13}{840} (n\pi)^4 \frac{b^2}{\ell^3} + \frac{1}{10} (n\pi)^2 \frac{1}{\ell} + 3 \frac{\ell}{b^2} \right],
$$

$$k_{e44} = k_{e88} := D \left[\frac{1}{210} (n\pi)^4 \left(\frac{b}{\ell} \right)^3 + \frac{2}{15} (n\pi)^2 \frac{b}{\ell} + 2\frac{\ell}{b} \right],$$

$$k_{e48} := D \left[-\frac{1}{280} (n\pi)^4 \left(\frac{b}{\ell} \right)^3 - \frac{1}{30} (n\pi)^2 \frac{b}{\ell} + \frac{\ell}{b} \right],$$

and:

$$k_{g11} = k_{g22} = k_{g55} = k_{g66} := -\frac{Ph}{A} (n\pi)^2 \frac{1}{6} \frac{b}{\ell}, \qquad k_{g12} = k_{g56} := 0,$$

$$k_{g15} = k_{g26} := -\frac{Ph}{A} (n\pi)^2 \frac{1}{12} \frac{b}{\ell}, \qquad k_{g16} = k_{g25} := 0,$$

$$k_{g33} = k_{g77} := -\frac{Ph}{A} (n\pi)^2 \frac{13}{70} \frac{b}{\ell}, \qquad k_{g34} = -k_{g78} := -\frac{Ph}{A} (n\pi)^2 \frac{11}{420} \frac{b^2}{\ell},$$

$$k_{g37} := -\frac{Ph}{A} (n\pi)^2 \frac{9}{140} \frac{b}{\ell}, \qquad k_{g38} = -k_{g47} := \frac{Ph}{A} (n\pi)^2 \frac{13}{840} \frac{b^2}{\ell},$$

$$k_{g44} = k_{g88} := -\frac{Ph}{A} (n\pi)^2 \frac{1}{210} \frac{b^3}{\ell}, \qquad k_{g48} := \frac{Ph}{A} (n\pi)^2 \frac{1}{280} \frac{b^3}{\ell}.$$

$$(10.142)$$

$\square$

Eigenvalue Problem

The element stiffness matrices $\mathbf{K}_{\alpha n}^{(e)}$ ($\alpha = e, g$), similarly to what done for planar frames, must (i) be expressed in the global basis, oriented as the X, Y, and Z axes, by taking into account the orientation of the strips, and (ii) be assembled, taking into account the topology of the cross-section of the shell. Since the local basis $x^{(e)}$, $y^{(e)}$, $z^{(e)}$ is identified by the angle $\varphi^{(e)}$ that the $y^{(e)}$ axis forms with the Y axis (Fig. 10.23b), the matrices must be transformed as follows:

$$\tilde{\mathbf{K}}_{\alpha n}^{(e)} = \mathbf{T}^{(e)^T} \mathbf{K}_{\alpha n}^{(e)} \mathbf{T}^{(e)}, \qquad (10.143)$$

where:

$$\mathbf{T}^{(e)} := \begin{bmatrix} \bar{\mathbf{T}}^{(e)} & \mathbf{0} \\ \mathbf{0} & \bar{\mathbf{T}}^{(e)} \end{bmatrix}, \qquad \bar{\mathbf{T}}^{(e)} := \begin{bmatrix} 1 & \cdot & \cdot & \cdot \\ \cdot & \cos \varphi^{(e)} & \sin \varphi^{(e)} & \cdot \\ \cdot & -\sin \varphi^{(e)} & \cos \varphi^{(e)} & \cdot \\ \cdot & \cdot & \cdot & 1 \end{bmatrix}. \qquad (10.144)$$

The matrices thus obtained must be assembled, and any (uniform) constraint imposed at the nodal lines.[36] Thus, the eigenvalue problem follows:

$$\left(\mathbf{K}_{en} + \mu \mathbf{K}_{gn} \right) \mathbf{q}_n = \mathbf{0}, \qquad n = 1, 2, \cdots, N, \qquad (10.145)$$

---

[36] It should be observed, that the constraints at the bases of the cylindrical shell have already been satisfied by the choice of the harmonic functions. Any other constraints concern the nodal lines. A frequent example is offered by the symmetry/antisymmetry conditions of the directrix, which allow to model only one of the two halves of Γ.

where $\mathbf{q}_n$ is a vector listing the nodal displacements in the global reference (four per node, if all free) and $\mathbf{K}_{en}, \mathbf{K}_{gn}$ the global matrices, all referenced to the harmonic n. The smallest eigenvalue of the problem, $\mu_c(n)$, represents the critical load associated with the harmonic n. The critical load is $\mu_c = \min_n ([\mu_c(n)])$, to be determined by trial and error.

10.10.2 Finite Element Sectional Model

When the interest is confined to the *local instability* of a TWM, the finite strip method, illustrated in the previous section, can be further simplified. As it was observed, indeed, the displacements, in this form of instability, are exclusively normal to the surface of the cylinder, i.e., no membrane elastic energy is stored in the plate but only flexural. By assuming that the deflections is sinusoidal in the longitudinal direction, one has:

$$w(x, s) = W(s) \sin\left(\frac{n\pi x}{\ell}\right), \tag{10.146}$$

where s is a curvilinear abscissa spanning the Γ line. The unknown function $W(s)$ must satisfy the indefinite equilibrium equations of the plate, Eq. 10.40, which, after the separation of variables, becomes:

$$W'''' - 2\left(\frac{n\pi}{\ell}\right)^2 W'' + \left(\frac{n\pi}{\ell}\right)^2 \left[\left(\frac{n\pi}{\ell}\right)^2 - \frac{p_x}{D}\right] W = 0. \tag{10.147}$$

This is *identical to the equation of the transverse elastic line*, Eq. 10.54, which was written for the single plate simply supported on the two opposite sides. As the only difference, Eq. 10.147 must be integrated on the polygonal line Γ, and not on a single segment of the y axis.[37] The geometric boundary conditions require that $W = 0$ at the internal nodes of the line (while it is generally $W \neq 0$ at the end nodes, if the section is open). Figure 10.24 exemplifies two cases.

Remark 10.7 According with the interpretation of Eq. 10.147, as analogous to that of a prestressed beam on Winkler soil, it can be affirmed, that the prismatic shell reduces to a frame formed by these beams, whose domain is the directrix Γ. Hence, the denomination of *sectional model*. The internal nodes of the frame are hinged at the ground, given the hypothesis of immobility of the edges of the shell, so that the constituent beams interact exclusively through the bending moments they exchange at the nodes.

---

[37] Hence, the notation $W(s)$ instead of $Y(y)$.

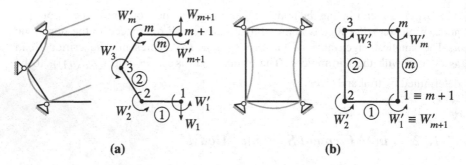

Fig. 10.24 Sectional model of a TWM with fixed internal nodes: (**a**) open, (**b**) closed cross-section

Exact One-Dimensional Finite Element Model

Equation 10.147 can be integrated using the exact finite element already introduced in the Sect. 10.6.2. For a generic open cross-section, here assumed for simplicity without branches[38] (Fig. 10.24a), consisting of m elements and $m + 1$ nodes, the global vector of configuration variables $\mathbf{q}$ is made up of $m + 3$ displacement components, the $m + 1$ rotations of all the nodes and the two transverse displacements of the end nodes:

$$\mathbf{q} = \left(W_1, W_1'; W_2', \cdots, W_m'; W_{m+1}, W_{m+1}' \right)^T . \tag{10.148}$$

For a generic closed cross-section (Fig. 10.24b), consisting of m elements and m nodes, the vector of the global configuration variables consists of only m rotations of the nodes:

$$\mathbf{q} = \left(W_1', W_2', \cdots, W_{m-1}', W_m' \right)^T . \tag{10.149}$$

The element local stiffness matrix, of dimension 4×4, is defined by Eqs. 10.74. In this matrix, it is necessary to delete rows and columns associated with the translations of the nodes, so that these reduce their dimensions to 2×2. The terminal elements $1, m + 1$ of open cross-sections, however, are exceptions, since their dimensions are 3×3.

The global matrix $\mathbf{K}$ of the structure is obtained according to the assembly schemes illustrated in Fig. 10.25. The critical load relative to the harmonic n is computed by requiring that $\det \mathbf{K} = 0$. This value must be minimized by repeating the analysis for several n.

Remark 10.8 The local critical load determined with the sectional model is generally very close to that offered by the finite strip model. The latter is more accurate, because it takes into account the displacements, although small, of the nodal lines,

---

[38] The presence of branches does not imply any substantial difficulty.

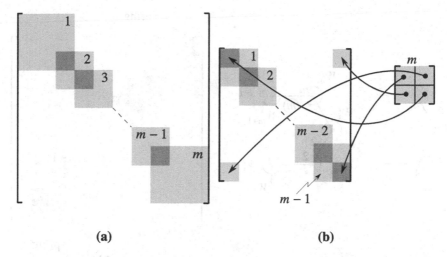

(a) **(b)**

Fig. 10.25 Assembly of stiffness matrices of the sectional model: (**a**) open, (**b**) closed cross-section

generally occurring in a local mode.[39] On the other hand, while the sectional model is exact, the finite strip model makes use of cubic interpolation functions, which therefore require a finer discretization than the "natural" one, used here.

An Artifice for Determining Distortional Critical Loads

The sectional model is valid only if the internal nodes do not move, to avoid inflections of the plates in their own plane. When the mode is distortional, therefore, it cannot be used. However, there are cases in which only a few plates, of small width, bend in their own plane, while all the other plates buckles exclusively out-of-plane.

This is the case, for example, of cross-sections with "lips" of extremity (Fig. 10.26a), for which, with a certain approximation, the sectional model can still be used, assimilating the lips to *stiffening ribs* (and therefore to beams), as already done in the Sect. 10.8. After having established an energy equivalence between the sinusoidal bending of the stiffening beam (ignoring the torsion, given the small thickness) and the elongation of a spring, it is possible to determine the stiffness k of the latter, to be introduced in the sectional model (Fig. 10.26a).[40] Equation 10.94a returns k as proportional to the bending stiffness EI of the lip.[41]

---

[39] Indeed, an out-of-plane shear force at the edge of a plate becomes a membrane force for the orthogonal plate, implying displacement of the common nodal line.

[40] The geometric stiffness of the compressed stiffeners, here ignored, could be also accounted for.

[41] The eccentricity of the rib is ignored. Moreover, it should be necessary to take into account a "collaborating width" of the wall, which contributes to the bending of the rib, and therefore to its inertia moment I. For such *effective width*, there exist approximate formulas, mostly developed in technical standards.

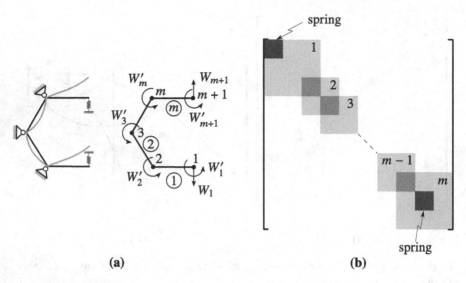

Fig. 10.26 Sectional model of a lipped cross-section: (a) equivalent springs, (b) assembly (1×1 spring matrix shaded dark)

The equivalent springs contribute to the global stiffness matrix of the sectional model (Fig. 10.26b). This is achieved from the matrix of the unstiffened model (Fig. 10.25a), by adding the stiffness k to the coefficients in the entries $(1, 1)$ and $(m + 2, m + 2)$, consistently with the order adopted in Eq. 10.148 for the configuration variables.

10.10.3 Illustrative Examples of Local and Distortional Buckling

The mechanical behavior of uniformly compressed TWM is discussed, with regard to the phenomenon of local (or distortional) instability. To this end, some numerical results of literature [12], obtained through the finite strip method (Sect. 10.10.1), are illustrated. They are partially reproduced here, with excellent approximation, with the sectional model described in the Sect. 10.10.2.

Reference is made to the profiles shown in Fig. 10.27, of uniform thickness, whose geometry is defined by the following nondimensional parameters:

$$\alpha := \frac{\ell}{a}, \quad \beta =: \frac{b}{a}, \quad \gamma := \frac{h}{a}, \quad \delta := \frac{d}{a}, \tag{10.150}$$

where α is a measure of slenderness (ratio between the length ℓ and the height a), β is the aspect ratio of the cross-section (ratio of the width b to height a), γ is a

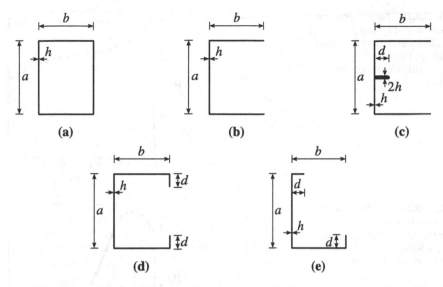

Fig. 10.27 Thin-walled members: (**a**) rectangular tube, (**b**) channel, (**c**) channel with stiffened web, (**d**) lipped channel, (**e**) lipped L-profile

measure of thinness (ratio between the thickness h and the height), and, finally, δ is a measure of the stiffening (ratio of the height d of the stiffener and that of the profile). Moreover, said p_x the compression load per unit of length of the directrix, a nondimensional load is defined as:

$$\mu := \frac{p_x a^2}{\pi^2 D}, \tag{10.151}$$

in analogy with what has been done for the single plate. In all the numerical simulation, the thickness is fixed, but its influence is qualitatively discussed; furthermore, the Poisson's ratio is taken as $\nu = 0.3$. The influence of the geometric parameters on the nondimensional critical load $\mu_c = \frac{p_{xc} a^2}{\pi^2 D}$ is investigated.

Rectangular Tube

The critical load μ_c of the rectangular tube *vs* its aspect ratio β is plotted in Fig. 10.28.

It has been obtained with either the finite strip method or the sectional model, drawing almost coincident results. The values of μ_c, as it occurs for the single plate, *are independent of γ*, as the thickness effect is already included in the bending stiffness D, which appears in Eq. 10.151. It is observed that the critical load is always larger than that relevant to the single plate hinged on all four sides, for which $\mu_c = 4$, and it is equal to this value only for $\beta = 1$. Indeed, it happens that the shorter (horizontal) walls exert a stiffening action on the longer (vertical) walls, by the way of the bending moment at the nodes, thus increasing the critical

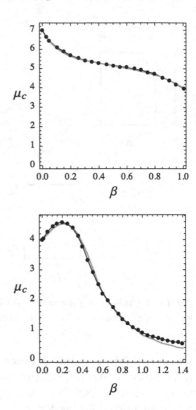

Fig. 10.28 Nondimensional critical load μ_c of a rectangular tube *vs* the aspect ratio β, for any γ; solid line: finite strips, dots: sectional model

Fig. 10.29 Nondimensional critical load μ_c of a channel *vs* the aspect ratio β, for any γ; solid line: finite strips, dots: sectional model

load. When, however, the tube is square, the four walls buckle in the same way, so that the bending moment at the nodes vanishes, together with the stiffening effect.

In the Problem 14.4.4, a rectangular tube with wings is studied by the Finite Strip method.

Channel

By carrying out a similar analysis on the channel (C-profile), the results shown in Fig. 10.29 are determined. Here, results from the two methods are slightly different, but, however, in good agreement. As in the previous case, the critical load is independent of the thickness γ.

Regarding the mechanical behavior, it is observed that (i) when $\beta = 0$, it is $\mu_c = 4$, as for the supported plate on four sides;[42] (ii) when $0 < \beta < 0.37$, it is $\mu_c > 4$; and (iii) when $\beta > 0.37$, it is $\mu_c < 4$. The phenomenon is usually interpreted by saying that, when the flanges (or wings, i.e., the horizontal walls) are sufficiently short, these exert a stiffening effect on the web (the vertical wall), by increasing its critical load. When, on the other hand, the flanges are long, they have a weakening effect on the assembly, as they "trigger" the phenomenon of instability.

---

[42] This is an extreme case, physically impossible, unless there are constraints along the edges.

Fig. 10.30 Ratio between the critical load μ_c of a web-stiffened channel and the critical load μ_{c0} of the unstiffened profile, vs the length δ of the stiffener and for different values of the aspect ratio β; $\gamma = 0.01$

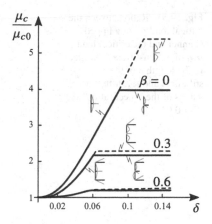

In other words (as well in use in technical language), the critical mode is "of web" or "of flanges" type, for short and long flanges, respectively. Although these claims are improper, since the critical mode is the eigenvector of a singular stiffness matrix, and therefore involves *all* the constituent elements, the explanation is suggestive and facilitates the interpretation of the results.

Channel with Stiffened Web

Now, the effect of a rib, located on the longitudinal axis of the web, is investigated. It has a double thickness, to simulate a cold fold of a sheet of uniform thickness.[43] Figure 10.30 reports the results obtained (with the finite strip method) as the length δ of the stiffener varies, for several values of the aspect ratio β and for a fixed γ. In ordinate, the nondimensional critical load μ_c is divided by μ_{c0}, i.e., the critical value of the unstiffened cross-section (which, in turn, depends on β according to the graph in Fig. 10.29). The ratio, therefore, expresses the "gain" offered by the stiffener.

The limiting case $\beta = 0$ refers to the web, hinged on four sides. It can be seen that, as the length of the stiffener increases, the critical load increases too, up to four times. Reached this value, it remains constant, since the critical mode, from initially symmetric, becomes antisymmetric of half-wave-length (similarly to what observed in Fig. 10.14). The horizontal dashed line refers to the symmetric mode of the web, which behaves as it were clamped along its own longitudinal axis, and simply supported on the opposite side. This value of the critical load cannot be reached in a natural way, and it is well separated from the antisymmetric critical load.

The cases $\beta = 0.3$, 0.6 show a similar qualitative behavior but with a gain that gradually decreases with the length of the flanges. Moreover, the antisymmetric and symmetric modes that occur at the horizontal straight lines are associated with closer critical loads (i.e., they are quasi-simultaneous).

---

[43] The error on the (small) torsional stiffness of the rib is negligible.

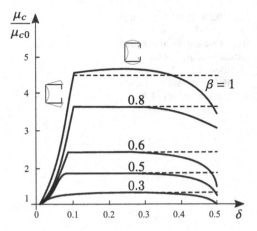

Fig. 10.31 Ratio between the critical load μ_c of a lipped channel and the critical load μ_{c0} of the unstiffened profile, vs the length δ of the stiffeners and for different values of the aspect ratio β; $\gamma = 0.01$

The mechanical behavior can be explained by arguments analogous to those developed for the unstiffened profile. When the flanges are short, so that the phenomenon is guided by the web, a stiffener of suitable size can significantly magnify the local critical load. However, when the flanges are long, being the phenomenon driven by these latter, the stiffener produces a small benefit, even negligible. In these cases, the shape of the mode of the web, antisymmetric or symmetric, slightly changes the critical load of the assembly, since it refers to the least important element.

If the thickness is changed, the ascending branches of the curves change, as the associate modes involve in-plane bending of the stiffener; the horizontal branches, on the other hand, are independent of the thickness, since related to out-of-plane deflection of all the walls.

Lipped Channel

A different way to stiffen the channel is to provide it with lips. It is of interest, then, to analyze how the length of the lips influences the local/distortional critical load, as the aspect ratio of profile changes. The results of the analysis, conducted with the finite strip method, are shown in Fig. 10.31, where the ratio μ_c/μ_{c0} between the critical loads of the stiffened and unstiffened cross-section (this latter plotted in Fig. 10.29) is shown.

It is observed that, for all the aspect ratios considered, two different forms of instability manifest themselves: (i) *distortional*, for short lips, and (ii) *local*, for long lips. It is necessary, therefore, that the length of the lips is large enough for the flange-lip nodes remain fixed. In the distortional phase, the critical load increases with the stiffness of the lips, which contribute to the elastic energy through bending in their own planes. Exceeded, however, a critical length value, which depends on the aspect ratio, the lips deflect only out-of-plane, like the other walls of the assembly. The longer the flanges, the longer must be the lips, in order the displacement of the common node is prevented. If, however, the lips are "excessively" long (e.g., $\delta > 0.3$), they have a negative influence on the

Fig. 10.32 Ratio, for different stiffened channels, between the maximum critical load $\mu_{c\infty}$, obtainable with an appropriate reinforcement δ, and the critical load μ_{c0} of the unstiffened profile, *vs* the aspect ratio β

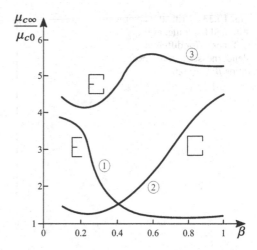

overall behavior, as they reduce the critical load (as for the channel of Fig. 10.29), representing the weak element of the assembly.

In the above analysis, the thickness γ is kept fixed. If it is varied, only the ascending branches, in which in-plane bending of the stiffeners occurs, are modified.

Comparative Analysis of the Effectiveness of the Stiffener Position

Referring to the channel, one wonders if it is more convenient to stiffen the web or the flanges. From the analysis of the previous results, it is evident that for small β (where the instability is triggered by the web), it is appropriate to stiffen the web, while for large β (where the instability is triggered by the flanges) it is more effective to stiffen the flanges. It is also of interest to examine the case in which both web and flanges are simultaneously reinforced, for which a good behavior is expected in all cases.

Figure 10.32 compares the critical load, for the three different arrangements of the reinforcements, *vs* the aspect ratio β. The figure shows the critical load $\mu_{c\infty} := \max_{\delta} (\mu_c)$, which corresponds, for an assigned β, to the *maximum critical load achievable with an appropriate δ*, divided the value μ_{c0} of the unstiffened cross-section.[44] The figure therefore shows the maximum gain that can be obtained for suitable lengths of the stiffeners.

The analysis quantifies what has already been observed on a qualitative basis. Stiffening the web is beneficial only for short flanges (e.g., $\beta < 0.25$); it magnifies the critical load by a factor of $2.5 \div 4$. Stiffening the flanges by lips is beneficial only for long flanges (e.g., $\beta > 0.7$); it magnifies the critical load by a factor of $2.5 \div 4.5$. For a large range of intermediate values of β, for example, $0.25 < \beta < 0.7$, both stiffeners are ineffective. In this interval, instead, the combination of the two

---

[44] The values of $\mu_{c\infty}/\mu_{c0}$ are taken from Figs. 10.30 and 10.31, as the ordinates of the horizontal lines, for each curve $\beta = \mathrm{cost}$.

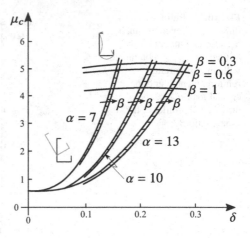

Fig. 10.33 Critical load of a stiffened L-profile, *vs* the stiffener δ, for different slendernesses α and aspect ratios β; $\gamma = 0.01$

stiffeners leads to a maximum gain, of the order of $4.5 \div 5.5$. The doubly stiffened profile, moreover, has a very good behavior in the whole $0 < \beta < 1$ field, at extremes of which it approaches, from above, the behavior of the cross-section individually stiffened.

Lipped L-Profile

Not all the TWM suffer from local instability: a clear example is represented by the lipped L-profile. For this beam, Fig. 10.33 returns the value of μ_c as a function of δ, for several values of α and β and for γ fixed. The solution is obtained with the finite strip method.

From the analysis of the results, it is observed that, as δ increases, the critical load initially increases with a monotonous law. The corresponding critical mode, however, has only one half-wave in the longitudinal direction, as in the case of a plate free on a longitudinal side and simply supported on the remaining three sides (case E in Fig. 10.9). The instability pattern, therefore, does not have a character of local mode (as the nodes move) nor distortional (as the loss of shape of the cross-section is weak); in contrast, it has a character of global mode, of flexural-flexural-torsional type. The critical load, indeed, can be determined with an excellent approximation (of the order of a few percentage units) by the Vlasov model. It strongly depends on the slenderness α but very little on the aspect ratio β (varied in the figure in the interval $(0.3, 1)$). If, however, the stiffener is long enough, the classic mode with several half-waves manifests, in which the nodes remain fixed. As already observed for the other profiles, the local critical load is practically independent of α, little dependent on δ, and strongly dependent on β.

References

1. Allen, H.G., Bulson, P.S.: Background to Buckling. McGraw-Hill, New York (1980)
2. Bazant, Z.P., Cedolin, L.: Stability of Structures: Elastic, Inelastic, Fracture, and Damage Theories. World Scientific Publishing, Singapore (2010)
3. Bleich, F.: Buckling Strength of Metal Structures. McGraw-Hill, New York (1952)
4. Brush, D.O., Almroth, B.O.: Buckling of Bars, Plates, and Shells. McGraw-Hill, New York (1975)
5. Chajes, A.: Principles of Structural Stability Theory. Prentice-Hall, New Jersey (1974)
6. Cheung, Y.K.: Finite strip method in structural analysis. Pergamon Press, Oxford (1976)
7. Como, M.: Theory of Elastic Stability (in Italian). Liguori, Napoli (1967)
8. Column Research Committee of Japan: Handbook of Structural Stability. Corona Publishing Company, Tokyo (1971)
9. Hancock, G.J.: Coupled instabilities in metal structures (CIMS) – What have we learned and where are we going? Thin Wall, Struct. **128**, 2–11 (2018)
10. Iyengar, N.G.R.: Elastic Stability of Structural Elements. Macmillan India, New Delhi (2007)
11. Pignataro, M., Luongo, A.: A finite strip static and dynamic analysis of thin-walled members. In: Pubblicazione II, Università di Roma **337** (1984)
12. Pignataro, M., Luongo, A., Rizzi, N.: Problemi di instabilità delle strutture (in Italian). In. Corso di aggiornamento AIMETA, Roma (1988)
13. Pignataro, M., Rizzi, N., Luongo, A.: Stability, Bifurcation and Postcritical Behaviour of Elastic Structures. Elsevier, Amsterdam (1990)
14. Southwell, R.V., Skan, S.W.: On the stability under shearing forces of a flat elastic strip. P. R. Soc. Lond. A-Conta. **105**(733), 582–607 (1924)
15. Timoshenko, S.P., Gere, J.M.: Theory of Elastic Stability. McGraw-Hill, New York (1961)
16. van Der Heijden, A.M.A.: W.T. Koiter's Elastic Stability of Solids and Structures. Cambridge University Press, Cambridge (2009)
17. Yoo, C.H., Lee, S.: Stability of Structures: Principles and Applications. Elsevier, Amsterdam (2011)

Chapter 11
Dynamic Bifurcations Induced by Follower Forces

11.1 Introduction

Follower forces are nonconservative forces of positional type, that is, they depend on the configuration assumed by the body to which they are applied; hence, they do not descend from a potential [11]. Examples of follower forces are: (a) the frictional forces, which are parallel and opposite in verse to the local velocity (e.g., the forces on the windshield wiper of a car), and (b) the reaction forces, caused by the emission of a fluid from a tank (such as those supporting the motion of a rocket).

When a compression follower force acts on a beam, or on a beam-like discrete system, the geometric stiffness matrix is non-symmetric, since it is not the Hessian of an energy. Hence, as discussed in Chap. 3, the system is circulatory and prone to a dynamic bifurcation (*flutter*, or circulatory Hopf bifurcation). If damping is added, this changes into a generic Hopf bifurcation, which generally manifests itself at a lower value of the circulatory critical load (except for special values of the parameters). This phenomena is called the *destabilizing effect of damping* (or *Ziegler paradox*). After the critical load has been exceeded, the system may execute periodic oscillations on a *limit cycle*, whose amplitude depends on the difference between the current and the critical value of the load.

In this chapter, the mentioned behavior is illustrated for: (a) a two degrees of freedom system (the *Ziegler column*, or *double pendulum*), and (b) a continuous system (the *Beck beam*). Exact, asymptotic and numerical solutions are provided, for both linear and nonlinear systems, and the role of different forms of damping on the mechanical behavior is discussed.

11.2 Nonconservative Nature of the Follower Forces

To verify the nonconservative character of the follower forces, it suffices to refer to the *Beck beam*, represented in Fig. 11.1a [1]. This is a fixed-free beam, loaded at its free end B by a follower force of intensity F, which, in each configuration, maintains the direction of the tangent to the centerline of the beam.[1] To prove that this force is nonconservative, it is necessary to show that the work W expended by the force, in passing from the reference to the current configuration, does depend on the path. Denoting the transverse displacement of the free end of the beam by v_B and its rotation by φ_B (both assumed to be infinitesimal), it is possible, among the infinite paths, to choose the following two: (I) the translation is assigned first, then the rotation follows (v_B, φ_B) (Fig. 11.1b), and (II) the rotation is assigned first, then the translation follows (φ_B, v_B) (Fig. 11.1c). Since the force does not expend work in the rotation (which leaves fixed the point of application), work is done in translation only. Along path I, since the force remains orthogonal to the displacement, it is $W_I = 0$; along path II, instead, the transverse component of the

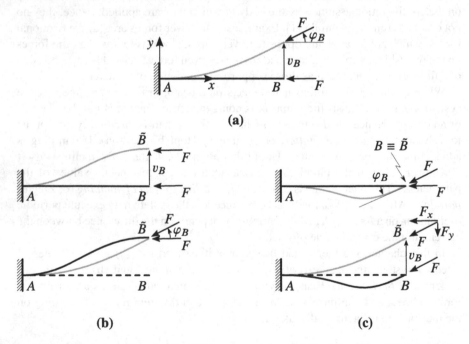

Fig. 11.1 Beck beam and work of the follower force: (**a**) model; (**b**) deformation path I, (v_B, φ_B); (**c**) deformation path II, (φ_B, v_B)

---

[1] For this reason, it is also called *tangential* force.

force, F_y, expends non-zero work, so that $W_{II} = -|F_y| v_B$. Since $W_I \neq W_{II}$, the follower force is nonconservative.

11.3 Ziegler Column

A paradigmatic two degrees of freedom system is introduced, known as the *Ziegler column* (Fig. 11.2) [12].[2] The system (namely, a double pendulum) consists of two massless and rigid rods, AB and BC, each of length ℓ, constrained mutually and to the ground by hinges. There are two concentrated masses on the pendulum: one, of intensity m_1, is located at the internal hinge B and the other, of intensity m_2, at the free end C. The two rods are connected, to each other and to the ground, by linear viscoelastic devices, of stiffness k_1 and k_2 and viscosity coefficients c_1 and c_2, respectively applied at A and B. The double pendulum is loaded at the free end by a compression follower force of intensity F, which maintains its direction parallel to the rod BC.

11.3.1 Linearized Equations of Motion

To build up the equations of motion, it needs to analyze the kinematics and the constitutive law, first. The equations of motion can successively be derived according to two alternatives: (a) the Extended Hamilton Principle, which reads (Eq. D.1 in the Appendix D):

Fig. 11.2 Ziegler column

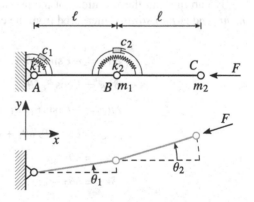

---

[2] A generalization of the Ziegler column, consisting of a triple pendulum, is dealt with in the Problem 14.5.1.

$$\delta H := \int_{t_1}^{t_2} (\delta T - \delta W_{int} + \delta W_{ext})\, dt = 0, \qquad \forall \delta \mathbf{q} \mid \delta \mathbf{q}\,(t_1) = \delta \mathbf{q}\,(t_2) = \mathbf{0},$$

$$(11.1)$$

where T is the kinetic energy, $\delta W_{int}, \delta W_{ext}$ is the virtual work of internal/external forces, and $\mathbf{q}$ are the Lagrangian parameters or (b) the balance of the forces in the current configuration. Both the procedures are illustrated ahead.

Kinematics

The system configuration is described by rotations of the two rods, $\theta_1\,(t)$ and $\theta_2\,(t)$, taken as Lagrangian parameters. From geometric considerations, the following *exact relations* are derived, which express the components of the longitudinal u and transverse v displacements of the points B, C:

$$u_B = -\ell\,(1 - \cos\theta_1)\,, \tag{11.2a}$$

$$v_B = \ell \sin\theta_1, \tag{11.2b}$$

$$u_C = -\ell\,(2 - \cos\theta_1 - \cos\theta_2)\,, \tag{11.2c}$$

$$v_C = \ell\,(\sin\theta_1 + \sin\theta_2)\,. \tag{11.2d}$$

The "curvatures" of the viscoelastic devices, expressing the "jump" of the rotations, are defined as follows:

$$\kappa_1 := \theta_1, \tag{11.3a}$$

$$\kappa_2 := \theta_2 - \theta_1. \tag{11.3b}$$

By carrying out the variations of the previous expressions, the *virtual displacements* and *virtual strains* (measured from the current configuration) are determined:

$$\delta u_B = -\ell \sin\theta_1\, \delta\theta_1, \tag{11.4a}$$

$$\delta v_B = \ell \cos\theta_1\, \delta\theta_1, \tag{11.4b}$$

$$\delta u_C = -\ell\,(\sin\theta_1\, \delta\theta_1 + \sin\theta_2\, \delta\theta_2)\,, \tag{11.4c}$$

$$\delta v_C = \ell\,(\cos\theta_1\, \delta\theta_1 + \cos\theta_2\, \delta\theta_2)\,, \tag{11.4d}$$

$$\delta\kappa_1 = \delta\theta_1, \tag{11.4e}$$

$$\delta\kappa_2 = \delta\theta_2 - \delta\theta_1. \tag{11.4f}$$

Limiting the analysis, for the moment, to *small motions* in the neighborhood of the rectilinear configuration,[3] one can expand in series Eq. 11.2 for small

---

[3] The nonlinear analysis will be developed later, in the Sect. 11.4.

displacements. By retaining up to the quadratic terms in the Lagrangian parameters, one finds:

$$u_B = -\frac{\ell\theta_1^2}{2}, \tag{11.5a}$$

$$v_B = \ell\,\theta_1, \tag{11.5b}$$

$$u_C = -\ell\left(\frac{\theta_1^2}{2} + \frac{\theta_2^2}{2}\right), \tag{11.5c}$$

$$v_C = \ell\,(\theta_1 + \theta_2). \tag{11.5d}$$

It should be noticed that while the transverse displacements are quantities of the first order in θ_i, the longitudinal displacements are of the second order. By performing an analogous series expansion of Eq. 11.4,[4] one gets:

$$\delta u_B = -\ell\,\theta_1\,\delta\theta_1, \tag{11.6a}$$

$$\delta v_B = \ell\,\delta\theta_1, \tag{11.6b}$$

$$\delta u_C = -\ell\,(\theta_1\,\delta\theta_1 + \theta_2\,\delta\theta_2), \tag{11.6c}$$

$$\delta v_C = \ell\,(\delta\theta_1 + \delta\theta_2), \tag{11.6d}$$

$$\delta\kappa_1 = \delta\theta_1, \tag{11.6e}$$

$$\delta\kappa_2 = \delta\theta_2 - \delta\theta_1. \tag{11.6f}$$

Constitutive Law

The constitutive law for the viscoelastic devices, according to the in-parallel functioning, is obtained by adding up elastic and viscous forces:

$$M_j = k_j\kappa_j + c_j\dot{\kappa}_j, \quad j = 1, 2, \tag{11.7}$$

where M_j ($j = 1, 2$) are the internal couples and, k_j, c_j ($j = 1, 2$) are constants. In terms of displacement and speed, they read:

$$M_1 = k_1\theta_1 + c_1\dot{\theta}_1, \tag{11.8a}$$

$$M_2 = k_2\,(\theta_2 - \theta_1) + c_2\,(\dot{\theta}_2 - \dot{\theta}_1). \tag{11.8b}$$

---

[4] Or, equivalently by performing the variation of Eqs. 11.5 and 11.3.

Variational Formulation

The extended Hamilton principle, Eq. 11.1, is applied, whose terms are calculated separately.

The kinetic energy of the system is:

$$T = \frac{1}{2}m_1\left(\dot{u}_B^2 + \dot{v}_B^2\right) + \frac{1}{2}m_2\left(\dot{u}_C^2 + \dot{v}_C^2\right) = \frac{\ell^2}{2}\left[m_1\dot{\theta}_1^2 + m_2\left(\dot{\theta}_1 + \dot{\theta}_2\right)^2\right],$$

(11.9)

in which the contributions of order higher than two have been ignored and Eqs. 11.5 have been used. By performing the variation and then integrating by parts, one has:

$$\int_{t_1}^{t_2} \delta T\,dt = \ell^2 \int_{t_1}^{t_2} \left[m_1\dot{\theta}_1\delta\dot{\theta}_1 + m_2\left(\dot{\theta}_1 + \dot{\theta}_2\right)\left(\delta\dot{\theta}_1 + \delta\dot{\theta}_2\right)\right]dt$$

$$= -\ell^2 \int_{t_1}^{t_2} \left[m_1\ddot{\theta}_1\delta\theta_1 + m_2\left(\ddot{\theta}_1 + \ddot{\theta}_2\right)\left(\delta\theta_1 + \delta\theta_2\right)\right]dt \qquad (11.10)$$

$$+ \left[m_1\dot{\theta}_1\delta\theta_1 + m_2\left(\dot{\theta}_1 + \dot{\theta}_2\right)\left(\delta\theta_1 + \delta\theta_2\right)\right]_{t_1}^{t_2},$$

where it has been taken into account that $\delta\theta_i = 0$ at the ends of the time interval.

The virtual work of the internal forces, taking into account Eqs. 11.6e,f, reads:

$$\delta W_{int} = M_1\delta\kappa_1 + M_2\delta\kappa_2 = M_1\delta\theta_1 + M_2\left(\delta\theta_2 - \delta\theta_1\right). \qquad (11.11)$$

The virtual work of the external forces, using Eqs. 11.6c,d, as well as $F_x = -F\cos\theta_2 \simeq -F$ and $F_y = -F\sin\theta_2 \simeq -F\theta_2$, is:

$$\delta W_{ext} = F_x\delta u_C + F_y\delta v_C = F\ell\left[\left(\theta_1\,\delta\theta_1 + \theta_2\,\delta\theta_2\right) - \theta_2\left(\delta\theta_1 + \delta\theta_2\right)\right]. \qquad (11.12)$$

Substituting the three contributions in the variational principle, Eq. 11.1, one has:

$$\delta H = -\int_{t_1}^{t_2} \left\{\left[(m_1 + m_2)\,\ell^2\ddot{\theta}_1 + m_2\ell^2\ddot{\theta}_2\right]\delta\theta_1 + m_2\ell^2\left(\ddot{\theta}_1 + \ddot{\theta}_2\right)\delta\theta_2\right\}dt$$

$$- \int_{t_1}^{t_2} \left[(M_1 - M_2 + F\ell\theta_2 - F\ell\theta_1)\,\delta\theta_1 + M_2\delta\theta_2\right]dt = 0, \qquad \forall\delta\theta_1, \delta\theta_2.$$

(11.13)

From this expression, the balance equations follow:

$$(m_1 + m_2)\,\ell^2\ddot{\theta}_1 + m_2\ell^2\ddot{\theta}_2 + M_1 - M_2 + F\ell\left(\theta_2 - \theta_1\right) = 0, \qquad (11.14a)$$

$$m_2\ell^2\ddot{\theta}_1 + m_2\ell^2\ddot{\theta}_2 + M_2 = 0. \qquad (11.14b)$$

Using the constitutive law in Eqs. 11.8, the equations of motion are finally derived:

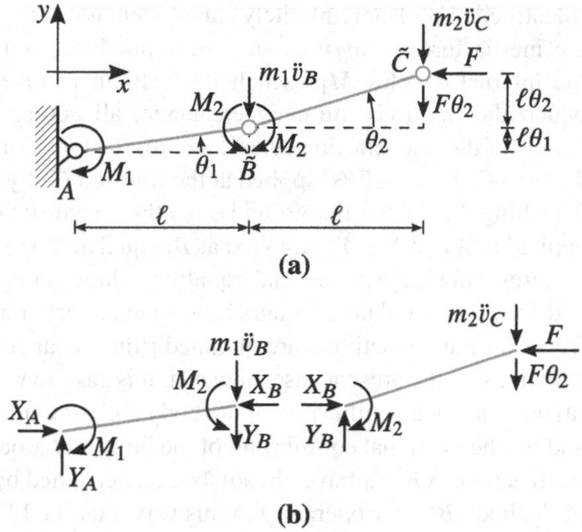

Fig. 11.3 Balance of forces acting on the Ziegler column: (a) inertial and active forces; (b) exploded view of the system, including the reactive forces

(a)

(b)

$$\ell^2 \begin{bmatrix} m_1 + m_2 & m_2 \\ m_2 & m_2 \end{bmatrix} \begin{pmatrix} \ddot{\theta}_1 \\ \ddot{\theta}_2 \end{pmatrix} + \begin{bmatrix} c_1 + c_2 & -c_2 \\ -c_2 & c_2 \end{bmatrix} \begin{pmatrix} \dot{\theta}_1 \\ \dot{\theta}_2 \end{pmatrix}$$
$$+ \left(\begin{bmatrix} k_1 + k_2 & -k_2 \\ -k_2 & k_2 \end{bmatrix} + F\ell \begin{bmatrix} -1 & 1 \\ 0 & 0 \end{bmatrix} \right) \begin{pmatrix} \theta_1 \\ \theta_2 \end{pmatrix} = \begin{pmatrix} 0 \\ 0 \end{pmatrix}. \tag{11.15}$$

They are of the type:

$$\mathbf{M}\ddot{\mathbf{q}} + \mathbf{C}\dot{\mathbf{q}} + \left(\mathbf{K}_e + \mathbf{K}_g \right) \mathbf{q} = \mathbf{0}, \tag{11.16}$$

where $\mathbf{q} := (\theta_1, \theta_2)^T$ and where the mass, damping, elastic, and geometric stiffness matrices appear in the order. It is worth noticing that, unlike the other matrices, $\mathbf{K}_g$ is *non-symmetric*, as an evidence of the nonconservativeness of the force F. Equation 11.15 shows that a viscoelastic system subject to follower forces is a *damped circulatory system*. The matrix $\mathbf{K}_g$ is the *circulatory matrix* of the system.

Remark 11.1 It is interesting to observe that the total stiffness matrix, $\mathbf{K} := \mathbf{K}_e + \mathbf{K}_g$, is non-singular for any value of F, being $\det \mathbf{K} = k_1 k_2 > 0$, independently of F. Hence, the Ziegler column *does not manifest static bifurcations*, however large is the intensity of the follower force. If the problem were approached from a purely static perspective, it should be concluded that the system is always stable! It will be immediately seen that this is wrong, as the loss of stability occurs through a dynamic bifurcation.

Supplement 11.1 (Balance of Forces) The equations of motion can alternatively be deduced from balance of forces, as illustrated below. Figure 11.3 shows the forces acting on the structure in a configuration adjacent to the reference one (identified, i.e., from infinitely small, but not zero, rotations θ_i). These are (a) the inertia forces $-m_1\ddot{v}_B$, $-m_2\ddot{v}_C$, applied respectively at the hinges B, C; (b) the internal couples M_1, which the column receives from the ground, and M_2, which the two rods mutually exchange, all acting in opposite directions to the strains of the viscoelastic devices; and (c) the two components of the external force, $F_x = -F$, $F_y = -F\theta_2$, applied at the free end C. By separating the two bodies and detaching them from the ground, also the constraint reactions appear, i.e., X_A, Y_A applied at A and X_B, Y_B, applied at B, equal and opposite on the two rods. Balance requires enforcing six cardinal equations, three for each body, in the four reactions and in the two unknown Lagrangian parameters. Lagrangian equations (i.e., free from constraint reactions) are obtained from linear combinations of the six cardinal equations. In the present case, however, it is easy to write them directly, by imposing (a) the rotational equilibrium of the body BC, around the point $\tilde{B}$, transformed of B, and (b) the rotational equilibrium of the body AB around the fixed point A, in which the reactions X_B, Y_B have already been determined by the translational equilibrium of the body BC. By operating in this way, Eqs. 11.14 are found.   □

Nondimensional Form of the Equations of Motion
Following Ziegler, reference is made to a particular family of systems, in which $m_1 =: 2m, m_2 =: m$, and $k_1 = k_2 =: k$. Furthermore, to facilitate the discussion, the equations of motion are made nondimensional. To this end, a new time $\tilde{t}$ is introduced, together with nondimensional damping coefficients ξ_1, ξ_2 and load μ, according to the following definitions:

$$\tilde{t} := t\sqrt{\frac{k}{m\ell^2}}, \qquad \xi_1 := \frac{c_1}{\ell\sqrt{km}}, \qquad \xi_2 := \frac{c_2}{\ell\sqrt{km}}, \qquad \mu := \frac{F\ell}{k}. \qquad (11.17)$$

The equations of motion, Eqs. 11.15, are thus rewritten in the simpler form:

$$\tilde{\mathbf{M}}\ddot{\mathbf{q}} + \tilde{\mathbf{C}}\dot{\mathbf{q}} + \left(\tilde{\mathbf{K}}_e + \mu\tilde{\mathbf{K}}_g\right)\mathbf{q} = \mathbf{0}, \qquad (11.18)$$

where:

$$\tilde{\mathbf{M}} := \begin{bmatrix} 3 & 1 \\ 1 & 1 \end{bmatrix}, \qquad \tilde{\mathbf{C}} := \begin{bmatrix} \xi_1 + \xi_2 & -\xi_2 \\ -\xi_2 & \xi_2 \end{bmatrix},$$

$$\tilde{\mathbf{K}}_e := \begin{bmatrix} 2 & -1 \\ -1 & 1 \end{bmatrix}, \qquad \tilde{\mathbf{K}}_g := \begin{bmatrix} -1 & 1 \\ 0 & 0 \end{bmatrix}, \qquad (11.19)$$

are nondimensional matrices and where the dot denotes the derivative with respect to the nondimensional time $\tilde{t}$.[5] In the following, the tildes will be omitted on the nondimensional quantities, matrices, and time.

---
[5] Therefore, $\dot{\mathbf{q}} \equiv \frac{d\mathbf{q}}{d\tilde{t}} = \frac{d\mathbf{q}}{dt}\frac{dt}{d\tilde{t}} = \frac{d\mathbf{q}}{dt}\sqrt{\frac{m\ell^2}{k}}$.

11.3.2 Undamped System

The circulatory system is examined first, which is drawn by Eq. 11.18 by vanishing the damping coefficients ξ_1, ξ_2. By looking for solutions of the type $\mathbf{q}\,(t) = \mathbf{u}\exp(\lambda t)$, an algebraic eigenvalue problem follows:

$$\left[\mathbf{M}\lambda^2 + \left(\mathbf{K}_e + \mu\mathbf{K}_g\right)\right]\mathbf{u} = \mathbf{0}, \tag{11.20}$$

or, in explicit form:

$$\begin{bmatrix} 3\lambda^2 + 2 - \mu & \lambda^2 - 1 + \mu \\ \lambda^2 - 1 & \lambda^2 + 1 \end{bmatrix}\begin{pmatrix} u_1 \\ u_2 \end{pmatrix} = \begin{pmatrix} 0 \\ 0 \end{pmatrix}. \tag{11.21}$$

The characteristic equation, whose roots λ decide on the stability of the position of trivial equilibrium, reads:

$$2\lambda^4 + (7 - 2\mu)\,\lambda^2 + 1 = 0. \tag{11.22}$$

As this is a biquadratic equation, it can be solved in closed form, to provide:

$$\lambda_{1,2}^2 = \frac{1}{4}\left(-(7 - 2\mu) \pm \sqrt{(7 - 2\mu)^2 - 8}\right). \tag{11.23}$$

When $\mu = 0$, the discriminant $\Delta := (7 - 2\mu)^2 - 8 > 0$, and therefore $\lambda_{1,2}^2 < 0$; by letting $\lambda_{1,2}^2 =: -\omega_{1,2}^2$, the four eigenvalues are purely imaginary, $\lambda_{1,2,3,4} = (\pm i\omega_1, \pm i\omega_2)$ (Fig. 11.4a). As μ increases, the discriminant Δ decreases, along with the difference $|\omega_1 - \omega_2|$. When $\Delta = 0$, the two pairs of eigenvalues collapse into a

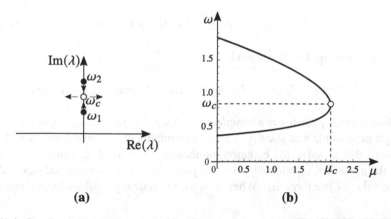

(a)   **(b)**

Fig. 11.4 Circulatory Hopf bifurcation of the Ziegler column: (a) collision of the eigenvalues; (b) imaginary parts of the eigenvalues, $\lambda_{1,2} = i\omega_1, i\omega_2$, as function of the load μ

unique pair, for which $\omega_1 = \omega_2 =: \omega_c$. For a further increase of μ, it is $\Delta < 0$, so that the four eigenvalues assume complex conjugates values of opposites in sign, of the type $\pm\delta \pm i\omega$. The bifurcation condition is therefore $\Delta = 0$, which occurs when the follower force assumes the critical value:

$$\mu_c := \frac{7}{2} - \sqrt{2} \simeq 2.09. \tag{11.24}$$

Correspondingly, the imaginary part of the coincident eigenvalues is $\omega_c = 2^{-1/4} \simeq 0.84$. This bifurcation is called *circulatory* or *reversible Hopf bifurcation*. The mechanism is often represented as in Fig. 11.4b, which shows the imaginary parts of the eigenvalues as a function of μ. When $\mu > \mu_c$ the eigenvalues are no longer purely imaginary and therefore no longer represented in the figure.

11.3.3   Damped System

The general case in which damping is different from zero is now examined. By letting again $\mathbf{q}(t) = \mathbf{u}\exp(\lambda t)$ in Eq. 11.18, the following eigenvalue problem is derived:

$$\left[\mathbf{M}\lambda^2 + \mathbf{C}\lambda + \left(\mathbf{K}_e + \mu\mathbf{K}_g\right)\right]\mathbf{u} = \mathbf{0}, \tag{11.25}$$

or:

$$\begin{bmatrix} 3\lambda^2 + \lambda(\xi_1 + \xi_2) + 2 - \mu & \lambda^2 - \lambda\xi_2 - 1 + \mu \\ \lambda^2 - \lambda\xi_2 - 1 & \lambda^2 + \lambda\xi_2 + 1 \end{bmatrix} \begin{pmatrix} u_1 \\ u_2 \end{pmatrix} = \begin{pmatrix} 0 \\ 0 \end{pmatrix}. \tag{11.26}$$

The characteristic equation is written as:

$$2\lambda^4 + I_1\lambda^3 + I_2(\mu)\lambda^2 + I_3\lambda + I_4 = 0, \tag{11.27}$$

where the following definitions hold:

$$I_1 := \xi_1 + 6\xi_2, \quad I_2(\mu) := 7 + \xi_1\xi_2 - 2\mu, \quad I_3 := \xi_1 + \xi_2, \quad I_4 := 1. \tag{11.28}$$

The characteristic equation is a *complete polynomial of fourth degree*, whose roots, although expressible analytically, assume extremely complicated form. It is therefore preferable to solve the equation numerically. For fixed damping coefficients ξ_1, ξ_2, the numerical solution supplies two pairs of complex conjugated eigenvalues for any value of the force μ.[6] When $\mu = 0$, the four eigenvalues have a negative

---

[6] Damping is assumed to be sufficiently small (sub-critical); for larger (super-critical) damping the eigenvalues are real.

real part, so that the trivial equilibrium position is asymptotically stable (Fig. 11.5a). When μ increases from zero, the eigenvalues approach the imaginary axis and *only one pair crosses it* at the critical value μ_d (d stands for "damped"); a further increase of the load brings this pair of eigenvalues to the right of the imaginary axis, making the equilibrium unstable. This mechanism is called *generic Hopf bifurcation*.

The Hopf critical load of the damped system, μ_d, is determined by imposing that the fourth degree Eq. 11.27 admits the root $\lambda = i\omega_d$ (and its complex conjugate, while the others roots are still to the left of the imaginary axis). By replacing this root in the equation, and separately equating to zero the real and imaginary parts of the equation, two real equations in the unknowns μ_d and ω_d are drawn, i.e.:

$$2\omega_d^4 - I_2(\mu_d)\omega_d^2 + I_4 = 0, \tag{11.29a}$$

$$\omega_d\left(I_3 - I_1\omega_d^2\right) = 0. \tag{11.29b}$$

From these equations, $\omega_d^2 = \frac{I_3}{I_1}$ and $I_2(\mu_d) = 2\frac{I_3}{I_1} + \frac{I_1 I_4}{I_3}$ follow. By making use of Eqs. 11.28, and solving, the critical load is finally obtained:

$$\mu_d = \mu_c + \frac{\xi_1 \xi_2}{2} - \frac{3 - 2\sqrt{2}}{2}\frac{\left[\xi_1 - \left(4 + 5\sqrt{2}\right)\xi_2\right]^2}{(\xi_1 + \xi_2)(\xi_1 + 6\xi_2)}; \tag{11.30}$$

moreover, $\omega_d^2 = \frac{\xi_1 + \xi_2}{\xi_1 + 6\xi_2}$.

Equation 11.30 is the Cartesian equation of a surface $\mu_d = \mu_d(\xi_1, \xi_2)$, which is the locus of a system family, parameterized by (μ, ξ_1, ξ_2), in a state of incipient bifurcation. This surface, represented in Fig. 11.5b, is known as the *Whitney umbrella* [4]; it separates the stable states of the system (points below the surface) from the unstable ones (points above). The following results are drawn [5]:

1. The critical load μ_d of the damped system is "almost everywhere" (i.e., for nearly all the weakly damped systems) smaller than the critical load μ_c of the undamped system. The damping, therefore, although dissipative,[7] has a detrimental effect on the mechanical behavior of the circulatory system. The $\mu_d < \mu_c$ property is known as *Ziegler's paradox* or the *destabilizing effect of damping*.
2. The damped system *is not in continuity* with the undamped one, in the sense that μ_d *does not* tend to μ_c when the damping modulus $\rho := \sqrt{\xi_1^2 + \xi_2^2}$ tends to zero. This limit, indeed, depends on *how* the operation is carried out, more precisely on the direction along which the origin is approached in the damping plane.

Property 2 can be easily checked by observing that, when in Eq. 11.30 $\xi_1 \to 0$ and $\xi_2 \to 0$, the second term to the right-hand member tends to zero, while the third

---

[7] Indeed, the matrix $\mathbf{C}$ is positive definite, since, as it is easy to check, its eigenvalues are positive $\forall(\xi_1, \xi_2) > 0$.

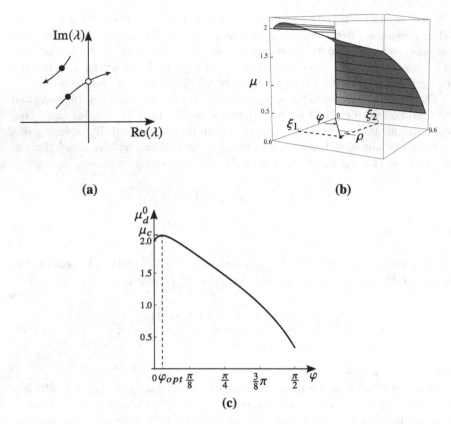

(a)   **(b)**

(c)

Fig. 11.5 Generic Hopf bifurcation of the Ziegler column: (**a**) path of the eigenvalues on the complex plane; (**b**) locus of critical states in the parameter space (μ, ξ_1, ξ_2); (**c**) critical value μ_d^0 of the follower force when damping is evanescent, *vs* $\varphi := \arctan\left(\frac{\xi_2}{\xi_1}\right)$

term tends to a definite and finite value.[8] This limit, however, depends on the $\frac{\xi_2}{\xi_1}$ ratio; for example, (i) if $\xi_2 = 0$, it is $\lim\limits_{\xi_1 \to 0} \mu_d = 2$; (ii) if $\xi_1 = 0$, it is $\lim\limits_{\xi_2 \to 0} \mu_d = \frac{1}{3}$.

To study, more generally, how the limit depends on the $\frac{\xi_2}{\xi_1}$ ratio, it is convenient to set $\xi_1 = \rho \cos \varphi$, $\xi_2 = \rho \sin \varphi$ (Fig. 11.5b) and to execute, for a fixed φ, the limit:

$$\mu_d^0 := \lim_{\rho \to 0} \mu_d = \frac{1}{2} \frac{33 \sin(2\varphi) + 8}{7 \sin(2\varphi) - 5 \cos(2\varphi) + 7}. \tag{11.31}$$

---

[8] In fact, both the numerator and denominator of the ratio are quadratic functions of the damping parameters.

The graph of μ_d^0 as a function of $\varphi \in \left(0, \frac{\pi}{2}\right)$ is represented in Fig. 11.5c. It is interesting to note that $\mu_d^0 < \mu_c$ for any $\varphi \neq \varphi_{opt}$, with $\varphi_{opt} := \arctan\left(\frac{1}{4+5\sqrt{2}}\right) \simeq 0.09$ rad. This angle identifies the *optimal direction*, along which $\mu_d^0 = \mu_c$, and therefore for which the damped system is in continuity with the undamped one. If one increases damping along the optimal direction, i.e., if $\xi_1 = \rho \cos\left(\varphi_{opt}\right)$, $\xi_2 = \rho \sin\left(\varphi_{opt}\right)$ in Eq. 11.30, then $\mu_d\left(\rho; \varphi_{opt}\right) \simeq 2.09 + 0.045\rho^2$ is found, i.e., damping has a very weak beneficial effect on the critical load.[9]

Remark 11.2 It can be proved that (i) if $\varphi > \varphi_{opt}$, the critical mode is that of lowest frequency; (ii) if $\varphi < \varphi_{opt}$, the critical mode is that of highest frequency. This circumstance, somewhat unusual, is often summarized by saying that "the first, or the second, mode becomes unstable." The locution is correct only if it is referred to the modes of the circulatory system, ordered according to their frequency, and not to the natural modes (whose frequencies and forms are modified by the follower force and damping).

11.4 Limit Cycles of the Ziegler Column

A damped circulatory system, such as the Ziegler column, suffers from a generic Hopf bifurcation when the load reaches the critical value μ_d. As already commented in the Chap. 3, this bifurcation causes the birth of periodic solutions, represented by a family of limit cycles in the phase space. To each μ close to μ_d (larger, if the bifurcation is super-critical, or smaller, if sub-critical), a specific closed orbit is associated. If this orbit is stable (as happens in the super-critical case), it is attractive of all the surrounding trajectories, so that the system, after having exhausted a transient, executes periodic oscillations. The amplitude and frequency of the cycle depend on the difference $|\mu - \mu_d|$, i.e., on the distance from the bifurcation point.

To build up the limit cycle, it is first necessary to formulate a *nonlinear model*; subsequently, periodic solutions admitted by the nonlinear equations of motion must be sought. This last operation can be performed by specific numerical algorithms, or by a perturbation method, if the interest is limited to small values of $|\mu - \mu_d|$. Among the different methods available in literature, here the *Lindstedt-Poincaré method* is illustrated,[10] since it appears to be the natural extension to the dynamic case of the asymptotic method that was used systematically in the Chap. 5 for static systems.[11] Further details can be found in [3, 7].

---

[9] For high damping, for example $\rho = 0.5$, it is $\mu_d \simeq 2.10$.

[10] The method belongs to the family of *Strained Parameter Methods* [8].

[11] Accordingly, this latter is also known as 'static perturbation method'.

11.4.1 Nonlinear Model

To obtain the nonlinear equations of motion of the Ziegler column, it is necessary to repeat the procedure illustrated in the Sect. 11.3, using the *exact* kinematic relations (Eqs. 11.2, 11.4), in place of those expanded in series (Eqs. 11.5, 11.6). Moreover, it has to be taken into account that, in finite kinematics, $F_x = -F \cos \theta_2$, $F_y = -F \sin \theta_2$.

Exact Equations of Motion

The first variation of the kinetic energy and the internal and external virtual works are:

$$\delta T = m_1 \left(\dot{u}_B \, \delta \dot{u}_B + \dot{v}_B \, \delta \dot{v}_B \right) + m_2 \left(\dot{u}_C \, \delta \dot{u}_C + \dot{v}_C \, \delta \dot{v}_C \right), \tag{11.32a}$$

$$\delta W_{int} = M_1 \, \delta \kappa_1 + M_2 \, \delta \kappa_2, \tag{11.32b}$$

$$\delta W_{ext} = -F \left(\cos \theta_2 \, \delta u_C + \sin \theta_2 \, \delta v_C \right). \tag{11.32c}$$

Substituting in these expressions the kinematic relationships, Eqs. 11.2 and 11.4, making use of the extended Hamilton principle (Eq. 11.1), and integrating by parts, one obtains the following exact balance equations:

$$(m_1 + m_2) \, \ell^2 \ddot{\theta}_1 + m_2 \ell^2 \ddot{\theta}_2 \cos (\theta_1 - \theta_2) + m_2 \ell^2 \dot{\theta}_2^2 \sin (\theta_1 - \theta_2) \tag{11.33a}$$

$$+ M_1 - M_2 - F \ell \sin (\theta_1 - \theta_2) = 0,$$

$$m_2 \ell^2 \ddot{\theta}_2 + m_2 \ell^2 \ddot{\theta}_1 \cos (\theta_1 - \theta_2) - m_2 \ell^2 \dot{\theta}_1^2 \sin (\theta_1 - \theta_2) + M_2 = 0. \tag{11.33b}$$

The same equations are drawn through direct equilibrium in the current configuration, proceeding as illustrated in the Supplement 11.1 for the linearized problem, but making use of the exact kinematics.

To obtain the equations of motion, it is necessary to introduce the viscoelastic law expressed in terms of displacement (Eqs. 11.8). Particularizing the equations of motion to the case examined by Ziegler, in which $m_1 = 2m$, $m_2 = m$, $k_1 = k_2 = k$, and rendering the equation nondimensional through the parameters already defined in Eqs. 11.17, one finally gets:

$$3 \ddot{\theta}_1 + \ddot{\theta}_2 \cos (\theta_1 - \theta_2) + \dot{\theta}_2^2 \sin (\theta_1 - \theta_2) + (\xi_1 + \xi_2) \, \dot{\theta}_1 - \xi_2 \dot{\theta}_2 \tag{11.34a}$$

$$+ 2\theta_1 - \theta_2 - \mu \sin (\theta_1 - \theta_2) = 0,$$

$$\ddot{\theta}_2 + \ddot{\theta}_1 \cos (\theta_1 - \theta_2) - \dot{\theta}_1^2 \sin (\theta_1 - \theta_2) + \xi_2 \left(\dot{\theta}_2 - \dot{\theta}_1 \right) + \theta_2 - \theta_1 = 0, \tag{11.34b}$$

where $\theta_i = \theta_i \left(\tilde{t} \right)$, with $\tilde{t}$ the nondimensional time.

Remark 11.3 In exact kinematics, and differently from the linearized theory, the longitudinal motion, $u_B (t), u_C (t)$, does contribute to the kinetic energy. The nonlinearities present in Eqs. 11.33 (or Eqs. 11.34) are (a) of inertial type, related to the motion of the masses, and (b) induced by the external (circulatory) force. The internal viscoelastic forces are instead linear.

Nondimensional Equations Expanded in Series
In view of a perturbation analysis, the Eqs. 11.34 are expanded in series, up to the terms of the third degree, thus obtaining:

$$\mathbf{M}\ddot{\mathbf{q}} + \mathbf{C}\dot{\mathbf{q}} + \left(\mathbf{K}_e + \mu\mathbf{K}_g\right)\mathbf{q} = \mathbf{F}\left(\mathbf{q}, \dot{\mathbf{q}}, \ddot{\mathbf{q}}\right) + \mu\mathbf{G}\left(\mathbf{q}\right), \tag{11.35}$$

where the matrices are defined in Eqs. 11.19. Moreover, $\mathbf{F}\left(\mathbf{q}, \dot{\mathbf{q}}, \ddot{\mathbf{q}}\right)$ is the vector of the inertial nonlinearities and $\mathbf{G}\left(\mathbf{q}\right)$ the vector of the circulatory nonlinearities, both cubic in theirs arguments and defined as follows:

$$\mathbf{F}\left(\mathbf{q}, \dot{\mathbf{q}}, \ddot{\mathbf{q}}\right) := \begin{pmatrix} \frac{1}{2}\ddot{\theta}_2 \left(\theta_1 - \theta_2\right)^2 - \dot{\theta}_2^2 \left(\theta_1 - \theta_2\right) \\ \frac{1}{2}\ddot{\theta}_1 \left(\theta_1 - \theta_2\right)^2 + \dot{\theta}_1^2 \left(\theta_1 - \theta_2\right) \end{pmatrix}, \quad \mathbf{G}\left(\mathbf{q}\right) := \begin{pmatrix} -\frac{1}{6} \left(\theta_1 - \theta_2\right)^3 \\ 0 \end{pmatrix}. \tag{11.36}$$

The linear part of these equations, of course, coincides with that studied in the Sect. 11.3.

11.4.2 Lindstedt-Poincaré Method

To determine the limit cycle, periodic solutions must be found for Eq. 11.35. The simpler perturbation algorithm to achieve this result is the *Lindstedt-Poincaré method*, discussed here[12] [2, 9]. Most importantly, it is strictly analogous to the static perturbation method, already used in the analysis of postbuckling.

Perturbation Equations
The method consists of the following steps:

- the time scale is stretched, by introducing an appropriate parameter Ω which transforms the limit cycle *unknown* period T (on the natural scale t) to a *known* period on the stretched scale τ;
- a family of periodic solutions $\mathbf{q} = \mathbf{q}\left(\tau; \epsilon\right)$, $\Omega = \Omega\left(\epsilon\right)$, $\mu = \mu\left(\epsilon\right)$ is sought for, parameterized by a perturbation parameter ϵ; via a suitable normalization, the perturbation parameter assumes the meaning of *amplitude of the limit cycle*;
- $\mathbf{q}$, Ω, μ are expanded in series of ϵ, and the perturbation equations are obtained;
- the perturbation equations are solved in sequence, by requiring that the solutions to the different orders fulfill the periodicity; from these conditions, the unknown coefficients of the Ω and μ series are computed.

---

[12] Actually, there are more sophisticated methods, such as the *multiple scale method* [8, 9] which allow one to analyze also the stability of the limit cycle and to describe the transient motions. Here, however, such a method is not discussed, to reduce the complexity of the analysis to a lower level.

To deform the time scale, the new variable $\tau := \Omega t$ is introduced, with $\Omega = \frac{2\pi}{T}$ being the unknown circular frequency of the periodic solution; hence, when $t = T$, it is $\tau = 2\pi$ (whatever ω). Moreover, it turns out that $\frac{d}{dt} = \frac{d}{d\tau}\frac{d\tau}{dt} = \Omega \frac{d}{d\tau}$ and, consequently, $\frac{d^2}{dt^2} = \Omega^2 \frac{d^2}{d\tau^2}$. In the new time, Eq. 11.35 is written as:

$$\Omega^2 \mathbf{M}\ddot{\mathbf{q}} + \Omega\, \mathbf{C}\dot{\mathbf{q}} + \left(\mathbf{K}_e + \mu \mathbf{K}_g\right) \mathbf{q} = \Omega^2 \mathbf{F}\left(\mathbf{q}, \dot{\mathbf{q}}, \ddot{\mathbf{q}}\right) + \mu \mathbf{G}\left(\mathbf{q}\right), \tag{11.37}$$

where the derivative with respect to τ is again indicated by a dot.

The dependent variable and the parameters are successively expanded in series as:

$$\mathbf{q}\left(\tau; \epsilon\right) = \epsilon \mathbf{q}_1\left(\tau\right) + \epsilon^3 \mathbf{q}_3\left(\tau\right) + \cdots, \tag{11.38a}$$

$$\Omega\left(\epsilon\right) = \Omega_0 + \epsilon^2 \Omega_2 + \cdots, \tag{11.38b}$$

$$\mu\left(\epsilon\right) = \mu_0 + \epsilon^2 \mu_2 + \cdots. \tag{11.38c}$$

in which the symmetry of the system has been accounted for.[13]

Substituting the series in the equation of motion, and equating separately to zero the terms with the same powers of ϵ, the following perturbation equations are obtained:

$$\epsilon^1: \ \Omega_0^2 \mathbf{M}\ddot{\mathbf{q}}_1 + \Omega_0 \mathbf{C}\dot{\mathbf{q}}_1 + \left(\mathbf{K}_e + \mu_0 \mathbf{K}_g\right) \mathbf{q}_1 = \mathbf{0}, \tag{11.39a}$$

$$\epsilon^3: \ \Omega_0^2 \mathbf{M}\ddot{\mathbf{q}}_3 + \Omega_0 \mathbf{C}\dot{\mathbf{q}}_3 + \left(\mathbf{K}_e + \mu_0 \mathbf{K}_g\right) \mathbf{q}_3 = -2\Omega_0 \Omega_2 \mathbf{M}\ddot{\mathbf{q}}_1 - \Omega_2 \mathbf{C}\dot{\mathbf{q}}_1 \tag{11.39b}$$

$$- \mu_2 \mathbf{K}_g \mathbf{q}_1 + \mu_0 \mathbf{G}\left(\mathbf{q}_1\right)$$

$$+ \Omega_0^2 \mathbf{F}\left(\mathbf{q}_1, \dot{\mathbf{q}}_1, \ddot{\mathbf{q}}_1\right).$$

These equations must be accompanied by the *periodicity conditions*, $\mathbf{q}\left(2\pi\right) = \mathbf{q}\left(0\right), \dot{\mathbf{q}}\left(2\pi\right) = \dot{\mathbf{q}}\left(0\right)$, which, due to the series expansions in Eq. 11.38a, imply:

$$\mathbf{q}_k\left(2\pi\right) = \mathbf{q}_k\left(0\right), \qquad \dot{\mathbf{q}}_k\left(2\pi\right) = \dot{\mathbf{q}}_k\left(0\right), \qquad k = 1, 3, \cdots. \tag{11.40}$$

To assign a geometric meaning to the perturbation parameter, the *normalization condition* $\theta_2\left(0\right) = \epsilon, \dot{\theta}_2\left(0\right) = 0$, is introduced.[14] Consequently, ϵ is the maximum

---

[13] Given the exclusive presence of odd nonlinearities, if the equation is satisfied by $(\mathbf{q}, \mu, \Omega)$, it is also satisfied by $(-\mathbf{q}, \mu, \Omega)$. Hence, the series Eqs 11.38 must contain only odd powers of ϵ in $\mathbf{q}$, and only even powers of ϵ in Ω, μ, in such a way that the two solutions are described by ϵ and $-\epsilon$.

[14] The origin of the times, being the system autonomous, is not relevant. Therefore, it has been chosen for the sake of simplicity, to take the origin at the instant in which the speed of θ_2 zeroes. Other choices would lead to describe the same orbit with different origins.

rotation (because it corresponds to a point of inversion of motion) of the upper rod of the double pendulum, taken as a measure of the *amplitude of the limit cycle*. Considering the series expansions in Eq. 11.38a, the normalization involves:

$$\theta_{21}(0) = 1, \qquad \dot{\theta}_{21}(0) = 0, \tag{11.41a}$$

$$\theta_{2k}(0) = 0, \qquad \dot{\theta}_{2k}(0) = 0, \qquad k = 3, 5, \cdots. \tag{11.41b}$$

Generating Solution

The lowest-order perturbation, Eq. 11.39a (also called generating equation), admits the solution $\mathbf{q}_1 = \mathbf{u} \exp(\lambda \tau)$, with $\mathbf{u} \in \mathbb{C}^2$. Because one is interested in periodic solutions of period 2π (i.e., of frequency 1), the eigenvalue must be purely imaginary, namely, $\lambda = \pm i$. Taking $\mathbf{q}_1 = \mathbf{u} \exp(i\tau)$, an algebraic problem is drawn:

$$\left[-\Omega_0^2 \mathbf{M} + i\Omega_0 \mathbf{C} + \left(\mathbf{K}_e + \mu_0 \mathbf{K}_g \right) \right] \mathbf{u} = \mathbf{0}. \tag{11.42}$$

This equation is identical to Eq. 11.25, already encountered in the linearized analysis, if one takes $(\mathbf{u}, \Omega_0, \mu_0) = (\mathbf{u}_d, \omega_d, \mu_d)$. Therefore, *the initial values of the series expansions* in Eqs. 11.38b,c *identify the Hopf bifurcation condition of the damped system*. In other words, the limit cycles originate from the Hopf point $\mu = \mu_d$; at this point, they assume the frequency ω_d and a "pattern" described by the eigenvector $\mathbf{u}_d$.

The first order motion is expressed by:[15]

$$\mathbf{q}_1(\tau) = \mathbf{u}_d \exp(i\tau) + \bar{\mathbf{u}}_d \exp(-i\tau), \tag{11.43}$$

where the bar indicates the complex conjugate. Given the normalization, it is $\mathbf{u}_d = (u_1, u_2)^T = (u_1, 1/2)^T$, with $u_1 = u_1(\xi_1, \xi_2)$. The solution in Eq. 11.43, of complex form, can also be written in real form:[16]

$$\mathbf{q}_1(\tau) = 2\operatorname{Re}(\mathbf{u}_d) \cos \tau - 2\operatorname{Im}(\mathbf{u}_d) \sin \tau. \tag{11.44}$$

This is the parametric equation of an ellipse in the plane (θ_1, θ_2), which is the first approximation of the limit cycle $\mathbf{q}(\tau; \epsilon) \simeq \epsilon \mathbf{q}_1(\tau)$. However, the amplitude of the ellipse, ϵ, is indeterminate at this order, and not related to the bifurcation parameter μ. To establish how μ depends on ϵ, it is necessary to proceed one step further.

Adjoint Problem

Before moving on to the perturbation equation of order ϵ^3, it is worth studying the *adjoint problem* to Eq. 11.42, which will be used soon. This problem is defined as

---

[15] For the motion to be real, the complex conjugate solution must be added.

[16] Remember Eq. 3.23, which (with a slight change of symbols) expresses the complex modes.

the *transpose of the original problem*.[17] Given that $(\Omega_0, \mu_0) = (\omega_d, \mu_d)$ and due to the symmetry of $\mathbf{M}, \mathbf{C}, \mathbf{K}_e$, but not of $\mathbf{K}_g$, the problem reads:

$$\left[-\omega_d^2 \mathbf{M} + i\omega_d \mathbf{C} + \left(\mathbf{K}_e + \mu_d \mathbf{K}_g^T \right) \right] \mathbf{v}_d = \mathbf{0}. \tag{11.45}$$

The eigenvector $\mathbf{v}_d$ (called the left eigenvector[18]) is in general different from $\mathbf{u}_d$ (called the right eigenvector), and *non* orthogonal to it.[19] The left eigenvector can be arbitrarily normalized, as will be discussed ahead in Note 24.

Perturbation Equation of Order ϵ^3

Equation 11.39b is now considered. Since, at this step, $\mathbf{q}_1(\tau)$ is known from Eq. 11.43,[20] the terms on the right-hand member of the equation are also known, to within the two indeterminate parameters Ω_2, μ_2. The linear terms are written immediately:

$$\mathbf{M}\ddot{\mathbf{q}}_1 = -\mathbf{M} \left(\mathbf{u}_d e^{i\tau} + \bar{\mathbf{u}}_d e^{-i\tau} \right), \tag{11.46a}$$

$$\mathbf{C}\dot{\mathbf{q}}_1 = i\mathbf{C} \left(\mathbf{u}_d e^{i\tau} + \bar{\mathbf{u}}_d e^{-i\tau} \right), \tag{11.46b}$$

$$\mathbf{K}_g \mathbf{q}_1 = \mathbf{K}_g \left(\mathbf{u}_d e^{i\tau} + \bar{\mathbf{u}}_d e^{-i\tau} \right). \tag{11.46c}$$

Nonlinear terms are more laborious to compute; shortly, since they are cubic forms of $\mathbf{q}_1(\tau)$ and its derivatives, they are of the type:

$$\mathbf{F}(\mathbf{q}_1, \dot{\mathbf{q}}_1, \ddot{\mathbf{q}}_1) = \mathbf{f}_3 e^{3i\tau} + \mathbf{f}_1 e^{i\tau} + \bar{\mathbf{f}}_3 e^{-3i\tau} + \bar{\mathbf{f}}_1 e^{-i\tau}, \tag{11.47a}$$

$$\mathbf{G}(\mathbf{q}) = \mathbf{g}_3 e^{3i\tau} + \mathbf{g}_1 e^{i\tau} + \bar{\mathbf{g}}_3 e^{-3i\tau} + \bar{\mathbf{g}}_1 e^{-i\tau}, \tag{11.47b}$$

where $\mathbf{f}_k = \mathbf{f}_k \left(u_j, \bar{u}_j \right)$, $\mathbf{g}_k = \mathbf{g}_k \left(u_j, \bar{u}_j \right)$ are complex-valued vectors. The detailed calculation of $\mathbf{f}_1$ and $\mathbf{g}_1$ is illustrated in the Supplement 11.2 (Eq. 11.58); $\mathbf{f}_3, \mathbf{g}_3$, are instead omitted, because they are irrelevant at the order in question, as it will appear clear soon.

With the results acquired, Eq. 11.39b is written as:

---

[17] Strictly speaking, the adjoint problem in complex field is the *transpose conjugate*. Conjugation, however, is inessential for practical purposes, as "balanced" by the definition of the scalar product in complex field, which will be introduced shortly (Eq. 11.49) and which would also require the use of the conjugate. Here, to simplify the discussion, and consistently with some authors, formal rigor is renounced, without prejudice to the exactness of the results.

[18] It is called left because the adjoint problem of $\mathbf{Au} = \mathbf{0}$ (with $\mathbf{u}$ to the right of $\mathbf{A}$) can also be written as $\mathbf{v}^T \mathbf{A} = \mathbf{0}$ (with $\mathbf{v}$ to the left of $\mathbf{A}$).

[19] Excluding cases of eigenvalues of multiplicity 2, for which the two eigenvectors are, in fact, orthogonal

[20] The complex form is preferred, which reduces writing. Examples of use of the real form will be presented in the following chapters.

$$\omega_d^2 \mathbf{M} \ddot{\mathbf{q}}_3 + \omega_d \mathbf{C} \dot{\mathbf{q}}_3 + \left(\mathbf{K}_e + \mu_d \mathbf{K}_g \right) \mathbf{q}_3 = \left(2\omega_d \Omega_2 \mathbf{M} - i\Omega_2 \mathbf{C} - \mu_2 \mathbf{K}_g \right) \mathbf{u}_d e^{i\tau}$$

$$+ \left(\omega_d^2 \mathbf{f}_1 + \mu_d \mathbf{g}_1 \right) e^{i\tau} + \left(\omega_d^2 \mathbf{f}_3 + \mu_d \mathbf{g}_3 \right) e^{3i\tau}$$

$$+ \text{c.c.,}$$

$$(11.48)$$

where c.c. indicates the complex conjugate of all the terms preceding it. This equation governs the motion of an oscillator with two degrees of freedom, subject to harmonic excitation, of frequencies 1 and 3 (on the τ time scale). Since the frequency 1 is also one of the natural frequencies of the oscillator, the excitation frequency 1 is resonant.[21] For generic values of the two free parameters, Ω_2, μ_2, the response $\mathbf{q}_3(\tau)$ diverges to infinity, and therefore the periodicity condition cannot be fulfilled. Periodicity, therefore, requires that an appropriate *solvability* (or compatibility) *condition* is satisfied (discussed in Supplement 11.3), namely, that the *resonant excitation is orthogonal to the left eigenvector* $\mathbf{v}_d$ associated with the same eigenvalue (i.e., to the solution of Eq. 11.45).[22] In formulas:

$$\mathbf{v}_d^T \left[\left(\Omega_2 \left(2\omega_d \mathbf{M} - i\mathbf{C} \right) - \mu_2 \mathbf{K}_g \right) \mathbf{u}_d + \omega_d^2 \mathbf{f}_1 + \mu_d \mathbf{g}_1 \right] = 0. \qquad (11.49)$$

The previous equation can also be written as:

$$\Omega_2 \left(2 m \omega_d - i c \right) - \mu_2 k_g + \omega_d^2 f_1 + \mu_d g_1 = 0, \qquad (11.50)$$

where complex scalar quantities have been introduced (named modal parameters and modal forces[23]):

$$m := \mathbf{v}_d^T \mathbf{M} \mathbf{u}_d, \quad c := \mathbf{v}_d^T \mathbf{C} \mathbf{u}_d, \quad k_g := \mathbf{v}_d^T \mathbf{K}_g \mathbf{u}_d, \quad f_1 := \mathbf{v}_d^T \mathbf{f}_1, \quad g_1 := \mathbf{v}_d^T \mathbf{g}_1. \qquad (11.51)$$

By separating the real and imaginary parts in Eq. 11.50, and using the notation $(\cdot)_R := \mathrm{Re}\,(\cdot)$ and $(\cdot)_I := \mathrm{Im}\,(\cdot)$, two real equations are obtained:

$$\begin{bmatrix} -2\omega_d m_R - c_I \ k_{gR} \\ -2\omega_d m_I + c_R \ k_{gI} \end{bmatrix} \begin{pmatrix} \Omega_2 \\ \mu_2 \end{pmatrix} = \begin{pmatrix} \omega_d^2 f_{1R} + \mu_d g_{1R} \\ \omega_d^2 f_{1I} + \mu_d g_{1I} \end{pmatrix}, \qquad (11.52)$$

from which Ω_2 and μ_2 are computed.[24]

---

[21] If 3 were also a natural frequency, a degenerate case would occur, called of *internal resonance*, which is excluded in the present discussion.

[22] The operation is also called "elimination of secular terms," with reference to the divergent contributions that, in celestial mechanics studied by Poincaré, significantly manifest themselves only after centuries (*siécles* in French, *saecula* in Latin).

[23] Here m should not be confused with the concentrated mass of the Ziegler column.

[24] Equation 11.49 shows that normalization of the left eigenvector is arbitrary, as $\mathbf{v}_d$ is a factor. A careful choice of normalization, however, can simplify Eq. 11.52. In fact, requesting that $\mathbf{v}_d^T \left(2\omega_d \mathbf{M} - i\mathbf{C} \right) \mathbf{u}_d = 1$ (or equal to i), the matrix in Eq. 11.52 becomes triangular, and the equations can be solved in cascade.

Remark 11.4 The similarities between the Lindstedt-Poincaré method and the static perturbation method, illustrated in Chap. 4, should be noticed. In both the procedures, the solvability condition, imposed at the various orders k, allows one to evaluate the free parameters. However, in statics, where all the equations are real, solvability determines only the μ_k parameters; in dynamics, where the equations are complex, solvability separates into *two* real equations, allowing to determine both the μ_k and the Ω_k parameters.

First-Order Limit Cycle
Truncating the procedure to this order, from Eqs. 11.38 and from the results obtained, the family of limit cycles is determined, expressed in parametric form:

$$\mathbf{q}(\tau; \epsilon) = \epsilon \left(2 \operatorname{Re}(\mathbf{u}_d) \cos \tau - 2 \operatorname{Im}(\mathbf{u}_d) \sin \tau\right), \tag{11.53a}$$

$$\Omega(\epsilon) = \omega_d + \epsilon^2 \Omega_2, \tag{11.53b}$$

$$\mu(\epsilon) = \mu_d + \epsilon^2 \mu_2, \tag{11.53c}$$

where use has been made of Eq. 11.44 and where Ω_2, μ_2 are provided by Eq. 11.52. The previous equations describe a family of ellipses on the plane θ_1, θ_2, parameterized by ϵ. The system performs harmonic motions on these ellipses. When $\epsilon \to 0$, the ellipse becomes evanescent; the motion assumes frequency $\Omega \to \omega_d$ and the load tends to the bifurcation value $\mu \to \mu_d$. As ϵ increases, the cycle widens, entailing changes in frequency and load.

Remembering that $\tau = \Omega t$, one can go back to the original time:

$$\mathbf{q}(t; \epsilon) = \epsilon \left\{2 \operatorname{Re}(\mathbf{u}_d) \cos \left[\left(\omega_d + \epsilon^2 \Omega_2\right) t\right] - 2 \operatorname{Im}(\mathbf{u}_d) \sin \left[(\omega_d + \Omega_2) t\right]\right\}, \tag{11.54a}$$

$$\mu(\epsilon) = \mu_d + \epsilon^2 \mu_2. \tag{11.54b}$$

Finally, one can eliminate ϵ and express the solution as parametrized by the load:

$$\mathbf{q}(t; \mu) = \sqrt{\frac{\mu - \mu_d}{\mu_2}} \left\{2 \operatorname{Re}(\mathbf{u}_d) \cos \left[\left(\omega_d + \frac{\mu - \mu_d}{\mu_2} \Omega_2\right) t\right] \right.$$
$$\left. - 2 \operatorname{Im}(\mathbf{u}_d) \sin \left[\left(\omega_d + \frac{\mu - \mu_d}{\mu_2} \Omega_2\right) t\right]\right\}. \tag{11.55}$$

Remark 11.5 The perturbation solution thus found is also called "of first order," because only one solvability condition has been used. If one pushes the procedure further on, the ϵ^3 order perturbation Eq. 11.48 must be solved. With the values of Ω_2, μ_2 now determined, the equation provides a periodic solution $q_3(\tau)$ in which both harmonics, of frequency 1 and 3, appear. The indeterminacy on the response of frequency 1, relative to the resonant excitation (however respectful of solvability), introduces an arbitrariness, which is removed by the normalization condition, Eq. 11.41b. The $q_3(\tau)$ component *modifies the shape of the cycle* (which is no longer an ellipse), as well as the time law with which the cycle is traveled. Successively, the solvability condition applied to the perturbation equation of order ϵ^5 yields Ω_4, μ_4. It is therefore understood that the first-order solution is unable to capture these changes, as it will emerge shortly from the comparison with numerical results.

Supplement 11.2 (Nonlinear Terms in the Perturbation Eq. 11.39b) To evaluate the nonlinear terms in Eq. 11.39b, some typical ones are calculated in advance. Taking into account the definitions in Eqs. 11.36, and the first-order solution in Eq. 11.43, one has:

$$\ddot{\theta}_k \theta_j \theta_h = -\epsilon^3 \left(u_k e^{i\tau} + \bar{u}_k e^{-i\tau} \right) \left(u_j e^{i\tau} + \bar{u}_j e^{-i\tau} \right) \left(u_h e^{i\tau} + \bar{u}_h e^{-i\tau} \right)$$

$$\tag{11.56a}$$

$$= -\epsilon^3 \left(\bar{u}_k u_j u_h + u_k \bar{u}_j u_h + u_k u_j \bar{u}_h \right) e^{i\tau} + (\cdots) e^{3i\tau} + \text{c.c.},$$

$$\dot{\theta}_k^2 \theta_j = -\epsilon^3 \left(u_k e^{i\tau} - \bar{u}_k e^{-i\tau} \right)^2 \left(u_j e^{i\tau} + \bar{u}_j e^{-i\tau} \right)$$

$$\tag{11.56b}$$

$$= -\epsilon^3 \left(u_k^2 \bar{u}_j - 2 u_k \bar{u}_k u_j \right) e^{i\tau} + (\cdots) e^{3i\tau} + \text{c.c.},$$

$$\theta_k \theta_j \theta_h = \epsilon^3 \left(u_k e^{i\tau} + \bar{u}_k e^{-i\tau} \right) \left(u_j e^{i\tau} + \bar{u}_j e^{-i\tau} \right) \left(u_h e^{i\tau} + \bar{u}_h e^{-i\tau} \right) \tag{11.56c}$$

$$= \epsilon^3 \left(\bar{u}_k u_j u_h + u_k \bar{u}_j u_h + u_k u_j \bar{u}_h \right) e^{i\tau} + (\cdots) e^{3i\tau} + \text{c.c.},$$

where the harmonics of order 3 have not made explicit. From Eqs. 11.36, it follows that:

$$\mathbf{F}(\mathbf{q}_1, \dot{\mathbf{q}}_1, \ddot{\mathbf{q}}_1) = \epsilon^3 \begin{pmatrix} -\frac{1}{2} u_1^2 \bar{u}_2 + 2 u_2^2 \bar{u}_1 - u_1 u_2 \bar{u}_1 - \frac{1}{2} u_2^2 \bar{u}_2 \\ -\frac{1}{2} u_1^2 \bar{u}_1 + 2 u_1^2 \bar{u}_2 - u_2 u_1 \bar{u}_2 - \frac{1}{2} u_2^2 \bar{u}_1 \end{pmatrix} e^{i\tau} + (\cdots) e^{3i\tau} + \text{c.c.},$$

$$\tag{11.57a}$$

$$\mathbf{G}(\mathbf{q}_1) = \epsilon^3 \begin{pmatrix} -\frac{1}{2}(u_1 - u_2)^2 (\bar{u}_1 - \bar{u}_2) \\ 0 \end{pmatrix} e^{i\tau} + (\cdots) e^{3i\tau} + \text{c.c..} \tag{11.57b}$$

From these latter, the vectors $\mathbf{f}_1$, $\mathbf{g}_1$ already introduced in Eq. 11.47 are drawn:

$$\mathbf{f}_1 := \begin{pmatrix} -\frac{1}{2} u_1^2 \bar{u}_2 + 2 u_2^2 \bar{u}_1 - u_1 u_2 \bar{u}_1 - \frac{1}{2} u_2^2 \bar{u}_2 \\ -\frac{1}{2} u_1^2 \bar{u}_1 + 2 u_1^2 \bar{u}_2 - u_2 u_1 \bar{u}_2 - \frac{1}{2} u_2^2 \bar{u}_1 \end{pmatrix}, \quad \mathbf{g}_1 := \begin{pmatrix} -\frac{1}{2}(u_1 - u_2)^2 (\bar{u}_1 - \bar{u}_2) \\ 0 \end{pmatrix}.$$

$$\tag{11.58}$$

$\square$

Supplement 11.3 (Solvability Condition in Complex Field) A linear oscillator subject to a harmonic excitation is considered:

$$\mathbf{M}\ddot{\mathbf{q}} + \mathbf{C}\dot{\mathbf{q}} + \mathbf{K}\mathbf{q} = \mathbf{f}e^{i\alpha t}, \tag{11.59}$$

where $\alpha \in \mathbb{R}$ is the excitation frequency. To determine a particular solution, $\mathbf{q} = \mathbf{w}e^{i\alpha t}$ is set, which leads to the algebraic problem:

$$\left(-\alpha^2 \mathbf{M} + i\alpha \mathbf{C} + \mathbf{K}\right)\mathbf{w} = \mathbf{f}. \tag{11.60}$$

If $i\alpha$ is an eigenvalue of the homogeneous system (i.e., if the frequency of the excitation coincides with a natural one), the matrix in the left member is singular, so that the algebraic problem cannot be solved (confirming that, for a generic $\mathbf{f}$, no harmonic solutions exist in resonance conditions). However, according to the Rouché-Capelli theorem, the algebraic problem equally admits solution (and therefore the response is harmonic of frequency α) if and only if the following *compatibility condition* is met (the discussion carried out in the Sect. 4.5.2 should be remembered):

$$\mathbf{v}^T \mathbf{f} = 0, \tag{11.61}$$

where $\mathbf{v}$ is the left eigenvector associated with α, solution to:

$$\left(-\alpha^2 \mathbf{M} + i\alpha \mathbf{C} + \mathbf{K}\right)^T \mathbf{v} = \mathbf{0}. \tag{11.62}$$

With solvability satisfied, the solution $\mathbf{w}$ to Eq. 11.60 is determined to within an additive contribution parallel to the eigenvector $\mathbf{u}$ associated with α. In fact, if $\hat{\mathbf{w}}$ is a solution, even $\hat{\mathbf{w}} + \beta\mathbf{u}$ is solution, whatever β. The response of the system in the resonant mode $\mathbf{u}$ is therefore indeterminate, as the system is "labile" in that direction. The indeterminate constant β can be chosen according to a normalization condition.   □

11.4.3   Numerical Results

Numerical applications, taken from [3], are illustrated, in which the analytical solution provided by the Lindstedt-Poincaré method is compared with the exact one, obtained through numerical integrations of the equations of motion, Eqs. 11.34.

Preliminary, an analysis is carried out about the influence of the damping parameters, ξ_1, ξ_2, on the coefficient μ_2; this latter governs the curvature of the graph of the function $\mu = \mu(\epsilon)$, according to Eq. 11.54b. The greater the value of μ_2 and, with the same deviation $|\mu - \mu_d|$ from the bifurcation, the smaller the amplitude of the cycle. Therefore, small μ_2 in modulus involve large amplitude

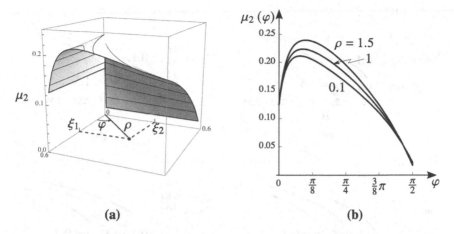

Fig. 11.6 Values of the μ_2 parameter in Eq. 11.54b *vs* the damping coefficients ξ_1, ξ_2: (a) 3D representation, (b) graph of $\mu_2\,(\varphi)$ for assigned values of the damping modulus ρ

oscillations of the double pendulum. Figure 11.6a illustrates the graph of the function $\mu_2 = \mu_2\,(\xi_1, \xi_2)$, obtained by solving Eq. 11.52. All values of μ_2 are found to be positive, so that *the Hopf bifurcation of the Ziegler column is super-critical* (i.e., the limit cycles exist for $\mu > \mu_d$ and are stable, as it can be proved by analytical arguments omitted here). The surface, around the origin, has a shape similar to the Whitney umbrella (graph of $\mu_d = \mu_d\,(\xi_1, \xi_2)$ in Fig. 11.5b). Fixed damping modulus $\rho := \sqrt{\xi_1^2 + \xi_2^2}$, μ_2 depends on the angle $\varphi := \arctan\left(\frac{\xi_2}{\xi_1}\right)$, as shown in Fig. 11.6b. It emerges that, far away from $\varphi \simeq 0.24$,[25] the coefficient μ_2 decreases to very small values, denoting the existence of limit cycles of great amplitude. The damping, therefore, produces detrimental effects not only in the linear field, where, as it has been seen, it generally reduces the critical load, but also in the nonlinear field, where it implies large oscillations.

For a certain pair of damping coefficients ξ_1, ξ_2, the graph in Fig. 11.6a gives the corresponding value of μ_2, which makes it is possible to draw the bifurcation diagram $\mu = \mu\,(\epsilon)$, according to Eq. 11.54b. Remembering that, due to the normal-ization adopted, $(\theta_{1max}, \theta_{2max}) = \epsilon\,(2\,\|u_1\|, 1)$, it is possible to represent the two rotations as a function of the load μ, such as $(\theta_{1max}, \theta_{2max}) = \sqrt{\frac{\mu - \mu_d}{\mu_2}}\,(2\,\|u_1\|, 1)$. Figure 11.7 shows the diagrams relevant to two differently damped systems: (a) with small damping angle $\varphi = 0.05$, whose destabilizing effect, according to the linear analysis (Fig. 11.5), is weak, and (b) with medium-large damping angle $\varphi = \frac{\pi}{4}$, whose destabilizing effect is relevant.

---

[25] The original paper [3] contains a mere printing error, removed here.

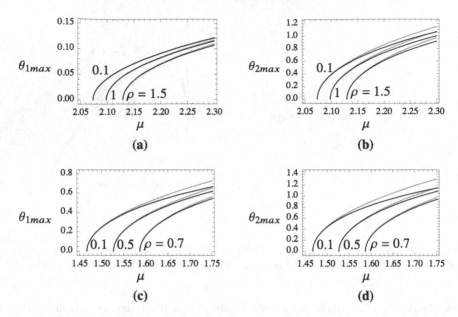

Fig. 11.7 Bifurcation diagrams for the Ziegler columns. Rotations θ_{1max}, θ_{2max}, for several values of ρ and for (**a**), (**b**) $\varphi = 0.05$; (**c**), (**d**) $\varphi = \frac{\pi}{4}$; thick gray lines, asymptotic Solution; thin black lines, numerical solution

From the examination of these figures, the following conclusions are drawn:

- the curves bifurcate from the fundamental path at values of μ_d slightly lower/higher than $\mu_c \simeq 2.09$, in case (a) (Fig. 11.7a,b), and at much smaller values than μ_c, in case (b) (Fig. 11.7c,d);
- the amplitudes of the rotations are very large, of the order of 1 rad, even for load increment of only 15% with respect to the critical value;
- the rotation of the free-end rod, θ_{2max}, is greater than that of the grounded rod, θ_{1max}, of the order of 10 times in case (a) (for which $\xi_2 \ll \xi_1$), and only twice in case (b) (for which $\xi_2 = \xi_1$);
- the agreement between asymptotic and numerical solutions is excellent, in spite of the fact that the exact equations have been expanded only to the third degree and the perturbation method applied at the lowest level.

Finally, the evolution of the response is observed for fixed damping coefficients and load. This can be done, for example, in terms of configurations variables (θ_1, θ_2), by tracing the ellipse described by Eq. 11.55, or of velocities $(\dot{\theta}_1, \dot{\theta}_2)$, by evaluating the time derivative of Eq. 11.55.[26] Figure 11.8 shows the limit cycle relevant to the two systems already analyzed in Fig. 11.7, for fixed values of ρ, φ, and μ.

---

[26] The cycle develops in the four-dimensional state-space $(\theta_1, \theta_2, \dot{\theta}_1, \dot{\theta}_2)$; the diagrams shown here are therefore two-dimensional projections.

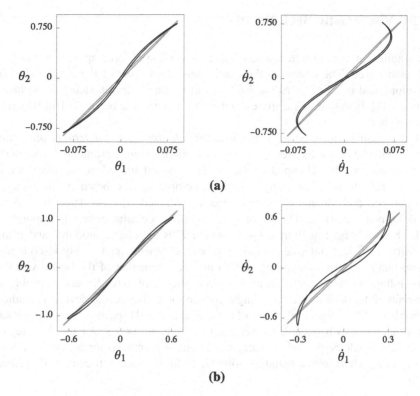

Fig. 11.8 Limit cycles of the Ziegler column for (**a**) $\varphi = 0.05$, $\rho = 0.1$, $\mu = 2.2$; (**b**) $\varphi = \frac{\pi}{4}$, $\rho = 0.1$, $\mu = 1.7$; thick gray lines, asymptotic solution; thin black lines, numerical solution

The following observations are made:

- ellipses (plotted by thick gray curves) are very flattened on their major axis, to denote a small contribution of the imaginary part of the eigenvector $\mathbf{u}_d$;[27]
- the numerical solution (plotted by thin black curves) shows the distortional effect of the harmonic of frequency 3, which was not considered in the perturbation analysis, as anticipated in the Remark 11.5;
- the distortion caused by the harmonic 3 is more pronounced when the motion is depicted in the velocity space, as the differentiation introduces a factor equal to 3;
- despite of the small differences, the asymptotic solution yields excellent qualitative results, also quantitatively satisfactory.

---

[27] Indeed, when an eigenvector is real, the trajectory in space of the phases is a straight line, because all the material points move in-phase. This is easily deduced from Eq. 11.44, when $\text{Im}(\mathbf{u}_d) = \mathbf{0}$..

11.5 Viscoelastic Beck Beam

The stability of a fixed-free beam, subject to a follower force applied at the free end, is analyzed. The system is the continuous counterpart of the Ziegler double pendulum, and it is known as the *Beck beam*; originally formulated for an elastic beam [1, 11], it was subsequently extended to a viscoelastic beam [5, 10], which is analyzed here.

The system is modeled as a planar *extensible and shear undeformable beam* (Fig. 11.9). The beam has length ℓ, cross-sectional inertia moment I and mass linear density m. It's clamped at the extreme A and loaded at the free end B by a follower force of intensity F, which compresses the beam in the straight configuration and maintains its own direction aligned with the tangent at B to the deformed centerline. The beam is made of viscoelastic material, responding to the Kelvin-Voigt constitutive law,[28] where E is the elastic modulus and η the viscosity coefficient (internal damping); moreover, it rests on a purely viscous soil of constant c (external damping), simulating the interaction of the beam with the surrounding fluid (e.g., air). The analysis is limited here to the linearized problem. A variant of the Beck beam, including a concentrated viscoelastic device, is studied in Problem 14.5.2. A generalization of the model to the 3D space, in which torsional vibrations are also triggered, is presented in Problem 14.5.3. As for the Ziegler columns, it would be possible to carry out a nonlinear analysis for the evaluation of the limit cycle. The relevant study, omitted here for the sake of brevity, is described in [6].

Fig. 11.9 Viscoelastic Beck beam

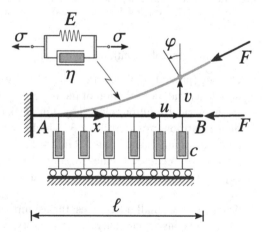

---

[28] The rheological model of Kelvin-Voigt consists of an elastic spring arranged in parallel to a viscous dashpot, in which the relative forces add up.

11.5.1 Linearized Model

A linearized model of extensible and shear undeformable beam is formulated in the framework of the theory of prestressed bodies (Chap. 6). The model generalizes, to the dynamic field and viscoelastic bodies, the Euler beam developed in Chap. 7.[29] As a major difference, since the system dealt with here is nonconservative, a virtual work approach is followed, as discussed in Sect. 6.3. Accordingly, the equations of motion are derived through the Extended Hamilton Principle; then, they are interpreted on the basis of the balance of forces.

Kinematics

The displacement field of the beam is described by the three scalar components $u(x, t)$, $v(x, t)$, $\varphi(x, t)$, which respectively represent the longitudinal and transverse translation and the cross-sectional rotation. The condition of shear undeformability, however, reduces the independent displacements to two, e.g., u, v, being $\varphi = v'$ in linear kinematics. Remembering the congruence Eqs. 7.5, it holds:

$$\varepsilon = u' + \frac{1}{2}v'^2 + \cdots, \tag{11.63a}$$

$$\kappa = v'' + \cdots, \tag{11.63b}$$

where ε is the unit extension and κ is the flexural curvature. In the former strain, a first-order part $\varepsilon_1 := u'$ and a second-order part $\varepsilon_2 := \frac{1}{2}v'^2$ are recognized; in the latter strain, the first-order part $\kappa_1 := v''$ is sufficient to a linearized analysis (since κ only appears, squared, in the elastic energy).

Viscoelastic Constitutive Law

The viscoelastic behavior of the beam is expressed by the Kelvin-Voigt model, which makes the solicitation depending both on strains and strain rates. Taking into account the axial prestress, and the linear part only of the strains, it reads:[30,31]

$$\begin{pmatrix} N \\ M \end{pmatrix} = \begin{pmatrix} N_0 \\ 0 \end{pmatrix} + \begin{pmatrix} \tilde{N} \\ \tilde{M} \end{pmatrix} = \begin{pmatrix} N_0 \\ 0 \end{pmatrix} + \begin{pmatrix} EA\varepsilon_1 + \eta A\dot{\varepsilon}_1 \\ EI\kappa_1 + \eta I\dot{\kappa}_1 \end{pmatrix}. \tag{11.64}$$

---

[29] The nonlinear model, as developed, for example, in [6], is instead more easily formulated in the hypothesis of inextensibility, as done in this book in Chap. 5, with reference to the elastic Euler beam.

[30] The second in Eq. 11.64 is formally identical to Eq. 11.7, valid for the (discrete) Ziegler column.

[31] Equation 11.64 highlights a *Principle of Similarity* existing between the elastic and the viscoelastic law, for which the second one is obtained from the first one by replacing the elastic modulus E by the viscoelastic "modulus" $E + \eta \frac{\partial}{\partial t}$.

The internal force soliciting the viscous soil is:[32]

$$f_{int}^v = c\,\dot{v}. \tag{11.65}$$

Variational Principle and Equations of Motion

The equations of motion are derived from the extended Hamilton principle (Eq. D.3 of the Appendix D) that reads:

$$\delta H\,[u, v] := \int_{t_1}^{t_2} (\delta T - \delta W_{int} + \delta W_{ext})\,\mathrm{d}t = 0, \tag{11.66}$$

$$\forall\,(\delta u, \delta v)\,|\,\delta u\,(x, t_i) = \delta v\,(x, t_i) = 0,\ i = 1, 2,\quad \forall x,$$

which is the continuous counterpart of Eq. 11.1, holding for discrete systems.

The kinetic energy of the beam is $T = \frac{1}{2}\int_0^\ell m\,(\dot{u}^2 + \dot{v}^2)\,\mathrm{d}x$, from which:

$$
\begin{aligned}
\int_{t_1}^{t_2} \delta T\,\mathrm{d}t &= \int_{t_1}^{t_2} \mathrm{d}t \int_0^\ell m\,(\dot{u}\delta\dot{u} + \dot{v}\delta\dot{v})\,\mathrm{d}x \\
&= \int_0^\ell \mathrm{d}x \int_{t_1}^{t_2} m\,(\dot{u}\delta\dot{u} + \dot{v}\delta\dot{v})\,\mathrm{d}t \\
&= \int_0^\ell \mathrm{d}x \left(-\int_{t_1}^{t_2} m\,(\ddot{u}\delta u + \ddot{v}\delta v)\,\mathrm{d}t + [m\dot{u}\delta u + m\dot{v}\delta v]_{t_1}^{t_2} \right),
\end{aligned}
\tag{11.67}
$$

having integrated by parts over the time and taken into account that the variations cancel out at the ends of the interval (t_1, t_2).

The internal virtual work is sum of the works expended by forces soliciting the beam and the viscous soil, that is, $\delta W_{int} := \delta W_{int}^b + \delta W_{int}^v$. Regarding the first contribution, it is necessary to consider that, in the spirit of linearized theory, $N\delta\varepsilon = \left(N_0 + \tilde{N}\right)(\delta\varepsilon_1 + \delta\varepsilon_2) \simeq \left(N_0 + \tilde{N}\right)\delta\varepsilon_1 + N_0\delta\varepsilon_2$, having neglected the product between the increments (Sect. 6.3). Taking into account the kinematics (Eq. 11.63), and that $N_0 = -F$, it follows:

---

[32] Here, the viscous force $f^v = f_{int}^v$ is considered as *internal* to the beam-soil system; consistently, it will be considered to contribute to the internal virtual work δW_{int} (Eq. 11.66). Alternatively, the system can be considered as consisting of the only beam, on which the environment (and no longer the soil) exerts an *external* viscous force $f_{ext}^v = -f_{int}^v = -c\,\dot{v}$, which should therefore be considered contributing to external virtual work δW_{ext}. In other words, in the first case, the force is the one acting on the dissipators, in the second one on the beam. The two procedures obviously lead to the same result.

$$\delta W_{int}^b = \int_0^\ell \left[\left(-F + \tilde{N} \right) \delta \varepsilon_1 - F \delta \varepsilon_2 + M \delta \kappa \right] dx$$

$$= \int_0^\ell \left[\left(-F + \tilde{N} \right) \delta u' - F v' \delta v' + M \delta v'' \right] dx$$

$$= \int_0^\ell \left[-\tilde{N}' \delta u + \left(F v'' + M'' \right) \delta v \right] dx \tag{11.68}$$

$$+ \left[\left(-F + \tilde{N} \right) \delta u + M \delta v' + \left(-F v' - M' \right) \delta v \right]_0^\ell,$$

having integrated twice for parts in space. The second contribution is:

$$\delta W_{int}^v = \int_0^\ell f_{int}^v \delta v dx. \tag{11.69}$$

The follower force, in the adjacent configuration, has components $F_x \simeq -F$, $F_y \simeq -F v_B'$; consequently, it does a virtual work equal to:

$$\delta W_{ext} = -F \delta u_B - F v_B' \delta v_B \quad . \tag{11.70}$$

By summarizing, the Principle in Eq. 11.66 requires that:

$$\delta H [u, v] = \int_{t_1}^{t_2} dt \int_0^\ell \left\{ -m \left(\ddot{u} \delta u + \ddot{v} \delta v \right) + \left[\tilde{N}' \delta u - \left(F v'' + M'' \right) \delta v \right] - f_{int}^v \delta v \right\} dx$$

$$- \int_{t_1}^{t_2} \left\{ \left[\left(-F + \tilde{N} \right) \delta u + M \delta v' + \left(-F v' - M' \right) \delta v \right]_0^\ell \right.$$

$$+ F \delta u_B + F v_B' \delta v_B \bigg\} dt = 0, \qquad \forall \, (\delta u, \delta v) \, . \tag{11.71}$$

From this latter, one gets the field equations:

$$m \ddot{u} - \tilde{N}' = 0, \tag{11.72a}$$

$$m \ddot{v} + M'' + F v'' + f_{int}^v = 0. \tag{11.72b}$$

Then, by introducing the geometric conditions $u_A = v_A = v_A' = 0$, and observing that the terms $\pm F \delta u_B$ and $\pm F v_B' \delta v_B$ cancel each other out, the mechanical boundary conditions follow:

$$\tilde{N}_B = 0, \tag{11.73a}$$

$$-M_B' = 0, \tag{11.73b}$$

$$M_B = 0. \tag{11.73c}$$

Using the constitutive law in Eqs. 11.64 and 11.65, as well as kinematics, the equations of motion, accompanied by the geometric and mechanical conditions, are finally found as:

$$m\ddot{u} - EAu'' - \eta A\dot{u}'' = 0, \tag{11.74a}$$

$$u_A = 0, \tag{11.74b}$$

$$EAu'_B + \eta A\dot{u}'_B = 0, \tag{11.74c}$$

together with:

$$m\ddot{v} + EIv'''' + \eta I\dot{v}'''' + Fv'' + c\dot{v} = 0, \tag{11.75a}$$

$$v_A = 0, \tag{11.75b}$$

$$v'_A = 0, \tag{11.75c}$$

$$-EIv'''_B - \eta I\dot{v}'''_B = 0, \tag{11.75d}$$

$$EIv''_B + \eta I\dot{v}''_B = 0. \tag{11.75e}$$

Equations 11.74 and 11.75 describe the longitudinal motion $u(x, t)$ and the transverse motion $v(x, t)$ of the beam, respectively. The two problems are uncoupled, similarly to what happens for the extensible Euler beam. Furthermore, since the longitudinal motion *does not* depend on the intensity of the follower force, it decays over time, due to internal damping. Therefore, the stability problem is governed by Eqs. 11.75 alone, which govern the transverse motion.

Supplement 11.4 (Balance of Forces) Equations 11.75, ruling the transverse motion, can be interpreted, or directly deducted, as a balance of forces, according to d'Alembert principle. The equilibrium equation of the slightly inflected elastic beam is:

$$EIv'''' - N_0v'' = f^{in} + f^v_{ext}, \tag{11.76}$$

where $N_0 = -F$ is the normal force in the prestressed configuration, $f^{in} = -m\ddot{v}$ is the inertia force, and $f^v_{ext} = -c\dot{v}$ is the external viscous force (exerted by the environment on the beam). By invoking the viscoelastic similarity principle (Note 31), the substitution $E \to E + \eta\frac{\partial}{\partial t}$ must be performed, from which the field Eq. 11.75a is recovered.

The geometric boundary conditions in Eq. 11.75b,c are obvious; however, the mechanical conditions in Eq. 11.75d,e require more attention. They establish that the viscoelastic shear force $T_B := -I\left(E + \eta\frac{\partial}{\partial t}\right)v'''_B$ and the viscoelastic bending moment $M_B := I\left(E + \eta\frac{\partial}{\partial t}\right)v''_B$, evaluated at the free end B of the beam, are equal to zero. While the condition on the bending moment coincides with that of the Euler problem, relevant to a beam subject to a gravitational force (Eq. 7.12c), the condition on the shear force differs from that problem (Eq. 7.12b). This happens because the geometric effect, already discussed for the Euler problem in the Remark 7.3, is now missing, due to the rotation of the follower force; hence, the shear force at the boundary is equal to zero in each configuration.

An alternative reasoning, in line with Remark 7.3, is the following. It is based on equilibrium at the boundary in the y direction (as given from the variational principle, and not in the direction normal to the deformed axis). The transverse component of the internal force in the adjacent configuration is $V = T \cos \varphi + N_0 \sin \varphi = T - F v'$. Since, for the equilibrium at boundary, $V_B = F_y$ must hold, with $F_y = -F v'_B$, it follows that $T_B = 0$. □

Remark 11.6 The condition of null shear force at the boundary embodies the essence of the nonconservativeness of the problem. Equations 11.75, in fact, cannot be derived from a potential, because of this boundary condition. In mathematical terms, it can be said that while the field equations are self-adjoint, the boundary conditions are not self-adjoint, so that the problem, as a whole, is *non-self-adjoint*.

Nondimensional Form of the Equations of Motion
Also in this case, in order to facilitate the discussion, the equations of motion are recast in nondimensional form. The following definitions are introduced:

$$\tilde{x} := \frac{x}{\ell}, \qquad \tilde{t} := t\sqrt{\frac{EI}{m\ell^4}}, \qquad \tilde{v} := \frac{v}{\ell},$$

$$\mu := \frac{F\ell^2}{EI}, \qquad \tilde{\eta} := \eta\sqrt{\frac{I}{Em\ell^4}}, \qquad \tilde{c} := \frac{c\ell^2}{\sqrt{EIm}}, \tag{11.77}$$

where $\tilde{x}, \tilde{t}$ and $\tilde{v}$ are, in the order, the nondimensional abscissa, time, and transverse displacement; μ is the nondimensional external force; $\tilde{c}$ and $\tilde{\eta}$ are nondimensional damping parameters, respectively, external and internal. The equations of motion, Eqs. 11.75, are accordingly rewritten in the form:

$$\ddot{\tilde{v}} + \tilde{v}'''' + \tilde{\eta}\,\dot{\tilde{v}}'''' + \mu\tilde{v}'' + c\,\dot{\tilde{v}} = 0, \tag{11.78a}$$

$$\tilde{v}_A = 0, \tag{11.78b}$$

$$\tilde{v}'_A = 0, \tag{11.78c}$$

$$-\tilde{v}'''_B - \tilde{\eta}\,\dot{\tilde{v}}'''_B = 0, \tag{11.78d}$$

$$\tilde{v}''_B + \tilde{\eta}\,\dot{\tilde{v}}''_B = 0, \tag{11.78e}$$

where the dot denotes differentiation with respect to the nondimensional time $\tilde{t}$ and the dash with respect to the nondimensional abscissa $\tilde{x}$. In the following, for brevity of notation, the tildes on nondimensional quantities will be omitted.

11.5.2 Undamped Beam

First, the stability analysis of the undamped beam is carried out. By letting $\eta = c = 0$, the equations of motion, Eqs. 11.78, reduce to:

$$\ddot{v} + v'''' + \mu\, v'' = 0,$$ (11.79a)

$$v_A = 0,$$ (11.79b)

$$v'_A = 0,$$ (11.79c)

$$-v'''_B = 0,$$ (11.79d)

$$v''_B = 0.$$ (11.79e)

By separating the variables, according to $v(x, t) = \phi(x) \exp(\lambda t)$, a boundary value problem in the space only follows, i.e.,

$$\phi'''' + \mu\, \phi'' + \lambda^2 \phi = 0,$$ (11.80a)

$$\phi_A = 0,$$ (11.80b)

$$\phi'_A = 0,$$ (11.80c)

$$-\phi'''_B = 0,$$ (11.80d)

$$\phi''_B = 0.$$ (11.80e)

The field Eq. 11.80a admits the solution:

$$v(x) = c_1 \cos(\alpha x) + c_2 \sin(\alpha x) + c_3 \cosh(\beta x) + c_4 \sinh(\beta x),$$ (11.81)

where c_j ($j = 1, \ldots 4$) are arbitrary constants and α, β are defined as:

$$\alpha^2 := \frac{\mu + \sqrt{\mu^2 - 4\lambda^2}}{2}, \quad \beta^2 := \frac{-\mu + \sqrt{\mu^2 - 4\lambda^2}}{2}.$$ (11.82)

Substitution of the general solution, Eq. 11.81, in the boundary conditions leads to the algebraic problem:

$$\mathbf{Sc} = \mathbf{0},$$ (11.83)

where $\mathbf{c} = \left(c_j\right)^T$ is a column matrix and:

Fig. 11.10 Frequencies of the undamped Beck beam *vs* the nondimensional follower force $\mu := \frac{F\ell^2}{EI}$; the critical values μ_{ci} of the force correspond to points with vertical tangent, at which a circulatory Hopf bifurcation occurs

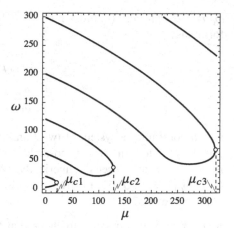

$$
\mathbf{S} := \begin{bmatrix} 1 & 0 & 1 & 0 \\ 0 & \alpha & 0 & \beta \\ -\alpha^3 \sin\alpha & \alpha^3 \cos\alpha & -\beta^3 \sinh\beta & -\beta^3 \cosh\beta \\ -\alpha^2 \cos\alpha & -\alpha^2 \sin\alpha & \beta^2 \cosh\beta & \beta^2 \sinh\beta \end{bmatrix}. \tag{11.84}
$$

The matrix $\mathbf{S}$ depends, through the wave-numbers α and β, on the eigenvalue λ and the bifurcation parameter μ; therefore, $\mathbf{S} = \mathbf{S}(\lambda, \mu)$. The roots of the characteristic equation $\det \mathbf{S} = 0$ are the eigenvalues λ, function of μ. The equation reads:

$$
\alpha^4 + \beta^4 + \alpha\beta \left(\alpha^2 - \beta^2\right) \sin\alpha \sinh\beta + 2\alpha^2\beta^2 \cos\alpha \cosh\beta = 0. \tag{11.85}
$$

When $\mu = 0$, Eq. 11.85 admits an infinite number of purely imaginary eigenvalues $\lambda = \pm i\,\omega_j$, coinciding with the natural frequencies of the beam. When the follower force increases from zero and reaches the first critical value $\mu = \mu_{c1}$, the first two pairs of eigenvalues collide (as it happens for the Ziegler column) and then separate, one on the Right and the other on the left of the imaginary axis (therefore, a *circulatory Hopf bifurcation* occurs). All the other (infinitely many) eigenvalues not affected by the collision remain on the imaginary axis, until, for $\mu = \mu_{c2}$, the third and fourth pair of eigenvalues collide, to give rise to a successive bifurcation. For larger values of μ, it is observed the collision between the fifth and sixth pair and so on. Figure 11.10 describes the locus of the ω frequencies as the nondimensional force μ increases. The bifurcation values μ_{ci} are the abscissas of points of the locus at which the tangent is vertical.

The critical loads are determined as follows. By letting $\lambda = i\omega_c$, $\mu = \mu_c$, and making use of the definitions in Eqs. 11.82, the characteristic equation becomes a real function of the two unknown parameters ω_c, μ_c. By denoting the equation as $f(\omega_c, \mu_c) = 0$, with $f \in \mathbb{R}$, it must be enforced (i) that this is satisfied and (ii)

ω_c *is a* *double root* (in order for the coalescence between two roots occurs).[33] In formulas:

$$f(\omega_c, \mu_c) = 0, \tag{11.86a}$$

$$\frac{\partial f(\omega_c, \mu_c)}{\partial \omega_c} = 0, \tag{11.86b}$$

which constitutes a system of two transcendental equations in the two unknowns. For this system it is not possible to draw a closed form solution, but it is necessary to proceed numerically. With reference to Fig. 11.10, one finds $(\mu_{c1}, \omega_{c1}) = (20.051, 11.016)$, $(\mu_{c2}, \omega_{c2}) = (127.811, 37.055)$, $(\mu_{c3}, \omega_{c3}) = (317.981, 67.567)$.

Remark 11.7 It is possible to check that the Beck beam (like the Ziegler column) *cannot buckle statically*. The equation $\det \mathbf{S} = 0$, in fact, as it can be seen from Fig. 11.10, does not admit the root $\omega_c = 0$, however great is the magnitude of the force.

Remark 11.8 The critical value of the follower force is $F_c = 20.051 \frac{EI}{\ell^2}$; the Eulerian critical load is instead $P_c = \pi^2 \frac{EI}{4\ell^2} \simeq 2.467 \frac{EI}{\ell^2}$. Hence, $\frac{F_c}{P_c} \simeq 8.13$, i.e., the dynamic instability manifests at significantly larger follower force than gravitational force.

11.5.3   Damped Beam

The more general case, in which both the internal η and external c damping coefficients are different from zero, is dealt with. By letting again $v(x,t) = \phi(x) \exp(\lambda t)$ in the equations of motion, Eqs. 11.78, a spatial boundary value problem follows, i.e.,

$$(1 + \lambda \eta) \phi'''' + \mu \phi'' + \left(\lambda^2 + \lambda c\right) \phi = 0, \tag{11.87a}$$

$$\phi_A = 0, \tag{11.87b}$$

$$\phi'_A = 0, \tag{11.87c}$$

$$-\phi'''_B = 0, \tag{11.87d}$$

$$\phi''_B = 0, \tag{11.87e}$$

---

[33] This second condition, therefore, replaces the simpler zeroing of the discriminant, used for the Ziegler column; in fact, an explicit expression of the solution, known in the discrete problem, is not available in the continuous problem.

having simplified the factor $1 + \lambda \eta$ in the mechanical boundary conditions. The field Eq. 11.87a admits a solution formally analogous to that of the undamped beam (Eq. 11.81), in which, however, the wave-numbers are redefined as follows:

$$\alpha^2 := \frac{\mu + \sqrt{\mu^2 - 4\lambda\,(c + \lambda)\,(1 + \lambda\,\eta)}}{2\,(1 + \lambda\,\eta)}, \quad \beta^2 := \frac{-\mu + \sqrt{\mu^2 - 4\lambda\,(c + \lambda)\,(1 + \lambda\,\eta)}}{2\,(1 + \lambda\,\eta)}.$$

$$(11.88)$$

Since the homogeneous boundary conditions in Eqs. 11.87b,c are not changed by damping, they are still expressed by Eqs. 11.83 and 11.84, in which the new definitions of α, β hold. The matrix $\mathbf{S}$ depends, through α and β, on the eigenvalue λ and the bifurcation parameter μ, as well as the damping coefficients; it is therefore $\mathbf{S} = \mathbf{S}\,(\lambda, \mu; c, \eta)$. The characteristic equation $\det \mathbf{S} = 0$ is also formally analogous to Eq. 11.85. For each fixed set of (μ, c, η), by letting $\lambda = \delta + i\omega$ in the characteristic equation and separating the real and imaginary parts, two equations in the unknowns δ, ω are obtained. The bifurcation condition corresponds to the crossing of the imaginary axis of only one pair of complex conjugated eigenvalues (generic Hopf bifurcation), for which $\lambda = i\omega_d$, $\mu = \mu_d$. Substitution in the characteristic equation, and separation of the real and imaginary part, leads to:

$$\mathrm{Re}\,[\det\,(\mathbf{S}\,(i\omega_d, \mu_d; c, \eta))] = 0, \qquad (11.89a)$$

$$\mathrm{Im}\,[\det\,(\mathbf{S}\,(i\omega_d, \mu_d; c, \eta))] = 0, \qquad (11.89b)$$

which, for a given damping (c, η), is a system of real equations in the unknowns (ω_d, μ_d). Among the infinite pairs of roots, one is interested in that admitting the smallest μ_d.

These equations can only be solved numerically. For each assigned damping pair, a critical point is found, lying on a surface of the (c, η, μ) space, illustrated in Fig. 11.11a. This surface is known as the *Whitney umbrella* for the Beck viscoelastic beam (similar to the surface of Fig. 11.5b) [4, 5, 10]. Therefore, the same considerations made for the Ziegler column apply to the Beck beam, that is, (a) damping generally produces an *unstable effect* on the undamped system and (b) the damped system is not in continuity with the undamped one. If one puts $c = \rho \cos \varphi$, $\eta = \rho \sin \varphi$ in Eqs. 11.89, and evaluates the limit as ρ tends to zero, he obtains the critical load μ_d^0 of a system affected by an infinitely small damping. The graph of μ_d^0 vs the angle $\varphi := \arctan\left(\frac{\eta}{c}\right)$, is represented in Fig. 11.11b. Unlike the Ziegler column, the optimal direction of the damping (for which, exceptionally, the two systems are in continuity), is $\varphi_{opt} = 0$, that is, $\eta = 0$. It is concluded that, for the Beck beam, the *internal damping is destabilizing*, while the *external damping is stabilizing*.

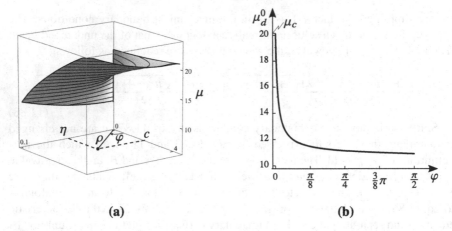

(a)   **(b)**

Fig. 11.11 Generic Hopf bifurcation of the viscoelastic Beck beam: (a) locus of the critical states in the parameter space (μ, c, η); (b) critical value μ_d^0 of the follower force for evanescent damping, $vs\ \varphi := \arctan\left(\frac{\eta}{c}\right)$

References

1. Bolotin, V.V.: Nonconservative Problems of the Theory of Elastic Stability. Macmillan, London (1963)
2. Cesari, L.: Asymptotic Behavior and Stability Problems in Ordinary Differential Equations. Springer, Berlin (1971)
3. D'Annibale, F., Ferretti, M.: On the effects of linear damping on the nonlinear Ziegler's column. Nonlinear Dynam. **103**(4), 3149–3164 (2021)
4. Kirillov, O.N.: Nonconservative Stability Problems of Modern Physics. De Gruyter, Berlin (2013)
5. Luongo, A., D'Annibale, F.: On the destabilizing effect of damping on discrete and continuous circulatory systems. J. Sound Vib. **333**(24), 6723–6741 (2014)
6. Luongo, A., D'Annibale, F.: Nonlinear hysteretic damping effects on the post-critical behaviour of the visco-elastic Beck's beam. Math. Mech. Solids **22**(6), 1347–1365 (2017)
7. Luongo, A., D'Annibale, F., Ferretti, M.: Hard loss of stability of Ziegler's column with nonlinear damping. Meccanica **51**(11), 2647–2663 (2016)
8. Nayfeh, A.H.: Perturbation Methods. Wiley, New York (1973)
9. Nayfeh, A.H., Mook, D.T.: Nonlinear Oscillations. Wiley, New York (1995)
10. Seyranian, A.P., Mailyaev, A.A.: Multiparameter Stability Theory with Mechanical Applications. World Scientific, Singapore (2003)
11. Sugiyama, Y., Langthjem, M.A., Katayama, K.: Dynamic Stability of Columns Under Nonconservative Forces. Springer, Cham (2019)
12. Ziegler, H.: Principles of Structural Stability. Blaisdell Publishing, Waltham (1968)

Chapter 12
Aeroelastic Stability

12.1 Introduction

Elastic structures under wind flow are subjected to forces that, under the strongly simplifying assumptions of the *quasi-steady theory*, can be considered as dependent only on the state of the system (i.e., from its position and velocity). Therefore, the forces contribute, in their linear part, to the geometric stiffness and damping of the structure. If these effects are instabilizing and the wind velocity is sufficiently high, they prevail on the elastic and dissipative forces, causing loss of stability. The bifurcation can either be of static type (similarly to that experienced by the Euler beam), which brings the system to a different equilibrium position, or, more frequently, of dynamic type (as for the damped Beck beam), which leads the system to periodically oscillate on a limit cycle.

The simplest form of aeroelastic instability, said *galloping*, is exhibited by a single degree of freedom oscillator. In this case, since the wind flow alters the damping coefficient only (since the structural velocity affects the relative wind velocity, as "seen" by the structure), when the total damping vanishes and incipiently becomes negative, the equilibrium loses stability, and the oscillator executes harmonic vibrations of amplitude exponentially diverging (possibly stabilizing on a stable limit cycle, if any). The zeroing of the total damping, as condition of instability, is known as the *Den Hartog criterion*. Such a simple criterion can be successfully applied to strings and beams, when they are reduced to single degree of freedom systems via the Galerkin method.

Rigid cylinders elastically constrained, allowed to translate in the two transverse directions and to rotate around their own axis, exhibit additional forms of instability (a) *rotational divergence*, (b) *rotational galloping*, (c) *translational galloping*, and (d) *roto-translational galloping* (more frequently said *flutter*). The first two forms involve just one degree of freedom, the last two forms, two degrees of freedom.

All these phenomena and methods of analysis are illustrated in this chapter. Moreover, the limits of the quasi-steady theory of the aerodynamic forces are discussed.

12.2   Aerodynamic Forces

Bodies, when hit by wind, are subject to pressures. To evaluate them, a *cylinder rigid and fixed to the ground* is considered, of axis $\mathbf{a}_z$, immersed in laminar wind flow of velocity $\mathbf{U} = U\mathbf{a}_d$, orthogonal ad $\mathbf{a}_z$ (Fig. 12.1a). If the cylinder is long enough, edge effects can be neglected, so that the velocity field of the fluid is assumed planar, orthogonal to $\mathbf{a}_z$. If the cylinder cross-section has a contour curve free of corners, with slowly variable curvature (i.e., if the figure has an "aerodynamic shape"), the fluid threads remain attached to the body, without breaking. In this case, by using the equations of motion of fluids, it is possible to determine the field of normal and tangential pressures acting on the body and, by integration on the cylinder lateral surface, to compute the resulting force $\mathbf{f}^a$ and couple $\mathbf{c}^a$ per unit length of cylinder. These are called *aerodynamic forces* and have physical dimensions $[\mathrm{MT}^{-2}]$ and $[\mathrm{MLT}^{-2}]$, respectively. By projecting the forces onto the basis $(\mathbf{a}_d, \mathbf{a}_l, \mathbf{a}_z)$, with $\mathbf{a}_l := \mathbf{a}_z \times \mathbf{a}_d$ which completes the left-handed triad, one has:

$$\mathbf{f}^a := f_d\mathbf{a}_d + f_l\mathbf{a}_l, \qquad \mathbf{c}^a := c_z\mathbf{a}_z. \tag{12.1}$$

The components f_d, f_l, c_z are respectively called:

- *drag force* (or resistance force) f_d, acting in the direction of the flow, and positive if oriented as $\mathbf{a}_d$;
- *lift force* (or carrying force) f_l, acting in the direction orthogonal to the flow, and positive if oriented as $\mathbf{a}_z \times \mathbf{a}_d$;
- *aerodynamic couple* c_z, defined with respect to a predefined center, and positive if oriented as $\mathbf{a}_z$.

The three components of the aerodynamic force are expressed as:

$$f_d = \frac{1}{2}\rho_a U^2 D C_d, \tag{12.2a}$$

$$f_l = \frac{1}{2}\rho_a U^2 D C_l, \tag{12.2b}$$

$$c_z = \frac{1}{2}\rho_a U^2 D^2 C_m, \tag{12.2c}$$

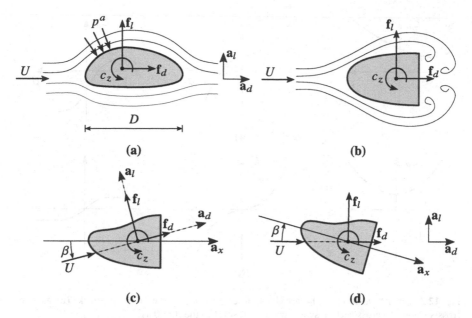

Fig. 12.1 Aerodynamic forces on fixed rigid cylinder: (**a**) aerodynamic cross-section, (**b**) squat cross-section, (**c**) cross-section subject to a flow rotated by an angle $\beta > 0$ (counterclockwise), (**d**) cross-section rotated by an angle $-\beta$, subject to horizontal wind

where ρ_a is the density of the air,[1] D is a characteristic dimension of the cross-section, and $\frac{1}{2}\rho_a U^2$ is the *kinetic pressure*, equal to the kinetic energy per unit of fluid volume. Moreover, C_d is the drag coefficient, C_l the lift coefficient, and C_m the moment coefficient, all nondimensional and called *aerodynamic coefficients*. These depend on the shape of the cross-section and, of course, on the incident direction of wind.

The aerodynamic coefficients can be determined analytically only in very particular conditions, for example, for idealized airfoils. In almost all cases, it is necessary to determine them experimentally by wind tunnel tests. This is all the more true, when considering *bluff bodies*, having sharp edges and highly variable curvatures (Fig. 12.1b). In these cases, indeed, the fluid threads break into vortices, rising strong pressure differences between the windward and leeward parts of the body.

Angle of Attack

As mentioned, the coefficients C_h ($h = d, l, m$) depend on the incident wind direction. Denoted by β the angle formed by $\mathbf{a}_d$ with a chosen material direction $\mathbf{a}_x$ of the cross-section, positive if $\mathbf{a}_x$ overlaps $\mathbf{a}_d$ rotating *counterclockwise* by an angle

---

[1] It is $\rho_a = 1.225$ Kg/m^3 in standard conditions of temperature and pressure.

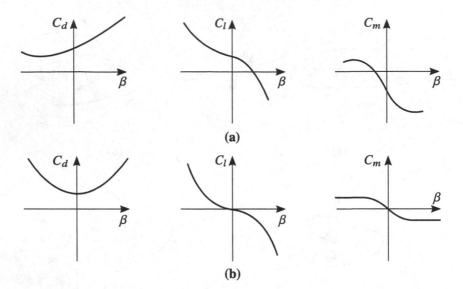

Fig. 12.2 Qualitative form of the aerodynamic coefficients *vs* the angle of attack: (**a**) generic cross-section, (**b**) cross-section symmetric with respect to the $\beta = 0$ axis

smaller than π (Fig. 12.1c), it is of interest to determine the laws $C_h(\beta)$. The angle β is called *angle of attack*. The laws can be determined in the wind tunnel, more easily by keeping fixed the direction of the flow and rotating the cylinder around its axis by a clockwise angle β (Fig. 12.1d).[2] Proceeding by small increments of β, and measuring the forces in Eq. 12.2 for each of the angles, the diagrams of $C_h(\beta)$ are constructed by points (and possibly interpolated by polynomials), as qualitatively shown in Fig. 12.2.

The aerodynamic coefficients are of the order of 1 for the most common cross-sections. It is observed that $C_d > 0$ for any β, as the medium (the air) always opposes to penetration by the body. The other two coefficients, instead, can be positive or negative. In particular, if the wind is horizontal and oriented as $\mathbf{a}_x$, a positive lift force indicates that the aerodynamic action is upward (as is the case of the aircraft wing profiles, whose weight is balanced by this force). If, on the other hand, the lift is negative (also called *downforce*), the aerodynamic action is downward (sometimes used in automotive sports to increase the grip of the tires on the road).

If the material axis $\mathbf{a}_x$ is of symmetry for the cross-section, then the $C_h(\beta)$ law also possess symmetries/antisymmetries. Indeed, it happens that, if the forces corresponding to an angle of attack β are considered (Fig. 12.3a) and the drawing is overturned around the symmetry axis (Fig. 12.3b), since the shape of the cross-

---

[2] In this way the relative rotation of the wind with respect to the body is, as wished, counterclockwise.

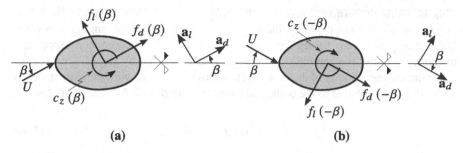

Fig. 12.3 Aerodynamic forces on a symmetric cross-section: (**a**) generic condition, (**b**) overturning around the symmetry axis

section remains unchanged, the forces so obtained correspond to the angle of attack $-\beta$. With reference to the basis $(\mathbf{a}_d, \mathbf{a}_l)$, which is *intrinsic* to the direction of the flow,[3] it is observed that $f_d(-\beta) = f_d(\beta)$ and also $f_l(-\beta) = -f_l(\beta)$, $c_z(-\beta) = -c_z(\beta)$. It is concluded that $C_d(\beta)$ is a symmetric function and that $C_l(\beta)$, $C_m(\beta)$ are antisymmetric functions (Fig. 12.2b). If, as a particular case, the section is circular, it is $C_l(\beta) = C_m(\beta) \equiv 0$, as all the axes are of symmetry.

Quasi-Steady Theory

The evaluation of the aerodynamic forces is much more complex when the cylinder does not remain fixed under the action of the wind, but vibrates, due to its elasticity. This can be diffused in the body or, more simply, concentrated in the constraints at the ground. As a matter of fact, it happens that the oscillations of the body modify the flow; consequently the forces, no longer steady, vary over the time as a function of the dynamic response of the body. For this reason, forces are called *aeroelastic*.[4] The determination of unsteady forces is a very complex problem, not entirely resolved, yet; later on (in the Sect. 12.9), a short mention will be given on them, referring to an approximate method widely used in applications. Here the attention is limited to the illustration of an extremely simple (and rough) theory, called *quasi-steady*,[5] which allows to use the (already described) results achieved in wind tunnel experimental tests [1, 3, 4, 8].

The quasi-steady theory is based on the following conjecture: if the body moves slowly, relatively to the material particles of the fluid that invest it, the aeroelastic forces at instant t are about the same that would be measured on the *fixed body*, under an incident flow having a constant angle of attack $\beta(t)$, frozen at time t.

---

[3] The base is not overturned, but follows the flow.

[4] More specifically, aeroelastic forces are generated by the interaction between the action of the fluid and the elastic response of the structure. In the following, the terms "aerodynamic coefficients" and "aerodynamic matrices," will be used, since they relate to measurements made on fixed cylinders; in contrast, "aeroelastic forces" and "aeroelastic actions" will be employed, since they relate to moving cylinders. However, the terms "aerodynamic" and "aeroelastic" are often confused in the common language.

[5] In technical jargon, the less precise wording "quasi-static" theory is also used.

This is called the *instantaneous attack angle* $\beta(t)$. It should be underlined that, due to the vibration of the body, the angle of attack is actually changing over time; hence, the quasi-steady theory ignores the dynamics of the phenomenon. According to this conjecture, the aeroelastic forces are still expressed by Eq. 12.2, except for (a) replacing the velocity of the flow U relative to the ground, with the instantaneous velocity $U_r(t)$, relative to the body, and (b) replacing β with $\beta(t)$; in formulas:

$$f_d = \frac{1}{2}\rho_a U_r^2(t)\, D\, C_d(\beta(t)), \tag{12.3a}$$

$$f_l = \frac{1}{2}\rho_a U_r^2(t)\, D\, C_l(\beta(t)), \tag{12.3b}$$

$$c_z = \frac{1}{2}\rho_a U_r^2(t)\, D^2 C_m(\beta(t)). \tag{12.3c}$$

The Limits of Applicability of the Quasi-steady Theory
One wonders when the quasi-steady theory is applicable. Since the motion of the body must be slow compared to that of the fluid, in order for the body to appear "almost steady" on the time scale in which the motion of the fluid develops, the natural period of the structure, $T_s = \frac{1}{\nu_s}$, with ν_s its natural frequency, must be large compared to the percurrence time $T_p := \frac{D}{U}$ that a fluid particle, moving with a velocity U, employs to cross a body of width D. Hence, it must be $\frac{T_s}{T_p} \gg 1$, that is, $U_{red} := \frac{U}{\nu_s D} \gg 1$. The nondimensional quantity U_{red} is called the *reduced velocity*, and it must be large enough for the quasi-steady theory to be applicable. Empirical rules, dictated in the technical field, indicate as $20 \div 30$ its minimum value. Unfortunately it happens that, in civil engineering, the cross-sections (e.g., of bridges) are wide, so the theory in question is of dubious application.[6] The wind velocity, in these cases, must be very large and the structure extremely flexible (e.g., a long bridge) in order to satisfy the hypotheses of the theory. The towers and the antennas, on the other hand, usually fall within these limits.

12.3 Galloping of Single Degree of Freedom Systems

A single degree of freedom (DOF) aeroelastic system is analyzed, whose loss of stability manifests itself through a phenomenon called *galloping* in the technical literature. The name is suggested by the large vertical oscillations caused, in cold areas, by a horizontal wind acting on iced overhead electric lines.[7]

---

[6] For example, if $D = 10$ m, $U = 60$ m/s (i.e., a hurricane of 216 Km/h) and $T_s = 4s$, it turns out $U_{rid} = 24$.

[7] The ice concretions modify the original circular cross-section of the cables, thus triggering lift forces, which otherwise would be zero.

12.3.1 Model

A horizontal rigid cylinder is considered, of length ℓ, constrained in such a way it can only translate vertically, to which a viscoelastic organ is applied. The cylinder is subject to horizontal steady wind of velocity U (Fig. 12.4), flowing orthogonally to its axis. The equation of motion of the body reads:

$$M\ddot{v} = F_y^{el}(v) + F_y^{v}(\dot{v}) + F_y^{a}(\dot{v}; U),\tag{12.4}$$

where $v(t)$ is the vertical displacement; M is the mass of the body; $F_y^{el}(v)$ is the elastic force exerted by the constraint, which depends on displacement; $F_y^{v}(\dot{v})$ is the viscous force, which depends on the structural velocity; $F_y^{a}(\dot{v}; U)$ is the vertical component of the aeroelastic force, which depends on the structural velocity, as it will be immediately seen, and, parametrically, on the velocity of the flow U; finally, a dot denotes differentiation with respect to time t.

The spring is assumed nonlinear, with symmetric behavior, of constitutive law:

$$F_y^{el} = -\left(k_1 v + k_3 v^3\right),\tag{12.5}$$

where $k_1 > 0$ is the linear stiffness and $k_3 \gtrless 0$ is the cubic stiffness. The spring is called "hardening" if $k_3 > 0$ (i.e., if the nonlinear restoring force is greater in modulus than the linear part) and "softening" if $k_3 < 0$. The viscous damper is instead assumed linear, of (structural) coefficient c_s, so that:

$$F_y^{v} = -c_s \dot{v}.\tag{12.6}$$

The equation of motion is therefore:

$$\ddot{v} + 2\xi_s \omega_s \dot{v} + \omega_s^2 v + \kappa_s v^3 = \frac{1}{M} F_y^{a}(\dot{v}; U),\tag{12.7}$$

Fig. 12.4 Single degree of freedom system, subject to wind

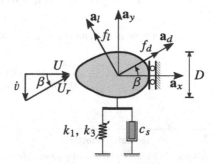

where $\omega_s := \sqrt{\frac{k_1}{M}}$ is the circular frequency of the undamped structure, $\xi_s := \dfrac{c_s}{2\omega_s M}$ is the structural damping factor, and $\kappa_s := \frac{k_3}{M}$ is a structural nonlinearity coefficient.

Aeroelastic Force

To determine the aeroelastic force according to the quasi-steady theory, it needs to determine the *wind relative to the body*, $U_r(t)$, and the instantaneous attack angle $\beta(t)$, which the relative flow forms, at time t, with the material direction $\mathbf{a}_x$. The wind velocity relative to the body, $U_r \mathbf{a}_d$, is the vector difference between the velocity relative to the ground,[8] $U \mathbf{a}_x$, and the body velocity $\dot{v}\,\mathbf{a}_y$, i.e., $U_r \mathbf{a}_d = U \mathbf{a}_x - \dot{v}\,\mathbf{a}_y$ (time t understood). Since $\mathbf{a}_d = \cos\beta\,\mathbf{a}_x + \sin\beta\,\mathbf{a}_y$, the previous relation is rewritten as $(U - U_r \cos\beta)\,\mathbf{a}_x - (U_r \sin\beta + \dot{v})\,\mathbf{a}_y = 0$, from which it follows:[9]

$$U_r = \frac{U}{\cos\beta}, \tag{12.8a}$$

$$\tan\beta = -\frac{\dot{v}}{U}. \tag{12.8b}$$

The relative velocity U_r and the instantaneous angle of attack β are represented in Fig. 12.4 for a negative structural velocity $-\dot{v}\,\mathbf{a}_y$, so that, according to Eq. 12.8b, β appears positive in the figure. The movement of the body has, therefore, two effects: (a) it modifies the magnitude of the velocity (Eq. 12.8a), and (b) it modifies the angle of attack (Eq. 12.8b), rendering both of them depending on the structural velocity, and therefore implicit functions of time. This explains the reason for which the aeroelastic forces, even in a simplified theory such as the quasi-steady, are dependent on the velocity of the body. Hence, by using Eq. 12.3a, b and 12.8 to evaluate the drag and lift forces (per unit of length), and then projecting them onto $\mathbf{a}_y$, the vertical force component (acting over the entire cylinder) is finally found to be:[10]

$$F_y^a = f_d \ell \sin\beta + f_l \ell \cos\beta = \frac{1}{2}\rho_a U^2 D\ell C_y(\beta), \tag{12.9}$$

in which a new aerodynamic coefficient has been defined:

$$C_y(\beta) := \frac{1}{\cos^2\beta}\,[C_d(\beta)\sin\beta + C_l(\beta)\cos\beta]. \tag{12.10}$$

---

[8] Also said "absolute" velocity.

[9] It should be noticed that many authors (e.g., [1, 8]) take the $\mathbf{a}_y$ axis positive downward. It entails a change of sign in the attack angle β, defined in Eq. 12.8b, with obvious consequences on Table 12.1.

[10] The horizontal force and the couple are absorbed by the constraints.

Remark 12.1 Since the aeroelastic forces depend on the velocity $\dot{v}$, they are *not* known forces, such as those usually considered in structural analysis to evaluate the static or dynamic effects of the wind. They, indeed, must be placed in the *left member* of the equations of motion, as a geometric effect related to the change of state of the elastic system. These geometric forces modify the intrinsic characteristics of the system, eventually making it unstable.

12.3.2 Linear Stability Analysis

A preliminary linear analysis is performed, sufficient to determine the onset of the instability phenomenon. Under the assumption that the structural velocity $\dot{v}$ is small with respect to the wind velocity U, from Eq. 12.8b, it follows that the angle of attack β is also small. By expanding in series the trigonometric functions, one finds:

$$\beta = -\frac{\dot{v}}{U} + \cdots , \qquad (12.11)$$

together with $C_y(\beta) = C_l(\beta) + C_d(\beta)\beta$. By also expanding the aerodynamic coefficients around $\beta = 0$, one has:

$$C_h(\beta) = C_{h_0} + C'_{h_0}\beta + \cdots , \quad h = d,l, \qquad (12.12)$$

where the dash denotes differentiation with respect to β and the sub-index 0 evaluation at $\beta = 0$. Thus, C'_{h_0} is the slope at the origin of the $C_h(\beta)$ curve. Ultimately, linearizing in β and taking into account Eq. 12.11, one gets:

$$C_y(\beta) = C_{l_0} - \left(C_{d_0} + C'_{l_0}\right)\frac{\dot{v}}{U} . \qquad (12.13)$$

The equation of motion, Eq. 12.7, once the elastic and aeroelastic nonlinearities have been ignored, and Eqs. 12.9, 12.13 used become:

$$\ddot{v} + 2\xi_s\omega_s\dot{v} + \omega_s^2 v = \frac{1}{2}\frac{\rho_a U^2 D\ell}{M}\left[C_{l_0} - \left(C_{d_0} + C'_{l_0}\right)\frac{\dot{v}}{U}\right]. \qquad (12.14)$$

The presence of a static force (independent of $\dot{v}$) and a time-variable force (proportional to $\dot{v}$) is observed in the equation. The first force coincides with that exerted on the body at rest. It doesn't contribute to stability, since it just modifies the equilibrium position;[11] accordingly, it will be ignored ahead. The second force, on the other hand, being proportional to $\dot{v}$, is *formally* a viscous force, so that it can be merged with the structural damping. The equation of motion is therefore rewritten as:

$$\ddot{v} + 2\omega_s \left(\xi_s + U\zeta_1\right)\dot{v} + \omega_s^2 v = 0, \tag{12.15}$$

in which:

$$\zeta_1 := \frac{\rho_a D\ell}{4\omega_s M}\left(C_{d_0} + C'_{l_0}\right). \tag{12.16}$$

The product $\xi_a := U\zeta_1$ is called the *aerodynamic damping factor*, which is proportional to the wind velocity and depends on ζ_1, whose sign is now investigated. This coefficient contains two contributions: the drag coefficient $C_{d_0} > 0$, which measures the resistance of the medium when the body is at rest, and $C'_{l_0} \gtrless 0$, which measures the slope of $C_l\left(\beta\right)$ at the origin. Two cases can occur:

- $C_{d_0} + C'_{l_0} \geq 0$ (i.e., (i) $C'_{l_0} \geq 0$, or, (ii) $C'_{l_0} < 0$, but $\left|C'_{l_0}\right| \leq C_{d_0}$); in this case, it is $\zeta_1 \geq 0$, i.e., the *aerodynamic damping is stabilizing* (or neutral) and adds itself to the structural damping; the section of the cylinder is called *aerodynamically stable*;
- $C_{d_0} + C'_{l_0} < 0$ (i.e., $C'_{l_0} < 0$, and, moreover $\left|C'_{l_0}\right| > C_{d_0}$); in this case, the *aerodynamic damping is instabilizing*, since it subtracts itself from the structural damping; the section is called *aerodynamically unstable*.

From the previous discussion, the following result emerges. In order for aeroelastic instability to occur, it needs that the first derivative of the lift coefficient, evaluated at $\beta = 0$, is (a) negative and (b) sufficiently large in modulus. In such occurrence, since the aerodynamic damping is proportional to U, *there exists a critical velocity* U_c *such that the total damping* $\xi_t := \xi_s + U\zeta_1$ *zeroes*; this is $U_c := -\frac{\xi_s}{\zeta_1}$, or:

$$U_c = \frac{4\omega_s M}{\rho_a D\ell}\frac{\xi_s}{\left|C_{d_0} + C'_{l_0}\right|}, \quad \text{if } C_{d_0} + C'_{l_0} < 0, \tag{12.17}$$

and it is called the *critical galloping velocity* or the Den Hartog velocity.[12] When $U = U_c$, the system has a pair of purely imaginary eigenvalues $\lambda = \pm i\omega_s$, with ω_s

---

[11] The equation of motion $\ddot{v} + 2\xi_s\omega_s\dot{v} + \omega_s^2 v = \frac{F_{st}}{M}$, with $F_{st} = \text{cost}$, can be rewritten as $\ddot{w} + 2\xi_s\omega_s\dot{w} + \omega_s^2 w = 0$, where $w := v - v_{st}$ measures the deviation from the equilibrium position $v_{st} := \frac{F_{st}}{M\omega_s^2} = \frac{F_{st}}{k_1}$.

[12] The stability criterion $\xi_t > 0$ is also called *the criterion by Den Hartog* (1901–1989).

the natural frequency of the oscillator, which is about to cross the imaginary axis.[13] Therefore a dynamic, or *Hopf*, bifurcation occurs. When $U < U_c$ the equilibrium is asymptotically stable, when $U > U_c$ the equilibrium is unstable.

Remark 12.2 Since $\omega_s M = \sqrt{k_1 M}$, from Eq. 12.17, it follows that the critical galloping velocity is low in systems: (a) slightly damped, (b) flexible, and (c) not very massive. An engineering technique to improve the mechanical performance of light and flexible structures consists in increasing their structural damping, for example, via added energy dissipation devices. Examples of such techniques are presented in Problems 14.6.1 and 14.6.2.

Remark 12.3 If the cylinder has uniformly distributed mass $m := \frac{M}{\ell}$, the critical velocity in Eq. 12.17 is independent of the length of the cylinder.

Remark 12.4 The reduced critical velocity $U_{c,red} := \frac{U_c}{v_s D}$, by using Eq. 12.17, reads:

$$U_{c,red} = \frac{M}{\rho_a D^2 \ell} \frac{8\pi \xi_s}{\left| C_{d_0} + C'_{l_0} \right|}. \tag{12.18}$$

It is *independent of* ω_s and directly proportional to the *mass ratio* $\frac{M}{\rho_a D^2 \ell}$. This latter, for a homogeneous cylinder with a square cross-section of side D, coincides with the ratio between the density of the material $\rho_m := \frac{M}{D^2 \ell}$ and that of the air, ρ_a. In order for the quasi-steady theory to hold, it must be $U_{c,red} > 20$. All other things being equal, the mass ratio must be large enough.[14]

Numerical Values of the Galloping Aerodynamic Coefficient

The values of the aerodynamic coefficients of long cylinders were determined, for different shapes of the cross-section, by experimental or numerical methods, and are available in the literature [1]. The results are not always in agreement, as they are affected by unavoidable measurement errors or numerical modeling inaccuracies, for example, due to (a) the nature of the flow, laminar, or turbulent, (b) the length

---

[13] The eigenvalues λ of Eq. 12.15, which rule the motion $v = \exp(\lambda t)$, satisfy the characteristic equation $\lambda^2 + 2\omega_s \xi_t \lambda + \omega_s^2 = 0$, and are equal to:

$$\lambda_{1,2} = -\omega_s \xi_t \pm i\omega_s \sqrt{1 - \xi_t^2}$$

For ξ_t small in modulus, $\lambda_{1,2}$ is (a) complex, with negative real part (entailing asymptotic stability), when $\xi_t > 0$; (b) purely imaginary, $\lambda = \pm i\omega_s$, when $\xi_t = 0$ (i.e., at the bifurcation); (c) complex, with real part positive (entailing instability), when $\xi_t < 0$.

[14] For example, if $\xi_s = 0.01$, $\left| C_{d_0} + C'_{l_0} \right| = 3$ (square cross-section), the mass ratio must be at least equal to 250. Since the density of air is $1.225 \, kg/m^3$, the average density of the cylinder, ρ_m, must be at least equal to $300 \, kg/m^3$.

Table 12.1 Values of
$C'_{y_0} := C_{d_0} + C'_{l_0}$ for
different cross-section shapes
of a long cylinder, subject to
laminar flow, horizontally
incident; negative values
indicate aerodynamically
unstable cross-sections

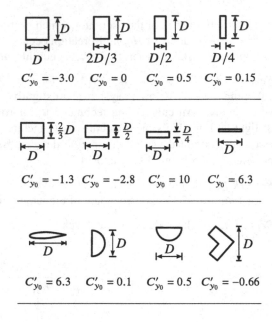

of the cylinder, and (c) the sharp or rounded corners of polygonal cross-sections, which strongly influence the results. Generally, the numerical values, rather than separately to C_{d_0} and C'_{l_0}, refer to their sum $C'_{y_0} := C_{d_0} + C'_{l_0}$. Table 12.1 (taken from [1], with signs adapted to the conventions used here) reports the values of C'_{y_0} for different cross-sectional shapes. It is seen that the square is unstable; the "high" rectangles (i.e., having transverse dimension to the wind larger than longitudinal dimension) are stable; and the "low" rectangles (with transverse dimension smaller than longitudinal) are unstable (like the square) for moderate thicknesses, but they become stable when thin (e.g., when the thickness-to-width ratio is less than 1/4). The airfoil is stable, with high coefficient $C'_{y_0} \simeq 2\pi$,[15] the D-section is stable; the L-profile is unstable.

Example 12.1 (Critical Galloping Velocity of a Square Box Cylinder) As an example, a steel cylinder is considered, of density $\rho_m = 7850 \, \text{kg/m}^3$, having square hollow cross-section, of width $D = 20 \, \text{cm}$ and uniform thickness $b = 0.2 \, \text{cm}$, natural period $T_s = 0.4 \, s$, damping factor $\xi_s = 0.01$. The critical wind velocity is given by Eq. 12.17. In it, the mass per unit length of the cylinder, $m := M/\ell$, is equal to $4Db\rho_m = 12.56 \, \text{kg/m}$. The aerodynamic coefficient of the square cross-section is drawn from Table 12.1, as $\left| C_{d_0} + C'_{l_0} \right| = 3.0$. The air density is $\rho = 1.225 \, \text{Kg/m}^3$. From Eq. 12.17 one draws:

---

[15] This theoretical value, relating to the so-called airfoil, was determined by the aerodynamicist T. Theodorsen (1897–1978). Hence, it turns out that $C'_{y_0} \simeq C'_{l_0}$, being $C_{d_0} \simeq 0.01$.

$$U_c = \frac{8\pi m \xi_s}{T_s \rho_a D \left| C_{d_0} + C'_{l_0} \right|} = \frac{8\pi \cdot 12.56 \cdot 0.01}{0.4 \cdot 1.225 \cdot 0.2 \cdot 3.0} = 10.74 \, \text{m/s}. \tag{12.19}$$

For the validity of the quasi-steady theory, it is necessary to check that the reduced critical velocity, given by Eq. 12.18, is much greater than 1 (e.g., greater than 20). In the case at hand, it is:

$$U_{c,red} = \frac{U_c}{v_s D} = \frac{T_s U_c}{D} = \frac{0.4 \cdot 10.74}{0.2} = 21.48 > 20. \tag{12.20}$$

The result obtained, although based on an empirical rule, is reliable.   □

Influence of the Orientation of the Cross-Section with Respect to the Flow

In the previous analysis, it was assumed that the material direction, taken as a reference to establish the laws $C_h (\beta)$ of the aerodynamic coefficients, coincides with the flow direction $\mathbf{a}_x$, orthogonal to motion (Fig. 12.5a). However, it is of interest to consider the case in which such a material direction, now denoted with $\mathbf{a}_{x_0}$, is distinct from the flow direction $-\mathbf{a}_x$, and it forms with it a finite angle $-\beta_0$ (i.e., clockwise, Fig. 12.5b). This situation occurs, for example, when one wants to analyze the dynamics in the vertical plane of a frozen electric duct, hit by a horizontal wind, the concretion of which (taken for simplicity of an assigned shape) can assume free orientation with respect to the flow.

The aeroelastic force acting on the cylinder can, in this case, immediately be determined by using the results of the previous analysis. Since the force depends exclusively on the value of the drag coefficient and on the derivative of the lift coefficient, both *evaluated at the flow direction when the body is fixed*, it is sufficient to replace the angle $\beta = 0$ by the new angle $\beta = \beta_0$ (which is the angle with which the cylinder, rotated by $-\beta_0$, "sees" the flow, Fig. 12.1c, d). Thus, $C_{d_0} + C'_{l_0} \equiv C_d (0) + C'_l (0)$ appearing in Eq. 12.16 must be replaced by $C_d (\beta_0) + C'_l (\beta_0)$. A more formal derivation of this result is illustrated in Supplement 12.1.

Remark 12.5 From the diagrams of $C_d (\beta)$ and $C_l (\beta)$, one can also find the value of β_0 for which the sum $C_d (\beta_0) + C'_l (\beta_0)$ is negative and maximum in absolute value. Thus, the *most dangerous exposure to wind* is identified for the cross-section, in correspondence of which the critical galloping velocity assumes the lowest possible value. Figure 12.5c illustrates an example, in which the cross-section is aerodynamically stable when $\beta_0 = 0$ (as $C'_l (0) > 0$), but it is unstable for a certain angle of attack β_0 (i.e., for an clockwise rotation $-\beta_0$ of the cross-section), as $C'_l (\beta_0) < 0$ is larger, in absolute value, than $C_d (\beta_0)$.

Supplement 12.1 (Aerodynamic Galloping Coefficient of a Cross-Section Generically Oriented) Said γ the angle that the relative velocity forms with the horizontal direction $\mathbf{a}_x$, consistently with Eqs. 12.8, one has $U_r = \frac{U}{\cos \gamma}$,

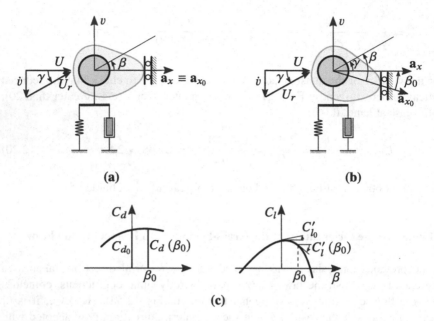

Fig. 12.5 Orientation of the cross-section with respect to the horizontal flow: (**a**) material axis $\mathbf{a}_{x_0}$ aligned with $\mathbf{a}_x$; (**b**) material axis $\mathbf{a}_{x_0}$ deviated from $\mathbf{a}_x$ by a (clockwise) angle $-\beta_0$; (**c**) aerodynamic coefficients evaluated at β_0 (i.e., relative to the orientation $-\beta_0$ of the cross-section)

$\tan \gamma = -\frac{\dot{v}}{U}$, or, for small structural velocities, $U_r = U$, $\gamma = -\frac{\dot{v}}{U}$. The instantaneous angle of attack that the relative velocity forms with the reference line $\mathbf{a}_{x_0}$ is instead $\beta = \gamma + \beta_0$ (which becomes $\beta = \beta_0$ when $\dot{v} = 0$). The aeroelastic forces, therefore, depend on β, by the way of the aerodynamic coefficient, but their projection on the vertical axis is ruled by γ. Hence:

$$F_y = f_d (\beta) \ell \sin \gamma + f_l (\beta) \ell \cos \gamma \simeq \frac{1}{2}\rho U^2 D \ell C_y (\beta, \gamma) , \qquad (12.21)$$

where, for small γ:

$$C_y (\beta, \gamma) := \frac{1}{\cos^2 \gamma} [C_d (\beta) \sin \gamma + C_l (\beta) \cos \gamma] \simeq \gamma C_d (\beta) + C_l (\beta) . \qquad (12.22)$$

By expanding in series the aerodynamic coefficients around β_0, one gets $C_h (\beta) = C_h (\beta_0) + (\beta - \beta_0) C'_h (\beta_0)$ $(h = d, l)$, from which it follows:

$$C_y (\beta, \gamma) = C_y \left(\beta_0, \frac{\dot{v}}{U}\right) = C_l (\beta_0) - [C_d (\beta_0) + C'_l (\beta_0)] \frac{\dot{v}}{U} , \qquad (12.23)$$

in accordance with what was previously stated. Equation 12.23 must be compared with the Eq.12.13.   □

12.3.3 Nonlinear Analysis: The Limit Cycle

The linear analysis supplies the critical value U_c at which the equilibrium position loses stability. It predicts, in the unstable region, exponentially divergent oscillations, without however giving any information about the existence of limit cycles (super-critical or sub-critical) which bound the oscillation amplitudes and how these cycles depend on U. To analyze the problem, it is necessary to carry out a nonlinear analysis, which takes into account both structural and aeroelastic nonlinearities. To this end, the equation of motion, Eq. 12.7, is considered again, in which the cubic contribution of the elastic spring, previously ignored, is retained, and, moreover, the terms of order higher than the first one are taken into account in the Taylor series expansion of the aeroelastic force.

Nonlinear Aeroelastic Forces

Equation 12.10 gives the exact expression of the vertical aerodynamic coefficient $C_y (\beta)$, in which β is given by Eq. 12.8b. By expanding this latter in series of small, but not infinitesimal, $\frac{\dot{v}}{U}$, it needs to consider, in addition to the linear term, also the quadratic, cubic, etc. The same holds for drag and lift coefficients. Therefore:

$$\beta = -\frac{\dot{v}}{U} + \frac{1}{3}\left(\frac{\dot{v}}{U}\right)^3 + \cdots , \tag{12.24a}$$

$$C_h (\beta) = C_{h_0} + C'_{h_0}\beta + \frac{1}{2}C''_{h_0}\beta^2 + \frac{1}{6}C'''_{h_0}\beta^3 + \cdots , \quad h = d, l. \tag{12.24b}$$

Substituting these expressions in Eq. 12.10, where the circular functions have also been expanded in series, one obtains, up to the cubic terms:

$$\begin{aligned}
C_y (\beta) =& C_{l_0} - \left(\frac{\dot{v}}{U}\right)(C_{d_0} + C'_{l_0}) + \left(\frac{\dot{v}}{U}\right)^2\left(\frac{1}{2}C_{l_0} + C'_{d_0} + \frac{1}{2}C''_{l_0}\right) \\
& - \left(\frac{\dot{v}}{U}\right)^3\left(\frac{1}{2}C_{d_0} + \frac{1}{6}C'_{l_0} + \frac{1}{2}C''_{d_0} + \frac{1}{6}C'''_{l_0}\right).
\end{aligned} \tag{12.25}$$

Consequently, the vertical aeroelastic force in Eq. 12.9 reads:

$$F_y = \frac{1}{2}\rho_a U^2 D\ell\left[A_0 - A_1\frac{\dot{v}}{U} + A_2\left(\frac{\dot{v}}{U}\right)^2 - A_3\left(\frac{\dot{v}}{U}\right)^3 + \cdots\right], \tag{12.26}$$

where:[16]

---

[16] The coefficients A_0, A_1 are consistent with those introduced in the linear analysis, as it can be seen from Eq. 12.13; in particular, $A_1 \equiv C'_{y_0}$.

$$A_0 := C_{l_0}, \qquad\qquad\qquad A_1 := C_{d_0} + C'_{l_0},$$

$$A_2 := \frac{1}{2}C_{l_0} + C'_{d_0} + \frac{1}{2}C''_{l_0}, \quad A_3 := \frac{1}{2}C_{d_0} + \frac{1}{6}C'_{l_0} + \frac{1}{2}C''_{d_0} + \frac{1}{6}C'''_{l_0}. \tag{12.27}$$

The nondimensional coefficients A_i are also called *aerodynamic coefficients*, and they are directly available in the literature for various cross-section shapes.

In case of symmetric cross-section, being, as mentioned above, $C_d(\beta)$ symmetric and $C_l(\beta)$ antisymmetric, it is $C'_{d_0} = C'''_{d_0} = \cdots = 0$, $C_{l_0} = C''_{l_0} = \cdots = 0$. Therefore, $A_0 = A_2 = \cdots = 0$, in agreement with the fact that F_y must be an odd function of $\dfrac{\dot{v}}{U}$; hence:

$$F_y = -\frac{1}{2}\rho_a U^2 D\ell \left[A_1 \frac{\dot{v}}{U} + A_3 \left(\frac{\dot{v}}{U} \right)^3 + \cdots \right]. \tag{12.28}$$

Nonlinear Equation of Motion

The equation of motion, Eq. 12.7, in the hypothesis of cross-section symmetric with respect to the flow, contains only cubic terms (and odd higher powers, neglected here). It can be written in the form;

$$\ddot{v} + 2\omega_s \left(\xi_s + U\zeta_1 \right) \dot{v} + \omega_s^2 v + \kappa_s v^3 + \frac{1}{U}\zeta_3 \dot{v}^3 = 0, \tag{12.29}$$

called the *Rayleigh-Duffing equation*. In it, for an aerodynamically unstable cross-section, the following definitions hold:[17]

$$\zeta_1 := \frac{\rho_a D\ell}{4\omega_s M} A_1 < 0, \qquad \zeta_3 := \frac{\rho_a D\ell}{2M} A_3. \tag{12.30}$$

Equation 12.29 contains two types of nonlinearities: (a) of structural (Duffing-like) type, expressed by the cubic term in the displacement, and (b) of aeroelastic (Rayleigh-like) type, expressed by the cubic term in the structural velocity.

Lindstedt-Poincaré Method

The Rayleigh-Duffing equation is nonlinear, not solvable in closed form. However, when the amplitude of motion is small (but finite), the nonlinearities are also small, so that the equation is weakly nonlinear, and it can be solved by a perturbation method. Here the *Lindstedt-Poincaré method* is illustrated, already discussed in

---

[17] The coefficient ζ_1 has already been introduced in Eq. 12.16.

detail in Sect. 11.4.2 for a circulatory system (to which reference is made for what is omitted here). The method consists in finding *periodic solutions* of the equation of motion, Eq. 12.29, and requires performing the following steps.

A nondimensional time $\tau := \Omega t$ is introduced, with Ω the unknown circular frequency. Equation 12.29 modifies into:

$$\Omega^2 \ddot{v} + 2\Omega \omega_s \left(\xi_s + U \zeta_1 \right) \dot{v} + \omega_s^2 v + \kappa_s v^3 + \frac{\Omega^3}{U} \zeta_3 \dot{v}^3 = 0, \tag{12.31}$$

in which the derivative with respect to the new time τ is still indicated with a dot. A one-parameter family of periodic solutions is sought for this equation, of the form $v = v(\tau; \epsilon)$, $\Omega = \Omega(\epsilon)$, $U = U(\epsilon)$, where ϵ is a perturbation parameter. Taking into account the symmetry of the system, the previous expressions are expanded in Taylor series, such as:

$$v(\tau; \epsilon) = \epsilon v_1(\tau) + \epsilon^3 v_3(\tau) + \cdots, \tag{12.32a}$$

$$\Omega(\epsilon) = \Omega_0 + \epsilon^2 \Omega_2 + \cdots, \tag{12.32b}$$

$$U(\epsilon) = U_0 + \epsilon^2 U_2 + \cdots, \tag{12.32c}$$

where all the coefficients are unknown. By replacing Eq. 12.32 into the equation of motion, Eq. 12.31, and separately equating to zero the terms with the same powers of ϵ, the following perturbation equations are obtained:

$$\epsilon^1 : \Omega_0^2 \ddot{v}_1 + 2\Omega_0 \omega_s \left(\xi_s + U_0 \zeta_1 \right) \dot{v}_1 + \omega_s^2 v_1 = 0, \tag{12.33a}$$

$$\epsilon^3 : \Omega_0^2 \ddot{v}_3 + 2\Omega_0 \omega_s \left(\xi_s + U_0 \zeta_1 \right) \dot{v}_3 + \omega_s^2 v_3 = -2\Omega_0 \Omega_2 \ddot{v}_1 - \kappa_s v_1^3 \tag{12.33b}$$

$$-2\Omega_0 \omega_s U_2 \zeta_1 \dot{v}_1$$

$$-2\Omega_2 \omega_s \left(\xi_s + U_0 \zeta_1 \right) \dot{v}_1$$

$$-\frac{\Omega_0^3}{U_0} \zeta_3 \dot{v}_1^3.$$

Since, on the τ scale, the solution is periodic with a known period 2π, the solution must respect the periodicity conditions $v(2\pi) = v(0)$, $\dot{v}(2\pi) = \dot{v}(0)$. From these, and from the series expansions in Eq. 12.32a, the periodicity conditions at the different orders are drawn:

$$v_k(2\pi) = v_k(0), \qquad \dot{v}_k(2\pi) = \dot{v}_k(0), \qquad k = 1, 3, \ldots. \tag{12.34}$$

Introduced the normalization $v(0; \epsilon) = \epsilon$, $\dot{v}(0; \epsilon) = 0$, which attributes to ϵ the meaning of *amplitude of limit* cycle, from Eq. 12.32a, it follows that:

$$v_1(0) = 1, \quad \dot{v}_1(0) = 0,$$

$$v_k(0) = 0, \quad \dot{v}_k(0) = 0, \quad k = 3, 5, \ldots. \tag{12.35}$$

Solution to the Perturbation Equations

The perturbation equation of order ϵ admits periodic solutions if, and only if, the total damping $\xi_s + U_0 \zeta_1$ vanishes. It is therefore $U_0 = -\frac{\xi_s}{\zeta_1} \equiv U_c$, that is, the critical galloping velocity is found. Also, for the period to be equal to 2π, it must be $\omega_0 = \omega_s$. Hence, the initial values of the series expansions, Eq. 12.32b, c, identify the bifurcation point; this implies that the family of limit cycles bifurcates from the equilibrium position at the Hopf bifurcation. With these results, taking into account the normalization, the solution is:

$$v_1 = \cos \tau. \tag{12.36}$$

Substituting it into the perturbation equation of order ϵ^3, and using known trigonometry formulas,[18] Eq. 12.33b becomes:

$$\omega_s^2 (\ddot{v}_3 + v_3) = 2\omega_s \Omega_2 \cos \tau + 2\omega_s^2 U_2 \zeta_1 \sin \tau$$
$$- \frac{1}{4} \kappa_s (3\cos \tau + \cos(3\tau)) + \frac{1}{4} \frac{\omega_s^3}{U_c} \zeta_3 (3\sin \tau - \sin(3\tau)). \tag{12.37}$$

It is observed that the cubic nonlinearities, in addition to frequency 1, "create" frequency 3 in the known term of the equation. Now, while the latter frequency generates a response frequency 3, which therefore respects the periodicity condition, the excitation frequency 1 is *in resonance* with the natural frequency 1 of the single DOF oscillator. If these resonant terms were not removed from the right member of the equation, the solution would diverge in time with linearly increasing amplitude, thus violating the periodicity.[19] Thus, by collecting the coefficients of $\cos \tau$ and $\sin \tau$, and canceling them separately, one obtains:[20]

$$2\omega_s \Omega_2 - \frac{3}{4} \kappa_s = 0, \tag{12.38a}$$

$$2\omega_s^2 U_2 \zeta_1 + \frac{3}{4} \frac{\omega_s^3}{U_c} \zeta_3 = 0. \tag{12.38b}$$

---

[18] It should be remembered that $\cos^3 \tau = \frac{1}{4}(3\cos \tau + \cos(3\tau))$ and that $\sin^3 \tau = \frac{1}{4}(3\sin \tau - \sin(3\tau))$.

[19] For example, a particular solution of $\ddot{q} + q = \sin \tau$ is $q = -\frac{1}{2}\tau \cos \tau$. These divergent term, as already mentioned in Chap. 11, is called "secular."

[20] It should be observed that, in case of an oscillator with a single DOF, the compatibility condition, already introduced in Sect. 11.4.2 (calling for the orthogonality of the known term to the solution of the transposed homogeneous problem), merely reduces to vanishing the resonant terms.

These are two algebraic equations in the two unknown parameters, which provide:

$$\Omega_2 = \frac{3\kappa_s}{8\omega_s}, \quad U_2 = -\frac{3\omega_s\zeta_3}{8U_c\zeta_1}. \tag{12.39}$$

With these results, the solution to Eq. 12.32 reads:

$$v(\tau; \epsilon) = \epsilon \cos(\tau), \tag{12.40a}$$

$$\Omega(\epsilon) = \omega_s\left(1 + \epsilon^2 \frac{3\kappa_s}{8\omega_s^2}\right), \tag{12.40b}$$

$$U(\epsilon) = U_c\left(1 - \epsilon^2 \frac{3\omega_s\zeta_3}{8U_c^2\zeta_1}\right). \tag{12.40c}$$

Remembering that $\tau = \Omega t$, it is possible to come back to the original time:

$$v(t; \epsilon) = \epsilon \cos\left(\left(1 + \epsilon^2 \frac{3\kappa_s}{8\omega_s^2}\right)\omega_s t\right), \tag{12.41a}$$

$$U(\epsilon) = U_c\left(1 + \epsilon^2 \frac{3\omega_s\zeta_3}{8U_c^2|\zeta_1|}\right), \tag{12.41b}$$

where it has taken into account that $\zeta_1 < 0$. Finally, by eliminating ϵ, one has:[21]

$$v(t; U) = \sqrt{\frac{8U_c|\zeta_1|(U - U_c)}{3\omega_s\zeta_3}} \cos\left(\left(1 + \frac{\kappa_s U_c|\zeta_1|(U - U_c)}{\omega_s^3\zeta_3}\right)\omega_s t\right). \tag{12.42}$$

Equations 12.40 describe a family of limit cycles; to any amplitude ϵ, a frequency $\Omega = \Omega(\epsilon)$ and a wind velocity $U = U(\epsilon)$ are associated. Alternatively, Eq. 12.42 describes the cycle directly as a function of U.

Figure 12.6 shows the graphs (bifurcation diagrams) of the frequency and wind velocity vs the amplitude. It is noticed that:

- if $\kappa_s > 0$ (i.e., if the spring is hardening), it is $\Omega > \omega_s$, that is, the cycle is performed at a higher frequency than the natural one; if $\kappa_s < 0$ (i.e., if the spring is softening), it is $\Omega < \omega_s$;
- if $\zeta_3 > 0$ (i.e., if $A_3 > 0$ in Eq. 12.27), the limit cycle is *super-critical*, that is, it exists for $U > U_c$; if $\zeta_3 < 0$ (i.e., if $A_3 < 0$ in Eq. 12.27), the limit cycle is *sub-critical*, i.e., it exists for $U < U_c$.

---

[21] Only the positive value of the square root has been taken, consistently with the symmetry properties of the solution. In fact, the sign change of ϵ corresponds to change $\cos(\Omega t)$ into $\cos(\Omega t + \pi)$ and therefore to translate, in an inessential way, the origin of time.

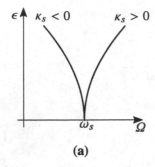 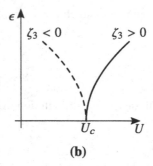

Fig. 12.6 Bifurcation diagrams for the Rayleigh-Duffing equation: (**a**) limit cycle frequency-amplitude relationship; (**b**) amplitude of the limit cycle *vs* the wind velocity (solid line: stable, dashed line: unstable)

- the structural nonlinearity has no effect (at this order) on the amplitude of the limit cycle, but only on the frequency; the aeroelastic nonlinearity has no effect (at this order) on the frequency, but only on the amplitude.

For fixed physical parameters of the system, and an assigned ϵ, the closed orbit described by the system is an ellipse in the $(v, \dot{v})$ phase plan, of parametric Eq. 12.40, traveled clockwise[22] (Fig. 12.7a, b). It is possible to prove (Supplement 12.2) that the orbit is *stable if super-critical, and unstable if sub-critical*, similarly to what happens for static fork bifurcations (Chap. 5).[23] The trajectories are attracted by the stable limit cycle (Fig. 12.7a), which therefore bounds the exponential growth of motion predicted by the linear theory. On the contrary, the trajectories are repelled from the unstable cycle (Fig. 12.7b). If, in this second case, the initial perturbation is large, in magnitude greater than ϵ, the motion diverges, although the equilibrium position is stable. Figure 12.7c,d, finally, gives a three-dimensional view of the family of limit cycles: for each fixed velocity U, respectively, super-critical or sub-critical, there exist a limit cycle, the amplitude of which increases with $|U - U_c|$, as described by Eq. 12.40c.

Remark 12.6 Structural nonlinearities play a secondary role in the galloping phenomenon, since, at the first order, they do not affect the amplitude of the limit cycle, but only its frequency. Therefore, no significant error is made if, in modeling the phenomenon, these nonlinearities are ignored (linear structural model) and only aeroelastic nonlinearities are taken into account (nonlinear aeroelastic model).

---

[22] In fact, if at time t is, for example, $v(t) = 0$, $\dot{v}(t) > 0$, then, at $t + dt$, it is $v(t + dt) \simeq \dot{v}(t)\,dt > 0$.

[23] An alternative proof requires the use of more sophisticated perturbation methods, such as that of Multiple Scales (not discussed in this book) or of the Floquet theory (to be illustrated in the Chap. 13).

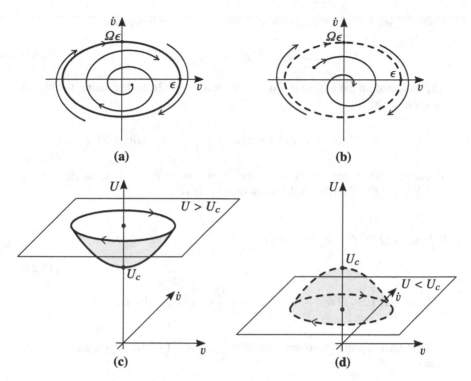

Fig. 12.7 Limit cycles for the Rayleigh-Duffing equation: (**a**) stable (super-critical) limit cycle; (**b**) unstable (sub-critical) limit cycle; (**c**) family of stable limit cycles, parameterized by $U > U_c$; (**d**) family of unstable limit cycle, parameterized by $U < U_c$

Supplement 12.2 (Energy-Based Interpretation of the Existence of a Limit Cycle and Its Stability Analysis) It is possible to give a meaningful energy-based interpretation about the existence of a limit cycle, as a periodic solution surrounded by non-periodic solutions. Assuming that, for a given wind velocity, the system describes an elliptical orbit Γ, of Eq. 12.40a, a generic amplitude ϵ of the cycle is chosen. If the work $W_\Gamma := W_\Gamma^c + W_\Gamma^{nc}$ that all forces, conservative or not, do along the orbit is computed, this is generally found to be non-zero. Since the conservative forces f^c, by definition, do null work on Γ, only the nonconservative forces f^{nc} contribute to this. Now, while viscous forces are dissipative, and therefore they do negative work, the aeroelastic forces can do either positive work (i.e., they put energy into the system) or negative (i.e., they subtract energy from the system). If it happens that $W_\Gamma^{nc} < 0$, the motion does not develop on the chosen orbit, but decays; if $W_\Gamma^{nc} > 0$, the motion does not develop on this orbit, but diverges; if $W_\Gamma^{nc} = 0$, the motion remains on Γ, i.e., the system executes the cycle indefinitely. Ultimately, among the family of elliptical orbits parameterized by the amplitude ϵ, the system chooses that one (dependent on U) on which, in one cycle, the dissipation equates the input of energy.

In the Rayleigh-Duffing equation, the nonconservative forces are:

$$f^{nc}(\dot{v}) := -2\omega_s \left(\xi_s - U|\zeta_1|\right)\dot{v} - \frac{1}{U}\zeta_3 \dot{v}^3, \tag{12.43}$$

which, evaluated on the ellipse in Eq. 12.40a, with $\tau = \Omega t$, and ignoring the higher harmonics, reads:

$$f^{nc}(\dot{v}) := 2\omega_s \Omega \left(\xi_s - U|\zeta_1|\right)\epsilon \sin(\Omega t) + \frac{3}{4}\frac{\Omega^3}{U}\zeta_3 \epsilon^3 \left(\sin(\Omega t) + \cdots\right). \tag{12.44}$$

In a time interval dt, the forces do, in the displacement $dv = \dot{v}\,dt$, an elementary work $dW_\Gamma^{nc} := f^{nc}(\dot{v})\,\dot{v}\,dt$, so the work on a cycle is:

$$W_\Gamma^{nc} = -\epsilon\Omega \int_0^{2\pi/\Omega} f^{nc}(\dot{v})\sin(\Omega t)\,dt \tag{12.45}$$

$$= -\epsilon\Omega \left[2\omega_s\Omega\left(\xi_s - U|\zeta_1|\right)\epsilon + \frac{3}{4}\frac{\Omega^3}{U}\zeta_3\epsilon^3\right]\int_0^{2\pi/\Omega}\sin^2(\Omega t)\,dt.$$

Since the time integral is non-zero, it is $W_\Gamma^{nc} = 0$ only if the term in square brackets vanishes, i.e., if:

$$U = \frac{\xi_s}{|\zeta_1|} + \frac{3}{8}\frac{\Omega^2}{U\omega_s}\frac{\zeta_3}{|\zeta_1|}\epsilon^2. \tag{12.46}$$

This is an implicit equation for U; however, by taking into account that $U = U_c + O\left(\epsilon^2\right)$, for asymptotic consistency, it is necessary to replace $U = U_c = \frac{\xi_s}{|\zeta_1|}$ in the second member; the same holds for $\Omega \simeq \omega_s$. The previous result is asymptotically equivalent to Eq. 12.41b.

Equation 12.45 also permits to justify the stability of the cycles. In fact, by calling ϵ_0 the amplitude of the cycle for which $W_\Gamma^{nc}(\epsilon_0) = 0$, it needs to examine the sign of:

$$\frac{dW_\Gamma^{nc}(\epsilon_0)}{d\epsilon} = -\left[2\omega_s\Omega\left(\xi_s - U|\zeta_1|\right) + \frac{9}{4}\frac{\Omega^3}{U}\zeta_3\epsilon_0^2\right] = -\frac{3}{2}\frac{\Omega^3}{U}\zeta_3\epsilon_0^2, \tag{12.47}$$

in which an inessential positive factor has been ignored and the definition of ϵ_0 replaced.[24] The cycle is stable when $\frac{dW_\Gamma^{nc}(\epsilon_0)}{d\epsilon} < 0$, since (a) $\Delta\epsilon > 0$ entails an energy dissipation (which reduces the perturbation) and (b) $\Delta\epsilon < 0$ an energy input (which increases the perturbation). The cycle is therefore stable if $\zeta_3 > 0$ (supercritical) and unstable if $\zeta_3 < 0$ (sub-critical).   □

---

[24] That is, the square bracket that appears in Eq. 12.45, with $\epsilon = \epsilon_0$.

12.4 Galloping of Strings and Beams

In the previous section, the motion of a *rigid* cylinder, subject to steady flow, was analyzed. The instantaneous action of wind was assumed to be identical on all the cross-sections (except those close to the ends of the cylinder). If the cylinder is instead deformable, for example, consisting of a taut string or a beam, the aeroelastic forces f_y^a are *function of the abscissa z*, as the $v(z,t)$ motion of the cross-sections is asynchronous and of different amplitude. Most importantly, the forces depend not only on the deflection at z but also on the deflections in a neighborhood of z, whose motion influences the flow, making it no longer planar. A drastic simplification of the problem is achieved by renouncing to consider the *non-local* aspect of the phenomenon and limiting to consider the effect of the *local motion* of the string/beam, i.e., assuming that $f_y^a(z,t) = f_y^a(\dot{v}(z,t))$.[25] Based on this assumption, the following hypothesis is introduced:

• the aeroelastic forces, instantaneously acting at the abscissa z of the beam, coincide with those exerted on an infinitely long and rigid cylinder that experiences the same translational velocity of the cross-section at that specific abscissa.

Accordingly, the curvature of the beam and the interaction between the structural velocities of close cross-sections are ignored.

Taking into account the considerations made about the role of the structural nonlinearities (Remark 12.6), *linear structural* models under *nonlinear aeroelastic forces* can be considered. The relevant equations of motion are of the type:

$$m\ddot{v} + c_e\dot{v} + \left(1 + \hat{\eta}\frac{\partial}{\partial t}\right)\mathcal{K}v = f_y^a(z,t), \qquad (12.48)$$

accompanied by appropriate boundary conditions. In these equations: $v = v(z,t)$ is the transverse displacement; m the mass linear density; c_e the *external damping* (which models the interaction between the solid and the still air); $\mathcal{K}$ the total stiffness operator (elastic plus geometric); and $\hat{\eta}$ a parameter of *internal damping*,[26] which describes the internal viscous force according to the rheological Kelvin-Voigt model; finally, $f_y^a(z,t)$ are the aeroelastic forces per unit length, given, in conditions of symmetry, by Eq. 12.28, i.e.,

$$f_y(z,t) = -\frac{1}{2}\rho_a U^2 D\left[A_1\frac{\dot{v}(z,t)}{U} + A_3\left(\frac{\dot{v}(z,t)}{U}\right)^3 + \cdots\right]. \qquad (12.49)$$

---

[25] The hypothesis is identical to that adopted in statics for the Winkler soil model. As a matter of fact, in that model, the reaction of the soil depends only by the displacement of the point at which the force is evaluated, and not by the displacements of points in the neighborhood.

[26] Usually $\hat{\eta}$ includes a geometric parameter, as it will be clarified by Eqs. 12.51 and 12.56.

The partial differential Eq. 12.48 requires the use of numerical algorithms (e.g., finite elements or finite differences). However, the problem can be strongly simplified if the Galerkin procedure is adopted,[27] in which the form $\phi(z)$ of the deflection is assigned, and its time-dependent amplitude $q(t)$ is free; accordingly:

$$v(z, t) = q(t) \phi(z). \tag{12.50}$$

The weak formulation of equilibrium (i.e., the Virtual Work Principle) then provides an ordinary differential equation in the unknown Lagrangian parameter q. Thus, the results already obtained for galloping of a single DOF system directly apply. The analysis is detailed ahead for strings and beams. More complex problems, relevant to base-isolated beams and suspension bridges, are discussed in Problems 14.6.1, 14.6.2, 14.6.3.

12.4.1 Strings

A taut string is considered, fixed at the supports A, B ($z = 0, \ell$), subject to a wind flow of velocity $U\mathbf{a}_x$. The string has a non-circular cross-section (because, e.g., iced), symmetric of axis $\mathbf{a}_x$, free to oscillate in the transverse direction $\mathbf{a}_y$. The equation of motion and the boundary conditions are:[28]

$$m\ddot{v} + c_e\dot{v} - \eta A_s \dot{v}'' - T_0 v'' = f_y^a(z, t), \tag{12.51a}$$

$$v(0, t) = 0, \tag{12.51b}$$

$$v(\ell, t) = 0, \tag{12.51c}$$

where $v = v(z, t)$ is the transverse displacement, m the mass linear density (including ice), c_e the external damping, η the internal damping, A_s the cross-section structural area of the string (i.e., without ice), $T_0 := \sigma_0 A_s$ the tension in the string, and f_y^a the aeroelastic forces per unit length given by Eq. 12.49.

If the string is constrained to oscillate in its first natural mode (which exactly satisfies the boundary conditions also of the aero-viscoelastic problem), it is:

---

[27] The Galerkin method has the same conceptual basis as the Ritz method, but it is founded on virtual work, not on potential energy. For this reason, it can also be applied to nonconservative systems, in which the forces do not derive from a potential.

[28] The equilibrium equation of the string, as known from Mechanics of Structures, is $T_0 v'' + p_y = 0$. Expressing the distributed force p_y as the sum of the inertia force $p_y^{in} = -m\ddot{v}$, of the external viscous force $p_y^{ve} = -c_e\dot{v}$ and of the aeroelastic force f_y^a, and using the Similarity Principle between elastic and viscous internal forces (where $\eta A_s \frac{\partial}{\partial t}$ is similar to $\sigma_0 A_s$), the Eq. 12.51a is obtained.

$$v(z, t) = q(t) \sin\left(\frac{\pi z}{\ell}\right). \tag{12.52}$$

It should be noticed that this is an exact solution for the linear part of the equations of motion, but it is not exact for the nonlinear part. The weak formulation of equilibrium consists in imposing that the *residual of the field equation* is orthogonal to the chosen shape function (i.e., in mechanical terms, that the unbalanced distributed force, consequent to the choice of Eq. 12.52, do null work in the admissible displacement field), i.e.,

$$\int_0^\ell \left[m\ddot{v} + c_e\dot{v} - \eta A_s \dot{v}'' - T_0 v'' - f_y^a(z, t) \right] \sin\left(\frac{\pi z}{\ell}\right) dz = 0. \tag{12.53}$$

By replacing Eq. 12.52 and performing the integrations, an ordinary differential equation (called the *Rayleigh oscillator*) is obtained:

$$M\ddot{q} + (c_s + Ub_1)\dot{q} + k_1 q + \frac{1}{U}b_3\dot{q}^3 = 0, \tag{12.54}$$

whose coefficients are defined as follows:

$$M := m \int_0^\ell \sin^2\left(\frac{\pi z}{\ell}\right) dz = \frac{1}{2}m\ell,$$

$$k_1 := T_0\frac{\pi^2}{\ell^2} \int_0^\ell \sin^2\left(\frac{\pi z}{\ell}\right) dz = T_0\frac{\pi^2}{2\ell},$$

$$c_s := \left(c_e + \eta A_s\frac{\pi^2}{\ell^2}\right) \int_0^\ell \sin^2\left(\frac{\pi z}{\ell}\right) dz = \frac{1}{2}\left(c_e\ell + \eta A_s\frac{\pi^2}{\ell}\right), \tag{12.55}$$

$$b_1 := \frac{1}{2}\rho_a D A_1 \int_0^\ell \sin^2\left(\frac{\pi z}{\ell}\right) dz = \frac{1}{4}\rho_a D\ell A_1,$$

$$b_3 := \frac{1}{2}\rho_a D A_3 \int_0^\ell \sin^4\left(\frac{\pi z}{\ell}\right) dz = \frac{3}{16}\rho_a D\ell A_3.$$

Equation 12.54 coincides with Eq. 12.29, when the parameters are thus identified:
$\omega_s^2 = \frac{k_1}{M}, \xi_s = \frac{c_s}{2\omega_s M}, \kappa_s = 0, \zeta_1 := \frac{b_1}{2\omega_s M}, \zeta_3 := \frac{b_3}{M}.$

12.4.2 Euler-Bernoulli Beams

The equation governing the dynamics transverse to the wind of an Euler-Bernoulli beam, subject to the flow $U\mathbf{a}_x$, is:[29]

$$m\ddot{v} + c_e \dot{v} + \eta I \dot{v}'''' + E I v'''' = f_y^a (z, t), \tag{12.56}$$

where: $v = v(z, t)$ is the displacement field, m the mass linear density, c_e the external damping, η the internal damping, I the moment of inertia of the cross-section with respect to $\mathbf{a}_x$, E the elastic modulus and f_y^a the aeroelastic forces for unit of length, given by Eq. 12.49. If the beam is clamped at A ($z = 0$) and free at B ($z = \ell$), the boundary conditions are:

$$v(0, t) = 0, \tag{12.57a}$$

$$v'(0, t) = 0, \tag{12.57b}$$

$$\left(E + \eta \frac{\partial}{\partial t} \right) I \, v''(\ell, t) = 0, \tag{12.57c}$$

$$-\left(E + \eta \frac{\partial}{\partial t} \right) I \, v'''(\ell, t) = 0. \tag{12.57d}$$

The beam is assumed to oscillate with an assigned shape, i.e., $v(z, t) = q(t)\phi(z)$. One possible choice is to take $\phi(z)$ as the first natural mode of the undamped beam, solution to the differential eigenvalue problem:

$$E I \phi'''' - m\omega_1^2 \phi = 0, \tag{12.58a}$$

$$\phi(0) = 0, \tag{12.58b}$$

$$\phi'(0) = 0, \tag{12.58c}$$

$$\phi''(\ell) = 0, \tag{12.58d}$$

$$\phi'''(\ell) = 0. \tag{12.58e}$$

The frequency $\omega_1 = \frac{1.875^2}{\ell^2} \sqrt{\frac{EI}{m}}$ is found, together with the eigenvector:

---

[29] The equilibrium equation of the Euler-Bernoulli beam is $E I v'''' - p_y = 0$. Expressing the distributed force p_y as the sum of the inertia force $p_y^{in} = -m\ddot{v}$, of the external viscous force $p_y^{ve} = -c_e \dot{v}$, of the aeroelastic force f_y^a, and using the Principle of Similarity between elastic and viscous internal forces (where $\eta I \frac{\partial}{\partial t}$ is similar to EI), Eq. 12.56 is obtained. The same principle leads to the boundary conditions in Eq. 12.57.

$$\phi(z) = 0.367 \left[\sin\left(1.875\,\frac{z}{\ell}\right) - \sinh\left(1.875\,\frac{z}{\ell}\right) \right]$$
$$+ \frac{1}{2} \left[\cosh\left(1.875\,\frac{z}{\ell}\right) - \cos\left(1.875\,\frac{z}{\ell}\right) \right], \tag{12.59}$$

normalized with the condition $\phi(\ell) = 1$.[30]

The weak formulation of equilibrium (virtual work principle) requires that:

$$\int_0^\ell \left[m\ddot{v} + c_e\dot{v} + \eta I \dot{v}'''' + EI v'''' - f_y^a(z,t) \right] \phi(z)\,dz + \cancel{C_B}\phi_B' + \cancel{F_B}\phi_B = 0, \tag{12.60}$$

where $C_B := EI v_B'' + \eta I \dot{v}_B'' = (EI q + \eta I \dot{q})\,\phi''(\ell)$, $F_B := -\left(EI v_B''' + \eta I \dot{v}_B'''\right) = -(EI q + \eta I \dot{q})\,\phi'''(\ell)$ are, respectively, the residual couple and force acting at the free end B. These, however, by virtue of Eq. 12.58, cancel in the problem under consideration, being $\phi''(\ell) = \phi'''(\ell) = 0$, so the only non-zero residual is in the field, whose equation is not satisfied by the shape function. From the virtual work principle, and from Eq. 12.50, Eq. 12.54 of the Rayleigh oscillator follows again, whose coefficients are defined as follows:

$$M := m \int_0^\ell \phi^2\,dz = \frac{1}{4}m\ell,$$

$$k_1 := EI \int_0^\ell \phi''''\phi\,dz = 3.091\,\frac{EI}{\ell^3},$$

$$c_s := c_e \int_0^\ell \phi^2\,dz + \eta I \int_0^\ell \phi''''\phi\,dz = \frac{1}{4}c_e\ell + 3.091\,\frac{\eta I}{\ell^3}, \tag{12.61}$$

$$b_1 := \frac{1}{2}\rho_a D A_1 \int_0^\ell \phi^2\,dz = \frac{1}{8}\rho_a D\ell A_1,$$

$$b_3 := \frac{1}{2}\rho_a D A_3 \int_0^\ell \phi^4\,dz = \frac{0.147}{2}\rho_a D\ell A_3.$$

---

[30] Normalization is arbitrary; to each normalization, a different meaning of the amplitude q corresponds.

As in the case of the string, by letting $\omega_s^2 = \frac{k_1}{M}$, $\xi_s = \frac{c_s}{2\omega_s M}$, $\kappa_s = 0$, $\zeta_1 := \frac{b_1}{2\omega_s M}$, $\zeta_3 := \frac{b_3}{M}$, Eq. 12.29 is recovered.

12.5   Planar Aeroelastic Systems

A three degrees of freedom aeroelastic model is formulated. It consists of a rigid cylinder, of axis $\mathbf{a}_z$, constrained by springs and viscous dampers, free to translate in the two traverse directions $\mathbf{a}_x$, $\mathbf{a}_y$ and to rotate around $\mathbf{a}_z$ (Fig. 12.8). The body is immersed in a wind flow of velocity $U\mathbf{a}_x$. The $\mathbf{a}_x$ axis is taken as the material direction of the section, with respect to which the angle of attack β is measured.

12.5.1   Three Degrees of Freedom Model

The cylinder is constrained by linear elastic springs k_x, k_y, k_θ and linear viscous dampers c_x, c_y, c_θ, applied in a particular cross-section, at a point C called *center of torsion*, the extensional devices being oriented as the transverse directions $\mathbf{a}_x$, $\mathbf{a}_y$ and the torsional organ as $\mathbf{a}_z$. The translations u_C, v_C of the point C, respectively, parallel to the directions $\mathbf{a}_x$, $\mathbf{a}_y$, are taken as Lagrangian parameters, together with the twist angle θ, expressing the rotation of the cylinder around $\mathbf{a}_z$. It is therefore $\mathbf{q} := (u_C, v_C, \theta)^T$. The equations of motion of the cylinder read:

$$\mathbf{M}\ddot{\mathbf{q}} + \mathbf{C}_s\dot{\mathbf{q}} + \mathbf{K}_s\mathbf{q} = \mathbf{f}^a, \qquad (12.62)$$

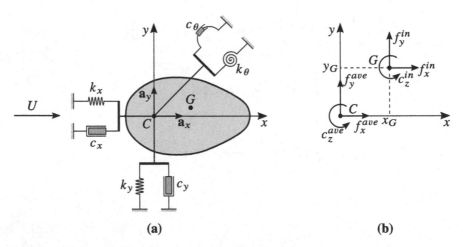

(a)   (b)

Fig. 12.8 Cross-section of an elastically constrained rigid cylinder: (a) model, (b) free body diagram

where $\mathbf{M}$ is the mass matrix, $\mathbf{C}_s$ the structural damping matrix, and $\mathbf{K}_s$ the structural stiffness matrix, defined as follows:

$$\mathbf{M} = M \begin{bmatrix} 1 & 0 & -y_G \\ 0 & 1 & x_G \\ -y_G & x_G & r_C^2 \end{bmatrix}, \quad \mathbf{C}_s = \begin{bmatrix} c_x & 0 & 0 \\ 0 & c_y & 0 \\ 0 & 0 & c_\theta \end{bmatrix}, \quad \mathbf{K}_s = \begin{bmatrix} k_x & 0 & 0 \\ 0 & k_y & 0 \\ 0 & 0 & k_\theta \end{bmatrix}.$$
(12.63)

In these expressions, M is the mass of the cylinder; r_C^2 is the squared mass polar inertia radius of the cross-section, computed with respect to the center of torsion; (x_G, y_G) are the coordinates of the center of mass G, measured in the coordinate system (C, x, y); finally, $\mathbf{f}^a := (F_x, F_y, C_m)^T$ is the aeroelastic force column matrix, to be evaluated later. Equation 12.62 is derived in Supplement 12.3, as an application of the Lagrange equations of motion (Appendix D). If, as a special case, $G \equiv C$, the equations uncouple in their structural part.

Supplement 12.3 (Derivation of the Lagrange Equations of Motion of a Rigid Cylinder with Viscoelastic Constraints) The velocity of a generic point $P = (x, y)$, belonging to the generic cross-section of abscissa z, has components:

$$\dot{u}(x, y, z) = \dot{u}_C - \dot{\theta}y, \quad \forall z,$$
(12.64a)

$$\dot{v}(x, y, z) = \dot{v}_C + \dot{\theta}x, \quad \forall z.$$
(12.64b)

By denoting with $\rho_m \, [\mathrm{ML}^{-3}]$ the mass volume density of the material, the kinetic energy of the cylinder reads:

$$\begin{aligned} T &= \frac{1}{2} \int_0^\ell dz \int_A \rho_m \left(\dot{u}^2 + \dot{v}^2 \right) dA \\ &= \frac{1}{2} \ell \int_A \rho_m \left[\dot{u}_C^2 + \dot{v}_C^2 + 2\dot{\theta} \left(x \dot{v}_C - y \dot{u}_C \right) + \dot{\theta}^2 \left(x^2 + y^2 \right) \right] dA \\ &= \frac{1}{2} \ell \left[m \left(\dot{u}_C^2 + \dot{v}_C^2 \right) + 2\dot{\theta} \dot{v}_C S_y - 2\dot{\theta} \dot{u}_C S_x + \dot{\theta}^2 I_C \right], \end{aligned}$$
(12.65)

where $m := \int_A \rho_m \, dA$, $[\mathrm{ML}^{-1}]$ is the mass per unit length of the cylinder; $S_x := \int_A \rho_m y \, dA =: m y_G$, $S_y := \int_A \rho_m x \, dA =: m x_G$ are the static mass moment of the cross-section with respect to the axes x, y; and $I_C = \int_A \rho_m \left(x^2 + y^2 \right) dA =: m r_C^2$ is the mass polar inertia moment of the cross-section with respect to the center of torsion C. Given that $M = m\ell$, it turns out that $T = \frac{1}{2} \dot{\mathbf{q}}^T \mathbf{M} \dot{\mathbf{q}}$, with $\mathbf{M}$ given by Eq. 12.63a.

The elastic potential energy of the system is:

$$U = \frac{1}{2} \left(k_x u_C^2 + k_y v_C^2 + k_\theta \theta^2 \right),$$
(12.66)

so that $U = \frac{1}{2}\mathbf{q}^T\mathbf{K}_s\mathbf{q}$, with $\mathbf{K}_s$ given by Eq. 12.63c. Because conservative external forces are absent, it is $V = 0$, hence $\Pi = U + V = U$.

The virtual work of nonconservative external forces is the sum of work expended by the viscous and aeroelastic forces, i.e.,

$$\delta W^{nc} = -c_x\dot{u}_C\delta u_C - c_y\dot{v}_C\delta v_C - c_\theta\dot{\theta}\delta\theta + \mathbf{f}^a\delta\mathbf{q}. \tag{12.67}$$

By letting $\delta W^{nc} =: \mathbf{Q}^T\delta\mathbf{q}$, it turns out that $\mathbf{Q} = -\mathbf{C}_s\mathbf{q} + \mathbf{f}^a$, with $\mathbf{C}_s$ given by Eq. 12.63b.[31] After having built up the Lagrangian $L := T - \Pi$, the Lagrange equations of motion, $\frac{\mathrm{d}}{\mathrm{d}t}\left(\frac{\partial L}{\partial\dot{\mathbf{q}}}\right) - \frac{\partial L}{\partial\mathbf{q}} = \mathbf{Q}$, supply Eq. 12.62.

The equations of motion can also be deduced from direct equilibrium (Fig. 12.8b). The following forces contribute to the balance: the forces of inertia $f_x^{in} := -M\left(\ddot{u}_C - y_G\ddot{\theta}\right)$, $f_y^{in} := -M\left(\ddot{v}_C + x_G\ddot{\theta}\right)$, applied at G; the inertia $c_z^{in} := -I_G\ddot{\theta}$ couple; the aero-viscoelastic forces $f_x^{ave} := -k_xu_C - c_x\dot{u}_C + f_x^a$, $f_y^{ave} := -k_yv_C - c_y\dot{v}_C + f_y^a$, $c_z^{vae} := -k_\theta\theta - c_\theta\dot{\theta} + c_z^a$, applied at C. The balance of forces and moments with respect to the torsion center is then imposed. Accounting for the Huygens theorem, for which $I_C = I_G + M\left(x_G^2 + y_G^2\right)$, Eqs. 12.62, 12.63 are finally obtained.   □

12.5.2 Aeroelastic Forces

To complete the aeroelastic model, the forces exerted by the wind on the mobile cylinder must be determined. According to the literature [3], two levels of approximation are possible, described and commented below.[32]

1. *Steady aeroelasticity* . The aeroelastic forces are called *steady* when the translational velocity $\dot{\mathbf{u}}_C = \dot{u}_C\mathbf{a}_x + \dot{v}_C\mathbf{a}_y$ and the angular velocity $\dot{\theta}\mathbf{a}_z$ are so small they can be neglected; the fluid-structure interaction is therefore entirely expressed by the twist θ; this angle, by modifying the wind exposure of the body, modifies the aeroelastic action. This is a low-level approximation, rarely adopted in applications.

2. *Quasi-steady aeroelasticity*. The aeroelastic forces are called *quasi-steady* when, in addition to the θ rotation, the velocity of the cylinder is also taken into account (although it is assumed small, compared to that of the fluid, as discussed in Sect. 12.2). This second approximation appears more significant. However, two variants of the theory are possible, in which (I) only $\dot{\mathbf{u}}_C$ is taken into account, but

---

[31] The structural damping matrix, being dissipative, cannot be deduced from an energy, but only from the virtual work. Sometimes reference is made to a pseudo-energy, called "Rayleigh dissipative potential." The procedure is justified by the formal analogy existing between elastic and viscous forces.

[32] The distinction does not arise when the cylinder can only translate, as in Sect. 12.3.

not $\dot{\theta}\mathbf{a}_z$, or, (II) both contributions are retained. In this book, the former will be referred as the the *first-level quasi-steady theory* and the latter the *second-level quasi-steady theory*.

A further and more advanced model, called *unsteady*, will be shortly discussed later (Sect. 12.9).

First-Level Quasi-steady Theory

At the generic time t, the cylinder possesses the translational velocity $\dot{\mathbf{u}}_C = \dot{u}_C \mathbf{a}_x + \dot{v}_C \mathbf{a}_y$ and the angular velocity $\dot{\theta}\mathbf{a}_z$ (Fig. 12.9). If, however, it is $\dot{\theta} D \ll |\dot{\mathbf{u}}_C|$, with D a characteristic dimension of the cross-section, it is possible to neglect the angular velocity and to assume that the body is just translating with velocity $\dot{\mathbf{u}}_C$. Denoted by $\mathbf{U} = U\mathbf{a}_x$ the wind velocity relative to the ground, the velocity relative to the body is:

$$\mathbf{U}_r := \mathbf{U} - \dot{\mathbf{u}}_C = (U - \dot{u}_C)\,\mathbf{a}_x - \dot{v}_C \mathbf{a}_y. \tag{12.68}$$

Because of the twist θ,[33] the material axis $\mathbf{a}_x$ rotates by the same angle, moving to $\hat{\mathbf{a}}_x$; hence, the instantaneous angle of attack β is the angle formed by $\mathbf{U}_r$ with the *current position* $\hat{\mathbf{a}}_x$ of the material axis. Consequently, $\mathbf{U}_r$ forms an angle $\beta + \theta$ with the original direction $\mathbf{a}_x$. From these considerations, and from Eq. 12.68, it follows that:

$$\tan(\beta + \theta) = \frac{\mathbf{U}_r \cdot \mathbf{a}_y}{\mathbf{U}_r \cdot \mathbf{a}_x} = -\frac{\dot{v}_C}{U - \dot{u}_C}, \tag{12.69}$$

so that the instantaneous angle of attack is:

$$\beta = -\theta - \arctan\left(\frac{\dot{v}_C}{U - \dot{u}_C}\right). \tag{12.70}$$

Fig. 12.9 Aeroelastic forces on the three degree of freedom cylinder

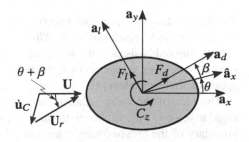

---

[33] Note that $\dot{\theta}$ is being neglected, but not θ.

This formula generalizes the definition in Eq. 12.8b, which is valid only in the case $\dot{u}_C = 0$, $\theta = 0$.

The aeroelastic force is $\mathbf{F} = F_d \mathbf{a}_d + F_l \mathbf{a}_l$, where $\mathbf{a}_d = \text{vers}(\mathbf{U}_r)$ and $\mathbf{a}_l = \mathbf{a}_z \times \mathbf{a}_d$; hence:

$$\mathbf{F} = \frac{1}{2}\rho_a U_r D\ell \left(\mathbf{U}_r C_d\left(\beta\right) + \mathbf{a}_z \times \mathbf{U}_r C_l\left(\beta\right)\right). \tag{12.71}$$

This force must be projected onto the $\left(\mathbf{a}_x, \mathbf{a}_y\right)$ base, as $F_x := \mathbf{F}\cdot\mathbf{a}_x$, $F_y := \mathbf{F}\cdot\mathbf{a}_y$; the force components and the couple read:[34]

$$F_x = \frac{1}{2}\rho_a U_r D\ell\left[C_d\left(\beta\right)\left(U - \dot{u}_C\right) + C_l\left(\beta\right)\dot{v}_C\right], \tag{12.72a}$$

$$F_y = \frac{1}{2}\rho_a U_r D\ell\left[-C_d\left(\beta\right)\dot{v}_C + C_l\left(\beta\right)\left(U - \dot{u}_C\right)\right], \tag{12.72b}$$

$$C_z = \frac{1}{2}\rho_a U_r^2 D^2\ell C_m\left(\beta\right). \tag{12.72c}$$

Remark 12.7 According to the first-level quasi-steady theory, the aeroelastic forces depend on the translational velocity $\dot{\mathbf{u}}_C$ of the cross-section (which modifies the relative velocity and therefore the angle of attack) together with the rotation θ (which modifies only the angle of attack, as the body, rotating, exposes a different face to the wind). Forces, on the other hand, *do not depend* on the angular velocity $\dot{\theta}$. As a consequence, while the aeroelastic action modifies the translational damping of the structure, it leaves the rotational damping unchanged. The theory, therefore, *cannot* explain the phenomenon of the *torsional galloping*, which was indeed observed in wind tunnel tests carried out on isosceles triangular prisms, constrained to rotate around their own axes [8] (as it will be commented in Sect. 12.6). To overcome these conceptual difficulties, a second-level theory was proposed.

Second-Level Quasi-steady Theory: The Mean Radius Conjecture

The theory of second level, also known as *the mean* (or characteristic) *radius theory*, extends the model discussed so far. Although this theory, as it will be seen soon, has a weak rational basis (and the topic, far from being exhaustively explained, is still subject of debate), it represents, at the moment, the only tool available for a simple approach to the problem [8].

The idea of the mean radius theory is the following (Fig. 12.10a). Due to the angular velocity $\dot{\theta}\mathbf{a}_z$, applied to the axis of the torsional centers C, a point P on the boundary of the cross-section is animated by the structural velocity $\dot{\mathbf{u}}_P = \dot{\theta}\mathbf{a}_z \times$

---

[34] In fact, taking into account Eq. 12.68, one has $\mathbf{a}_z \times \mathbf{U}_r \cdot \mathbf{a}_x = -\dot{v}_C \mathbf{a}_z \times \mathbf{a}_y \cdot \mathbf{a}_x = \dot{v}_C$ and $\mathbf{a}_z \times \mathbf{U}_r \cdot \mathbf{a}_y = \left(U - \dot{u}_C\right)\mathbf{a}_z \times \mathbf{a}_x \cdot \mathbf{a}_y = U - \dot{u}_C$.

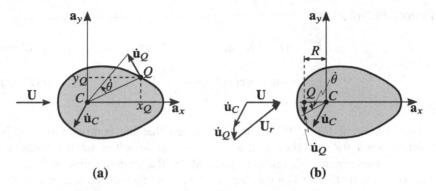

Fig. 12.10 Characteristic radius theory: (a) characteristic point Q on the boundary of the cross-section, (b) characteristic point Q on the $\mathbf{a}_x$ axis, upwind

$\overrightarrow{CP}$, which differs from point to point. It happens, therefore, that *the wind velocity relative to P*, i.e., $\mathbf{U} - \dot{\mathbf{u}}_P$, *is a function of point*, together with the instantaneous angle of attack. As a consequence, the quasi-steady theory (in which the angle of attack is unique for the whole cross-section) cannot be applied. It is then postulated that there exists a *characteristic point* $Q = (x_Q, y_Q)$, such that its velocity $\dot{\mathbf{u}}_Q = \dot{\theta}\mathbf{a}_z \times \overrightarrow{CQ} = \dot{\theta}\left(-y_Q\mathbf{a}_x + x_Q\mathbf{a}_y\right)$ is representative of the entire velocity field at the boundary, so that it can be taken to evaluate, in average, the relative velocity of the wind with respect to the body.[35] This is a pure conjecture, which leaves open the choice of the characteristic point, which should be determined on the basis of experimental tests.

Assuming the conjecture is valid, the relative velocity in Eq. 12.68 should be corrected as follows:

$$\mathbf{U}_r := \mathbf{U} - \left(\dot{\mathbf{u}}_C + \dot{\mathbf{u}}_Q\right) = \left(U - \dot{u}_C + y_Q\dot{\theta}\right)\mathbf{a}_x - \left(\dot{v} + x_Q\dot{\theta}\right)\mathbf{a}_y. \qquad (12.73)$$

Consequently, the angle of attack in Eq. 12.70 should also be corrected into:

$$\beta = -\theta - \arctan\left(\frac{\dot{v}_C + x_Q\dot{\theta}}{U - \dot{u}_C + y_Q\dot{\theta}}\right). \qquad (12.74)$$

The choice of the point Q, as mentioned above, is arbitrary. By referring to a compact cross-section, whose center of torsion is close to the center of gravity, it has been proposed in the literature, to assume $Q \in \mathbf{a}_x$, at a distance from the center $x_Q = -R$ (i.e., upwind with respect to C), with R the *mean radius*, of the cross-section (Fig. 12.10b), to be properly selected.[36] With this choice, the previous

---

[35] The point Q can fall on the boundary, but also inside the domain.

[36] A frequent choice is $R = D/2$, with D the cross-section width in the direction $\mathbf{a}_x$.

equations are written as:

$$\mathbf{U}_r = (U - \dot{u}_C)\,\mathbf{a}_x - \left(\dot{v}_C - R\dot{\theta}\right)\mathbf{a}_y, \tag{12.75a}$$

$$\beta = -\theta - \arctan\left(\frac{\dot{v}_C - R\dot{\theta}}{U - \dot{u}_C}\right). \tag{12.75b}$$

Remark 12.8 From Eq. 12.75, it can be deduced that the contribution of $\dot{\theta}$ is important when $\dot{\theta}R = O(\dot{v}_C)$, that is, when the velocities on the boundary of the cross-section are significantly affected by the angular velocity (i.e., for quasi-harmonic motions, when the displacements are significantly affected by the rotation). Indications in this regard can be drawn from the natural modes of the cylinder, and precisely from the ratio between the translational and rotational components, which establishes the extent of coupling.

Linearized Aeroelastic Forces

With reference to the quasi-steady theory of the second level, the aeroelastic forces are linearized, assuming that $|\dot{u}_C| \ll U$ and $R\dot{\theta} \ll U$. By Taylor expanding the module of the relative velocity (Eq. 12.75a), one obtains:

$$U_r = \sqrt{(U - \dot{u}_C)^2 + \left(\dot{v}_C - R\dot{\theta}\right)^2} = U - \dot{u}_C + \cdots. \tag{12.76}$$

Analogously, the angle of attack (12.75b) is:[37]

$$\beta = -\theta - \frac{\dot{v}_C}{U} + \frac{R\dot{\theta}}{U} + \cdots, \tag{12.77}$$

and the aerodynamic coefficients $C_h(\beta)$ are:

$$C_h(\alpha) = C_{h0} + C'_{h0}\beta + \cdots, \quad h = d, l, m. \tag{12.78}$$

Substituting these expressions in Eq. 12.72, and retaining the linear terms only, one finds:

$$\mathbf{f}^a = \mathbf{f}_0^a - \mathbf{C}_a\dot{\mathbf{q}} - \mathbf{K}_a\mathbf{q}. \tag{12.79}$$

---

[37] Many authors (e.g., [1, 8]), in addition to take v positive downward, also take θ positive clockwise, so that β is opposite to that defined here. This has consequences on Table 12.2 and on the discussion carried out in Remark 12.11.

Here, $\mathbf{f}_0^a := \frac{1}{2}\rho_a U^2 D\ell \left(C_{d_0} \; C_{l_0} \; C_{m_0} D \right)^T$ are the forces acting on the fixed cylinder (which will be ignored ahead); moreover:

$$
\mathbf{C}_a := \frac{1}{2}\rho_a U D\ell \begin{bmatrix} 2C_{d_0} & -C_{l_0} + C'_{d_0} & -C'_{d_0} R \\ 2C_{l_0} & C_{d_0} + C'_{l_0} & -C'_{l_0} R \\ 2C_{m_0}D & C'_{m_0}D & -C'_{m_0} D R \end{bmatrix},
\tag{12.80a}
$$

$$
\mathbf{K}_a := \frac{1}{2}\rho_a U^2 D\ell \begin{bmatrix} 0 & 0 & C'_{d_0} \\ 0 & 0 & C'_{l_0} \\ 0 & 0 & C'_{m_0}D \end{bmatrix}.
\tag{12.80b}
$$

In the previous definitions, $\mathbf{C}_a$ is the *aerodynamic damping matrix* and $\mathbf{K}_a$ is the *aerodynamic stiffness matrix*. If the effect of the angular velocity is ignored (first-level theory), $R = 0$ must be taken in the third column of $\mathbf{C}_a$.

12.5.3 Linear Stability Analysis

The stability of the equilibrium position of the cylinder is now analyzed, by linearizing the motion in the neighborhood of $\mathbf{q} = \mathbf{0}$. Making use of linearized expressions of the forces (Eq. 12.79), and ignoring the constant known term, the equation of motion, Eq. 12.62, reads:

$$
\mathbf{M}\ddot{\mathbf{q}} + (\mathbf{C}_s + U\mathbf{B})\,\dot{\mathbf{q}} + \left(\mathbf{K}_s + U^2\mathbf{H} \right)\mathbf{q} = \mathbf{0},
\tag{12.81}
$$

in which the following definitions have been introduced, to make explicit the dependence on U, i.e.,

$$
\mathbf{C}_a =: U\mathbf{B}, \qquad \mathbf{K}_a =: U^2\mathbf{H}.
\tag{12.82}
$$

In full, one has:

$$
M \begin{bmatrix} 1 & 0 & -y_G \\ 0 & 1 & x_G \\ -y_G & x_G & r_C^2 \end{bmatrix} \begin{pmatrix} \ddot{u}_C \\ \ddot{v}_C \\ \ddot{\theta} \end{pmatrix} + \begin{bmatrix} c_x + Ub_{11} & Ub_{12} & Ub_{13} \\ Ub_{21} & c_y + Ub_{22} & Ub_{23} \\ Ub_{31} & Ub_{32} & c_\theta + Ub_{33} \end{bmatrix} \begin{pmatrix} \dot{u}_C \\ \dot{v}_C \\ \dot{\theta} \end{pmatrix}
$$

$$
+ \begin{bmatrix} k_x & 0 & U^2 h_{13} \\ 0 & k_y & U^2 h_{23} \\ 0 & 0 & k_\theta + U^2 h_{33} \end{bmatrix} \begin{pmatrix} u_C \\ v_C \\ \theta \end{pmatrix} = \begin{pmatrix} 0 \\ 0 \\ 0 \end{pmatrix},
$$

$$
\tag{12.83}
$$

where:

$$b_{11} := \rho_a D\ell C_{d_0}, \qquad\qquad b_{21} := \rho_a D\ell C_{l_0}, \qquad\qquad b_{31} := \rho_a D^2 \ell C_{m_0},$$

$$b_{12} := \frac{1}{2}\rho_a D\ell\left(-C_{l_0} + C'_{d_0}\right), \; b_{22} := \frac{1}{2}\rho_a D\ell\left(C_{d_0} + C'_{l_0}\right), \; b_{32} := \frac{1}{2}\rho_a D^2 \ell C'_{m_0},$$

$$b_{13} := -\frac{1}{2}\rho_a D\ell C'_{d_0} R, \qquad b_{23} := -\frac{1}{2}\rho_a D\ell C'_{l_0} R, \qquad b_{33} := -\frac{1}{2}\rho_a D^2 \ell C'_{m_0} R,$$

$$h_{13} := \frac{1}{2}\rho_a D\ell C'_{d_0}, \qquad\qquad h_{23} := \frac{1}{2}\rho_a D\ell C'_{l_0}, \qquad\qquad h_{33} := \frac{1}{2}\rho_a D^2 \ell C'_{m_0}.$$
$$\tag{12.84}$$

Remark 12.9 The aeroelastic forces depend (a) on the structural (translational and angular) velocities and (b) the rotation of the cylinder. The former, $-\mathbf{C}_a\dot{\mathbf{q}}$, add themselves to the structural damping and contribute (as a geometric effect) to the total damping; the latter, $-\mathbf{K}_a\mathbf{q}$, add themselves to the elastic forces and contribute (as a geometric effect) to the total stiffness.

Cross-Sections Symmetric with Respect to the Flow Direction

In the particular case of a symmetric cross-section with respect to the $\mathbf{a}_x$ axis, it is $y_G = 0$ in the mass matrix; moreover it is $C_{l_0} = C_{m_0} = C'_{d_0} = 0$ in the aerodynamic matrices. The equations of motion, therefore, simplify as follows:

$$M\begin{bmatrix} 1 & 0 & 0 \\ 0 & 1 & x_G \\ 0 & x_G & r_C^2 \end{bmatrix}\begin{pmatrix} \ddot{u}_C \\ \ddot{v}_C \\ \ddot{\theta} \end{pmatrix} + \begin{bmatrix} c_x + Ub_{11} & 0 & 0 \\ 0 & c_y + Ub_{22} & Ub_{23} \\ 0 & Ub_{32} & c_\theta + Ub_{33} \end{bmatrix}\begin{pmatrix} \dot{u}_C \\ \dot{v}_C \\ \dot{\theta} \end{pmatrix}$$

$$+ \begin{bmatrix} k_x & 0 & 0 \\ 0 & k_y & U^2 h_{23} \\ 0 & 0 & k_\theta + U^2 h_{33} \end{bmatrix}\begin{pmatrix} u_C \\ v_C \\ \theta \end{pmatrix} = \begin{pmatrix} 0 \\ 0 \\ 0 \end{pmatrix}.$$
$$\tag{12.85}$$

Remarkably, the motion $u_C(t)$ along the symmetry direction $\mathbf{a}_x$ is uncoupled from the roto-translational motion, $v_C(t), \theta(t)$.

Dynamic Bifurcations

The equation of motion, Eq. 12.81, admits the solution $\mathbf{q}(t) = \mathbf{u}\exp(\lambda t)$, which leads to the algebraic eigenvalue problem:

$$\left[\lambda^2 \mathbf{M} + \lambda\left(\mathbf{C}_s + U\mathbf{B}\right) + \left(\mathbf{K}_s + U^2\mathbf{H}\right)\right]\mathbf{u} = \mathbf{0}. \tag{12.86}$$

When $U \to 0$, and since the elastic system is damped, the six eigenvalues λ are all to the left of the imaginary axis; they consist of three pairs of complex conjugate numbers, if damping is small. The equilibrium position is therefore asymptotically stable. For increasing U, if appropriate conditions about the aerodynamic coefficients exist (i.e., if the cross-section is aerodynamically unstable), a pair of eigenvalues approaches the imaginary axis, and for $U = U_c$ passes through it, by taking the value $\lambda_c = \pm i\omega_c$. Thus, a *dynamic* (or Hopf) *bifurcation* occurs, in correspondence of which the system loses stability, oscillating with exponentially increasing amplitude. The critical eigenvalue $i\omega_c$, the critical velocity U_c, and the critical mode $\mathbf{u}_c$ satisfy the equation:

$$\left[-\omega_c^2 \mathbf{M} + i\omega_c \left(\mathbf{C}_s + U_c \mathbf{B} \right) + \left(\mathbf{K}_s + U_c^2 \mathbf{H} \right) \right] \mathbf{u}_c = \mathbf{0}. \tag{12.87}$$

The algebraic problem, being of sixth degree in λ, cannot be solved in closed form, but only numerically (or with perturbation methods). To determine the critical condition, one can proceed in two alternative ways:

1. A sufficiently small value of U is taken, and the six eigenvalues of the problem in Eq. 12.86 are determined. Making U to increase in small increments, the corresponding eigenvalues are computed, and the locus of the roots of the characteristic polynomial is plotted, until the condition $\mathrm{Re}\,(\lambda) = 0$ is satisfied by the couple with the smallest real part in modulus.
2. By letting $\lambda = i\omega_c$, the characteristic polynomial of Eq. 12.87, equated to zero, reads $\sum_{k=0}^{6} (i\omega_c)^{6-k} \, p_k \, (U_c) = 0$, where p_k are polynomials in U_c. The real and imaginary parts are zeroed separately, thus obtaining two real equations in the unknowns ω_c, U_c :

$$p_0 \omega_c^6 - p_2 \, (U_c) \, \omega_c^4 + p_4 \, (U_c) \, \omega_c^2 - p_6 \, (U_c) = 0, \tag{12.88a}$$

$$\omega_c \left[p_1 \omega_c^4 - p_3 \, (U_c) \, \omega_c^2 + p_5 \, (U_c) \right] = 0. \tag{12.88b}$$

Solving (in closed form) the second equation with respect to ω_c^2 and substituting the result in the first one, an equation in U_c only is drawn. This can be solved by Newton-Raphson techniques, or simply represented in a graph, to determine, if it exists, the smallest real root $U_c > 0$ (whose associate ω_c is real).

In the following sections, systems with one or two degrees of freedom will be examined, deduced from the three degrees of freedom model.

12.6   Unidirectional Motions: Galloping and Rotational Divergence

If the cylinder in Fig. 12.8 has only one DOF, its equation of motion is drawn from Eq. 12.81 by deleting the columns and rows corresponding to the suppressed displacements. Three possible cases are examined.

Along-Flow Motion

If $v_C = \theta = 0$, the cylinder can only move parallel to the flow, i.e., in the $\mathbf{a}_x$ direction. The equation of motion is:

$$M\ddot{u}_C + (c_x + Ub_{11})\,\dot{u}_C + k_x u_C = 0, \qquad (12.89)$$

where $b_{11} := \rho_a D\ell C_{d_0}$. Since $C_{d_0} > 0$, wind is stabilizing. Therefore no aeroelastic bifurcations can occur. The oscillations consequent to an initial perturbation exponentially decay over time.

Cross-Flow Galloping

If $u_C = \theta = 0$, the cylinder can only move transversely to the wind. This problem has already been studied in Sect. 12.3.2; it is governed by the following equation of motion:

$$M\ddot{v}_C + \left(c_y + Ub_{22}\right)\dot{v}_C + k_y v_C = 0, \qquad (12.90)$$

where $b_{22} := \dfrac{1}{2}\rho_a D\ell\left(C_{d_0} + C_{l_0}'\right)$. The bifurcation condition (galloping) occurs when $U = U_c := -\dfrac{b_{22}}{c_y}$, according to the Den Hartog criterion. Bifurcation, however, as it was observed, takes place only if $b_{22} < 0$.

Torsional Galloping and Torsional Divergence

If $u_C = v_C = 0$, that is, if the cylinder can only rotate around to its axis, the equation of motion becomes:

$$M r_C^2 \ddot{\theta} + (c_\theta + Ub_{33})\,\dot{\theta} + \left(k_\theta + U^2 h_{33}\right)\theta = 0, \qquad (12.91)$$

where $b_{33} := -\dfrac{1}{2}\rho_a D^2 R\ell C_{m_0}'$ and $h_{33} := \dfrac{1}{2}\rho_a D^2\ell C_{m_0}'$. Two cases must be discussed, according to the first- or second-level theory adopted.

- The mean radius R is taken equal to zero (quasi-steady theory of first level). Since the aeroelastic action does not alter the structural damping (being $b_{33} = 0$), no dynamic bifurcation can occur. On the other hand, since the flow modifies the structural stiffness ($h_{33} \neq 0$), a static bifurcation can occur, called *rotational* (o torsional[38]) *divergence*, provided the aeroelastic forces are instabilizing (i.e., $h_{33} < 0$). The phenomenon is very similar to that exhibited by the compressed inverted pendulum, where, for a sufficiently large axial load, the geometric action prevails on the elastic stiffness. Therefore, a torsional divergence occurs when the total stiffness vanishes, i.e., at the critical velocity $U_c := \sqrt{-\dfrac{k_\theta}{h_{33}}}$, or:

---

[38] The term "torsional" is referred to the torsion of a beam, of which the elastically constrained cylinder represents a rough model.

$$U_c = \sqrt{\frac{2k_\theta}{\rho_u D^2 \ell \, |C'_{m_0}|}}, \qquad C'_{m_0} < 0. \tag{12.92}$$

If, instead, it is $C'_{m_0} > 0$, the *static* aeroelastic action is stabilizing.
- The mean radius R is taken as non-zero (quasi-steady theory of second level). Since the aeroelastic forces modify the structural damping, dynamic instability occurs when the total torsional damping vanishes; the circumstance is formally analogous to that causing transverse galloping. The phenomenon is called *rotational* (or torsional) *galloping*; it occurs at the critical velocity $U_c := -\frac{c_\theta}{b_{33}}$, or:

$$U_c := \frac{4\xi_\theta \sqrt{k_\theta M r_C^2}}{\rho_a D^2 R \ell C'_{m_0}}, \qquad C'_{m_0} > 0. \tag{12.93}$$

If, instead, it is $C'_{m_0} < 0$, the aeroelastic *dynamic* action is stabilizing.

Remark 12.10 The nature of the instability, static or dynamic, is governed by the sign of C'_{m_0}: (a) if $C'_{m_0} < 0$, only divergence can occur, (b) if $C'_{m_0} > 0$, only galloping can manifest, at the corresponding critical velocities.

Remark 12.11 Regarding the torsional divergence, the following considerations apply. The condition $C'_{m_0} < 0$ indicates that, in the neighborhood of $\beta = 0$, the aeroelastic couple $C_z(\beta)$ decreases with the angle of attack, that is, $\frac{dC_z}{d\beta}\Big|_{\beta=0} < 0$. Since, from Eq. 12.77, it is $\beta = -\theta$ in static conditions, this entails that $\frac{dC_z}{d\theta}\Big|_{\theta=0} > 0$. This is a potentially unstable situation, as *the geometric couple is concordant in sign with the angular deviation*, differently from what occurs for a spring. The opposite happens about the torsional galloping, since, from the same Eq. 12.77, but under dynamic conditions, it follows that $\beta = \frac{R\dot\theta}{U}$. The difference between the two phenomena can be understood by noticing that (a) a counterclockwise rotation of the cross-section implies a clockwise rotation of the incident wind and (b) a counterclockwise rotational motion (with Q upwind) implies a counterclockwise rotation of the incident wind.

Numerical Values of the Moment Coefficients
The numerical values of the derivative of the aerodynamic moment coefficient, C'_{m_0}, can be found in the literature for a number of cross-section shapes. Table 12.2 (taken from [1] and adapted to the conventions used here), refers to rectangular cross-sections and to the airfoil.[39] All the rectangular cross-sections considered are prone

---

[39] Regarding the airfoil, a is the distance between the center of torsion and the point of application of the resultant of the aeroelastic forced, assumed to be located at $1/4$ of the chord. Therefore, the moment with respect to the center of torsion is equal to the lift force (proportional to $C'_{y_0} \simeq 2\pi$, as seen in Table 12.1) multiplied by the distance a.

Table 12.2   Values of C'_{m_0} for different cross-sections of a long cylinder, subject to horizontal incident flow; positive values indicate dynamically unstable cross-sections (i.e., subject to galloping); negative values denote statically unstable cross-sections (i.e., subject to divergence)

$C'_{m_0} = 0.18$ $C'_{m_0} = 0.16$ $C'_{m_0} = 1.125$

$C'_{m_0} = 1.04$ $C'_{m_0} = -\frac{2\pi a}{D}$

to torsional galloping only, whose critical velocity is lower, the more the cross-section is narrow. The airfoil, on the other hand, is stable to galloping, but it is prone to torsional divergence.

12.7   Two Degrees of Freedom Translational Galloping

If the cylinder in Fig. 12.8 can translate in both directions, but cannot rotate, the system has two degrees of freedom, $u_C(t)$, $v_C(t)$. It is governed by two equations of motion deduced from Eq. 12.81, i.e.,

$$
\begin{bmatrix} 1 & 0 \\ 0 & 1 \end{bmatrix} \begin{pmatrix} \ddot{u}_C \\ \ddot{v}_C \end{pmatrix} + \begin{bmatrix} 2\xi_x \omega_x + U\frac{b_{11}}{M} & U\frac{b_{12}}{M} \\ U\frac{b_{21}}{M} & 2\xi_y \omega_y + U\frac{b_{22}}{M} \end{bmatrix} \begin{pmatrix} \dot{u}_C \\ \dot{v}_C \end{pmatrix}
$$
$$
+ \begin{bmatrix} \omega_x^2 & 0 \\ 0 & \omega_y^2 \end{bmatrix} \begin{pmatrix} u_C \\ v_C \end{pmatrix} = \begin{pmatrix} 0 \\ 0 \end{pmatrix},
$$

(12.94)

where the following natural frequencies and damping factors have been introduced:

$$
\omega_x := \sqrt{\frac{k_x}{M}}, \quad \omega_y := \sqrt{\frac{k_y}{M}}, \quad \xi_x := \frac{c_x}{2M\omega_x}, \quad \xi_y := \frac{c_y}{2M\omega_y}.
$$

(12.95)

To limit algebra, it is assumed that the structural damping is the same in both the directions, $\xi_x = \xi_y =: \xi_s$. The critical velocity at which the system loses stability is said of *translational* (or flexural) *galloping with two degrees of freedom.*[40]

---

[40] The locution "flexural," less precise but widely used in the technical field, alludes to the fact that the rigid cylinder with elastic constraints could represent the motion of a beam in bending.

Remark 12.12 The problem is of interest only if the $\mathbf{a}_x$ direction is *not of symmetry* for the cross-section. Indeed, in the symmetric case, the equations uncouple, because $b_{12} = b_{21} = 0$, and, since the longitudinal motion is always stable, they describe the simpler occurrence of transverse galloping. In the non-symmetric case, on the other hand, it is of interest investigating if, due to the coupling induced by $b_{12} \neq 0$, $b_{21} \neq 0$, the longitudinal motion alters, and to what extent, the transverse galloping velocity.

Exact Solution
By letting $(u_C, v_C) = (\hat{u}_C, \hat{v}_C) \exp(\lambda t)$, an eigenvalue problem is drawn from Eq. 12.94, i.e.,

$$\begin{bmatrix} \lambda^2 + \left(2\xi_s\omega_x + U\frac{b_{11}}{M}\right)\lambda + \omega_x^2 & \lambda U\frac{b_{12}}{M} \\ \lambda U\frac{b_{21}}{M} & \lambda^2 + \left(2\xi_s\omega_y + U\frac{b_{22}}{M}\right)\lambda + \omega_y^2 \end{bmatrix} \begin{pmatrix} \hat{u}_C \\ \hat{v}_C \end{pmatrix} = \begin{pmatrix} 0 \\ 0 \end{pmatrix}. \tag{12.96}$$

The associated characteristic equation is:

$$\lambda^4 + p_1\lambda^3 + p_2\lambda^2 + p_3\lambda + p_4 = 0, \tag{12.97}$$

where the coefficients $p_i = p_i(U)$ are polynomials, defined as follows:

$$p_1 := 2\xi_s(\omega_x + \omega_y) + U\left(\frac{b_{11}}{M} + \frac{b_{22}}{M}\right),$$

$$p_2 := \omega_x^2 + \omega_y^2 + \left(2\xi_s\omega_x + U\frac{b_{11}}{M}\right)\left(2\xi_s\omega_y + U\frac{b_{22}}{M}\right) - U^2\frac{b_{12}}{M}\frac{b_{21}}{M},$$

$$p_3 := 2\xi_s\left(\omega_x\omega_y^2 + \omega_y\omega_x^2\right) + U\left(\frac{b_{11}}{M}\omega_y^2 + \frac{b_{22}}{M}\omega_x^2\right),$$

$$p_4 := \omega_x^2\omega_y^2. \tag{12.98}$$

In the critical state, it is $\lambda = i\omega_c$ and $U = U_c$. The characteristic equation splits into two real equations, which, in explicit form, read:

$$\omega_c^4 - \left(\omega_x^2 + \omega_y^2 + f(U_c)\right)\omega_c^2 + \omega_x^2\omega_y^2 = 0, \tag{12.99a}$$

$$\omega_c\left[\left(\omega_y^2 - \omega_c^2\right)\left(2\xi_s\omega_x + U_c\frac{b_{11}}{M}\right) + \left(\omega_x^2 - \omega_c^2\right)\left(2\xi_s\omega_y + U_c\frac{b_{22}}{M}\right)\right] = 0, \tag{12.99b}$$

where:

$$f(U_c) := \left(2\xi_s\omega_x + U_c\frac{b_{11}}{M}\right)\left(2\xi_s\omega_y + U_c\frac{b_{22}}{M}\right) - U_c^2\frac{b_{12}}{M}\frac{b_{21}}{M}. \tag{12.100}$$

The pair of real roots U_c, ω_c with the smallest U_c, identifies the critical state.[41]

Resonant Case

A notable particular case is first considered, in which the two natural frequencies are equal: $\omega_x = \omega_y =: \omega_s$ (*resonant* case). In this condition, it is possible to get a closed form solution (detailed later in Supplement 12.4). The solution does not depend on the individual aerodynamic coefficients, but on the *invariants of the aerodynamic matrix* [7]:

$$\text{tr}\mathbf{B} := b_{11} + b_{22}, \qquad \det \mathbf{B} := b_{11}b_{22} - b_{12}b_{21}. \qquad (12.101)$$

Figure 12.11 illustrates the different types of bifurcations. In each region of the invariant plane $(\det \mathbf{B}, \text{tr}\mathbf{B})$, the (stylized) λ root loci are represented, as the wind velocity increases from zero. The plot depicts the following scenario:

- In the first quadrant the equilibrium is stable, whatever the velocity of the wind.
- A single Hopf bifurcation occurs in the second and third quadrants, at the critical velocity:

$$U_{c,1} := -\xi_s M \omega_s \frac{\text{tr}\mathbf{B} + \sqrt{\text{tr}^2\mathbf{B} - 4 \det \mathbf{B}}}{\det \mathbf{B}}. \qquad (12.102)$$

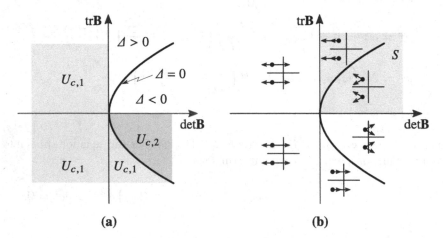

(a)   (b)

Fig. 12.11 Linear stability diagram for translational galloping, on the aerodynamic invariant plane, in the resonant case $\omega_x = \omega_y =: \omega_s$: (**a**) domains of validity of the critical velocities $U_{c,1}$, $U_{c,2}$; (**b**) stylized loci of the eigenvalues, as U increases, coalescing at $\lambda = -\xi_s \omega_s \pm i \omega_s \sqrt{1 - \xi_s^2}$, when $U = 0$; the parabola has equation $\text{tr}^2\mathbf{B} - 4 \det \mathbf{B} = 0$

---

[41] By solving Eq. 12.99b for ω_c^2 and substituting it in Eq. 12.99a, a single equation for U_c is found.

- In the fourth quadrant it is necessary to distinguish: outside the parabola of equation $\mathrm{tr}^2 \mathbf{B} - 4 \det \mathbf{B} = 0$, a single Hopf bifurcation manifests at the velocity $U_{c,1}$ (since the second bifurcation occurs at a higher velocity); inside the parabola, a double Hopf bifurcation occurs, at the critical velocity:

$$U_{c,2} := -\frac{4\xi_s M \omega_s}{\mathrm{tr}\mathbf{B}}. \tag{12.103}$$

Remarkable differences emerge between *resonant two DOF galloping* and transverse galloping, commented ahead.

- The condition $b_{22} < 0$, which is necessary and sufficient for causing transverse galloping, is neither necessary nor sufficient to trigger galloping with two degrees of freedom. In fact, if b_{11}, that is always positive, is large enough to make $\mathrm{tr}\mathbf{B} > 0$, and, moreover, if $\det \mathbf{B} > 0$, the system is stable for any wind velocity. In this case, *coupling with longitudinal motion restabilizes the system*, which otherwise would be unstable.
- Conversely, $b_{22} > 0$ is not sufficient to ensure stability, since, although $\mathrm{tr}\mathbf{B} > 0$, if $\det \mathbf{B} < 0$, the system falls into the second quadrant, in which, at $U = U_{c,1}$, the equilibrium is unstable. This is the most striking aspect of coupling, capable of *making unstable an otherwise stable system*.
- In the second and third quadrant of the invariant plane, the velocity $U_{c,1}$ can be *either smaller or larger than that of transverse galloping*; coupling, therefore, can be either beneficial or detrimental.
- In the fourth quadrant of the invariant plane, both velocities $U_{c,1}, U_{c,2}$ (which, respectively, hold outside and inside the parabola), are larger than the transverse galloping velocity, so that *the effect of the longitudinal motion is beneficial*.
- In a small region of the invariant plane, as a result of the resonance, a *double Hopf bifurcation* occurs, in spite of the fact the longitudinal motion is aerodynamically stable.

The previous results are justified in Supplement 12.4.

Supplement 12.4 (Solution to the Eigenvalue Problem in the Resonant Case)
In the resonant case $\omega_x = \omega_y =: \omega_s$, Eq. 12.99 simplifies into:

$$\omega_c^4 - \left(2\omega_s^2 + f_r(U_c)\right)\omega_c^2 + \omega_s^4 = 0, \tag{12.104a}$$

$$\omega_c \left(\omega_s^2 - \omega_c^2\right)\left(4\xi_s\omega_s + \mathrm{tr}\mathbf{B}\frac{U_c}{M}\right) = 0, \tag{12.104b}$$

where:

$$f_r(U_c) := \det\mathbf{B}\frac{U_c^2}{M^2} + 2\xi_s\omega_s\mathrm{tr}\mathbf{B}\frac{U_c}{M} + 4\omega_s^2\xi_s^2. \tag{12.105}$$

Equation 12.104b admits two solutions:

- $\omega_c = \omega_s$; by replacing it in Eq. 12.104a, $f_r(U_c) = 0$ is drawn, from which $U = U_{c,1}$ follows.[42] This is real and positive only if (i) $\det \mathbf{B} < 0$ or if (ii) $\det \mathbf{B} > 0$, but $\Delta := \text{tr}^2\mathbf{B} - 4\det \mathbf{B} > 0$, together with $\text{tr}\mathbf{B} < 0$. The domain of existence of $U_{c,1}$ is indicated in Fig. 12.11a on the plane of invariants $(\text{tr}\mathbf{B}, \det \mathbf{B})$; it is bounded by the positive semi-axis $\text{tr}\mathbf{B}$ and the lower branch of the parabola $\Delta = 0$. In the second and third quadrant, the condition (i) is satisfied; in the fourth quadrant, outside the parabola, the conditions (ii) are fulfilled. Since the critical frequency is unique, $\omega_c = \omega_s$, a single pair of eigenvalues crosses the imaginary axis, thus giving rise to a single Hopf bifurcation.

- $\omega_c \neq \omega_s$; Eq. 12.104b directly provides the velocity $U_{c,2}$. This is positive if and only if $\text{tr}\mathbf{B} < 0$; however, since $U_{c,2} > U_{c,1}$,[43] the domain of validity of $U_{c,2}$ is delimited by the positive semi-axis $\det \mathbf{B}$ and the lower branch of the parabola $\Delta = 0$ (Fig. 12.11a). Substituting $U_c = U_{c,2}$ in Eq. 12.104a, two critical frequencies, $\omega_{c,1} \neq \omega_{c,2}$, real and distinct are determined,[44] corresponding to the same velocity $U_{c,2}$; therefore a double Hopf bifurcation occurs.   □

Asymptotic Analysis of the Nonresonant Case

The analysis previously carried out clarifies the behavior of the system under the resonant condition $\omega_x = \omega_y$; however, in the *nonresonant* $\omega_x \neq \omega_y$ case, it needs to resort to a numerical solution of Eq. 12.99. Alternatively, to draw qualitative indications, one can perform an asymptotic analysis, along the following lines. By considering all the aeroelastic and damping forces as small of order ϵ, that is, by letting $\xi_s = O(\epsilon)$, $b_{ij} = O(\epsilon)$, the eigenvalue problem Eq. 12.96 can be regarded as a ϵ-perturbation of the free oscillation problem. Equation 12.99a, being $f(U_c) = O(\epsilon^2)$, gives $\omega_c = \omega_x + O(\epsilon^2)$, $\omega_c = \omega_y + O(\epsilon^2)$. When these results are replaced in Eq. 12.99b and only terms $O(1)$ are kept, the following critical velocities are found:

$$U_c \simeq -\frac{2\xi_s M\omega_x}{b_{11}} < 0, \qquad U_c \simeq -\frac{2\xi_s M\omega_y}{b_{22}} \gtrless 0, \tag{12.106}$$

that is, the velocities relevant to the uncoupled motions. It is concluded that, in the nonresonant case, the *longitudinal motion produces only a small modification of the critical transverse galloping velocity*.

Numerical Analysis of the Quasi-Resonant Case

From the above findings and from the continuity of the solution, it follows that, when the two natural frequencies are not exactly equal, but they are close to

---

[42] The positive sign in front of the square root is discarded, because it provides either negative or positive solutions of larger modulus.

[43] Indeed, it can also be written as $U_{c,1} = -\frac{4\xi_s M\omega_s}{\text{tr}\mathbf{B} - \sqrt{\text{tr}^2\mathbf{B} - 4\det \mathbf{B}}} < U_{c,2}$.

[44] Indeed, $f_r(U_{c,2}) = \frac{4\xi_s^2\omega_s^2}{\text{tr}^2\mathbf{B}}(4\det \mathbf{B} - \text{tr}^2\mathbf{B}) > 0$, so that the discriminant of Eq. 12.104a is also positive.

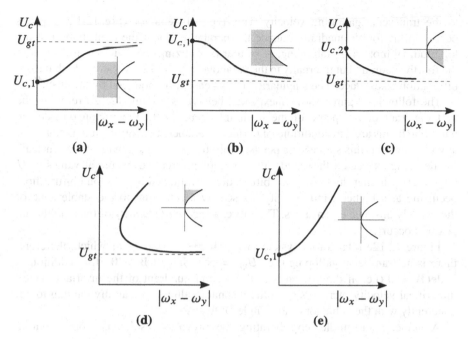

Fig. 12.12 Bifurcation diagrams for translational galloping, plotting U_c vs $|\omega_x - \omega_y|$, for different regions of the aerodynamic invariant plane: (**a**)–(**d**) $b_{22} < 0$, (**e**) $b_{22} > 0$

each other, the intensity of coupling between longitudinal and transverse motions decreases with $|\omega_x - \omega_y|$. The analysis of the quasi-resonant system necessarily requires employment of numerical solutions.[45] The relevant (qualitative) results are illustrated in the bifurcation diagrams shown in Fig. 12.12, in which the critical velocity is expressed *vs* the difference $|\omega_x - \omega_y|$ (detuning parameter), for different values of the aerodynamic coefficients. These latter fall into the regions of the plan of the invariants indicated in the figure, with $b_{22} < 0$ (Fig. 12.12a–d) or $b_{22} > 0$ (Fig. 12.12e). The U_{gt} values denotes the transverse galloping velocity, as provided by Den Hartog criterion.

Figure 12.12a relates to a cross-section for which tr$\mathbf{B} < 0$. In this case, the critical velocity at resonance, $U_{c,1}$, is lower than the transverse galloping velocity U_{gt} (i.e., coupling exerts a detrimental effect). When, however, the detuning increases, $U_{c,1} \rightarrow U_{gt}^-$, i.e., it approaches monotonously from below the critical velocity of the single DOF system. In the same region, but for different values of the aerodynamic coefficients, the opposite case can occur (Fig. 12.12b), in which $U_{c,1} > U_{gt}$ (i.e., coupling brings a beneficial effect). When the detuning increases, it is seen that $U_{c,1} \rightarrow U_{gt}^+$, that is, the critical velocity tends monotonously from above

---

[45] A perturbation method has also been employed in [7], where $|\omega_x - \omega_y|$ is considered as an imperfection of the resonant system.

to the transverse galloping velocity. This type of behavior (Fig. 12.12b) always occurs in the fourth quadrant, outside the parabola, where the contribution of the longitudinal motion is stabilizing. A similar (stabilizing) behavior is still found in the fourth quadrant, but internally to the parabola (Fig. 12.12c), where a double Hopf bifurcation takes place. The singularity, in this case, is denounced by the cusp.[46]

The following figures show unexpected behaviors. Figure 12.12d refers to the first quadrant of the plane of the invariants, namely, to the region outside the parabola. In this area, in conditions of perfect resonance, the equilibrium is stable for any U. However, this stable state persists only for $|\omega_x - \omega_y|$ small. When, instead, the detuning exceeds a threshold value, a Hopf bifurcation occurs at values of U only slightly higher than U_{gt}. This bifurcation is followed by a second bifurcation, occurring at a higher wind velocity, caused by a returning to the stable zone of the initially unstable eigenvalues. Therefore, a *regain of stability* of the equilibrium position occurs.

Figure 12.12e relates to a system in which $b_{22} > 0$, and for which, therefore, there is no transverse galloping (i.e., $U_{gt} = \infty$). Since $\mathrm{tr}\mathbf{B} > 0$, if, in addition, it is $\det \mathbf{B} < 0$ (i.e., if the system is in the second quadrant of the invariant plane), the critical velocity increases from the resonant value $U_{c,1}$, rapidly tending to ∞, coherently with the behavior of the single DOF system.

A numerical example, corroborating the previous results, is worked out in Problem 14.6.4.

12.8   Roto-translational Flutter and Galloping

If the cylinder in Fig. 12.8 is free to rotate and to translate in the direction $\mathbf{a}_y$ and transverse to the flow, but it cannot translate in the direction $\mathbf{a}_x$, parallel to the flow, the system has two degrees of freedom, $v_C(t)$, $\theta(t)$. The equations of motion, Eq. 12.81, reduces to[47] [2, 6, 10]:

$$\begin{bmatrix} 1 & x_G \\ \frac{x_G}{r_C^2} & 1 \end{bmatrix} \begin{pmatrix} \ddot{v}_C \\ \ddot{\theta} \end{pmatrix} + \begin{bmatrix} 2\xi_y\omega_y + U\frac{b_{22}}{M} & U\frac{b_{23}}{M} \\ U\frac{b_{32}}{Mr_C^2} & 2\xi_\theta\omega_\theta + U\frac{b_{33}}{Mr_C^2} \end{bmatrix} \begin{pmatrix} \dot{v}_C \\ \dot{\theta} \end{pmatrix}$$
$$+ \begin{bmatrix} \omega_y^2 & U^2\frac{h_{23}}{M} \\ 0 & \omega_\theta^2 + U^2\frac{h_{33}}{Mr_C^2} \end{bmatrix} \begin{pmatrix} v_C \\ \theta \end{pmatrix} = \begin{pmatrix} 0 \\ 0 \end{pmatrix},$$

(12.107)

where:

---

[46] This is most noticeable when the graph is symmetrically extended to account for positive and negative $\omega_x - \omega_y$.

[47] The model is suitable for analyzing, for example, the motion of a bridge, whose stiffness in the horizontal plane is much larger than that in the vertical plane, although within the limits of the quasi-steady theory, already mentioned.

$$\omega_y := \sqrt{\frac{k_y}{M}}, \quad \omega_\theta := \sqrt{\frac{k_\theta}{Mr_C^2}}, \quad \xi_y := \frac{c_y}{2M\omega_y}, \quad \xi_\theta := \frac{c_\theta}{2Mr_C^2\omega_\theta}. \tag{12.108}$$

The same equations hold when no constraints are present in the $\mathbf{a}_x$ direction, but the cross-section is symmetric with respect to this axis, in that, as it was observed for Eq. 12.85, the motion $u_C(t)$ is uncoupled from $v_C(t)$, $\theta(t)$, and decays over time.

The critical velocity at which the system loses stability is said of *roto-translational* (or flexural-torsional) *galloping*. This velocity is determined ahead by first referring to a reduced model, consistent with the steady theory of the aeroelastic forces, and then to the complete model, according to the quasi-steady theory.

12.8.1 Steady Aeroelasticity

As a first approximation of the problem, *structural damping is ignored*, and reference is made to the *steady theory of the aeroelastic forces*. According to this model, the change in position of the cylinder is taken into account (which therefore translates and rotates by v_C, θ), but the translational and rotational velocities $\dot{v}_C, \dot{\theta}$ are ignored. Said in other terms, for the purpose of determining the aeroelastic action, the cylinder is "frozen" at time t, and *not* animated by any motion, as it is instead in the quasi-steady theory [3]. The equations of motion, therefore, reduce to the contributions of mass and stiffness, only, and they read:

$$\begin{bmatrix} 1 & x_G \\ \frac{x_G}{r_C^2} & 1 \end{bmatrix} \begin{pmatrix} \ddot{v}_C \\ \ddot{\theta} \end{pmatrix} + \begin{bmatrix} \omega_y^2 & U^2\frac{h_{23}}{M} \\ 0 & \omega_\theta^2 + U^2\frac{h_{33}}{Mr_C^2} \end{bmatrix} \begin{pmatrix} v_C \\ \theta \end{pmatrix} = \begin{pmatrix} 0 \\ 0 \end{pmatrix}. \tag{12.109}$$

It should be observed that due to the aeroelastic forces proportional to h_{23}, the stiffness matrix *is non-symmetric*. The system is therefore *circulatory*, formally similar to the Ziegler column (studied in Chap. 11), although the physical nature of the nonconservative force is different. The coefficient h_{33}, moreover, modifies the torsional stiffness of the elastic system, through a geometric effect already discussed for the torsional divergence.

By letting the solution as $(v_C, \theta) = \left(\hat{v}_C, \hat{\theta}\right)\exp(\lambda t)$, the equations of motion lead to the following algebraic eigenvalue problem:

$$\begin{bmatrix} \lambda^2 + \omega_y^2 & \lambda^2 x_G + U^2\frac{h_{23}}{M} \\ \lambda^2\frac{x_G}{r_C^2} & \lambda^2 + \omega_\theta^2 + U^2\frac{h_{33}}{Mr_C^2} \end{bmatrix} \begin{pmatrix} \hat{v}_C \\ \hat{\theta} \end{pmatrix} = \begin{pmatrix} 0 \\ 0 \end{pmatrix}. \tag{12.110}$$

The associate characteristic equation is:

$$p_0\lambda^4 + p_2\lambda^2 + p_4 = 0, \tag{12.111}$$

where:

$$p_0 := 1 - \frac{x_G^2}{r_C^2},$$

$$p_2 := \omega_y^2 + \omega_\theta^2 + \frac{U^2}{Mr_C^2}(h_{33} - x_G h_{23}), \tag{12.112}$$

$$p_4 := \omega_y^2 \left(\omega_\theta^2 + U^2 \frac{h_{33}}{Mr_C^2} \right).$$

When $U = 0$, as the mass and elastic stiffness matrices are positive definite, there exist four purely imaginary eigenvalues, complex conjugate in pairs, for which the system is marginally stable. As U increases, the roots approach two by two, collide and then separate, two on the left (stable eigenvalues) and two on the right (unstable eigenvalues). At the coalescence of the eigenvalues, a *circulatory Hopf bifurcation* (also said *flutter*) occurs. The corresponding wind velocity U_c is found by zeroing the discriminant of the characteristic equation, $\Delta := p_2^2 - 4p_0 p_4 = 0$, which reads:

$$\left[\omega_y^2 + \omega_\theta^2 + \frac{U_c^2}{Mr_C^2}(h_{33} - x_G h_{23}) \right]^2 - 4\omega_y^2 \left(1 - \frac{x_G^2}{r_C^2} \right) \left(\omega_\theta^2 + U_c^2 \frac{h_{33}}{Mr_C^2} \right) = 0. \tag{12.113}$$

This is a biquadratic equation for U_c, which can be solved in closed form.

Remark 12.13 If $x_G = 0$, as it happens in bi-symmetric cross-sections, the eigenvalue problem in Eqs. 12.110 triangularizes and admits the roots:

$$\lambda = \pm i\omega_y, \pm i \sqrt{\omega_\theta^2 + U^2 \frac{h_{33}}{Mr_C^2}}. \tag{12.114}$$

These, by assuming that the cross-section is not prone to divergence (i.e., $C_{m_0}' > 0$, and therefore $h_{33} > 0$), are *purely imaginary for any wind velocity*. The coalescence, therefore, does not imply the birth of a real part of the eigenvalue (as the discriminant is always positive), and therefore flutter is not triggered. Of course, this circumstance is purely ideal, because a small imperfection in the system implies $x_G \neq 0$; however, the critical velocity of flutter is high.

Case $C_{m_0}' = 0$

The critical velocity assumes a particularly simple expression when $C_{m_0}' \simeq 0$, which implies $h_{33} \simeq 0$. This circumstance occurs when the aeroelastic couple (proportional to $C_{m_0}' U^2$) is small, compared to the elastic couple. In this case, the solution U_c^0 of Eq. 12.113 is:

$$U_c^0 = \sqrt{\frac{Mr_C}{x_G h_{23}} \left[r_C \left(\omega_y^2 + \omega_\theta^2 \right) - 2\omega_y \omega_\theta \sqrt{r_C^2 - x_G^2} \right]}, \qquad x_G \neq 0. \qquad (12.115)$$

Since $r_C^2 - x_G^2 = \frac{I_G}{M} > 0$, and the term in square bracket is also positive,[48] it turns out that U_c^0 is real *if and only if* $x_G h_{23} > 0$. Now, if the section is aerodynamically unstable to transverse galloping, it is $C_{l_0}' < 0$, which implies $h_{23} < 0$. For these cross-sections, therefore, in order for flutter to occur, it must be $x_G < 0$, that is, *the center of mass must be upwind*, with respect to the torsional center; in the downwind case, the cross-section is stable for any flow rate. If instead the section is aerodynamically stable to transverse galloping, as $C_{l_0}' > 0$ entails $h_{23} > 0$, the reverse is true.

The critical velocity U_c^0 depends on the natural frequencies. By fixing, e.g., ω_y, one wonders which frequency ω_θ minimizes U_c^0. Solving $\frac{dU_c^0}{d\omega_\theta} = 0$, $\omega_\theta = \omega_y \sqrt{1 - \left(\frac{x_G}{r_C} \right)^2}$ is found, which turns out to be a minimum. It is concluded that, for small eccentricities $x_G \ll r_C$, the minimum velocity is close to the resonance $\omega_\theta = \omega_y$; however, if the eccentricity of G is large, it remarkably deviates from resonance.

A numerical example, relevant to the roto-translational flutter of a two DOF system, is illustrated in Problem 14.6.5.

12.8.2 Quasi-steady Aeroelasticity

A higher level of approximation, consistent with the quasi-steady theory of the aeroelastic forces, is obtained by considering all the terms in Eq. 12.107 [3]. The associated eigenvalue problem becomes:

$$\begin{bmatrix} \lambda^2 + \lambda \left(2\xi_y \omega_y + U \frac{b_{22}}{M} \right) + \omega_y^2 & \lambda^2 x_G + \lambda U \frac{b_{23}}{M} + U^2 \frac{h_{23}}{M} \\ \lambda^2 \frac{x_G}{r_C^2} + \lambda U \frac{b_{32}}{Mr_C^2} & \lambda^2 + \lambda \left(2\xi_\theta \omega_\theta + U \frac{b_{33}}{Mr_C^2} \right) + \omega_\theta^2 + U^2 \frac{h_{33}}{Mr_C^2} \end{bmatrix}$$
$$\begin{pmatrix} \hat{v}_C \\ \hat{\theta} \end{pmatrix} = \begin{pmatrix} 0 \\ 0 \end{pmatrix}.$$

$$(12.116)$$

Zeroing the determinant of the coefficient matrix, the characteristic equation follows:

$$p_0 \lambda^4 + p_1 \lambda^3 + p_2 \lambda^2 + p_3 \lambda + p_4 = 0, \qquad (12.117)$$

---

[48] Indeed, a lower bound obtained by ignoring x_G^2 is $r_C \left(\omega_y - \omega_\theta \right)^2 > 0$.

whose coefficients are thus defined:

$$p_0 := 1 - \frac{x_G^2}{r_C^2},$$

$$p_1 := 2\left(\xi_y\omega_y + \xi_\theta\omega_\theta\right) + \frac{U}{M}\left(b_{22} + \frac{b_{33}}{r_C^2}\right) - \frac{Ux_G}{Mr_C^2}\left(b_{23} + b_{32}\right),$$

$$p_2 := \omega_y^2 + \omega_\theta^2 + \frac{U^2}{Mr_C^2}\left(h_{33} - x_Gh_{23}\right) + \left(2\xi_y\omega_y + U\frac{b_{22}}{M}\right)\left(2\xi_\theta\omega_\theta + U\frac{b_{33}}{Mr_C^2}\right)$$

$$- U^2\frac{b_{23}b_{32}}{M^2r_C^2},$$

$$p_3 := 2\omega_y\omega_\theta\left(\xi_\theta\omega_y + \xi_y\omega_\theta\right) + U\left(\frac{b_{22}}{M}\omega_\theta^2 + \frac{b_{33}}{Mr_C^2}\omega_y^2\right) + 2\xi_y\omega_y\frac{U^2h_{33}}{Mr_C^2}$$

$$+ \frac{U^3}{M^2r_C^2}\left(h_{33}b_{22} - h_{23}b_{32}\right),$$

$$p_4 := \omega_y^2\left(\omega_\theta^2 + U^2\frac{h_{33}}{Mr_C^2}\right).$$

$$(12.118)$$

The presence of odd powers of λ changes the mechanism of bifurcation. When $U = 0$, due to the structural damping, the eigenvalues are two pairs of complex conjugate numbers, having real negative part; that is, the system is asymptotically stable. When U increases, it can happen that a pair of eigenvalues crosses the imaginary axis, giving rise to a generic *Hopf bifurcation*.[49]

To determine the critical velocity at which the bifurcation occurs, $\lambda = i\omega_c$, $U = U_c$ are substituted in the characteristic equation, and its real and imaginary parts are separated, thus obtaining:

$$p_0\omega_c^4 - p_2\omega_c^2 + p_4 = 0, \qquad (12.119a)$$

$$\omega_c\left(p_3 - p_1\omega_c^2\right) = 0. \qquad (12.119b)$$

---

[49] Although this phenomenon is identical to that observed in the two DOF translational galloping, it is (improperly) said of flutter in the technical literature, alluding to the simultaneous presence, in the critical mode, of a translation and a rotation, as it happens in the properly said flutter. The physical, rather than the mathematical aspect, is thus preferred. Here, the locution "roto-translational galloping" is, in fact, used.

By evaluating $\omega_c^2 = \frac{p_3}{p_1}$ from the second equation, a polynomial equation in U_c is obtained from the first one:

$$p_0 p_3^2 - p_1 p_2 p_3 + p_1^2 p_4 = 0. \tag{12.120}$$

A numerical example, relevant to the roto-translational galloping of a two DOF system, is discussed in Problem 14.6.6.

Remark 12.14 The critical wind velocity, determined with the quasi-steady approximation, is generally smaller than the velocity determined with the steady approximation. This result is consistent with what has been discussed about the *Ziegler paradox* (Chap. 11).

12.9 Unsteady Aeroelasticity

The quasi-steady theory is not applicable when the reduced velocity, $U_{red} = \dfrac{U}{v_s D} < 20$. In this case, indeed, the time that a fluid particle of velocity U takes to cross the cylinder cross-section of width D is not small compared to the natural period of the cylinder, $T_s = \frac{1}{v_s}$. Consequently, the motion of the body modifies that of the flow, so that the results obtained in the wind tunnel, with fixed body, are inapplicable. This circumstance frequently occurs in civil engineering, given that the width D is generally large, a few meters [5, 6, 8, 9].

Scanlan Model
In cases in which the quasi-steady theory fails, the *unsteady theory* must be used. Far from having today a satisfactory mathematical model, the approach to the problem is semi-experimental and mostly limited to the linear field [3, 9]. It consists in carrying out experiments in which the aeroelastic forces are measured with the body in motion. The tests that are actually feasible and meaningful are those in which the body moves with harmonic motion, of frequency ω assigned. If one admits that the aeroelastic forces depend exclusively on the position $\mathbf{q}$ and velocity $\dot{\mathbf{q}}$ of the cylinder, it must hold:

$$\mathbf{f}^a = -\mathbf{C}_a(\omega)\,\dot{\mathbf{q}} - \mathbf{K}_a(\omega)\,\mathbf{q}, \tag{12.121}$$

where the aerodynamic damping and stiffness matrices are functions of the assigned frequency. Equation 12.121 generalizes, in the unsteady case, Eq. 12.79, valid in the quasi-steady case, making the previously constants coefficients now dependent on ω.[50] The coefficients appearing in $\mathbf{C}_a(\omega)$, $\mathbf{K}_a(\omega)$ are called *aerodynamic* (or flutter)

---

[50] In principle, i.e., to within experimental uncertainties, the coefficients of the unsteady theory tend to those of the quasi-steady theory when the frequency tends to zero.

derivatives, because they measure the first order part of the aeroelastic forces. The resulting theory is known as the *Scanlan model*.

The experimental determination of the aerodynamic derivatives requires carrying out a large number of tests, in which, for each frequency ω with which the body is moved, the in-phase forces, $\mathbf{K}_a (\omega) \, \mathbf{q}$, and those in quadrature (out of phase), $\mathbf{C}_a (\omega) \, \dot{\mathbf{q}}$, are measured.[51] It is therefore possible to build up some graphs, partly available in the literature, which give, for a given cross-section and for a given wind direction, the coefficients of the two matrices vs ω. More conveniently, these diagrams are provided in nondimensional form, in function of the *reduced frequency* $k := \dfrac{D \, \omega}{U} \equiv \dfrac{1}{U_{red}}$, equal to the inverse of the reduced velocity. The equations of motion, therefore, are written as follows:

$$\mathbf{M}\ddot{\mathbf{q}} + [\mathbf{C}_s + \mathbf{C}_a (k)] \, \dot{\mathbf{q}} + [\mathbf{K}_s + \mathbf{K}_a (k)] \, \mathbf{q} = \mathbf{0}. \tag{12.122}$$

Aerodynamic Instability

At the dynamic bifurcation, the system executes harmonic oscillations, of unknown frequency ω_c. Therefore, by letting $\mathbf{q} = \hat{\mathbf{q}} \exp(i\omega_c t)$, Eq. 12.122 gives:

$$\left(-\omega_c^2 \mathbf{M} + i\omega_c \, [\mathbf{C}_s + \mathbf{C}_a (k_c)] + [\mathbf{K}_s + \mathbf{K}_a (k_c)] \right) \hat{\mathbf{q}} = \mathbf{0}, \tag{12.123}$$

where $k_c := \dfrac{D\omega_c}{U_c}$. This is an algebraic eigenvalue problem with complex coefficients. In order for it to admit non-trivial solutions, the following equation has to be satisfied:

$$f (U_c, k_c) := \det \left[-\left(\frac{U_c k_c}{D} \right)^2 \mathbf{M} + i \left(\frac{U_c k_c}{D} \right) [\mathbf{C}_s + \mathbf{C}_a (k_c)] + [\mathbf{K}_s + \mathbf{K}_a (k_c)] \right]$$

$$= 0,$$

$$\tag{12.124}$$

where ω_c has been expressed as a function of U_c and k_c. By separating the real and imaginary parts, two real equations are obtained, $\mathrm{Re}\,[f (U_c, k_c)] = 0$, $\mathrm{Im}\,[f (U_c, k_c)] = 0$ in the two unknowns U_c, k_c. If analytic expressions of the aerodynamic derivatives were available, these last two equations would allow to determine the unknowns, analytically or numerically. As generally this is not the case, graphs or tables should be used, expressing the coefficients as a function of k_c. One, therefore, proceeds in iterative way, by performing the following steps: (i) a "guess" value of k_c is chosen, and the values of the aerodynamic derivatives are read in the graph; (ii) with these now known values, one checks that a value of U_c exists, which, with a predetermined tolerance, zeroes *both* the real and imaginary

---

[51] By letting, indeed, $\mathbf{q} = \hat{\mathbf{q}} \exp(i\omega t)$, $\mathbf{f} = \hat{\mathbf{f}} \exp(i\omega t)$, from Eq. 12.121 it follows, that $\hat{\mathbf{f}} = -(\mathbf{K}_a (\omega) + i\omega \mathbf{C}_a (\omega)) \, \hat{\mathbf{q}}$.

parts of the characteristic equation; and (iii) if this does not happen, the procedure is repeated with a new value of k_c, until convergence.

References

1. Blevins, R.D.: Flow-Induced Vibration. Van Nostrand Reinhold, New York (1977)
2. Bolotin, V.V.: Nonconservative Problems of the Theory of Elastic Stability. Macmillan, London (1963)
3. Dowell, E.H. et al.: A Modern Course in Aeroelasticity. Kluwer Academic Publishers, Dordrecht (2004)
4. Fung, Y.C.: An Introduction to the Theory of Aeroelasticity. Dover, New York (1969)
5. Jurado, J.A., Hernández, S., Nieto, F., Mosquera, A.: Bridge Aeroelasticity: Sensitivity Analysis and Optimal Design. WIT Press, Southampton (2011)
6. Lacarbonara, W.: Nonlinear Structural Mechanics: Theory, Dynamical Phenomena and Modeling. Springer Science & Business Media, New York (2013)
7. Luongo, A., Piccardo, G.: Linear instability mechanisms for coupled translational galloping. J. Sound Vib. **288**(4–5), 1027–1047 (2005)
8. Païdoussis, M.P., Price, S.J., de Langre, E.: Fluid-Structure Interactions: Cross-Flow-Induced Instabilities. Cambridge University Press, New York (2011)
9. Simiu, E., Scanlan, R.H.: Wind Effects on Structures: An Introduction to Wind Engineering. Wiley, New York (1986)
10. Ziegler, H.: Principles of Structural Stability. Blaisdell Publishing, Waltham (1968)

Chapter 13
Parametric Excitation

13.1 Introduction

The parametric excitation of a mechanical system is a solicitation causing an assigned, and periodic, variation of the parameters. The loading, therefore, does not consist of an external forcing, described by a known function of time, to be placed in the right side of the equation, but of an "excitation" of the parameters that appear in the left side of the equation itself. The resulting system is non-autonomous, as it explicitly depends on time, but it is homogeneous, because the known term (the external excitation) is zero. The dependence on time makes the stability problem quite different from all those discussed so far (relating to autonomous systems), although some important formal analogies between the two classes of problems can be identified.

The mechanical origin of the parametric excitation can be of various kind. For example, motion impressed at the constraints, motion relative to a rotating reference, geometry variable over time, mass moving on an oscillating system, and pulsating axial loads acting on a beam. Parametric excitation, in all these systems, is characterized by an amplitude and a frequency. When the parametric frequency is "in resonance" (i.e., it is in particular ratios, to be determined) with one or more of the natural frequencies of the unexcited system, the trivial equilibrium position becomes unstable, as soon as the excitation amplitude exceeds a threshold value. Instability manifests itself with oscillations of increasing amplitude, theoretically tending to infinity, as a result of the linearization of the equations. The presence of suitable nonlinearities can limit this growth, so that the oscillations stabilize on a limit cycle.

To analyze stability of the equilibrium position of a parametrically excited finite dimensional systems, the Floquet theory of the linear differential equations with periodic coefficients must be used. Equivalently, one can resort to a discrete-time vision of the system, obtained by sampling the states at time intervals equal to the period of the excitation.

© The Author(s), under exclusive license to Springer Nature Switzerland AG 2023   493
A. Luongo et al., *Stability and Bifurcation of Structures*,
https://doi.org/10.1007/978-3-031-27572-2_13

A representative mathematical equation, encompassing many of the character-istics of parametrically excited systems, is the *Mathieu equation*, governing the dynamics of a class of single degree of freedom systems. For it, the linear stability diagram is known (i.e., the *Strutt diagram*), showing the occurrence of *divergence* and *flip* bifurcations, alternately occurring at the boundaries of certain instability tongues. On these boundaries, the natural frequency of the system is in *direct resonance* with the parametric frequency (e.g., 1:1 in *divergence*, and 1:2 in the *flip bifurcations*). However, a more complex form of instability, namely the *Neimark-Sacker* bifurcation, which involves two modes in *combination resonance* with the parametric excitation, requires at least two degrees of freedom to exist. Among the physical systems, the paradigm is expressed by the *Bolotin beam*, i.e., an Euler beam in which the axial load is pulsating.

In this chapter, an overview of these problems is given. Moreover, perturbation methods are illustrated, able to evaluate the postcritical behavior of systems with one or two degrees of freedom.

13.2 Introductory Examples

Some sample mechanical systems, excited parametrically, are now browsed (Fig. 13.1).

Pendulum with Motion Impressed at the Base

A planar pendulum is considered, of mass m and constant length ℓ, whose suspension point is subject to an assigned periodic motion $y(t)$ (Fig. 13.1a). The acceleration a_r (taken positive upward) relative to the suspension point is the difference between the acceleration of gravity, $-g$, and that of the constraint, $\ddot{y}$, and therefore it is $a_r := -g - \ddot{y}$. The linearized equation of motion in the rotation $\theta(t)$ is, therefore:

$$\ddot{\theta} + \frac{1}{\ell}(g + \ddot{y})\theta = 0. \tag{13.1}$$

In this example, the parametric excitation "modifies the gravity" and consequently induces a periodic variation of the geometric stiffness of the pendulum, which depends on the apparent weight of the mass.

Point Mass Moving on a Rotating Disk

A point mass m is considered, resting without friction on a disk rotating with periodic angular speed $\Omega(t)$, constrained by two viscoelastic devices of stiffness k_i and damping c_i, $i = 1, 2$ (Fig. 13.1b). The linearized equations of motion, in the coordinates $x(t)$, $y(t)$ of the point instantaneously occupied by the mass, are:

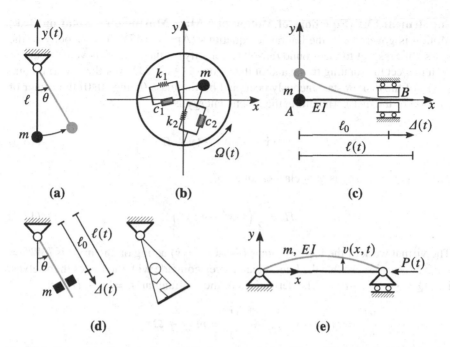

Fig. 13.1 Parametrically excited mechanical systems: (**a**) pendulum with motion impressed at the base; (**b**) point mass moving on a rotating disk; (**c**) beam of variable length; (**d**) pendulum with moving mass; (**e**) beam subjected to a pulsating axial force (Bolotin beam)

$$
\begin{bmatrix} m & 0 \\ 0 & m \end{bmatrix}\begin{pmatrix} \ddot{x} \\ \ddot{y} \end{pmatrix} + \begin{bmatrix} c_1 & -2m\Omega \\ 2m\Omega & c_2 \end{bmatrix}\begin{pmatrix} \dot{x} \\ \dot{y} \end{pmatrix} + \begin{bmatrix} k_1 - m\Omega^2 & -m\dot{\Omega} \\ m\dot{\Omega} & k_2 - m\Omega^2 \end{bmatrix}\begin{pmatrix} x \\ y \end{pmatrix} = \begin{pmatrix} 0 \\ 0 \end{pmatrix},
$$
(13.2)

as derived in Supplement 13.1. In these equations, the following forces are recognized,[1] due to $\Omega(t)$:

- the Coriolis force $(-2m\Omega\dot{y}, 2m\Omega\dot{x})$, which gives rise to a (antisymmetric) gyroscopic matrix;[2]
- the radial (centripetal) inertia force, $-m\Omega^2 r$, with $r = \sqrt{x^2 + y^2}$, of components $(-m\Omega^2 x, -m\Omega^2 y)$, which gives rise to a (symmetric) matrix of geometric stiffness;
- the circumferential inertia force, $m\dot{\Omega} r$, of components $(-m\dot{\Omega} y, m\dot{\Omega} x)$, which results in a circulatory (antisymmetric) stiffness matrix.

Since the angular speed is periodic, all these matrices are periodic.

---

[1] Here reference is made to the apparent forces, but changed in sign, consistently with the inertia forces $(m\ddot{x}, m\ddot{y})$.

[2] The Coriolis force is given by the expression $\vec{F}_c = 2m\,\vec{\omega} \times \vec{v}_r$ where $\vec{\omega}$ is the angular velocity of the reference and $\vec{v}_r$ the velocity relative to the reference itself.

Supplement 13.1 (Equations of Motion of a Mass Moving on a Rotating Disk)
Motion is governed by the Lagrange equations (Appendix D). The velocity of the
mass with respect to the ground (absolute velocity) is the sum of its velocity $(\dot{x}, \dot{y})$
with respect the rotating frame and of the velocity $(-\Omega y, \Omega x)$ of the material point
(x, y) of the disk, instantaneously occupied by the mass (being $(0, 0)$ the center of
rotation of the disk). Therefore, the kinetic energy is:[3]

$$T = \frac{1}{2} m \left[(\dot{x} - \Omega y)^2 + (\dot{y} + \Omega x)^2 \right]. \tag{13.3}$$

The potential energy is only elastic, equal to:

$$\Pi = \frac{1}{2} \left(k_1 x^2 + k_2 y^2 \right). \tag{13.4}$$

The virtual work of the viscous forces $(-c_1 \dot{x}, -c_2 \dot{y})$ acting on the mass is $\delta W^{nc} = -c_1 \dot{x} \delta x - c_2 \dot{y} \delta y$, so that the Lagrangian forces coincide with the forces themselves,
i.e., $\mathbf{Q} = (-c_1 \dot{x}, -c_2 \dot{y})$. The derivatives of the Lagrangian $L = T - \Pi$ are:

$$\frac{\partial L}{\partial \dot{x}} = m (\dot{x} - \Omega y), \qquad \frac{\partial L}{\partial \dot{y}} = m (\dot{y} + \Omega x),$$

$$\frac{\partial L}{\partial x} = m \Omega (\dot{y} + \Omega x) - k_1 x, \qquad \frac{\partial L}{\partial y} = -m \Omega (\dot{x} - \Omega y) - k_2 y, \tag{13.5}$$

from which the equations of motion follow:

$$m \left(\ddot{x} - \dot{\Omega} y - \Omega \dot{y} \right) - m \left(\Omega \dot{y} + \Omega^2 x \right) + k_1 x = -c_1 \dot{x}, \tag{13.6a}$$

$$m \left(\ddot{y} + \dot{\Omega} x + \Omega \dot{x} \right) + m \left(\Omega \dot{x} - \Omega^2 y \right) + k_2 y = -c_2 \dot{y}. \tag{13.6b}$$

They assume the matrix form in Eq. 13.2.   □

Beam of Variable Length
A beam is considered, of bending stiffness EI, massless, free at the end A, where a
point mass m is fixed and clamped at an intermediate point B, whose position varies
along the beam with periodic law $\Delta(t)$ (Fig. 13.1c). By indicating with ℓ_0 the length
of the AB beam at time $t = 0$, the instantaneous length is $\ell(t) := \ell_0 + \Delta(t)$. The
system is equivalent to a single DOF oscillator of stiffness $k(t) = \frac{3EI}{\ell(t)^3}$, governed
by the following equation of motion:

---

[3] The same symbol T is used in this chapter to denote the kinetic energy and the period of the
parametric excitation, which, however, should not create ambiguity for the reader.

$$my + \frac{3EI}{(\ell_0 + \Delta(t))^3} y = 0, \tag{13.7}$$

where $y(t)$ denotes the displacement of the point mass, transverse to the beam. The periodic variation of the length involves a periodic variation of the elastic stiffness of the system.

Pendulum with Moving Mass: A Swing Model

A planar pendulum is considered, under the gravity g, whose end mass m is subject to an assigned periodic motion $\Delta(t)$ along the wire, which makes the length of the pendulum variable over time as $\ell(t) := \ell_0 + \Delta(t)$ (Fig. 13.1d). The model simulates the behavior of a swing, in which the user (a) by flexing the knees, if standing, or (b) by extending and folding the legs, when sitting, periodically changes the center of mass of the system. It is common experience that if this periodic movement occurs with an appropriate frequency (as it will be seen, equal to the double of the natural frequency of the swing), the system performs oscillations of large amplitude.

The equation governing the small rotations $\theta(t)$ results to be:

$$(\ell_0 + \Delta(t))^2 \ddot{\theta} + 2(\ell_0 + \Delta(t)) \dot{\Delta}(t) \dot{\theta} + g(\ell_0 + \Delta(t)) \theta = 0, \tag{13.8}$$

as deduced in Supplement 13.2. The radial movement of the point mass therefore has the following effects:

- it modifies the inertia, through the variation of the arm (measured from the suspension point of pendulum) of the tangent acceleration $\ddot{\theta}\ell$;
- it introduces a tangent Coriolis acceleration $2\dot{\theta}\dot{\ell}$, induced by the radial velocity $\dot{\ell}$;
- it modifies the geometric stiffness, due to the variation of the arm of the weight force.

Remark 13.1 It is worth noticing that the pendulum with moving mass (Eq. 13.8) and the mass on rotating disk (Eq. 13.2) are systems very different from the pendulum with movable constraint (Eq. 13.1) and from the beam with variable length (Eq. 13.7). As a matter of fact, in the first two systems, the excitation affects the inertia, gyroscopic, and geometric forces; in the last two systems, the excitation only affects the stiffness.

Supplement 13.2 (Equation of Motion of the Pendulum with Moving Mass) To derive the equations of motion of the pendulum with moving mass, the Lagrange equations are used (Appendix D). The instantaneous velocity of the mass has components $(\dot{\ell}, \dot{\theta}\ell)$, radial and circumferential, respectively, so that the kinetic energy reads:

$$T = \frac{1}{2}m \left(\dot{\ell}^2 + \dot{\theta}^2 \ell^2 \right). \tag{13.9}$$

The total potential energy coincides with the gravitational energy possessed by the mass, i.e.,

$$\Pi = mg\ell\,(1 - \cos\theta). \tag{13.10}$$

Defined the Lagrangian $L = T - \Pi$, it results:

$$\frac{\partial L}{\partial\dot{\theta}} = m\ell^2\dot{\theta}, \qquad \frac{\partial L}{\partial\theta} = -mg\ell\sin\theta. \tag{13.11}$$

Since nonconservative forces are absent, the equation of motion is:

$$\frac{\mathrm{d}}{\mathrm{d}t}\left(\ell^2\dot{\theta}\right) + g\ell\sin\theta = 0. \tag{13.12}$$

For small rotations, the sine is nearly equal to the angle; moreover, by making explicit $\ell\,(t) = \ell_0 + \Delta\,(t)$, Eq. 13.8 is finally drawn.   □

Simply Supported Bolotin Beam, as Paradigmatic Continuous System

As an example of a parametrically excited continuous system, an Euler-Bernoulli beam is considered, supported at the ends, and subjected to an axial force $P\,(t) = P_0 + P_1\cos\,(\Omega t)$, partly constant and partly pulsating with harmonic law (Fig. 13.1e). The system is known as the *Bolotin beam* [1]. The relevant equations of motion are a generalization to the dynamic case of the equations governing the buckling of the Euler beam, namely:

$$m\ddot{v} + EIv'''' + (P_0 + P_1\cos\,(\Omega t))\,v'' = 0, \tag{13.13a}$$

$$v\,(0, t) = v''\,(0, t) = 0, \tag{13.13b}$$

$$v\,(\ell, t) = v''\,(\ell, t) = 0, \tag{13.13c}$$

where $v\,(x, t)$ is the transverse displacement at the abscissa x at time t; moreover, a dash indicates differentiation with respect to x and a dot with respect to t. Furthermore, m is the mass linear density, EI the bending stiffness, and ℓ the length of the beam, all constant over time.

Since the boundary conditions are of simple support, a solution is tried in the form:

$$v\,(x, t) = \sum_{k=1}^{\infty} q_k\,(t)\sin\left(\frac{k\pi x}{\ell}\right). \tag{13.14}$$

Substituting it into the field equation, and by requiring this satisfied for any x, an infinite number of uncoupled equations in the Lagrangian parameters $q_k\,(t)$ follow, i.e.,

$$m\ddot{q}_k + \left[\left(\frac{k\pi}{\ell}\right)^4 EI - \left(\frac{k\pi}{\ell}\right)^2 (P_0 + P_1 \cos(\Omega t))\right] q_k = 0. \qquad k = 1, 2, \cdots.$$

$$(13.15)$$

These are the equations of motion of infinite in number and *independent* single DOF oscillators, which can be rewritten as:

$$\ddot{q}_k + \omega_k^2 \left(1 - \mu_k \cos(\Omega t)\right) q_k = 0, \qquad k = 1, 2, \cdots, \tag{13.16}$$

in which:

$$\omega_k^2 := \left(\frac{k\pi}{\ell}\right)^4 \frac{EI}{m}\left(1 - \frac{P_0}{P_{ck}}\right), \qquad \mu_k := \left(\frac{k\pi}{\ell}\right)^2 \frac{P_1}{m\omega_k^2}, \tag{13.17}$$

with $P_{ck} := \left(\frac{k\pi}{\ell}\right)^2 EI$ being the kth Eulerian critical load of the beam. In Eq. 13.17, ω_k is the kth natural frequency of the beam axially compressed by the static component of the force;[4] μ_k is a nondimensional parameter, which measures the intensity of the parametric excitation of the kth mode.

The pulsating component of the force, therefore, enters the geometric stiffness of the beam and excites it parametrically. Since the system has infinite DOF, *excitation can make unstable any of the modes*.

Bolotin Beam Under Constraint Conditions Other Than Simple Support
If the boundary conditions of the Bolotin beam are not of simple support, the system could be discretized by invoking the extended Hamilton principle. However, when the "strong form" (i.e., the field equation) of a problem is known, it is more convenient to apply the *Galerkin method*, which leads to the same results.

In the spirit of the Galerkin method, it is assumed that:

$$v(x, t) = \sum_{k=1}^{\infty} q_k(t)\, \phi_k(x), \tag{13.18}$$

where the $\phi_k(x)$ are assigned trial functions. Here, $\phi_k(x)$ is taken as the kth natural mode of oscillation of the beam prestressed by the static component of the axial force. Since these functions are orthogonal, after normalization, they satisfy the conditions $\int_0^\ell \phi_j \phi_k \mathrm{d}x = \delta_{jk}$, with δ_{jk} the Kronecker delta.[5] The series, however, does not exactly satisfy the field equation and the boundary conditions, but gives rise to residues, which represent unbalanced forces, in the field and at the boundary. By requiring that such forces expend zero virtual work on any trial functions, an infinite number of *averaged* equilibrium conditions are obtained.

---

[4] The compression force P_0 reduces the stiffness of the beam, and therefore also its natural frequencies.

[5] It is $\delta_{jk} = 0$, if $j \neq k$, and $\delta_{jk} = 1$, if $j = k$.

Exemplifying the procedure in the case of the clamped-free beam, the averaged equilibrium conditions are written as:

$$\int_0^\ell \sum_{j=1}^\infty \left\{ m\phi_j \ddot{q}_j + \left[EI\phi_j'''' + (P_0 + P_1 \cos(\Omega t)) \phi_j'' \right] q_j \right\} \phi_k \mathrm{d}x$$

$$+ \sum_{j=1}^\infty \left[\left(-EI\phi_j''' (\ell) - P_0\phi_j' (\ell) - P_1 \cos(\Omega t) \phi_j' (\ell) \right) \right] q_j \phi_k (\ell) \qquad (13.19)$$

$$+ \sum_{j=1}^\infty EI\phi_j'' (\ell) q_j \phi_k' (\ell) = 0, \quad k = 1, 2, \cdots,$$

where the barred boundary terms are zero, given the choice of the trial functions. The preceding conditions leads to a system of infinite coupled equations, such as:

$$\ddot{q}_k + \omega_k^2 q_k - \mu \cos(\Omega t) \sum_{j=1}^\infty h_{kj} q_j = 0 \qquad k = 1, 2, \cdots, \qquad (13.20)$$

where ω_k is the kth natural frequency of the beam compressed by the force P_0 and μ is a nondimensional multiplier of P_1, which measures the intensity of the parametric excitation; moreover:

$$h_{kj} := \frac{P_1}{m} \int_0^\ell \phi_j' \phi_k' \mathrm{d}x = h_{jk}, \qquad (13.21)$$

having integrated by parts Eq. 13.19.[6] Uncoupling, therefore, is a peculiarity of the supported beam [9]. The procedure illustrated can be applied in a similar way to other constraint conditions.

---

[6] Indeed, by accounting for $\phi_k (0) = 0$, it is:

$$\int_0^\ell \phi_j'' \phi_k \mathrm{d}x - \phi_j' (\ell) \phi_k (\ell) = -\int_0^\ell \phi_j' \phi_k' \mathrm{d}x + \left[\phi_j' (\ell) \phi_k (\ell) - \phi_j' (0) \phi_k (0) \right] - \phi_j' (\ell) \phi_k (\ell)$$

13.3 Theory of Linear Ordinary Differential Equations with Periodic Coefficients

As it has been shown by examples in the previous section, the equations of the motion of a general linear system with n DOF, parametrically excited, are of the following type:

$$\mathbf{M}(t)\,\ddot{\mathbf{q}} + \mathbf{C}(t)\,\dot{\mathbf{q}} + \mathbf{K}(t)\,\mathbf{q} = 0, \tag{13.22}$$

where $\mathbf{M}(t+T) = \mathbf{M}(t)$ is the (symmetric) mass matrix; $\mathbf{C}(t+T) = \mathbf{C}(t)$ is the damping matrix, which includes possible gyroscopic effects, and therefore generally non-symmetric; $\mathbf{K}(t+T) = \mathbf{K}(t)$ is the stiffness matrix, elastic plus geometric, which includes possible circulatory effects, and therefore non-symmetric; and finally, T is the *known period* of the parametric excitation.

The previous n scalar equations of the second order can more conveniently be rewritten as $N := 2n$ differential equations of first order, i.e.,

$$\dot{\mathbf{x}} = \mathbf{A}(t)\,\mathbf{x}, \qquad \mathbf{A}(t+T) = \mathbf{A}(t), \tag{13.23}$$

where $\mathbf{x} = (\mathbf{q}, \dot{\mathbf{q}})^T$ is the *state variable* vector and $\mathbf{A}$ is the state matrix:

$$\mathbf{A} := \begin{bmatrix} \mathbf{0} & \mathbf{I} \\ -\mathbf{M}^{-1}\mathbf{K} & -\mathbf{M}^{-1}\mathbf{C} \end{bmatrix}. \tag{13.24}$$

The behavior of the system in Eq. 13.23 is described by the theory of linear ordinary differential equations with periodic coefficients, due to Floquet (1847–1920). Here, the main results are only stated, without any proof, which can be easily found in other books (e.g., [7]). Successively, an equivalent theory is developed, based on a discrete-time view of the system.

13.3.1 Floquet Theorem

For the linear system of periodic ordinary differential equations, Eq. 13.23, the following theorem holds [2, 7, 8]:

> **Floquet Theorem**: *The T-periodic system $\dot{\mathbf{x}} = \mathbf{A}(t)\,\mathbf{x}$ admits a fundamental system of N solutions, called "normal," of the type:*[7]
>
> $$\mathbf{y}_k(t) = \exp(\gamma_k t)\,\boldsymbol{\phi}_k(t), \qquad k = 1, 2, \cdots, N, \tag{13.25}$$

---

[7] The analogy with the normal solutions of a system of linear equations with constant coefficients should be noticed, for which the $\boldsymbol{\phi}_k$ assume constant values and represent the modes of the system.

where $\boldsymbol{\phi}_k (t + T) = \boldsymbol{\phi}_k (t)$ are complex-valued periodic functions of period T (or sub-multiples integers of T). The numbers $\gamma_k \in \mathbb{C}$ are the *characteristic exponents* of the periodic system.

Given the periodicity of $\boldsymbol{\phi}_k (t)$, the multipliers γ_k decide about the stability. In particular, from Eq. 13.25, it follows, that the origin $\mathbf{x} = \mathbf{0}$:

- is asymptotically stable, if Re $(\gamma_k) < 0$ for any $k \in \{1, 2, \cdots, N\}$;
- is unstable, if Re $(\gamma_j) > 0$ for at least one $j \in \{1, 2, \cdots, N\}$.

Remark 13.2 Floquet Theorem establishes that $\boldsymbol{\phi}_k (t + T) = \boldsymbol{\phi}_k (t)$, where T is the period of the excitation. This does *not* imply that $\boldsymbol{\phi}_k (t)$ has *minimum period T*, since this latter could be an integer sub-multiple of T, that is, $\frac{T}{p}$ $(p = 1, 2, \cdots)$. However the term "period T" is almost universally adopted in the literature in the context of the parametric excitation and therefore will also be used here, except to specify, when necessary, the true period of the response.

Computation of the Characteristic Exponents

The characteristics exponents are computed through the algorithm illustrated below. An alternative method, based on the Fourier series, is discussed in the Supplement 13.3.

1. A *fundamental system* of (real) solutions is determined *in the* $[0, T]$ *interval*, by numerical or approximate analytical methods:

$$\mathbf{X}(t) := \left[\mathbf{x}^{(1)}(t), \mathbf{x}^{(2)}(t), \cdots, \mathbf{x}^{(N)}(t) \right], \tag{13.26}$$

where $\mathbf{x}^{(k)}(t)$ are solutions satisfying the initial conditions:

$$\mathbf{x}^{(k)}(0) = (0, \cdots, 0, 1, 0, \cdots, 0)^T, \tag{13.27}$$

where only the kth element is non-null; consequently, is $\mathbf{X}(0) = \mathbf{I}$.

2. The fundamental system is evaluated at the instant $t = T$, by getting the *monodromy matrix* (or Floquet matrix):[8]

$$\mathbf{F} := \mathbf{X}(T). \tag{13.28}$$

3. The following algebraic eigenvalue problem is solved:

$$(\mathbf{F} - \lambda_k \mathbf{I}) \mathbf{u}_k = \mathbf{0}, \tag{13.29}$$

to compute the N *characteristic multipliers* λ_k $(k = 1, 2, \cdots, N)$. Being $\mathbf{F}$ real and generic, multipliers are real or complex conjugated in pairs.

---

[8] More often denoted in literature by $\mathbf{C}$, a symbol that is avoided here, in order not to create ambiguity with the damping matrix.

4. By letting $\lambda_k := \exp(\gamma_k T)$, the *characteristic exponents* γ_k are evaluated as:

$$\gamma_k = \frac{1}{T}\ln\lambda_k. \tag{13.30}$$

Remark 13.3 As already pointed out, it is not necessary to integrate the system of equations, for example, numerically, over a large period of time, which would nullify the use of the Floquet theory, but it is sufficient to do this in an interval equal to the (known) period of the coefficients. This circumstance allows using approximate analytical solutions, such as provided by asymptotic methods (not discussed in this book), which, even if they diverge for large times, give instead a reasonable approximation in $[0, T]$.

Stability in Terms of Characteristic Multipliers
The stability conditions, by virtue of the link expressed by Eq. 13.30, can be reformulated in terms of multipliers, instead of exponents. To this end, by setting $\lambda_k = \rho_k \exp(i\varphi_k)$, where $\rho_k := |\lambda_k|$ is the module and $\varphi_k \in (-\pi, \pi)$ the phase, from Eq. 13.30, it results:

$$\gamma_k = \frac{1}{T}(\ln\rho_k + i\varphi_k), \tag{13.31}$$

so that $\mathrm{Re}(\gamma_k) = \frac{1}{T}\ln\rho_k$ and $\mathrm{Im}(\gamma_k) = \frac{\varphi_k}{T}$. It therefore happens that the origin $\mathbf{x} = \mathbf{0}$:

- is asymptotically stable, if $\rho_k < 1$ for any $k \in (1, 2, \cdots, N)$;
- is unstable, if $\rho_j > 1$ is for at least one $j \in (1, 2, \cdots, N)$.

It is concluded that:

- if all the characteristic multipliers are contained in the circle of unit radius of the complex plane, the origin is asymptotically stable;
- if at least one multiplier is outside the circle of unit radius of the complex plane, the origin is unstable.

Remark 13.4 A multiplier on the unit circle indicates a condition of *marginal stability*, potentially of incipient instability, if the "speed" of the multiplier, as a parameter varies, is directed outside the circle. Marginal stability, on the other hand, persists if the velocity is tangent to the unit circle.

Remark 13.5 Since the characteristic multipliers are defined to within $\exp(2ip\pi)$ $(p = 1, 2, \cdots)$, the characteristic exponents, from Eq. 13.30, are defined to within $ip\frac{2\pi}{T} = ip\Omega$, with $\Omega := \frac{2\pi}{T}$ the fundamental circular frequency of the parametric excitation. By limiting the angle φ to the range $(-\pi, \pi)$, and taking into account that $\mathrm{Im}(\gamma_k) = \frac{\varphi_k}{T}$, it follows that $\mathrm{Im}(\gamma) \in \left(-\frac{\pi}{T}, \frac{\pi}{T}\right) = \left(-\frac{\Omega}{2}, \frac{\Omega}{2}\right)$. The relationship between multipliers and characteristic exponents is represented in Fig. 13.2. The circle of unit radius, belonging to the plane $(\mathrm{Re}(\lambda), \mathrm{Im}(\lambda))$, maps into the

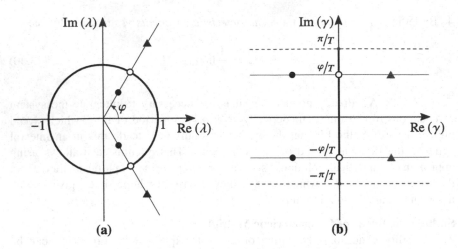

Fig. 13.2   Relation between (**a**) characteristic multipliers and (**b**) characteristic exponents; *filled circle* stable, *open circle* critical, *filled right triangle* unstable

segment $\left(-\frac{\pi}{T}, \frac{\pi}{T}\right)$ of the imaginary axis of the plane $(\text{Re}\,(\gamma)\,, \text{Im}\,(\gamma))$. Therefore, to multipliers internal to the unit circle correspond exponents to the left of the imaginary axis; to multipliers external to the circle correspond exponents to the right of the imaginary axis.

Supplement 13.3 (Computation of the Characteristic Exponents via the Hill Infinite Determinant Method)   The characteristic exponents of the system in Eq. 13.23 can also be calculated by the method illustrated below. By letting, according to Eq. 13.25:

$$\mathbf{x}\,(t) = \exp\,(\gamma t)\,\boldsymbol{\phi}\,(t)\,, \tag{13.32}$$

and substituting it in the equations of the problem, it follows:

$$\dot{\boldsymbol{\phi}} + \gamma \boldsymbol{\phi} = \mathbf{A}\,(t)\,\boldsymbol{\phi}. \tag{13.33}$$

Since $\boldsymbol{\phi}$ is periodic of period T, it admits a Fourier series, which, in complex form, reads:

$$\boldsymbol{\phi} = \sum_{m=-\infty}^{\infty} \mathbf{c}_m \exp\,(im\Omega t)\,, \tag{13.34}$$

where $\Omega = \frac{2\pi}{T}$ and $\mathbf{c}_m$ is a vector of dimension N (with $\mathbf{c}_m \neq \bar{\mathbf{c}}_{-m}$, being $\boldsymbol{\phi}$ complex). Substituting Eq. 13.34 into Eq. 13.33, and separately equating to zero the harmonics $m = 0, \pm 1, \pm 2, \cdots$, an algebraic eigenvalue problem is drawn:

$$(\mathbf{L} - \gamma\mathbf{I})\,\mathbf{c} = \mathbf{0}, \tag{13.35}$$

where $\mathbf{L}$ is a matrix and $\mathbf{c} := (\cdots, \mathbf{c}_{m-1}, \mathbf{c}_m, \mathbf{c}_{m+1}, \cdots)$ a vector, both of infinite dimension. The determinant of $\mathbf{L}$ is called the *Hill infinite determinant*. Truncating the series at the harmonic Mth, its dimension is equal to $D := (1 + 2M)\,N$. There exist, therefore, D eigensolutions $(\gamma_k, \mathbf{c}_k)$ which provide as many normal solutions $\mathbf{x} = \mathbf{y}_k$:

$$\mathbf{y}_k\,(t) = \exp(\gamma_k t) \sum_{m=-M}^{M} \mathbf{c}_m \exp(im\Omega t), \qquad k = 1, \cdots, D. \tag{13.36}$$

However, since the Floquet Theorem states that there exist only $N < D$ normal solutions, it is evident that the procedure provides "spurious" solutions. This is due to the fact that, as said in Remark 13.5, γ is defined to within multiples of $i\Omega$, so that *only the eigenvalues whose imaginary part falls in the $\left(-\frac{\Omega}{2}, \frac{\Omega}{2}\right)$ interval are significant.*

To understand the reason for the appearance of spurious solutions, it is observed that, if no limits are imposed on γ, the representation of Eq. 13.36 (with $M = \infty$) is ambiguous, since it can also be written as:

$$\mathbf{y}_k\,(t) = \exp[(\gamma_k + ip\Omega)\,t] \sum_{m=-\infty}^{\infty} \mathbf{c}_m \exp[i\,(m - p)\,\Omega t], \qquad k = 1, \cdots, D. \tag{13.37}$$

It is therefore expected that for M sufficiently large, the eigenvalue problem in Eq. 13.35 provides an approximation of the characteristic exponents $\mathrm{Im}\,(\gamma_{k0}) \in \left(-\frac{\Omega}{2}, \frac{\Omega}{2}\right)$ $(k = 1, 2, \cdots, N)$, together with $\gamma_k = \gamma_{k0} \pm ip\Omega$ $(p = 1, 2, \cdots, M)$.

As an example, the case in which $\mathbf{A}\,(t) = \mathbf{A}_0 + \mathbf{B}\cos(\Omega t)$ is analyzed, with $\mathbf{A}_0$ and $\mathbf{B}$ constant arrays. From Eq. 13.33, it follows that:

$$\dot{\boldsymbol{\phi}} + \gamma\boldsymbol{\phi} = \sum_{m=-\infty}^{\infty} (\gamma + im\Omega)\,\mathbf{c}_m e^{im\Omega t}, \tag{13.38a}$$

$$\mathbf{A}\,(t)\,\boldsymbol{\phi} = \sum_{m=-\infty}^{\infty} \left[\mathbf{A}_0\mathbf{c}_m e^{im\Omega t} + \frac{1}{2}\mathbf{B}\mathbf{c}_m e^{i(m+1)\Omega t} + \frac{1}{2}\mathbf{B}\mathbf{c}_m e^{i(m-1)\Omega t} \right], \tag{13.38b}$$

having written $\cos(\Omega t) = \frac{1}{2}\,(\exp(i\Omega t) + \exp(-i\Omega t))$. By separating the harmonics in Eq. 13.33, one gets:

$$\frac{1}{2}\mathbf{B}\mathbf{c}_{m-1} + [\mathbf{A}_0 - (\gamma + im\Omega)\,\mathbf{I}]\,\mathbf{c}_m + \frac{1}{2}\mathbf{B}\mathbf{c}_{m+1} = \mathbf{0}, \tag{13.39}$$

with $m \in (-\infty, \infty)$; hence, the **L** matrix is three-banded in blocks of dimension $N \times N$.

Now, if $(\gamma, \mathbf{c}_m)$ is a solution of Eq. 13.39, by letting $\gamma^* := \gamma + i\Omega$, $\mathbf{c}_m^* := \mathbf{c}_{m+1}$, also $(\gamma^*, \mathbf{c}_m^*)$ is a solution of the same equations. This is easily deduced by writing the equation relative to the $(m + 1)$th harmonic:

$$\frac{1}{2}\mathbf{B}\mathbf{c}_m + [\mathbf{A}_0 - (\gamma + i\,(m + 1)\,\Omega)\,\mathbf{I}]\,\mathbf{c}_{m+1} + \frac{1}{2}\mathbf{B}\mathbf{c}_{m+2} = \mathbf{0}, \qquad (13.40)$$

which, with the positions introduced, becomes identical to Eq. 13.39, with the starred quantities. With the same reasoning, it is shown that $(\gamma \pm pi\Omega, \mathbf{c}_{m\pm p})$ are also solutions. Given what has been observed about Eq. 13.37, *all the spurious solutions replicate the fundamental N*, not only in the characteristic exponents, but also in the response. This is strictly true if the Fourier series is infinite, and it is approximately true if it is truncated. These aspects are further discussed in Problem 14.7.2, with reference to a numerical example.   □

13.3.2 Periodic Systems as Discrete-Time Systems: The Poincaré Map

An alternative (but equivalent) approach to the Floquet theory is viable for Eq. 13.23. This is based on a discrete-time vision of the system, made possible by the periodicity of the state matrix $\mathbf{A}(t)$. Accordingly, instead of observing the evolution $\mathbf{x}(t)$ of the continuous-time system, the response is sampled at discrete-time instants, $t = 0, T, 2T, \cdots, hT, \cdots$, in order to build up the sequence $\mathbf{x}_0, \mathbf{x}_1, \mathbf{x}_2, \cdots, \mathbf{x}_h, \cdots$, where $\mathbf{x}_h := \mathbf{x}(hT)$. From a geometrical point of view, by referring to the $N + 1$-dimensional extended state-space $(\mathbf{x}, t)$, the succession can be considered as made by the coordinates of points $\mathbf{x}_h$ intersected by the trajectories of the continuous-time system on the hyper-planes $t = hT$ (Fig. 13.3a). When these points are brought back to the same N-dimensional plane, a plot is obtained, called a *Poincaré* section. The goal of the analysis is to determine the relationship existing between two successive elements of the sequence, $\mathbf{x}_{h+1} = \mathcal{F}(\mathbf{x}_h)$, which takes the name of *Poincaré map* [4].

To this end, the following fundamental observation is introduced. Since *the continuous-time system assumes at the time $t_h := hT$ the same characteristics that it had at the time $t = 0$*, being $\mathbf{A}(t_h) = \mathbf{A}(0)$, its response at the time $t_h + t$ under initial conditions imposed a t_h is identical to that the system would have at the time t, if the same initial conditions were imposed at the time 0. Said in other terms, the

Fig. 13.3 Discrete-time system: (a) sampling of the response $\mathbf{x}(t) \in \mathbb{R}^N$ at discrete-times $t = 0, T, 2T, \cdots$ and transfer matrix $\mathbf{F}$; (b) eigenvectors $\mathbf{u}_k \in \mathbb{C}^N$ and eigenvalues λ_k of matrix $\mathbf{F}$

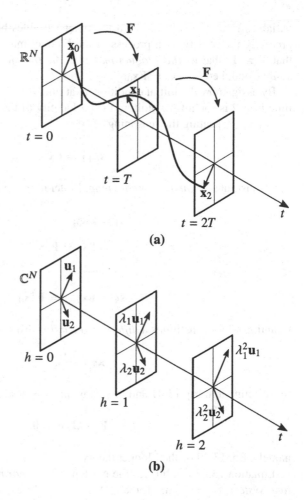

(a)

(b)

response depends only on the time $t - t_h$, elapsed from the application of the initial conditions to the time t_h, and not on the value of h.[9]

One can imagine, therefore, to assign to the system the initial condition $\mathbf{x}_0$ and to record its response $\mathbf{x}_1$. Between the two vectors, belonging to the same state-space, there exist the linear relationship $\mathbf{x}_1 = \mathbf{T}\mathbf{x}_0$, where $\mathbf{T}$ is a matrix of dimension $N \times N$, called *transfer matrix* (as it transfers the state from time $h = 0$ to time $h = 1$). To build $\mathbf{T}$ columnwise, it needs to evaluate N responses at time $h = 1$, each corresponding to a condition at time $h = 0$ in which only one of the state

---

[9] If the system were aperiodically dependent on time, the response at $t > t_h$ to initial conditions $\mathbf{x}_h$ assigned at time t_h, would depend on t *and* on t_h, that is, $\mathbf{x} = \mathbf{x}(t, t_h; \mathbf{x}_h)$. This is the case of systems subject to aging (e.g., rubber, which loses its elastic properties, or fresh concrete, that hardens over time). If, on the other hand, the system is periodic, that is, it has a *null average* aging over a period T, the response does not depend on t_h, but only on the difference $t - t_h$, that is, $\mathbf{x} = \mathbf{x}(t - t_h; \mathbf{x}_h)$.

variables has a unitary value, all the other variables being set to zero. But this is precisely the construction process of the Floquet matrix! It is therefore concluded that $\mathbf{T} \equiv \mathbf{F}$, that is, *the Floquet matrix is the transfer matrix of the discrete-time system*, and therefore $\mathbf{x}_1 = \mathbf{F}\mathbf{x}_0$.

By assign now the initial condition $\mathbf{x}_1$ at time $h = 1$, the response at the discrete-time $h = 2$ is sought. Exploiting the periodicity of the system, it is concluded that $\mathbf{x}_2 = \mathbf{F}\mathbf{x}_1$. Repeating the reasoning:

$$\mathbf{x}_{h+1} = \mathbf{F}\mathbf{x}_h. \tag{13.41}$$

This constitutes a *linear Poincaré map*. Reiterating the map, one has:

$$\mathbf{x}_1 = \mathbf{F}\mathbf{x}_0, \tag{13.42a}$$

$$\mathbf{x}_2 = \mathbf{F}\mathbf{x}_1 = \mathbf{F}^2\mathbf{x}_0, \tag{13.42b}$$

$$\cdots = \cdots \tag{13.42c}$$

$$\mathbf{x}_h = \mathbf{F}\mathbf{x}_{h-1} = \mathbf{F}^h\mathbf{x}_0. \tag{13.42d}$$

Equation 13.41 constitutes a *difference equation*. To solve it, one tries the solution:

$$\mathbf{x}_h = \lambda^h \mathbf{u}. \tag{13.43}$$

Substituting it in Eq. 13.41 and imposing that this is satisfied for any h, one obtains:

$$(\mathbf{F} - \lambda\mathbf{I})\,\mathbf{u} = \mathbf{0}, \tag{13.44}$$

namely, Eq. 13.29 of the Floquet theory.

Equation 13.43 establishes the existence of N *normal solutions of the discrete-time system*, $\mathbf{y}_{k_h} = \lambda_k^h \mathbf{u}_k$, for $k = 1, 2 \cdots, N$. In each of these, an eigenvector $\mathbf{u}_k$ of the Floquet matrix, when reiterated under the map $\mathbf{F}$, *transforms being affected by a complex scalar* (Fig. 13.3b).[10] Therefore, given the initial condition $\mathbf{y}_{k_0} = \mathbf{u}_k$, the result is $\mathbf{y}_{k_1} = \lambda_k \mathbf{u}_k$, $\mathbf{y}_{k_2} = \lambda_k^2 \mathbf{u}_k$, $\cdots$, $\mathbf{y}_{k_h} = \lambda_k^h \mathbf{u}_k$. If a certain λ_k is in modulus smaller than 1, the corresponding eigenvector decays in the discrete-time; if it has modulus larger than 1, it diverges in the discrete-time. If, on the other hand, λ_k has unit modulus, it needs to distinguish:

- if $\lambda_k = 1$, the result is $\mathbf{y}_{k_h} = \mathbf{u}_k$ for $h = 0, 1, \cdots$, i.e., $\mathbf{u}_k$ is a *fixed point* (or of equilibrium) of the map; the associated continuous-time system admits a T-periodic solution, of the same period (or sub-multiple $\frac{T}{p}$, $p = 1, 2, \cdots$) of the parametric excitation (Fig. 13.4a).
- if $\lambda_k = -1$, the result is $\mathbf{y}_{k_h} = \mathbf{u}_k$ for $h = 0, 2, \cdots$ and $\mathbf{y}_{k_h} = -\mathbf{u}_k$ for $h = 1, 3, , \cdots$, i.e., the response changes sign at each discrete-time; the

---

[10] If the scalar is real, the vector remains parallel to itself, as illustrated in the figure.

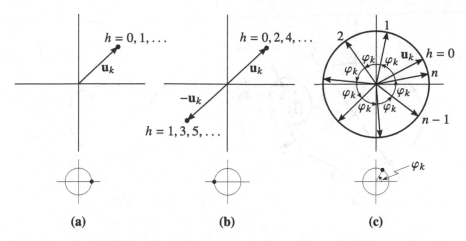

Fig. 13.4 Iterations of an eigenvector $\mathbf{u}_k$ of the Floquet matrix associated with an eigenvalue of unit modulus: (a) $\lambda_k = 1$, (b) $\lambda_k = -1$, (c) $\lambda_k = \exp(i\varphi_k)$

associated continuous-time system admits a $2T$-periodic solution (or odd sub-multiple $\frac{2T}{p}$, $p = 1, 3, 5, \cdots,$[11]), with a period double than the period of the parametric excitation (Fig. 13.4b).

- if $\lambda_k = \exp(i\varphi_k)$, the result is $\mathbf{y}_{k_h} = \mathbf{u}_k \exp(ih\varphi_k)$ for $h = 0, 1, \cdots$, i.e., $\mathbf{u}_k$ is rotated by an angle φ_k at each discrete-time; it is necessary to further distinguish:

  (a) the *degenerate case*, in which $\varphi_k = \frac{2\pi}{q}$, $q = 1, 2, \cdots$; in this case, the trajectory intersects the Poincaré section in q points on the circle, which are visited in sequence; the associated continuous-time system admits a periodic solution of period qT (or sub-multiples $\frac{q}{p}T$ (with $p, q = 1, 2, \cdots$);
  (b) the *generic case*, in which φ_k is incommensurable to 2π; in this case, the trajectory intersects the Poincaré section filling the circle in a dense and discrete way; the continuous-time system admits a *quasi-periodic* solution (Fig. 13.4c).

Periodic and quasi-periodic motions of the continuous-time system are susceptible of an interesting geometric representation, discussed in Supplement 13.4 and illustrated in Fig. 13.5. These motions, in a three-dimensional state-space (which can be understood as a projection of a larger space), develop on a torus. Periodic motions consist of closed trajectories (Fig. 13.5b) and quasi-periodic motions of open trajectories, which densely cover the torus (Fig. 13.5c,d).

Remark 13.6 The N normal solutions of the discrete-time system, $\mathbf{y}_{k_h} = \lambda_k^h \mathbf{u}_k$, coincide with the normal solutions of the continuous-time system (Eq. 13.25), when

---

[11] The even sub-multiples, indeed, do not respect the change of sign at $t = T$, reproducing the T-periodic motions.

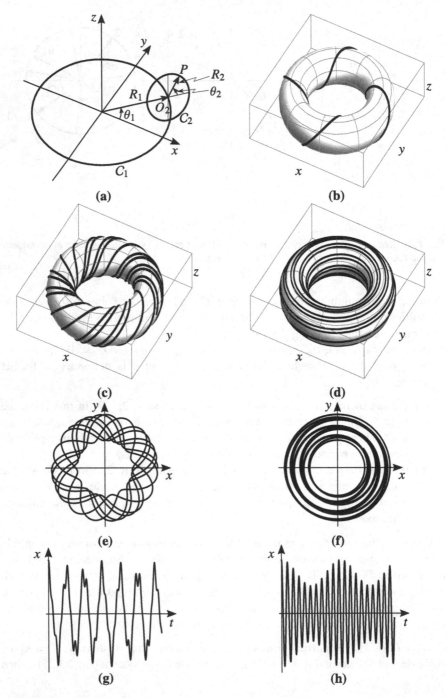

Fig. 13.5 Periodic and quasi-periodic motions on a torus: (**a**) canonical representation of the torus in parametric form; (**b**) periodic motion on the torus ($\omega_2 = 4\,\omega_1$); (**c,d**) quasi-periodic motions on the torus: (**c**) $\omega_2 = 3.7\,\omega_1$, (**d**) $\omega_2 = 0.123\,\omega_1$; (**e,f**) projections of the trajectories on the (x, y) plane; (**g,h**) time-laws relevant to cases (**c**) and (**d**), respectively

these latter are sampled with step T. Indeed, from Eq. 13.25, it follows:

$$\mathbf{y}_k(hT) = \exp(\gamma_k hT)\,\boldsymbol{\phi}_k(hT) = \lambda_k^h \boldsymbol{\phi}_k(0), \tag{13.45}$$

where Eq. 13.30 has been used, together with the periodicity of $\boldsymbol{\phi}_k$. Ultimately, $\mathbf{u}_k = \boldsymbol{\phi}_k(0)$ and $\mathbf{y}_{k_h} = \mathbf{y}_k(hT)$.

Supplement 13.4 (Periodic and Quasi-Periodic Motions on a Torus) The torus is a surface generated by the revolution of a circle C_2, of radius R_2 (Fig. 13.5a), around a coplanar and external axis z. In the motion of revolution, the center O_2 of C_2 describes in the plane (x, y) a circle C_1 of radius R_1, so that it can also be said that the torus is the Cartesian product $C_1 \times C_2$. Taken a radial section of the torus, identified by the angle θ_1 that the section forms with the axis x, and taken on this section a point P, whose radius vector $\overrightarrow{O_2P}$ forms an angle θ_2 with the radial direction, the coordinates of P are:

$$x(\varphi_1, \varphi_2) = (R_1 + R_2 \cos \theta_2) \cos \theta_1, \tag{13.46a}$$

$$y(\varphi_1, \varphi_2) = (R_1 + R_2 \cos \theta_2) \sin \theta_1, \tag{13.46b}$$

$$z(\varphi_1, \varphi_2) = R_2 \sin \theta_2. \tag{13.46c}$$

where $\theta_1 \in (0, 2\pi)$, $\theta_2 \in (0, 2\pi)$. Equation 13.46 is the parametric equation of the surface represented in Fig. 13.5b–d.

A *bi-periodic motion* is considered, now, obtained from the composition of two uniform circular motions, of angular velocities ω_1, ω_2, respectively, developing on the circles C_1, C_2. Accordingly, while the radial section performs a revolution around z, of angular velocity ω_1, the vector radius O_2P performs a rotational motion, of angular velocity ω_2, around the instantaneous tangent to C_1. The resulting trajectory has parametric equations that are deduced from Eq. 13.46 by expressing the values assumed by the parameters at time t, i.e., $\theta_1 = \omega_1 t, \theta_2 = \omega_2 t$:

$$x(t) = [R_1 + R_2 \cos(\omega_2 t)] \cos(\omega_1 t), \tag{13.47a}$$

$$y(t) = [R_1 + R_2 \cos(\omega_2 t)] \sin(\omega_1 t), \tag{13.47b}$$

$$z(t) = R_2 \sin(\omega_2 t). \tag{13.47c}$$

If ω_1 and ω_2 are in a rational ratio, the motion in Eq. 13.47 is periodic and develops on a closed trajectory (Fig. 13.5b). If the ratio is irrational, the motion in Eq. 13.47 is quasi-periodic and takes place on an open trajectory, densely covering the torus (Fig. 13.5c,d). If $\omega_2 > \omega_1$ (Fig. 13.5c,e,g), the slow motion ω_1 is "disturbed" by a higher frequency; if $\omega_2 < \omega_1$ (Fig. 13.5d,f,h), the fast motion ω_1 is modulated in amplitude by a lower frequency. The case of modulation is that of major interest here. It occurs when $\lambda_k = \exp(i\varphi_k)$, $|\varphi_k| < \pi$; indeed, taken $\omega_1 = \Omega$, it is $\omega_2 = \mathrm{Im}(\gamma_k) = \frac{1}{T}\varphi_k < \frac{\Omega}{2}$. □

13.4   Characteristic Multipliers of Single Degree of Freedom Systems

Parametrically excited single DOF systems are studied and the properties of their characteristic multipliers investigated. The general case is considered first, in which both damping and stiffness are parametrically excited; then, the excitation is particularized to the stiffness only. For these latter class of systems, the bifurcation mechanisms are illustrated [7].

13.4.1   General Systems

A general single DOF system, parametrically excited, possesses mass $m(t)$, damping $c(t)$ and stiffness $k(t)$, all variable over time by periodic laws of the same period T. It falls into this class of problems, the pendulum with movable mass, already introduced in Sect. 13.2. The equation of motion of these systems is:

$$m(t)\ddot{q} + c(t)\dot{q} + k(t)q = 0. \tag{13.48}$$

By admitting $m(t) > 0 \ \forall t \in [0, T]$, the equation in recast the form:

$$\ddot{q} + f_1(t)\dot{q} + f_2(t)q = 0, \qquad f_i(t+T) = f_i(t), \tag{13.49}$$

where: $f_1(t) := \frac{c(t)}{m(t)}$, $f_2(t) := \frac{k(t)}{m(t)}$.

Equation 13.49 is equivalent to two equations of the first order in the state variables $\mathbf{x} := (q, \dot{q})^T$. It is assumed that two solutions, $\mathbf{x}^{(1)}(t)$, $\mathbf{x}^{(2)}(t)$, satisfying the initial conditions $\mathbf{x}^{(1)}(0) = (1, 0)^T$ and $\mathbf{x}^{(2)}(0) = (0, 1)^T$, have been determined in the interval $[0, T]$, possibly by numerical integration. Hence, the matrix of the fundamental system of solutions is built up as:

$$\mathbf{X}(t) := \left[\mathbf{x}^{(1)}(t)\ \mathbf{x}^{(2)}(t) \right] \equiv \begin{bmatrix} q^{(1)}(t)\ q^{(2)}(t) \\ \dot{q}^{(1)}(t)\ \dot{q}^{(2)}(t) \end{bmatrix}, \qquad \mathbf{X}(0) = \mathbf{I}. \tag{13.50}$$

By evaluating the matrix at $t = T$, the Floquet matrix is drawn:

$$\mathbf{F} := \begin{bmatrix} q^{(1)}(T)\ q^{(2)}(T) \\ \dot{q}^{(1)}(T)\ \dot{q}^{(2)}(T) \end{bmatrix}. \tag{13.51}$$

The characteristic multipliers are the eigenvalues of $\mathbf{F}$, i.e., the roots of the characteristic equation:

$$\lambda^2 - \lambda \, \text{tr} \mathbf{F} + \det \mathbf{F} = 0. \tag{13.52}$$

The two roots of Eq. 13.52 are real or complex conjugate and satisfy the condition $\lambda_1 \lambda_2 = \det \mathbf{F}$. It should be remembered, from mathematical analysis, that $W(T) := \det(\mathbf{X}(T))$ is the *Wronskyan* of the differential equation, evaluated at time T, for which the *Liouville theorem* holds:[12]

$$W(T) = W(0) \exp\left(-\int_0^T f_1(t)\,dt\right). \tag{13.53}$$

By taking into account that $W(0) = 1$, it follows that:

$$\lambda_1 \lambda_2 = \exp\left(-\int_0^t f_1(t)\,dt\right). \tag{13.54}$$

13.4.2 Undamped and Damped Hill Equation

A notable case is considered, in which the single DOF system, Eq. 13.49, is parametrically excited in the stiffness only. For example, they fall into this class of problems: the pendulum with motion enforced at the base, and the variable length beam, already introduced in Sect. 13.2. The equation of motion governing such systems is called the *Hill equation*. The case of undamped system is examined first, and then the case of damping (kept constant over time) is addressed.

Undamped System
The undamped Hill equation is:

$$\ddot{q} + f_2(t)\,q = 0, \qquad f_2(t+T) = f_2(t), \tag{13.55}$$

---

[12] The Wronskian of the system of linear differential equations $\dot{\mathbf{x}} = \mathbf{A}(t)\mathbf{x}$ is defined as the determinant of the matrix of fundamental solutions, $W(t) := \det(\mathbf{X}(t))$. For it, the Liouville theorem holds:

$$W(t) = W(0) \exp\left(\int_0^t \text{tr}\mathbf{A}(t)\,dt\right)$$

If the matrix equation is the first-order form of a scalar equation of order n:

$$q^{(n)} + a_1(t)\,q^{(n-1)} + \cdots + a_{n-1}(t)\,\dot{q} + a_n(t)\,q = 0$$

then $\text{tr}\mathbf{A}(t) = -a_1(t)$.

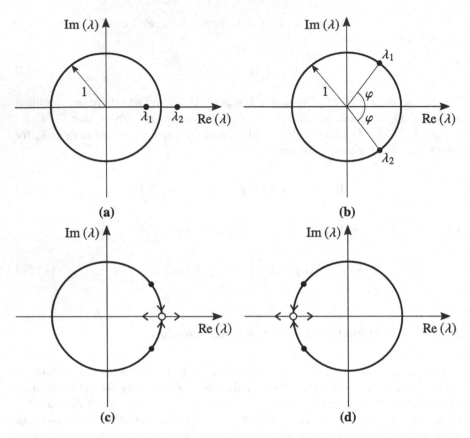

Fig. 13.6 Characteristic multipliers and bifurcations for the undamped Hill equation: (**a**) real multipliers, (**b**) complex conjugated multipliers of unit modulus; (**c**) bifurcation of *divergence* type, (**d**) bifurcation of *flip* type

which is a special case of Eq. 13.49, where $f_1(t) \equiv 0$. Its characteristic multipliers satisfy Eq. 13.54, which, in the case in question, reads:

$$\lambda_1 \lambda_2 = 1. \tag{13.56}$$

Only two cases are, therefore, possible:

- the characteristic multipliers are real and reciprocal, one internal to the unit circle, the other external to it (Fig. 13.6a);
- the characteristic multipliers are complex conjugate, of modulus unitary, $\lambda_{1,2} = \exp(\pm i\varphi)$, i.e., they lie on the unit circle (Fig. 13.6b).

It follows that if the multipliers are real, the equilibrium is unstable; if they are complex conjugates, the equilibrium is marginally stable. If the multipliers depend on a parameter, there are only two bifurcation mechanisms:

- bifurcation of *divergence* type , in which the multipliers run on the unit circle, collide at $\lambda = 1$, and separate, one inside, the other outside (remaining reciprocal) (Fig. 13.6c);
- bifurcation of *flip* type, in which the multipliers run on the unit circle, collide at $\lambda = -1$, and separate, one inside, the other outside (remaining reciprocal) (Fig. 13.6d).

At the divergence bifurcation, the system performs periodic oscillations of period T (Fig. 13.4a); at the flip bifurcation, the system executes oscillations of period $2T$ (Fig. 13.4b).

Remark 13.7 Remembering the Remark 13.2, it can be said that at the divergence bifurcation, the period of the response is $T_r = T, \frac{T}{2}, \frac{T}{3}, \cdots$, while, at the flip bifurcation, it is $T_r = 2T, \frac{2T}{3}, \frac{2T}{5}, \cdots$. Overall, in the two types of bifurcation, the minimum period is $T_r = \frac{2T}{p}$ ($p = 1, 2, \cdots$), in the sequence of which flip and divergence bifurcations alternate.

Damped System
The effect of an added damping to the Hill equation is now analyzed. The equation of motion, Eq. 13.55, changes into:

$$\ddot{q} + \zeta \dot{q} + f_2(t) q = 0, \qquad f_2(t + T) = f_2(t), \tag{13.57}$$

where $\zeta > 0$ is a damping coefficient. With reference to the general case of Eq. 13.49, since $f_1(t) = \zeta = \text{cost}$, the relationship between the characteristic multipliers (Eq. 13.54) becomes:

$$\lambda_1 \lambda_2 = \exp(-\zeta T) < 1. \tag{13.58}$$

The following occurrences are therefore possible:

- the characteristic multipliers are both real (Fig. 13.7a), at least one of which is inside the unit circle (the other being either internal or external);
- the characteristic multipliers are complex conjugates (Fig. 13.7b), of type $\lambda_{1,2} = \rho \exp(\pm i\varphi)$ and lie on a circle of radius $\rho = \exp\left(-\frac{\zeta T}{2}\right) < 1$.

It is concluded that it is not enough that the multipliers are real to make the equilibrium unstable, since the multipliers could both be smaller than 1 in modulus. If the multipliers depend on a parameter, there exist the following two bifurcation mechanisms:

- bifurcation of *divergence* type, , in which the multipliers run on the circle of radius $\rho < 1$, collide at $\lambda = \rho$, and then separate, one toward the center of the circle, the other outward; when this latter crosses the unit circle at $\lambda = 1$, the divergence bifurcation occurs (Fig. 13.7c);
- bifurcation of *flip* type, in which the multipliers run on the circle of radius $\rho < 1$, collide at $\lambda = -\rho$, and then separate, one toward the center of the circle, the other

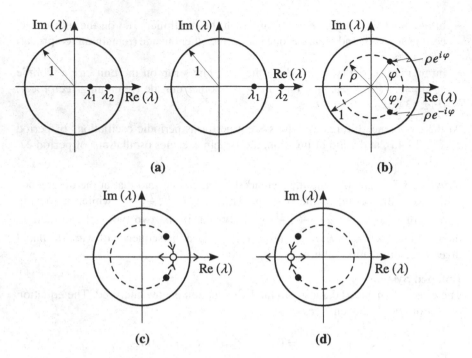

Fig. 13.7 Characteristic multipliers and bifurcations for the damped Hill equation: (**a**) real multipliers, (**b**) complex conjugated multipliers of modulus smaller than one; (**c**) bifurcation of *divergence* type, (**d**) bifurcation of *flip* type

outward; when this latter crosses the unit circle at $\lambda = -1$, the flip bifurcation occurs (Fig. 13.7d).

Remark 13.8 Similarities and differences should be noticed between the bifurcation mechanisms of parametrically excited non-autonomous systems and self-excited autonomous systems. The divergence bifurcation of the undamped Hill system (Fig. 13.6c) is similar to the static divergence of Hamiltonian systems (where the bifurcation follows the collision in the origin of the purely imaginary eigenvalues of the Jacobian matrix). Damping has the same effect on the two classes of systems, anticipating the collision in the stable zone (Fig. 13.7c). The flip bifurcations, undamped (Fig. 13.6d) or damped (Fig. 13.7d), instead, are peculiar to periodic systems. It will be seen later, dealing with systems with several DOF (Sect. 13.8), that there exists a further mechanism, similar to the Hopf bifurcation of autonomous systems.

13.5 Mathieu Equation

The *Mathieu equation* is, by far, the most studied equation concerning parametric excitation, of which it constitutes a paradigm.[13] It is a particular case of the Hill equation, in which the stiffness varies with harmonic law around a constant value. Its canonical form is[14] [7]:

$$\ddot{q} + (\delta + 2\epsilon \cos(2t)) q = 0, \qquad (13.59)$$

where the nondimensional parameters δ and 2ϵ measure the constant part of the stiffness and the amplitude of the parametric excitation, respectively. The *period of the parametric excitation is instead kept constant and equal to $T = \pi$*. Further on Sect. 13.6.1, an example will be shown that explains how to switch between physical and nondimensional parameters.

Remark 13.9 In analyzing the stability of the trivial solution of Eq. 13.59, one has to assign the period of the forcing, $T = \pi$, and to scan all the members of a family of systems, characterized by the parameter δ. The canonical form, therefore, overturns the usual perspective of the structural mechanics, according to which the system is fixed and the excitation is variable. However, it will be seen shortly that the two points of view are equivalent, although that relevant to Eq. 13.59 provides more readable results.

13.5.1 Strutt Diagram

The analysis of the Mathieu equation is aimed at identifying the combinations (δ, ϵ) for which the origin is stable and those for which it is unstable. The answer to this problem is supplied by the *Strutt diagram*, which constitutes a "behavior chart" of the system.[15] It identifies, on the parameter plane (δ, ϵ), the stable and unstable regions. A portion of the diagram of Strutt is represented in Fig. 13.8, where δ

---

[13] A review on the subject, and on its generalizations, is presented in [3].

[14] Another form, also used in literature, is:

$$\frac{d^2 q}{d\tilde{t}^2} + \left(\tilde{\delta} + \tilde{\epsilon} \cos \tilde{t}\right) q = 0$$

One can switch from this equation to Eq. 13.59 by introducing the change of the time scale, $\tilde{t} = 2t$. Since $\frac{d}{dt} = \frac{d}{d\tilde{t}} \frac{d\tilde{t}}{dt} = \frac{1}{2} \frac{d}{d\tilde{t}}$, it is also $\frac{d^2}{dt^2} = \frac{1}{4} \frac{d^2}{d\tilde{t}^2}$. Therefore, from comparison, it is $\tilde{\delta} = \frac{1}{4}\delta$ and $\tilde{\epsilon} = \frac{1}{2}\epsilon$.

[15] Although the diagram was published by M.J.O. Strutt in 1928, it had already been obtained by E. Ince, in 1927 [3]. For this reason, it is also called the *Ince-Strutt diagram*.

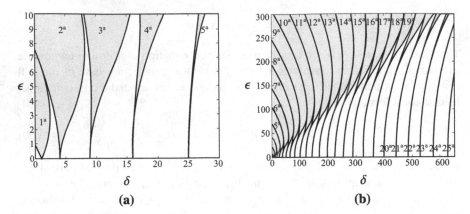

Fig. 13.8 Strutt diagram for the Mathieu equation: (**a**) first 5 resonance zones, (**b**) first 25 resonance zones; gray areas, unstable; white areas, stable; transition curves, odd order, of period $2T$; even order, of period T

and ϵ are limited to positive values[16, 17] The figure shows the so-called tongues of instability that emanate from the δ axis, at the integer values $\delta = p^2$, $p = 1, 2, \cdots$. The regions are named in the order: *first-order resonance* (or *principal resonance*) zone, emanating from $\delta = 1$; *second order resonance* zone, which originates from $\delta = 4$, and so on; the regions following the first one are also called, cumulatively, *secondary resonance* zones. It is worth noticing that the regions of instability gradually decrease in width, becoming more and more "narrow" and "high." However, for ϵ large enough, they touch each other. This means that, at low levels of parametric excitation (ϵ small), instability occurs only near the resonant values $\delta = p^2$ ($p = 1, 2, \cdots$). When, however, the excitation is strong enough (ϵ large), almost all values of δ are unstable. Fortunately, in applications to mechanical systems, the excitation is of small intensity, for which it is meaningful to identify the intervals of δ for which the equilibrium is unstable.

The Strutt diagram cannot be determined in closed form, and therefore a numerical approach is required.[18] However, there are perturbation solutions holding for ϵ small, which will be briefly illustrated later. Analytical solutions valid for larger ϵ

---

[16] Negative values of δ are also of interest, as they refer to statically unstable systems (e.g., an inverted mathematical pendulum, or the Bolotin beam under a static force larger than the critical one), which can eventually be restabilized by parametric excitation. An interesting physical example is offered by a juggler, who keeps in equilibrium a vertical rod resting on the palm of his hand, simply impressing a suitable motion to the base.

[17] The negative values of ϵ are obtained overturning the diagram around the horizontal axis δ, which therefore is an axis of symmetry. Indeed, a change of sign of ϵ in the Mathieu Eq. 13.59, is equivalent to add a phase π to the argument of the cosine function; it therefore corresponds to a mere translation of the origin of time, which does not affect the stability.

[18] An example of application of the numerical procedure is illustrated in Problem 14.7.1.

are also available in the literature (e.g., in [3]), where polynomial expressions for $\delta = \delta\,(\epsilon)$ are provided, up to the power ϵ^{12}.

Origin of the Unstable Regions

Since the Mathieu equation is a Hill equation, the bifurcation mechanisms illustrated in Fig. 13.6c,d hold for it. This implies that the loci of incipient bifurcation of Strutt diagram (i.e., the curves that separate stable from unstable states) identify systems admitting *periodic solutions of period T* (divergence bifurcations) *or of period 2T* (flip bifurcations), with $T = \pi$. These loci are called *transition curves*. To find them, their intersections δ_0 with the δ axis are first examined.

When $\epsilon \to 0$, the Mathieu equation becomes a constant coefficient equations, admitting the solution:

$$q_0\,(t) = a \cos\left(\sqrt{\delta_0}t\right) + b \sin\left(\sqrt{\delta_0}t\right), \tag{13.60}$$

where the index 0 remembers that it relates to $\epsilon = 0$. This solution is periodic, of period $T_r = \frac{2\pi}{\sqrt{\delta_0}} \equiv \frac{2T}{\sqrt{\delta_0}}$, for any δ_0. However, T_r is equal T or $2T$, or their sub-multiples, *if and only if* $\sqrt{\delta_0}$ *is an integer number* p. It is therefore $T_r = \frac{2T}{p}$ ($p = 1, 2, \cdots$), in accordance with the Remark 13.7. In particular, by denoting by $\mathbf{x}_0 := (q_0, \dot{q}_0)^T$ the response in the state-space, it happens that:

- if $\sqrt{\delta_0} = 1, 3, 5, \cdots$, then $T_r = 2T, \frac{1}{3}2T, \frac{1}{5}2T, \cdots$, i.e., $\mathbf{x}_0\,(2T) = \mathbf{x}_0\,(0)$;[19]
- if $\sqrt{\delta_0} = 2, 4, 6, \cdots$, then $T_r = T, \frac{1}{2}T, \frac{1}{3}T, \cdots$, i.e., $\mathbf{x}_0\,(T) = \mathbf{x}_0\,(0)$.[20]

It is concluded that:

- the resonance regions of *odd* order emanates from $\delta_0 = 1, 9, 25, \cdots$ and are generated by *flip* bifurcations, which give rise to periodic solutions of period $2T$ (or odd sub-multiples);
- the resonance regions of *even* order emanates from $\delta_0 = 4, 16, 36, \cdots$ and are generated by *divergence* bifurcations, which give rise to periodic solutions of period T (or even and odd sub-multiples).

Remark 13.10 The regions of instability are called *of resonance*, as in them the frequency $\sqrt{\delta_0}$ of the unexcited system and the frequency 2 of the parametric excitation are in a ratio next to a rational number ($\frac{1}{2}, \frac{3}{2}, \frac{5}{2}, \cdots$ or $1, 2, 3, \cdots$).

---

[19] For example, if $\sqrt{\delta_0} = 1$, it is $\mathbf{x}_0\,(t) = \begin{pmatrix} a \cos t + b \sin t \\ -a \sin t + b \cos t \end{pmatrix}$, so that: $\mathbf{x}_0\,(0) = \begin{pmatrix} a \\ b \end{pmatrix}$, $\mathbf{x}_0\,(\pi) = \begin{pmatrix} -a \\ -b \end{pmatrix}$, $\mathbf{x}_0\,(2\pi) = \begin{pmatrix} a \\ b \end{pmatrix}$, $\cdots$.

[20] For example, if $\sqrt{\delta_0} = 2$, it is $\mathbf{x}_0\,(t) = \begin{pmatrix} a \cos\,(2t) + b \sin\,(2t) \\ -2a \sin\,(2t) + 2b \cos\,(2t) \end{pmatrix}$, so that: $\mathbf{x}_0\,(0) = \begin{pmatrix} a \\ 2b \end{pmatrix}$, $\mathbf{x}_0\,(\pi) = \begin{pmatrix} a \\ 2b \end{pmatrix}$, $\mathbf{x}_0\,(2\pi) = \begin{pmatrix} a \\ 2b \end{pmatrix}$, $\cdots$.

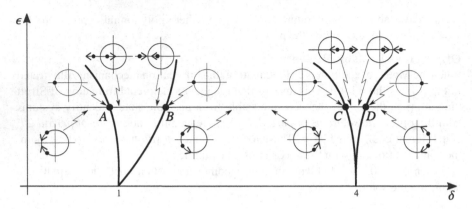

Fig. 13.9 Movement of the characteristic multipliers of the Mathieu equation, as the parameter δ is changed, while ϵ is fixed

Movement of the Characteristic Multipliers

From the previous results, it is possible to obtain information on the movement of the characteristic multipliers, when a parameters is quasi-statically varied. To investigate the topic, the excitation level ϵ is fixed, and the frequency-like parameter δ is increased from zero (Fig. 13.9).

When δ is small, the characteristic multipliers $\lambda_{1,2}$ are complex conjugates and lie on the unit circle (i.e., the system is in the stable region). As δ increases, the multipliers travel around the circle, and, at the transition curve, they collide at $\lambda = -1$ (point A in the figure, where a flip bifurcation occurs). Upon entering the unstable region, the multipliers first separate on the real axis, and then they *invert the motion* and collide again at $\lambda = -1$ (point B in the figure, of flip bifurcation), where the system regains stability. For further increments of δ, the characteristic multipliers travel the unit circle and collide at $\lambda = 1$ (point C in the figure, of divergence bifurcation). Entering the region of instability, they first separate on the real axis, then they invert the motion and collide again at $\lambda = 1$ (point D in the figure, where another divergence bifurcation occurs), regaining stability. The process goes on in the same way, alternating flip and divergence bifurcations.

13.5.2 Asymptotic Construction of the Transition Curves

The transition curves of the Strutt diagram for the Mathieu equation can be evaluated, for small ϵ, through a perturbation method. This is based on the search of the combinations of two parameters δ and ϵ, for which the response is periodic of assigned period, π or 2π. Among these methods, the simplest is the *strained parameter method*, already used for autonomous systems in the Chaps. 11 and 12, in the form of the Lindstedt-Poincaré method.[21]

The unknown link between the two parameters is expressed in the form $\delta = \delta(\epsilon)$, which entails $q = q(t; \delta(\epsilon), \epsilon) = q(t; \epsilon)$, where ϵ is the perturbation parameter.[22] A family of solutions is sought, satisfying the periodicity conditions:

$$q(t + 2\pi; \epsilon) = q(t; \epsilon), \qquad q(t + \pi; \epsilon) = q(t; \epsilon),$$
$$\text{or} \qquad\qquad\qquad\qquad\qquad\qquad (13.61)$$
$$\dot{q}(t + 2\pi; \epsilon) = \dot{q}(t; \epsilon), \qquad \dot{q}(t + \pi; \epsilon) = \dot{q}(t; \epsilon),$$

on the boundaries $\delta = \delta(\epsilon)$ of the resonance zones, of odd or even order, respectively.

Expanding in series of ϵ both the dependent variable $q(t; \epsilon)$, and the parameter δ, it follows:

$$q(t; \epsilon) = q_0(t) + \epsilon q_1(t) + \epsilon^2 q_2(t) + \cdots, \qquad (13.62a)$$

$$\delta(\epsilon) = \delta_0 + \epsilon \delta_1 + \epsilon^2 \delta_2 + \cdots. \qquad (13.62b)$$

Substituting these series in Mathieu Eq. 13.59 and separately equating to zero the terms with the same powers of ϵ, the perturbation equations are obtained:

$$\epsilon^0: \quad \ddot{q}_0 + \delta_0 q_0 = 0, \qquad (13.63a)$$

$$\epsilon^1: \quad \ddot{q}_1 + \delta_0 q_1 = -\delta_1 q_0 - 2q_0 \cos(2t), \qquad (13.63b)$$

$$\epsilon^2: \quad \ddot{q}_2 + \delta_0 q_2 = -\delta_2 q_0 - \delta_1 q_1 - 2q_1 \cos(2t). \qquad (13.63c)$$

Performing similar steps on the periodicity conditions, Eq. 13.61, it turns out that:

$$q_k(t + 2\pi) = q_k(t), \qquad q_k(t + \pi) = q_k(t),$$
$$\qquad\qquad\qquad \text{or} \qquad\qquad\qquad\qquad\qquad k = 0, 1, \cdots. \qquad (13.64)$$
$$\dot{q}_k(t + 2\pi) = \dot{q}_k(t), \qquad \dot{q}_k(t + \pi) = \dot{q}_k(t),$$

The solution to Eq. 13.63a is given by Eq. 13.60; with the 2π-periodicity condition, it gives $\delta_0 = 1, 9, \cdots$; with the π-periodicity condition, it supplies $\delta_0 = 4, 16, \cdots$. The case of principal resonance, $\delta_0 = 1$, is the unique treated in detail ahead.

Principal Resonance
With $\delta_0 = 1$, the generating solution reads:

---

[21] The substantial difference with the Lindstedt-Poincaré method lies in the fact that the period of the response is now known, and not unknown. This circumstance implies that it needs to expand in series a single parameter, instead of two.

[22] Unlike all the other problems tackled in this book, ϵ has now a physical meaning, as it appears in the equation of motion. For this reason, no normalization conditions are required to identify it.

$$q_0 = a \cos t + b \sin t. \tag{13.65}$$

With the help of well-known formulas of trigonometry,[23] Eq. 13.63b is rewritten as:

$$\ddot{q}_1 + q_1 = -a (1 + \delta_1) \cos t + b(1 - \delta_1) \sin t$$
$$- a \cos (3t) - b \sin (3t) . \tag{13.66}$$

The frequency 1 terms appearing on the right side of the equation, since they repeat the same natural frequency 1 of the oscillator, are resonant. Therefore, they must be removed from the equation, since they would make the solution aperiodic. By vanishing them for any t, one obtains:

$$\begin{bmatrix} 1 + \delta_1 & 0 \\ 0 & 1 - \delta_1 \end{bmatrix} \begin{pmatrix} a \\ b \end{pmatrix} = \begin{pmatrix} 0 \\ 0 \end{pmatrix}. \tag{13.67}$$

This equation admits two solutions, (i) $\delta_1 = -1$, $b = 0$, $\forall a$ and (ii) $\delta_1 = 1$, $a = 0$, $\forall b$, which represent as many branches of the transition curve. The first-order approximation of the curve is therefore (Fig. 13.10a):

$$\delta = 1 \mp \epsilon. \tag{13.68}$$

The approximation can be improved by proceeding to the second order. However, one has to repeat the process *separately* for each of the two branches found, i.e.,

- on branch (i), Eq. 13.63b admits the solution[24] $q_1 = \frac{a}{8} \cos (3t)$; when this is introduced in Eq. 13.63c, one obtains:

$$\ddot{q}_2 + q_2 = -a \left(\delta_2 + \frac{1}{8} \right) \cos t + \text{NRT}, \tag{13.69}$$

where NRT denotes "nonresonant terms," i.e., harmonics of frequency other than 1. The periodicity condition, then, provides $\delta_2 = -\frac{1}{8}$;

---

[23] It should be remembered that:

$$\cos \alpha \, \cos \beta = \frac{1}{2} [\cos (\alpha - \beta) + \cos (\alpha + \beta)]$$

$$\sin \alpha \, \cos \beta = \frac{1}{2} [\sin (\alpha - \beta) + \sin (\alpha + \beta)]$$

$$\sin \alpha \sin \beta = \frac{1}{2} [\cos (\alpha - \beta) - \cos (\alpha + \beta)]$$

[24] The complementary solution is ignored, since it repeats the generating solution.

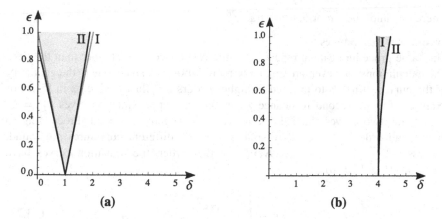

Fig. 13.10 Asymptotic approximation of the transition curves for the (undamped) Mathieu equation: (**a**) first resonance zone, (**b**) second resonance zone; (I) first-order approximation, (II) second-order approximation

- on branch (ii), Eq. 13.63b admits the solution $q_1 = \frac{b}{8} \sin(3t)$; when this is replaced in Eq. 13.63c, one gets:

$$\ddot{q}_2 + q_2 = -b\left(\delta_2 + \frac{1}{8}\right)\sin t + \text{NRT}, \tag{13.70}$$

The periodicity condition, then, still yields $\delta_2 = -\frac{1}{8}$.

Collecting all the results, the transition curves originating from $\delta = 1$, at the second-order approximation, are (Fig. 13.10a):[25]

$$\delta = 1 \mp \epsilon - \frac{1}{8}\epsilon^2. \tag{13.71}$$

The periodic motion on the two branches is expressed by:

$$q(t) = \begin{cases} a\left[\cos t + \epsilon\frac{1}{8}\cos(3t)\right], \\ b\left[\sin t + \epsilon\frac{1}{8}\sin(3t)\right], \end{cases} \tag{13.72}$$

---

[25] The transition curves in Fig. 13.10a, as well as in Fig. 13.10b, are extended up to $\epsilon = 1$, for which, in principle, the asymptotic approximation is no longer valid. In spite of this, it can be checked, by comparison with numerical or higher-order asymptotic solutions [3] that the approximation is still excellent for this value of ϵ.

where the amplitude, a or b, is arbitrary.[26]

Secondary Resonances

The same procedure can be repeated for the resonance zones higher than the first. The calculations, however, are gradually more laborious, since, due to the geometry of the curves, it needs to proceed to higher orders (as illustrated, e.g., in [5]). For example, for the second resonance zone, the first approximation gives $\delta_1 = 0$, that is, the two curves share the same vertical tangent at $\delta = 4$. Proceeding to the second order, however, they split, since two different curvatures are found, expressed by $\delta_2 = -\frac{1}{12}, \frac{5}{12}$. Overall, at this order, the transition curves have equations (Fig. 13.10b):

$$\delta = \begin{cases} 4 - \frac{1}{12}\epsilon^2, \\ 4 + \frac{5}{12}\epsilon^2. \end{cases} \tag{13.73}$$

13.5.3 Influence of Damping

By introducing a viscous damping in Mathieu Eq. 13.59, this latter becomes:

$$\ddot{q} + \zeta\,\dot{q} + (\delta + 2\epsilon \cos{(2t)})\,q = 0. \tag{13.74}$$

To determine the relevant transition curves, the same perturbation method, already used for the undamped system, is applied. If damping is small, it can be rescaled as $\zeta = \epsilon\hat{\zeta}$ in the first region or as $\zeta = \epsilon^2\hat{\zeta}$ in the second region, thus ordering it at the first non-trivial perturbation equation. The procedure is identical to that adopted in statics to take into account the imperfections (Sect. 4.6).

Principal Resonance

Limiting to consider the principal region of resonance, the first two perturbation Eq. 13.63a,b change as follows:

$$\epsilon^0: \quad \ddot{q}_0 + q_0 = 0, \tag{13.75a}$$

$$\epsilon^1: \quad \ddot{q}_1 + q_1 = -\delta_1 q_0 - 2q_0 \cos{(2t)} - \hat{\zeta}\dot{q}_0, \tag{13.75b}$$

---

[26] It is worth noticing that Eq. 13.72 *does not* represent the general solution of the motion on each of the two branches of the transition curve. Indeed, it depends on just one, and not two, arbitrary constants, as required by a differential equation of the second order. Actually, the coincidence of the characteristic multipliers at $\lambda = 1$ generates *also* an aperiodic motion, of the type $t \sin t$, or $t \cos t$, similarly to what happens for systems of differential equations with constant coefficients (as explained, e.g., in [7]). For generic initial conditions, therefore, the motion on the boundary of the instability zone diverges. This circumstance, however, does not have practical relevance, because the existence on the curve of a periodic solution, although particular, denotes the desired stability limit.

where $\delta_0 = 1$ has been taken. The solution to the ϵ^0 order equation is given by Eq. 13.65, which, after replacing in the ϵ^1 order equation, gives (Note 23 should be remembered):

$$\ddot{q}_1 + q_1 = -\left[a\,(1+\delta_1) + b\hat{\zeta}\right]\cos t + \left[b(1-\delta_1) + a\hat{\zeta}\right]\sin t$$
$$- a\cos(3t) - b\sin(3t). \tag{13.76}$$

Zeroing the resonant terms leads to:

$$\begin{bmatrix} 1+\delta_1 & \hat{\zeta} \\ \hat{\zeta} & 1-\delta_1 \end{bmatrix} \begin{pmatrix} a \\ b \end{pmatrix} = \begin{pmatrix} 0 \\ 0 \end{pmatrix}, \tag{13.77}$$

which generalizes Eq. 13.67, valid for the undamped system. The coupling terms (absent in the undamped case) entail that the transition curve is made up of *a single branch*, whose equation is obtained by vanishing the determinant of the matrix. From this equation the unknown $\delta_1 = \pm\sqrt{1-\hat{\zeta}^2}$ is drawn. By substituting it in the series, Eq. 13.62b, and coming back to the unrescaled physical parameter, $\zeta = \epsilon\hat{\zeta}$, one finally obtains:

$$\delta = 1 \pm \sqrt{\epsilon^2 - \zeta^2}. \tag{13.78}$$

The curve is plotted in Fig. 13.11a. It is noticed that damping has a *stabilizing effect*, producing an "uplift" of the instability region. There exists therefore a threshold of the intensity of the parametric excitation, $\epsilon_{min} := \zeta$, below which the trivial equilibrium is stable, even when $\delta \simeq 1$.

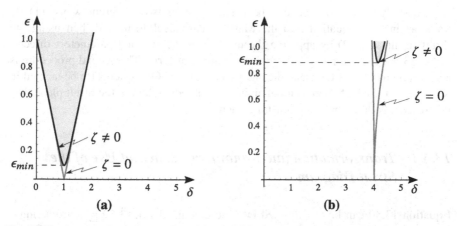

Fig. 13.11 Asymptotic approximation of the transition curves for the damped Mathieu equation: (a) first resonance zone, (b) second resonance zone; $\zeta = 0.1$

Secondary Resonances

Proceeding as illustrated above for the second resonance zone, the following transition curve is obtained [7]:

$$\delta = 4 + \frac{1}{6}\epsilon^2 \pm \sqrt{\frac{1}{16}\epsilon^4 - 4\zeta^2}, \tag{13.79}$$

represented in Fig. 13.11b. It appears that the threshold value, $\epsilon_{min}^2 := 8\zeta$, is significantly larger than that of the principal resonance, to prove that a small damping stabilizes secondary resonances to a greater extent.

13.6   Instability Regions of a Physical System: The Bolotin Beam

A physical system is considered as an example, namely, the simply supported beam, axially loaded by a pulsating load $P(t) = P_0 + P_1 \cos(\Omega t)$ (Bolotin beam), already introduced in Sect. 13.2 (Fig. 13.1e). By assimilating the beam to a single DOF system, i.e., by considering the response in its first mode only, the equation of motion, as deduced from Eq. 13.16, is:

$$\ddot{q} + \omega^2 (1 - \mu \cos(\Omega t)) q = 0, \tag{13.80}$$

where, remembering Eq. 13.17, it is:

$$\omega^2 := \left(\frac{\pi}{\ell}\right)^4 \frac{EI}{m} \left(1 - \frac{P_0}{P_c}\right), \qquad \mu := \left(\frac{\pi}{\ell}\right)^2 \frac{P_1}{m\omega^2}. \tag{13.81}$$

The study of the motion can be carried out in two different ways: (1) by transforming the equation into the Mathieu canonical form and then using the Strutt results and (2) by applying the method of the strained parameters directly to Eq. 13.80, without passing through the canonical form. The second procedure is a useful preparatory exercise to the analysis of multi-DOF systems (to be tackled in Sect. 13.8), for which there is no of an instrument equivalent to the Strutt plane. For this reason, both methods are illustrated here.

13.6.1   Transformation into Canonical Form and Use of the Strutt Diagram

Equation 13.80 can be transformed into the canonical Eq. 13.59 by introducing a new time τ, such that $\Omega t = 2\tau$. Since $\frac{d}{dt} = \frac{d}{d\tau}\frac{d\tau}{dt} = \frac{\Omega}{2}\frac{d}{d\tau}$, and $\frac{d^2}{dt^2} = \frac{\Omega^2}{4}\frac{d^2}{d\tau^2}$, Eq. 13.80 becomes:

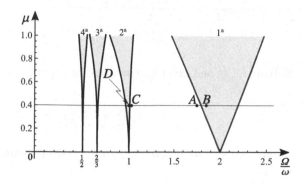

Fig. 13.12   Strutt diagram for the Bolotin beam, in terms of physical parameters

$$\frac{d^2 q}{d\tau^2} + \frac{4\omega^2}{\Omega^2}\left(1 - \mu \cos{(2\tau)}\right) q = 0. \tag{13.82}$$

Comparing it with Eq. 13.59, it follows that:[27]

$$\delta = 4\frac{\omega^2}{\Omega^2}, \qquad |\epsilon| = 2\frac{\omega^2}{\Omega^2}\mu. \tag{13.83}$$

The results of the Strutt diagram can therefore directly be applied. Since the regions of instability emanate from the points $\delta_0 = p^2$, these correspond, in terms of physical parameters, to:

$$\Omega = \frac{2\omega}{p}, \qquad p = 1, 2, \cdots, \tag{13.84}$$

that is, $\Omega = \left(2, 1, \frac{2}{3}, \frac{1}{2}, \frac{2}{5}, \cdots\right)\omega$. It happens that, unlike the Strutt diagram, the region of principal resonance is the rightmost one, so that all the secondary resonance regions are more and more to the left (Fig. 13.12).[28]

It is concluded that the *principal parametric excitation occurs when the excitation frequency Ω is (approximately) double the natural frequency ω;*[29] smaller excitation frequencies can only trigger secondary resonances; higher frequencies do not induce instability.

The principal resonance region, recalling Eq. 13.71 and the definitions in Eq. 13.83, is delimited by curves of the following equations:

---

[27] The sign of ϵ is irrelevant, as discussed in Note 17.

[28] The representation on the plane of the physical parameters, therefore, is less effective than that of Strutt, as the infinite regions of resonance are "squashed" near the origin of Ω.

[29] This result is also valid for the frequency of the excitation that instabilizes the swing, as anticipated in Sect. 13.2.

$$4\frac{\omega^2}{\Omega^2} = 1 \pm 2\frac{\omega^2}{\Omega^2}\mu - \frac{1}{2}\left(\frac{\omega^2}{\Omega^2}\right)^2\mu^2. \tag{13.85}$$

Solving for $\frac{\Omega}{\omega}$ and expanding in series for small μ, one gets:

$$\Omega = \omega\left(2 \mp \frac{1}{2}\mu - \frac{1}{32}\mu^2\right). \tag{13.86}$$

Similarly, by recalling Eq. 13.73, the second resonance region is bounded by the curves:

$$4\frac{\omega^2}{\Omega^2} = \begin{cases} 4 - \frac{1}{3}\left(\frac{\omega^2}{\Omega^2}\right)^2\mu^2, \\ 4 + \frac{5}{3}\left(\frac{\omega^2}{\Omega^2}\right)^2\mu^2. \end{cases} \tag{13.87}$$

Solving for $\frac{\Omega}{\omega}$ and expanding in series for small μ, one finds:

$$\Omega = \begin{cases} \omega\left(1 + \frac{1}{24}\mu^2\right), \\ \omega\left(1 - \frac{5}{24}\mu^2\right). \end{cases} \tag{13.88}$$

Figure 13.13 reports some time histories, resulting from the same initial conditions, obtained by numerical integration of Eq. 13.80. They relate to points close to the transition curves of the first and second resonance regions, marked in Fig. 13.12. The solutions are divergent in the unstable zones (at B and D points) and limited in amplitude in the stable zones (at A and C points). The period of the response is approximately double or equal to the period of the forcing $T = \frac{2\pi}{\Omega}$, respectively close to the first or second region.

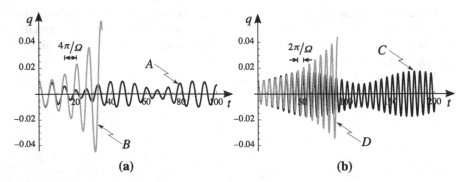

Fig. 13.13 Numerical responses of Eq. 13.80, relevant to the points marked in Fig. 13.12: (a) $\frac{\Omega}{\omega} = 1.75, 1.85$, (b) $\frac{\Omega}{\omega} = 1.02, 0.99$; $\mu = 0.4$ and $\omega = 1$

Additional examples of physical systems (namely, the pendula in Sect. 13.2), whose equations are recast in the canonical Mathieu form, are presented in Problems 14.7.3, 14.7.4.

13.6.2   Direct Construction of the Transition Curves

The transition curves can also be built up by directly tackling the original Eq. 13.80, without recurring to the Strutt diagram. As a first step, and following [7], a *straightforward series expansion* of the equation of motion is performed. It consists in (a) assuming μ as the perturbation parameter and (b) expanding in series only the Lagrangian coordinate, by leaving unchanged the parameter Ω. The straightforward expansion brings out the *singular values of* Ω, in correspondence of which the response tends to infinity. From these singular points on the Ω axis, the resonance regions emanate. In addition to providing a useful result, the procedure helps to understand the mechanism which generates the resonances at the different orders, as illustrated in Supplement 13.5.

The straightforward expansion, however, does not allow to determine the transition curves. For these, it needs to stretch the time axis and to expand also Ω in series of μ. The unknown coefficients of the Ω series permit to remove the singularities at the different orders. In this way, the law $\Omega = \Omega(\mu)$ is drawn, which describes the locus on which the periodic solutions exist in the (Ω, μ) parameter plane. The procedure is described in Supplement 13.6, only for the principal resonance zone.

Supplement 13.5 (Straightforward Expansion of Eq. 13.80) Taking μ as the perturbation parameter, the dependent variable $q(t; \mu)$ is expanded as:

$$q(t, \mu) = q_0(t) + \mu q_1(t) + \mu^2 q_2(t) + \cdots, \tag{13.89}$$

while Ω is left unchanged. The perturbation equations follow:

$$\mu^0: \quad \ddot{q}_0 + \omega^2 q_0 = 0, \tag{13.90a}$$

$$\mu^1: \quad \ddot{q}_1 + \omega^2 q_1 = \omega^2 q_0 \cos(\Omega t), \tag{13.90b}$$

$$\mu^2: \quad \ddot{q}_2 + \omega^2 q_2 = \omega^2 q_1 \cos(\Omega t), \tag{13.90c}$$

$$\cdots \tag{13.90d}$$

$$\mu^p: \quad \ddot{q}_p + \omega^2 q_p = \omega^2 q_{p-1} \cos(\Omega t). \tag{13.90e}$$

From the perturbation equation of order μ^0, $q_0 = a \cos(\omega t) + b \sin(\omega t)$ is drawn, i.e., q_0 contains the unique frequency ω. When this solution is substituted in the order μ equation, the product $q_0 \cos(\Omega t)$ generates the new frequencies $\Omega \pm \omega$. These are in resonance with the natural frequency ω if $\Omega \pm \omega = \pm \omega$ (independent signs), i.e., if:

$$\Omega = 2\omega. \tag{13.91}$$

In this circumstance, the series is singular, because q_1 diverges to infinity; the singularity point is the principal resonance point. However, if $\Omega \neq 2\omega$, q_1 is limited; taking the particular solution only, it contains the frequencies $\Omega \pm \omega$. When q_1 is, in turn, substituted in the μ^2 order equation, the term $q_1 \cos(\Omega t)$ generates the new frequencies $2\Omega \pm \omega$ and again ω. The latter one is due to a resonance which, not involving the frequency Ω, is not related to the parametric excitation (but it is a peculiarity of all the straightforward expansions). The frequencies $2\Omega \pm \omega$, instead, are resonant with ω, if $2\Omega \pm \omega = \pm\omega$, i.e., if:

$$\Omega = \omega. \tag{13.92}$$

This is a singularity caused by the parametric excitation, which denotes secondary resonance. If $\Omega \neq 2\omega$ and $\Omega \neq \omega$, the procedure can be carried out to the next orders, where one finds other resonances linked to Ω. In particular, by reasoning by induction, the resonance condition $p\Omega \pm \omega = \pm\omega$ is met at the order μ^p, or:

$$\Omega = \frac{2}{p}\omega, \qquad p = 1, 2, \cdots, \tag{13.93}$$

which, for $p = 1, 2$, includes the previous results. Equation 13.84 is thus recovered.

□

Supplement 13.6 (Strained Parameter Method Applied to Eq. 13.80) The transition curves for Eq. 13.80, limited to the principal zone, are now determined with the strained parameter method. The first step consists in "stretching" the time scale[30] in order to bring the period of the forcing to a known one. By letting, for example, $\tau := \Omega t$, one gets:[31]

$$\Omega^2 \ddot{q} + \omega^2 (1 - \mu \cos \tau) q = 0, \tag{13.94}$$

with $()^\bullet = \frac{d()}{d\tau}$. Taking μ as the perturbation parameter, the second step call for expanding in series *both* the dependent variable $q(\tau; \mu)$ and the Ω parameter (in order to establish the link $\Omega = \Omega(\mu)$):

$$q(\tau; \mu) = q_0(\tau) + \mu q_1(\tau) + \mu^2 q_2(\tau) + \cdots, \tag{13.95a}$$

$$\Omega(\mu) = \omega\left(2 + \mu\sigma_1 + \mu^2\sigma_2 + \cdots\right). \tag{13.95b}$$

---

[30] Hence, the denomination of the method of strained parameters, or coordinates

[31] Other authors put $\Omega t = 2\tau$, as done in this book for the Mathieu equation; the final result, of course, is independent of this choice.

Here, the series of Ω has a starting point 2ω (principal resonance) and $\sigma_1, \sigma_2, \cdots = O\,(1)$ are detuning parameters, measuring the deviation of the excitation frequency from the resonant value. The following perturbation equations are obtained:

$$\mu^0: \quad 4\ddot{q}_0 + q_0 = 0, \tag{13.96a}$$

$$\mu^1: \quad 4\ddot{q}_1 + q_1 = -4\sigma_1 \ddot{q}_0 + \cos\tau\, q_0, \tag{13.96b}$$

$$\mu^2: \quad 4\ddot{q}_2 + q_2 = -\left(4\sigma_2 + \sigma_1^2\right)\ddot{q}_0 - 4\sigma_1\ddot{q}_1 + \cos\tau\, q_1. \tag{13.96c}$$

The solution to the μ^0 order equation is $q_0 = a\cos\left(\frac{\tau}{2}\right) + b\sin\left(\frac{\tau}{2}\right)$, which, substituted in the equation of order μ^1, gives:

$$\begin{aligned}
4\ddot{q}_1 + q_1 = &\, a\left(\sigma_1 + \frac{1}{2}\right)\cos\left(\frac{\tau}{2}\right) + b\left(\sigma_1 - \frac{1}{2}\right)\sin\left(\frac{\tau}{2}\right) \\
&+ \frac{1}{2}a\cos\left(\frac{3}{2}\tau\right) + \frac{1}{2}b\sin\left(\frac{3}{2}\tau\right).
\end{aligned} \tag{13.97}$$

By eliminating the resonant terms, one gets $\sigma_1 = -\frac{1}{2}$, $b = 0$ or $\sigma_1 = \frac{1}{2}$, $a = 0$. By solving the previous equation, $q_1 = -\frac{a}{16}\cos\left(\frac{3}{2}\tau\right)$ or $q_1 = -\frac{b}{16}\sin\left(\frac{3}{2}\tau\right)$ are drawn, which, substituted in the next perturbation equation, lead to:

$$4\ddot{q}_2 + q_2 = a\left(\sigma_2 + \frac{1}{32}\right)\cos\left(\frac{\tau}{2}\right) + \text{NRT}, \tag{13.98}$$

or:

$$4\ddot{q}_2 + q_2 = b\left(\sigma_2 + \frac{1}{32}\right)\sin\left(\frac{\tau}{2}\right) + \text{NRT}. \tag{13.99}$$

Zeroing the resonant terms, one finds, in both cases, $\sigma_2 = -\frac{1}{32}$. With these results, Eq. 13.86 is recovered. $\qquad\qquad\square$

13.7 Nonlinear Single Degree of Freedom Systems: The Mathieu-Duffing Oscillator

When a single DOF system is in a zone of instability, the linear Floquet theory predicts a response diverging over time. The result is obviously fallacious, since, as the amplitude of the motion increases, the nonlinearities gradually become relevant, so that the linear model loses its validity. It is therefore important to determine whether, in the unstable zone, there exist stable motions (possibly of large but finite amplitude), able to limit the response. As it will be seen shortly, such motions do exist and are periodic, so that the system, once the unstable equilibrium position has been left, moves, after a transient, to a stable *limit cycle*, in an analogous manner to what happens for self-excited autonomous systems (e.g., loaded by follower forces, in Chap. 11, or velocity-dependent forces, in Chap. 12). However, it happens that such limit cycles, which originate from the boundaries of the instability regions, *extend their existence outside the unstable zone*, thus affecting the dynamics also in the stable area. Here, the phenomenon is illustrated with reference to a sample system in principal resonance. Attention is paid to the description of the analytical tools necessary to perform the nonlinear analysis [7].

Mathieu-Duffing Oscillator as Paradigmatic System
A *nonlinear* single DOF system is considered, damped and parametrically excited in stiffness. Assuming that the nonlinearity is cubic, the equation of motion reads:

$$\ddot{q} + 2\xi\omega\dot{q} + \omega^2\left(1 - \mu\cos\left(\Omega t\right)\right)q + \kappa\omega^2 q^3 = 0, \tag{13.100}$$

where ξ is the damping ratio and κ is the stiffness ratio.[32]

This equation, called the *Mathieu-Duffing damped equation*, models, for example, the Bolotin beam (already described by Eq. 13.80), in which damping and nonlinear curvature effects have been added. In this specific case, since the behavior of the Euler-Bernoulli beam is *hardening*,[33] it turns out that $\kappa > 0$; however, to account also for systems with softening behavior, it is assumed that it may also be $\kappa < 0$.

13.7.1   Principal Resonance

Periodic solutions to Eq. 13.100 are sought, close to the principal resonance $\Omega = 2\omega$. The perturbation method of strained parameters is employed, as done in Supplement 13.6 for the linear undamped version of the system.

By straining the independent variable according to $\tau = \Omega t$, the equation becomes:

$$\Omega^2\ddot{q} + 2\xi\omega\Omega\dot{q} + \omega^2\left(1 - \mu\cos\tau\right)q + \kappa\omega^2 q^3 = 0, \tag{13.101}$$

with $()^{\bullet} = \frac{d()}{d\tau}$. Since the equation is nonlinear, the dependent variable must be rescaled, to make $q^3 \ll q$. Moreover, it is convenient to order the parametric

---

[32] The constitutive law of a cubic spring is $f = -\left(k_1 q + k_3 q^3\right) = -k_1\left(q + \kappa q^3\right)$, where $\kappa := \frac{k_3}{k_1}$ is a parameter of physical dimensions $\left[L^{-2}\right]$.

[33] Indeed, as seen in Sect. 5.4, the beam exhibits super-critical fork bifurcations.

excitation at the same level of the nonlinearity, by requiring that $O\,(\mu q) = O\,(q^3)$, i.e., $q = O\,(\sqrt{\mu})$; accordingly, the rescaling $q = \sqrt{\mu}\hat{q}$ is introduced.[34] By ordering the viscous force at the same level, damping is rescaled as $\xi = \mu\hat{\xi}$. Instead, the coefficient of nonlinearity κ is taken of order 1 (since the "smallness" of the force is already expressed by q^3). With these transformations, after division by $\sqrt{\mu}\omega^2$, the rescaled equation becomes:

$$\left(\frac{\Omega}{\omega}\right)^2 \ddot{q} + \omega^2 q + \mu \left(2\xi\frac{\Omega}{\omega}\dot{q} + \cos\tau\,q + \kappa q^3\right) = 0, \qquad (13.102)$$

where the hat on q has been omitted. Introduced the series expansions:

$$q = q_0 + \mu q_1 + \cdots, \qquad (13.103a)$$
$$\Omega = \omega\,(2 + \mu\sigma_1 + \cdots), \qquad (13.103b)$$

the perturbation equations follow:

$$\mu^0:\quad 4\ddot{q}_0 + q_0 = 0, \qquad (13.104a)$$
$$\mu^1:\quad 4\ddot{q}_1 + q_1 = -4\sigma_1\ddot{q}_0 - 4\hat{\xi}\dot{q}_0 - \cos\tau\,q_0 - \kappa\,q_0^3, \qquad (13.104b)$$

to be supplemented by the periodicity conditions.

The solution at the μ^0 order is:

$$q_0 = a\cos\left(\frac{\tau}{2}\right) + b\sin\left(\frac{\tau}{2}\right), \qquad (13.105)$$

or, equivalently:

$$q_0 = A\cos\left(\frac{\tau}{2} + \theta\right), \qquad (13.106)$$

where $A := \sqrt{a^2 + b^2}$ is the *amplitude* of the response and $\theta = -\arctan\left(\frac{b}{a}\right)$ is the phase. When Eq. 13.105 is substituted in the next perturbation equation, and use is made of trigonometric transformations, one gets:

$$4\ddot{q}_1 + q = \left\{\frac{1}{4}a\left[-3\kappa\left(a^2 + b^2\right) + 2 + 4\sigma_1\right] - 2b\hat{\xi}\right\}\cos\left(\frac{\tau}{2}\right)$$
$$+ \left\{\frac{1}{4}b\left[-3\kappa\left(a^2 + b^2\right) - 2 + 4\sigma_1\right] + 2a\hat{\xi}\right\}\sin\left(\frac{\tau}{2}\right) \qquad (13.107)$$
$$+ NRT.$$

---

[34] This rescaling can alternatively be incorporated into Eq. 13.103a, as done for static systems in Chap. 4.

The condition of periodicity (or the removal of the secular terms) requires that:

$$a\left[-3\kappa\left(a^2+b^2\right)+2+4\sigma_1\right]-8b\hat{\xi}=0,\qquad(13.108a)$$

$$b\left[-3\kappa\left(a^2+b^2\right)-2+4\sigma_1\right]+8a\hat{\xi}=0.\qquad(13.108b)$$

This is a nonlinear algebraic system of *two bifurcation equations* into the two unknowns a, b.

Before studying Eq. 13.108, it is convenient to analyze the undamped case.

13.7.2 Undamped System

In absence of damping, Eqs. 13.108 can *formally* be written as a linear system of equations, whose coefficients depend on the unknowns:

$$\begin{bmatrix} -3\kappa\left(a^2+b^2\right)+2+4\sigma_1 & 0 \\ 0 & -3\kappa\left(a^2+b^2\right)-2+4\sigma_1 \end{bmatrix}\begin{pmatrix} a \\ b \end{pmatrix}=\begin{pmatrix} 0 \\ 0 \end{pmatrix}.$$
$$(13.109)$$

Vanishing the determinant of the matrix, a relationship follows between σ_1 and the amplitude $A := \sqrt{a^2+b^2}$, i.e.,

$$\left(4\sigma_1-3\kappa A^2\right)^2-4=0,\qquad(13.110)$$

which, solved with respect to σ_1, reads:

$$\sigma_1=\frac{3}{4}\kappa A^2\mp\frac{1}{2}.\qquad(13.111)$$

Multiplying both members of this equation by μ, and taking into account that $\mu\sigma_1 = \frac{\Omega}{\omega}-2$, and $\sqrt{\mu}A \to A$, one returns to the original variables:[35]

$$\Omega=\omega\left(2\mp\frac{\mu}{2}+\frac{3}{4}\kappa A^2\right).\qquad(13.112)$$

This equation provides, for each point (Ω, μ) of the bifurcation parameter plane, the finite amplitude A of the periodic solution existing there. From a geometric point of view, the equation represents a two-branch surface, $A = A^{\pm}(\Omega, \mu)$, which intersects the plane of the bifurcation parameters at the transition curves.[36] The

---

[35] Indeed A, being the amplitude of $\hat{q}$, suffers from the same rescaling of this latter.

[36] Equation 13.86, linearized in μ, should be remembered.

two branches collapse on the same parabola $\frac{\Omega}{\omega} = 2 + \frac{3}{4}\kappa A^2$ belonging to the plane $\mu = 0$. The transition curves are therefore loci of incipient bifurcation, from which a family of periodic solutions emanates, of amplitude vanishingly small on the locus, but increasing with the distance from it. The surface is represented in Fig. 13.14 for the two cases $\kappa > 0$ (hardening system) and $\kappa < 0$ (softening system). In addition to the three-dimensional plot, a section $\mu = $ const is represented on the (Ω, A) plane and a section $\Omega = $ const in the plane (μ, A). The $\mu = $ const section is commented first. It is seen that the two branches bend to the right (hardening behavior) or to the left (softening behavior). It is possible to prove that the lower branch A^- is unstable, while the upper branch A^+ is stable.[37] Both branches, as anticipated, extend themselves to the stable zone. With reference to the hardening case, the following behavior is observed:

- in the unstable zone, the periodic solution A^+ limits the amplitude of the response;
- in the stable zone, to the left of the resonance, no periodic solutions exist, so that the stable equilibrium position $A = 0$ does not suffer from the presence of coexisting steady states;[38]
- in the stable region, to the right of the resonance, there exist periodic motions, one unstable, A^-, and one stable, A^+, in addition to the stable equilibrium position $A = 0$. For small perturbations, the equilibrium $A = 0$ is attractive; for large disturbances, however, the state is attracted by A^+, at which points the system performs periodic oscillations of large amplitude, albeit the trivial equilibrium is stable.

As a consequence of the previous results, if the frequency of the parametric excitation is quasi-statically increased from a value to the left of the unstable zone, once the system instabilizes, it runs along the curve A^+ *also to the right of the unstable zone*. Therefore, the recovery of stability of the trivial equilibrium position does not imply any reduction in the amplitude of oscillation. If, on the other hand, the frequency of the parametric excitation is quasi-statically decreased from a value to the right of the instability zone, when the bifurcation occurs, the system experiences a *jump*, which takes it instantly (actually, after a transient) on the curve A^+; this phenomenon is called *hard loss of stability*, as relating to a sudden change of state. Subsequently, the system runs through A^+, until returning, after exiting

---

[37] The proof requires the study of the variational equation:

$$\Omega^2 \delta\ddot{q} + \left[\omega^2 \left(1 - \mu\cos\tau\right) + 3\kappa\omega^2 q_0^2\left(\tau\right)\right]\delta q = 0$$

obtained by letting $q = q_0(\tau) + \delta q$ in Eq. 13.101, where q_0 is given by Eq. 13.105 and δq is a disturbance. Since $q_0(\tau)$ is periodic of period 4π, $q_0^2(\tau)$ is periodic of period 2π, for which the variational equation is a Hill equation, of period 2π. Its characteristic multipliers decide stability.

[38] Only the neighborhood of the principal resonance zone is being explored, consistently with the perturbation solution. Hypothetically, could there exist other solutions originating from the secondary resonance tongue, to the left of the principal one, not examined here.

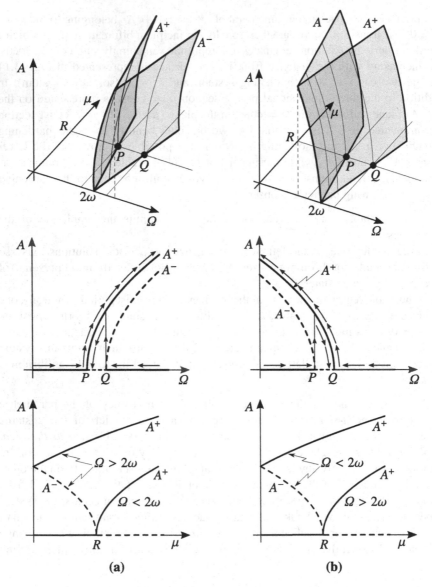

Fig. 13.14 Bifurcation diagram of the undamped Mathieu-Duffing equation: (**a**) hardening system ($\kappa > 0$); (**b**) softening system ($\kappa < 0$); 3D views, sections at $\mu =$ cost and $\Omega =$ cost

the resonance, to the equilibrium point $A = 0$. Similar considerations apply to the softening case.

A different behavior is observed in the $\Omega =$ const plane, where the cases $\Omega < 2\omega$ and $\Omega > 2\omega$ must be distinguished. Still referring to the hardening case (Fig. 13.14a), it is seen that , when $\Omega > 2\omega$, the lower (unstable) branch A^- and the upper (stable) branch A^+ are both crossed. However, they match each other at a cusp point when $\mu = 0$. Also in this case a hard loss of stability occurs, when μ is increased from zero at $\Omega = $ const. When, instead, $\Omega < 2\omega$, only the stable branch A^+ is crossed. In the softening case (Fig. 13.14b), a similar behavior manifests itself, but with the two cases $\Omega \gtrless 2\omega$ exchanged.

13.7.3   Damped System

In presence of damping, the bifurcation Eq. 13.108 can *formally* be written, again, as two linear equations whose coefficients depend on the unknowns, i.e.,

$$
\begin{bmatrix} -3\kappa \left(a^2 + b^2\right) + 2 + 4\sigma_1 & -8\hat{\xi} \\ 8\hat{\xi} & -3\kappa \left(a^2 + b^2\right) - 2 + 4\sigma_1 \end{bmatrix} \begin{pmatrix} a \\ b \end{pmatrix} = \begin{pmatrix} 0 \\ 0 \end{pmatrix}.
$$
(13.113)

Zeroing the determinant of the matrix, and letting $A = \sqrt{a^2 + b^2}$, one gets:

$$
\left(4\sigma_1 - 3\kappa A^2\right)^2 - 4 + 64\hat{\xi}^2 = 0,
$$
(13.114)

from which:

$$
\sigma_1 = \frac{1}{4}\left(3\kappa A^2 \mp 2\sqrt{1 - 16\hat{\xi}^2}\right).
$$
(13.115)

Returning to the original variables, the following result is found:

$$
\Omega = \omega\left(2 \mp \frac{1}{2}\sqrt{\mu^2 - 16\xi^2} + \frac{3}{4}\kappa A^2\right).
$$
(13.116)

This is the equation of a surface in the (A, Ω, μ) space, whose existence requires $\mu > \mu_{min} := 4\xi$.[39] The surface intersects the plane of the bifurcation parameters at the transition curve of the damped linear system; moreover it is tangent to the plane $\mu = \mu_{min}$, with which it shares the parabola $\frac{\Omega}{\omega} = 2 + \frac{3}{4}\kappa A^2$ (Fig. 13.15). Ultimately, not only the system is stable when $\mu < \mu_{min}$ (as predicted by the linear analysis), but there are no limit cycles in this range (as determined by the nonlinear analysis). When $\mu > \mu_{min}$, the behavior of the damped system is analogous to that of the undamped system, described by the 2D diagrams in Fig. 13.14.

   As the unique qualitative difference, the cusp point appearing in the $\Omega = $ const diagram transforms into a regular point with vertical tangent and occurs no more at $\mu = 0$, but at $\mu = \mu_{min}$.

---

[39] This result is consistent with that found in Sect. 13.5.3, where $\zeta = 2\xi\omega$.

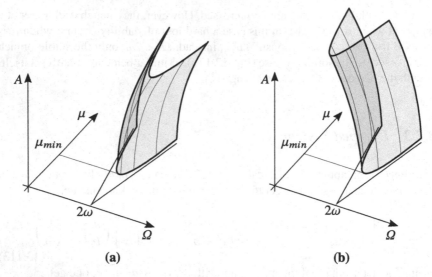

Fig. 13.15 Bifurcation diagram of the damped Mathieu-Duffing equation: (**a**) hardening system ($\kappa > 0$); (**b**) softening system ($\kappa < 0$)

13.8   Linear Systems with Multiple Degrees of Freedom

Undamped *linear* systems with multiple DOF, when harmonically excited in the their stiffness, are governed by fully coupled Mathieu equations. By expressing the motion in the modal basis of the unexcited system, it is possible to uncouple the part of the equation which is independent of time (not that depending on time), thus obtaining:

$$\ddot{q}_k + \omega_k^2 q_k - \mu \cos{(\Omega t)} \sum_{j=1}^{n} h_{kj} q_j = 0, \qquad k = 1, 2, \cdots, n, \qquad (13.117)$$

where, generally, $h_{kj} \neq h_{jk}$. Equation 13.20, relevant to the fixed-free Bolotin beam (where, however, $h_{kj} = h_{jk}$), is an example of this class of problems.

Mechanical systems governed by Eq. 13.117 exhibit different instability zones in the (Ω, μ) parameter plane. Some of these, relevant to flip and divergence bifurcations, replicate those already encountered in single DOF systems; a further type of bifurcation, said of Neimark-Sacker, which is peculiar to multiple DOF systems, is discussed below.

13.8.1 Flip and Divergence Bifurcations

A special case is first analyzed, in which the natural frequencies ω_k are *incommeasurable*, i.e., they are in irrational ratios $\frac{\omega_r}{\omega_s} \neq \frac{p}{q}$, $\forall r, s$, with p, q integers. If $\Omega = 2\omega_r$, the excitation is in principal resonance with the rth mode (which undergoes a flip bifurcation). The system essentially responds as a single DOF system in this mode, with higher-order contributions of the nonresonant modes. The same happens for resonances of higher order (alternatively of divergence or flip type), if $\Omega = \omega_r, \frac{2}{3}\omega_r, \frac{1}{2}\omega_r, \cdots$. As Ω varies, however, all modes, separately, can be excited. Thus, for example, in a two DOF system, there are two principal resonances, $\Omega = 2\omega_1, 2\omega_2$, two second-order resonances, $\Omega = \omega_1, \omega_2$, two third-order resonances $\Omega = \frac{2}{3}\omega_1, \frac{2}{3}\omega_2$, and so on. The zones originating from these frequencies (Fig. 13.16a), given the incommensurability of the natural frequencies, are separate from each other, at least for small values of μ; however, they merge for larger values of μ, resulting in more complex interactions [7].

The number of zones of instability becomes very large as the number of DOF of the system increases; on the other hand, the regions relevant to the higher

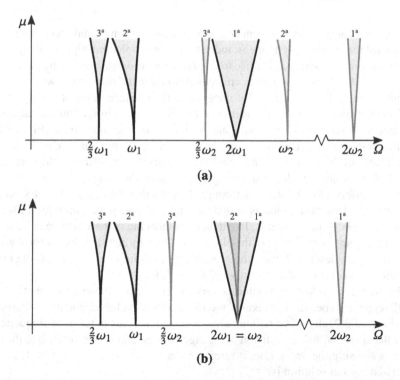

Fig. 13.16 Qualitative resonance zones for a system with two DOF: (**a**) incommensurable frequencies ($\omega_2 = 2.54\,\omega_1$ in the example), (**b**) commensurable frequencies ($\omega_2 = 2\omega_1$ in the example), but with uncoupled equations; transition curves: black 1st mode, gray 2nd mode

modes are progressively narrower, and this is more true in presence of damping, which generally increases with the mode number. Therefore, if one is interested in excitations of small amplitude, only low-order resonances are relevant.

Another simple case to investigate is that in which the natural frequencies ω_r are commensurable, but the equations are uncoupled (i.e., $h_{kj} = 0$ when $k \neq j$). This circumstance occurs, for example, for the simply supported Bolotin beam (Eq. 13.16), in absence of static force P_0, for which the frequencies are multiple of the fundamental one, $\frac{\omega_r}{\omega_1} = r^2$, or for a taut string subject to pulsating tension, for which $\frac{\omega_r}{\omega_1} = r$. In these systems, the *zones of instability can overlap*. Indeed, if, for example, it is $\omega_2 = 2\omega_1$, the principal resonance zone relative to the first mode, $\Omega = 2\omega_1$, originates from the same abscissa of the secondary resonance zone relative to the second mode, $\Omega = \omega_2$ (Fig. 13.16b). However, since the former is wider than the latter, only the principal resonance is significant, in the region of which the first mode is everywhere unstable. Close to the perfect resonance, also the second mode becomes unstable.

13.8.2 Neimark-Sacker Bifurcation

The principal and secondary resonance zones related to each modal coordinate q_k do not exhaust the casuistry. It happens, indeed, that, in coupled multiple DOF systems, few modes can combine, to give rise to *combination resonances* with the parametric excitation [7, 9]. These resonances create *additional instability zones*, which are not present in single DOF systems. To understand their nature, a two DOF system is considered, which, for small intensities of the excitation, admits four characteristic multipliers, in pairs complex conjugate, which lie on the unit circle (Fig. 13.17). As the solicitation increases, the multipliers move along the circle, and, at a critical value of the parameter, they coalesce in pairs. Upon collision, they separate, two of them going inside, and two going outside the circle. The equilibrium, therefore, suffers a loss of stability through a mechanism known as *Neimark-Sacker bifurcation*.[40] This bifurcation is similar to that of Hopf, which manifests itself in autonomous circulatory systems; like that, it requires the system to have at least two DOF. If these are more than two, the Neimark-Sacker bifurcation occurs through the mechanism just described, with the remaining multipliers remaining on the circle and being distinct (i.e., in marginally stable condition).

The Neimark-Sacker bifurcation is very different from those so far met, of flip and divergence type. In this regard, the discussion developed about the Poincaré map (Sect. 13.3.2) should be remembered, here summarized in terms of frequencies, rather than periods. At the collision of the eigenvalues, and with reference to the pair which is crossing the unit circle, it turns out that $\lambda_k = \exp(\pm i\varphi_k)$ ($k = 1, 2$); the associated normal solution is:

---

[40] Also called *secondary Hopf bifurcation*, for the reasons discussed ahead in Supplement 13.7.

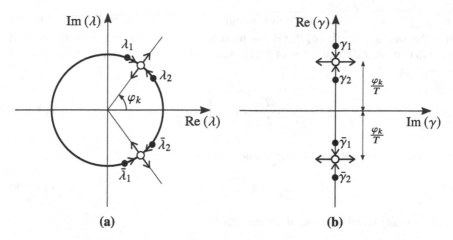

Fig. 13.17 Neimark-Sacker bifurcation: (**a**) characteristic multipliers, (**b**) characteristic exponents

- periodic, of frequency Ω (or multiples $\Omega_r = p\Omega$, $p = 1, 2, \cdots$), when $\varphi_k = 0$, i.e., at the divergence bifurcation;
- periodic of frequency $\frac{\Omega}{2}$ (or multiples $\Omega_r = p\frac{\Omega}{2}$, $p = 1, 3, 5, \cdots$), when $\varphi_k = \pi$, i.e., at the flip bifurcation;
- still periodic, of frequency $\frac{\Omega}{q}$ (or multiples $\Omega_r = \frac{p}{q}\frac{\Omega}{2}$, $p = 1, 2, \cdots$), when $\varphi_k = \varphi_k^* := \frac{2\pi}{q}$, $q = 3, 4, \cdots$ (*resonant Neimark-Sacker bifurcation*);
- *quasi-periodic*, when $\varphi_k \neq (0, \pi)$ and also $\varphi_k \neq \varphi_k^*$ (*nonresonant* or generic *Neimark-Sacker bifurcation*).

The geometric locus which, on the plane (Ω, μ), describes the states of incipient (generic) Neimark-Sacker bifurcation, and which separates the stable states from the unstable ones, is therefore constituted from *transition curves on which the system experiences quasi-periodic motions* (unlike the divergence and flip loci, where the motion is periodic). The algorithms for the determination of the stability boundaries are based on this property.

Origin of the Stability Boundaries

Quasi-periodic motions, according to the Floquet theorem, arise because the normal solution (Eq. 13.25) has two frequencies: (a) the frequency of the periodic function $\phi_k(t)$, equal to Ω (or its multiples $p\Omega$), and (b) the frequency of a characteristic exponent $\gamma_k = \frac{1}{T}\ln\lambda_k = \frac{1}{T}i\varphi_k := i\alpha_k$, with $\alpha_k := \Omega\frac{\varphi_k}{2\pi}$ and $\varphi_k \in (0, \pi)$. As already said, when $\varphi_k = 0$, the motion is periodic of frequency $\Omega_r = p\Omega$, $(p = 1, 2, \cdots)$; when $\varphi_k = \pi$, the motion is periodic of frequency $\Omega_r = p\frac{\Omega}{2}$, $(p = 1, 3, \cdots)$. Therefore, in order to put in evidence the modulating frequency, it is convenient to measure the angle φ_k from 0, when $0 < \varphi_k \leq \frac{\pi}{2}$, and from π, when $\frac{\pi}{2} < \varphi_k < \pi$. Accordingly, it is $\alpha_k := \Omega\frac{\varphi_k}{2\pi}$ or $\alpha_k := \Omega\frac{\pi - \varphi_k}{2\pi}$, in the two cases, respectively. The frequency of the periodic motion,

$p\frac{\Omega}{2}$, $(p = 1, 2, 3, \cdots)$ and that of modulation, α_k, combine and give rise to two new frequencies, $\pm p\frac{\Omega}{2} + \alpha_k$. Since, as the amplitude tends to zero, the solution must tend to two natural frequencies of the unexcited system, for example, $\pm\omega_j$, $\pm\omega_h$, it must hold:

$$\pm p\frac{\Omega}{2} + \alpha_k = \begin{cases} \omega_j, \\ \mp\omega_h. \end{cases} \tag{13.118}$$

By subtracting member by member, one obtains:

$$p\Omega = |\omega_j \pm \omega_h|, \qquad p = 1, 2, \cdots, \tag{13.119}$$

and summing member to member, one gets:[41]

$$\alpha_k = \frac{1}{2}|\omega_j \mp \omega_h|. \tag{13.120}$$

Equation 13.119 determines the points of the Ω axis from which the combination resonance regions (sum or difference) emanate. It generalizes Eq. 13.84, already determined for single DOF systems; Eq. 13.120 describes the modulation frequency.

An alternative procedure for the determination of Eq. 13.119, based on a perturbation method, is illustrated in Sect. 13.8.3.

Remark 13.11 In a single DOF system, in order for the solution to be monofrequent, the frequency of the parametric excitation Ω and the natural frequency ω must be rationally dependent (e.g., $\Omega = 2\omega$, ω, $\frac{2}{3}\omega$). In a two DOF system, for the solution to be bi-frequent, the three frequencies $(\Omega, \omega_1, \omega_2)$ must be rationally dependent, according to Eq. 13.119.

Bifurcation of a Torus from an Equilibrium Position
The Neimark-Sacker bifurcation admits an interesting geometric interpretation. As it was commented in Supplement 13.4, while discussing quasi-periodic motions, the bifurcation establishes that the *motion develops on an invariant manifold*[42] of the state-space, *taking the shape of a torus*.[43] If the Neimark-Sacker bifurcation is generic, the trajectories are open and densely fill the torus (quasi-periodic motions).

---

[41] The absolute value is used to make expressions independent of the relative magnitude of the two natural frequencies.

[42] The invariant manifold is a hyper-surface of the state-space, such that if the system is placed in one of its points (as an initial condition), it travels indefinitely on this surface. The trajectories, therefore, belong to the manifold.

[43] The linear analysis conducted here, however, only allows to ascertain the birth of the torus. To observe its evolution *vs* a bifurcation parameters, and in order to identify a *family of tori*, it should be necessary to carry out a nonlinear analysis, which is beyond the limits of this book.

If the bifurcation is resonant, the trajectories are closed, to replicate themselves over time (periodic motion).

Remark 13.12 The torus assumes in the Neimark-Sacker bifurcation the same role that the limit cycle has in the Hopf bifurcation. The limit cycle is visited periodically, the torus by bi-periodic motions. Both the limit cycle and the torus belong to a family, parameterized by a bifurcation parameter.

Supplement 13.7 (Secondary Hopf Bifurcation) The Neimark-Sacker bifurcation is also known as *secondary Hopf bifurcation* for the reason discussed soon [6]. In the *non-autonomous* system under study, the bifurcation gives rise to quasi-periodic solutions stemming from equilibrium. It, however, can also occur in *autonomous systems*, where the bifurcation creates quasi-periodic solutions from a cycle limit.[44] Accordingly, a limit cycle arising from an equilibrium position after a Hopf bifurcation can in turn post-bifurcate into a quasi-periodic motion after a "secondary" Hopf bifurcation. As a matter of fact, *the variational equation governing the stability of a limit cycle takes the same form as the equations governing the parametric excitation.* To verify this, an autonomous dynamical system is considered, of equation $\dot{\mathbf{x}} = \mathbf{F}(\mathbf{x})$. Once a limit cycle $\mathbf{x}_0(t) = \mathbf{x}_0(t + T)$ of known period T has been determined, its stability is analyzed. By letting $\mathbf{x} = \mathbf{x}_0 + \delta\mathbf{x}$, linearizing in the $\delta\mathbf{x}$ perturbation and expanding the right member, it is $\mathbf{F}(\mathbf{x}_0 + \delta\mathbf{x}) = \mathbf{F}(\mathbf{x}_0) + \mathbf{F}_{,\mathbf{x}}(\mathbf{x}_0)\,\delta\mathbf{x}$. Given that $\dot{\mathbf{x}}_0 = \mathbf{F}(\mathbf{x}_0)$, the variational equation $\delta\dot{\mathbf{x}} = \mathbf{F}_{,\mathbf{x}}(\mathbf{x}_0(t))\,\delta\mathbf{x}$ is obtained. This is a linear system with T-periodic coefficients (formally identical to that describing a parametric excitation), whose solution is governed by Floquet theorem. If a Neimark-Sacker bifurcation occurs, the perturbation $\delta\mathbf{x}$ is bi-periodic; hence such is the perturbed motion $\mathbf{x}_0 + \delta\mathbf{x}$. $\square$

13.8.3 Evaluation of the Combination Resonances by Straightforward Expansions

Equation 13.119, provided by the Floquet theory, establishes a relationship among the frequencies $(\Omega, \omega_1, \omega_2)$ involved in the combination resonance. It can also be determined through a straightforward expansion of the equations of motion, similarly to what was done in Sect. 13.6.2 for a system with a single DOF. Supplement 13.8 describes the procedure.

As already said in Sect. 13.6.2, the straightforward expansion does not allow to identify the transition curves, separating the stable from the unstable states, but only the points of the Ω axis from which they emanate. To determine the curves, the method of the strained parameters must be used, which generalizes the procedure illustrated in Supplement 13.6 for a single DOF. The method is illustrated in detail in

---

[44] Circumstances not tackled in this book.

the Sect. 13.8.4 for a two DOF system, mainly referring to the principal combination resonance.

Supplement 13.8 (Straightforward Expansion of Eq. 13.117) Introduced the series:

$$q_k(t; \mu) = q_{k0}(t) + \mu q_{k1}(t) + \mu^2 q_{k2}(t) + \cdots, \qquad k = 1, 2, \cdots, n, \qquad (13.121)$$

in Eq. 13.117, the following perturbation equations are derived:

$$\mu^0: \quad \ddot{q}_{k0} + \omega_k^2 q_{k0} = 0, \tag{13.122a}$$

$$\mu^1: \quad \ddot{q}_{k1} + \omega_k^2 q_{k1} = \sum_{j=1}^{n} h_{kj} q_{j0} \cos(\Omega t), \tag{13.122b}$$

$$\mu^2: \quad \ddot{q}_{k2} + \omega_k^2 q_{k2} = \sum_{j=1}^{n} h_{kj} q_{j1} \cos(\Omega t), \tag{13.122c}$$

$$\cdots \tag{13.122d}$$

$$\mu^p: \quad \ddot{q}_{kp} + \omega_k^2 q_{kp} = \sum_{j=1}^{n} h_{kj} q_{j,p-1} \cos(\Omega t). \tag{13.122e}$$

with $k = 1, 2, \cdots, n$. The solution to the μ^0 order equations is $q_{k0} = a_k \cos(\omega_k t) + b_k \sin(\omega_k t)$, with a_k, b_k arbitrary constants. Substituting it in the perturbation equations of order μ, the parametric excitation terms generate, in the kth equation, the frequencies $\Omega \pm \omega_j$, $j = 1, 2, \cdots, n$. If $\Omega \pm \omega_j = \pm \omega_k$ (independent signs), the harmonic forcing is in resonance with the natural frequency ω_k. There are therefore two forms of resonance:

$$\Omega = \begin{cases} 2\omega_k, & j = k, \\ |\omega_j \pm \omega_k|, & j \neq k, \end{cases} \tag{13.123}$$

said of the *first order*, or principal ones. The first of these has been already encountered in single DOF systems and involves only the coordinate q_k; the second, in contrast, affects two coordinates, q_j and q_k. Indeed, it happens that if, for example, $\Omega + \omega_j = \omega_k$, it is also $\Omega - \omega_k = -\omega_j$, so that the excitation frequency $\Omega - \omega_k$ in the jth equation is in resonance with the natural frequency ω_j. Said in others terms, the resonance manifests itself *simultaneously* in two equations. This type of resonance, which involves the three frequencies $(\Omega, \omega_j, \omega_k)$, is called *combination resonance of the first order* (of sum or difference type).

When the perturbation equations of order μ are solved, q_{k1} contains the frequencies $\Omega \pm \omega_j$. Substituting it in the equations of order μ^2, the new frequencies $2\Omega \pm \omega_j$ are generated; when these are equal $\pm \omega_k$, the following resonances occur:

$$\Omega = \begin{cases} \omega_k, & j = k, \\ \frac{1}{2}\left|\omega_j \pm \omega_k\right|, & j \neq k, \end{cases} \tag{13.124}$$

called of the *second order*, or secondary. Again, the first of this has already been encountered in single DOF systems and affects only the q_k coordinate. The second, on the other hand, is peculiar of multiple DOF systems and involves the two coordinates q_j and q_k; it is called *combination resonance of the second order*.

Proceeding to the successive orders, and reasoning by induction, the resonance conditions manifested at the order μ^p are $p\Omega \pm \omega_j = \pm\omega_k$, or:

$$\Omega = \begin{cases} \frac{2}{p}\omega_k, & j = k, \\ \frac{1}{p}\left|\omega_j \pm \omega_k\right|, & j \neq k. \end{cases} \tag{13.125}$$

The first of these coincides with Eq. 13.84; the second is Eq. 13.119, which describes the *combination resonance of order p*.   □

13.8.4   Combination Resonance and Transition Curves in a Two Degree of Freedom System

It will be shown, by an example, how to compute the principal combination resonance zones of a two DOF system, by employing the strained parameter perturbation method. Equation 13.117 specializes into:

$$\ddot{q}_1 + \omega_1^2 q_1 - \mu\left(h_{11}q_1 + h_{12}q_2\right)\cos\left(\Omega t\right) = 0, \tag{13.126a}$$

$$\ddot{q}_2 + \omega_2^2 q_1 - \mu\left(h_{21}q_1 + h_{22}q_2\right)\cos\left(\Omega t\right) = 0. \tag{13.126b}$$

Proceeding as for a single DOF system (Sect. 13.6, Supplement 13.6), the time scale is stretched to make it independent of the Ω parameter. Here the nondimensional time $\tau := \Omega t$ is chosen,[45] which transforms the equations into:

$$\Omega^2\ddot{q}_1 + \omega_1^2 q_1 - \mu\left(h_{11}q_1 + h_{12}q_2\right)\cos\tau = 0, \tag{13.127a}$$

$$\Omega^2\ddot{q}_2 + \omega_2^2 q_2 - \mu\left(h_{21}q_1 + h_{22}q_2\right)\cos\tau = 0, \tag{13.127b}$$

where the dot indicates differentiation with respect to τ. Then, the dependent variables q_k and the control parameter Ω are expanded in series of the small parameter μ:

---

[45] The factor 2, used in Sect. 13.6, is not essential.

$$q_1 = q_{10} + \mu q_{11} + \mu^2 q_{12} + \cdots, \tag{13.128a}$$

$$q_2 = q_{20} + \mu q_{21} + \mu^2 q_{22} + \cdots, \tag{13.128b}$$

$$\Omega = \Omega_0 \left(1 + \mu \sigma_1 + \mu^2 \sigma_2 + \cdots \right), \tag{13.128c}$$

where Ω_0 is the (unknown) frequency Ω from which the instability region emanates and $\sigma_1, \sigma_2, \cdots$ are (unknown) detuning parameters. Substituting Eq. 13.128 in Eq. 13.127, and equating to zero the terms with the same power of μ, the following perturbation equations are obtained:

- Order μ^0:

$$\ddot{q}_{10} + \beta_1^2 q_{10} = 0, \tag{13.129a}$$

$$\ddot{q}_{20} + \beta_2^2 q_{20} = 0, \tag{13.129b}$$

- Order μ^1:

$$\ddot{q}_{11} + \beta_1^2 q_{11} = -2\sigma_1 \ddot{q}_{10} + \frac{1}{\Omega_0^2} \cos \tau \, (h_{11} q_{10} + h_{12} q_{20}), \tag{13.130a}$$

$$\ddot{q}_{21} + \beta_2^2 q_{21} = -2\sigma_1 \ddot{q}_{20} + \frac{1}{\Omega_0^2} \beta_2^2 \cos \tau \, (h_{21} q_{10} + h_{22} q_{20}), \tag{13.130b}$$

- Order μ^2:

$$\ddot{q}_{12} + \beta_1^2 q_{12} = - \left(2\sigma_2 + \sigma_1^2 \right) \ddot{q}_{10} - 2\sigma_1 \ddot{q}_{11} + \frac{1}{\Omega_0^2} \cos \tau \, (h_{11} q_{11} + h_{12} q_{21}),$$
$$\tag{13.131a}$$

$$\ddot{q}_{22} + \beta_2^2 q_{22} = - \left(2\sigma_2 + \sigma_1^2 \right) \ddot{q}_{20} - 2\sigma_1 \ddot{q}_{21} + \frac{1}{\Omega_0^2} \cos \tau \, (h_{21} q_{11} + h_{22} q_{21}),$$
$$\tag{13.131b}$$

in which the following *dimensionless natural frequencies* have been introduced:

$$\beta_1 := \frac{\omega_1}{\Omega_0}, \qquad \beta_2 := \frac{\omega_2}{\Omega_0}. \tag{13.132}$$

The solution to Eq. 13.129 is:

$$q_{10} = a_1 \cos (\beta_1 \tau) + b_1 \sin (\beta_1 \tau), \tag{13.133a}$$

$$q_{20} = a_2 \cos{(\beta_2 \tau)} + b_2 \sin{(\beta_2 \tau)}, \tag{13.133b}$$

where a_1, b_1, a_2, b_2 are arbitrary constants. The first-order motion is therefore bi-periodic, of frequencies β_1, β_2. Substituting these expressions in Eq. 13.130, and taking into account known trigonometry formulas,[46] one has:

$$\ddot{q}_{11} + \beta_1^2 q_{11} = 2\sigma_1 \beta_1^2 \left[a_1 \cos{(\beta_1 \tau)} + b_1 \sin{(\beta_1 \tau)} \right] \tag{13.134a}$$

$$+ \frac{1}{2\Omega_0^2} h_{11} a_1 \left[\cos{((1 - \beta_1)\tau)} + \cos{((1 + \beta_1)\tau)} \right]$$

$$+ \frac{1}{2\Omega_0^2} h_{11} b_1 \left[-\sin{((1 - \beta_1)\tau)} + \sin{((1 + \beta_1)\tau)} \right]$$

$$+ \frac{1}{2\Omega_0^2} h_{12} a_2 \left[\cos{((1 - \beta_2)\tau)} + \cos{((1 + \beta_2)\tau)} \right]$$

$$+ \frac{1}{2\Omega_0^2} h_{12} b_2 \left[-\sin{((1 - \beta_2)\tau)} + \sin{((1 + \beta_2)\tau)} \right],$$

$$\ddot{q}_{21} + \beta_2^2 q_{21} = 2\sigma_1 \beta_2^2 \left[a_2 \cos{(\beta_2 \tau)} + b_2 \sin{(\beta_2 \tau)} \right] \tag{13.134b}$$

$$+ \frac{1}{2\Omega_0^2} h_{21} a_1 \left[\cos{((1 - \beta_1)\tau)} + \cos{((1 + \beta_1)\tau)} \right]$$

$$+ \frac{1}{2\Omega_0^2} h_{21} b_1 \left[-\sin{((1 - \beta_1)\tau)} + \sin{((1 + \beta_1)\tau)} \right]$$

$$+ \frac{1}{2\Omega_0^2} h_{22} a_2 \left[\cos{((1 - \beta_2)\tau)} + \cos{((1 + \beta_2)\tau)} \right]$$

$$+ \frac{1}{2\Omega_0^2} h_{22} b_2 \left[-\sin{((1 - \beta_2)\tau)} + \sin{((1 + \beta_2)\tau)} \right].$$

The frequencies of the forcing are now inspected, to detect possible resonances with the natural frequencies. Concerning Eq. 13.134a, the frequencies to be investigated

---

[46] It should be remembered that:

$$\cos{\tau}\, \cos{(\beta_i \tau)} = \frac{1}{2} \left[\cos{((1 - \beta_i)\tau)} + \cos{((1 + \beta_i)\tau)} \right]$$

$$\cos{\tau}\, \sin{(\beta_i \tau)} = \frac{1}{2} \left[-\sin{((1 - \beta_i)\tau)} + \sin{((1 + \beta_i)\tau)} \right]$$

are $1 \pm \beta_1$, $1 \pm \beta_2$; when one of these equates $\pm\beta_1$, a resonance occurs. Since $1 + \beta_1 \neq \pm\beta_1$, $1 - \beta_1 \neq -\beta_1$, $1 + \beta_2 \neq -\beta_1$, the following four cases, only, must be examined.

- The frequency $1 - \beta_1$ is equal to β_1 if and only if $2\beta_1 = 1$, or, in terms of dimensional frequencies, if $2\omega_1 = \Omega_0$. This is the principal resonance condition involving the first mode only, because, under this condition, all the terms in the right member of Eq. 13.134b are nonresonant. Hence, the second Lagrangian coordinate participates only passively to the motion (i.e., it brings contributions of higher order). An identical circumstance occurs when modes are exchanged, i.e., when $2\beta_2 = 1$, i.e., when $2\omega_2 = \Omega_0$, for which the excitation is in principal resonance with the second mode.
- The frequency $1 - \beta_2$ is equal to β_1 if and only if $\beta_1 + \beta_2 = 1$; simultaneously, in Eq. 13.134b, the frequency $1 - \beta_1$ is equal to β_2, so that all terms underlined in the two equations are resonant. This implies that *both the modes* are actively involved in the response. This is the *combination resonance of sum type*, which, in terms of dimensional frequencies, requires:

$$\Omega_0 = \omega_1 + \omega_2, \tag{13.135}$$

  already found in Eq. 13.123b.
- The frequency $1 + \beta_2$, which appears in Eq. 13.134a, is equal to β_1 if and only if $\beta_1 - \beta_2 = 1$; simultaneously, in Eq. 13.134b, the frequency $1 - \beta_1$ is equal to $-\beta_2$, and therefore still in resonance. Similarly, $1 - \beta_2 = -\beta_1$ if and only if $\beta_2 - \beta_1 = 1$; simultaneously, in the second equation, $1 + \beta_1 = \beta_2$. The two cases are summarized by the resonance condition $1 = |\beta_1 - \beta_2|$, that is, in dimensional form:

$$\Omega_0 = |\omega_1 - \omega_2|. \tag{13.136}$$

This is the *combination resonance* of *difference type*.

It is worth noticing that, at this order, no other resonances appear. These, indeed, will manifest themselves only at higher orders, as already discussed.

Principal Resonance of Sum Type
The principal combination resonance of sum type is examined, for which Eq. 13.135, i.e., $1 = \beta_1 + \beta_2$, holds. By writing $1 - \beta_2 = \beta_1$ in Eq. 13.134a, and $1 - \beta_1 = \beta_2$ in Eq. 13.134b, and making explicit only the (underlined) resonant terms, the perturbation equations of order μ become:

$$\ddot{q}_{11} + \beta_1^2 q_{11} = 2\sigma_1 \beta_1^2 \left[a_1 \cos(\beta_1\tau) + b_1 \sin(\beta_1\tau) \right] \tag{13.137a}$$

$$+ \frac{1}{2\Omega_0^2} h_{12} \left(a_2 \cos(\beta_1\tau) - b_2 \sin(\beta_1\tau) \right) + \text{NRT},$$

$$\ddot{q}_{21} + \beta_2^2 q_{21} = 2\sigma_1 \beta_2^2 \left[a_2 \cos (\beta_2 \tau) + b_2 \sin (\beta_2 \tau) \right] \tag{13.137b}$$

$$+ \frac{1}{2\Omega_0^2} h_{21} \left(a_1 \cos (\beta_2 \tau) - b_1 \sin (\beta_2 \tau) \right) + \text{NRT},$$

where NRT denotes nonresonant terms. In order for the q_{k1} to be bi-periodic too, the resonant terms in the second member must be removed. To this end, the coefficients of $\cos (\beta_i \tau)$, $\sin (\beta_i \tau)$ must be separately canceled in each equation, thus obtaining:[47]

$$
\begin{bmatrix}
2\beta_1^2 \sigma_1 & \frac{h_{12}}{2\Omega_0^2} & 0 & 0 \\
\frac{h_{21}}{2\Omega_0^2} & 2\beta_2^2 \sigma_1 & 0 & 0 \\
0 & 0 & 2\beta_1^2 \sigma_1 & -\frac{h_{12}}{2\Omega_0^2} \\
0 & 0 & -\frac{h_{21}}{2\Omega_0^2} & 2\beta_2^2 \sigma_1
\end{bmatrix}
\begin{pmatrix}
a_1 \\ a_2 \\ b_1 \\ b_2
\end{pmatrix}
=
\begin{pmatrix}
0 \\ 0 \\ 0 \\ 0
\end{pmatrix}.
\tag{13.138}
$$

Zeroing the determinant of the matrix leads to find four roots, two by two coincident:

$$\sigma_1 = \pm \frac{1}{4\beta_1 \beta_2 \Omega_0^2} \sqrt{h_{12} h_{21}} = \pm \frac{1}{4\omega_1 \omega_2} \sqrt{h_{12} h_{21}}. \tag{13.139}$$

This solution is real if and only if $h_{12} h_{21} > 0$; if, instead, $h_{12} h_{21} \leq 0$, the combination resonance of sum type does not give rise to any unstable zone (at the first order). It should be noticed that h_{11}, h_{22} do not contribute to the first-order solution. With the above results, by using Eqs. 13.128c and 13.135, the transition curve is found in the form:

$$\Omega = (\omega_1 + \omega_2) \left(1 \pm \frac{\mu}{4\omega_1 \omega_2} \sqrt{h_{12} h_{21}} \right). \tag{13.140}$$

It consists of two branches, namely, two straight lines outgoing from $\Omega = \omega_1 + \omega_2$.

The *motion on the transition curves* is described, at this order, by $\mathbf{q} = \mathbf{q}_0$, which, using Eqs. 13.133, 13.138, reads:

$$\mathbf{q} = a_1 \begin{pmatrix} \cos (\tilde{\omega}_1 t) \\ \mp \frac{\omega_1}{\omega_2} \sqrt{\frac{h_{21}}{h_{12}}} \cos (\tilde{\omega}_2 t) \end{pmatrix} + b_1 \begin{pmatrix} \sin (\tilde{\omega}_1 t) \\ \pm \frac{\omega_1}{\omega_2} \sqrt{\frac{h_{21}}{h_{12}}} \sin (\tilde{\omega}_2 t) \end{pmatrix}. \tag{13.141}$$

In this expression, the original time has been reintroduced via $\beta_i \tau = \beta_i \Omega t = \beta_i \Omega_0 \left(1 \pm \frac{\mu}{4\omega_1 \omega_2} \sqrt{h_{12} h_{21}} \right) t = \tilde{\omega}_i t$, having used Eqs. 13.128c and 13.139 and having indicated with $\tilde{\omega}_i := \omega_i \left(1 \pm \frac{\mu}{4\omega_1 \omega_2} \sqrt{h_{12} h_{21}} \right)$ the ith natural frequency

---

[47] The equations have been rearranged to highlights the block structure of the matrix.

modified by the parametric excitation. The sign $\pm$ in $\tilde{\omega}_i$ must be taken consistently with Eq. 13.140, that is, according to which of the two transition branches the system belongs.

Remark 13.13 The two frequencies $\tilde{\omega}_i$ that appear in the response, Eq. 13.141, can also be written as $\tilde{\omega}_{1,2} = \frac{1}{2} (\tilde{\omega}_1 + \tilde{\omega}_2) \pm \frac{1}{2} (\tilde{\omega}_1 - \tilde{\omega}_2)$, to denote a bi-periodic motion of higher frequency $\frac{1}{2} (\tilde{\omega}_1 + \tilde{\omega}_2)$, modulated by the lower frequency $\frac{1}{2} (\tilde{\omega}_1 - \tilde{\omega}_2)$. For μ small, the higher frequency is close to $\frac{\Omega_0}{2}$, that is, to the half the frequency of the parametric excitation, such as it occurs in the principal resonance for a single DOF system. However, unlike this system, and according to the discussion previously carried out on the Neimark-Sacker bifurcation, this frequency is slowly modulated by another frequency, which is much smaller than the former, the closer the natural frequencies are.

Secondary Resonance of Sum Type

A secondary combination resonance of sum type occurs when, according to Eq. 13.124b, it is $\beta_1 + \beta_2 = 2$. Since the frequencies $1 \pm \beta_i$ which appear on the right member of the perturbation Eq. 13.134 are nonresonant, the suppression of the secular terms at the order μ leads to $\sigma_1 = 0$. This implies that the two transition curves that bound the resonance region have the same vertical tangent at the branching point. To determine the curvatures σ_2, one must first solve Eq. 13.134 and then replace the results in Eq. 13.131 and finally cancel the resonant terms. Equations of the type of Eq. 13.138 are still found, from which σ_2 is evaluated, by requiring that the determinant of the coefficient matrix is zero. The transition curve is finally written as:

$$\Omega = \frac{1}{2} (\omega_1 + \omega_2) \left(1 \pm \mu^2 \sigma_2 \right), \tag{13.142}$$

that is, it consists of two tangent parabolas. Here, for the sake of brevity, the derivation of σ_2 is omitted.

References

1. Bolotin, V.V.: The Dynamic Stability of Elastic Systems. Holden Day, San Francisco (1964)
2. Cesari, L.: Asymptotic Behavior and Stability Problems in Ordinary Differential Equations. Springer, Berlin (1971)
3. Kovacic, I., Rand, R., Mohamed Sah, S.: Mathieu's equation and its generalizations: overview of stability charts and their features. Appl. Mech. Rev. **70**(2), 22 p. (2018)
4. Lacarbonara, W.: Nonlinear Structural Mechanics: Theory, Dynamical Phenomena and Modeling. Springer Science & Business Media, New York (2013)
5. Nayfeh, A.H.: Perturbation Methods. Wiley, New York (1973)
6. Nayfeh, A.H., Balachandran, B.: Applied Nonlinear Dynamics: Analytical, Computational, and Experimental Methods. Wiley, New York (2008)
7. Nayfeh, A.H., Mook, D.T.: Nonlinear Oscillations. Wiley, New York (1995)

8. Seyranian, A.P., Mailybaev, A.A.: Multiparameter Stability Theory with Mechanical Applications. World Scientific, Singapore (2003)
9. Sugiyama, Y., Langthjem, M.A., Katayama, K.: Dynamic Stability of Columns Under Nonconservative Forces. Springer, Cham (2019)

Chapter 14
Solved Problems

14.1 Introduction

Solved problems are proposed in this chapter, aimed to stimulate the reader in applying the algorithms illustrated in the book, and in interpreting the mechanical behavior of the structures at the bifurcation. Problems concern: (a) buckling of conservative systems (planar beams, open thin-walled beams and plates, in various loading and constraint conditions), and (b) nonconservative systems (rigid rods or beams under follower forces, beams and cables manifesting aeroelastic instability, and parametrically excited systems). Exact, Ritz and Finite Element solutions are worked out, and results commented.

14.2 Elastic Buckling of Planar Beam Systems

The critical load of few systems, made of elastic beams, with or without springs, is determined.

14.2.1 Stepped Beam

A clamped-free beam, having two different stiffnesses, EI_1 and EI_2, in sub-intervals of length ℓ_1 and ℓ_2, respectively, is compressed by a load P (Fig. 14.1). (a) Define the total potential energy. (b) Determine the critical load by means of the Ritz method, by using polynomial trial functions defined on the whole domain $(0, \ell_1 + \ell_2)$ (Case 1). (c) Repeat the Ritz analysis by using two sets of trial functions, one for each sub-interval, after having enforced geometric compatibility at the

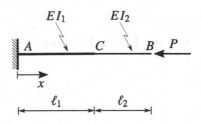

Fig. 14.1 Clamped-free
stepped beam, uniformly
compressed

singular point (Case 2). (d) Plot the critical load vs the stiffness ratio $\eta := \sqrt{\frac{EI_1}{EI_2}}$,
and compare the Ritz results with the exact findings of Sect. 7.5.1.

(a) The total potential energy (TPE) is defined as in Eq. 7.53, i.e.:

$$
\Pi = \int_{0}^{\ell_1} \frac{1}{2} EI_1 v''^2 \, dx + \int_{\ell_1}^{\ell_1+\ell_2} \frac{1}{2} EI_2 v''^2 \, dx - \int_{0}^{\ell_1+\ell_2} \frac{1}{2} Pv'^2 \, dx. \tag{14.1}
$$

(b) A large number ($n = 12$) of polynomial trial functions $\phi_i(x) = x^{i+1}$,
$i = 1, 2, \cdots, n$ (which satisfy the boundary conditions at the clamp), defined
on the whole interval, is used. Accordingly, $v(x) = \sum_{i=1}^{n} x^{i+1} a_i$, with $0 <
x \le \ell_1 + \ell_2$ and a_i unknown constants. Operating as in Sect. 7.5.2, algebraic
equilibrium equations are obtained in the form of Eq. 7.57. The elastic and
geometric stiffness matrices are found to be similar to that in Eq. 7.59, but of
larger dimensions 12×12. By zeroing the determinant of the total matrix, the
critical load P_c is evaluated numerically.

(c) Aimed at using a smaller number of DOF, polynomial trial functions are
considered again, but *different in each of the two sub-intervals*, able to capture
the jump of the curvature at the singular point. By still taking $v_1(x) =
\sum_{i=1}^{n} x^{i+1} a_i$ in the sub-interval 1 (i.e., for $0 < x \le \ell_1$), one can take
$v_2(x) = c_0 + c_1(x - \ell_1) + \sum_{i=1}^{n} (x - \ell_1)^{i+1} b_i$ in the sub-interval 2 (i.e.,
for $\ell_1 \le x < \ell_2$). Here, b_i are unknown constants (additional Lagrangian
parameters) independent of a_i, while the constants c_0, c_1 must be chosen in
order to satisfy the compatibility at $x = \ell_1$. This latter calls for $v_1(\ell_1) =
v_2(\ell_1)$, $v_1'(\ell_1) = v_2'(\ell_1)$, i.e., $v_1(\ell_1) = c_0$, $v_1'(\ell_1) = c_1$. Therefore:

$$
v_2(x) = \sum_{i=1}^{n} \ell_1^{i+1} a_i + (x - \ell_1) \sum_{i=1}^{n} (i+1) \ell_1^{i} a_i + \sum_{i=1}^{n} (x - \ell_1)^{i+1} b_i. \tag{14.2}
$$

The first two terms describe the tangent to the deformed beam at the singular
point $x = \ell_1$; the third term, the deviation of the beam from such straight
line. By taking $n = 2$ (i.e., by using four independent trial functions),
equilibrium Eqs. 7.57 are derived in the unknown Lagrangian parameters $\mathbf{q} =
(a_1, b_1, a_2, b_2)^T$, where:

$$\mathbf{K}_e = EI_1\ell_1 \begin{pmatrix} 4 & 0 & 6\ell_1 & 0 \\ 0 & 0 & 0 & 0 \\ 6\ell_1 & 0 & 12\ell_1^2 & 0 \\ 0 & 0 & 0 & 0 \end{pmatrix} + EI_2\ell_2 \begin{pmatrix} 0 & 0 & 0 & 0 \\ 0 & 4 & 0 & 6\ell_2 \\ 0 & 0 & 0 & 0 \\ 0 & 6\ell_2 & 0 & 12\ell_2^2 \end{pmatrix},$$

$$\mathbf{K}_g = -P\ell_1 \begin{pmatrix} \frac{4\ell_1(\ell_1+3\ell_2)}{3} & 2\ell_2^2 & \frac{3\ell_1^2(\ell_1+4\ell_2)}{2} & 2\ell_2^3 \\ 2\ell_2^2 & \frac{4\ell_2^3}{3\ell_1} & 3\ell_1\ell_2^2 & \frac{3\ell_2^4}{2\ell_1} \\ \frac{3\ell_1^2(\ell_1+4\ell_2)}{2} & 3\ell_1\ell_2^2 & \frac{9\ell_1^3(\ell_1+5\ell_2)}{5} & 3\ell_1\ell_2^3 \\ 2\ell_2^3 & \frac{3\ell_2^4}{2\ell_1} & 3\ell_1\ell_2^3 & \frac{9\ell_2^5}{5\ell_1} \end{pmatrix}.$$

$$(14.3)$$

The relevant characteristic equation takes a cumbersome expression, here omitted. However, if $n = 1$ is taken, the (nondimensional) critical load assumes a simple expression, i.e.:

$$\frac{P_c \ell_1^2}{\pi^2 EI_1} = \frac{6\gamma \left(\gamma(\gamma + 3) + \eta^2 - \sqrt{\gamma^2(\gamma + 3)^2 + \gamma(3 - 2\gamma)\eta^2 + \eta^4} \right)}{\pi^2 \eta^2 (4\gamma + 3)},$$

$$(14.4)$$

where $\eta := \sqrt{\frac{EI_1}{EI_2}}$ and $\gamma := \frac{\ell_1}{\ell_2}$ (Eq. 7.52) are the stiffness ratio and the length ratio, respectively.

(d) Figure 14.2 shows the dimensionless critical load vs η, for different γ, both for (a) trial functions defined on the whole domain (Case 1) and (b) trial functions defined in sub-domains (Case 2). A comparison with the exact solution of the problem (Sect. 7.5.1) is also made. It is observed that in Case 1, several ($n = 12$) polynomial functions are needed to fit the exact solutions in a satisfactory way; on the other hand, in Case 2, the agreement between Ritz and exact solutions is good with only two ($n = 1$) trial functions and excellent with four ($n = 2$).

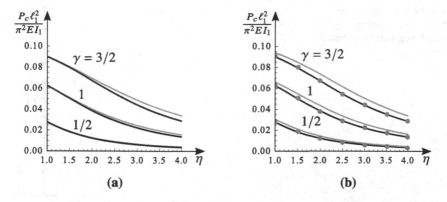

Fig. 14.2 Critical load of the stepped beam in Fig. 14.1; comparison between the Ritz (in gray) and the exact solution (in black) in (a) Case 1 ($n = 12$ trial functions defined on the whole domain) and (b) Case 2, for $n = 1$ (solid lines) and $n = 2$ (dots) functions per sub-interval

14.2.2   Clamped-Free Beam Under Distributed and Concentrated Axial Loads

A clamped-free beam, of length ℓ and bending stiffness EI, is subject to uniformly distributed longitudinal compression loads $p(x) = p = $ const and to a concentrated longitudinal load P applied at the free end (Fig. 14.3).

(a) Define the total potential energy of the system. (b) Determine the critical combinations of the longitudinal loads (pl and P) by using the Ritz method, by adopting the following trial functions:

- *Case 1: a single trial function, describing the buckling mode of the compressed clamped-free beam under the concentrated load P alone, as given in Eq. B.23;*
- *Case 2a: a single polynomial trial function, as given in Eq. 7.92, with $i = 1$;*
- *Case 2b: two polynomial trial functions, as given in Eq. 7.92, with $i = 1, 2$.*

(c) By assuming pl and P as independent bifurcation parameters, plot the stability domain for each of the above cases, and compare the results with finite element analyses.

(a) The total potential energy (TPE), according to Eq. 7.81, reads:

$$\Pi = \int_0^\ell \left(\frac{1}{2} EI v''^2 + \frac{1}{2} N_0 v'^2 \right) dx, \qquad (14.5)$$

where $N_0 = -(P + p(\ell - x))$.

(b) Trial functions are used to approximate the transverse displacement $v(x)$, leading to the following results:

- Case 1. By letting $v(x) = \left(1 - \cos\left(\frac{\pi x}{2\ell}\right)\right) q_1$ in Eq. 14.5, the TPE becomes:

Fig. 14.3 Clamped-free beam compressed by distributed and concentrated loads

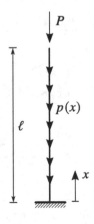

$$\Pi = \left[\frac{EI\pi^4}{64\ell^3} - \frac{P\pi^2}{16\ell} - \frac{p}{32} \left(\pi^2 - 4 \right) \right] q_1^2. \tag{14.6}$$

By imposing stationary, the equilibrium equation follows:

$$\left[\frac{EI\pi^4}{32\ell^3} - \frac{P\pi^2}{8\ell} - \frac{p}{16} \left(\pi^2 - 4 \right) \right] q_1 = 0, \tag{14.7}$$

from which the critical combination of the axial loads is derived as:

$$\frac{EI\pi^4}{32\ell^3} - \frac{P\pi^2}{8\ell} - \frac{p}{16} \left(\pi^2 - 4 \right) = 0. \tag{14.8}$$

- Case 2a. By letting $v(x) = x^2 q_1$ in Eq. 14.5, the TPE reads:

$$\Pi = \left(2EI\ell - \frac{2}{3} P\ell^3 - \frac{1}{6} p\ell^4 \right) q_1^2. \tag{14.9}$$

By imposing stationary,

$$\left(4EI\ell - \frac{4}{3} P\ell^3 - \frac{1}{3} p\ell^4 \right) q_1 = 0 \tag{14.10}$$

follows, from which the critical load combination is drawn:

$$4EI\ell - \frac{4}{3} P\ell^3 - \frac{1}{3} p\ell^4 = 0. \tag{14.11}$$

- Case 2b. By letting $v(x) = \sum_{i=1}^{2} x^{i+1} q_i$ in Eq. 14.5, the TPE takes the form:

$$\Pi = \left(2EI\ell - \frac{2}{3} P\ell^3 - \frac{1}{6} p\ell^4 \right) q_1^2 + \left(6EI\ell^3 - \frac{9}{10} P\ell^5 - \frac{3}{20} p\ell^6 \right) q_2^2$$
$$+ \left(6EI\ell^2 - \frac{3}{2} P\ell^4 - \frac{3}{10} p\ell^5 \right) q_1 q_2. \tag{14.12}$$

From stationary, the equilibrium conditions follow:

$$\begin{pmatrix} 4EI\ell - \frac{4}{3} P\ell^3 - \frac{1}{3} \ell^4 p & 6EI\ell^2 - \frac{3}{2} P\ell^4 - \frac{3}{10} p\ell^5 \\ 6EI\ell^2 - \frac{3}{2} P\ell^4 - \frac{3}{10} p\ell^5 & 12EI\ell^3 - \frac{9}{5} P\ell^5 - \frac{3}{10} p\ell^6 \end{pmatrix} \begin{pmatrix} q_1 \\ q_2 \end{pmatrix} = \begin{pmatrix} 0 \\ 0 \end{pmatrix}. \tag{14.13}$$

The system admits a nontrivial solution if the determinant of the matrix of the coefficients is equal to zero, i.e., if:

Fig. 14.4 Stability domains
for the beam in Fig. 14.3, for
different discretizations: Case
1 (solid gray line), Case 2a
(solid black line), Case 2b
(dashed black line), FE
results (black dots)

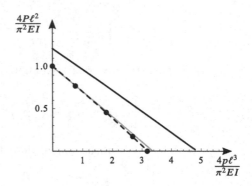

$$60EI^2 - 26EIP\ell^2 - 8EIp\ell^3 + \frac{3}{4}P^2\ell^4 + \frac{1}{2}pP\ell^5 + \frac{1}{20}p^2\ell^6 = 0. \quad (14.14)$$

(c) The critical combinations of the nondimensional distributed and concentrated
loads are plotted in Fig. 14.4 for each examined case. A comparison with FE
(exact) results is also provided. It is observed that the Ritz solution in Case 2a
(in which a single polynomial trial function is used) leads to a stability domain
larger than the exact one. On the other hand, the curves obtained in Case 1 and
Case 2b are in a very good agreement with the FE results.

14.2.3   Clamped-Sliding Beam on Partial Elastic Soil

*A compressed clamped-clamped beam, of length ℓ, in which one of the two
constraints permits longitudinal sliding, partially rests on a Winkler soil of constant
k_f (Fig. 14.5). (a) Define the total potential energy. (b) Determine the critical
load by using the Ritz method, in which the trial functions are taken coincident
with the first three, symmetric or antisymmetric, buckling modes of the compressed
clamped-clamped beam. (c) Plot the critical load vs the soil-to-beam stiffness ratio
$\eta := \frac{k_f}{EI}\left(\frac{\ell}{\pi}\right)^4$, together with a sketch of the corresponding buckling modes. (d)
Compare the results with finite element analyses.*

(a) The TPE is defined in Eq. 7.116 and reads:

$$\Pi = \int_0^\ell \left(\frac{1}{2}EIv''^2 - \frac{1}{2}Pv'^2 \right) dx + \int_{\ell/3}^{2\ell/3} \frac{1}{2}k_f v^2 \, dx. \quad (14.15)$$

(b) The transverse displacement is expressed according to Eq. 7.91, with $n = 3$, as:

Fig. 14.5 Compressed clamped-sliding beam, partially resting on Winkler soil

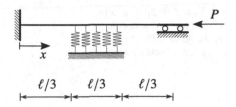

$$v^h(x) = \sum_{i=1}^{3} \phi_i^h(x) q_i, \quad h = S, A, \tag{14.16}$$

where $\phi_i^S(x)$ and $\phi_i^A(x)$ are given in Eqs. B.21a and B.21b, with reference to the symmetric ($h = S$) and antisymmetric ($h = A$) buckling modes of a compressed clamped-clamped beam, respectively. By substituting Eq. 14.16 in Eq. 14.15, and by imposing stationary of the TPE, the equilibrium equations are obtained:

$$\begin{pmatrix} k_{11}^h & k_{12}^h & k_{13}^h \\ & k_{22}^h & k_{23}^h \\ \text{SYM} & & k_{33}^h \end{pmatrix} \begin{pmatrix} q_1 \\ q_2 \\ q_3 \end{pmatrix} = \begin{pmatrix} 0 \\ 0 \\ 0 \end{pmatrix}, \quad h = S, A, \tag{14.17}$$

where:

$$k_{11}^S := \frac{8EI\pi^4}{\ell^3} - \frac{2P\pi^2}{\ell} + k_f\ell\left(\frac{4\pi+9\sqrt{3}}{8\pi}\right), \quad k_{11}^A := \frac{3257.9EI}{\ell^3} + 0.27k_f\ell - \frac{40.4P}{\ell},$$

$$k_{12}^S := \frac{k_f\ell}{3}, \quad k_{12}^A := \frac{13.6EI}{\ell^3} - 0.16k_f\ell - \frac{0.07P}{\ell},$$

$$k_{13}^S := k_f\ell\left(\frac{16\pi+27\sqrt{3}}{48\pi}\right), \quad k_{13}^A := -\frac{10.2EI}{\ell^3} + 0.18k_f\ell + \frac{0.005P}{\ell},$$

$$k_{22}^S := \frac{128EI\pi^4}{\ell^3} - \frac{8P\pi^2}{\ell} + k_f\ell\left(\frac{8\pi-9\sqrt{3}}{16\pi}\right), \quad k_{22}^A := \frac{28549.7EI}{\ell^3} + 0.12k_f\ell - \frac{119.4P}{\ell},$$

$$k_{23}^S := k_f\ell\left(\frac{20\pi-27\sqrt{3}}{60\pi}\right), \quad k_{23}^A := -\frac{104.1EI}{\ell^3} - 0.15k_f\ell + \frac{0.3P}{\ell},$$

$$k_{33}^S := \frac{648EI\pi^4}{\ell^3} - \frac{18P\pi^2}{\ell} + \frac{k_f\ell}{2}, \quad k_{33}^A := \frac{112952.6EI}{\ell^3} + 0.21k_f\ell - \frac{237.7P}{\ell}. \tag{14.18}$$

Equations 14.17 admit nontrivial solution if the determinant of the matrix is equal to zero, from which the critical load P_c is evaluated for both symmetric and antisymmetric cases; here, the explicit expressions are omitted.

(c) Figure 14.6 shows the nondimensional critical load $\frac{P_c\ell^2}{4\pi^2EI}$, relative to both symmetric and antisymmetric modes, vs the soil-to-beam stiffness ratio $\eta := \frac{k_f}{EI}\left(\frac{\ell}{\pi}\right)^4$. The envelope of the lowest values is indicated with a thicker line, together with the corresponding critical modes depicted in Fig. 14.7 for selected values of η. It is seen that the critical mode is alternately symmetric or antisymmetric, with number of half-waves increasing with η.

(d) FE results are provided, denoted by black dots in Figs. 14.6 and 14.7. An excellent agreement between analytical and numerical solutions is detected.

Fig. 14.6 Critical load *vs* the
soil-to-beam stiffness ratio η,
for the beam in Fig. 14.5.
Black dots denote FE results

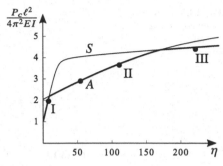

Fig. 14.7 Buckling modes of the beam in Fig. 14.5, for selected values of η, referring to the different branches in Fig. 14.6: (**a**) I, (**b**) II, (**c**) III. Black dots denote FE results

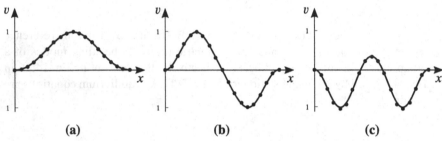

(**a**)   (**b**)   (**c**)

14.2.4   Free-Free Beam on Elastic Soil

A compressed beam of length ℓ, unconstrained at the ends, rests on elastic soil of Winkler constant k_f (Fig. 14.8). (a) Determine the exact critical load, by referring to the general theory developed in Sect. 7.9.3. (b) Plot the critical load vs the soil-to-beam stiffness ratio $\gamma \ell := \sqrt[4]{\frac{k_f \ell^4}{EI}}$, together with the corresponding buckling mode.

(a) The general theory of beam on elastic soil, arbitrarily constrained at the ends (Sect. 7.9.3), is applied to a free-free beam. By placing the origin of the coordinates at midspan, the boundary conditions read:

$$v''' \left(-\ell/2\right) + 2\mu\gamma^2 v' \left(-\ell/2\right) = 0, \quad v'' \left(-\ell/2\right) = 0,$$
$$v''' \left(\ell/2\right) + 2\mu\gamma^2 v' \left(\ell/2\right) = 0, \quad v'' \left(\ell/2\right) = 0,$$

$$(14.19)$$

where the definitions $\gamma^4 := \frac{k_f}{EI}$ (Eq. 7.129) and $\mu := \frac{1}{2}\frac{P}{\sqrt{EI\,k_f}}$ (Eq. 7.130) have been used. In order to determine the critical load, it is necessary to investigate in what of the three cases discussed in the Sect. 7.9.3 the solution falls. By assuming, as a first attempt, that the nondimensional critical load μ_c is less than 1, case III occurs. Substituting the relevant general solution (Eq. 7.142) in the boundary conditions, four algebraic equations in the four unknown constants c_i, $i = 1, \ldots, 4$ are drawn:

Fig. 14.8 Free-free beam on elastic soil, uniformly compressed

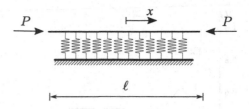

$$\begin{pmatrix} A_{11} & A_{12} & A_{13} & A_{14} \\ A_{21} & A_{22} & A_{23} & A_{24} \\ -A_{13} & A_{14} & -A_{11} & A_{12} \\ A_{23} & -A_{24} & A_{21} & -A_{22} \end{pmatrix} \begin{pmatrix} c_1 \\ c_2 \\ c_3 \\ c_4 \end{pmatrix} = \begin{pmatrix} 0 \\ 0 \\ 0 \\ 0 \end{pmatrix}, \tag{14.20}$$

where α and ω are defined in Eq. 7.141, and:

$$\begin{aligned}
A_{11} &:= e^{-\frac{\alpha\ell}{2}} \gamma^2 \left(\omega \sin\left(\frac{\omega\ell}{2}\right) - \alpha \cos\left(\frac{\omega\ell}{2}\right) \right), \\
A_{12} &:= e^{-\frac{\alpha\ell}{2}} \gamma^2 \left(\alpha \sin\left(\frac{\omega\ell}{2}\right) + \omega \cos\left(\frac{\omega\ell}{2}\right) \right), \\
A_{13} &:= e^{\frac{\alpha\ell}{2}} \gamma^2 \left(\omega \sin\left(\frac{\omega\ell}{2}\right) + \alpha \cos\left(\frac{\omega\ell}{2}\right) \right), \\
A_{14} &:= -e^{\frac{\alpha\ell}{2}} \gamma^2 \left(\alpha \sin\left(\frac{\omega\ell}{2}\right) - \omega \cos\left(\frac{\omega\ell}{2}\right) \right), \\
A_{21} &:= e^{-\frac{\alpha\ell}{2}} \left(2\alpha\omega \sin\left(\frac{\omega\ell}{2}\right) - \gamma^2\mu \cos\left(\frac{\omega\ell}{2}\right) \right), \\
A_{22} &:= e^{-\frac{\alpha\ell}{2}} \left(\gamma^2\mu \sin\left(\frac{\omega\ell}{2}\right) + 2\alpha\omega \cos\left(\frac{\omega\ell}{2}\right) \right), \\
A_{23} &:= -e^{\frac{\alpha\ell}{2}} \left(2\alpha\omega \sin\left(\frac{\omega\ell}{2}\right) + \gamma^2\mu \cos\left(\frac{\omega\ell}{2}\right) \right), \\
A_{24} &:= e^{\frac{\alpha\ell}{2}} \left(\gamma^2\mu \sin\left(\frac{\omega\ell}{2}\right) - 2\alpha\omega \cos\left(\frac{\omega\ell}{2}\right) \right).
\end{aligned} \tag{14.21}$$

This system admits nontrivial solution if the determinant of the coefficient matrix is zero, so that the problem is governed by the following characteristic equation:

$$4\mu^3 - 3\mu + (\mu+1)(1-2\mu)^2 e^{4\ell\alpha} + 2e^{2\ell\alpha}\left(\left(-4\mu^3 + 3\mu + 1\right)\cos(2\ell\omega) - 2 \right) + 1 = 0. \tag{14.22}$$

Substitution of Eqs. 7.141 for ω and α in the previous equation leads to a transcendental equation in the (nondimensional) critical load μ and the soil-to-beam stiffness ratio $\gamma\ell$. The relevant graph is shown in Fig. 14.9, where the envelope of the lowest values is indicated with a thicker line. The corresponding critical modes are depicted in Fig. 14.10 for selected values of $\gamma\ell$ (each referring to the branches I, II, and III in Fig. 14.9a and IV, V, and VI in Fig. 14.9b). It is seen that the critical mode is alternately antisymmetric or symmetric, with number of half-waves increasing with $\gamma\ell$.

If the analysis is repeated for the cases I and II discussed in Sect. 7.9.3, different algebraic systems in the unknowns c_i are found, whose characteristic equations admit solutions corresponding to higher critical loads. The only significant solution, therefore, is that falling in case III.

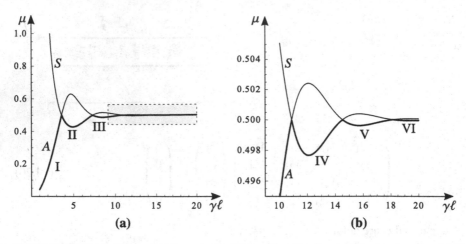

Fig. 14.9 (a) Critical load of the beam in Fig. 14.8 *vs* the soil-to-beam stiffness ratio $\gamma \ell$ and (b) a zoom of the highlighted region

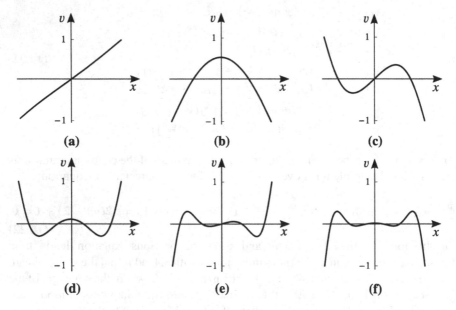

Fig. 14.10 Buckling modes of the beam in Fig. 14.8 for selected values of $\gamma \ell$, relevant to the different branches in Fig. 14.9: (a) I, (b) II, (c) III, (d) IV, (e) V, (f) VI

14.2.5   *Beam Elastically Restrained Against Rotation*

A clamped-supported beam of flexural stiffness EI and length ℓ, elastically restrained against rotation by a torsional spring of stiffness k_t, placed at the supported end, subject to uniform compression P, is considered (Fig. 14.11). (a)

Fig. 14.11
Clamped-supported beam
with torsional spring at one
end, uniformly compressed

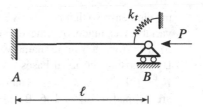

Evaluate the exact nondimensional critical load $\mu_c := \frac{P_c \ell^2}{4\pi^2 EI}$ *vs the spring-to-stiffness ratio* $\eta_t := \frac{k_t \ell}{EI}$ *by using a single exact finite element. (b) Perform the same computation by using a single polynomial finite element. (c) Repeat the latter analysis by using two polynomial finite elements of equal length. (d) Plot* μ_c *vs* η_t *for the three cases examined above, and comment the results.*

(a) By using the exact finite element formulated in Sect. 7.12.2, and accounting for the constraints, which reduce to one the number of DOF of the system, the equilibrium equation (to the rotation of the node B) reads:

$$(k_{33}(P) + k_t)\,\varphi_B = 0. \tag{14.23}$$

Here, the total (elastic plus geometric) stiffness coefficient k_{33}, which depends on the prestress, is given in Eqs. 7.182. By introducing the nondimensional quantities defined above, and zeroing the total stiffness, the following characteristic equation is derived:

$$\frac{2\pi^2\mu \cos\left(2\pi\sqrt{\mu}\right) - \pi\sqrt{\mu}\sin\left(2\pi\sqrt{\mu}\right)}{\pi\sqrt{\mu}\sin\left(2\pi\sqrt{\mu}\right) + \cos\left(2\pi\sqrt{\mu}\right) - 1} + \eta_t = 0. \tag{14.24}$$

This is a transcendental equation for the unknown critical load $\mu = \mu_c(\eta_t)$, depending on η_t.

(b) By using only one polynomial finite element, as formulated in Sect. 7.12.1, and remembering Eqs. 7.166 with $N_0 = -P$, the equilibrium condition reads:

$$\left(\frac{4EI}{\ell} + k_t - P\frac{2\ell}{15}\right)\varphi_B = 0. \tag{14.25}$$

The characteristic equation, written in terms of nondimensional quantities and solved with respect to the load, is:

$$\mu_c = \frac{15}{8\pi^2}(4 + \eta_t). \tag{14.26}$$

(c) To discretize the beam by two polynomial finite elements of equal length $\frac{\ell}{2}$, a node C is introduced at midspan. The load induces axial compression in both the elements, so that $N_0^{(1)} = N_0^{(2)} = -P$. The global vector of the configuration variables is made of nine displacement components, two translation, and a rotation for each node, namely, $\mathbf{q} = (u_A, v_A, \varphi_A, u_C, v_C, \varphi_C, u_B, v_B, \varphi_B)^T$. By choosing the local bases coincident with the global one, one gets $\mathbf{K}^{(e)} = \mathbf{K}_e^{(e)} + \mathbf{K}_g^{(e)}$, $e = 1, 2$, where the elastic and geometric stiffness matrices are defined in Eqs. 7.166. By assembling the matrices in sequence and adding the torsional spring k_t in the position $(9, 9)$, the global stiffness matrices are obtained. By introducing the constraints $u_A = v_A = \varphi_A = u_C = u_B = v_B = 0$, in which it has been taken into account, that $u \equiv 0$ along the beam, the problem reduces to three DOF, namely:

$$\left[\begin{pmatrix} \frac{192EI}{\ell^3} & 0 & \frac{24EI}{\ell^2} \\ 0 & \frac{16EI}{\ell} & \frac{4EI}{\ell} \\ \frac{24EI}{\ell^2} & \frac{4EI}{\ell} & \frac{8EI}{\ell} + k_t \end{pmatrix} + P \begin{pmatrix} -\frac{24}{5\ell} & 0 & -\frac{1}{10} \\ 0 & -\frac{2\ell}{15} & \frac{\ell}{60} \\ -\frac{1}{10} & \frac{\ell}{60} & -\frac{\ell}{15} \end{pmatrix} \right] \begin{pmatrix} v_C \\ \varphi_C \\ \varphi_B \end{pmatrix} = \begin{pmatrix} 0 \\ 0 \\ 0 \end{pmatrix}.$$

$$(14.27)$$

By zeroing the determinant of the matrix coefficients, the characteristic equation follows, which, in nondimensional form, reads:

$$3\pi^6 \mu^3 - 4\pi^4 (3\eta_t + 55)\mu^2 + 480\pi^2 (\eta_t + 8)\mu - 3600(\eta_t + 4) = 0. \quad (14.28)$$

This is a cubic polynomial equation for μ, which is more conveniently solved numerically. For any assigned η_t, the smallest root of the cubic equation is the critical load $\mu_c (\eta_t)$ sought for.

(d) The results achieved are plotted in Fig. 14.12, which shows the nondimensional critical load $\mu_c := \frac{P_c \ell^2}{4\pi^2 EI}$ vs the spring-to-stiffness ratio $\eta_t := \frac{k_t \ell}{EI}$, according to (i) the exact solution (Eq. 14.24, gray curve), (ii) the solution provided by a single polynomial finite element (Eq. 14.26, continuous black line), and (iii) the solution provided by two polynomial finite elements (Eq. 14.27, dashed black line). From the plot emerges what follows.

- The exact critical load increases with the stiffness of the torsional spring. It passes (Table 7.1 should be remembered) from the critical value relevant to the clamped-supported beam ($P_c = 2.05\frac{\pi^2 EI}{\ell^2}$, i.e., $\mu_c = 0.51$) when $\eta_t = 0$ (infinitely soft spring) to the critical value of the clamped-sliding beam ($P_c = 4\frac{\pi^2 EI}{\ell^2}$, i.e., $\mu_c = 1$) when $\eta_t \to \infty$ (infinitely stiff spring).
- The model with just one finite element is poorly approximated. It gives a reasonable approximation for soft spring (e.g., $\eta_t \simeq 1$) but completely wrong results for stiffer springs. This is due to the fact that the unique trial function adopted in the discretization (namely, $\psi_6 (x)$ in Eq. 7.164, plotted in Fig. 7.19), possesses *just one flex point* and therefore is unable to capture the

Fig. 14.12 Nondimensional critical load of the beam in Fig. 14.11 *vs* the spring-to-beam stiffness ratio $\eta_t := \frac{k_t \ell}{EI}$; exact solution (gray curve), one polynomial finite element (continuous black line), two polynomial finite elements (dashed black line)

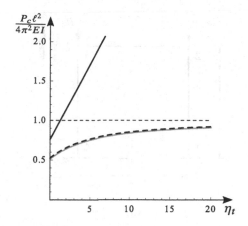

occurrence of a second flex point close to the spring, when this is sufficiently stiff.[1]

- When two finite elements are employed, the critical mode can be more accurately described. As a matter of fact, the FE solution is in excellent agreement with the exact solution, representing an upper bound.

14.2.6 Braced Frame

A frame, elastically braced by a horizontal spring of stiffness k, is considered (Fig. 14.13a). It consists of an assembly of beams, all of length ℓ, having mechanical characteristics EA and EI. Two vertical concentrated loads P are applied at the nodes 2 and 3. (a) Using the polynomial finite elements, define the global stiffness matrix of the system. (b) Determine and plot the critical load vs the spring stiffness. (c) Check the convergence of the results with respect to the number of the finite elements used, also accounting for the findings of Problem 14.2.5.

(a) Based on the linear equilibrium analysis, the loads induce compressive axial forces only in the beams 1 and 3 (Fig. 14.13b), equal to $N_0^{(1)} = N_0^{(3)} = -P$. In what follows, each beam is modeled with a single polynomial finite element. The (global) vector of configuration variables $\mathbf{q}$ consists of twelve displacement components, three $(u_i, v_i, \varphi_i,)$ for each node $(i = 1, \cdots, 4)$, expressed in the global basis:

$$\mathbf{q} = (u_1, v_1, \varphi_1, u_2, v_2, \varphi_2, u_3, v_3, \varphi_3, u_4, v_4, \varphi_4)^T. \tag{14.29}$$

---

[1] As a limit case, when $\eta_t = \infty$, the single polynomial FE cannot be used, for the reasons discussed in Example 10.1.

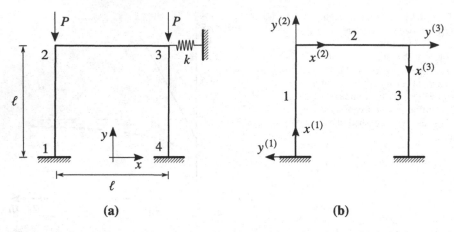

Fig. 14.13 Braced frame with compressed columns: (a) static scheme, (b) local bases and coordinate systems

Choosing a local basis for each beam (as illustrated in Fig. 14.13b), the total (elastic plus geometric) element stiffness matrices are:

$$
\begin{aligned}
\mathbf{K}^{(1)} &= \mathbf{K}_e^{(1)} + \mu \mathbf{K}_g^{(1)}, \\
\mathbf{K}^{(2)} &= \mathbf{K}_e^{(2)}, \\
\mathbf{K}^{(3)} &= \mathbf{K}_e^{(3)} + \mu \mathbf{K}_g^{(3)},
\end{aligned}
\tag{14.30}
$$

where μ is a load multiplier and $\mathbf{K}_e^{(e)}$ and $\mathbf{K}_g^{(e)}$ (for $e = 1, \cdots, 3$) are the elastic and geometric matrices, respectively, defined in Eq. 7.166. The stiffness matrices in the global basis $\tilde{\mathbf{K}}^{(e)}$ (for $e = 1, \cdots, 3$) are obtained according to Eq. 7.170, where the element rotation matrices are defined in Eq. 7.171, with $\theta_1 = \pi/2$, $\theta_2 = 0$ and $\theta_3 = -\pi/2$. The brace spring has a stiffness matrix of dimension 1×1, indicated as $\mathbf{K}^s := [k]$. By assembling the latter and the arrays $\tilde{\mathbf{K}}^{(e)}$, according to the scheme in Fig. 14.14, and imposing the constraints $u_1 = v_1 = \varphi_1 = u_4 = v_4 = \varphi_4 = 0$ (i.e., by deleting rows and columns corresponding to the suppressed displacements), a system of algebraic equations is obtained, of the type $\bar{\mathbf{K}}(\mu)\,\bar{\mathbf{q}} = \mathbf{0}$. The global stiffness matrix reads:

$$
\bar{\mathbf{K}}(\mu) =
\begin{pmatrix}
k_{11} & 0 & k_{13} & k_{14} & 0 & 0 \\
 & k_{22} & k_{23} & 0 & k_{25} & k_{23} \\
 & & k_{33} & 0 & -k_{23} & k_{36} \\
 & & & k_{44} & 0 & k_{13} \\
 & & & & k_{22} & -k_{23} \\
\text{SYM} & & & & -k_{23} & k_{33}
\end{pmatrix},
\tag{14.31}
$$

where:

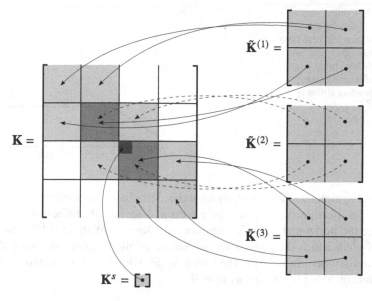

Fig. 14.14 Assembly scheme of the global stiffness matrix of the braced frame

$$
\begin{aligned}
k_{11} &:= \tfrac{12EI}{\ell^3} + \tfrac{EA}{\ell} - \tfrac{6\mu P}{5\ell}, & k_{25} &:= -\tfrac{12EI}{\ell^3}, \\
k_{13} &:= -\tfrac{\mu P}{10} + \tfrac{6EI}{\ell^2}, & k_{33} &:= \tfrac{8EI}{\ell} - \tfrac{2\mu P \ell}{15}, \\
k_{14} &:= -\tfrac{EA}{\ell}, & k_{36} &:= \tfrac{2EI}{\ell}, \\
k_{22} &:= \tfrac{12EI}{\ell^3} + \tfrac{EA}{\ell}, & k_{44} &:= k_{22} - \tfrac{6\mu P}{5\ell} + k. \\
k_{23} &:= \tfrac{6EI}{\ell^2},
\end{aligned}
\qquad (14.32)
$$

(b) By requiring that the determinant of the global stiffness matrix is equal to zero, μ_c is determined, from which a closed-form expression of the critical load P_c, as function of the spring stiffness, is obtained (here omitted). The latter is plotted in Fig. 14.15 (gray line), in the nondimensional form $\frac{P_c \ell^2}{\pi^2 EI}$, vs the spring-to-beam stiffness ratio $\eta := k\frac{\ell^3}{EI}$ (for a given axial-to-bending stiffness ratio, $\frac{EA\ell^2}{EI} = 5500$). It is seen that the bracing spring produces a linear increase of the critical load up to a certain value, after which P_c remains constant; consequently, further stiffening does not entail any advantage on stability. The critical mode is antisymmetric, with movable nodes, along the ascendent branch; it is symmetric, with (almost) fixed nodes, along the horizontal branch.

(c) The previous procedure is repeated by increasing the number of the finite elements for each beam. It has been found that using two FE per each beams (black line in Fig. 14.15), while the antisymmetric critical load remains almost unaffected, the symmetric critical load drastically reduces. By increasing the number of the FE, these results are confirmed, so that the model with two FE per beam is sufficiently accurate. As a matter of fact, when the frame buckles

Fig. 14.15 Critical load of the braced frame *vs* the spring-to-beam stiffness ratio η, using one (gray line) and two (black line) finite elements for each beam; axial-to-bending stiffness ratio $\frac{EA\ell^2}{EI} = 5500$

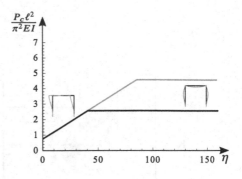

in the symmetric mode with fixed nodes, it behaves as a clamped-supported column, restrained at the top by a torsional spring of stiffness $k_t = \frac{2EI}{\ell}$,[2] since the horizontal beam is unprestressed. Results of Probl. 14.2.5 therefore apply,[3] in which the reasons for the failure of the rough model have already been discussed. In contrast, the rough model well captures the antisymmetric deflection along the ascendent branch.

14.3   Buckling of Open Thin-Walled Beams

The critical load of thin-walled beams (TWB), under different constraint and load conditions, is determined.

14.3.1   Uniformly Compressed Clamped-Free Beam

A clamped-free TWB, with a warping constraint, is presolicited by a compression load P, applied at the centroid of the free end (Fig. 14.16a). (a) Define the total potential energy of the system. (b) Write the bifurcation condition using the Ritz method, with reference to a generic cross-section. (c) Determine the critical load for the T-section of Fig. 14.16b, and compare the results with finite element analyses. The following numerical data are given: $\ell = 1$ m, $a = b = 100$ mm, $s_w = s_f = 2$ mm, $E = 207,000$ N/mm^2, and $G = 79,300$ N/mm^2.

---

[2] Indeed, the moment at the end A of a clamped-clamped AB beam is $m_A = \frac{EI}{\ell}(4\varphi_A + 2\varphi_B)$; since $\varphi_B = -\varphi_A$, then $m_A = 2\frac{EI}{\ell}\varphi_A$.

[3] As a check of this occurrence, by letting $\eta_t = 2$ in the exact solution plotted in Fig. 14.12, one gets $\frac{P_c \ell^2}{4\pi^2 EI} = 0.6378$, and therefore $P_c = 2.55\frac{\pi^2 EI}{\ell^2}$ which is almost identical to that in Fig. 14.15.

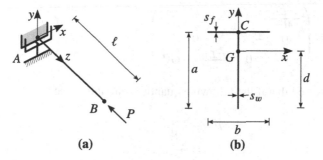

Fig. 14.16 Uniformly compressed clamped-free TWB: (a) structure, (b) T-section

(a) The total potential energy is written as sum of the elastic contributions (Eqs. 9.9 and 9.10) and that of the prestress (Eq. 9.35); it reads:

$$
\Pi = \frac{1}{2} \int_0^\ell \left(EI_y u_C''^2 + EI_x v_C''^2 + EI_\omega \theta''^2 + GJ\theta'^2 \right) dz +
$$

$$
\hspace{6cm} (14.33)
$$

$$
- \frac{P}{2} \int_0^\ell \left(u_C'^2 + v_C'^2 + r_C^2 \theta'^2 + 2y_C u_C' \theta' - 2x_C v_C' \theta' \right) dz.
$$

(b) In order to apply the Ritz method, the displacements are approximate as follows, by taking as shape functions the buckling modes of the clamped-free beam (Eq. B.23):

$$
\begin{pmatrix} u_C\,(z) \\ v_C\,(z) \\ \theta\,(z) \end{pmatrix} = \begin{pmatrix} \hat{u}_C \\ \hat{v}_C \\ \hat{\theta} \end{pmatrix} \left(1 - \cos\left(\frac{\pi}{2\ell} z \right) \right). \tag{14.34}
$$

It should be noticed that, with this choice, $\theta'(0) = 0$, so that the geometric boundary condition requested by the warping constraint is satisfied. By substituting Eq. 14.34 in Eq. 14.33, and by imposing stationarity of the TPE, the equilibrium equations are obtained:

$$
\left(\mathbf{K}_e + \mathbf{K}_g \right) \mathbf{q} = \mathbf{0}, \tag{14.35}
$$

where $\mathbf{q} = \left(\hat{u}_C, \hat{v}_C, \hat{\theta} \right)$ and the stiffness matrices, respectively, elastic and geometric, are:

$$\mathbf{K}_e = \begin{pmatrix} \frac{\pi^2 E I_y}{4\ell^2} & 0 & 0 \\ 0 & \frac{\pi^2 E I_x}{4\ell^2} & 0 \\ 0 & 0 & \frac{\pi^2 E I_\omega}{4\ell^2} + GJ \end{pmatrix}, \quad \mathbf{K}_g = -P \begin{pmatrix} 1 & 0 & y_C \\ 0 & 1 & -x_C \\ y_C & -x_C & r_C^2 \end{pmatrix}.$$

$$(14.36)$$

It is observed that if the following quantities are introduced:

$$\begin{aligned} P_x &:= \frac{\pi^2 E I_x}{4\ell^2}, \\ P_y &:= \frac{\pi^2 E I_y}{4\ell^2}, \\ P_\theta &:= \frac{1}{r_C^2}\left(\frac{\pi^2 E I_\omega}{4\ell^2} + GJ\right), \end{aligned} \quad (14.37)$$

Equation 14.35 assumes the form of Eq. 9.48, derived for an uniformly compressed beam, simply resting on warping-unrestrained torsional supports. Accordingly, the bifurcation condition in Eq. 9.50 holds.

(c) All the terms of the latter equations are evaluated with reference to the T-section in Fig. 14.16b, where y is axis of symmetry, containing the center of torsion; therefore, $x_C = 0$, $y_C = a - d$, and the geometric characteristics are:[4]

$$\begin{aligned} d &= \frac{a\,bs_f + \frac{1}{2}a^2 s_w}{bs_f + as_w}, \\ A &= bs_f + as_w, \\ I_x &= a^2 bs_f + \frac{1}{3}a^3 s_w - d^2\left(bs_f + as_w\right), \\ I_y &= \frac{1}{12}b^3 s_f, \\ I_\omega &= 0, \\ J &= \frac{1}{3}bs_f^3 + \frac{1}{3}as_w^3, \end{aligned} \quad (14.38)$$

from which $r_C^2 = \frac{I_x + I_y}{A} + y_C^2$ is calculated. Then, the critical load is determined according to the discussion developed in Sect. 9.4.2 for the mono-symmetric cross-sections, i.e., by evaluating the minimum between the Eulerian P_x and the flexural-torsional critical load (Eq. 9.55). Finally, the Ritz solution to this problem yields $P_c = 18.718$ kN, nearly coincident with that supplied by the finite element analysis, performed according to the theory illustrated in the Sect. 9.8.

---

[4] Note 9 of Chap. 9 and Rmrk. 9.7 should be remembered.

14.3.2 Uniformly Bent Clamped-Free Beam

A clamped-free TWB, with a warping constraint, is prestressed by bi-axial uniform bending, induced by couples $\pm\mathbf{C}$ applied at the ends, with $\mathbf{C} = C_x\mathbf{a}_x + C_y\mathbf{a}_y$ (Fig. 14.17a). (a) Define the total potential energy of the system. (b) Write the bifurcation condition using the Ritz method for a generic cross-section, and then, specialize it to mono-symmetric and bi-symmetric cross-sections. (c) Plot the stability domains for both T-section (Fig. 14.17b) and I-section (Fig. 14.17c), and compare the results with finite element analyses. The same numerical data of Exercise 14.3.1 are considered.

(a) The total potential energy is written as sum of the elastic contributions (Eqs. 9.9 and 9.10) and that of prestress (Eq. 9.66); it reads:

$$
\Pi = \frac{1}{2}\int_0^\ell \left(EI_y u_C''^2 + EI_x v_C''^2 + EI_\omega\theta''^2 + GJ\theta'^2 \right)dz+
$$

$$(14.39)$$

$$
- \int_0^\ell \left[C_x\left(u_C' + y_H\theta'\right)\theta' + C_y\left(v_C' - x_H\theta'\right)\theta' \right]dz.
$$

(b) Operating as in the previous problem, the displacements are taken as in Eq. 14.34 and, then, substituted in Eq. 14.39. By imposing stationarity of the TPE, the equilibrium equations, Eq. 14.35, are obtained, where the elastic stiffness matrix is the same of Eq. 14.36a and the geometric one becomes:

$$
\mathbf{K}_g = \begin{pmatrix} 0 & 0 & -C_x \\ 0 & 0 & -C_y \\ -C_x & -C_y & -2\left(y_H C_x - x_H C_y\right) \end{pmatrix}.
$$

$$(14.40)$$

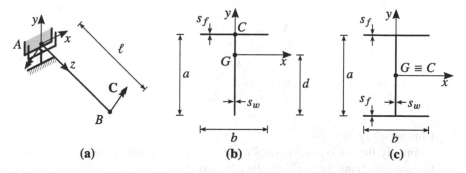

(a)   (b)   (c)

Fig. 14.17 Uniformly bent clamped-free TWB: (a) structure, (b) T-section, (c) I-section

By introducing the definitions given in Eq. 14.37, the equilibrium equations assume the form of Eq. 9.74, derived for a uniformly bent beam, simply resting on warping-unrestrained torsional supports. Accordingly, by vanishing the determinant of the coefficient matrix, the characteristic equation, Eq. 9.75, is obtained. In the case of mono-symmetric cross-sections, with y the axis being of symmetry (i.e., $x_H = 0$), the latter simplifies into:

$$P_x P_y \left(r_C^2 P_\theta - 2y_H C_x \right) - C_x^2 P_x - C_y^2 P_y = 0. \tag{14.41}$$

Also, for bi-symmetric cross-sections, with x and y axes of symmetry (i.e., $x_H = y_H = 0$), Eq. 9.75 becomes:

$$r_C^2 P_x P_y P_\theta - C_x^2 P_x - C_y^2 P_y = 0. \tag{14.42}$$

(c) Equations 14.41 and 14.42 each describe an ellipse in the (C_x, C_y) plane, which represents the locus of the critical states (i.e., the combinations between the two components of the external couple that induce incipient instability of the TWB).

With reference to the T-section in Fig. 14.17b, for which Eq. 14.41 holds, the geometric characteristics are defined in Eq. 14.38; moreover:

$$x_C = 0,$$
$$y_C = a - d,$$
$$r_C^2 = \frac{I_x + I_y}{A} + y_C^2, \tag{14.43}$$
$$y_H = y_C - \frac{1}{2I_x} \left(s_f y_C \left(\frac{b^3}{12} + b y_C^2 \right) + \frac{s_w}{4} \left(y_C^4 - d^4 \right) \right).$$

Concerning the I-section in Fig. 14.17c, for which Eq. 14.42 applies, the geometric characteristics are:

$$\begin{aligned}
A &= 2bs_f + as_w, \\
I_x &= \tfrac{1}{2}s_f b a^2 + \tfrac{1}{12}a^3 s_w, \\
I_y &= \tfrac{1}{6}b^3 s_f, \\
I_\omega &= \tfrac{1}{24}s_f a^2 b^3, \\
J &= \tfrac{2}{3}bs_f^3 + \tfrac{1}{3}as_w^3,
\end{aligned} \tag{14.44}$$

from which $r_C^2 = \frac{I_x + I_y}{A}$.

Finally, the corresponding stability domains are plotted in Fig. 14.18 for both the cross-sections; here, FE results are also provided (according to the theory illustrated in the Sect. 9.8), indicated by black dots. An excellent agreement between Ritz and the FE is detected. It is observed that the critical condition for

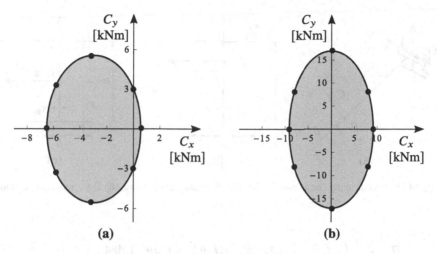

Fig. 14.18 Stability domains of the TWB in Fig. 14.17 for (**a**) T-section and (**b**) I-section; black dots denote FE results

the T-section is sensitive to the change in the sign of C_x; in particular, positive couples (which compress the web) induce an early instability. In contrast, the stability domain of the I-section is symmetric, such that the TWB is insensitive to the sign of the moment.

14.3.3 Compressed and Bent Clamped-Free Beam

A clamped-free TWB, with a warping constraint, is prestressed by a compressive axial force P and by the couples $\pm \mathbf{C}$ applied at the ends, being $\mathbf{C} = C_x \mathbf{a}_x + C_y \mathbf{a}_y$ (Fig. 14.19a). (a) Define the total potential energy of the system. (b) Write the bifurcation condition using the Ritz method for a generic cross-section, and then, specialize it to mono-symmetric and bi-symmetric cross-sections. (c) Referring to the T-section in Fig. 14.19b and to the I-section in Fig. 14.19c, plot the three-dimensional stability domains and their cross-sections $P = $ const, by taking P, C_x, C_y as independent bifurcation parameters; then compare the results with finite element analyses. The same numerical data of Exercise 14.3.1 are given.

(a) The TPE is written as sum of the elastic contributions (Eqs. 9.9 and 9.10) and that of prestress (Eqs. 9.35 and 9.66); it reads:

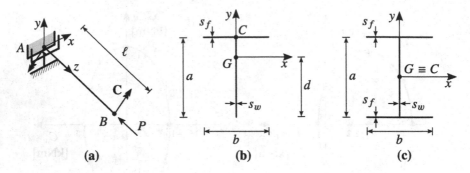

Fig. 14.19 Eccentrically compressed clamped-free beam: (**a**) structure, (**b**) T-section, (**c**) I-section

$$\Pi = \frac{1}{2} \int_0^\ell \left(EI_y u_C''^2 + EI_x v_C''^2 + EI_\omega \theta''^2 + GJ\theta'^2 \right) dz +$$

$$- \frac{P}{2} \int_0^\ell \left(u_C'^2 + v_C'^2 + r_C^2 \theta'^2 + 2y_C u_C'\theta' - 2x_C v_C'\theta' \right) dz + \qquad (14.45)$$

$$- \int_0^\ell \left[C_x \left(u_C' + y_H \theta' \right) \theta' + C_y \left(v_C' - x_H \theta' \right) \theta' \right] dz.$$

(b) Operating as in Problems 14.3.1 and 14.3.2, the displacement field is approximated as in Eq. 14.34. By substituting it into the TPE and by imposing stationarity, equilibrium Eqs. 14.35 are obtained. Here, the elastic stiffness matrix is the same defined in Eq. 14.36, while the geometric stiffness matrix is:

$$\mathbf{K}_g = \begin{pmatrix} -P & 0 & -C_x - Py_C \\ 0 & -P & -C_y + Px_C \\ -C_x - Py_C & -C_y + Px_C & -Pr_C^2 - 2\left(y_H C_x - x_H C_y\right) \end{pmatrix}.$$
$$(14.46)$$

The latter can also be derived from the superimposition of the compression and bending effects, exploiting the linearity of the problem (i.e., by summing Eqs. 14.36b and 14.40).

Then, by using the definitions in Eq. 14.37 and by vanishing the determinant of the coefficient matrix, the characteristic equation follows:

$$\left(P - P_y\right)\left(C_y - Px_C\right)^2 + \left(P - P_x\right)\left(C_x + Py_C\right)^2$$

$$- r_C^2\left(P - P_x\right)\left(P - P_y\right)\left(P - P_\theta\right) \tag{14.47}$$

$$+ 2\left(P - P_x\right)\left(P - P_y\right)\left(C_y x_H - C_x y_H\right) = 0.$$

The latter is the same derived for an eccentrically compressed beam, simply resting on warping-unrestrained torsional supports (Eq. 9.83), when $C_x = -Py_S$ and $C_y = Px_S$ are taken. For mono-symmetric cross-sections, with y being the axis of symmetry (entailing $x_C = 0$ and $x_H = 0$), Eq. 14.47 simplifies into:

$$\left(P - P_y\right)C_y^2 + \left(P - P_x\right)\left(C_x + Py_C\right)^2 - r_C^2\left(P - P_x\right)\left(P - P_y\right)\left(P - P_\theta\right)$$

$$- 2y_H\left(P - P_x\right)\left(P - P_y\right)C_x = 0.$$

$$\tag{14.48}$$

Also, for bi-symmetric cross-sections, with x and y the axes of symmetry (implying $x_C = y_C = 0$ and $x_H = y_H = 0$), the latter becomes:

$$\left(P - P_y\right)C_y^2 + \left(P - P_x\right)C_x^2 - r_C^2\left(P - P_x\right)\left(P - P_y\right)\left(P - P_\theta\right) = 0.$$

$$\tag{14.49}$$

(c) Equations 14.48 and 14.49 represent the locus of the critical states in the $\left(C_x,\ C_y,\ P\right)$ space; in particular, for a given value of the load P, they describe an ellipse in the $\left(C_x,\ C_y\right)$ plane. Figure 14.20 shows the stability domains, referred to as the T-section and I-section in Fig. 14.19b,c, respectively. FE results are also provided (using the theory of Sect. 9.8), as indicated by black dots. An excellent agreement between Ritz and FE solutions is observed. Moreover, it is seen that the stability domain in the $\left(C_x,\ C_y\right)$ plane shrinks with P when positive (i.e., of compression) and expands with P when negative (i.e., of tension).

14.3.4 Simply Supported Beam, Bent by a Uniformly Distributed Load

A TWB, constrained at the ends by simple supports and torsional clamps that leave the warping free, is prestressed by a distributed load of intensity $\mathbf{p} = -p\mathbf{a}_y$, *applied on a line parallel to the z axis, of trace* $x_Q = 0,\ y_Q \gtrless 0$ *(Fig. 14.21a). The cross-section is bi-symmetric, with x and y the axes of symmetry. (a) Determine the critical load, as a function of the two nondimensional parameters* $\chi := \frac{GJ\ell^2}{EI_\omega}$ *and* $\eta := \frac{y_Q}{\ell}\sqrt{\frac{EI_y}{GJ}}$, *using the Ritz method. (b) Plot the critical load as χ varies, for different values of η. (c) With reference to an I-section (Fig. 14.21b), analyze the influence of the load application line.*

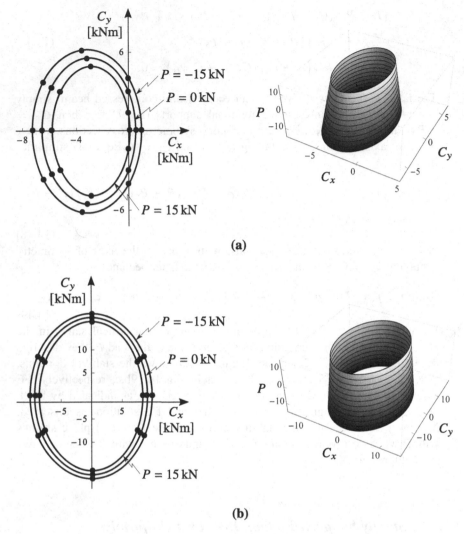

Fig. 14.20 Stability domains of the TWB in Fig. 14.19 (three-dimensional view and contour lines $P = $ const) for (**a**) T-section and (**b**) I-section; black dots denote FE results

(a) Operating like in Sect. 9.7.3, the bifurcation condition requires that the equilibrium equation $\left(\mathbf{K}_e + \mathbf{K}_g\right)\mathbf{q} = \mathbf{0}$ admits a nontrivial solution; the generic stiffness matrices, respectively, elastic and geometric, are defined in Eqs. 9.140 and 9.141. Here, the shape functions, respectful of the geometric boundary conditions, are taken as:

Fig. 14.21 Simply supported TWB non-uniformly bent by distributed transverse loads: (a) structure, (b) I-section

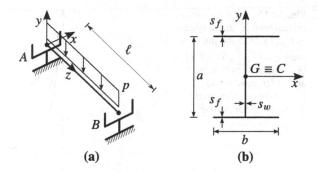

$$\phi_u(z) = \sin\left(\tfrac{\pi}{\ell}z\right),$$
$$\phi_v(z) = 0,$$
$$\phi_\theta(z) = \sin\left(\tfrac{\pi}{\ell}z\right),$$
(14.50)

while the bending moments in the prestressed state are:

$$M_x^0(z) = -p\tfrac{\ell}{2}z + p\tfrac{z^2}{2},$$
$$M_y^0(z) = 0.$$
(14.51)

Referring to a bi-symmetric cross-section, in which point H coincides with the centroid $G \equiv C$ (i.e., $x_H = y_H = 0$), the following stiffness matrices are obtained:

$$\mathbf{K}_e = \begin{pmatrix} \dfrac{\pi^4 EI_y}{2\ell^3} & 0 \\ 0 & \dfrac{\pi^2 GJ}{2\ell} + \dfrac{\pi^4 EI_\omega}{2\ell^3} \end{pmatrix},$$
$$\mathbf{K}_g = \begin{pmatrix} 0 & \tfrac{1}{24}\left(3+\pi^2\right)p\ell \\ \tfrac{1}{24}\left(3+\pi^2\right)p\ell & -\tfrac{1}{2}y_Q p\ell \end{pmatrix},$$
(14.52)

where rows and columns corresponding to the suppressed displacements have been deleted. Concerning the $(2, 2)$ element of the matrix $\mathbf{K}_g$, it has been taken into account that $e_Q = |y_Q|$ and that the downward load is negative when $y_Q > 0$ (since entering in the torsion center) and positive when $y_Q < 0$ (since outgoing from the torsion center). Therefore, the characteristic equation $(\det(\mathbf{K}_e + \mathbf{K}_g) = 0)$ reads:

$$\frac{\pi^8 EI_y EI_\omega}{4\ell^6} + \frac{\pi^6 EI_y GJ}{4\ell^4} - p\frac{\pi^4 EI_y y_Q}{4\ell^2} - \frac{p^2\ell^2}{576}\left(3+\pi^2\right)^2 = 0. \quad (14.53)$$

The latter is recast in the nondimensional form:

$$\frac{\left(3+\pi^2\right)^2}{144}\mu^2 + \pi^4\eta\,\mu - \frac{\pi^6\left(\pi^2+\chi\right)}{\chi} = 0, \quad (14.54)$$

Fig. 14.22 Nondimensional critical load μ_c of the system in Fig. 14.21 vs the stiffness ratio $\chi := \frac{GJ\ell^2}{EI_\omega}$, for a generic bi-symmetric cross-section and different values of η ($\eta > 0$ denotes downward loads applied above the torsion center)

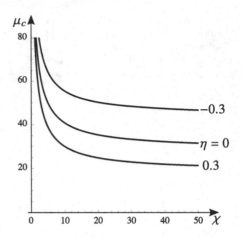

where $\mu := \frac{p\ell^3}{\sqrt{EI_y GJ}}$ is the nondimensional load and the definitions $\chi := \frac{GJ\ell^2}{EI_\omega}$ and $\eta := \frac{y_Q}{\ell}\sqrt{\frac{EI_y}{GJ}}$ have been introduced. The nondimensional critical load is the smallest root of Eq. 14.54, from which the following closed-form expression is found:

$$\mu_c = \frac{12\pi^3}{\left(3 + \pi^2\right)^2} \left(-6\pi\eta + \sqrt{\frac{\pi^2\left(3 + \pi^2\right)^2}{\chi} + 6\pi^2\left(1 + 6\eta^2\right) + \pi^4 + 9}\right).$$
(14.55)

It is observed that when $\chi \to \infty$ (i.e., when the beam is torsionally stiff or long), the critical load tends to:

$$\mu_{c\infty} := \lim_{\chi \to \infty}\left(\mu_c\right) = \frac{12\pi^3}{\left(3 + \pi^2\right)^2}\left(-6\pi\eta + \sqrt{6\pi^2\left(1 + 6\eta^2\right) + \pi^4 + 9}\right),$$
(14.56)

which is independent of χ.

(b) Figure 14.22 shows the critical load vs the χ parameter, for few values of η. It is seen that, for a fixed η, the increase of χ produces a decrease of the critical load, which, consistently with above, tends to an asymptotic value for high values of χ. On the contrary, for a fixed χ, negative values of η lead to higher critical loads and positive values of η to lower critical loads.

(c) When reference is made to the I-section in Fig. 14.21b, whose geometric characteristics are defined in Eq. 14.44, the η parameter further simplifies into $\eta = \frac{y_Q}{a/2}\frac{1}{\sqrt{\chi}}$, so that, remarkably, the two independent parameters η, χ reduce to just one. The following three case studies are investigated: (C) load applied to the torsion center axis ($y_Q = 0$); (I) load applied at the intrados ($y_Q = -\frac{a}{2}$); and (E) load applied at the extrados ($y_Q = \frac{a}{2}$).

Fig. 14.23 Nondimensional critical load μ_c of the system in Fig. 14.21 *vs* $\chi := \frac{GJ\ell^2}{EI_\omega}$, for an I-section, when the load is applied at the intrados (I), torsion center (C), and extrados (E)

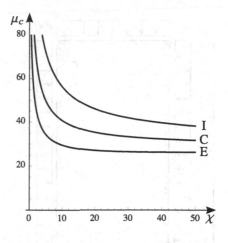

The laws $\mu_c(\chi)$, for each case study, are shown in Fig. 14.23. Curves I and E are representative of any I-section, and curve C is valid for any bi-symmetric cross-section. The latter, indeed, coincides with the curve depicted in Fig. 14.22 for $\eta = 0$. It is observed that the load eccentricity affects the critical value, which increases when the (downward) load is at the bottom and decreases when the load is at the top.

14.4 Buckling of Plates and Prismatic Shells

The critical load of sample systems, made of elastic plates and springs, is determined.

14.4.1 Plate Simply Supported on Four Sides and Subject to Bi-axial Stress

Three rectangular plates (having different aspect ratios), simply supported on the entire contour, are bi-axially prestressed by forces p_x, p_y, uniformly distributed and normal to the edges, positive when they generate compression (Fig. 14.24).[5] (a) Determine the critical load as a function of the ratio between the magnitudes of the two loads. (b) Determine the critical combinations of the two loads, and plot the stability domain.

---

[5] Differently from what is done in Sect. 10.5, here the two loads are free to take both signs.

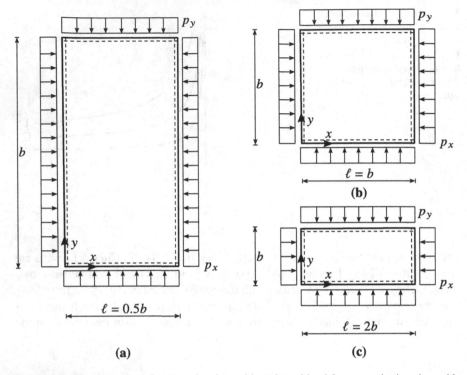

Fig. 14.24 Simply supported rectangular plate subjected to a bi-axial compression/traction, with aspect ratio (**a**) $\alpha = 0.5$, (**b**) $\alpha = 1$, (**c**) $\alpha = 2$

(a) The bifurcation condition of a rectangular plate, simply supported on the entire contour, prestressed by forces p_x, p_y, uniformly distributed and normal to the edges, is defined in Eq. 10.48. By letting $\alpha := \frac{\ell}{b}$ be the aspect ratio of the plate, $\rho := \frac{p_y}{p_x}$ the ratio (positive or negative) between the magnitudes of the two loads, and $p_{xc} := \mu_c \dfrac{\pi^2 D}{b^2}$ the critical nondimensional load, $\mu_c \gtrless 0$, is determined by Eq. 10.50. The values of n, m (number of half-waves along x and y, respectively) which minimize μ_c are determined, by trial and error, for different load ratios ρ, as reported in Table 14.1. It is observed that when $\rho < 0$, both positive and negative values of the critical load are found; they refer to the load combinations $p_x > 0$, $p_y < 0$ (when $\mu_c > 0$) and $p_x < 0$, $p_y > 0$ (when $\mu_c < 0$), respectively. Moreover, when $p_x > 0$, it is seen that as the load ratio increases (from negative to positive values), the critical load decreases.

(b) By letting $\bar{p}_{xc} := 4\dfrac{\pi^2 D}{b^2}$ and $\bar{p}_{yc} := 4\dfrac{\pi^2 D}{\ell^2}$ (i.e., the critical values for uniaxial compression in two directions), Eq. 10.48 is rewritten as:

Table 14.1 Nondimensional critical load of simply supported plates, compressed/taut in two directions, for different values of the load ratio ρ

$\alpha = 0.5$				$\alpha = 1$				$\alpha = 2$			
ρ	n	m	μ_c	ρ	n	m	μ_c	ρ	n	m	μ_c
-2	1	1	12.5	-2	2	1	12.5	-2	5	1	12.37
	1	3	-12.07		1	2	-3.57		1	1	-0.89
-1	1	1	8.33	-1	2	1	8.33	-1	4	1	8.33
	1	4	-33.3		1	2	-8.33		1	1	-2.08
0	1	1	6.25	0	1	1	4	0	2	1	4
1	1	1	5	1	1	1	2	1	1	1	1.25
2	1	1	4.17	2	1	1	1.33	2	1	1	0.69

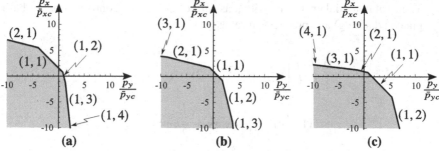

Fig. 14.25 Stability domain of the simply supported rectangular plate, subject to bi-axial compression/traction, with aspect ratios (**a**) $\alpha = 0.5$, (**b**) $\alpha = 1$, and (**c**) $\alpha = 2$; the labels denote the pairs (n, m) relevant to each boundary segment

$$\frac{1}{\alpha^2}\left[\left(\frac{p_x}{\bar{p}_{xc}}\right)n^2 + \left(\frac{p_y}{\bar{p}_{yc}}\right)m^2\right] = \frac{1}{4}\left[\left(\frac{n}{\alpha}\right)^2 + m^2\right]^2, \qquad (14.57)$$

which defines a straight line in the $\left(\frac{p_x}{\bar{p}_{xc}}, \frac{p_y}{\bar{p}_{yc}}\right)$ plane, depending on the two integers n, m; the segments closest to the origin constitute a piecewise straight line that delimits the domain of stability. The latter is represented in Fig. 14.25 for some (n, m) pairs. It is seen that when $\alpha = 1$ (square plate), the stability domain is symmetric with respect to the bisector of the first and third quadrant. Moreover, the points belonging to the two stability boundaries relevant to $\alpha = 0.5$ and $\alpha = 2$ satisfy the following relationship: $\left(\frac{p_y}{\bar{p}_{yc}}, \frac{p_x}{\bar{p}_{xc}}\right)_{\alpha=0.5} = \left(\frac{p_x}{\bar{p}_{xc}}, \frac{p_y}{\bar{p}_{yc}}\right)_{\alpha=2}$, corresponding to a rotation of the structure of $\frac{\pi}{2}$ rad.

14.4.2   Clamped-Free Plate Elastically Supported at a Vertex, Equally Compressed in Two Directions

A rectangular plate of dimensions $\ell \times b$, clamped at the two sides $x = 0$ and $y = b$ and free at the remaining sides, is elastically braced by a spring of stiffness k, located at the vertex $A = (\ell, 0)$ (Fig. 14.26). The plate is prestressed by normal forces $p_x = p_y = p$, uniformly distributed on the free edges (positive when they generate compression). (a) Define the total potential energy. (b) Determine the critical load using the Ritz method. (c) Plot the critical load vs the spring-to-plate stiffness ratio $\eta := \frac{b^2 k}{\pi^2 D}$, together with the corresponding buckling modes. (d) Compare the results with finite element analyses.

(a) The total potential energy is written as sum of the elastic contributions (of both plate and spring) and prestress; it reads:

$$
\Pi = \frac{D}{2} \int_0^\ell \int_0^b \left[w_{,xx}^2 + w_{,yy}^2 + 2\nu w_{,xx} w_{,yy} \right.
$$

$$
\left. + 2(1-\nu) w_{,xy}^2 \right] dx \, dy + \frac{1}{2} k w^2 (\ell, 0)
$$

$$
+ \frac{1}{2} \int_0^\ell \int_0^b \left[N_x^0 w_{,x}^2 + N_y^0 w_{,y}^2 + 2N_{xy}^0 w_{,x} w_{,y} \right] dx \, dy, \quad (14.58)
$$

where use has been made of Eq. 10.9 and of Eq. 10.31, with $N_x^0 = -p$, $N_y^0 = -p$, $N_{xy}^0 = 0$.

(b) Ritz solutions are derived by approximating the transverse deflection $w(x, y)$ according to Eq. B.14, by taking $N = 2$, in which the buckling modes of a clamped-free beam (Eq. B.23) are used as trial functions, namely:

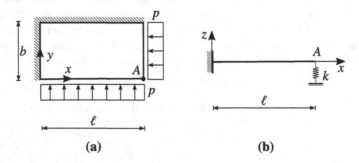

Fig. 14.26 Braced plate equally compressed in two directions: (a) structure, (b) cross-section at $y = 0$

$$X(x) = 1 - \cos\left[\frac{\pi}{\ell}\left(n - \frac{1}{2}\right)x\right], \qquad n = 1, 2,$$

$$Y(y) = 1 - \cos\left[\frac{\pi}{b}\left(n - \frac{1}{2}\right)(y - b)\right].$$

(14.59)

By substituting Eq. B.14 in Eq. 14.58, and imposing stationary of the TPE, the equilibrium equations are obtained:

$$\left(\mathbf{K}_e + \mathbf{K}_g\right)\mathbf{q} = \mathbf{0}.$$

(14.60)

Here, $\mathbf{q} = (a_{11}, a_{12}, a_{21}, a_{22})^T$ and:

$$\mathbf{K}_e = \begin{pmatrix} k_{11}^e & k_{12}^e & k_{13}^e & k_{14}^e \\ & k_{22}^e & k_{23}^e & k_{24}^e \\ & & k_{33}^e & k_{34}^e \\ \text{SYM} & & & k_{44}^e \end{pmatrix} + \eta \begin{pmatrix} 1 & 1 & 1 & 1 \\ & 1 & 1 & 1 \\ & & 1 & 1 \\ \text{SYM} & & & 1 \end{pmatrix},$$

(14.61)

$$\mathbf{K}_g = \mu \begin{pmatrix} k_{11}^g & k_{12}^g & k_{13}^g & 0 \\ & k_{22}^g & 0 & k_{24}^g \\ & & k_{33}^g & k_{34}^g \\ \text{SYM} & & & k_{44}^g \end{pmatrix},$$

where $\eta := \frac{b^2 k}{\pi^2 D}$ is the spring-to plate stiffness ratio, $\mu := \frac{b^2 p}{\pi^2 D}$ is the nondimensional load, and moreover:

$$k_{11}^e := \frac{32\alpha^2 v - 8\pi\left(\alpha^4 + 2\alpha^2 v + 1\right) + \pi^2\left(3\alpha^4 + 2\alpha^2 + 3\right)}{64\alpha^3},$$

$$k_{12}^e := \frac{12(\pi - 4)\alpha^2 v + 3\pi^2 - 4\pi}{96\alpha^3},$$

$$k_{13}^e := \frac{\pi(3\pi - 4)\alpha^2 + 12(\pi - 4)v}{96\alpha},$$

$$k_{14}^e := -\frac{3v}{2\alpha},$$

$$k_{22}^e := \frac{-288\alpha^2 v - 8\pi\left(243\alpha^4 + 18\alpha^2 v - 1\right) + 9\pi^2\left(81\alpha^4 + 6\alpha^2 + 1\right)}{192\alpha^3},$$

$$k_{23}^e := \frac{5v}{2\alpha},$$

$$k_{24}^e := \frac{27\pi(3\pi - 4)\alpha^2 + 12(4 + 3\pi)v}{32\alpha},$$

$$k_{33}^e := \frac{-288\alpha^2 v + 8\pi\left(\alpha^4 - 18\alpha^2 v - 243\right) + 9\pi^2\left(\alpha^4 + 6\alpha^2 + 81\right)}{192\alpha^3},$$

$$k_{34}^e := \frac{48\alpha^2 v + 9\pi\left(4\alpha^2 v + 9\pi - 12\right)}{32\alpha^3},$$

$$k_{44}^e := \frac{288\alpha^2 v + 216\pi\left(\alpha^4 + 2\alpha^2 v + 1\right) + 81\pi^2\left(3\alpha^4 + 2\alpha^2 + 3\right)}{64\alpha^3},$$

$$k_{11}^g := -\frac{\pi(3\pi - 8)\left(\alpha^2 + 1\right)}{16\alpha},$$

$$k_{12}^g := \frac{(4 - 3\pi)\pi}{24\alpha},$$

$$k_{13}^g := \frac{1}{24}(4 - 3\pi)\pi\alpha,$$

$$k_{22}^g := -\frac{\pi\left(27(3\pi - 8)\alpha^2 + 9\pi + 8\right)}{48\alpha},$$

$$k_{24}^g := \frac{3}{8}(4 - 3\pi)\pi\alpha,$$

$$k_{33}^g := -\frac{\pi\left(8\left(\alpha^2 - 27\right) + 9\pi\left(\alpha^2 + 9\right)\right)}{48\alpha},$$

$$k_{34}^g := \frac{3(4 - 3\pi)\pi}{8\alpha},$$

$$k_{44}^g := -\frac{3\pi(8 + 9\pi)\left(\alpha^2 + 1\right)}{16\alpha},$$

(14.62)

Fig. 14.27 Critical load of
the braced plate in Fig. 14.26
vs the spring-to-plate stiffness
ratio η, for different aspect
ratios α ($\nu = 0.2$); black dots
denote FE results

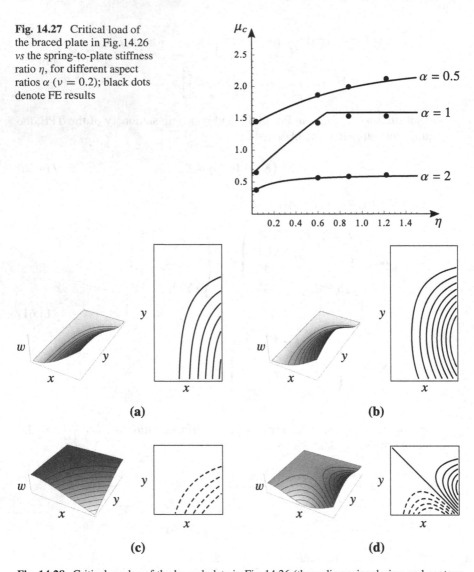

Fig. 14.28 Critical modes of the braced plate in Fig. 14.26 (three-dimensional view and contour
lines), when (**a**) $\alpha = 0.5$, $\eta = 0.2$, (**b**) $\alpha = 0.5$, $\eta = 1$, (**c**) $\alpha = 1$, $\eta = 0.2$, and (**d**) $\alpha = 1$, $\eta = 1$

with $\alpha := \frac{\ell}{b}$ the aspect ratio of the plate. Equations 14.60 admit nontrivial
solution if the determinant of the matrix is equal to zero, from which the critical
load μ_c is evaluated, as a function of α and η.

(c) Figure 14.27 shows the nondimensional critical load μ_c *vs* the spring-to-plate
    stiffness ratio η, for different values of the aspect ratio. A plot of the critical
    modes, corresponding to two different values of η, is also provided in Fig. 14.28
    for $\alpha = 0.5$ and $\alpha = 1$. It is observed that as η increases, the critical load

also increases until to a certain value, after which it remains constant. For the square plate ($\alpha = 1$) to reach this value, the critical mode abruptly changes from symmetric, with respect to a diagonal of the plate, to antisymmetric. On the other hand, in rectangular plates, the change in the buckling mode is smooth; it involves variations in the tangent at the vertex A.

(d) Comparisons with FE results (black dots in Fig. 14.27) are illustrated in terms of critical load. A very good agreement between analytical and numerical solutions is detected.

14.4.3 Square Plate on Elastic Soil, Simply Supported on Four Sides and Subject to Bi-axial Stress

A square plate of side ℓ, simply supported on the entire contour, prestressed by forces p_x, p_y uniformly distributed and normal to the edges (positive when they generate compression), rests on elastic soil of Winkler constant k_f (Fig. 14.29). (a) Write the equilibrium equations and boundary conditions. (b) Determine the critical load as a function of the ratio between the magnitudes of the two loads, and plot it vs the soil-to-plate stiffness $\eta := \frac{k_f}{D} \left(\frac{\ell}{\pi}\right)^4$. (c) Determine the critical combinations of the loads, and plot the stability domain. (d) Represent the corresponding critical modes.

(a) The state of prestress is $N_x^0 = -p_x$, $N_y^0 = -p_y$, $N_{xy}^0 = 0$; the equilibrium equations are derived by Eqs. 10.47 by adding the soil contribution:

$$D \left(w_{,xxxx} + 2w_{,xxyy} + w_{,yyyy}\right) + p_x w_{,xx} + p_y w_{,yy} + k_f w = 0,$$
(14.63a)

$$w = 0, \text{ at } x = 0, \ell,$$
(14.63b)

$$w_{,xx} = 0, \text{ at } x = 0, \ell,$$
(14.63c)

$$w = 0, \text{ at } y = 0, \ell,$$
(14.63d)

$$w_{,yy} = 0, \text{ at } y = 0, \ell,$$
(14.63e)

(b) Still assuming the solution as in Eq. 10.41, and proceeding as in Sect. 10.4, the bifurcation condition is obtained:

$$p_x \left(\frac{n\pi}{\ell}\right)^2 + p_y \left(\frac{m\pi}{\ell}\right)^2 = k_f + D \left[\left(\frac{n\pi}{\ell}\right)^2 + \left(\frac{m\pi}{\ell}\right)^2\right]^2.$$
(14.64)

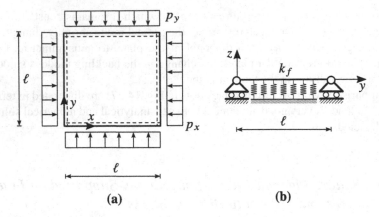

Fig. 14.29 Simply supported square plate resting on elastic soil, subjected to a bi-axial compression/traction: (**a**) structure, (**b**) cross-section

By letting $\rho := \frac{p_y}{p_x}$ (and taking $p_x > 0$, $p_y \gtrless 0$) be the ratio between the magnitudes of the two loads, and defining $p_{xc}(n, m) := \mu_c \frac{\pi^2 D}{\ell^2}$, a nondimensional critical load $\mu_c > 0$ is obtained as:

$$\mu_c = \frac{(n^2 + m^2)^2 + \eta}{n^2 + \rho m^2}, \tag{14.65}$$

where $\eta := \frac{k_f}{D}\left(\frac{\ell}{\pi}\right)^4$ is the soil-to-plate stiffness ratio. By fixing $\rho \gtrless 0$, the values of n, m which minimize the critical load μ_c are sought, by trial and error. Figure 14.30 shows the nondimensional critical load μ_c vs the soil-to-plate stiffness ratio η, for different values of ρ. It is observed that as η increases, the critical load increases according to a piecewise straight line; each segment of this line corresponds to a certain (n, m) pair.

Furthermore, by fixing the soil characteristics, an increase of the ratio ρ entails a decrease of the critical load. Such phenomenon is sometimes accompanied by a change of buckling mode, expressed by the (n, m) pair. As an example, some results are reported in Table 14.2, obtained for $\eta = 25$ and $\eta = 50$ (to be compared with results of Table 10.1, referred to $\eta = 0$).

(c) By denoting by $\bar{p}_{xc} = \bar{p}_{yc} = 4\frac{\pi^2 D}{\ell^2}$ the critical loads for uniaxial compression, Eq. 14.64 is rewritten as:

$$\left(\frac{p_x}{\bar{p}_{xc}}\right)n^2 + \left(\frac{p_y}{\bar{p}_{yc}}\right)m^2 = \frac{1}{4}\left(n^2 + m^2\right)^2 + \frac{1}{4}\eta, \tag{14.66}$$

which establishes a linear relationship between the two (independent) loads p_x, p_y. In particular, it defines a straight line in $\left(\frac{p_x}{\bar{p}_{xc}}, \frac{p_y}{\bar{p}_{yc}}\right)$ plane, depending on

Fig. 14.30 Critical load of the square plate on Winkler foundation vs the soil-to-plate stiffness ratio η, for different values of the load ratio ρ

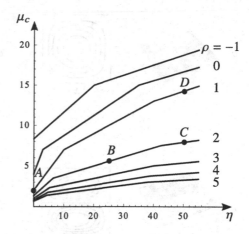

Table 14.2 Nondimensional critical load μ_c of the square plate on Winkler foundation and (n, m) pairs, for some values of the load ratio ρ: (a) $\eta = 25$, (b) $\eta = 50$

(a)

ρ	n	m	μ_c
-1	3	1	15.6
0	2	1	12.5
1	2	1	10
	1	2	

(b)

ρ	n	m	μ_c
-1	3	1	18.8
0	3	1	16.7
1	2	2	14.3

Fig. 14.31 Stability domain of the square plate on Winkler foundation, for different soil-to-plate stiffness ratios η

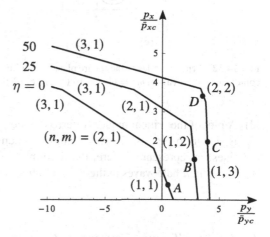

the two integers n, m; the segments closest to the origin constitute a piecewise straight line that delimits the domain of stability. The latter is shown in Fig. 14.31 for different values of the soil-to-plate stiffness ratio. It is observed that the stability domain expands as the soil stiffness increases. Moreover, a correspondence with Fig. 14.30 is shown via points A, B, C, and D, indicated in both plots.

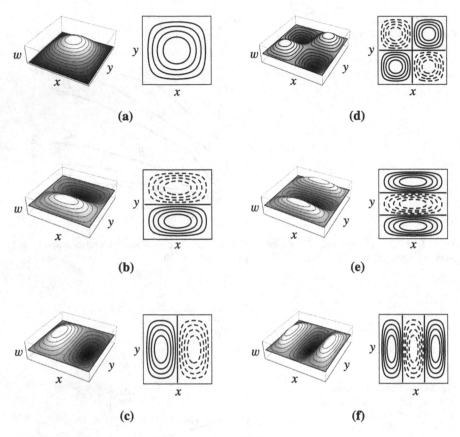

Fig. 14.32 Critical modes of the square plate on Winkler foundation (three-dimensional view and contour lines), for (**a**) $(n, m) = (1, 1)$, (**b**) $(1, 2)$, (**c**) $(2, 1)$, (**d**) $(2, 2)$, (**e**) $(1, 3)$, and (**f**) $(3, 1)$

(d) A plot of the critical modes (referred to the (n, m) pairs indicated in Fig. 14.31) is shown in Fig. 14.32, where a three-dimensional view, together with contour lines, is represented. Here, the deformed configurations describe a plate that bends in n half-waves in the x direction and in m half-waves in the y direction.

14.4.4   Uniformly Compressed Rectangular Tube with Wings

A rectangular tube with wings, constrained at the ends by simple supports and torsional clamps that leave the warping free, is solicited by a compression load P (Fig. 14.33). (a) By using the finite strip method, determine the local critical load. (b) Plot the local critical load as a function of the wing length, for different aspect

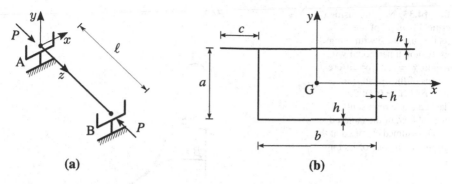

Fig. 14.33 Uniformly compressed rectangular tube with wings: (**a**) structure, (**b**) cross-section

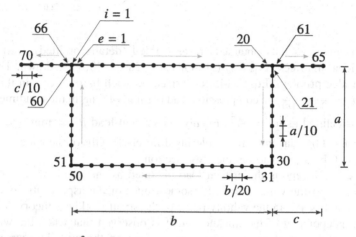

Fig. 14.34 Cross-section of the rectangular tube with wings: numbering of the finite strips e (gray arrows indicate the sense of numbering)

ratios of the cross-section. (c) Represent a sketch of the corresponding buckling modes.

(a) In order to use the finite strip method, the domain occupied by the body is divided into 70 finite strips e, delimited by 70 lines parallel to the longitudinal axis of the structure and by the end cross-section. A scheme is reported in Fig. 14.34, where the numbering of the finite strips is also indicated. By operating as described in the Sect. 10.10.1, the element stiffness matrices $\mathbf{K}_{\alpha n}^{(e)}$ ($\alpha = e, g$), defined in Eqs. 10.140, are expressed in the global basis, by taking into account the orientation of the strips, i.e., according the transformation in Eq. 10.143 and Eq. 10.144. In the latter, $\varphi^{(e)} = 0$ for $e = 1, \cdots, 20$, $\varphi^{(e)} = -\pi/2$ for $e = 21, \cdots, 30$, $\varphi^{(e)} = -\pi$ for $e = 31, \cdots, 50$, $\varphi^{(e)} = -3\pi/2$ for $e = 51, \cdots, 60$, $\varphi^{(e)} = 0$ for $e = 61, \cdots, 65$ and $\varphi^{(e)} = \pi$ for $e = 66, \cdots, 70$.

Fig. 14.35 Nondimensional critical load μ_c of the rectangular tube with wings vs the nondimensional wing length $\chi = c/a$, for different values of the aspect ratio $\beta = b/a$ ($\nu = 0.3$); the dashed curve represents the critical load of the isolated wing, assumed clamped at the ground along one of its long sides

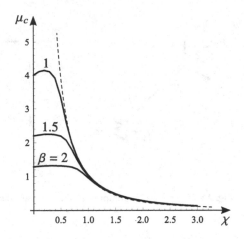

The matrices thus obtained are assembled (details omitted), from which the global matrices $\mathbf{K}_{en}$, $\mathbf{K}_{gn}$ (of dimension 280×280) are drawn. Then, the eigenvalue problem, Eq. 10.145, is derived for each harmonic n; in this latter, said $p := \frac{P}{2(a+b+c)}$ the compression load per unit of length, the nondimensional load is defined as $\mu := \frac{pa^2}{\pi^2 D}$. Finally, the critical load $\mu_c = \min_n ([\mu_c(n)])$ is determined by trial and error, exploring different lengths of the wings and some values of the aspect ratio of the cross-section.

(b) Figure 14.35 reports the critical load obtained as the nondimensional length $\chi := c/a$ of the wings varies, for some values of the aspect ratio $\beta := b/a$. The values of μ_c, in the validity range of the small thickness theory, are found to be independent of the thickness h. It is observed that when the wings are sufficiently short, they exert a stiffening effect on the tube, by increasing its critical load. This gain is less evident in cross-sections with a high aspect ratio β. On the other hand, the wings have a weakening effect on the assembly, by reducing the critical load, when they are sufficiently long. Moreover, when they are very long ($\chi \geq 1$), the critical load of the assembly tends to that of a clamped-free plate (scheme D in Fig. 10.9) of dimensions $\ell \times c$, indicated by the dashed curve in Fig. 14.35, having equation $\mu_c = 1.53/\chi^2$.[6]

(c) Figure 14.36 shows a sketch of the local buckling modes of cross-sections with short ($\chi = 0.2$) and long ($\chi = 1$) wings, for aspect ratios $\beta = 1$ and $\beta = 2$. It is confirmed that when the wings are short, the buckling phenomenon is guided by the tube; when the wings are long, they drive the buckling.

---

[6] This equation is obtained by accounting that $p_c \simeq 1.53\frac{\pi^2 D}{c^2}$ when $\alpha := \frac{\ell}{c} \geq 3$, as appears in Fig. 10.10 (case D), together with the definition in Eq. 10.45. Its successive substitution in the definition used here, $\mu_c := \frac{p_c a^2}{\pi^2 D}$, furnishes the sought law.

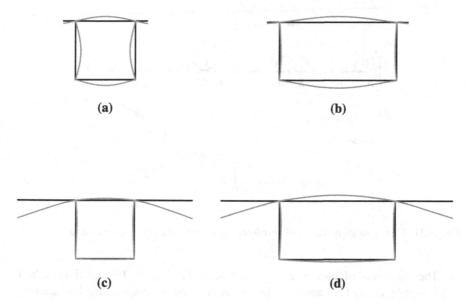

(a)   **(b)**

(c)   **(d)**

Fig. 14.36 Sketch of the local buckling modes of the rectangular tube with (**a, b**) short wings ($\chi = 0.2$) and (**c,d**) long wings ($\chi = 1$); (**a,c**) $\beta = 1$, (**b,d**) $\beta = 2$

14.5 Dynamic Bifurcations Induced by Follower Forces

The dynamic bifurcations induced by follower forces, acting on rigid rods or beams, are investigated.

14.5.1 Triple Pendulum Subjected to Follower Forces

A triple pendulum, made of three massless and rigid rods, AB, BC, and CD, each of which of length ℓ, is loaded by two compressive follower forces of intensity F, applied at the points C and D, which maintain their directions parallel to the rods BC and CD, respectively (Fig. 14.37). There are three lumped masses on the pendulum: two, of intensity m_1, are located at the internal hinges B and C and the other one, of intensity m_2, at D. The rods are connected, to each other and to the ground, by linear viscoelastic devices applied at A, B, and C, of stiffness k_1, k_2, and k_3 and viscosity coefficients c_1, c_2, and c_3, respectively. (a) Determine the equations of motion of the system under the hypothesis of small displacements. (b) Determine the critical value of the forces for the undamped case, when $m_1 = 2m$, $m_2 = m$, and $k_1 = k_2 = k_3 = k$. (c) Detect the possible occurrence of static bifurcations.

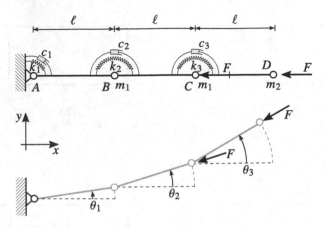

Fig. 14.37 Triple pendulum under follower forces: structure and current configuration

(a) The equations of motion are derived according to the Extended Hamilton Principle, Eq. D.1; to apply it, the kinematics and the constitutive law are first analyzed.

The configuration of the system is described by the rotations of the rods, $\theta_1(t)$, $\theta_2(t)$, and $\theta_3(t)$, taken as Lagrangian parameters (Fig. 14.37). From geometric considerations, analogous to those developed for the Ziegler double pendulum (Sect. 11.3), the following relationships are derived, which express the longitudinal u and transverse v displacement components of points B, C, and D:

$$u_B = -\frac{\ell\theta_1^2}{2}, \tag{14.67a}$$

$$v_B = \ell\,\theta_1, \tag{14.67b}$$

$$u_C = -\ell\left(\frac{\theta_1^2}{2} + \frac{\theta_2^2}{2}\right), \tag{14.67c}$$

$$v_C = \ell\,(\theta_1 + \theta_2), \tag{14.67d}$$

$$u_D = -\ell\left(\frac{\theta_1^2}{2} + \frac{\theta_2^2}{2} + \frac{\theta_3^2}{2}\right), \tag{14.67e}$$

$$v_D = \ell\,(\theta_1 + \theta_2 + \theta_3). \tag{14.67f}$$

The "curvatures" of the viscoelastic device are defined as follows:

$$\kappa_1 := \theta_1, \tag{14.68a}$$

$$\kappa_2 := \theta_2 - \theta_1, \tag{14.68b}$$

$$\kappa_3 := \theta_3 - \theta_2, \tag{14.68c}$$

and the constitutive laws read:

$$M_j = k_j \kappa_j + c_j \dot{\kappa}_j, \quad j = 1, 2, 3, \tag{14.69}$$

where M_j ($j = 1, 2, 3$) are the internal couples, k_j, c_j ($j = 1, 2, 3$) are constants, and the dot denotes time differentiation.

The Extended Hamilton Principle, Eq. D.1, is applied. *Virtual displacements* and *virtual strains* are determined by taking the variations of Eqs. 14.67 and 14.68. The kinetic energy of the system is:

$$T = \frac{1}{2} m_1 \left(\dot{u}_B^2 + \dot{v}_B^2 \right) + \frac{1}{2} m_1 \left(\dot{u}_C^2 + \dot{v}_C^2 \right) + \frac{1}{2} m_2 \left(\dot{u}_D^2 + \dot{v}_D^2 \right), \tag{14.70}$$

in which the contributions of order higher than two have been ignored. The virtual work of the internal forces reads:

$$\delta W_{int} = M_1 \delta \kappa_1 + M_2 \delta \kappa_2 + M_3 \delta \kappa_3 \quad . \tag{14.71}$$

The virtual work of the external forces, accounting for $F_{xC} = -F \cos \theta_2 \simeq -F$, $F_{xD} = -F \cos \theta_3 \simeq -F$, $F_{yC} = -F \sin \theta_2 \simeq -F\theta_2$, and $F_{yD} = -F \sin \theta_3 \simeq -F\theta_3$, is:

$$\delta W_{ext} = F_{xC} \delta u_C + F_{yC} \delta v_C + F_{xD} \delta u_D + F_{yD} \delta v_D. \tag{14.72}$$

After having performed the variation of the kinetic energy, substitution of the three contributions in Eq. D.1 and integration by parts lead to the following balance equations:

$$(2m_1 + m_2) \ell^2 \ddot{\theta}_1 + (m_1 + m_2) \ell^2 \ddot{\theta}_2 + m_2 \ell^2 \ddot{\theta}_3 + M_1 - M_2 \tag{14.73a}$$

$$+ F\ell (\theta_3 - \theta_1) + F\ell (\theta_2 - \theta_1) = 0,$$

$$(m_1 + m_2) \ell^2 \ddot{\theta}_1 + (m_1 + m_2) \ell^2 \ddot{\theta}_2 + m_2 \ell^2 \ddot{\theta}_3 + M_2 - M_3 \tag{14.73b}$$

$$+ F\ell (\theta_3 - \theta_2) = 0,$$

$$m_2 \ell^2 \ddot{\theta}_1 + m_2 \ell^2 \ddot{\theta}_2 + m_2 \ell^2 \ddot{\theta}_3 + M_3 = 0. \tag{14.73c}$$

Using Eqs. 14.68 and 14.69, the equations of motion are finally derived:

$$
\ell^2 \begin{bmatrix} 2m_1 + m_2 & m_1 + m_2 & m_2 \\ m_1 + m_2 & m_1 + m_2 & m_2 \\ m_2 & m_2 & m_2 \end{bmatrix} \begin{pmatrix} \ddot{\theta}_1 \\ \ddot{\theta}_2 \\ \ddot{\theta}_3 \end{pmatrix} + \begin{bmatrix} c_1 + c_2 & -c_2 & 0 \\ -c_2 & c_2 + c_3 & -c_3 \\ 0 & -c_3 & c_3 \end{bmatrix} \begin{pmatrix} \dot{\theta}_1 \\ \dot{\theta}_2 \\ \dot{\theta}_3 \end{pmatrix}
$$

$$
+ \left(\begin{bmatrix} k_1 + k_2 & -k_2 & 0 \\ -k_2 & k_2 + k_3 & -k_3 \\ 0 & -k_3 & k_3 \end{bmatrix} + F\ell \begin{bmatrix} -2 & 1 & 1 \\ 0 & -1 & 1 \\ 0 & 0 & 0 \end{bmatrix} \right) \begin{pmatrix} \theta_1 \\ \theta_2 \\ \theta_3 \end{pmatrix} = \begin{pmatrix} 0 \\ 0 \\ 0 \end{pmatrix}.
$$

$$(14.74)$$

(b) In the undamped case, when $m_1 = 2m$, $m_2 = m$, and $k_1 = k_2 = k_3 = k$, the equations of motion, Eqs. 14.74, read:

$$
m\ell^2 \begin{bmatrix} 5 & 3 & 1 \\ 3 & 3 & 1 \\ 1 & 1 & 1 \end{bmatrix} \begin{pmatrix} \ddot{\theta}_1 \\ \ddot{\theta}_2 \\ \ddot{\theta}_3 \end{pmatrix} + \left(k \begin{bmatrix} 2 & -1 & 0 \\ -1 & 2 & -1 \\ 0 & -1 & 1 \end{bmatrix} + F\ell \begin{bmatrix} -2 & 1 & 1 \\ 0 & -1 & 1 \\ 0 & 0 & 0 \end{bmatrix} \right) \begin{pmatrix} \theta_1 \\ \theta_2 \\ \theta_3 \end{pmatrix} = \begin{pmatrix} 0 \\ 0 \\ 0 \end{pmatrix}.
$$

$$(14.75)$$

They are recast in the following nondimensional form:

$$
\begin{bmatrix} 5 & 3 & 1 \\ 3 & 3 & 1 \\ 1 & 1 & 1 \end{bmatrix} \begin{pmatrix} \ddot{\theta}_1 \\ \ddot{\theta}_2 \\ \ddot{\theta}_3 \end{pmatrix} + \left(\begin{bmatrix} 2 & -1 & 0 \\ -1 & 2 & -1 \\ 0 & -1 & 1 \end{bmatrix} + \mu \begin{bmatrix} -2 & 1 & 1 \\ 0 & -1 & 1 \\ 0 & 0 & 0 \end{bmatrix} \right) \begin{pmatrix} \theta_1 \\ \theta_2 \\ \theta_3 \end{pmatrix} = \begin{pmatrix} 0 \\ 0 \\ 0 \end{pmatrix},
$$

$$(14.76)$$

according to the definitions:

$$
\tilde{t} := t \sqrt{\frac{k}{m\ell^2}}, \quad \mu := \frac{F\ell}{k}. \tag{14.77}
$$

The solutions of Eq. 14.76 are of the type:

$$
\begin{pmatrix} \theta_1 \\ \theta_2 \\ \theta_3 \end{pmatrix} = \begin{pmatrix} u_1 \\ u_2 \\ u_3 \end{pmatrix} e^{\lambda t}, \tag{14.78}
$$

which lead to the following algebraic eigenvalue problem:

$$
\begin{bmatrix} 5\lambda^2 - 2\mu + 2 & 3\lambda^2 + \mu - 1 & \lambda^2 + \mu \\ 3\lambda^2 - 1 & 3\lambda^2 - \mu + 2 & \lambda^2 + \mu - 1 \\ \lambda^2 & \lambda^2 - 1 & \lambda^2 + 1 \end{bmatrix} \begin{pmatrix} u_1 \\ u_2 \\ u_3 \end{pmatrix} = \begin{pmatrix} 0 \\ 0 \\ 0 \end{pmatrix}. \tag{14.79}
$$

The relevant characteristic equation decides on the stability of trivial equilibrium position; it reads:

$$
\lambda^6 + \left(\frac{13}{2} - 3\mu \right) \lambda^4 + \left(\frac{3\mu^2}{2} - 7\mu + \frac{13}{2} \right) \lambda^2 + \frac{1}{4} = 0. \tag{14.80}
$$

Fig. 14.38 Circulatory Hopf bifurcation of the triple pendulum: imaginary parts of the eigenvalues, $\lambda_{1,2,3} = i\omega_1, i\omega_2, i\omega_3$, as function of the follower force multiplier μ

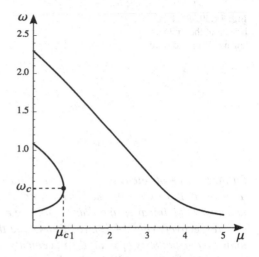

When $\mu = 0$, Eq. 14.80 admits six purely imaginary eigenvalues, $\lambda_{1,\cdots,6} = (\pm i\omega_1, \pm i\omega_2, \pm i\omega_3)$. As μ increases, the difference between the first two frequencies decreases, and when $\mu = \mu_c$, the two pairs of eigenvalues collapse into a unique pair, for which $\omega_1 = \omega_2 =: \omega_c$. The mechanism is represented in Fig. 14.38, which shows the imaginary parts of the eigenvalues as a function of μ (when $\mu > \mu_c$, the first two eigenvalues are no longer purely imaginary and then no longer represented in the figure). The critical values of the force and of the frequency are detected numerically, by increasing μ in small steps until the condition $\omega_1 = \omega_2 =: \omega_c$ is satisfied, to within a prefixed tolerance;[7] they are:

$$\mu_c \simeq 0.774, \qquad \omega_c \simeq 0.511. \qquad (14.81)$$

(c) The system cannot experience static bifurcation. Indeed, the determinant of the stiffness matrix is different from zero (equal to 1), independently of the intensity of follower forces. As a confirmation, the frequencies never vanish in Fig. 14.38.

14.5.2 Planar Beam Braced at the Tip, Subjected to a Follower Force

A planar beam of length ℓ, axial stiffness EA, flexural stiffness EI, and linear mass density m are clamped at the end A and loaded at the free end B by a

---

[7] Alternatively, the characteristic equation $f(\omega_c; \mu_c) = 0$, obtained by letting $\lambda = i\omega_c$ in Eq. 14.80, is solved for the two unknowns, by making system with the coalescence condition $\frac{df(\omega_c; \mu_c)}{d\omega_c} = 0$.

Fig. 14.39 Beck beam braced at the tip by a spring-damper device

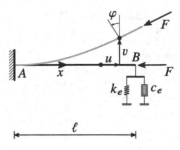

follower force of intensity F, which maintains its own direction aligned with the tangent at B to the deformed centerline. A viscoelastic device is applied at B, composed of a linear spring-damper having elastic and viscosity coefficients k_e and c_e,[8] respectively (Fig. 14.39). (a) Determine the equations of motion of the system under the hypothesis of small displacements and in the framework of the theory of prestressed bodies. (b) Write the characteristic equation. (c) Discuss the effects of the coefficients k_e and c_e on the critical load of the system. (d) Plot the stability domain in the undamped and damped cases.

(a) The system is similar to the Beck beam studied in the Sect. 11.5 but differs (i) for the absence of internal and external distributed damping (of coefficients η and c, respectively) and (ii) for the presence of a viscoelastic device applied at B. The equations of motion can be derived by following the lines of Sect. 11.5, by putting $\eta = c = 0$ and adding the contribution δW_{int}^{br} of the bracing force f_{int}^{br} to the internal virtual work. This is:

$$\delta W_{int}^{br} = f_{int}^{br} \delta v_B, \tag{14.82}$$

with:

$$f_{int}^{br} = k_e\, v_B + c_e\, \dot{v}_B. \tag{14.83}$$

By performing the same steps of Sect. 11.5, Eq. 11.74 (with $\eta = 0$) follows for the longitudinal motion. On the other hand, the equations governing the transverse motion are found to be:

$$m\,\ddot{v} + EI v'''' + F v'' = 0, \tag{14.84a}$$

$$v_A = 0, \tag{14.84b}$$

$$v'_A = 0, \tag{14.84c}$$

$$-EI v'''_B + k_e\, v_B + c_e\, \dot{v}_B = 0, \tag{14.84d}$$

$$EI v''_B = 0. \tag{14.84e}$$

---

[8] The subscript e stands for "extensional."

in which the restoring force due to the bracing appears in the boundary condition Eq. 14.84d. Since the axial and transverse problems are uncoupled, the stability is determined by the latter.[9] The governing equations are recast in the nondimensional form:

$$\ddot{v} + v'''' + \mu \, v'' = 0, \tag{14.85a}$$

$$v_A = 0, \tag{14.85b}$$

$$v'_A = 0, \tag{14.85c}$$

$$-v'''_B + k_e \, v_B + c_e \, \dot{v}_B = 0, \tag{14.85d}$$

$$v''_B = 0, \tag{14.85e}$$

having introduced the following definitions:

$$\tilde{x} := \frac{x}{\ell}, \qquad \tilde{t} := t \sqrt{\frac{EI}{m\ell^4}}, \qquad \tilde{v} := \frac{v}{\ell},$$

$$\mu := \frac{F\ell^2}{EI}, \qquad \tilde{k}_e := \frac{k_e \ell^3}{EI}, \qquad \tilde{c}_e := \frac{c_e \, \ell}{\sqrt{EI \, m}}, \tag{14.86}$$

and omitted the tilde.

(b) Equation 14.85 admit solutions of the form $v(x, t) = \phi(x) \exp(\lambda t)$. A boundary value problem in the space follows:

$$\phi'''' + \mu \, \phi'' + \lambda^2 \phi = 0, \tag{14.87a}$$

$$\phi_A = 0, \tag{14.87b}$$

$$\phi'_A = 0, \tag{14.87c}$$

$$-\phi'''_B + (k_e + c_e \, \lambda) \, \phi_B = 0, \tag{14.87d}$$

$$\phi''_B = 0. \tag{14.87e}$$

The general solution to the field Eq. 14.87a is:

$$\phi(x) = c_1 \cos(\alpha x) + c_2 \sin(\alpha x) + c_3 \cosh(\beta x) + \frac{c_4}{\beta} \sinh(\beta x), \tag{14.88}$$

---

[9] In absence of damping, the longitudinal motion is periodic, but however limited, so that it does not affect stability.

where c_j $(j = 1, \ldots 4)$ are arbitrary constants and α, β are defined in Eq. 11.82.[10] Substitution of the general solution, Eq. 14.88, in the boundary conditions, Eqs. 14.87b–e, leads to the algebraic problem $\mathbf{Sc} = \mathbf{0}$, where $\mathbf{c} = \left(c_j\right)^T$ is a column matrix and:

$$
\mathbf{S} := \begin{bmatrix} 1 & 0 & 1 & 0 \\ 0 & \alpha & 0 & 1 \\ S_{31} & S_{32} & S_{33} & S_{34} \\ -\alpha^2 \cos\alpha & -\alpha^2 \sin\alpha & \beta^2 \cosh\beta & \beta \sinh\beta \end{bmatrix}, \tag{14.89}
$$

with:

$$
\begin{aligned}
& S_{31} := -\alpha^3 \sin\alpha + (k_e + c_e \lambda)\cos\alpha, \quad S_{32} := \alpha^3 \cos\alpha + (k_e + c_e \lambda)\sin\alpha, \\
& S_{33} := -\beta^3 \sinh\beta + (k_e + c_e \lambda)\cosh\beta, \quad S_{34} := -\beta^2 \cosh\beta + (k_e + c_e \lambda)\frac{\sinh\beta}{\beta}.
\end{aligned}
$$
$$\tag{14.90}$$

The characteristic equation ($\det \mathbf{S} = 0$) reads:

$$
\alpha \left(\alpha^4 + \beta^4 + \alpha\beta \left(\alpha^2 - \beta^2\right)\sin\alpha \sinh\beta + 2\alpha^2\beta^2 \cos\alpha \cosh\beta\right)
$$

$$
+ (k_e + c_e \lambda)\left(\alpha^2 + \beta^2\right)\left(\sin\alpha \cosh\beta - \frac{\alpha}{\beta}\cos\alpha \sinh\beta\right) = 0.
$$
$$\tag{14.91}$$

(c) The roots of Eq. 14.91 are the eigenvalues λ, functions of (μ, k_e, c_e), which decide on the stability of the trivial equilibrium position. Depending on the parameters of the system, different bifurcation mechanisms can occur, namely, (i) circulatory Hopf bifurcation, (ii) simple Hopf bifurcation, and (iii) divergence, as discussed ahead.

(i) In the undamped case $(c_e = 0)$, being the system circulatory, the loss of stability manifests via a circulatory Hopf bifurcation, for critical combinations of the parameters (μ, k_e). In this bifurcation, two pairs of purely imaginary eigenvalues coalesce. Denoting by $f_u\left(\omega, \mu_c; k_e\right) = 0$ Eq. 14.91, in which $\lambda = i\omega_c$ has been substituted, such condition calls for satisfying the following two equations:

$$
f_u\left(\omega_c, \mu_c; k_e\right) = 0, \tag{14.92a}
$$

---

[10] The form adopted in Eq. 14.88 differs from that in Eq. 11.81, because of an inessential constant. This writing, indeed, makes Eq. 14.88 valid also in the limit for $\lambda \to 0$ (a case of interest ahead). As a matter of fact, if $\lambda \to 0$, then $\beta \to 0$; therefore:

$$
\phi(x) \to c_1 \cos(\alpha x) + c_2 \sin(\alpha x) + c_3 + c_4 x,
$$

so that the general solution of the compressed beam (in the static regime) is recovered.

$$\frac{\partial f_u (\omega_c, \mu_c; k_e)}{\partial \omega_c} = 0. \qquad (14.92b)$$

(ii) In the damped case $c_e \neq 0$, being the system circulatory and damped, a simple Hopf bifurcation occurs for critical combinations of the parameters (μ, k_e, c_e). In it, only one pair of complex conjugate eigenvalues crosses the imaginary axis. To find the bifurcation condition, $\lambda = i\omega_d$ and $\mu = \mu_d$ are substituted in Eq. 14.91, and the real and imaginary parts are separated, thus obtaining two real equations: $\mathrm{Re}\,[\det(\mathbf{S})] = 0$, $\mathrm{Im}\,[\det(\mathbf{S})] = 0$; these are found to assume the following forms:

$$f_d (\omega_d, \mu_d; k_e) = 0, \qquad (14.93a)$$

$$\mathscr{C}_e\, g_d (\omega_d, \mu_d) = 0, \qquad (14.93b)$$

i.e., damping does appear neither in the real part of the equation nor (after factorization) in the imaginary part; moreover, the stiffness only contributes to the real part.

(iii) The scenario is completed by a third bifurcation mechanism, namely, the divergence (i.e., a static bifurcation). The relevant bifurcation condition is determined by calculating the limit of Eq. 14.91 for $\lambda \to 0$, which leads to the following relationship between load and stiffness of the device:

$$\mu^{3/2} + k_e \left(\sin\sqrt{\mu} - \sqrt{\mu}\cos\sqrt{\mu}\right) = 0. \qquad (14.94)$$

As expected, such a relationship is independent of the damping coefficient (and therefore it does exist both in undamped and damped systems). Remarkably, divergence cannot occur when $k_e = 0$, i.e., in the unbraced structure studied in the Sect. 11.5.

(d) The results of the stability analysis for the undamped case are represented by the stability diagram in the (μ, k_e)-plane, shown in Fig. 14.40a (stable zone in gray). The stability domain is delimited by two curves: (i) $\mathscr{H}^u$, which is the undamped Hopf locus, at which the beam loses stability via a circulatory Hopf; (ii) $\mathscr{D}$, which is the divergence locus, at which the straight configuration loses stability via a static bifurcation. When $k_e = 0$, the critical load of the undamped Beck beam is recovered ($\mu_c = 20.051$). When μ is increased, the beam can exploit two different mechanisms of bifurcation, depending on the magnitude of k_e, which are of dynamic type (curve $\mathscr{H}^u$) or of static type (curve $\mathscr{D}$). The two curves meet, by sharing their tangents, at a degenerate point, which is a *double-zero point* (denoted by DZ in Fig. 14.40a), at which the two imaginary eigenvalues of the circulatory Hopf bifurcation collide at the origin.

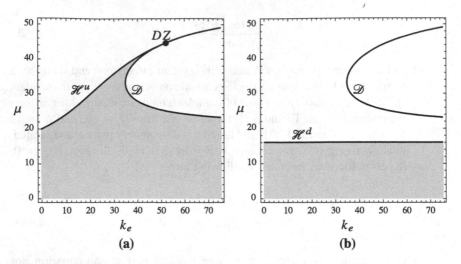

Fig. 14.40 Stability domain of the beam of Fig. 14.39 in (**a**) the undamped case, (**b**) the damped case; stable zone in gray; $\mathcal{H}^u$ circulatory Hopf locus, $\mathcal{D}$ divergence locus, DZ double zero point; $\mathcal{H}^d$ simple Hopf bifurcation locus (for any c_e)

The scenario discussed for the undamped case abruptly changes, when a dashpot is added at the tip of the beam (which, of course, leaves the $\mathcal{D}$-curve unchanged). It is found that, independently of the magnitude of the damping parameter c_e (as commented about Eqs. 14.93), the simple Hopf curve changes into the straight horizontal line $\mu = \mu_d := 16.05$, labelled $\mathcal{H}^d$ in Fig. 14.40b, which is below the divergence curve.[11] Therefore, the damped beam always loses stability via a dynamic bifurcation. Remarkably, the critical load is independent both of c_e and k_e.[12]

14.5.3   Foil Beam in 3D, Eccentrically Braced at the Tip, Subjected to a Follower Force

A spatial clamped-free foil beam of length ℓ, rectangular cross-section of width b much larger than the thickness h, eccentrically braced at the tip, is loaded by a tangential follower force of intensity F applied at the centroid. The bracing device is orthogonal to the axis line x and put to a distance e_Q with respect to it; it is composed of a linear spring-damper having elastic and viscosity coefficients k_e and

---

[11] The original paper [1] contained a miscalculation affecting $\mathcal{H}^d$, removed here.

[12] The independence of k_e is technically explained as follows. If $f_d(\omega_d, \mu_d; k_e) = 0$ is plotted on the (ω_d, μ_d) plane for several values of k_e, it is found that all the curves intersect at the same point $(\omega_d, \mu_d) = (7.06, 16.05)$; moreover, the (unique, for any k_e) curve $g_d(\omega_d, \mu_d) = 0$ also passes through this point, which is therefore the solution to Eqs. 14.93 for any k_e.

Fig. 14.41 Foil beam under a tangential follower force, eccentrically braced at the tip by a viscoelastic device

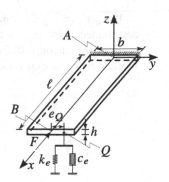

c_e, respectively (Fig. 14.41). *(a) Determine the linearized equations of motion of the system, in the framework of the theory of prestressed bodies, by assuming that the beam is unflexible in the xy-plane (the "strong" plane) and that the torsional moment is comparable with the bending moment in the xz-plane (the "weak" plane). (b) Write the characteristic equation. (c) Plot the stability domains in the undamped and damped cases, for an eccentricity $e_Q = b/2$ and for length-to-width ratios equal to $\ell/b = 5$ and $\ell/b = 10$, respectively.*

(a) The foil beam behaves as a flexural-torsional beam with flexural and torsional stiffnesses EI_y and GJ, respectively; moreover, m is the mass per unit length, and I_x is the mass moment of inertia around the x-axis. The deflection of the beam in the weak plane is denoted by $v = v(x, t)$ and the twist angle around the x-axis by $\theta = \theta(x, t)$. Similarly to Probl. 14.5.2, the equations of motion can be derived by following the lines of Sect. 11.5. In particular, the internal and the external distributed damping are neglected, namely, $\eta = c = 0$, the contribution of the torsion is added to the kinetic energy, i.e., $\frac{1}{2} \int_0^\ell I_x \dot{\theta}^2 dx$, and the contribution of the torsional moment and that of the bracing force are added to the internal virtual work. The latter ones, here denoted as δW_{int}^t and δW_{int}^{br}, respectively, assume the following expressions:

$$\delta W_{int}^t = \int_0^\ell M_t \delta \kappa_t dx, \tag{14.95a}$$

$$\delta W_{int}^{br} = f_{int}^{br} \left(\delta v_B + e_Q \delta \theta_B \right), \tag{14.95b}$$

where $\kappa_t = \theta'$ is the torsional curvature and M_t, f_{int}^{br} are the torsional moment and the bracing force, respectively. These internal forces are linked to the strains by the constitutive laws:

$$M_t = GJ\kappa_t, \tag{14.96a}$$

$$f_{int}^{br} = k_e \left(v_B + e_Q \theta_B \right) + c_e \left(\dot{v}_B + e_Q \dot{\theta}_B \right). \tag{14.96b}$$

By performing the same steps of Sect. 11.5, the equations of motion are determined. Also in this case (as in Probl. 14.5.2), the equations of the axial motion are uncoupled from those governing the transverse and rotational motion, which decide on stability. In nondimensional form, they read:

$$\ddot{v} + v'''' + \mu\, v'' = 0, \tag{14.97a}$$

$$-\rho^2\ddot{\theta} + \eta^2\,\theta'' = 0, \tag{14.97b}$$

$$v_A = 0, \tag{14.97c}$$

$$v'_A = 0, \tag{14.97d}$$

$$-v'''_B + k_e\left(v_B + e_Q\theta_B\right) + c_e\left(\dot{v}_B + e_Q\dot{\theta}_B\right) = 0, \tag{14.97e}$$

$$v''_B = 0, \tag{14.97f}$$

$$\theta_A = 0, \tag{14.97g}$$

$$\eta^2\,\theta'_B + k_e\,e_Q\left(v_B + e_Q\theta_B\right) + c_e\,e_Q\left(\dot{v}_B + e_Q\dot{\theta}_B\right) = 0, \tag{14.97h}$$

where the following definitions have been introduced:

$$\tilde{x} := \frac{x}{\ell}, \qquad \tilde{t} := t\sqrt{\frac{EI_y}{m\ell^4}}, \qquad \tilde{v} := \frac{v}{\ell},$$

$$\mu := \frac{F\ell^2}{EI_y}, \qquad \eta := \sqrt{\frac{GJ}{EI_y}}, \qquad \tilde{e}_Q := \frac{e_Q}{\ell}, \tag{14.98}$$

$$\rho := \sqrt{\frac{I_x}{m\ell^2}}, \qquad \tilde{k}_e := \frac{k_e\ell^3}{EI_y}, \qquad \tilde{c}_e := \frac{c_e\,\ell}{\sqrt{EI_y\,m}}.$$

and the tilde removed. For a rectangular $b \times h$ cross-section, it is:

$$I_y = \frac{1}{12}bh^3, \quad J = \frac{1}{3}bh^3, \quad m = \rho_m bh, \quad I_x = \frac{1}{12}b^3h\rho_m, \tag{14.99}$$

with ρ_m the mass density of the foil

It is worth noticing that (i) the equations of motion are coupled only in the boundary conditions, (ii) the external force enters only the translational equation, and (iii) if the bracing is non-eccentric, i.e., $e_Q = 0$, the equations for twist are uncoupled with respect to the translational equations, so that twist does not affect stability; in this case, the results of Probl. 14.5.2 still apply.

(b) The stability analysis of the trivial equilibrium position $v = \theta = 0$ is here addressed. The solution of Eq. 14.97 is sought in the form:

$$\begin{pmatrix} v \\ \theta \end{pmatrix} = \begin{pmatrix} \phi \\ \psi \end{pmatrix}\exp\left(\lambda t\right), \tag{14.100}$$

from which the following (spatial) boundary value problem is obtained:

$$\lambda^2 \phi + \phi'''' + \mu \phi'' = 0, \tag{14.101a}$$

$$-\rho^2 \lambda^2 \psi + \eta^2 \psi'' = 0, \tag{14.101b}$$

$$\phi_A = 0, \tag{14.101c}$$

$$\phi'_A = 0, \tag{14.101d}$$

$$-\phi'''_B + (k_e + c_e \lambda)\left(\phi_B + e_Q \psi_B\right) = 0, \tag{14.101e}$$

$$\phi''_B = 0, \tag{14.101f}$$

$$\psi_A = 0, \tag{14.101g}$$

$$\eta^2 \psi'_B + e_Q (k_e + c_e \lambda)\left(\phi_B + e_Q \psi_B\right) = 0. \tag{14.101h}$$

The uncoupled field equations, Eqs. 14.101a,b admit the solutions:[13]

$$\phi(x) = c_1 \cos(\alpha x) + c_2 \sin(\alpha x) + c_3 \cosh(\beta x) + \frac{c_4}{\beta} \sinh(\beta x), \tag{14.102a}$$

$$\psi(x) = c_5 \cos(\gamma x) + \frac{c_6}{\gamma} \sin(\gamma x), \tag{14.102b}$$

where c_j $(j = 1, \cdots, 6)$ are arbitrary constants, α, β are defined in Eq. 11.82, and

$$\gamma^2 := -\left(\frac{\lambda \rho}{\eta}\right)^2. \tag{14.103}$$

By substituting Eq. 14.102 in the boundary conditions, Eqs. 14.101c-h, the algebraic problem $\mathbf{Sc} = \mathbf{0}$ is obtained, where $\mathbf{c} = \left(c_j\right)^T$ is a column matrix and the 6×6 matrix $\mathbf{S}$ reads:

$$\mathbf{S} := \begin{bmatrix} 1 & 0 & 1 & 0 & 0 & 0 \\ 0 & \alpha & 0 & 1 & 0 & 0 \\ S_{31} & S_{32} & S_{33} & S_{34} & S_{35} & S_{36} \\ S_{41} & S_{42} & S_{43} & S_{44} & 0 & 0 \\ 0 & 0 & 0 & 0 & 1 & 0 \\ S_{61} & S_{62} & S_{63} & S_{64} & S_{65} & S_{66} \end{bmatrix}, \tag{14.104}$$

---

[13] The solution of Eq. 14.101b is valid also in the limit case $\lambda \to 0$. Indeed, if $\lambda \to 0$, then $\gamma \to 0$, so that

$$\psi(x) \to c_5 + c_6 x.$$

Therefore, the general static solution of the beam in uniform torsion is recovered.

whose coefficients are:

$$
\begin{aligned}
&S_{31} := -\alpha^3 \sin\alpha + (k_e + c_e\,\lambda)\cos\alpha, &&S_{32} := \alpha^3 \cos\alpha + (k_e + c_e\,\lambda)\sin\alpha,\\
&S_{33} := -\beta^3 \sinh\beta + (k_e + c_e\,\lambda)\cosh\beta, &&S_{34} := -\beta^2 \cosh\beta + (k_e + c_e\,\lambda)\frac{\sinh\beta}{\beta},\\
&S_{35} := e_Q\,(k_e + c_e\,\lambda)\cos\gamma, &&S_{36} := e_Q\,(k_e + c_e\,\lambda)\frac{\sin\gamma}{\gamma},\\
&S_{41} := -\alpha^2 \cos\alpha, &&S_{42} := -\alpha^2 \sin\alpha,\\
&S_{43} := \beta^2 \cosh\beta, &&S_{44} := \beta \sinh\beta,\\
&S_{61} := e_Q\,(k_e + c_e\,\lambda)\cos\alpha, &&S_{62} := e_Q\,(k_e + c_e\,\lambda)\sin\alpha,\\
&S_{63} := e_q\,(k_e + c_e\,\lambda)\cosh\beta, &&S_{64} := e_Q\,(k_e + c_e\,\lambda)\frac{\sinh\beta}{\beta},\\
&S_{65} := e_Q^2\,(k_e + c_e\,\lambda)\cos\gamma - \gamma\eta^2 \sin\gamma, &&S_{66} := e_Q^2\,(k_e + c_e\,\lambda)\frac{\sin\gamma}{\gamma} + \eta^2 \cos(\gamma).
\end{aligned}
$$
$$(14.105)$$

Then, the characteristic equation, $\det \mathbf{S} = 0$, reads:

$$
\alpha\left(\alpha^4 + \beta^4 + \alpha\beta\left(\alpha^2 - \beta^2\right)\sin\alpha \sinh\beta + 2\alpha^2\beta^2 \cos\alpha \cosh\beta\right)
$$
$$
\times\left(\eta^2 \cos\gamma + e_Q^2\,(k_e + c_e\,\lambda)\frac{\sin\gamma}{\gamma}\right) + (k_e + c_e\,\lambda)\left(\alpha^2 + \beta^2\right)
$$
$$
\times\left(\sin\alpha \cosh\beta - \frac{\alpha}{\beta}\cos\alpha \sinh\beta\right)\eta^2 \cos\gamma = 0. \qquad (14.106)
$$

(c) The eigenvalues λ, which decide on the stability of trivial equilibrium position, are the root of Eq. 14.106. Similarly to Probl. 14.5.2, depending on the parameters of the system, three different bifurcation mechanisms can occur, which are the circulatory and the simple Hopf bifurcation, as well as the divergence. The conditions under which the three bifurcations take place are the same previously discussed. In particular, the divergence locus is described by the following equation:

$$
\left(e_Q^2 k_e + \eta^2\right)\mu^{3/2} + k_e\eta^2\left(\sin\left(\sqrt{\mu}\right) - \sqrt{\mu}\cos\left(\sqrt{\mu}\right)\right) = 0, \qquad (14.107)
$$

while the dynamic bifurcation loci are evaluated numerically. Two cases are studied, in which the parameters in Eqs. 14.98 assume the following values:

- Case study I: $\ell/b = 5$, entailing:

$$
\rho = \frac{1}{10\sqrt{3}}, \quad \eta = \sqrt{\frac{2}{1+\nu}}. \qquad (14.108)
$$

- Case study II: $\ell/b = 10$, entailing:

$$
\rho = \frac{1}{20\sqrt{3}}, \quad \eta = \sqrt{\frac{2}{1+\nu}}, \qquad (14.109)
$$

with ν being the Poisson's ratio, which has been assumed equal to 0.3. Moreover, the dimensional eccentricity is taken equal to $b/2$, so that the nondimensional value is $e_Q = \frac{b}{2\ell}$.

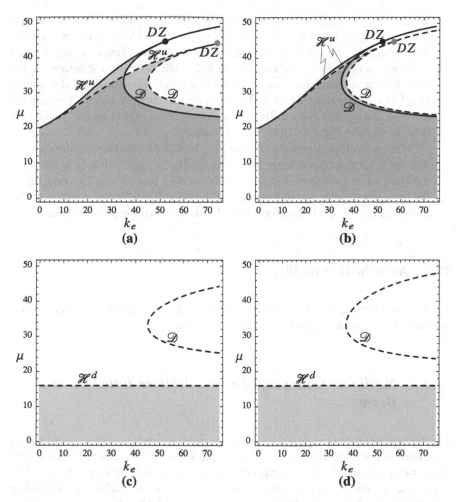

Fig. 14.42 Linear stability diagrams of the beam in Fig. 14.41: (**a**, **b**) undamped system, (**c,d**) damped system; (**a**, **c**) case study I; (**b**, **d**) case study II; solid and dashed curves refer to $e_Q = 0$ and $e_Q \neq 0$, respectively; stable zone in gray; $\mathscr{H}^u$ circulatory Hopf locus, $\mathscr{D}$ divergence locus, $\mathscr{H}^d$ simple Hopf bifurcation locus (for any c_e), DZ double zero point

The linear stability diagrams for the undamped case are displayed in Fig. 14.42a,b. There, the critical load μ is plotted versus the stiffness of the spring k_e, for the case studies I (Fig. 14.42a) and II (Fig. 14.42b), respectively. The dashed curves of the two case studies are compared with the solid curves of the non-eccentric bracing analyzed in Probl. 14.5.2; the shaded region represents the stable zone of the parameter plane.

Also in this case, the beam can lose stability through a static bifurcation ($\mathscr{D}$ curves in figure) or through a dynamic bifurcation ($\mathscr{H}^u$ curves in figure). It is seen that in both cases I and II, the flexural-torsional coupling, due to the eccentricity of the spring, is detrimental for the dynamic bifurcation; as a consequence, the best location for the spring is $e_Q = 0$. Conversely, the coupling is beneficial for the static bifurcation, since now the divergence occurs at larger k_e-values; consequently, the eccentricity $b/2$ is the optimum for increasing the static bifurcation load. It is worth noticing that the effect of the flexural-torsional coupling strongly depends on the length-to-width ratio ℓ/b; indeed, the larger is the ratio, the smaller is the difference with the case of non-eccentric bracing [7] (Fig. 14.42a and Fig. 14.42b should be compared).

When a dashpot is added to the tip (Fig. 14.42c,d), similar considerations to that made for the planar beam hold, namely, the $\mathscr{D}$ locus remains unchanged; the Hopf locus ($\mathscr{H}^d$ curve) is a horizontal straight line for any c_e, lying below the divergence curve and independent of the eccentricity (consistently with the independency of c_e). It therefore coincides with that of the planar beam.

14.6 Aeroelastic Stability

The dynamic bifurcations triggered by aeroelastic forces, acting on elastic beams, or elastically constrained rigid cylinders, are investigated.

14.6.1 Nonlinear Galloping of a Base-Isolated Euler-Bernoulli Beam

The dynamics of a base-isolated tower, of square cross-section, exposed to a steady wind flow, uniformly distributed along its height, is modeled as a planar viscoelastic Euler-Bernoulli beam. The latter is constrained at the bottom end A ($z = 0$) by a viscoelastic device, free at the top end B ($z = \ell$), and subject to a wind of velocity $U\mathbf{a}_x$, orthogonal to the plane of motion (Fig. 14.43). The relevant equations of motion and boundary conditions are [5, 6]:

$$m\ddot{v} + c_e \dot{v} + EI \left(1 + \eta \frac{\partial}{\partial t} \right) v'''' = f_y^a (z, t), \tag{14.110a}$$

$$v'_A = 0, \tag{14.110b}$$

$$EI \left(1 + \eta \frac{\partial}{\partial t} \right) v'''_A + k_d v_A + C_d \dot{v}_A = 0, \tag{14.110c}$$

$$-EI \left(1 + \eta \frac{\partial}{\partial t} \right) v'''_B = 0, \tag{14.110d}$$

$$EI \left(1 + \eta \frac{\partial}{\partial t} \right) v''_B = 0. \tag{14.110e}$$

Fig. 14.43 Base-isolated beam under transverse uniform steady wind flow

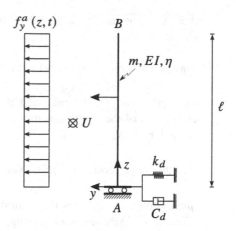

Here, $v(z,t)$ is the transverse displacement; EI is the flexural stiffness; m is the mass per unit length; η is the internal viscous damping coefficient;[14] c_e is the external damping coefficient; $f^a(z,t)$ are the nonlinear aerodynamic loads, defined by the odd power series in Eq. 12.49; and k_d and C_d are the spring stiffness and the viscosity coefficient, respectively, of the device at the base. The following numerical data are assigned: $\ell = 300$ m, $EI = 2.95 \times 10^{13}$ Nm^2, $m = 45000$ Kg/m, $\eta = 0.00091$ s, $c_e = 320$ Ns/m^2; $C_d = 80000$ Ns/m is fixed, while k_d is left as free parameter; $\rho_a = 1.225$ Kg/m^3; the cross-section is taken square, of side $D = 12$ m; the relevant aerodynamic coefficients are $A_1 = -0.9298$ and $A_3 = 7.677$. (a) Determine the first natural mode of the undamped system, and plot it for $k_d = (10.9, 17.5, 24.0) \times 10^6$ N/m. (b) By using the Galerkin method, reduce the problem to a single DOF (i.e., to the Rayleigh oscillator), by taking the fundamental undamped mode as shape function. (c) Determine the critical galloping velocity of the damped system , plot it vs the device stiffness and for different dashpots $C_d = (30000, 50000, 80000)$ Ns/m, and comment the influence of the base insulation on the mechanical performances of the beam. (d) Compute the limit cycle using a perturbation solution, and plot the bifurcation diagram (amplitude vs wind velocity) when $k_d = (10.9, 17.5, 24.0) \times 10^6$ N/m.

(a) The natural modes of the undamped isolated beam are solution to the following differential eigenvalue problem:

$$EI\phi'''' - m\omega^2\phi = 0, \tag{14.111a}$$

$$\phi'_A = 0, \tag{14.111b}$$

$$EI\phi'''_A + k_d\phi_A = 0, \tag{14.111c}$$

$$EI\phi'''_B = 0, \tag{14.111d}$$

$$EI\phi''_B = 0. \tag{14.111e}$$

---

[14] A slight notational difference with Eq. 12.56 exists, concerning η.

where ω is the unknown natural frequency. The general solution of the field equation is:

$$\phi(z) = c_1 \cos(\beta z) + c_2 \cosh(\beta z) + c_3 \sin(\beta z) + c_4 \sinh(\beta z), \qquad (14.112)$$

where $\beta^4 := \frac{m\omega^2}{EI}$. By enforcing the boundary conditions, and zeroing the determinant of the coefficient matrix, a transcendental characteristic equation is found, namely:

$$\cosh(\beta\ell)\left(k_d \cos(\beta\ell) - \beta^3 EI \sin(\beta\ell)\right) + k_d = \beta^3 EI \cos(\beta\ell) \sinh(\beta\ell), \qquad (14.113)$$

which implicitly expresses the natural frequencies as depending on the device stiffness. The smallest root β_1 supplies the fundamental frequency ω_1. By successively solving the homogeneous problem, the constants in Eq. 14.112 are found; by enforcing the normalization $\phi(\ell) = 1$, they are:

$$c_1 = \frac{2\beta^3 EI \sinh(\beta\ell) - k_d(\cos(\beta\ell) + \cosh(\beta\ell))}{k_d(\sin(\beta\ell) - \sinh(\beta\ell))} c_4, \qquad (14.114a)$$

$$c_2 = \frac{k_d(\cos(\beta\ell) + \cosh(\beta\ell)) - 2\beta^3 EI \sin(\beta\ell)}{k_d(\sin(\beta\ell) - \sinh(\beta\ell))} c_4, \qquad (14.114b)$$

$$c_3 = -c_4, \qquad (14.114c)$$

$$c_4 = \frac{k_d(\sinh(\beta\ell) - \sin(\beta\ell))}{2\beta^3 EI \sin(\beta\ell) \cosh(\beta\ell) - 2\sinh(\beta\ell)\left(\beta^3 EI \cos(\beta\ell) + k_d \sin(\beta\ell)\right)}. \qquad (14.114d)$$

The fundamental mode is plotted in Fig. 14.44 for three different values of the device stiffness. The magnitude of the displacement at the base, compared with the displacement at the tip, and how it increases for softer springs, should be noticed.

(b) To discretize the problem, $v(z, t) = \phi(z) q(t)$ is taken, with q the displacement at the tip of the beam (being $\phi(\ell) = 1$). The weak formulation of equilibrium is offered by the Virtual Work Principle (VWP) which requires that:

$$\int_0^\ell \left[m\ddot{v} + c_e\dot{v} + EIv'''' + EI\eta\dot{v}'''' - f_y^a(z,t)\right]\phi(z)\,dz - F_A\phi_A + \cancel{F_B}\phi_B + \cancel{C_B}\phi'_B = 0, \qquad (14.115)$$

where F_H, C_H $(H = A, B)$ are the residual forces and couples acting at the free ends. By using the constitutive law, and successively discretizing, it follows:

$$F_A := -EIv'''_A - EI\eta\dot{v}'''_A - k_d v_A - C_d\dot{v}_A \qquad (14.116a)$$

$$= -(EIq + EI\eta\dot{q})\phi'''_A - k_d\phi_A q - C_d\phi_A\dot{q},$$

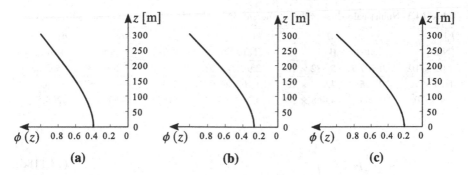

Fig. 14.44 Fundamental mode of the undamped isolated beam, for increasing $k_d =$ $(10.9, 17.5, 24.0) \times 10^6$ N/m

$$F_B := -EIv'''_B - EI\eta\dot{v}'''_B \tag{14.116b}$$

$$= -(EIq + EI\eta\dot{q})\,\phi'''_B,$$

$$C_B := EIv''_B + \eta EI\dot{v}''_B \tag{14.116c}$$

$$= (EIq + \eta EI\dot{q})\,\phi''_B.$$

Since, by Eq. 14.111, it is $\phi''_B = \phi'''_B = 0$, then F_B and C_B cancel in the problem under consideration.

From the VWP, the Rayleigh oscillator equation follows (Eq. 12.54):

$$M\ddot{q} + (c_s + Ub_1)\,\dot{q} + k_1 q + \frac{1}{U}b_3\dot{q}^3 = 0, \tag{14.117}$$

whose coefficients are:

$$M := m \int_0^\ell \phi^2 \mathrm{d}z, \tag{14.118a}$$

$$k_1 := EI \int_0^\ell \phi''''\phi\,\mathrm{d}z + EI\phi'''_A\phi_A + k_d\phi_A^2 = M\omega_1^2, \tag{14.118b}$$

$$c_s := c_e \int_0^\ell \phi^2 \mathrm{d}z + \eta EI \int_0^\ell \phi''''\phi\,\mathrm{d}z + \eta EI\phi'''_A\phi_A + C_d\phi_A^2 \tag{14.118c}$$

$$= \left(\frac{c_e}{m} + \eta\omega_1^2\right) M + (C_d - \eta k_d)\,\phi_A^2,$$

Table 14.3 Numerical values of the coefficients in Eq. 14.118 for different device stiffnesses

k_d	ω_1	M	k_1	c_s	b_1	b_3
[N/m]	[rad/s]	[kg]	[N/m]	[Ns/m]	[kg/m]	[kg/m]
1.09×10^7	0.707	5.882×10^6	2.941×10^6	55252.6	-893.33	4302.83
1.75×10^7	0.795	4.920×10^6	3.107×10^6	42582.6	-747.26	3526.
2.4×10^7	0.843	4.468×10^6	3.173×10^6	37167.9	-678.54	3204.57

$$b_1 := \frac{1}{2}\rho_a D A_1 \int_0^\ell \phi^2 dz, \tag{14.118d}$$

$$b_3 := \frac{1}{2}\rho_a D A_3 \int_0^\ell \phi^4 dz, \tag{14.118e}$$

where use has been made of Eq. 14.111a,c to recast the expressions of k_1, c_s in a more convenient form. By numerically evaluating the coefficients for different device stiffnesses k_d, the values in Table 14.3 are obtained. It is seen that the smaller k_d, the larger the equivalent total damping c_s, and the larger the absolute value of the aerodynamic coefficients b_1, b_3.

(c) The critical wind velocity, according to the discussion of Sect. 12.3.2, is that one which zeroes the total damping, therefore $U_c = -\frac{c_s}{b_1}$. Its values are plotted in Fig. 14.45 vs the device stiffness and for different dashpots at the base. In the figure, the limit case $k_d \to \infty$ (beam clamped at the ground) is also represented by a dashed line. When the stiffness is very high, the galloping velocity tends (from below) to that of the clamped beam. The generally beneficial effect of the base insulation is observed, since it increases the critical velocity of the clamped beam in a wide range of low stiffness values. This range is as much larger as the damping at the base, C_d, is larger. The reason of the good performance of the isolated beam is due to the contribution brought by the added device, C_d, to the global damping of the system, c_s (as shown by Eq. 14.118c). The added dashpot, indeed, dissipates energy so far the spring k_d permits translation of the clamp at the ground, so that its beneficial effect is large for soft springs, but decreases (up to zero) for stiff springs. In contrast, a detrimental effect manifests for large stiffness, entailing even a reduction of the critical velocity of the clamped beam. This is caused by the fact that the device stiffness k_d also influences the global internal damping (by the way of the modal frequency ω_1, the modal mass M, and the displacement at the base, as highlighted by Eq. 14.118c), being able to produce an overall detrimental effect which counterbalances that of C_d. More specifically, the internal damping η brings to c_s a negative contribution at the boundary of the beam and a (prevailing, due to the dissipative nature of damping) positive contribution in the domain. The combined effect, however, decreases when the stiffness k_d is decreased from infinity.

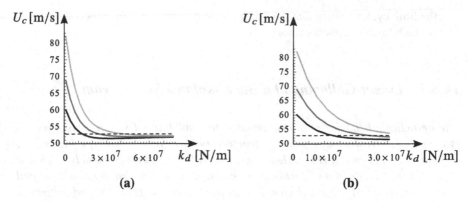

Fig. 14.45 Critical wind velocity of the base-isolated beam *vs* the device stiffness, for different dashpots at the base, $C_d = (30000, 50000, 80000)$ Ns/m (black, dark gray, and light gray lines, respectively): (a) large view showing the asymptotic trend, (b) zoom of the soft spring range

Fig. 14.46 Bifurcation diagrams of the base-isolated beam: amplitude of the limit cycle *vs* the wind velocity; $k_d = (10.9, 17.5, 24.0) \times 10^6$ N/m (black, dark gray, and light gray lines, respectively)

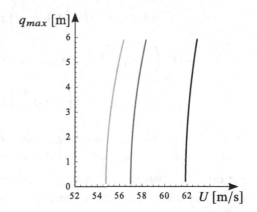

(d) To compute the limit cycle, use is made of the perturbation solution developed in Sect. 12.3.3. First, the Rayleigh Eq. 14.117 is put in the canonical form of Eq. 12.29, by performing the following identifications: $\omega_s^2 = \frac{k_1}{M}$, $\xi_s = \frac{c_s}{2\omega_s M}$, $\kappa_s = 0$, $\zeta_1 := \frac{b_1}{2\omega_s M}$, $\zeta_3 := \frac{b_3}{M}$. Then, Eq. 12.41b is used, with $\epsilon = q_{max}$ the amplitude of oscillation of the tip of the beam, to obtain:

$$U = U_c \left(1 + q_{max}^2 \frac{3\omega_s \xi_s}{8U_c^2 |\zeta_1|}\right) = U_c \left(1 + q_{max}^2 \frac{3\omega_s c_s}{8U_c^2 |b_1|}\right). \qquad (14.119)$$

Figure 14.46 shows the diagrams of the wind velocity *vs* the amplitude q_{max}, for $k_d = (10.9, 17.5, 24.0) \times 10^6$ N/m. It is observed that in the explored range of the amplitudes, the base device does not substantially influence the amplitude of the limit cycles, stating that the linear passive controller only affects the trigger of the bifurcation phenomenon.[15]

14.6.2   Linear Galloping of a Base-Isolated Shear Beam

The aeroelastic behavior of a base-isolated tall building, of square cross-section, is considered. By assuming that the structure behaves as a shear-type frame (i.e., a multi DOF system), it is modeled as an equivalent continuum, namely, a planar shear beam,[16] made of a viscoelastic material. It is constrained at the bottom end A ($z = 0$) by a viscoelastic device, free at the top end B ($z = \ell$), and subject to a steady wind of velocity $U\mathbf{a}_x$, orthogonal to the plane of motion (Fig. 14.47). The relevant equations of motion and boundary conditions are [2]:

$$m\ddot{v} + c_e\dot{v} - GA\left(1 + \eta\frac{\partial}{\partial t}\right)v'' = f_y^a(z,t), \qquad (14.120a)$$

$$-GA\left(1 + \eta\frac{\partial}{\partial t}\right)v_A' + k_d v_A + C_d \dot{v}_A = 0, \qquad (14.120b)$$

$$GA\left(1 + \eta\frac{\partial}{\partial t}\right)v_B' = 0. \qquad (14.120c)$$

Here, $v(z,t)$ is the transverse displacement; GA is the shear stiffness of the beam; m is the mass per unit length; η is the internal viscous damping coefficient; c_e is the external damping coefficient; $f^a(z,t) = -\frac{1}{2}\rho_a U D A_1 \dot{v}(z,t)$ are the linear aerodynamic loads (defined according to Eq. 12.49); and k_d and C_d are the device stiffness and viscosity coefficients, respectively. The following numerical data are given: $\ell = 36$ m, $GA = 8.65 \times 10^7$ N, $m = 4737$ Kg/m, $\eta = 0.0014859$ s, $c_e = 34.87$ Ns/m^2; $C_d = 19202.3$ Ns/m is fixed, while k_d is left as a free parameter; $\rho_a = 1.225$ Kg/m^3; the cross-section is square, of side $D = 16$ m, and the relevant aerodynamic coefficient is $A_1 = -0.9298$. (a) Evaluate the fundamental natural mode of the undamped system, and plot it for $k_d = (3.60, 7.20, 24.02) \times 10^6$ N/m. (b) By applying the Galerkin method, reduce the system to the Rayleigh oscillator, by using the undamped fundamental mode as shape function. (c) Determine the critical galloping velocity, and plot it vs the spring stiffness, for different dashpot damping values $C_d = (12801.5, 19202.3, 25603)$ Ns/m.

---

[15] To reduce the limit cycle, a *nonlinear* dashpot is needed, as discussed in [5].

[16] A planar shear beam is a 1D internally constrained polar continuum, whose material points are allowed only to experience transversal translations $v(z)$, without rotation. It can be derived by the inextensible Timoshenko beam, by zeroing the curvature, $\kappa = \varphi' = 0$. Therefore, the unique non-zero strain is the shear $\gamma = v'$. Accordingly, the bending moment is a reactive internal force, while the shear force is active.

Fig. 14.47 Base-isolated shear beam under transverse uniform steady wind flow

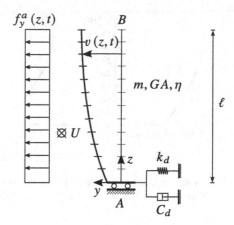

(a) The differential eigenvalue problem governing the undamped oscillations is:

$$- GA\phi'' - m\omega^2\phi = 0, \tag{14.121a}$$

$$-GA\phi'_A + k_d\phi_A = 0, \tag{14.121b}$$

$$GA\phi'_B = 0, \tag{14.121c}$$

whose field equation admits the general solution:

$$\phi(z) = c_1\cos(\beta z) + c_2\sin(\beta z), \tag{14.122}$$

where $\beta^2 := \frac{m\omega^2}{GA}$. By using the boundary conditions, the characteristic equation follows:

$$\cos(\beta\ell) = \frac{GA}{k_d}\beta\sin(\beta\ell), \tag{14.123}$$

whose smallest root determines the fundamental frequency ω_1 as a function of k_d. The associate eigenvector, such that $\phi(\ell) = 1$, is given by Eq. 14.122, with:

$$c_1 = \frac{\beta GA}{\beta GA\cos(\beta\ell) + k_d\sin(\beta\ell)}, \quad c_2 = \frac{k_d}{\beta GA\cos(\beta\ell) + k_d\sin(\beta\ell)}. \tag{14.124}$$

The fundamental mode is plotted in Fig. 14.48 for three different values of the device stiffness.

(b) The weak formulation of equilibrium (VWP) is adopted, with $v(z, t) = \phi(z)q(t)$; it reads:

$$\int_0^\ell \left[m\ddot{v} + c_e\dot{v} - GAv'' - GA\eta\dot{v}'' - f_y^a(z, t)\right]\phi(z)\,dz - F_A\phi_A + \cancel{F_B}\phi_B = 0, \tag{14.125}$$

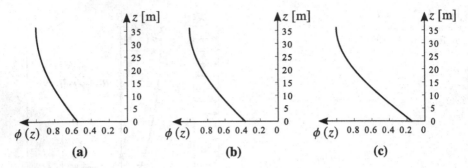

Fig. 14.48 Fundamental mode of the undamped base-isolated shear beam, when (a) $k_d = 3.60 \times 10^6$ N/m, (b) $k_d = 7.20 \times 10^6$ N/m, (c) $k_d = 24.02 \times 10^6$ N/m

where:

$$F_A := GAv'_A + GA\eta\dot{v}'_A - k_d v_A - C_d \dot{v}_A \tag{14.126a}$$

$$= (GAq + GA\eta\dot{q})\,\phi'_A - k_d\phi_A q - C_d\phi_A\dot{q},$$

$$F_B := GAv'_B + GA\eta\dot{v}'_B \tag{14.126b}$$

$$= (GAq + GA\eta\dot{q})\,\phi'_B,$$

are residual forces acting at the free ends; however, by virtue of Eq. 14.121, F_B cancels, being $\phi'_B = 0$. From the VWP, Eq. 14.117 of the Rayleigh oscillator follows, whose coefficients are:

$$M := m\int_0^\ell \phi^2 \, dz, \tag{14.127a}$$

$$k_1 := -GA\int_0^\ell \phi''\phi \, dz - \cancel{GA\phi'_A\phi_A} + \cancel{k_d\phi_A^2} = M\omega_1^2, \tag{14.127b}$$

$$c_s := c_e\int_0^\ell \phi^2 \, dz - \eta GA\int_0^\ell \phi''\phi \, dz - \eta GA\phi'_A\phi_A + C_d\phi_A^2 \tag{14.127c}$$

$$= \left(\frac{c_e}{m} + \eta\omega_1^2\right)M + (C_d - \eta k_d)\,\phi_A^2,$$

$$b_1 := \frac{1}{2}\rho_a DA_1\int_0^\ell \phi^2 \, dz, \tag{14.127d}$$

Table 14.4 Numerical values of the coefficients in Eq. 14.127 for different device stiffnesses

k_d	ω_1	M	k_1	c_s	b_1
[N/m]	[rad/s]	[kg]	[N/m]	[Ns/m]	[kg/m]
3.60×10^6	3.71	124900.	1.72×10^6	7664	-240.3
7.20×10^6	4.48	109800.	2.20×10^6	5236	-211.2
24.02×10^6	5.36	93620.	2.69×10^6	4360	-180.1

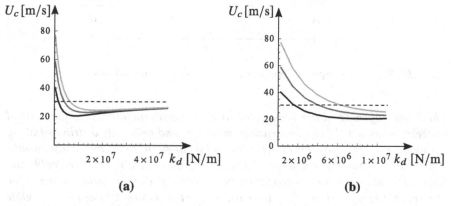

Fig. 14.49 Critical wind velocity of the base-isolated shear beam *vs* the device stiffness, for different dashpots at the base, $C_d = (12801.5, 19202.3, 25603)$ Ns/m (black, dark gray, and light gray lines, respectively): (**a**) large view showing the asymptotic trend, (**b**) zoom of the soft spring range

where use has been made of Eq. 14.121a,c. By numerically evaluating the coefficients for different device stiffnesses k_d, the values in Table 14.4 are obtained.

(c) Similarly to Problem 14.6.1, the critical galloping velocity is derived as $U_c = -\frac{c_s}{b_1}$, and shown in Fig. 14.49 *vs* the spring stiffness and for different damping at the base; the case $k \to \infty$ (clamped beam) is also represented by the dashed line. The previous considerations (reported in Problem 14.6.1), about the influence of the base insulation on the mechanical performances of the beam, still hold.

14.6.3 Galloping of a Pipeline Suspension Bridge

A single-span pipeline suspension bridge, subjected to a wind of velocity U, which blows orthogonally to the plane of the bridge, is considered. It consists of a pair of identical parallel cables (referred to ahead as "the cable"), a stiffening box-girder of rectangular cross-section, equispaced hangers, and two supporting towers. Its planar aeroelastic behavior can be described by a simplified model, according to

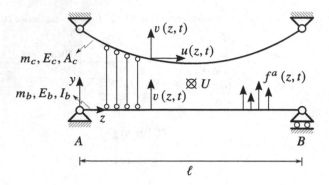

Fig. 14.50 Single-span suspension bridge under transverse uniform wind flow

which, the stiffening girder is modeled as an Euler-Bernoulli beam; the vertical hangers are assumed to be inextensible, massless, and uniformly distributed along the span (i.e., they work as an inextensible and shear-flexible curtain); consequently, cable and beam undergo the same vertical displacement field $v(z, t)$ (i.e., cable and beam are two in-parallel subsystems); the horizontal displacements of the cable are free, not being restrained by the curtain; the towers are assumed of negligible flexibility, so that the cables are fixed at the supports A, B ($z = 0, \ell$) (Fig. 14.50). The relevant equations of motion and boundary conditions are [3, 4]:[17]

$$m\ddot{v} + c_e\dot{v} + E_bI_b\left(1 + \eta_b\frac{\partial}{\partial t}\right)v'''' - T_0\left(1 + \eta_c\frac{\partial}{\partial t}\right)v'' \tag{14.128a}$$

$$+E_cA_c\left(1 + \eta_c\frac{\partial}{\partial t}\right)\frac{k^2}{\ell}\int_0^\ell v\,\mathrm{d}s = f_y^a(z, t),$$

$$v_A = 0, \tag{14.128b}$$

$$E_bI_b\left(1 + \eta_b\frac{\partial}{\partial t}\right)v_A'' = 0, \tag{14.128c}$$

$$v_B = 0, \tag{14.128d}$$

$$E_bI_b\left(1 + \eta_b\frac{\partial}{\partial t}\right)v_B'' = 0. \tag{14.128e}$$

Here, $m := m_b + m_c$ is the total mass (beam plus cable) per unit length; c_e is the external damping coefficient, accounting for dissipation of both girder and cable in

---

[17] The cable equation (obtained from Eq. 14.128a by zeroing EI_b) is consistent with the Irvine theory [9], according to which the longitudinal displacement $u(z, t)$ is *statically condensed*, leading to the integral term. This integral, however, which is proportional to the total elongation of the cable, disappears when the motion is antisymmetric, i.e., in the case which is of interest here.

motionless air; $E_b I_b$ *is the bending stiffness of the girder;* η_b, η_c *are internal viscous damping coefficients, for the cable and girder, respectively;* T_0 *is the pretension in the cable and* $k = \frac{8d}{\ell^2}$ *(with d the sag at midspan and* ℓ *the length) the curvature under self-weight (including the weight of the deck);* $E_c A_c$ *is the axial stiffness of the cable; and* $f^a(z, t) = -\frac{1}{2} \rho_a U D A_1 \dot{v}(z, t)$ *are the aerodynamic loads (defined according to Eq. 12.49), assumed to act on the girder only. The following numerical data are given:* $\ell = 195$ m, $d = 19.5$ m, $m = 1095.6$ Kg/m, $T_0 = 2.62 \times 10^6$ N, $E_c A_c = 4.75 \times 10^9$ N, $E_b I_b = 2.4 \times 10^9$ Nm^2, $\eta_b = 0.00027$ s, $\eta_c = 0.000014$ s, $c_e = 62.17$ Ns/m^2; $\rho_a = 1.225$ Kg/m^3; $A_1 = -3.47$, *referred to a rectangular cross-section with aspect ratio* 2 : 1 *and height* $D = 1$ m. (a) *Find the first antisymmetric mode of the system.*[18] (b) *Use it to derive the equivalent Rayleigh oscillator, by means of the Galerkin procedure.* (c) *Determine the critical galloping velocity, and comment the role of the beam in the collaborating system.*

(a) The undamped free vibrations of the bridge are governed by the following eigenvalue problem:

$$E_b I_b \phi'''' - T_0 \phi'' - m \omega^2 \phi + E_c A_c \frac{k^2}{\ell} \int_0^\ell \phi \, ds = 0, \qquad (14.129\text{a})$$

$$\phi_A = 0, \qquad (14.129\text{b})$$

$$E_b I_b \phi_A'' = 0, \qquad (14.129\text{c})$$

$$\phi_B = 0, \qquad (14.129\text{d})$$

$$E_b I_b \phi_B'' = 0. \qquad (14.129\text{e})$$

By inspection, it is easy to check that $\phi(z) = \sin\left(\frac{2\pi}{\ell} z\right)$ satisfies the boundary conditions, as well as the field equation (in which the integral is zero), provided the frequency ω assumes the expression:

$$\omega_1 = \frac{2\pi}{\ell} \sqrt{\frac{T_0}{m} \left(1 + 4\pi^2 \frac{E_b I_b}{\ell^2 T_0}\right)}. \qquad (14.130)$$

When $E_b I_b \to 0$, then ω_1 tends to the *second frequency* of the taut string (the first one not existing in sagged cables).

(b) The unknown displacement field is taken as $v(z, t) = \phi(z) q(t)$; the weak formulation of equilibrium (VWP) is adopted, which requires:

---

[18] As it has been established in the Irvine theory, while flat cables possess a symmetric fundamental mode, sagged cables have an antisymmetric mode. It can be checked [3, 4] that, for this system, the first mode is antisymmetric.

$$\int_0^\ell \left[m\ddot{v} + c_e\dot{v} + E_bI_bv'''' + E_bI_b\eta_b\dot{v}'''' - T_0v'' - T_0\eta_c\dot{v}'' + E_cA_c\frac{k^2}{\ell}\int_0^\ell vdz \right.$$

$$\left. + E_cA_c\eta_c\frac{k^2}{\ell}\int_0^\ell \dot{v}\,dz - f_y^a(z,t) \right] \phi(z)\,dz - \mathcal{C}_A\phi_A' + \mathcal{C}_B\phi_B' = 0,$$

$$(14.131)$$

where $C_H := E_bI_bv_H'' + \eta_b E_bI_b\dot{v}_H'' = (E_bI_bq + \eta_b E_bI_b\dot{q})\,\phi_H''$ (with $H = A, B$) are residual couples acting at the ends. These latter, however, being $\phi_A'' = \phi_B'' = 0$, cancel out in the problem under consideration. From the VWP, the Rayleigh oscillator follows (Eq. 14.117), whose coefficients are:

$$M := m\int_0^\ell \phi^2 dz = \frac{\ell m}{2}, \qquad (14.132a)$$

$$k_1 := E_bI_b\int_0^\ell \phi''''\phi\,dz - T_0\int_0^\ell \phi''\phi\,dz = \frac{8\pi^4 E_bI_b}{\ell^3} + \frac{2\pi^2 T_0}{\ell}, \qquad (14.132b)$$

$$c_s := c_e\int_0^\ell \phi^2 dz + \eta_b E_bI_b\int_0^\ell \phi''''\phi\,dz - \eta_c T_0\int_0^\ell \phi''\phi\,dz \qquad (14.132c)$$

$$= \frac{c_e\ell}{2} + \frac{8\pi^4\eta_b E_bI_b}{\ell^3} + \frac{2\pi^2\eta_c T_0}{\ell},$$

$$b_1 := \frac{1}{2}\rho_a DA_1\int_0^\ell \phi^2 dz = \frac{1}{4}\rho_a DA_1\ell. \qquad (14.132d)$$

in which it has been accounted for, that $\int_0^\ell \phi\,dz = 0$, due to the antisymmetry of ϕ.

(c) The critical velocity zeroes the total damping, i.e., $U_c = -\frac{c_s}{b_1}$, from which a closed-form expression is derived:

$$U_c = -\frac{2\left(c_e\ell^4 + 16\pi^4\eta_b E_bI_b + 4\pi^2\ell^2\eta_c T_0\right)}{DA_1\ell^4\rho_a} = 29.6\,\text{m/s}.$$

It can be seen, that the beam increases the critical velocity of the system by the way of the internal damping η_b. It is significant so far $\frac{E_b I_b}{T_0 \ell^2}$ is not small, i.e., the bridge is not too long.

14.6.4 Two Degrees of Freedom Translational Galloping

A two DOF rigid cylinder, free to translate but not to rotate, roughly representative of an iced string subject to a wind flow, is characterized by the following numerical data:[19] *length $\ell = 200\,m$, total mass $M = 500\,Kg$, damping factor $\xi_s = 0.01$, transverse to wind frequency $\omega_y = 2.5\,rad/s$, characteristic dimension of the cross-section $D = 0.025\,m$, air density $\rho_a = 1.225\,Kg/m^3$, aerodynamic coefficients $C_{d_0} = 0.96$, $C'_{d_0} = -0.49$, and $C'_{l_0} = -1.49$. Three case studies A, B, C are considered, in which the lift aerodynamic coefficient assumes the following values: $C_{l_0} = \left(C_{l_0}^A, C_{l_0}^B, C_{l_0}^C \right) = (-0.40, -0.63, -1.05)$, respectively. (a) In resonant conditions (i.e., for equal longitudinal and transverse frequencies $\omega_x = \omega_y =: \omega_s$), evaluate the critical wind velocity U_c, and discuss the effects of coupling on the transverse galloping velocity U_{gt} (i.e., relevant to the single DOF model). (b) In nonresonant conditions $\omega_x \neq \omega_y$, show the procedure to numerically find the critical velocity, by referring to the case study C (i.e., $C_{l_0} = -1.05$), by taking $\omega_x = 2.65, 2.9, 3.25\,rad/s$. (c) Repeat the analysis for the three case studies, and plot the critical wind velocity as a function of the detuning $|\omega_x - \omega_y|$.*

(a) In resonant conditions $\omega_x = \omega_y$, the problem is governed by the equations of motion Eq. 12.94, whose aerodynamic coefficients $(b_{11}, b_{22}, b_{12}, b_{21})$ must be evaluated according to Eqs. 12.84. The aerodynamic matrix, in the SI, and for the three cases considered, reads:

$$\mathbf{B} = \begin{pmatrix} 5.88 & -0.275625 \\ -2.45 & -1.62313 \end{pmatrix}, \begin{pmatrix} 5.88 & 0.42875 \\ -3.85875 & -1.62313 \end{pmatrix}, \begin{pmatrix} 5.88 & 1.715 \\ -6.43125 & -1.62313 \end{pmatrix}.$$

$$(14.133)$$

The corresponding invariants are:

$$\mathrm{tr}\mathbf{B} = 4.25688, 4.25688, 4.25688,$$

$$(14.134)$$

$$\det \mathbf{B} = -10.2193, -7.88954, 1.48562.$$

By referring to the linear stability diagram in Fig. 12.11, valid in the resonant case, it is seen that (i) case studies A, B fall into the second quadrant, where a single Hopf bifurcation occurs at the critical wind velocity $U_c = U_{c,1}$, defined in

---

[19] The problem has been built up by adapting the data contained in [11] and [8].

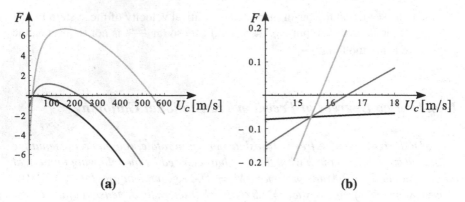

Fig. 14.51 (a) Graph of the left member F of the characteristic Eq. 12.97 after condensation of ω_c, and (b) its zoom; case study C (in the first quadrant of the invariant plane) for nonresonant systems with detuning $|\omega_x - \omega_y| = 0.15, 0.4, 0.75$ (black, dark gray, light gray lines, respectively)

Eq. 12.102; (ii) case C falls into the first quadrant (out of the parabola), where the equilibrium is stable. Accordingly, the critical wind velocities are:

$$U_{C,1} = 14.60, \ 17.91, \ \infty \text{ m/s}.$$

In all the cases considered, the transverse galloping velocity (Den Hartog, Eq. 12.17) is $U_{gt} = -\frac{2\xi_s M \omega_y}{b_{22}} = 15.4 \text{m/s}$. In system A (whose representative point is farther from the stable zone), coupling produces a detrimental effect, as it reduces the transverse galloping velocity by approximately 5.2%. On the other hand, in system B (whose representative point is closer to the stable zone), coupling has a beneficial effect, because it increases the transverse galloping velocity by about 16.3%.

(b) In nonresonant conditions $\omega_x \neq \omega_y$, it needs to numerically solve the nonlinear Eqs. 12.99 for the unknowns (U_c, ω_c). As discussed in Note 41 of Chap. 12, it is convenient to eliminate ω_c between the two equations, to obtain a unique equation of type $F(U_c; \omega_x, \omega_y) = 0$ in the U_c unknown, for any assigned pair of frequencies. The smallest intersection of the graph of the function with the axis of abscissas provides the critical velocity. Typical plots of $F(U_c; \omega_x, \omega_y)$, concerning system C for fixed ω_y and different ω_x, are represented in Fig. 14.51. It is seen that when the detuning is small, the system is stable for any wind velocity; when it is large, $U_c > U_{gt}$ of few percent units.

(c) By performing the previous analysis for each case study, the critical wind velocity U_c is found *vs* the difference $|\omega_x - \omega_y|$, as shown in Fig. 14.52 by solid black lines; there, the Den Hartog velocity is also indicated by dashed black lines. The plots are consistent with Fig. 12.12a,b,d, for systems A, B, C, respectively, according to the location of the representative points on the invariant plane.

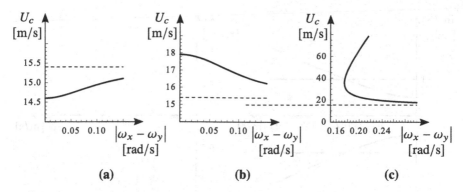

Fig. 14.52 Bifurcation diagrams of the suspension bridge, reporting U_c vs $|\omega_x - \omega_y|$, for **(a)** case A, **(b)** case B, and **(c)** case C; dashed lines denote the Den Hartog velocity

14.6.5 Flutter in the Steady Theory

A two degrees of freedom cylinder, roughly modeling a tower building,[20] free to twist and to translate transversally to the wind flow, has the following properties: length $\ell = 125$ m, characteristic dimension of the cross-section $D = 15$ m; total mass $M = 3 \times 10^6$ Kg; radius of inertia $r_C = 6$ m; damping factors $\xi_y = \xi_\theta = 0.01$; translational frequency $\omega_y = 1$ rad/s; rotational frequency $\omega_\theta = 0.5$ rad/s; air density $\rho_a = 1.225$ Kg/m³, and aerodynamic coefficients $C'_{l_0} = -4.38$ and $C'_{m_0} = 0.5$. (a) Write the characteristic equation using the steady theory, and discuss the nature of the eigenvalues in the case $x_G = 0$ (i.e., of no eccentricity of the center of mass with respect to the center of torsion) vs the wind velocity U. (b) By taking $x_G = -1$ m, plot the locus of the eigenvalues, and evaluate the critical velocity. (c) Determine and plot the critical wind velocity U_c as a function of x_G. (d) Specialize the critical wind velocity to the case $C'_{m_0} = 0$, and discuss the results according to the sign of C'_{l_0}. (e) With reference to the case $C'_{m_0} = 0$, detect the influence of the rotational frequency on the critical wind velocity, by exploring the range $x_G \in (-5 \div 0)$ m.

(a) The problem, based on the steady theory of aeroelastic forces, is governed by the equations of motion Eq. 12.109. There, the aerodynamic coefficients are evaluated according to Eq. 12.84, from which $h_{23} = -5030.16$ and $h_{33} = 8613.28$ (in the SI) follow. The associate characteristic equation is given by Eq. 12.111. If $x_G = 0$, the latter admits the roots in Eq. 12.114, which are purely imaginary eigenvalues for any wind velocity. Figure 14.53 shows the imaginary and real parts of the eigenvalues as the wind velocity varies. It is seen that the coalescence of the eigenvalues occurs at $U = 97$ m/s, without implying the birth of a real part of the eigenvalue, and therefore flutter is not triggered.

---

[20] A similar problem is dealt with in [10].

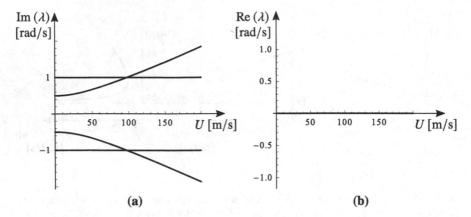

Fig. 14.53 Flutter avoidance in a two DOF system, when $x_G = 0$: (**a**) imaginary and (**b**) real parts of the eigenvalues *vs* the wind velocity

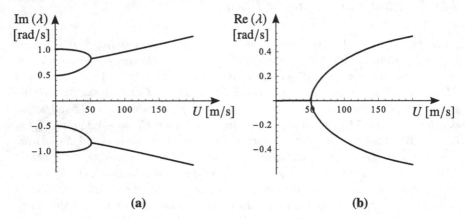

Fig. 14.54 Flutter of a two DOF: (**a**) imaginary and (**b**) real parts of the eigenvalues *vs* the wind velocity, when $x_G = -1$

(b) When $x_G \neq 0$, the locus of the eigenvalues changes, as ruled by the characteristic Eq. 12.111. As an example, when $x_G = -1$ m, the eigenvalues behave as illustrated in Fig. 14.54, entailing a circulatory Hopf bifurcation (flutter) at the coalescence, occurring, in this case, at $U_c = 51.3$ m/s.

(c) In the general case $C'_{m_0} \neq 0$, the critical wind velocity U_c is the smallest positive root of Eq. 12.113. This latter can be solved analytically, providing a closed-form expression of the critical wind velocity as a function of the eccentricity x_G of the center of mass (here omitted). The results are plotted in Fig. 14.55a. It is observed that (i) the critical velocity is real only when $x_G < 0$ and (II) when x_G is small in modulus, the critical velocity is high (consistently with Rmrk. 12.13).

(d) In the particular case $C'_{m_0} = 0$, the expression of the critical wind velocity assumes the simpler form in Eq. 12.115, denoted by U_c^0. It is plotted in

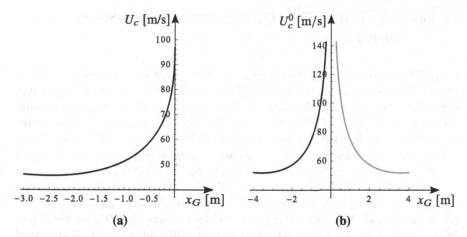

Fig. 14.55 Flutter in a two DOF system: critical wind velocity vs the position of the center of mass x_G, in the cases: (a) $C'_{m_0} \neq 0$; (b) $C'_{m_0} = 0$, with $C'_{l_0} < 0$ (black line) and $C'_{l_0} > 0$ (gray line)

Fig. 14.56 Flutter in a two DOF system: critical wind velocity vs the rotational frequency ω_θ, when $C'_{m_0} = 0$, for different values of the eccentricity of the center of mass $x_G \in (-5 \div 0)$ (lighter gray lines refer to increasing magnitudes of x_G); $\omega_\theta = 1$ denotes resonance

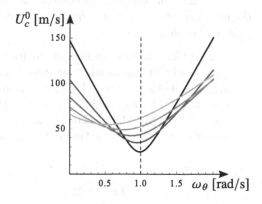

Fig. 14.55b vs x_G, for opposite signs of the aerodynamic coefficient $C'_{l_0} = \pm 4.38$. The results are consistent with what is discussed in Sect. 12.8.1, namely, (i) when $C'_{l_0} < 0$, the flutter occurs only if $x_G < 0$ (i.e., for *center of mass upwind* with respect to the torsion center); when $C'_{l_0} > 0$, only if $x_G > 0$ (downwind case).

(e) By fixing $\omega_y = 1$ rad/s, the critical velocity U_c^0 (Eq. 12.115) is shown in Fig. 14.56 as the rotational frequency ω_θ varies in the range $\omega_\theta \in (0 \div 2)$, for different values of the eccentricity of the center of mass $x_G \in (-5 \div 0)$ (lighter gray lines denote increasing modulus of the eccentricity. It is seen (according to the discussion of Sect. 12.8.1) that all the curves possess a minimum velocity U_c^0, different for any x_G. These minima are lower when the eccentricity is small and occur close to the resonance $\omega_\theta = \omega_y$; they are higher when the eccentricity is large and occur far from the resonance.

14.6.6 Roto-Translational Galloping in the Quasi-Steady Theory

A two DOF cylinder, roughly modeling a chimney,[21] free to rotate and to translate in the direction transverse to the wind flow, has the following properties: length $\ell = 100$ m, characteristic dimension of the cross-section $D = 1$ m; characteristic radius $R = 0.5$ m; total mass $M = 6 \times 10^4$ Kg; no eccentricity of the center of mass, $x_G = 0$; radius of inertia $r_C = 0.2$ m; translational and rotational coincident frequencies $\omega_y = \omega_\theta = 6$ rad/s; translational and rotational equal damping factors $\xi_y = \xi_\theta =: \xi_s$; air density $\rho_a = 1.225$ Kg/m^3; aerodynamic coefficients $C_{d_0} = 2.04$, $C'_{l_0} = -4.38$, and $C'_{m_0} = 0.496$. (a) According to the quasi-steady theory, write the characteristic equation (by leaving ξ_s and U as free parameters), and check its validity in the problem at hand. (b) Discuss the nature of the eigenvalues as the wind velocity varies, for different values of the structural damping. (c) Determine and represent the critical wind velocity vs the structural damping. (d) Repeat the previous analysis by modifying the mass eccentricity to $x_G = 0, \pm 0.1$ m, fixing the structural damping at $\xi_s = 0.008$, and investigate the influence of the rotational frequency on the critical wind velocity and on the galloping modes.

(a) The problem, when formulated in the framework of the quasi-steady theory, is governed by the equations of motion Eq. 12.107. The aerodynamic coefficients are evaluated according to Eq. 12.84, from which it follows: $b_{22} = -143.325$, $b_{23} = 134.138$, $b_{32} = 30.38$, $b_{33} = -15.19$, $h_{23} = -268.275$, and $h_{33} = 30.38$ (in the SI). The associate characteristic equation is given by Eq. 12.117, whose coefficients, defined in Eq. 12.118, assume the following expressions:

$$p_0 = 1,$$

$$p_1 = 24\xi_s - 0.00871792U,$$

$$p_2 = 144\xi_s^2 + 0.0126452U^2 - 0.104615\xi_s U + 72, \qquad (14.135)$$

$$p_3 = 864\xi_s + 0.000026361U^3 + 0.1519\xi_s U^2 - 0.313845U,$$

$$p_4 = 0.4557U^2 + 1296.$$

As explained in Sect. 12.2, the quasi-steady theory is applicable if the (dimensionless) reduced velocity U_{red} is larger than $20 \div 30$. Referring to the transverse galloping velocity (Den Hartog, Eq. 12.17), i.e., $U_{gt} = -\frac{2\xi_s M \omega_y}{b_{22}}$, the corresponding reduced velocity is $U_{gt,red} := \frac{2\pi U_{gt}}{\omega_y D}$. The latter linearly depends on the structural damping ξ_s, as plotted in Fig. 14.57. It is seen that $U_{gt,red} > 20$

---

[21] A further example is discussed in [10].

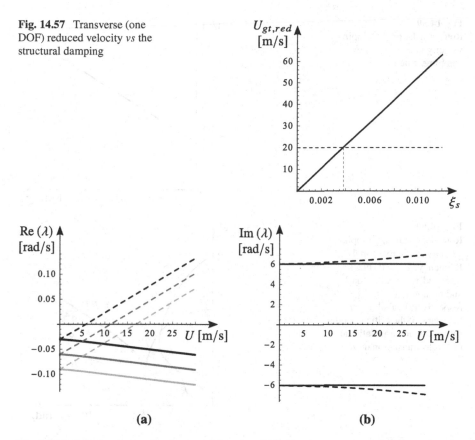

Fig. 14.57 Transverse (one DOF) reduced velocity *vs* the structural damping

Fig. 14.58 Roto-translational galloping: (**a**) real and (**b**) imaginary parts of the eigenvalues *vs* the wind velocity; $x_G = 0$ and $\xi_s = 0.005, 0.01, 0.015$ (black, dark gray and light gray lines, respectively)

when the damping $\xi_s > 0.0038$. Therefore, the quasi-steady theory can be used in this range of damping.

(b) The fourth degree characteristic Eq. 12.117 is numerically solved, and two pairs of complex conjugate eigenvalues are determined. Their real and imaginary parts are represented in Fig. 14.58 *vs* the wind velocity, for different values of the structural damping $\xi_s = 0.005, 0.01, 0.015$ (black, dark gray, and light gray lines, respectively). It is seen that when $U = 0$, the real part of the eigenvalues is negative, that is, the system is asymptotically stable. However, when U increases, the real part of a pair of eigenvalues increases, becoming zero at a certain velocity (depending on the magnitude of damping). That means that a pair of eigenvalues crosses the imaginary axis, giving rise to a generic Hopf bifurcation. The imaginary part of the four eigenvalues is found to be almost independent of damping (the relevant curves are coincident) and weakly variable in the range of velocity examined (nearly constant the stable pair).

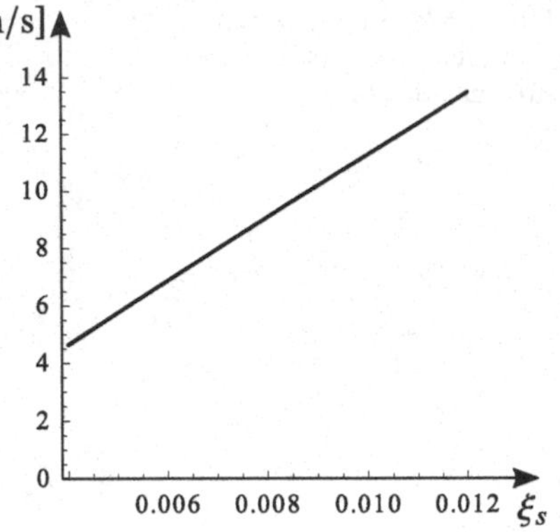

Fig. 14.59
Roto-translational galloping
velocity *vs* the structural
damping, when $x_G = 0$

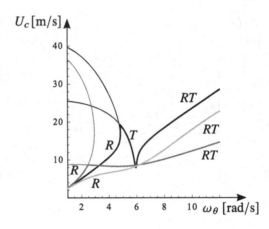

Fig. 14.60
Roto-translational galloping
velocity *vs* the rotational
frequency; $x_G = 0$ (black
line) and $x_G = \pm 0.1$ (dark
and light gray lines,
respectively); label denotes
the shape of the critical mode:
rotational (R), translational
(T), roto-translational (RT)

(c) For assigned value of damping, the critical wind velocity is determined by solving Eq. 12.120, with the coefficients p_k given in Eq. 14.135. The critical velocity, thus determined, is plotted in Fig. 14.59 *vs* the structural damping. It is observed that U_c increases with ξ_s, according to the physics of the phenomenon. The dependence, in this example, is almost linear, since due to the magnitude of the numerical values in Eq. 14.135 and to the smallness of ξ_s, the coefficients p_0, p_2, p_4 are nearly constant on U and p_3 nearly linear in the range of the U's of interest (while p_1 is strictly linear).

(d) Figure 14.60 shows the critical wind velocity as the rotational frequency varies in the range $\omega_\theta = (1 \div 12)$ rad/s, when $x_G = 0, \pm 0.1$ m and $\xi_s = 0.008$. Due to the occurrence of coincident critical modes at the crossing points, the lower critical velocity curve (denoted by a thicker line) presents some cusps. Here, abrupt changes of the modal form occur (rotational, translational or mixed), in a similar way to what is observed, in static bifurcations, for the beam on elastic foundation or for plates. The curves strongly depend on the eccentricity of the center of mass. When this is absent (black line), three branches are distinguishable, two ascendant and one descendant. When $x_G > 0$

Table 14.5 Normalized critical modes $\left(\hat{v}_C, \hat{\theta} r_C\right) = \left(1, \hat{\theta} r_C\right)$

ω_θ	$\hat{\theta} r_C$		
	$x_G = 0$	$x_G = +0.1$	$x_G = -0.1$
2.05	$49.45 + 10.39i$	$10.03 + 0.53i$	$-18.48 + 1.05i$
4.05	$6.58 + 1.39i$	$-0.71 + 0.0024i$	$-3.33 + 0.10i$
5.55	$0.00015 + 0.083i$	$-0.92 - 0.0025i$	$-1.31 + 0.017i$
8.2	$-4.23 - 0.89i$	$-1.3 - 0.014i$	$-0.41 - 0.002i$

(dark gray line), two branches exist; when $x_G < 0$ (light gray line), the curve is smooth. Generally, the critical wind velocity increases with the rotational frequency, when this is far from the resonant value $\omega_\theta = 6$, while it has a more complex behavior close to the resonance. Most importantly, when the critical wind velocity of the two DOF system is compared with that of the single DOF (namely, $U_{gt} = 40.2$ m/s, for the chosen damping), the detrimental effect of coupling with twist is observed, this being responsible of a strong reduction of the bifurcation value.

For some values of the rotational frequency ($\omega_\theta = 2.05, 4.05, 5.55, 8.2$), the normalized critical modes $\left(\hat{v}_C, \hat{\theta} r_C\right) = \left(1, \hat{\theta} r_C\right)$ (in which the rotation has been multiplied by a characteristic length, to make it comparable with $\hat{v}_C = 1$) are determined and shown in Table 14.5. It emerges that, for all the eccentricities examined, the twist prevails on the translation for small values of the rotational frequency, while the two components are comparable for higher values. An almost purely translational mode occurs close to the resonance, when $x_G = 0$. Since the imaginary part of the modes is generally much smaller than the real part, especially when $x_G \neq 0$, the orbits are very elongated ellipses.

14.7 Parametric Excitation

The Mathieu equation is numerically studied. Then, few parametrically excited systems are analyzed.

14.7.1 Exact Stability Analysis of the Mathieu Equation

The Mathieu equation, Eq. 13.59, is considered, in which $\epsilon = 0.4$ is fixed and δ varied, according to the following case studies:

$$A: \quad \delta = 0.4,$$
$$B: \quad \delta = 1,$$
$$C: \quad \delta = 3, \tag{14.136}$$
$$D: \quad \delta = 4.03,$$
$$E: \quad \delta = 4.5,$$

which are displayed in the Strutt diagram of Fig. 14.61. For each of the cases, (a) determine, by numerical integration, the fundamental system of solutions. (b) Build up the Floquet matrix, and compute the characteristic multipliers. (c) Discuss the stability, first using characteristic multipliers and then characteristic exponents.

(a) The fundamental system of solutions for Eq. 13.59 is composed of two solutions, namely, $q^{(1)}, q^{(2)}$ and their time derivatives, which satisfy the following initial conditions:

$$q^{(1)}(0) = 1, \qquad\qquad \dot{q}^{(1)}(0) = 0, \tag{14.137a}$$
$$q^{(2)}(0) = 0, \qquad\qquad \dot{q}^{(2)}(0) = 1. \tag{14.137b}$$

They are determined for each case study, by numerical integration in the interval $(0, \pi)$ of Eqs. 13.59, 14.137. They are plotted in Fig. 14.62, where the black curves represent $q^{(j)}(t)$, $j = 1, 2$ and the gray curves $\dot{q}^{(j)}(t)$, $j = 1, 2$.

(b) By evaluating the fundamental system of solutions at $t = T$, with $T = \pi$ the period of the parametric excitation of the Mathieu equation, the Floquet matrices are built up for each case study as:

$$\mathbf{F} := \begin{bmatrix} q^{(1)}(T) & q^{(2)}(T) \\ \dot{q}^{(1)}(T) & \dot{q}^{(2)}(T) \end{bmatrix}, \tag{14.138}$$

from which:

Fig. 14.61 Case studies represented on the Strutt diagram for the Mathieu equation

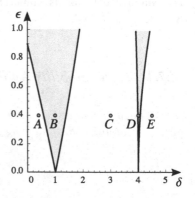

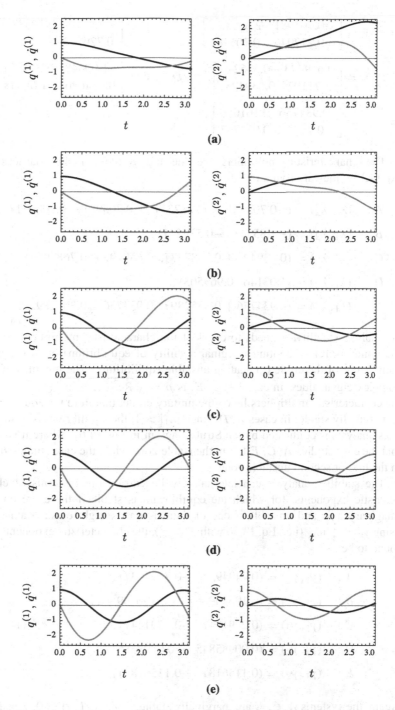

Fig. 14.62 Fundamental system of solutions for the Mathieu equation: (**a**) case study A; (**b**) case study B; (**c**) case study C; (**d**) case study D; (**e**) case study E. Black curves $q^{(j)}(t)$, $j = 1, 2$; gray curves $\dot{q}^{(j)}(t)$, $j = 1, 2$

$$A: \quad \mathbf{F} = \begin{bmatrix} -0.705981 & 2.35548 \\ -0.212946 & -0.705981 \end{bmatrix}, \quad B: \quad \mathbf{F} = \begin{bmatrix} -1.19648 & 0.593873 \\ 0.726712 & -1.19648 \end{bmatrix},$$

$$C: \quad \mathbf{F} = \begin{bmatrix} 0.639493 & -0.345245 \\ 1.71197 & 0.639493 \end{bmatrix}, \quad D: \quad \mathbf{F} = \begin{bmatrix} 1.00048 & 0.0148451 \\ 0.0646361 & 1.00048 \end{bmatrix},$$

$$E: \quad \mathbf{F} = \begin{bmatrix} 0.934756 & 0.161053 \\ -0.783778 & 0.934757 \end{bmatrix}.$$

$$(14.139)$$

The characteristic multipliers are the eigenvalues of the matrices in Eq. 14.139; they are found to be:

$A: \quad (\lambda_1, \lambda_2) = (-0.705981 + 0.708231\,i, \ -0.705981 - 0.708231\,i),$

$B: \quad (\lambda_1, \lambda_2) = (-1.85343, \ -0.539541),$

$C: \quad (\lambda_1, \lambda_2) = (0.639493 + 0.768797\,i, \ 0.639493 - 0.768797\,i),$

$D: \quad (\lambda_1, \lambda_2) = (1.03146, \ 0.969503),$

$E: \quad (\lambda_1, \lambda_2) = (0.934756 + 0.355289\,i, \ 0.934756 - 0.355289\,i).$

$$(14.140)$$

(c) To analyze stability, it needs to check if the characteristic multipliers are on the unitary circle (denoting marginal stability of equilibrium) or just one of them is out of the circle (indicating unstable equilibrium). The modulus of the complex eigenvalues, in cases A, C, E, is $\rho = \sqrt{\mathrm{Re}^2(\lambda) + \mathrm{Im}^2(\lambda)} = 1$, i.e., the characteristic multipliers lie on the unitary circle; therefore the equilibrium is marginally stable. In cases B, D, since $|\lambda_1| > 1$, the equilibrium is unstable. This analysis is confirmed by the Strutt diagram in Fig. 14.61, where it is seen that the case studies A, C, E fall in the stable zone while the case studies B, D in the first or second unstable zone.

The stability analysis can alternatively be carried out in terms of characteristic exponents, for which the equilibrium is stable if these are on the imaginary axis and unstable if one of them is on the right of such axis. By using $\gamma_k = \frac{1}{\pi}\ln\lambda_k$ (i.e., Eq. 13.30 with $T = \pi$), the characteristic exponents are found to be:

$A: \quad (\gamma_1, \gamma_2) = (0.749493\,i, \ -0.749493\,i),$

$B: \quad (\gamma_1, \gamma_2) = (0.196409 + i, \ -0.196409 + i),$

$C: \quad (\gamma_1, \gamma_2) = (0.279144\,i, \ -0.279144\,i),$ $\qquad\qquad (14.141)$

$D: \quad (\gamma_1, \gamma_2) = (0.00985845, \ -0.00985851),$

$E: \quad (\gamma_1, \gamma_2) = (0.115618\,i, \ -0.115618\,i).$

Again, the systems A, C, E are marginally stable, since $\mathrm{Re}\,(\gamma_j) = 0$, $j = 1, 2$; instead, the systems B, D, are unstable, since $\mathrm{Re}\,(\gamma_1) > 0$.

14.7.2 Computation of the Characteristic Exponents of the Mathieu Equation via the Hill Infinite Determinant

The Mathieu equation, Eq. 13.59, is considered. For case studies A (stable) and B (unstable) of Problem 14.7.1 (defined in Eq. 14.136), (a) determine the characteristic exponents via the Hill infinite determinant method. (b) Discuss the numerical convergence of the method and the occurrence of spurious solutions.

(a) By letting $\mathbf{x} = (q, \dot{q})^T$, the Mathieu equation is rewritten in the state-space form $\dot{\mathbf{x}} = \mathbf{A}(t)\,\mathbf{x}$, where $\mathbf{A}(t) = \mathbf{A}_0 + \mathbf{B}\cos(\Omega t)$, with $\Omega = 2$ and:

$$\mathbf{A}_0 = \begin{bmatrix} 0 & 1 \\ -\delta & 0 \end{bmatrix}, \quad \mathbf{B} = \begin{bmatrix} 0 & 0 \\ -2\epsilon & 0 \end{bmatrix}. \tag{14.142}$$

To determine the characteristic exponents, the Hill method discussed in the Suppl. 13.3 is applied, according to which the matrix $\mathbf{L}$ (appearing in Eq. 13.35) has to be built up. In particular, once the number M of the harmonic has been selected, the coefficients of the matrix $\mathbf{L}$ are explicitly defined by Eq. 13.39, where the definitions Eq. 14.142 are used. Below, such matrix is evaluated, for example, for $M = 1, 2, 3$, respectively:[22]

$$\mathbf{L} = \begin{pmatrix} 2i & 1 & 0 & 0 & 0 & 0 \\ -\delta & 2i & -\epsilon & 0 & 0 & 0 \\ 0 & 0 & 0 & 1 & 0 & 0 \\ -\epsilon & 0 & -\delta & 0 & -\epsilon & 0 \\ 0 & 0 & 0 & 0 & -2i & 1 \\ 0 & 0 & -\epsilon & 0 & -\delta & -2i \end{pmatrix}, \tag{14.143}$$

$$\mathbf{L} = \begin{pmatrix} 4i & 1 & 0 & 0 & 0 & 0 & 0 & 0 & 0 & 0 \\ -\delta & 4i & -\epsilon & 0 & 0 & 0 & 0 & 0 & 0 & 0 \\ 0 & 0 & 2i & 1 & 0 & 0 & 0 & 0 & 0 & 0 \\ -\epsilon & 0 & -\delta & 2i & -\epsilon & 0 & 0 & 0 & 0 & 0 \\ 0 & 0 & 0 & 0 & 0 & 1 & 0 & 0 & 0 & 0 \\ 0 & 0 & -\epsilon & 0 & -\delta & 0 & -\epsilon & 0 & 0 & 0 \\ 0 & 0 & 0 & 0 & 0 & 0 & -2i & 1 & 0 & 0 \\ 0 & 0 & 0 & 0 & -\epsilon & 0 & -\delta & -2i & -\epsilon & 0 \\ 0 & 0 & 0 & 0 & 0 & 0 & 0 & 0 & -4i & 1 \\ 0 & 0 & 0 & 0 & 0 & 0 & 0 & -\epsilon & 0 & -\delta & -4i \end{pmatrix}, \tag{14.144}$$

---

[22] The case $M = 4$ is also considered ahead, but the relevant matrix is omitted here.

$$
\mathbf{L} = \begin{pmatrix}
6i & 1 & 0 & 0 & 0 & 0 & 0 & 0 & 0 & 0 & 0 & 0 & 0 & 0 \\
-\delta & 6i & -\epsilon & 0 & 0 & 0 & 0 & 0 & 0 & 0 & 0 & 0 & 0 & 0 \\
0 & 0 & 4i & 1 & 0 & 0 & 0 & 0 & 0 & 0 & 0 & 0 & 0 & 0 \\
-\epsilon & 0 & -\delta & 4i & -\epsilon & 0 & 0 & 0 & 0 & 0 & 0 & 0 & 0 & 0 \\
0 & 0 & 0 & 0 & 2i & 1 & 0 & 0 & 0 & 0 & 0 & 0 & 0 & 0 \\
0 & 0 & -\epsilon & 0 & -\delta & 2i & -\epsilon & 0 & 0 & 0 & 0 & 0 & 0 & 0 \\
0 & 0 & 0 & 0 & 0 & 0 & 0 & 1 & 0 & 0 & 0 & 0 & 0 & 0 \\
0 & 0 & 0 & 0 & -\epsilon & 0 & -\delta & 0 & -\epsilon & 0 & 0 & 0 & 0 & 0 \\
0 & 0 & 0 & 0 & 0 & 0 & 0 & 0 & -2i & 1 & 0 & 0 & 0 & 0 \\
0 & 0 & 0 & 0 & 0 & 0 & -\epsilon & 0 & -\delta & -2i & -\epsilon & 0 & 0 & 0 \\
0 & 0 & 0 & 0 & 0 & 0 & 0 & 0 & 0 & 0 & -4i & 1 & 0 & 0 \\
0 & 0 & 0 & 0 & 0 & 0 & 0 & 0 & -\epsilon & 0 & -\delta & -4i & -\epsilon & 0 \\
0 & 0 & 0 & 0 & 0 & 0 & 0 & 0 & 0 & 0 & 0 & 0 & -6i & 1 \\
0 & 0 & 0 & 0 & 0 & 0 & 0 & 0 & 0 & 0 & -\epsilon & 0 & -\delta & -6i
\end{pmatrix}.
$$

$$(14.145)$$

Then, by computing the eigenvalues of the matrix $\mathbf{L}$, the characteristic exponents are obtained as a function of the number $M = 1, \cdots, 4$ of the selected harmonic. They are reported in Tables 14.6 and 14.7, for case studies A and B, respectively.

(b) By remembering, from Eqs. 14.141, that the exact values are $\gamma_{1,2} = \pm 0.749493\,i$ in case A and $\gamma_{1,2} = \pm 0.196409 + i$ in case B, it is seen, from Tables 14.6 and 14.7 that the characteristic exponents γ_1, γ_2 are well approximated (with two digits), in both case studies, also in the rough approximation $M = 1$. The others characteristic exponents are not so well approximated for small M but converge when M is increased. Moreover, as pointed out in the Supplement 13.3, the characteristic exponents successive to γ_1, γ_2 represent spurious solutions which can be obtained by summing $\pm i p \Omega$ ($p = 1, 2, \cdots, M$) to γ_1 and γ_2. This circumstance can be observed in Table 14.8 where, for each of the case studies, a chain of characteristic exponents (of infinite length) is generated starting from γ_1 and γ_2. By comparing the chains of Table 14.8 with the results of Tables 14.6 and 14.7, it is observed that the higher M, the better the approximation of the chains.

Table 14.6 Characteristic exponents in the case study A as a function of the number $M = 1, \cdots, 4$ of harmonics

M	γ_1, γ_2	γ_3, γ_4	γ_5, γ_6	γ_7, γ_8	γ_9, γ_{10}	γ_{11}, γ_{12}	γ_{13}, γ_{14}	γ_{15}, γ_{16}	γ_{17}, γ_{18}
1	$\pm 0.75\,i$	$\pm 1.27\,i$	$\pm 2.65\,i$	–	–	–	–	–	–
2	$\pm 0.75\,i$	$\pm 1.25\,i$	$\pm 2.75\,i$	$\pm 3.27\,i$	$\pm 4.65\,i$	–	–	–	–
3	$\pm 0.75\,i$	$\pm 1.25\,i$	$\pm 2.75\,i$	$\pm 3.25\,i$	$\pm 4.75\,i$	$\pm 5.27\,i$	$\pm 6.65\,i$	–	–
4	$\pm 0.75\,i$	$\pm 1.25\,i$	$\pm 2.75\,i$	$\pm 3.25\,i$	$\pm 4.75\,i$	$\pm 5.25\,i$	$\pm 6.75\,i$	$\pm 7.27\,i$	$\pm 8.65\,i$
$\cdots$	$\cdots$	$\cdots$	$\cdots$	$\cdots$	$\cdots$	$\cdots$	$\cdots$	$\cdots$	$\cdots$

Table 14.7 Characteristic exponents in the case study B as a function of the number $M = 1, \cdots, 4$ of harmonics

M	γ_1, γ_2	γ_3, γ_4	γ_5, γ_6	γ_7, γ_8	γ_9, γ_{10}	γ_{11}, γ_{12}	γ_{13}, γ_{14}	γ_{15}, γ_{16}	γ_{17}, γ_{18}
1	$\pm 0.2 + i$	$\pm 0.2 - i$	$\pm 3.01\,i$	–	–	–	–	–	–
2	$\pm 0.2 + i$	$\pm 0.2 - i$	$\pm 0.2 + 3i$	$\pm 0.2 - 3i$	$\pm 5.01\,i$	–	–	–	–
3	$\pm 0.2 + i$	$\pm 0.2 - i$	$\pm 0.2 + 3i$	$\pm 0.2 - 3i$	$\pm 0.2 + 5i$	$\pm 0.2 - 5i$	$\pm 7.01\,i$	–	–
4	$\pm 0.2 + i$	$\pm 0.2 - i$	$\pm 0.2 + 3i$	$\pm 0.2 - 3i$	$\pm 0.2 + 5i$	$\pm 0.2 - 5i$	$\pm 0.2 + 7i$	$\pm 0.2 - 7i$	$\pm 9.01\,i$
$\vdots$	$\vdots$	$\vdots$	$\vdots$	$\vdots$	$\vdots$	$\vdots$	$\vdots$	$\vdots$	$\vdots$

Table 14.8 Characteristic exponents and spurious solutions

Case study A		Case study B	
γ_1, γ_2	$\pm 0.75\,i$	γ_1, γ_2	$\pm 0.2 + i$
$\gamma_3 = \gamma_2 + i\Omega,\ \gamma_4 = \gamma_1 - i\Omega$	$\pm 1.25\,i$	$\gamma_3 = \gamma_1 - i\Omega,\ \gamma_4 = \gamma_2 - i\Omega$	$\pm 0.2 - i$
$\gamma_5 = \gamma_1 + i\Omega,\ \gamma_6 = \gamma_2 - i\Omega$	$\pm 2.75\,i$	$\gamma_5 = \gamma_1 + i\Omega,\ \gamma_6 = \gamma_2 + i\Omega$	$\pm 0.2 + 3i$
$\gamma_7 = \gamma_2 + 2i\Omega,\ \gamma_8 = \gamma_1 - 2i\Omega$	$\pm 3.25\,i$	$\gamma_7 = \gamma_1 - 2i\Omega,\ \gamma_8 = \gamma_2 - 2i\Omega$	$\pm 0.2 - 3i$
$\gamma_9 = \gamma_1 + 2i\Omega,\ \gamma_{10} = \gamma_2 - 2i\Omega$	$\pm 4.75\,i$	$\gamma_9 = \gamma_1 + 2i\Omega,\ \gamma_{10} = \gamma_2 + 2i\Omega$	$\pm 0.2 + 5i$
$\gamma_{11} = \gamma_2 + 3i\Omega,\ \gamma_{12} = \gamma_1 - 3i\Omega$	$\pm 5.25\,i$	$\gamma_{11} = \gamma_1 - 3i\Omega,\ \gamma_{12} = \gamma_2 - 3i\Omega$	$\pm 0.2 - 5i$
$\gamma_{13} = \gamma_1 + 3i\Omega,\ \gamma_{14} = \gamma_2 - 3i\Omega$	$\pm 6.75\,i$	$\gamma_{13} = \gamma_1 + 3i\Omega,\ \gamma_{14} = \gamma_2 + 3i\Omega$	$\pm 0.2 + 7i$
$\gamma_{15} = \gamma_2 + 4i\Omega,\ \gamma_{16} = \gamma_1 - 4i\Omega$	$\pm 7.25\,i$	$\gamma_{15} = \gamma_1 - 4i\Omega,\ \gamma_{16} = \gamma_2 - 4i\Omega$	$\pm 0.2 - 7i$
$\cdots$	$\cdots$	$\cdots$	$\cdots$

14.7.3 Pendulum with Motion Impressed at the Base

A planar pendulum, of mass m and length ℓ, is subject to an assigned periodic motion y (t) = a cos (Ωt) of its suspension point. (a) Transform the equation of motion governing the problem into the Mathieu canonical form. (b) Determine the first and second resonance regions.

(a) The system is governed by Eq. 13.1 with $\ddot{y} = -a\Omega^2 \cos(\Omega t)$, i.e.:

$$\ddot{\theta} + \frac{1}{\ell}\left(g - a\Omega^2 \cos(\Omega t)\right)\theta = 0. \tag{14.146}$$

It can be transformed in the canonical form of Eq. 13.59 by introducing a new time τ, such that $\Omega t = 2\tau$, in analogy with what has been done in Sect. 13.6.1. The following rules hold:

$$\frac{\mathrm{d}}{\mathrm{d}t} = \frac{\mathrm{d}}{\mathrm{d}\tau}\frac{\mathrm{d}\tau}{\mathrm{d}t} = \frac{\Omega}{2}\frac{\mathrm{d}}{\mathrm{d}t}, \quad \frac{\mathrm{d}^2}{\mathrm{d}t^2} = \frac{\Omega^2}{4}\frac{\mathrm{d}^2}{\mathrm{d}t^2}, \tag{14.147}$$

which, substituted in Eq. 14.146, lead to:

$$\frac{\mathrm{d}^2\theta}{\mathrm{d}\tau^2} + \left(4\frac{\omega^2}{\Omega^2} - 4\mu \cos(2\tau)\right)\theta = 0, \tag{14.148}$$

where $\omega^2 := g/\ell$ and $\mu := a/\ell$. By a direct comparison with Eq. 13.59, the following correspondence between the parameters is obtained:

$$\delta = 4\frac{\omega^2}{\Omega^2}, \quad |\epsilon| = 2\mu, \tag{14.149}$$

in which the sign of ϵ is irrelevant (as discussed in Note 17 of Chap. 13).

(b) The principal resonance region, recalling Eq. 13.71 and the definitions in Eq. 14.149, is delimited by curves of equations:

$$4\frac{\omega^2}{\Omega^2} = 1 \pm 2\mu - \frac{1}{2}\mu^2. \tag{14.150}$$

Solving the previous equation for $\frac{\Omega}{\omega}$ and expanding in series for small amplitude μ of the excitation, one gets:

$$\Omega = \omega\left(2 \mp 2\mu + \frac{7}{2}\mu^2\right). \tag{14.151}$$

Similarly, the second resonance region, by recalling Eqs. 13.73 and 14.149, is bounded by the curves:

$$4\frac{\omega^2}{\Omega^2} = \begin{cases} 4 - \frac{1}{3}\mu^2, \\ 4 + \frac{5}{3}\mu^2. \end{cases} \tag{14.152}$$

Solving for $\frac{\Omega}{\omega}$ and expanding in series for small μ, it follows:

$$\Omega = \begin{cases} \omega\left(1 + \frac{1}{24}\mu^2\right), \\ \omega\left(1 - \frac{5}{24}\mu^2\right). \end{cases} \tag{14.153}$$

14.7.4 Pendulum with Moving Mass

The mass m of a planar pendulum, under the gravity g, is subject to an assigned periodic motion $\Delta(t)$ which makes the length of the pendulum variable in time, as $\ell(t) := \ell_0 + \Delta(t)$. (a) Use the change of variable $q(t) = \theta(t)\ell(t)$, to transform the equation of small motions in an undamped Hill equation. (b) Show that when $\Delta(t) = a\cos(\Omega t)$, with $a \ll \ell$, the equation of motion is a Mathieu equation. (c) Transform such equation in the canonical form. (d) Determine the principal resonance region.

(a) The system is governed by Eq. 13.8 or, in a more compact form, by Eq. 13.12; for small θ it reads:

$$\frac{d}{dt}\left(\ell^2\dot{\theta}\right) + g\ell\theta = 0. \tag{14.154}$$

By letting $q = \theta \ell$, it is $\dot{q} = \dot{\theta}\ell + \theta \dot{\ell}$, and therefore $\ddot{\theta}\ell^2 = \dot{q}\ell - q\dot{\ell}$. The equation of motion, consequently, transforms into:

$$\ddot{q}\ell + \dot{q}\dot{\ell} - \dot{q}\dot{\ell} - q\ddot{\ell} + g\,q = 0. \tag{14.155}$$

Since $\ddot{\ell} = \ddot{\Delta}$, by dividing by ℓ, an undamped Hill equation is found, namely:

$$\ddot{q} + \frac{1}{\ell\,(t)}\left(g - \ddot{\Delta}\right)q = 0. \tag{14.156}$$

(b) In the case at hand, the equation of motion specializes into:

$$\ddot{q} + \frac{1}{\ell_0 + a\cos\left(\Omega t\right)}\left(g + a\Omega^2\cos\left(\Omega t\right)\right)q = 0. \tag{14.157}$$

By linearizing for small excitation amplitudes a, it finally reads:

$$\ddot{q} + \omega^2\left[1 - \mu\left(1 - \frac{\Omega^2}{\omega^2}\right)\cos(\Omega t)\right]q = 0, \tag{14.158}$$

where $\omega^2 := g/\ell_0$ and $\mu := a/\ell_0$; therefore, it is a Mathieu equation.

(c) Equation 14.158 is transformed into the canonical form Eq. 13.59, by introducing the new time τ, such that $\Omega t = 2\tau$, and proceeding as in Problem 14.7.3; accordingly, it becomes:

$$\frac{d^2q}{d\tau^2} + 4\frac{\omega^2}{\Omega^2}\left[1 - \mu\left(1 - \frac{\Omega^2}{\omega^2}\right)\cos(2\tau)\right]q = 0, \tag{14.159}$$

where the following correspondence with the parameters in Eq. 13.59 holds:

$$\delta = 4\frac{\omega^2}{\Omega^2}, \quad |\epsilon| = 2\mu\left(\frac{\omega^2}{\Omega^2} - 1\right). \tag{14.160}$$

(d) By substituting Eq. 14.160 in Eq. 13.71, one gets:

$$4\frac{\omega^2}{\Omega^2} = 1 \pm 2\left(\frac{\omega^2}{\Omega^2} - 1\right)\mu - \frac{1}{2}\left(\frac{\omega^2}{\Omega^2} - 1\right)^2\mu^2. \tag{14.161}$$

Solving the previous equation for $\frac{\Omega}{\omega}$ and expanding in series for small nondimensional excitation amplitudes μ, the curves bounding the principal resonance region are found:

$$\Omega = \omega\left(2 \pm \frac{3}{2}\mu + \frac{87}{32}\mu^2\right). \tag{14.162}$$

Acknowledgments Authors are grateful to the Ph.D. students Francesca Pancella and Yuri De Santis, who collaborated in carrying out the solved problems of Chap. 14.

References

1. Di Egidio, A., Luongo, A., Paolone, A.: Linear and non-linear interactions between static and dynamic bifurcations of damped planar beams. Int. J. Nonlinear Mech. **42**(1), 88–98 (2007)
2. Di Nino, S., Luongo, A.: Nonlinear aeroelastic behavior of a base-isolated beam under steady wind flow. Int. J. Nonlinear Mech. **119**, 10 pp. (2020)
3. Di Nino, S., Luongo, A.: Nonlinear aeroelastic in-plane behavior of suspension bridges under steady wind flow. Appl. Sci. **10**(5), 19 pp. (2020)
4. Di Nino, S., Luongo, A.: Nonlinear interaction between self-and parametrically excited wind-induced vibrations. Nonlinear Dyn. **103**(1), 79–101 (2021)
5. Di Nino, S., Luongo, A.: Nonlinear dynamics of a base-isolated beam under turbulent wind flow. Nonlinear Dyn. **107**(2), 1529–1544 (2022)
6. Di Nino, S., Zulli, D., Luongo, A.: Nonlinear dynamics of an internally resonant base-isolated beam under turbulent wind flow. Appl. Sci. **11**(7), 17 pp. (2021)
7. Ferretti, M., D'Annibale, F., Luongo, A.: Flexural-torsional flutter and buckling of braced foil beams under a follower force. Math. Probl. Eng. **2691963**, 10 pp. (2017)
8. Ferretti, M., Zulli, D., Luongo, A.: A continuum approach to the nonlinear in-plane galloping of shallow flexible cables. Adv. Math. Phys. **6865730**, 12 pp. (2019)
9. Irvine, M.: Cable Structures. Dover, New York (1992)
10. Luongo, A., Pancella, F., Piccardo, G.: Flexural-torsional galloping of prismatic structures with double-symmetric cross-section. J. Appl. Comput. Mech. **7**, 1049–1069 (2021)
11. Luongo, A., Piccardo, G.: Linear instability mechanisms for coupled translational galloping. J. Sound Vib. **288**(4-5), 1027–1047 (2005)

Appendix A
Calculus of Variations

In this appendix, a brief outline of the calculus of variations is provided. By using a structural example in the linearized context, the concepts of functional and its first variation are introduced. The Euler-Lagrange equations, accompanied by the natural boundary conditions, are derived. A more formal illustration of the matter can be found, for example, in [1–3].

A.1 The Concept of Functional via a Structural Example

An extensible and shear undeformable beam is considered, immersed in the 2D space, uniformly compressed by an axial force of intensity P, loaded by distributed transverse forces of intensity $p = p(x)$ (Fig. A.1). The beam is clamped at the end $x = 0$ and elastically constrained at the end $x = \ell$ by an extensional spring of stiffness k (Fig. A.1).

Using the results obtained in Chap. 7, concerning the linearized theory, the total potential energy (TPE) possessed by the system in the current configuration is:[1]

$$\Pi[v] = \int_0^\ell \left(\frac{1}{2} EI \, v''^2 - \frac{1}{2} P \, v'^2 - p \, v \right) dx + \frac{1}{2} k \, v(\ell)^2, \tag{A.1}$$

where $v = v(x)$ is the transverse displacement field.

It can be observed that the TPE associates with any kinematically admissible displacement field $v(x)$ (i.e., continuous, with continuous first derivative) a number $\Pi[v]$, defined by Eq. A.1. Differently from the TPE of a discrete system, which is a function of several scalar variables, the TPE of a continuous system depends on one

---

[1] The precritical deformations are neglected, and it is taken into account that, in the buckling mode, the longitudinal displacement is $u \equiv 0$.

Fig. A.1 Compressed beam
in the undeformed
configuration

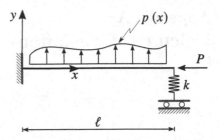

(or more) *functions*, appearing as its arguments. This application, which associates
a real number to each function, is called a *functional*.[2]

As for discrete systems, the equilibrium equations are obtained imposing the
stationary of the TPE. With reference to the problem of the beam, the goal is
determining the function $\bar{v}(x)$ (solution to the elastic problem), which makes the
TPE in Eq. A.1 "locally flat," in the sense that by perturbing $\bar{v}(x)$ by a quantity of
the first order, the TPE changes by smaller quantities, at least of second order.[3] The
stationary of a functional, however, cannot be imposed as for a function of several
variables. In what follows, information in this regard is provided.

A.2   First Variation of a Functional

By hypothesis, let $\bar{v}(x)$ be the displacement field which makes stationary the TPE.
A generic $v(x)$ function "close" to it can be written as:

$$v(x) = \bar{v}(x) + \eta\,\delta v(x), \tag{A.2}$$

where $\delta v(x)$ is an *arbitrary function*, which respects the geometric boundary
conditions, and $0 < \eta \ll 1$ is a scalar variable that governs its amplitude (Fig. A.2).
The function $\delta v(x)$ is called the *variation of the function* $\bar{v}(x)$. In the example
considered, the constraints prescribe the following geometric boundary conditions:

---

[2] A simple example of functional is the *arc length* of a curve:

$$L[y] := \int_{x_1}^{x_2} \sqrt{1 + y'(x)^2}\,\mathrm{d}x.$$

This associates each continuous function $y(x)$ of prescribed values $y(x_i) = y_i$ at the ends of the
interval (x_1, x_2), with a scalar $L[y]$, which represents the length of the curve $y(x)$ in this interval.

[3] Returning to the example of the arc length of a curve, the function that makes stationary (and in
that case minimum) the length of the arc of the curve, is the straight line which joins the two end
points (x_i, y_i). Perturbing the line with a function of amplitude $0 < \eta \ll 1$, the length the curve
changes (increases) by a quantity of order η^2.

Fig. A.2 Displacement field
$\bar{v}\,(x)$ and its variation

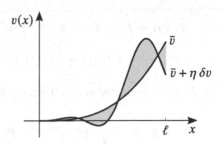

$$v\,(0) = v'\,(0) = 0. \tag{A.3}$$

Therefore, the $\delta v\,(x)$ field must meet the following restrictions:

$$
\begin{aligned}
v\,(0) &= \bar{v}\,(0) + \eta\,\delta v\,(0) = 0, \qquad \forall \eta \Rightarrow \delta v\,(0) = 0, \\
v'\,(0) &= \bar{v}'\,(0) + \eta\,\delta v'\,(0) = 0, \quad \forall \eta \Rightarrow \delta v'\,(0) = 0.
\end{aligned}
\tag{A.4}
$$

By fixing both $\bar{v}\,(x)$ and $\delta v\,(x)$, the functional $\varPi\,[\bar{v}\,(x) + \eta\,\delta v\,(x)]$ becomes a scalar function (no longer a functional!) of the variable η, that is:

$$F\,(\eta) := \varPi\,[\bar{v} + \eta\,\delta v]. \tag{A.5}$$

By virtue of the definition given for $\bar{v}\,(x)$, it follows immediately, that *the scalar function $F\,(\eta)$ admits a stationary point when $\eta = 0$.* Such an observation results in the fulfillment of the following condition:

$$\delta\varPi := \left.\frac{\mathrm{d}F}{\mathrm{d}\eta}\right|_{\eta=0} = 0, \tag{A.6}$$

which is called the *first variation* of the functional $\varPi$, and denoted by $\delta\varPi$. By this definition, *the TPE is stationary when its first variation vanishes.*

Hence, to enforce stationary of the TPE in Eq. A.1, the following algorithm must be applied:

1. The substitution $v \to v + \eta\,\delta v$ is performed, to obtain $F\,(\eta)$ (the overbar on v and the dependence on x are omitted for the sake of brevity):

$$
F\,(\eta) = \int_0^\ell \left[\frac{1}{2}EI\,\left(v'' + \eta\,\delta v''\right)^2 - \frac{1}{2}P\,\left(v' + \eta\,\delta v'\right)^2 - p\,\left(v + \eta\,\delta v\right)\right]\mathrm{d}x
$$

$$
+ \frac{1}{2}k\,\left(v\,(\ell) + \eta\,\delta v\,(\ell)\right)^2.
$$

$$\tag{A.7}$$

2. $F\,(\eta)$ is differentiated with respect to η:

$$F'(\eta) = \int_0^\ell \left[EI \left(v'' + \eta \, \delta v'' \right) \delta v'' - P \left(v' + \eta \, \delta v' \right) \delta v' - p \, \delta v \right] dx$$

$$+ k \left(v(\ell) + \eta \, \delta v(\ell) \right) \delta v(\ell) .$$

(A.8)

3. $F'(\eta)$ is evaluated at $\eta = 0$ and equated to zero:

$$\delta \Pi = F'(0) = \int_0^\ell \left(EI \, v'' \delta v'' - P \, v' \delta v' - p \, \delta v \right) dx + k v(\ell) \, \delta v(\ell) = 0, \ \forall \delta v,$$

(A.9)

where $\forall \delta v$ remembers the arbitrariness of the $\delta v(x)$ field which appears in Eq. A.5.

Remark A.1 In practice, the variation $\delta \Pi$ of the functional Π can be computed in a *formal way,* by operating as follows: (a) The δ operator is applied to Π, bringing it under the integral sign; (b) the variation of the integrand is performed, endowing δ with the same properties holding for the differentiation operator d, namely, $d(f(x)) = f'(x) dx$ becomes $\delta(f(v(x))) = f'(v(x)) \delta v(x)$. Thus, for example, $\delta(v''^2) = 2v'' \delta v''$. In this way, Eq. A.9 is more immediately obtained.

A.3   Euler-Lagrange Equations and Natural Conditions

Equation A.9 is an integral condition on $v(x)$ for the TPE to be stationary. It, however, as it will be seen soon, is equivalent to an ordinary differential equation, accompanied by suitable boundary conditions. Indeed, by integrating it twice by parts with respect to the variable x, in order to "free" the variation δv from any differentiations, Eq. A.9 is rewritten as:

$$\delta \Pi = \int_0^\ell \left(EI \, v'''' + P \, v'' - p \right) \delta v \, dx + \left[EI \, v'' \delta v' \right]_0^\ell + \left[\left(-EI \, v''' - P \, v' \right) \delta v \right]_0^\ell$$

$$+ k v(\ell) \, \delta v(\ell) = 0, \qquad \forall \delta v.$$

(A.10)

By introducing the geometric restrictions $\delta v(0) = 0$, $\delta v'(0) = 0$ required by the constraints on the field of the admissible variations, the former equation becomes:

$$\delta \Pi = \int_0^\ell \left(EI \, v'''' + P \, v'' - p \right) \delta v \, dx + EI \, v''(\ell) \, \delta v'(\ell)$$

$$+ \left(-EI \, v'''(\ell) - P \, v'(\ell) + k v(\ell) \right) \delta v(\ell) = 0, \qquad \forall \delta v.$$

(A.11)

Since δv is arbitrary in the range $(0, \ell)$, and since $\delta v(\ell)$ and $\delta v'(\ell)$ are arbitrary at the unconstrained boundary, it follows that the respective factors, in the domain and at the boundary, must cancel out separately. Hence:

$$EI\,v'''' + P\,v'' - p = 0, \qquad x \in (0, \ell), \tag{A.12}$$

together with:

$$EI\,v''(\ell) = 0, \tag{A.13a}$$

$$-EI\,v'''(\ell) - P\,v'(\ell) + k\,v(\ell) = 0. \tag{A.13b}$$

Equation A.12 is named the *Euler-Lagrange equation of the variational problem*; Eqs. (A.13) are called the *natural conditions* of the variational problem. The differential problem is completed by the geometric conditions $v(0) = 0$, $v'(0) = 0$, expressed by the constraints. The solution (possibly not unique) of the boundary value problem supplies the function $\bar{v}(x)$ sought for.

References

1. Gelfand, I.M., Fomin, S.V.: Calculus of Variations. Prentice-Hall, Englewood Cliffs (1963)
2. Lanczos, C.: The Variational Principles of Mechanics. Dover Publications, New York (1986)
3. Sokolnikoff, I.S.: Mathematical Theory of Elasticity. McGraw-Hill, New York (1956)

Appendix B
Ritz Method

The principle at the base of the Ritz method is illustrated [1–3]. The procedure is described with reference to a continuous meta-model (solid beam, thin-walled beam, plate, etc.), which admits a total potential energy which is quadratic and homogeneous in the displacements. The method is exemplified for a rectangular plate prestressed in its plan.

B.1 Discretization Method

The Ritz method, which belongs to the class of the *variational methods*, consists in expressing the unknown displacement field of the structure as a linear combination, with unknown coefficients, of known *trial functions* (also called "shape functions") , each of which satisfies *at least the geometric conditions*. In this way, the total potential energy (TPE) is transformed from a *functional* (i.e., dependent on one or several functions) to an *ordinary function* of a finite number of variables. For this reason, the method is also called of *discretization*. Making the TPE stationary with respect to the unknown variables, a system of algebraic (no longer differential) equations is found. When applied to a buckling problem, in the context of linearized theory (i.e., ruled by a quadratic and homogeneous TPE), the method returns a linear system of algebraic equations, also homogeneous, whose solution is an approximation of the critical load[1] and of the critical mode.

Remark B.1 The method can also be used in a *semi-variational* form, to reduce, for example, a two-dimensional problem, governed by partial differential equations, to

---

[1] The critical load supplied by the Ritz method is un upper bound for the true value, as it can be explained on a mechanical ground. Indeed, the choice of the approximate trial functions forces the structure to deform in a prescribed way, by making the system "more constrained" than it really is.

645
A. Luongo et al., *Stability and Bifurcation of Structures*,
https://doi.org/10.1007/978-3-031-27572-2

an ordinary differential problem. In this case, the known trial functions describe the spatial dependence in one direction, while the unknown coefficients depend continuously on the other spatial variable. This technique can also be used in dynamics, to transform the space-time partial differential equations into ordinary differential equations in time.

B.2   Algorithm

To illustrate the procedure, reference is made to a TPE expressed by a homogeneous quadratic form, in which derivatives up to the order K appear, precisely:

$$\varPi\,[\mathbf{u}] := \frac{1}{2} \sum_{k=0}^{K} \int_{\Omega} [\mathcal{D}_k \mathbf{u}\,(\mathbf{x})]^T \, \mathbf{A}_k\,(\mathbf{x};\,\mu)\,[\mathcal{D}_k \mathbf{u}\,(\mathbf{x})]\,\mathrm{d}\Omega, \tag{B.1}$$

where $\mathbf{u}\,(\mathbf{x})$ is a vector of M configuration variables, defined in the domain Ω, spanned by the coordinates $\mathbf{x}$; $\mathcal{D}_k$ is a formal matrix, whose coefficients are partial differential operators $\partial_{\mathbf{x}}^k$ of order k; and $\mathbf{A}_k$ is an array of functions (or constants) that describe the mechanical characteristics of the system and the state of prestress, governed by the multiplier μ.

According to the Ritz method, the displacement is expressed as:

$$\mathbf{u} = \boldsymbol{\Phi}\,(\mathbf{x})\,\mathbf{q}, \tag{B.2}$$

where $\boldsymbol{\Phi}\,(\mathbf{x})$ is a $M \times N$ matrix of known trial functions of the spatial coordinates, differentiable up to the order K, which satisfy the prescribed geometric conditions on the boundary of Ω, and $\mathbf{q}$ is a vector of N unknown constants (Lagrangian parameters). Substituting Eq. B.2 in Eq. B.1, and taking into account that $\mathbf{q}$ does not depend on $\mathbf{x}$, the TPE is written as:

$$\varPi\,(\mathbf{q}) := \frac{1}{2} \sum_{k=0}^{K} \mathbf{q}^T \int_{\Omega} [\mathcal{D}_k \boldsymbol{\Phi}\,(\mathbf{x})]^T \, \mathbf{A}_k\,(\mathbf{x};\,\mu)\,[\mathcal{D}_k \boldsymbol{\Phi}\,(\mathbf{x})]\,\mathrm{d}\Omega\,\,\mathbf{q}, \tag{B.3}$$

namely, it is an ordinary function of the N variables $\mathbf{q}$. By performing the integrations of the functions $\boldsymbol{\Phi}\,(\mathbf{x})$ and their derivatives on the domain Ω, one obtains:

$$\varPi\,(\mathbf{q}) = \frac{1}{2} \mathbf{q}^T \mathbf{K}\,(\mu)\,\mathbf{q}, \tag{B.4}$$

where:

$$\mathbf{K}(\mu) := \sum_{k=0}^{K} \int_{\Omega} [\mathcal{D}_k \mathbf{\Phi}(\mathbf{x})]^T \mathbf{A}_k(\mathbf{x}; \mu) [\mathcal{D}_k \mathbf{\Phi}(\mathbf{x})] \, d\Omega, \tag{B.5}$$

is a *stiffness matrix*, of dimension $N \times N$.

The stationary condition of TPE provides:

$$\delta \Pi(\mathbf{q}) = \delta \mathbf{q}^T \mathbf{K}(\mu)\mathbf{q} = 0, \quad \forall \delta \mathbf{q}, \tag{B.6}$$

or:

$$\mathbf{K}(\mu)\,\mathbf{q} = \mathbf{0}, \tag{B.7}$$

which constitutes the (approximate) equilibrium condition of the problem. The smallest value μ_c of the load multiplier for which $\det(\mathbf{K}(\mu_c)) = 0$ is the critical load sought for. The associate eigenvector $\mathbf{q}_c$ describes the critical mode via Eq. B.2.

B.3   Ritz Method for Rectangular Plates

The Ritz method is exemplified with reference to a rectangular plate prestressed in its own plane (Chap. 10). A possible choice for the trial functions is discussed.

B.3.1   Stiffness Matrices

For notational convenience, it is appropriate to rewrite the TPE of the plate as $\Pi = \tilde{U} + U^0$ and to use a matrix form, i.e.:

$$\tilde{U} = \frac{1}{2} \iint_{\Omega} (\mathcal{D}w)^T \mathbf{E}(\mathcal{D}w) \, dx dy, \tag{B.8a}$$

$$U^0 = \frac{1}{2}\mu \iint_{\Omega} (\nabla w)^T \mathbf{T}^0 (\nabla w) \, dx dy, \tag{B.8b}$$

where $\tilde{U}$ is the elastic contribution (integral on domain of Eq. 10.8) and U^0 is the prestress energy (Eq. 10.31); moreover:

$$\mathcal{D} := \begin{pmatrix} \partial_{xx}^2 \\ \partial_{yy}^2 \\ 2\partial_{xy}^2 \end{pmatrix}, \qquad \nabla = \begin{pmatrix} \partial_x \\ \partial_y \end{pmatrix}, \tag{B.9}$$

are two differential operators, which transform w in the curvature $\kappa = \mathcal{D}w$ (Eq. 10.2) and in the gradient ∇w, respectively; furthermore, $\mathbf{E}$ is the elastic matrix (Eq. 10.5) and $\mathbf{T}^0$ the stress tensor (Eq. 10.26), in which a load multiplier μ has been introduced.

According to the Ritz method, the deflection of the plate is taken as:

$$w(x, y) = \sum_{k=1}^{N} \phi_k(x, y) q_k \equiv \mathbf{\Phi}\mathbf{q}, \tag{B.10}$$

where $\phi_k(x, y)$ are known trial functions, respectful of the geometric boundary conditions, and q_k are unknown Lagrangian parameters; also, $\mathbf{\Phi} := (\phi_1 \ \phi_2 \ \cdots \ \phi_N)$ is a row matrix and $\mathbf{q} := (q_1 \ q_2 \ \cdots \ q_N)^T$ a column matrix. Substituting Eq. B.10 into the energies in Eqs. B.8, these latter take the form:

$$\tilde{U} = \frac{1}{2}\mathbf{q}^T \mathbf{K}_e \mathbf{q}, \qquad U^0 = \frac{1}{2}\mu\mathbf{q}^T \mathbf{K}_g \mathbf{q}, \tag{B.11}$$

where:

$$\mathbf{K}_e := \iint_{\Omega} (\mathcal{D}\mathbf{\Phi})^T \mathbf{E}(\mathcal{D}\mathbf{\Phi}) \, dx dy, \tag{B.12a}$$

$$\mathbf{K}_g := \iint_{\Omega} (\nabla\mathbf{\Phi})^T \mathbf{T}^0 (\nabla\mathbf{\Phi}) \, dx dy, \tag{B.12b}$$

are $N \times N$ stiffness matrices, respectively, elastic and geometric.[2] By requesting $\Pi = \tilde{U} + U^0$ is stationary with respect to $\mathbf{q}$, an algebraic eigenvalue problem is drawn:

$$\left(\mathbf{K}_e + \mu\mathbf{K}_g\right)\mathbf{q} = \mathbf{0}, \tag{B.13}$$

---

[2] The coefficients of the two matrices, $\mathbf{K}_e := \left[k_{ij}^e\right]$, $\mathbf{K}_g := \left[k_{ij}^g\right]$, consequently, are:

$$k_{ij}^e = \iint_{\Omega} \left(\phi_{i,xx} \ \phi_{i,yy} \ 2\phi_{i,xy}\right) \mathbf{E} \begin{pmatrix} \phi_{j,xx} \\ \phi_{j,yy} \\ 2\phi_{j,xy} \end{pmatrix} dx dy,$$

$$k_{ij}^g = \iint_{\Omega} \left(\phi_{i,x} \ \phi_{i,y}\right) \mathbf{T}^0 \begin{pmatrix} \phi_{j,x} \\ \phi_{j,y} \end{pmatrix} dx dy.$$

in the unknown multiplier μ. The smallest eigenvalue μ_c is the (approximate) critical load, according to Ritz; the corresponding eigenvector $\mathbf{q}_c$, when substituted in Eq. B.10, describes the associated buckling mode.

Remark B.2   The matrices in Eqs. B.12 can easily take into account of possible non-uniformity in the domain of the elastic characteristics and/or prestress.

B.3.2   Choice of the Trial Functions

The greatest difficulty in applying the Ritz method lies in the choice of the trial functions. Here, with reference to a rectangular domain and uniform constraint conditions on each side, an option is illustrated, according to which the functions are taken in a separable variable form, i.e., $\phi_{ij}(x, y) = X_i(x) Y_j(y)$. Correspondingly, Eq. B.10 is written as:

$$w(x, y) = \sum_{i=1}^{N} \sum_{j=1}^{N} a_{ij} X_i(x) Y_j(y), \tag{B.14}$$

that is, it appears as a *generalized double Fourier series*, whose coefficients $\mathbf{q} := \left(a_{ij} \right)$ are the Lagrangian parameters. The functions $X_i(x)$ (or $Y_j(y)$) can be taken as the ith eigenfunction of a boundary value problem, for example, the ith buckling (or vibrating) mode of a beam, whose boundary conditions are identical to those of the plate on the sides $x = 0, \ell$ (or $y = 0, b$).

With reference to the buckling modes in the x direction, the $X(x)$ functions are the eigenvectors of the problem:[3]

$$X'''' + \beta^2 X'' = 0, \tag{B.15}$$

i.e.:

$$X = c_1 \cos(\beta x) + c_2 \sin(\beta x) + c_3 x + c_4, \tag{B.16}$$

where the constants c_i (except one) and the wave-number β are derived by the boundary conditions. Example B.1 reports some of these functions.

Example B.1 (Buckling Modes of Various Constrained Beams)   The solutions to Eq. B.15, for several possible constraints, are:

- *Supported-supported.* By letting $X(0) = X''(0) = X(\ell) = X''(\ell)$, one gets:

---

[3] The buckling modes $Y(y)$ are obtained by the $X(x)$ modes by replacing x by y and ℓ by b.

$$X = \sin\left(\frac{n\pi x}{\ell}\right), \qquad n = 1, 2, \ldots . \tag{B.17}$$

- *Clamped-supported*. By letting $X(0) = X'(0) = X(\ell) = X''(\ell)$, the following characteristic equation is obtained:

$$\tan(\beta\ell) = \beta\ell, \tag{B.18}$$

with the associate eigenmode:

$$X = 1 - \cos(\beta x) - \frac{1 - \cos(\beta\ell)}{\sin(\beta\ell) - \beta\ell}(\sin(\beta x) - \beta x),$$

$$\frac{\beta\ell}{\pi} = 1.43, 2.46, 3.47, \ldots . \tag{B.19}$$

- *Clamped-clamped*. By letting $X(0) = X'(0) = X(\ell) = X'(\ell)$, the following characteristic equation is drawn:

$$\beta\ell \sin(\beta\ell) + 2\cos(\beta\ell) = 2, \tag{B.20}$$

together with:

$$X = \begin{cases} 1 - \cos(\beta x), & \frac{\beta\ell}{\pi} = 2, 4, 6, \ldots, \\ 1 - \cos(\beta x) - \frac{1-\cos(\beta\ell)}{\sin(\beta\ell)-\beta\ell}(\sin(\beta x) - \beta x), & \frac{\beta\ell}{\pi} = 2.86, 4.92, 6.94, \ldots, \end{cases} \tag{B.21}$$

for symmetric and antisymmetric modes, respectively.
- *Supported-free*. By letting $X(0) = X''(0) = X''(\ell) = 0$, $X'''(\ell) + \beta^2 X'(\ell) = 0$, one gets:[4]

$$X = \sin(\beta x), \qquad \frac{\beta\ell}{\pi} = n, \qquad n = 1, 2, \ldots . \tag{B.22}$$

- *Clamped-free*. By letting $X(0) = X'(0) = X''(\ell) = 0$, $X'''(\ell) + \beta^2 X'(\ell) = 0$, one finds:

$$X = 1 - \cos(\beta x), \qquad \frac{\beta\ell}{\pi} = n - \frac{1}{2}, \qquad n = 1, 2, \ldots . \tag{B.23}$$

□

---

[4] To these buckling modes, the rigid motion $X = x$, representing the eigenvector associated with the zero eigenvalue, must be added.

B.3.3   *Exploiting the Orthogonality Properties of the Buckling Modes*

When the buckling modes in Eqs. B.17–B.23 are used as trial functions for the plate, simpler expressions of the coefficients of the stiffness matrices are derived, by exploiting the orthogonality properties of modes. This properties read:[5]

$$\int_0^\ell X_i''(x)\,X_j''(x)\,\mathrm{d}x = 0, \quad \int_0^\ell X_i'(x)\,X_j'(x)\,\mathrm{d}x = 0, \quad \text{when } i \neq j, \tag{B.24}$$

as well as:

$$\int_0^b Y_i''(y)\,Y_j''(y)\,\mathrm{d}y = 0, \quad \int_0^b Y_i'(y)\,Y_j'(y)\,\mathrm{d}y = 0, \quad \text{when } i \neq j. \tag{B.25}$$

The variation of the TPE of the plate, under uniform bi-axial compression $N_x^0 = -p_x$, $N_y^0 = -p_y$ and shear $N_{xy}^0 = p_{xy}$, reads (Eqs. 10.10 and 10.32):

---

[5] The orthogonality conditions in Eqs. B.24 (or, similarly, in Eqs. B.25) are proved as follows. The TPE of the compressed beam, undergoing purely transverse displacement $v(x)$, is $\Pi = \int_0^\ell \frac{1}{2} EI\,v''^2 \mathrm{d}x - \int_0^\ell \frac{1}{2} P v'^2 \mathrm{d}x$. The stationary condition $\delta\Pi = 0$ provides the Virtual Work Principle:

$$\int_0^\ell EI\,v''\delta v''\mathrm{d}x = \int_0^\ell P\,v'\delta v'\mathrm{d}x,$$

which is consistent with Eq. 6.30. When $v = X_i$ is the ith critical mode and $P = P_i$ is the associated critical load, this equation is satisfied for any δv, in particular for $\delta v = X_j$; the same occurs when i and j are exchanged. Therefore:

$$EI \int_0^\ell X_i''X_j''\mathrm{d}x = P_i \int_0^\ell X_i'X_j'\mathrm{d}x$$

$$EI \int_0^\ell X_j''X_i''\mathrm{d}x = P_j \int_0^\ell X_j'X_i'\mathrm{d}x$$

By subtracting the previous equations member to member and assuming that $P_i \neq P_j$, the second in Eq. B.24 is drawn; hence, from the former equations, the first in Eq. B.24 also follows.

$$\delta\Pi = D \int_0^\ell \int_0^b \left[\left(w_{,xx} + v w_{,yy} \right) \delta w_{,xx} + \left(w_{,yy} + v w_{,xx} \right) \delta w_{,yy} \right.$$

$$+ 2 \left(1 - v \right) w_{,xy} \delta w_{,xy} \Big] dxdy + \int_0^\ell \int_0^b p_{xy} \left(w_{,y} \delta w_{,x} + w_{,x} \delta w_{,y} \right) dxdy$$

$$- \int_0^\ell \int_0^b p_x w_{,x} \delta w_{,x} dxdy - \int_0^\ell \int_0^b p_y w_{,y} \delta w_{,y} dxdy.$$

$$(B.26)$$

By using the double series in Eq. B.14 (with the extremes $1, N$ of the sum understood):

$$w = \sum_i \sum_j a_{ij} X_i (x) Y_j (y), \quad \delta w = \sum_h \sum_k \delta a_{hk} X_h (x) Y_k (y), \qquad (B.27)$$

it follows:

$$\delta\Pi = \sum_i \sum_j \sum_h \sum_k \left\{ a_{ij} \delta a_{hk} D \int_0^\ell \int_0^b \left[\left(X_i'' Y_j + v X_i Y_j'' \right) X_h'' Y_k \right. \right.$$

$$+ \left(X_i Y_j'' + v X_i'' Y_j \right) X_h Y_k'' + 2 \left(1 - v \right) X_i' Y_j' X_h' Y_k' \Big] dxdy$$

$$+ p_{xy} a_{ij} \delta a_{hk} \int_0^\ell \int_0^b \left(X_i Y_j' X_h' Y_k + X_i' Y_j X_h Y_k' \right) dxdy$$

$$(B.28)$$

$$- p_x a_{ij} \delta a_{hk} \int_0^\ell \int_0^b \left(X_i' Y_j X_h' Y_k \right) dxdy$$

$$- p_y a_{ij} \delta a_{hk} \int_0^\ell \int_0^b \left(X_i Y_j' X_h Y_k' \right) dxdy \Bigg\}.$$

By splitting the integrals, the previous expression is recast in the form:

$$\delta\Pi = \sum_i \sum_j \sum_h \sum_k a_{ij} \delta a_{hk} \left(K_{ij,hk}^e + K_{ij,hk}^g \right), \qquad (B.29)$$

where:[6]

$$
K_{ij,hk}^{e} := D \left[\delta_{ih} \int_0^{\ell} X_i'' X_h'' dx \int_0^b Y_j Y_k dy + \delta_{jk} \int_0^{\ell} X_i X_h dx \int_0^b Y_j'' Y_k'' dy \right.
$$

$$
+ v \left(\int_0^{\ell} X_i X_h'' dx \int_0^b Y_j'' Y_k dy + \int_0^{\ell} X_i'' X_h dx \int_0^b Y_j Y_k'' dy \right) \tag{B.30}
$$

$$
\left. + 2 \left(1 - v \right) \delta_{ih} \delta_{jk} \int_0^{\ell} X_i' X_h' dx \int_0^b Y_j' Y_k' dy \right]
$$

are the coefficients of the elastic stiffness matrix, and:

$$
K_{ij,hk}^{g} := p_{xy} \left(\int_0^{\ell} X_i X_h' dx \int_0^b Y_j' Y_k dy + \int_0^{\ell} X_i' X_h dx \int_0^b Y_j Y_k' dy \right)
$$

$$
- p_x \delta_{ih} \int_0^{\ell} X_i' X_h' dx \int_0^b Y_j Y_k dy - p_y \delta_{jk} \int_0^{\ell} X_i X_h dx \int_0^b Y_j' Y_k' dy
$$

$$
\tag{B.31}
$$

the coefficients of the geometric stiffness matrix. Here the Kronecker delta[7] accounts for the orthogonality condition expressed by Eqs. B.24 and B.25.

---

[6] As an example, when two trial functions are taken in each direction, and the four Lagrangian paremeters are ordered as $\mathbf{q} := (a_{11}, a_{12}, a_{21}, a_{22})^T$, the elastic stiffness matrix appears as (apex e suppressed):

$$
\left[K_{ij,hk}^{e} \right] = \begin{bmatrix} K_{11,11} & K_{12,11} & K_{21,11} & K_{22,11} \\ K_{11,12} & K_{12,12} & K_{21,12} & K_{22,12} \\ K_{11,21} & K_{12,21} & K_{21,21} & K_{22,21} \\ K_{11,22} & K_{12,22} & K_{21,22} & K_{22,22} \end{bmatrix}.
$$

[7] The Kronecker delta is defined as follows:

$$
\delta_{ih} = \begin{cases} 1 & i = h \\ 0 & i \neq h \end{cases}.
$$

References

1. Courant, R., Hilbert, D.: Methods of Mathematical Physics, vol. 1. Wiley, New York (1989)
2. El Naschie, M.S.: Stress Stability and Chaos in the Structural Engineering: An Energy Approach. McGraw-Hill, London (1990)
3. Sokolnikoff, I.S.: Mathematical Theory of Elasticity. McGraw-Hill, New York (1956)

Appendix C
Non-uniform Torsion of Open Thin-Walled Beams

The Vlasov theory of non-uniform torsion of open thin-walled beams is illustrated. It is based on two classical hypotheses: (a) *the cross-section is undeformable* in its plane, and (b) the middle surface of the beam is *shear undeformable*, this entailing *warping* of the cross-section. From the 3D continuum, by using energy equivalences, a *one-dimensional model of shaft* is derived, which takes into account of a warping variable along the axis of the beam. A one-dimensional finite element of open thin-walled shaft is formulated. Several illustrative examples are discussed, comparing the results supplied by the Vlasov and the de Saint-Venant derived 1D-models.

C.1 Mechanics of Torsion

This appendix is devoted to analyze the torsion of cylindrical thin-walled beams (TWB) with open cross-section. Such structures, for their peculiar mechanical behavior, occupy a specific chapter in the theory of beams, strongly distinguishing themselves from compact or thin closed-profile beams. As a matter of fact, TWB (understood open) possess a *low torsional stiffness*, for which they experience twist angles significantly larger than other beams. Moreover, the *de Saint-Venant Postulate*, according to which "the effects of self-equilibrated forces, applied at the bases, extinguish fast with the distance," *is not longer valid for TWB*. In contrast, the disturb induced by the constraints propagate to the whole TWB.

From a phenomenological point of view, two different types of deformation are observed in TWB: (1) the loss of shape of the cross-section in its plane and (2) the loss of planarity of the cross-section, also called *warping*. The in-plane loss of shape makes the beam theory inapplicable, so that modeling calls for considering the structure as a plate assembly. If, however, stiffening ribs, or thin *diaphragms*, are present, able to prevent (or significantly reduce) the loss of shape, without altering warping, it can be assumed that the cross-section remains undeformed in its own plane while being free to warp out of plane. In these cases, reference can still be made to a one-dimensional model, however suitably implemented with respect to that of compact cross-section beam (hereafter called classical, or de Saint-Venant shaft). The best known model in the literature, due to Vlasov [7], will be illustrated here (further details can be found in [1, 6]).

C.1.1   Effects of the Torsional Warping on the State of Stress

Based on the results of de Saint-Venant (DSV) theory, valid for uniform torsion, the effect of the torsional warping is first qualitatively discussed, comparing uniform and non-uniform torsion.

Uniform Torsion

The phenomenon of warping of the cross-sections of a beam in torsion is well known from the DSV theory [4, 5]. In this theory, since the cylinder is subject to forces exclusively applied at the bases, *torsion is uniform*; accordingly, the twist angle $\theta(z)$ varies linearly along the longitudinal axis z,[1] so that the torsional curvature (also said "torsion") is constant:

$$\kappa_t = \theta' = \text{cost.} \tag{C.1}$$

Furthermore, since the warping $w(x, y)$ is found to be independent of z, all sections of the beam warp in the same way, so that the longitudinal strain $\varepsilon_z = \frac{\partial w}{\partial z}$, as well as the normal stress σ_z, identically vanish.

This model of behavior is usually adopted in the technical theories of compact cross-section beams, even (a) when *the forces are distributed along the axis of the beam*, so that the torsional curvature is no longer constant along the axis, and (b) in the presence of constraints at ends of the beam, which prevent the free warping. The technical model, as is known, is based on the DSV Postulate [3, 5], which allows to extend the results of the theory to more complex situations, such as those discussed.

Non-uniform Torsion

When, however, the cross-section has no compact shape, the DSV Postulate loses its validity, so that (strictly speaking) it is no longer possible to apply the uniform

---

[1] In this appendix, and in Chap. 9, the beam axis is denoted by z (and not by x), in accordance with the common usage in 3D models.

torsion theory to TWB, being:

$$\kappa_t(z) = \theta' \neq \text{cost} \tag{C.2}$$

(i.e., the torsion is variable along the z axis). In that case, indeed, the warping $w(x, y, z)$ also depends (in an unknown manner) on z, so that $\varepsilon_z = \frac{\partial w}{\partial z} \neq 0$. From the elastic law, it follows that $\sigma_z = E\varepsilon_z \neq 0$ too, i.e., the non-uniform torsion induces *nonzero normal stresses*. However, if the external solicitation is only of torsional moment (i.e., the axial and shear forces and the bending moments are identically zero), the *normal stresses produced by the torsion must be self-equilibrated*.

If the normal stresses are variable along z, they, in turn, generate tangential stresses, as deduced from the indefinite equilibrium equation along z, i.e.:

$$\nabla \cdot \boldsymbol{\tau} + \frac{\partial \sigma_z}{\partial z} = 0 \Rightarrow \boldsymbol{\tau} \neq 0. \tag{C.3}$$

These tangential stresses (which can be assumed to be constant in the small thickness) *add themselves to those predicted by the DSV torsion theory* (which are linear and of opposite sign in the thickness); they both contribute to balance the torsional moment.

Remark C.1 Non-uniform torsion generates normal and tangential stresses, and this happens, indeed, for any type of cross-section. However, when the cross-section is compact, or it is thin but closed, the beam possesses a *high* DSV torsional stiffness, so that the phenomena qualitatively described here produce effects comparatively negligible. In contrast, since open TWB have *low* DSV torsional stiffness, the mechanisms induced by the variable warping have considerable relevance.

C.1.2   Introductory Examples: The I- and C-Cross-Sections

Examples of non-uniform torsion are qualitatively discussed, one relating to a bisymmetric, the other to a monosymmetric cross-section.

I-Cross-Section

An I-beam is considered, of small uniform thickness b, height h, and equal flanges of width a, subjected to torsion (Fig. C.1a). It is assumed that, due to symmetry, the cross-section rotates by an angle $\theta(z)$ around the centroid. The two flanges undergo equal and opposite horizontal displacements $u_{1,2} = \mp \theta(z)\frac{h}{2}$; therefore, they are bent in their own plane with a curvature $\kappa_y = \frac{\partial^2 u_{1,2}(z)}{\partial z^2} = \mp \frac{h}{2}\theta''(z)$

Fig. C.1 Non-uniform torsion of an I-beam: (**a**) twist angle, (**b**) flange deflections, (**c**) normal stresses and bimoment, (**d**) tangential stresses and bishear

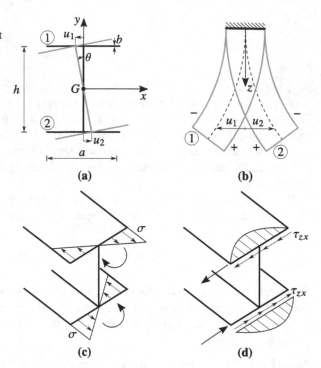

(Fig. C.1b).[2] Consequently, a field of unit extensions $\varepsilon_z = -\kappa_y x$ takes place, and normal stresses $\sigma_z = \pm \dfrac{Eh}{2} x \, \theta''(z)$ (Fig. C.1c), proportional to the second derivative of the twist angle. Given the symmetry, the equilibrium equations of the cross-section are automatically satisfied, namely:

$$N = \int_A \sigma \, dA = 0, \qquad M_y = -\int_A \sigma x \, dA = 0, \qquad M_x = \int_A \sigma y \, dA = 0.$$

$$\text{(C.4)}$$

Normal stresses are equivalent to a new quantity, which is not identifiable with any of the usual characteristics of solicitation (axial force, bending moment, shear force, torsional moment), referred to as the rigid cross-section. Vlasov called this internal action *bimoment*, as suggested by the normal stresses acting on the I-cross-section, which are equivalent to two equal and opposite moments. The bimoment will be more formally defined as an *integral mechanical action* of the normal stresses, dual of the torsion gradient $\kappa_t'(z)$.[3]

---

[2] The flanges, of course, as the web, undergo also torsion, which, however, due to the small thickness, is *energetically negligible* with respect to in-plane bending.

[3] In the same way, the axial force, bending moment, shear force and torsional moment are dual quantities of the extension, bending, average shear strain, and torsion, respectively.

Normal stresses, varying along the z axis, in turn generate tangential stresses, ruled by equilibrium. By assuming that the tangential stresses on the flanges are directed as x, it must be, $\tau_{zx,x} + \sigma_{z,z} = 0$,[4] by integrating it with the boundary condition $\tau_{zx}(-\frac{a}{2}, z) = 0$, it follows:

$$\tau_{zx}(x, z) = \mp \frac{Eh}{2}\theta'''(z) \int_{-\frac{a}{2}}^{x} x\,dx = \pm \frac{Eh}{4}\left(\frac{a^2}{4} - x^2\right)\theta'''(z). \tag{C.5}$$

The tangential stresses are therefore parabolic on the flanges, as in the Jourawsky theory (Fig. C.1d), and are proportional to the third derivative of the twist. These stresses have zero resultant force, but their *torque is different from zero* and equal to:

$$M_t^*(z) := -h \int_{-\frac{a}{2}}^{\frac{a}{2}} \tau_{zx}(x, z)b\,dx = -E\frac{bh^2a^3}{24}\theta'''(z). \tag{C.6}$$

The quantity M_t^* is sometimes called *bishear*, being a couple made up of two opposite shear forces solicitating the flanges; however, the prevalent diction, today, is *complementary torsional moment*.

Remark C.2 It should be observed that in the example in Fig. C.1d, the complementary torsional moment, M_t^*, has the same sign of the DSV torsional moment, M_t.[5] It follows that the external torsional moment is balanced by the internal action:

$$\bar{M}_t := M_t + M_t^*, \tag{C.7}$$

i.e., partly by M_t and partly by M_t^*. Since the tangential stresses induced by M_t have *small arm*, equal to $\frac{2b}{3}$, while those induced by M_t^* have *large arm*, equal to h, the result is a better mechanical performance in non-uniform torsion, compared to that in uniform torsion. However, this better behavior is counterbalanced by the presence of normal stresses, which are absent in uniform torsion.

C-Cross-Section

A channel beam is considered, subject to non-uniform torsion (Fig. C.2a). The cross-section rotates by an angle $\theta(z)$ around a point C, which, for reasons of symmetry (to be discussed later), is supposed to belong to the x axis, *but it is not coincident*

---

[4] The comma denotes partial differentiation with respect to the following variable.

[5] Indeed, as it is known from Jourawsky theory, the non-uniform bending of the upper (lower) flange, illustrated in Fig. C.1d, induces tangential stresses, which are opposite (concordant) to the x axis.

Fig. C.2 Non-uniform torsion of the channel beam: (a) twist angle, (b) normal stresses

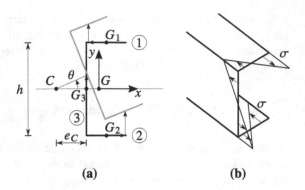

(a)   (b)

with the centroid, as in the case of doubly symmetry. By denoting the flanges by 1 and 2, and the web by 3, the displacements of the respective centroids G_i are $u_{1,2} = \mp \dfrac{h}{2}\theta\,(z)$ and $v_3 = e_C\theta\,(z)$, with h being the height of the cross-section and e_C the eccentricity of C with respect to the web. Correspondingly, the three walls undergo flexural curvatures $\kappa_{y1,2} = \mp\dfrac{h}{2}\theta''\,(z)$ and $\kappa_{y3} = e_C\theta''\,(z)$.

However, displacements corresponding to these inflections are not compatible, as they do not guarantee the continuity at the nodes, both in-plane and out of plane. For compatibility, it needs not only to normally translate and to rotate the segments 1,2,3 to ensure the continuity of the translations u, v in the plane (which happens with an almost zero energy expenditure), but it is also necessary to extend/contract the flanges along the z axis, in order to restore the continuity of warping w. Hence, the walls are *extended and inflected* and, therefore stressed by piecewise linear normal stresses, continue at nodes (Fig. C.2b). Also executing this operation, the field of normal stresses, it is not in general self-equilibrated: to set to zero the resultant force, $N = 0$, a further and suitable overall extension must be applied. Moreover, while the equilibrium condition $M_x = 0$ is satisfied by the symmetry, the other condition $M_y = 0$ requires an appropriate choice of the C pole, as it will be discussed later.

C.2   Vlasov Theory of Non-uniform Torsion

A beam is considered, subjected to distributed and concentrated torsional couples, which induce a solicitation of only torsional moment, arbitrarily variable along the axis z of the beam. One wants to solve the elastic problem of non-uniform torsion for the 3D continuum and, then, to formulate an equivalent 1D model of shaft.

The beam has an open cross-section of small thickness, with principal inertia axes x and y passing through the centroid G (Fig. C.3a). Γ is the *middle line* of the cross-section, i.e., the curve equidistant from its two long edges. On Γ, a curvilinear abscissa s is taken, originating from a point O, for the moment chosen arbitrarily. The thickness of the cross-section, generally variable, is described by

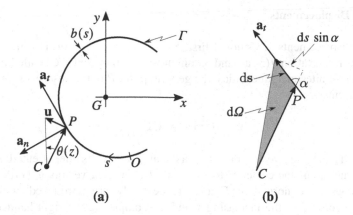

Fig. C.3   Open cross-section of small thickness: (**a**) displacements due to torsion, (**b**) sectorial area element

the $b(s)$ function. Since the thickness is, by hypothesis, small compared to the transverse dimensions of the cross-section, it is assumed that all the quantities of interest are constant on the chord normal to Γ, of length $b(s)$. So, for example, the displacement field is $\mathbf{u} = \mathbf{u}(s, z)$, that is, it refers to the *average surface S* of the profile, consisting of the cylinder of generatrix Γ and axis z. The TWB is therefore modeled as a (polar) 2D continuum,[6] immersed in a 3D space.

C.2.1   Kinematics

Two fundamental hypotheses are introduced:

1. The cross-section is rigid in its own plane, but warps out of plane;
2. The shear strains γ_{zs}, at the middle surface S of the TWB, are identically zero.

The first hypothesis is verified with a good approximation if the profile has thin diaphragms (or ribs), which contrast the in-plane loss of shape of the cross-section but leave this free of warp out of plane. The second hypothesis assumes membrane shear undeformability of the surface S, as suggested by the DSV theory, in which the torsion induces a field of shear strains linear on the chord, which zeroes at the midline Γ.[7]

---

[6] The model is polar to account for the DSV behavior. However, the Vlasov beam will be referred to as a 3D model, according to the literature.

[7] From the two hypotheses, it follows that, drawn a squared mesh on S, oriented according to the directrix Γ and the generatrices z, this transforms, after deformation, in a rectangular mesh, since only elongations along z occur.

In-plane Displacements

In-plane displacements are studied first. Since the cross-section is rigid, the displacement is a rotation $\theta(z)$ around an unknown rotation center C, called *center of torsion*. In infinitesimal kinematics, the generic point P at the abscissa s undergoes the displacement:

$$\mathbf{u} = \theta(z)\,\mathbf{a}_z \times \mathbf{CP}(s), \tag{C.8}$$

where $\mathbf{a}_z$ is the unit vector of the z axis and $\mathbf{CP}(s)$ is the oriented *position vector*, which joins the center C to the point P. The unit vectors $\mathbf{a}_t(s)\,and\,\mathbf{a}_n(s)$ of the tangent and normal to Γ at s, respectively, are introduced, with $\mathbf{a}_t(s)$ oriented according to the increasing s and $\mathbf{a}_n$ completing the right-handed basis. The displacement component tangent to the curve reads $u_t = \mathbf{u} \cdot \mathbf{a}_t$, i.e.:

$$u_t = \theta(z)\,\mathbf{a}_z \times \mathbf{CP}(s) \cdot \mathbf{a}_t(s). \tag{C.9}$$

Warping

From the second hypothesis, it is possible to determine the warping $w(z,s)$ as a function of the in-plane displacements. Indeed, since $\gamma_{zs} = w_{,s} + u_{t,z} = 0$, integrating with respect to s, one gets:

$$w = \cancel{w_0(z)} - \int_0^s u_{t,z}\,ds = -\theta'(z) \int_0^s \mathbf{CP}(s) \times d\mathbf{s}\cdot\mathbf{a}_z, \tag{C.10}$$

in which the factors of the mixed product have been permuted and the oriented segment on Γ has been denoted by $d\mathbf{s} := ds\,\mathbf{a}_t$; furthermore, $w_0(z) = 0$ has been taken.[8] In this expression, $\frac{1}{2}\mathbf{CP}(s) \times d\mathbf{s}\cdot\mathbf{a}_z = \pm\frac{1}{2}\overline{CP}\,ds\,\sin\alpha =: d\Omega$ is the area element, endowed with sign, spanned by the position vector $\mathbf{CP}(s)$, when its free end P runs along the arc element ds (Fig. C.3b); it is positive if the position vector rotates counterclockwise and negative if clockwise. Consequently:

$$\omega(s) := \int_0^s \mathbf{CP}(s) \times d\mathbf{s}\cdot\mathbf{a}_z, \tag{C.11}$$

---

[8] Indeed, $w_0(z)$ generates longitudinal strains, and therefore stresses, uniform on the cross-section, which, for the equilibrium, must cancel out.

is the double of the area spanned by the position vector when its free end travels the OP segment of Γ. The $\omega(s)$ function is called the *sectorial area* or *warping function*. It is a geometric property of the section, which depends on three independent parameters: the two coordinates of the torsion center C and the origin O; it, therefore, should be more correctly denoted by $\omega_C(s; O)$ (with C and O omitted when there is no ambiguity).

Ultimately, the warping field is written as:

$$w = -\theta'(z)\,\omega(s). \tag{C.12}$$

This entails unit extensions $\varepsilon_z = w_{,z}$ of the longitudinal fibers, equal to:

$$\varepsilon_z = -\theta''(z)\,\omega(s). \tag{C.13}$$

Remark C.3 The warping function $\omega(s)$ describes on Γ the *shape* of warping w, extension ε_z and normal stress σ_z (as stated in Eq. C.14 ahead), whose relevant amplitudes are ruled by the successive derivatives of $\theta(z)$.

Remark C.4 If the torsion is uniform, it is $\theta'(z) = \text{cost}$; consequently, warping depends on s but not on z, and the extension (and stress) cancels out identically.

Example C.1 (Some Warping Functions) As a few examples, the cross-sections examined in the Sect. C.1.2 are considered again.

For the I-cross-section, O and C are taken coincident with the centroid G (Fig. C.4a). Making the generic point P move from the origin O, the sectorial area spanned by the position vector **CP** is null on the web; it is different from zero on the flanges, on which it varies with a linear law, with a sign depending on the verse of the rotation of **CP**. The maximum absolute value occur at the extremes of the flanges, and it is equal to $\omega_A = 2\left(\frac{1}{2}\frac{h}{2}\frac{a}{2}\right)$.

For the C-cross-section, C and O are taken on the unique symmetry axis (Fig. C.4b). By making a generic point P move from the origin O, the warping function $\omega(s)$ is determined as piecewise linear. In absolute value, at the junction between flanges and web, it is $\omega_A = 2\left(\frac{1}{2}ec\frac{h}{2}\right)$; at the free end of the flanges, it is

$$\omega_B = \omega_A - 2\left(\frac{1}{2}\frac{h}{2}a\right). \qquad \square$$

C.2.2 Center of Torsion

To determine the center of torsion, it needs to enforce that the normal stresses induced by the non-uniform torsion are self-equilibrated.

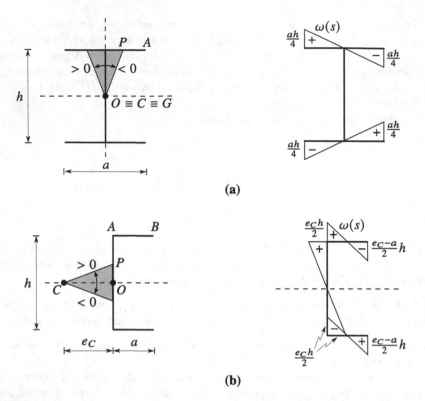

Fig. C.4 Examples of warping functions: (**a**) I-cross-section, (**b**) C-cross-section

Normal Stresses

By using the elastic law $\sigma_z = E\varepsilon_z$,[9] the unit extensions in Eq. C.13 trigger the normal stresses:

$$\sigma_z = -E\theta''(z)\,\omega(s). \tag{C.14}$$

Since, by hypothesis, the external forces induce torsional moment, but no axial force $N(z)$ and no bending moments $M_x(z)$, $M_y(z)$, the following conditions must hold:

$$N(z) = \int_A \sigma\,\mathrm{d}A = -E\theta''(z)\int_\Gamma \omega(s)\,b(s)\,\mathrm{d}s = 0, \tag{C.15a}$$

---

[9] Here, the same law valid in the DSV theory is used, admitting that, due to the Poisson effect, there is a contraction of the cross-section in its own plan, although this contradicts the hypothesis of transverse undeformability.

$$M_x(z) = \int_A \sigma \, y \mathrm{d}A = -E\theta''(z) \int_\Gamma \omega(s) \, y(s) \, b(s) \, \mathrm{d}s = 0, \qquad \text{(C.15b)}$$

$$M_y(z) = -\int_A \sigma \, x \mathrm{d}A = -E\theta''(z) \int_\Gamma \omega(s) \, x(s) \, b(s) \, \mathrm{d}s = 0. \qquad \text{(C.15c)}$$

These equations, having to be satisfied for any z, entail:

$$S_\omega := \int_\Gamma \omega(s) \, b(s) \, \mathrm{d}s = 0, \qquad \text{(C.16a)}$$

$$I_{\omega x} := \int_\Gamma \omega(s) \, y(s) \, b(s) \, \mathrm{d}s = 0, \qquad \text{(C.16b)}$$

$$I_{\omega y} := \int_\Gamma \omega(s) \, x(s) \, b(s) \, \mathrm{d}s = 0, \qquad \text{(C.16c)}$$

which are called the *sectorial area properties*. To satisfy them, it is necessary to look for an appropriate center of torsion C and an origin O of the sectorial area.

Remark C.5 Equations C.16 define the geometric quantities S_ω, $I_{\omega x}$, and $I_{\omega y}$ which are, respectively, called *sectorial static moment* (of dimension $[L^4]$) and *sectorial mixed inertia moments* (of dimensions $[L^5]$). These geometric characteristics follow the usual definitions of static moments and inertia moments of the areas, but with $\omega(s)$ in place of $x(s)$ and $y(s)$ (hence, the increased dimensions). The properties of the sectorial area, Eqs. C.16, are formally analogous to those possessed by the centroid and by the central inertia axes of a plane figure.

Remark C.6 Equations C.16 are also called *orthogonality properties*, as they state the vanishing of the scalar products between the functions $\omega(s)$ and 1, $x(s)$ ed $y(s)$.[10] They express the following geometric properties: In kinematics, 1, $x(s)$ and $y(s)$ represent three *modes* of displacement of the rigid section, which consist, respectively, in a translation along the z axis and two rotations around the axes y and x. The twist generates an additional $\omega(s)$ mode, describing the warping of the cross-section. The four modes are *mutually orthogonal*, due to the well-known properties of the centroid, to which the properties of the sectorial area in Eq. C.16 add themselves.

---

[10] In a space of function, and with an appropriate definition of scalar product, $f(x)$ and $g(x)$ are called orthogonal when $\int_a^b f(x)g(x)\mathrm{d}x = 0$.

Remark C.7 The same orthogonality properties apply to the normal stresses σ_z. Those deriving from extension and bending are proportional to 1, $x(s)$, and $y(s)$ and those descending from torsion to $\omega(s)$. It follows that, given the proportionality between σ_z and ε_z, the *mixed virtual work*, between stresses of one type and strains of another type, is zero. It means that extension, bending, and torsion are *energetically orthogonal*.

Coordinates of the Center of Torsion

The center of torsion C is determined by imposing the conditions in Eqs. C.16b, c (i.e., $I_{\omega x} = I_{\omega y} = 0$). Through elementary calculations, it can be proved (in the Supplement C.1) that its coordinates are:

$$x_C = x_D + \frac{1}{I_x} \int_\Gamma \omega_D(s)\, y(s)\, b(s)\, ds, \tag{C.17a}$$

$$y_C = y_D - \frac{1}{I_y} \int_\Gamma \omega_D(s)\, x(s)\, b(s)\, ds, \tag{C.17b}$$

where I_x, I_y are the central inertia moments of the cross-section and where $\omega_D(s)$ is the warping function evaluated with respect to an arbitrary pole D of coordinates x_D, y_D, and to an *arbitrary origin* O.[11] Therefore, the following operation must be performed: (a) a pole D and an origin O are arbitrarily chosen, and the $\omega_D(s)$ function is built-up; (b) from Eqs. C.17, the coordinates of the pole C are evaluated.

Remark C.8 Equations C.17 show that *the torsion center coincides with the shear center*, as determined by the Jourawsky approximate theory [4]. A concise demonstration of their coincidence will be given immediately.

Supplement C.1 (Proof of Eqs. C.17) To prove Eqs. C.17, the relation that links the sectorial areas $\omega_C(s)$ and $\omega_D(s)$, referred to two arbitrary poles C, D and to the same origin O, must be sought for (Fig. C.5a). Applying the definition in Eq. C.11, and taking into account that $\mathbf{CP}(s) = \mathbf{CD} + \mathbf{DP}(s)$, it follows:

$$\omega_C(s) = \int_0^s \mathbf{CP}(s) \times ds\cdot\mathbf{a}_z$$

$$= \mathbf{CD} \times \int_0^P ds \cdot \mathbf{a}_z + \int_0^P \mathbf{DP}(s) \times ds \cdot \mathbf{a}_z \tag{C.18}$$

$$= \mathbf{CD} \times \mathbf{OP} \cdot \mathbf{a}_z + \omega_D(s).$$

---

[11] Indeed, as it will be formally proven in the following Eq. C.25, if one translates the origin O, the sectorial area function changes by an additive constant, whose contribution to Eq. C.17 is irrelevant.

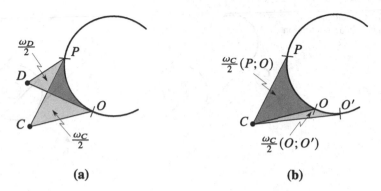

Fig. C.5 Dependence of the sectorial area: (**a**) from the pole, (**b**) from the origin

Since $\mathbf{CD} = \mathbf{x}_D - \mathbf{x}_C$ and $\mathbf{OP} = \mathbf{x} - \mathbf{x}_O$, the mixed product is equal to:

$$\mathbf{CD} \times \mathbf{OP} \cdot \mathbf{a}_z = \begin{vmatrix} x_D - x_C & y_D - y_C & 0 \\ x - x_O & y - y_O & 0 \\ 0 & 0 & 1 \end{vmatrix} = (x_D - x_C)(y - y_O) - (y_D - y_C)(x - x_O),$$

(C.19)

hence, the desired relationship reads:

$$\omega_C = \omega_D + (x_D - x_C)(y - y_O) - (y_D - y_C)(x - x_O).$$ (C.20)

By imposing that the sectorial mixed inertia moment $I_{\omega Cx}$, evaluated with respect to the C pole, vanishes, and taking into account that the axes x, y are central of of inertia, one gets:

$$\int_\Gamma \omega_C(s)\, y(s)\, b(s)\, ds = \int_\Gamma [\omega_D(s) + (x_D - x_C)(y - y_O) - (y_D - y_C)(x - x_O)]$$

$$\times y(s)\, b(s)\, ds = \int_\Gamma \omega_D(s)\, y(s)\, b(s)\, ds + (x_D - x_C) I_x = 0.$$

(C.21)

Proceeding in the same way for $I_{\omega Cy}$, one obtains:

$$\int_\Gamma \omega_C(s)\, x(s)\, b(s)\, ds = \int_\Gamma [\omega_D(s) + (x_D - x_C)(y - y_O) - (y_D - y_C)(x - x_O)]$$

$$\times x(s)\, b(s)\, ds = \int_\Gamma \omega_D(s)\, x(s)\, b(s)\, ds - (y_D - y_C) I_y = 0.$$

(C.22)

From these conditions, Eqs. C.17 follow.   □

Fig. C.6 Coincidence
between the shear and torsion
centers

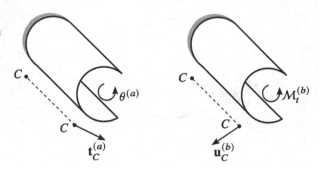

Coincidence Between Torsion and Shear Centers

The coincidence of the Vlasov torsion center with the Jourawsky shear center can
be easily proved by applying the Betti Theorem (or Reciprocity Theorem [3]).[12]

A TWB is considered, at an end of which the following systems of forces are
separately applied (Fig. C.6):

- a system (a), consisting of a shear force $\mathbf{t}_C^{(a)}$ applied at the *shear center* C; for the
  definition of shear center, the cross-section undergoes a displacement $\mathbf{u}_C^{(a)} \neq \mathbf{0}$
  and a rotation $\theta^{(a)} = 0$;
- a system (b), consisting of a torsional couple $\mathcal{M}_t^{(b)}$, which induces an unknown
  displacement $\mathbf{u}_C^{(b)}$ of the shear center C.

To determine $\mathbf{u}_C^{(b)}$, the Betti Theorem is applied, which is written as:

$$\mathbf{t}_C^{(a)} \cdot \mathbf{u}_C^{(b)} = \mathcal{M}_t^{(b)} \theta^{(a)}. \tag{C.23}$$

Since $\theta^{(a)} = 0$, it turns out that $\mathbf{u}_C^{(b)} = \mathbf{0}$. It is concluded that since the couple leaves
unchanged the position of the shear center, this is also the center of torsion.

---

[12] *Betti Theorem.* Given a system of forces $\mathbf{f}_a$ and a system of forces $\mathbf{f}_b$, acting on a linearly elastic
body, which, respectively, produce displacements $\mathbf{u}_a$, $\mathbf{u}_b$, the virtual work of the forces $\mathbf{f}_a$ expended
on the displacement $\mathbf{u}_b$ is equal to the virtual work of the forces $\mathbf{f}_b$ expended on the displacements
$\mathbf{u}_a$, i.e.:

$$\sum \mathbf{f}_a \cdot \mathbf{u}_b = \sum \mathbf{f}_b \cdot \mathbf{u}_a$$

The Theorem is an application of the Virtual Work Principle. Written the elastic law in the form
$\sigma = \mathbf{E}\varepsilon$, with $\mathbf{E} = \mathbf{E}^T$, it follows:

$$\sum \mathbf{f}_a \cdot \mathbf{u}_b = \int_V \sigma_a^T \cdot \varepsilon_b dV = \int_V \varepsilon_a^T \mathbf{E}\varepsilon_b dV = \int_V \varepsilon_a^T \sigma_b dV = \sum \mathbf{f}_b \cdot \mathbf{u}_a.$$

Principal Origin of the Sectorial Area

Once the pole C has been determined, the origin of the sectorial area must be evaluated by imposing that $S_\omega = 0$, i.e., by remembering Eq. C.16a, that the warping function has a *zero mean*. This property makes superfluous the actual determination of O (called *principal origin*), making it easier to proceed as follows: (a) an arbitrary origin O' (with respect to which the calculations are easier) is introduced, and the function $\omega_C\left(s; O'\right)$ is evaluated; (b) the main value, $\overline{\omega}_C\left(s; O'\right)$ is computed; (c) the mean value is subtracted from the function, thus obtaining:

$$\omega_C\left(s; O\right) = \omega_C\left(s; O'\right) - \overline{\omega}_C\left(s; O'\right). \tag{C.24}$$

A formal proof follows.

Supplement C.2 (Proof of Eq. C.24) The following relationship holds between the sectorial areas $\omega_C\left(P; O\right)$ and $\omega_C\left(P; O'\right)$, with $P = P\left(s\right)$ (Fig. C.5b):

$$\omega_C\left(P; O\right) = \omega_C\left(P; O'\right) - \omega_C\left(O; O'\right). \tag{C.25}$$

By enforcing $S_\omega = 0$, and using the previous equation:

$$\int_\Gamma \omega_C\left(P; O\right) b\left(s\right) ds = \int_\Gamma \omega_C\left(P; O'\right) b\left(s\right) ds - A\omega_C\left(O; O'\right) = 0, \tag{C.26}$$

where A is the cross-section area. Hence:

$$\omega_C\left(O; O'\right) = \frac{1}{A} \int_\Gamma \omega_C\left(P; O'\right) b\left(s\right) ds, \tag{C.27}$$

and consequently, by Eq. C.25:

$$\omega_C\left(P; O\right) = \omega_C\left(P; O'\right) - \frac{1}{A} \int_\Gamma \omega_C\left(P; O'\right) b\left(s\right) ds, \tag{C.28}$$

i.e., Eq. C.24.   □

C.3 One-Dimensional Shaft Model

The sectorial area law $\omega\left(s\right)$ defines the *shape* of the torsional warping, which, depending it only from the geometry of the cross-section, can be evaluated in advance. The *amplitude* of warping instead is unknown and, according to Eq. C.12, depends on the twist angle $\theta\left(z\right)$. This is therefore the only kinematic parameter; since it is defined in the domain $(0, \ell)$, the problem reduces to *one-dimensional*.

C.3.1   Formulation

To formulate a 1D model, starting from the 3D model, it is convenient to make use of a variational procedure, as described below.

Generalized Strains

The *generalized strains* of the shaft must describe (a) the DSV shear strains,[13] and (b) the longitudinal extensions. The former quantities depend exclusively on the torsion $\kappa_t (z)$; the latter, on the *torsion gradient* $\beta (z) := \kappa_t' (z)$. The generalized strains are related to the twist angle $\theta (z)$ from the congruence equations:

$$\kappa_t = \theta'(z), \tag{C.29a}$$

$$\beta = \theta''(z). \tag{C.29b}$$

These differential equations must be accompanied by geometric conditions at the boundary $\bar{z} := 0, \ell$, which are of two types:

- *Torsional clamp*, which implies:

$$\theta (\bar{z}) = 0; \tag{C.30}$$

- *Warping clamp*, $w (s, \bar{z}) = 0 \; \forall s$, which, by virtue of Eq. C.12, implies:

$$\theta' (\bar{z}) = 0. \tag{C.31}$$

Remark C.9 It is worth noticing that it is not possible to enforce a warping constraint only *on a portion of the cross-section*, as this is incompatible with the TWB model, generating, instead, a shell behavior.

Generalized Stresses

Generalized stresses (or solicitations) are defined as the dual quantities of the generalized strains. Generalized stresses and strains express the internal virtual work of the 3D beam, denoted as δW_i^{3D}, in terms of quantities pertinent to the 1D model, denoted by δW_i^{1D}.

     A first generalized stress is the DSV torsional moment, M_t, dual of the torsion κ_t. It represents the virtual work that the tangential stresses τ_{zs} expend on the shear

---

[13] Effect not made explicit in the Vlasov model, where the thickness has been condensed.

strains γ_{zs} occurring in the thickness of the profile, as a result of a unit torsion $\delta \kappa_t = 1$.

To determine the dual quantity of the torsion gradient, one proceeds as follows. Assigned, according to Eq. C.13, a field of virtual strains $\delta \varepsilon_z = -\omega \delta \theta'' = -\omega \delta \beta$, the internal virtual work expended by the stress field σ_z is:

$$\delta W_i^{3D} := \int_V \sigma_z \delta \varepsilon_z dV = -\int_0^\ell \delta \beta dz \int_\Gamma \sigma_z \omega b ds = \int_0^\ell B \delta \beta dz =: \delta W_i^{1D}, \qquad (C.32)$$

in which:

$$B(z) := -\int_\Gamma \sigma_z(s, z) \omega(s) b(s) ds. \qquad (C.33)$$

This quantity is called the *bimoment* and is the dual quantity of the torsion gradient β. It represents the virtual work that the normal stresses do on the longitudinal strains induced by a unit torsion gradient $\delta \beta = 1$; it is therefore an (internal) *Lagrangian force*.

Equilibrium Equations

The equilibrium of the 1D model is imposed by a variational approach, via the Virtual Work Principle, $\delta W_i^{1D} = \delta W_e^{1D}$. Let the shaft be loaded by distributed couples $c(z)$. By assigning virtual fields of displacements $\delta \theta$ and strains $\delta \kappa_t, \delta \beta$, the Principle reads:

$$\int_0^\ell M_t \delta \kappa_t dz + \int_0^\ell B \delta \beta dz = \int_0^\ell c \delta \theta dz. \qquad (C.34)$$

For the congruence, $\delta \kappa_t = \delta \theta'$ and $\delta \beta = \delta \theta''$ must hold, so that:

$$\int_0^\ell \left(M_t \delta \theta' + B \delta \theta'' - c \delta \theta \right) dz = 0, \qquad \forall \delta \theta, \qquad (C.35)$$

which, integrated by parts, becomes:

$$\int_0^\ell \left(B'' - M_t' - c \right) \delta \theta dz + \left[B \delta \theta' - B' \delta \theta + M_t \delta \theta \right]_0^\ell = 0, \qquad \forall \delta \theta. \qquad (C.36)$$

From this variational equation, the indefinite equilibrium equation is derived:

$$(M_t - B')' + c = 0, \tag{C.37}$$

together with the boundary conditions:

$$(M_t - B')\, \delta\theta = 0, \qquad\qquad \text{at } z = 0, \ell, \tag{C.38a}$$

$$B\, \delta\theta' = 0, \qquad\qquad \text{at } z = 0, \ell. \tag{C.38b}$$

From Eqs. C.38, taking into account the geometric conditions, the *mechanical conditions* follow, and precisely:

- If twist is free at the ends $\bar{z} = 0, \ell$, $\delta\theta$ is there arbitrary, and therefore it must be:

$$M_t(\bar{z}) - B'(\bar{z}) = 0; \tag{C.39}$$

- If warping is free at the ends $\bar{z} = 0, \ell$, $\delta\theta'$ is there arbitrary, and therefore it must be:

$$B(\bar{z}) = 0. \tag{C.40}$$

The latter condition is a consequence of the fact that, in the 3D model, the normal stress must vanish at the end cross-sections, entailing, by virtue of Eq. C.33, that also the bimoment vanishes.

Remark C.10 The field Eq. C.37 and the boundary condition in Eqs. C.39 are similar to those usually written for the compact cross-section shaft, i.e., $M_t' + c = 0$ and $M_t(\bar{z}) = 0$. However, these equations are additively modified by the new quantity $M_t^* := -B'$. This represents a torsional moment, called *complementary* (or also *bishear*), which adds itself to the DSV torsional moment M_t, to balance the external moment, so that the equilibrium equations become:

$$(M_t + M_t^*)' + c = 0, \tag{C.41a}$$

$$M_t(\bar{z}) + M_t^*(\bar{z}) = 0. \tag{C.41b}$$

Constitutive Law

The constitutive law that links the torsional moment and the torsion is known from the DSV theory, i.e.:

$$M_t = GJ\kappa_t, \tag{C.42}$$

where G is the tangential elastic modulus and $J := \frac{1}{3} \int_\Gamma b^3 (s) \, ds$ is the torsional inertia.

To determine the link between the bimoment and the torsion gradient, the elastic potential energy of the 3D model is equated to the elastic potential energy of the 1D model:

$$U^{3D} := \frac{1}{2} \int_V E \varepsilon_z^2 \, dV = \frac{1}{2} \int_0^\ell dz \int_A E \, (-\beta \omega)^2 \, dA = \frac{1}{2} E I_\omega \int_0^\ell \beta^2 dz =: U^{1D},$$

(C.43)

where a new geometric quantity has been introduced:

$$I_\omega := \int_\Gamma \omega^2 (s) \, b (s) \, ds,$$ (C.44)

called the *sectorial inertia moment*, having dimension $[L^6]$. The density of the elastic potential energy of the 1D model is defined as $e := \frac{dU^{1D}}{dz}$; therefore:

$$e := \frac{1}{2} E I_\omega \beta^2.$$ (C.45)

From the Green elastic law, valid for hyperelastic materials, one obtain $B = \dfrac{\partial e}{\partial \beta}$, i.e.:

$$B = E I_\omega \beta.$$ (C.46)

The bimoment is therefore proportional to the torsion gradient.[14]

Elastic Problem in Terms of Displacements

In summary, the elastic problem for open thin-walled shafts is governed by the following field equations:

- Congruence:

$$\begin{pmatrix} \kappa_t \\ \beta \end{pmatrix} = \begin{bmatrix} \partial_z \\ \partial_z^2 \end{bmatrix} \theta,$$ (C.47)

---

[14] The analogy with bending should be observed, for which $M_x = E I_x \kappa_x$, where $I_x = \int_A y^2 dA$.

where $\partial_z = \dfrac{d}{dz}$; *the system is kinematically impossible for internal constraints.*

- Equilibrium:

$$[-\partial_z \ \partial_z^2]\begin{pmatrix} M_t \\ B \end{pmatrix} = c, \tag{C.48}$$

in which the static operator is the adjunct of the kinematic one; *the system is hyperstatic for internal constraints.*

- Constitutive law:

$$\begin{pmatrix} M_t \\ B \end{pmatrix} = \begin{bmatrix} GJ & 0 \\ 0 & EI_\omega \end{bmatrix}\begin{pmatrix} \kappa_t \\ \beta \end{pmatrix}. \tag{C.49}$$

Combining the previous equations according to the displacement method, one gets:

$$EI_\omega \theta'''' - GJ\theta'' = c\,(z), \tag{C.50}$$

which constitutes the *elastic line equation of the shaft.* By substituting the elastic law in Eqs. C.38, the mechanical boundary conditions are recast in terms of displacement as:

$$[GJ\theta' - EI_\omega \theta''']\delta\theta = 0, \qquad\qquad \text{at } z = 0, \ell, \tag{C.51a}$$

$$EI_\omega \theta''\delta\theta' = 0, \qquad\qquad \text{at } z = 0, \ell. \tag{C.51b}$$

C.3.2   Solution to the Problem

The field Eq. C.50 is rewritten as:

$$\theta'''' - \alpha^2\theta'' = \frac{c\,(z)}{EI_\omega}, \tag{C.52}$$

where:

$$\alpha^2 := \frac{GJ}{EI_\omega}, \tag{C.53}$$

is a quantity of dimension $[\mathrm{L}^{-2}]$. The characteristic equation of the problem, $\lambda^2(\lambda^2 - \alpha^2) = 0$, admits the four real roots $\lambda_{1,2,3,4} = (0, 0, +\alpha, -\alpha)$. The general solution to the non-homogeneous problem, therefore, reads:

$$\theta = c_1 e^{\alpha(z-\ell)} + c_2 e^{-\alpha z} + c_3 z + c_4 + \theta_p\,(z), \tag{C.54}$$

where c_i are arbitrary constants and $\theta_p(z)$ is a particular solution to the non-homogeneous problem.[15] In the special case in which $c(z) = c = \text{cost}$, one can take:

$$\theta_p(z) = -\frac{1}{2}\frac{c}{GJ}z^2. \tag{C.55}$$

In the complementary part of the solution Eq. C.54, *exponential* contributions add themselves to the well-known linear part, which describes the uniform torsion of the DSV shaft. These extra terms allow to satisfy the increased number of boundary conditions, induced by the free or prevented warping at the end cross-sections. The exponential part decays from the extremes as quickly as the larger the nondimensional number $\alpha\ell$; therefore it happens that:

- If the beam has a high value of the DSV stiffness GJ, the $\alpha\ell$ ratio is large, so that *the exponential decay is fast;* the additional boundary conditions, therefore, produce only a local disturbance, confirming the validity of the DSV Postulate.[16]
- If the beam has a low value of the DSV stiffness GJ, the $\alpha\ell$ ratio is small, so that *the exponential decay is slow;* the additional boundary conditions produce significant effects on the whole length of the beam, confirming the invalidity of the DSV Postulate.[17]

C.3.3 Normal and Tangential Stresses

Once the elastic problem for the 1D model has been solved and therefore $\theta(z)$ has been evaluated, one can go back to the 3D model to determine the stresses, as described below.

Normal Stresses

By remembering that $\sigma_z = -E\omega\theta'' = -E\omega\beta$ and $\beta = \frac{B}{EI_\omega}$, one finds:

$$\sigma_z(s,z) = -\frac{B(z)}{I_\omega}\omega(s). \tag{C.56}$$

---

[15] The writing $c_1 e^{\alpha(z-\ell)}$ is equivalent to $\tilde{c}_1 e^{\alpha z}$, with $\tilde{c}_1 = c_1 e^{-\alpha\ell}$, as the constants are arbitrary. However, if α is large, the first form allows to avoid *numerical malconditioning*. Indeed, a boundary layer close to $z = \ell$ implies $c_1 = O(1)$, while it involves $\tilde{c}_1 \ll 1$ (with the other constants being $O(1)$). Thus, $c_2 e^{-\alpha z}$ describes the left boundary layer (and tends to zero to the right), and $c_1 e^{\alpha(z-\ell)}$ describes the right boundary layer (and tends to zero to the left).

[16] This is the case of compact beams or closed TWB, for which equations similar to Eq. C.50 hold, although with a different definition of I_ω.

[17] This is, generally, the case of open TWB.

This expression has the same formal structure as the Navier formula for one-axis bending: $\sigma_z (s, z) = \frac{M_x(z)}{I_x} y (s)$.

Tangential Stresses

Complementary tangential stresses $\boldsymbol{\tau}^* = \tau^* \mathbf{a}_t$ (to be superimposed to the DSV tangential stresses) are *reactive quantities* (in that they are associated with the undeformability constraint of the middle surface S). They must therefore be determined by the equilibrium equation $\nabla \cdot \boldsymbol{\tau}^* = -\frac{\partial \sigma_z}{\partial z}$. Taking into account Eq. C.56, this is rewritten as:

$$\nabla \cdot \boldsymbol{\tau}^* = -\frac{M_t^* (z)}{I_\omega} \omega (s) , \qquad (C.57)$$

having used $M_t^* = -B'$. Therefore, by proceeding as in the Jourawsky theory of non-uniform bending [4], one can integrate this equation on the portion $\mathcal{A}^\#$ of the cross-section, of midline $\Gamma^\#$, cut at the abscissa s from the chord C (Fig. C.7), to get:[18]

$$\int_{\mathcal{A}^\#} \nabla \cdot \boldsymbol{\tau}^* \mathrm{d}A = -\frac{M_t^* (z)}{I_\omega} \int_{\Gamma^\#} \omega (\bar{s}) b (\bar{s}) \mathrm{d}\bar{s} . \qquad (C.58)$$

By the Divergence Theorem, it follows:

$$q (s, z) = -\frac{M_t^* (z) S_\omega^\# (s)}{I_\omega} , \qquad (C.59)$$

where $q (s, z) := \int_C \tau_{zs}^* \mathrm{d}C$ is the *flow of tangential stresses through the chord C* and also:

$$S_\omega^\# (s) := \int_{\Gamma^\#} \omega (\bar{s}) b (\bar{s}) \mathrm{d}\bar{s} , \qquad (C.60)$$

is the *sectorial static moment extended to the portion* $\Gamma^\#$ (depending on s) of the middle line. Assuming the stress is uniform on the chord, the final expression is drawn:

$$\tau^* = -\frac{M_t^* (z) S_\omega^\# (s)}{I_\omega b (s)} . \qquad (C.61)$$

---

[18] Here the superscript # is used, instead of the more common asterisk, this latter having already been used to denote the "complementary" quantities.

Fig. C.7 Flow of the complementary tangential stresses through the chord C

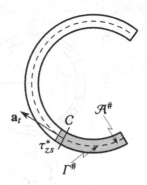

This has the same structure as the Jourawsky formula for bending and shear (non-uniform flexure). The tangential stresses τ^* are positive when outgoing from the $\mathcal{A}^{\#}$ area (Fig. C.7).

C.4 Illustrative Example: The Open Circular Tube

The results so far obtained are exemplified for a circular tube of average radius R and uniform thickness b, cut along a generatrix (Fig. C.8a). For such a cross-section, the geometric characteristics are calculated, and the distribution of normal and tangential stresses is evaluated. Finally, an application to a shaft is shown.

Geometric Characteristics of the Open Annular Cross-Section
The DSV torsional stiffness of the open annular cross-section is:

$$J = \frac{2}{3}\pi R b^3. \tag{C.62}$$

The shear center C, as it is known from the Jourawsky theory, lies on the symmetry axis, opposite to the cut, $2R$ apart from the centroid.

To evaluate the warping function ω_C with respect to the torsion center $C = (0, -2R)$, it is convenient to choose first a pole $D \equiv G = (0, 0)$ (Fig. C.8b). Taking the origin $O = (0, -R)$, it turns out that $\omega_D = R^2\varphi$, with $\varphi := \frac{s}{R}$ the angle at center, positive counterclockwise. Using the change of pole rule in Eq. C.20, and taking into account that $x = R\sin\varphi$, $y = -R\cos\varphi$, one obtains (suppressing the subscript C):[19]

$$\omega(\varphi) = R^2(\varphi - 2\sin\varphi), \tag{C.63}$$

---

[19] Alternatively, by observing from Fig. C.8b that $\omega_D - \omega_C$ is equal to double the area of the triangle with base $2R$ and height $R\sin\varphi$, Eq. C.63 follows.

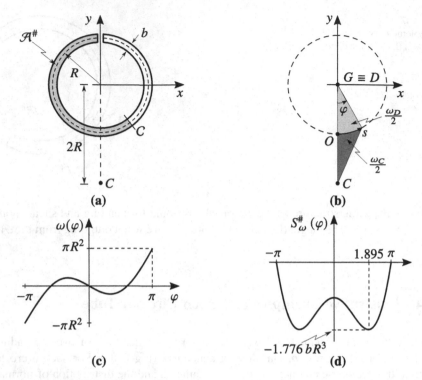

Fig. C.8 Open annular cross-section: (**a**) dimensions, (**b**) sectorial area, (**c**) warping function $\omega(\varphi)$, (**d**) sectorial static moment $S_\omega^{\#}(\varphi)$

which is plotted in Fig. C.8c. Since this function has zero mean, it is already referred to as the principal origin. Integrating $\omega^2(\varphi)$ according to Eq. C.44, the sectorial inertia moment is determined:

$$I_\omega = bR \int_{-\pi}^{\pi} \omega^2(\varphi)\, d\varphi = \frac{2}{3}\pi \left(\pi^2 - 6\right) b\, R^5. \qquad (C.64)$$

Using Eq. C.60, the sectorial static moment, function of the $\mathscr{A}^{\#}$ area, is derived (Fig. C.8a):

$$S_\omega^{\#}(\varphi) := bR \int_{-\pi}^{\varphi} \omega(\bar{\varphi})\, d\bar{\varphi} = \frac{1}{2} bR^3 \left(4 - \pi^2 + \varphi^2 + 4\cos\varphi\right), \qquad (C.65)$$

which is diagrammed in Fig. C.8d.

Normal and Tangential Stresses

Normal stresses are expressed by Eq. C.56 and are proportional to $\omega(\varphi)$. The function takes the maximum value in modulus, πR^2, at $\varphi = \pm\pi$; there, the maximum normal stress on the cross-section is:

$$\sigma_{z,\max}(z) = \frac{3B(z)}{2(\pi^2 - 6)bR^3}. \tag{C.66}$$

The complementary tangential stresses are calculated with Eq. C.61. Taking in this equation $\varphi = \pm 1.895$, where $S_\omega^\#(\varphi)$ is maximum in modulus and equal to $-1.776\,bR^3$, the maximum complementary tangential stress on the cross-section is evaluated:

$$\tau_{\max}^* = 0.219\,\frac{M_t^*(z)}{bR^2}. \tag{C.67}$$

To this stress, the maximum DSV shear stress must be added:

$$\tau_{\max} = \frac{M_t b}{J} = \frac{3M_t}{2\pi Rb^2}. \tag{C.68}$$

Remark C.11 If M_t and M_t^* are of the same order of magnitude (as in the examples below), it is $\tau_{\max}^* \ll \tau_{\max}$, being $R \gg b$. If, in contrast, $M_t^* \gg M_t$, τ^* can be of the same order as τ (consistently with what discussed in Fig. C.1d).

Example C.2 (Non-uniform Torsion Induced by a Constraint) An example is worked out, concerning a shaft clamped at the end $z = 0$, both to torsion and to warping, loaded by a torsional moment $C_B = 30\,\mathrm{N\,m}$ applied at the free end $z = \ell$ (Fig. C.9a). If warping were free at the clamp, the torsion would be uniform; since, however, warping is prevented there, torsion is non-uniform. Goal of the analysis is to study the effect induced by the constraint, comparing the Vlasov and DSV models of shaft. Numerical results refer to the open annular cross-section studied above, by taking: $E = 207,000\,\mathrm{N/mm^2}$, $G = 79,300\,\mathrm{N/mm^2}$, $b = 4\,\mathrm{mm}$, $R = 40\,\mathrm{mm}$, $\ell = 4\,\mathrm{m}$.

The boundary conditions at the two ends are:

$$\theta(0) = 0, \qquad\qquad \theta'(0) = 0, \tag{C.69a}$$

$$GJ\theta'(\ell) - EI_\omega\theta'''(\ell) = C_B, \qquad\qquad \theta''(\ell) = 0, \tag{C.69b}$$

which, imposed to the general solution in Eq. C.54 (in which $\theta_p \equiv 0$), lead to:

$$\theta = \frac{C_B}{GJ}\left[z + \frac{1}{\alpha}\left(\operatorname{sech}(\alpha\ell)\sinh(\alpha(\ell - z)) - \tanh(\alpha\ell)\right)\right]. \tag{C.70}$$

This is plotted in Fig. C.9b and compared with the solution of the DSV shaft. It is seen that the warping, although prevented in a single section, *changes the uniform*

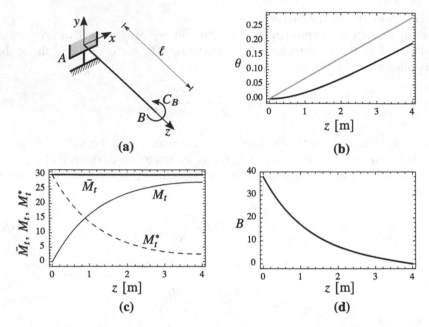

Fig. C.9 Clamped-free shaft, with prevented warping at the clamp, subject to a torsional moment at the free end: (**a**) structure, (**b**) twist angle, (**c**) torsional moments [Nm], (**d**) bimoment $\left[\text{Nm}^2\right]$. Black line, Vlasov shaft; gray line, DSV shaft

torsion regime into non-uniform, modifying the structural response significantly, unfolding its effects even at a great distance from the constraint.

The solicitations are calculated by differentiating $\theta\,(z)$, as:

$$M_t = GJ\theta' = C_B\left(1 - \operatorname{sech}(\alpha\ell)\cosh(\alpha\,(\ell - z))\right), \tag{C.71a}$$

$$B = EI_\omega\theta'' = \frac{C_B}{\alpha}\operatorname{sech}(\alpha\ell)\sinh(\alpha\,(\ell - z)), \tag{C.71b}$$

$$M_t^* = -EI_\omega\theta''' = C_B\operatorname{sech}(\alpha\ell)\cosh(\alpha\,(\ell - z)), \tag{C.71c}$$

and are plotted in the Figs. C.9c, d. It is observed that the total moment, $\bar{M}_t = M_t + M_t^*$, is constant, as required by equilibrium; however, since the prevented warping imposes at the clamp that $\kappa_t = \theta' = 0$, in this section, it is $M_t = 0$, i.e., the external action is entirely balanced by the complementary torsional moment.

Normal stresses are deduced from Eq. C.56 and plotted in Fig. C.10a; their maximum value occurs at the clamp $z = 0$, where the bimoment is maximum, and at the abscissa $\varphi = \pm\pi$, where the warping function is maximum. The complementary tangential stresses are calculated with Eq. C.61; they are maximum at the clamp $z = 0$, where the complementary torsional moment is maximum, and at the abscissa $\varphi = \pm1.895$, where $S_\omega^{\#}$ attains its maximum absolute value. Since $\tau^* > 0$, the tangential stress leaves the $\mathcal{A}^{\#}$ area and therefore produces a counterclockwise

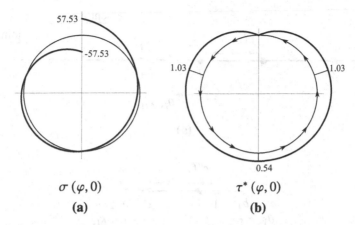

$$\sigma\,(\varphi,0) \qquad\qquad\qquad \tau^*\,(\varphi,0)$$

(a) **(b)**

Fig. C.10 Clamped-free shaft, with prevented warping at the clamp, subject to a torsional moment at the free end: (a) polar diagram of normal stresses $\sigma\,(\varphi,0)$ at the clamp; (b) polar diagram of the complementary tangential stresses $\tau^*\,(\varphi,0)$ at the clamp. Stresses in [N/mm²]

twisting moment M_t^*. The DSV tangential stresses are calculated with Eq. C.68; at the free end, they are equal to 20.46 N/mm². $\qquad\qquad\qquad\qquad\qquad\qquad\square$

C.5 Finite Element Analysis

A finite element of shaft, in non-uniform torsion, is developed here [2]. The beam is straight, subject to generic distributed couples $c\,(z)$, and concentrated couple C_j applied at points z_j (Fig. C.11a). It is assumed that the torsional stiffness GJ and the warping stiffness EI_ω are piecewise constant, with discontinuity points z_k. The points z_j and z_k are singularity points, which are taken as *nodes* of the structure. The system is thus described by n nodes of coordinates z_i ($i = 1, \ldots, n$) and $m = n - 1$ elements ($e = 1, \ldots, m$). Each node has two degrees of freedom: the twist θ_i and the torsional curvature $\kappa_{ti} = \theta_i'$, so that the system has $2n$ degrees of freedom.

C.5.1 Exact Finite Element

A generic element e of the system is considered (Fig. C.11b), of end nodes 1 and 2 (in local numbering). Let $z^e \in (0, \ell^e)$ a local abscissa with origin at node 1 and ℓ^e the length of the element. Let also $\theta_1^e, \kappa_{t1}^e$ and $\theta_2^e, \kappa_{t2}^e$ be the generalized displacements of nodes 1 and 2; $c^e\,(z^e)$ the external distributed couples applied to the element, referred to the local abscissa; m_{t1}^e, b_1^e and m_{t2}^e, b_2^e the generalized internal forces (couples and *bicouples*, the meaning of which will appear clear from the next Eqs. C.81).

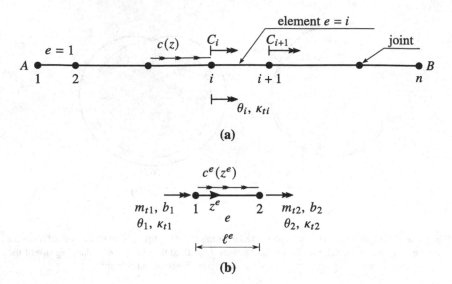

Fig. C.11 Shaft subject to distributed $c\,(x)$ and concentrated C_j couples: (**a**) nodes and elements, (**b**) eth element: nodal forces m_{ti}, b_i and displacements θ_i, κ_{ti} ($i = 1, 2$)

The first step consists in searching for the (linear) relationship existing between the nodal forces m_{ti}^e, b_i^e and the nodal displacements θ_i^e, κ_{ti}^e, once the distributed couples $c^e\,(z)$ have been assigned. To this end, it is needs to integrate the following differential problem:[20]

$$EI_\omega\,\theta'''' - GJ\,\theta'' = c\,(z)\,, \tag{C.72a}$$

$$\theta\,(0) = \theta_1\,, \tag{C.72b}$$

$$\theta'\,(0) = \kappa_{t1}\,, \tag{C.72c}$$

$$\theta\,(\ell) = \theta_2\,, \tag{C.72d}$$

$$\theta'\,(\ell) = \kappa_{t2}\,. \tag{C.72e}$$

This system describes the shaft element subject to distributed couples and assigned nodal displacement. Solved this problem, the nodal forces will be evaluated as the reactions exerted by the constraints.

---

[20] For simplicity of writing, the index e is omitted, when no ambiguity exists.

Displacement Field

Given the linearity of Eqs. C.72, it is possible to express the solution as the sum of two solutions:

$$\theta(z) = \theta_f(z) + \theta_m(z),\tag{C.73}$$

where (a) $\theta_f(z)$ is related to the problem (called with *fixed nodes*) where the known terms are different from zero in the domain and zero at the boundary and (b) $\theta_m(z)$ is related to the problem (called with *movable nodes*) in which the known terms are nonzero at the boundary and zero in the domain. The two sub-problems read:

(a) Fixed node problem:

$$EI_\omega \theta_f'''' - GJ\theta_f'' = c(x),\tag{C.74a}$$

$$\theta_f(0) = \theta_f'(0) = 0,\tag{C.74b}$$

$$\theta_f(\ell) = \theta_f'(\ell) = 0.\tag{C.74c}$$

(b) Movable node problem:

$$EI_\omega \theta_m'''' - GJ\theta_m'' = 0,\tag{C.75a}$$

$$\theta_m(0) = \theta_1,\tag{C.75b}$$

$$\theta_m'(0) = \kappa_{t1},\tag{C.75c}$$

$$\theta_m(\ell) = \theta_2,\tag{C.75d}$$

$$\theta_m'(\ell) = \kappa_{t2}.\tag{C.75e}$$

The fixed node solution must be determined case by case; if $c(x) = c = $ cost, it is equal to:

$$\theta_f = \frac{c\,e^{-\alpha z}}{2\alpha^3 EI_\omega\left(e^{\alpha\ell} - 1\right)}\left\{\ell e^{2\alpha z} + e^{\alpha z}\left[\alpha z\left(z - \ell\right) - \ell\right]\right.$$
$$\left. + e^{\alpha(z+\ell)}\left[\alpha z\left(\ell - z\right) - \ell\right] + \ell e^{\alpha\ell}\right\}.\tag{C.76}$$

The movable node solution, on the other hand, must be determined once and for all; it takes the form:

$$\theta_m(z) = \hat{\psi}_1(z)\,\theta_1 + \hat{\psi}_2(z)\,\kappa_{t1} + \hat{\psi}_3(z)\,\theta_2 + \hat{\psi}_4(z)\,\kappa_{t2},\tag{C.77}$$

where, remembering Eq. C.54:

$$\hat{\psi}_i\,(z) = c_{i1}\,e^{\alpha(z-\ell)} + c_{i2}\,e^{-\alpha z} + c_{i3}\,z + c_{i4}, \tag{C.78}$$

are the *interpolation functions*. In them, $\alpha^2 := GJ/EI_\omega$, and also:

$$
\begin{aligned}
c_{11} &:= \frac{1}{\alpha\ell + e^{-\alpha\ell}\,(\alpha\ell + 2) - 2}\,, & c_{12} &= -c_{11}, \\[2mm]
c_{13} &:= -\frac{\alpha\,(1 + e^{\alpha\ell})}{\alpha\ell + e^{\alpha\ell}\,(\alpha\ell - 2) + 2}\,, & c_{31} &= -c_{11}, \\[2mm]
c_{14} &:= \frac{\alpha\ell + e^{\alpha\ell}\,(\alpha\ell - 1) + 1}{\alpha\ell + e^{\alpha\ell}\,(\alpha\ell - 2) + 2}\,, & c_{32} &= c_{11}, \\[2mm]
c_{21} &:= \frac{e^{\alpha\ell}\,(e^{\alpha\ell} - \alpha\ell - 1)}{\alpha\,(e^{\alpha\ell} - 1)\,[\alpha\ell + e^{\alpha\ell}\,(\alpha\ell - 2) + 2]}\,, & c_{33} &= -c_{13}, \\[2mm]
c_{22} &:= \frac{e^{\alpha\ell}\,[e^{\alpha\ell}\,(\alpha\ell - 1) + 1]}{\alpha\,(1 - e^{\alpha\ell})\,[\alpha\ell + e^{\alpha\ell}\,(\alpha\ell - 2) + 2]}\,, & c_{34} &= c_{23}, \\[2mm]
c_{23} &:= \frac{1 - e^{\alpha\ell}}{\alpha\ell + e^{\alpha\ell}\,(\alpha\ell - 2) + 2}\,, & c_{41} &= -c_{22}, \\[2mm]
c_{24} &:= \frac{\alpha\ell + e^{2\alpha\ell}\,(\alpha\ell - 1) + 1}{\alpha\,(e^{\alpha\ell} - 1)\,[\alpha\ell + e^{\alpha\ell}\,(\alpha\ell - 2) + 2]}\,, & c_{42} &= -c_{21}, \\[2mm]
c_{44} &:= \frac{e^{\alpha\ell}\,(e^{\alpha\ell} - 2\alpha\ell) - 1}{\alpha\,(e^{\alpha\ell} - 1)\,[\alpha\ell + e^{\alpha\ell}\,(\alpha\ell - 2) + 2]}\,, & c_{43} &= c_{23}.
\end{aligned}
\tag{C.79}
$$

The interpolation functions are plotted in Fig. C.12 for different values of the nondimensional ratio $\alpha\ell = \sqrt{\frac{GJ\ell^2}{EI_\omega}}$. It emerges that:

- when $\alpha\ell \to \infty$ (i.e., when the DSV torsional stiffness GJ is large, or the beam is long): (a) the functions $\hat{\psi}_1, \hat{\psi}_3$ tend to straight lines, characteristics of the classical shaft; (b) the functions $\hat{\psi}_2, \hat{\psi}_4$ also tend to straight lines, however connected to zero by rapidly variable functions at the ends of the beam (where boundary layers occur);
- when $\alpha\ell = O\,(1)$ (i.e., when the DSV torsional stiffness GJ is small, or the beam is short), the functions $\hat{\psi}_2, \hat{\psi}_4$ significantly deviate from straight lines.

Nodal Forces

The displacement field in Eq. C.73 induces the following generalized stresses:

$$\bar{M}_t\,(z) := M_t\,(z) + M_t^\star\,(z) = GJ\,\theta'\,(z) - EI_\omega\,\theta'''\,(z), \tag{C.80a}$$

$$B\,(z) = EI_\omega\,\theta''\,(z), \tag{C.80b}$$

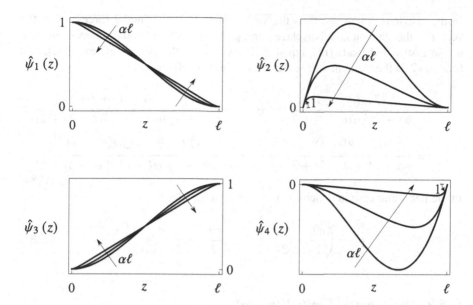

Fig. C.12 Exact interpolation functions for the shaft in non-uniform torsion, for different values of the nondimensional parameter $\alpha\ell = \sqrt{\frac{GJ\ell^2}{EI_\omega}}$ ($\alpha\ell = 1, 10, 50$)

from which the forces at the ends of the beam are computed as:

$$\begin{pmatrix} m_{t1} \\ b_1 \\ m_{t2} \\ b_2 \end{pmatrix} = \begin{pmatrix} -\bar{M}_t(0) \\ -B(0) \\ \bar{M}_t(\ell) \\ B(\ell) \end{pmatrix} = GJ \begin{pmatrix} -\theta'(0) \\ 0 \\ \theta'(\ell) \\ 0 \end{pmatrix} + EI_\omega \begin{pmatrix} \theta'''(0) \\ -\theta''(0) \\ -\theta'''(\ell) \\ \theta''(\ell) \end{pmatrix}. \tag{C.81}$$

By using Eq. C.77, and recasting Eq. C.81 in matrix form, the desired force-displacement relationship is finally determined:

$$\mathbf{f}^e = \mathbf{K}^e \mathbf{q}^e - \mathbf{p}^e, \tag{C.82}$$

in which:

$$\mathbf{f}^e := \begin{pmatrix} m_{t1} & b_1 & m_{t2} & b_2 \end{pmatrix}^T, \qquad \mathbf{q}^e := \begin{pmatrix} \theta_1 & \kappa_{t1} & \theta_2 & \kappa_{t2} \end{pmatrix}^T,$$

$$\mathbf{K}^e := \begin{pmatrix} k_{11} & k_{12} & -k_{11} & k_{12} \\ k_{12} & k_{22} & -k_{12} & k_{24} \\ -k_{11} & -k_{12} & k_{11} & -k_{12} \\ k_{12} & k_{24} & -k_{12} & k_{22} \end{pmatrix}, \qquad \mathbf{p}^e := \begin{pmatrix} -m_{t1p} \\ -b_{1p} \\ -m_{t2p} \\ -b_{2p} \end{pmatrix}. \tag{C.83}$$

In the previous equations: $\mathbf{f}^e$ is the vector of the element nodal forces; $\mathbf{q}^e$ is the vector of the element nodal displacement; $\mathbf{p}^e$ is the vector of nodal forces equivalent to the distributed loads (i.e., equal and opposite to the constraint reactions); and finally, $\mathbf{K}^e$ is the element stiffness matrix, whose coefficients are:

$$k_{11} := \frac{GJ\,\alpha\left(e^{\alpha\ell}+1\right)}{\alpha\ell + e^{\alpha\ell}\left(\alpha\ell-2\right)+2}, \quad k_{22} := \frac{GJ\left(e^{2\alpha\ell}(\alpha\ell-1)+1+\alpha\ell\right)}{\alpha\left(e^{\alpha\ell}-1\right)\left(\alpha\ell + e^{\alpha\ell}\left(\alpha\ell-2\right)+2\right)},$$

$$k_{12} := \frac{GJ\left(e^{\alpha\ell}-1\right)}{\alpha\ell + e^{\alpha\ell}\left(\alpha\ell-2\right)+2}, \quad k_{24} := \frac{GJ\left(e^{2\alpha\ell}-2\alpha\ell e^{\alpha\ell}-1\right)}{\alpha\left(e^{\alpha\ell}-1\right)\left(\alpha\ell + e^{\alpha\ell}\left(\alpha\ell-2\right)+2\right)}.$$

$$\text{(C.84)}$$

In the particular case in which $c(z) = c = \text{cost}$, one has:

$$\mathbf{p}^e = \left(\frac{c\ell}{2}, \quad \frac{c}{2\alpha^2 k_{12}}, \quad \frac{c\ell}{2}, \quad -\frac{c}{2\alpha^2 k_{12}}\right)^T. \tag{C.85}$$

C.5.2   Polynomial Finite Element

The exact interpolation functions, as it has been seen, are transcendental and depend on the ratio between the two torsional stiffnesses. To avoid this dependence, much simpler polynomial functions can be used, which, however, being approximate, require a finer discretization of the beam. To derive the force-displacement relationship, moreover, one cannot make use of the equilibrium equations (*strong formulation*), but he needs to use a variational principle, which minimizes the error (*weak formulation*).

The total potential energy of the shaft in non-uniform torsion, subject to distributed couples of intensity c (assumed constant) and nodal forces m_{ti}, b_i ($i = 1, 2$), reads:

$$\Pi = \int_0^\ell \left(\frac{1}{2} GJ\,\theta'^2 + \frac{1}{2} EI_\omega\,\theta''^2 - c\,\theta\right) dx$$

$$- \left(m_{t1}\,\theta(0) + b_1\,\theta'(0) + m_{t2}\,\theta(\ell) + b_2\,\theta'(\ell)\right). \tag{C.86}$$

In the same spirit of the Ritz method (but applied to the element), the twist field $\theta(z)$ is approximated as:

$$\theta(z) = \psi_1(z)\,\theta_1 + \psi_2(z)\,\kappa_{t1} + \psi_3(z)\,\theta_2 + \psi_4(z)\,\kappa_{t2}, \tag{C.87}$$

where $\psi_i(x)$, $i = 1, \ldots, 4$, are polynomial interpolation functions, which take the form:[21]

---

[21] The interpolation functions (here renamed) have already been used for the polynomial finite element of a prestressed extensible beam (Sect. 7.12).

$$\psi_1(z) = 1 - 3\left(\frac{z}{\ell}\right)^2 + 2\left(\frac{z}{\ell}\right)^3, \quad \psi_2(z) = \left[1 - 2\frac{z}{\ell} + \left(\frac{z}{\ell}\right)^2\right]x,$$

$$\psi_3(z) = 3\left(\frac{z}{\ell}\right)^2 - 2\left(\frac{z}{\ell}\right)^3, \quad \psi_4(z) = \left[\left(\frac{z}{\ell}\right)^2 - \frac{z}{\ell}\right]z.$$

(C.88)

Substituting Eqs. C.87 and C.88 in the functional, Eq. C.86, and imposing the stationary with respect to the displacement parameters, Eqs. C.83 and C.85 are obtained, where the quantities are redefined as follows:

$$\mathbf{K}^e := \begin{pmatrix} \frac{12EI_\omega}{\ell^3} + \frac{6GJ}{5\ell} & \frac{6EI_\omega}{\ell^2} + \frac{GJ}{10} & -\frac{12EI_\omega}{\ell^3} - \frac{6GJ}{5\ell} & \frac{6EI_\omega}{\ell^2} + \frac{GJ}{10} \\ & \frac{4EI_\omega}{\ell} + \frac{2\ell GJ}{15} & -\frac{6EI_\omega}{\ell^2} - \frac{GJ}{10} & \frac{2EI_\omega}{\ell} - \frac{\ell GJ}{30} \\ & & \frac{12EI_\omega}{\ell^3} + \frac{6GJ}{5\ell} & -\frac{6EI_\omega}{\ell^2} - \frac{GJ}{10} \\ \text{SYM} & & & \frac{4EI_\omega}{\ell} + \frac{2\ell GJ}{15} \end{pmatrix},$$

$$\mathbf{p}^e := \left(\frac{c\ell}{2} \quad \frac{c\ell^2}{12} \quad \frac{c\ell}{2} \quad -\frac{c\ell^2}{12}\right)^T.$$

(C.89)

C.5.3 Numerical Examples

Some numerical applications are developed, related to shaft in non-uniform torsion, for various loads and boundary conditions. All the results concern the cut annular cross-section, studied in the Sect. C.4; they are compared with those supplied by the classical shaft model. The numerical values adopted are: $E = 207,000\,\text{N/mm}^2$, $G = 79,300\,\text{N/mm}^2$, $b = 4\,\text{mm}$, $R = 40\,\text{mm}$; $\ell = 4\,\text{m}$ or $\ell = 10\,\text{m}$.

Example C.3 (Clamped-Clamped Shaft, Free to Warp, Loaded by Distributed Couples) The structure illustrated in Fig. C.13a is considered, consisting of a shaft of length ℓ, subject to distributed couples of intensity $c = 5\,\text{N}$, constrained at the ends by devices which prevent the twist of the cross-section but leave the warping free. First, the exact and then the polynomial finite elements are used.

When using the exact finite element (Sect. C.5.1), it is sufficient to consider just one element; hence, the (global) vector of the configuration variables $\mathbf{q}$ consists of four displacement components, two for each node: $\mathbf{q} = (\theta_1, \kappa_{t1}, \theta_2, \kappa_{t2})^T$. The global stiffness matrix and the vector of the equivalent nodal loads coincide with those defined in Eqs. C.83 and C.85. By imposing the constraints $\theta_1 = \theta_2 = 0$, that is, by deleting in the global stiffness matrix the rows and columns corresponding to the suppressed displacements and eliminating from the vector of the equivalent

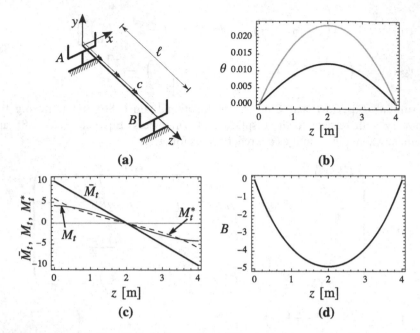

Fig. C.13 Clamped-clamped shaft, free to warp ($\ell = 4$ m): (**a**) structure; (**b**) twist angle, (**c**) torsional moments [Nm], (**d**) bimoment $[\text{Nm}^2]$. Black line, Vlasov shaft; gray line, DSV shaft; exact solution

nodal loads the corresponding rows, one arrives to a non-homogeneous algebraic system of two equations in the unknowns κ_{t1}, κ_{t2}. By solving that system and using Eqs. C.73 and C.80, the results illustrated in Fig. C.13 are achieved. It is observed that the two models of shaft provide very different twist angles (Fig. C.13b), the DSV model being much more flexible than the Vlasov model. The total torsional moment $\bar{M}_t\,(z)$ (Fig. C.13c) is linear and antisymmetric and therefore coincides (for reasons of symmetry) with that of the classical model; however, it is the sum of two contributions, $M_t\,(z)$ and $M_t^*\,(z)$, which appear to be close to each other and of the same sign.

If one changes the length of the shaft, increasing it to $\ell = 10$ m, the results of Fig. C.14 are obtained. From these, it is observed that the effect of non-uniform torsion decreases, consistently with the fact that the characteristic exponent $\alpha\ell$ in the general solution, Eq. C.54, increases.

The same problem is now solved using the polynomial finite element developed in the Sect. C.5.2. The shaft is divided in three elements of length $\ell/3$, so that the (global) vector of configuration variables $\mathbf{q}$ consists of eight displacement components, two for each node: $\mathbf{q} = (\theta_1, \kappa_{t1}, \theta_2, \kappa_{t2}, \theta_3, \kappa_{t3}, \theta_4, \kappa_{t4})^T$. The local stiffness matrix and the vector of equivalent nodal loads of the generic element are obtained by putting $\ell/3$ in place of ℓ in Eqs. C.89a and C.89b. The global stiffness matrix and the global load vector are built up by assembling the local stiffness

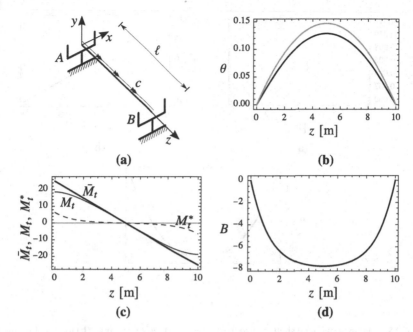

Fig. C.14   Clamped-clamped shaft, free to warp ($\ell = 10$ m): (**a**) structure; (**b**) twist angle, (**c**) torsional moments [Nm], (**d**) bimoment $\left[\text{Nm}^2\right]$. Black line, Vlasov shaft; gray line, DSV shaft; exact solution

matrices and the vectors of the nodal loads through a sequential scheme (analogous to that used for extensible beam systems in Chap. 7, Fig. 7.21). By introducing the external constraints $\theta_1 = \theta_4 = 0$, one gets a system of non-homogeneous algebraic equations in the six unknown Lagrangian parameters. Solving this system, and using the Eqs. C.87 and C.80, the results shown in Fig. C.15 are obtained. These are compared with those supplied by exact finite element. From comparison, it is possible to observe an excellent agreement between the two solutions in terms of displacements $\theta(z)$ and $\kappa_t(z)$ and a slight discrepancy of the solicitation fields, which are approximated only on average by the polynomial finite element. However, by increasing the number of finite elements, the approximate solution converges fast to the exact one, also in terms of solicitations.   □

Example C.4 (Clamped-Clamped Shaft, with Warping Prevented at One End, Solicited by Distributed Couples) The clamped-clamped shaft, already considered in the Example C.3, is analyzed, in which, however, a warping constraint is introduced at the right end only (Fig. C.16a). Using a single exact finite element, and imposing the constraints $\theta_1 = k_{t1} = \theta_2 = 0$, the results displayed in Fig. C.16 are obtained. It is observed that the additional constraint considerably increases the torsional stiffness of the beam, since the maximum twist angle reduces to about one fourth (Fig. C.16b). Furthermore, since the warping constraint destroys the symmetry of the structure, the torsional moments at the extremes are no longer

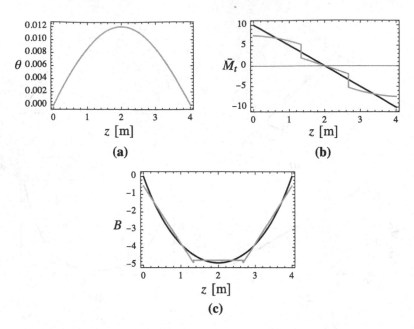

Fig. C.15 Clamped-clamped shaft, free to warp ($\ell = 4$ m): (**a**) twist angle, (**b**) torsional moments [Nm], (**c**) bimoment $[\text{Nm}^2]$. Black line, finite exact element; gray line, polynomial finite element

equal, so that the diagram of $\bar{M}_t(z)$ is found to be translated (Fig. C.16c). The two contributions, $M_t(z)$ and $M_t^*(z)$, are of the same sign, with $M_t(\ell) = 0$, due to the prevented warping at the right end. The bimoment, also, is strongly non-symmetrized (Fig. C.16d), since it must vanish at the left (where the normal stresses are zero, because the cross-section warps freely), but not at right (where the normal stresses must prevent warping).   □

Example C.5 (Clamped-Free Shaft, with Warping Prevented at the Clamp, Solicited by Distributed Couples) A shaft is considered, constrained at the left end only, where both twist and warping are prevented (Fig. C.17a). The shaft is solicited by uniformly distributed couples of intensity $c = 5$ N. Still using a single exact finite element, and imposing the constraints $\theta_1 = k_{t1} = 0$, one obtains the results illustrated in Fig. C.17. The most relevant aspect is that, close to the free end, *the two contributions to the total torsional moment are discordant in sign*, so that the DSV torsional moment is, in this region, larger than that predicted by the classical model.   □

Example C.6 (Clamped-Clamped Shaft, Free to Warp, Solicited by a Concentrated Couple) The shaft of the Example C.3 is considered again but solicited by a couple concentrated at midspan, of intensity $C_C = 20$ N m (Fig. C.18a). To solve the problem, two exact finite elements are employed, so that the Lagrangian parameter

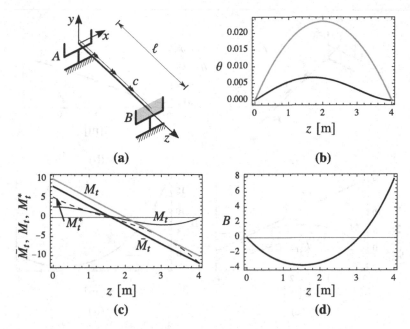

Fig. C.16 Clamped-clamped shaft, with prevented/free warping ($\ell = 4$ m): (**a**) structure; (**b**) twist angle, (**c**) torsional moments [Nm], (**d**) bimoment $\left[\text{Nm}^2\right]$. Black line, Vlasov shaft; gray line, DSV shaft; exact solution

vector is $\mathbf{q} = (\theta_1, \kappa_{t1}, \theta_2, \kappa_{t2}, \theta_3, \kappa_{t3})^T$, having denoted with 1,2,3 the nodes A, C, B, respectively. After having assembled two element stiffness matrices and imposed the constraints $\theta_1 = \theta_3 = 0$, a four DOF system is obtained. Its solution leads to the results illustrated in Fig. C.18, when $\ell = 4$ m, and Fig. C.19, when $\ell = 10$ m.

Looking first at the shorter beam (Fig. C.18), it emerges that the solutions provided by the two models, DSV and Vlasov, are very different, especially near the point at which the couple is applied, where warping must vanish for symmetry reasons. The two contributions to the moment are equally important, except near midspan, where, being $M_t = 0$ for the symmetry, the contribution of M_t^* prevails. However, when the beam becomes longer (Fig. C.19), the two solutions are close each other, and the effect of warping is appreciated only around the midspan. □

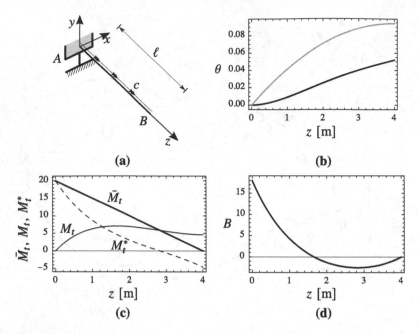

Fig. C.17 Clamped.free shaft, with warping prevented at the clamp ($\ell = 4$ m): (**a**) structure; (**b**) twist angle, (**c**) torsional moments [Nm], (**d**) bimoment $[\text{Nm}^2]$. Black line, Vlasov shaft; gray line, DSV shaft; exact solution

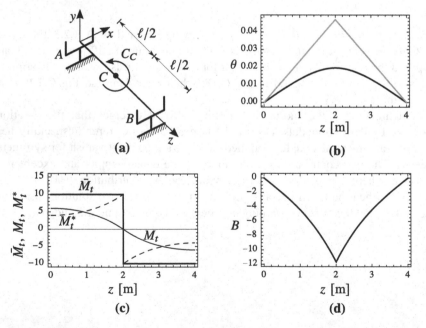

Fig. C.18 Clamped-clamped shaft, free to warp ($\ell = 4$ m): (**a**) structure; (**b**) twist angle, (**c**) torsional moments [Nm], (**d**) bimoment $[\text{Nm}^2]$. Black line, Vlasov shaft; gray line, DSV shaft; exact solution

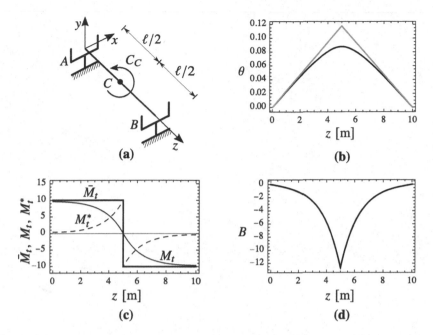

Fig. C.19 Clamped-clamped shaft, free to warp ($\ell = 10$ m): (**a**) structure; (**b**) twist angle, (**c**) torsional moments [Nm], (**d**) bimoment $\left[\text{Nm}^2\right]$. Black line, Vlasov shaft; gray line, DSV shaft; exact solution

References

1. Corradi Dell' Acqua, L.: Meccanica delle strutture: Le teorie strutturali e il metodo degli elementi finiti (in Italian), vol. 2. McGraw-Hill, Milano (1994)
2. Hoogenboom, P.C.J., Borgart, A.: Method for including restrained warping in traditional frame analyses. Heron **50**(1), 55–68 (2005)
3. Luongo, A., Paolone, A.: Scienza delle costruzioni: Il continuo di Cauchy (in Italian), vol. 1. Casa Editrice Ambrosiana, Milano (2005)
4. Luongo, A., Paolone, A.: Scienza delle costruzioni: Il problema di De Saint Venant (in Italian), vol. 2. Casa Editrice Ambrosiana, Milano (2005)
5. Sokolnikoff, I.S.: Mathematical Theory of Elasticity. McGraw-Hill, New York (1956)
6. Viola, E.: Teoria delle strutture: Stati tensionali e piastre (in Italian), vol. 2. Pitagora, Bologna (2010)
7. Vlasov, V.Z.: Thin-Walled Elastic Beams (English translation of the 2nd Russian edition of 1959). Israel Program for Scientific Translation, Jerusalem (1961)

Appendix D
Extended Hamilton Principle and Lagrange Equations of Motion

The *Extended Hamilton Principle* is used as a suitable tool to establish the equations of motion of nonconservative systems. From this Principle, the *Lagrange equations of motion* are derived for discrete systems [1–3].

D.1 Variational Principles for Nonconservative Systems

An elastic system is considered, subject to nonconservative forces. The nonconservativeness of these latter has important consequences on the analysis of stability, namely:

- Since the Lagrange-Dirichlet Theorem is no longer valid, neither the energy criterion nor the static criterion is applicable to ascertain stability; therefore, it needs to resort to the *dynamic criterion*, which requires writing the equations of motion of the system.
- The equations of motion *cannot* be deduced from an energy principle (such as the total potential energy, used for static systems) but must be deducted, either (a) from the balance of forces (by applying the d'Alembert Principle) or (b) through a variational principle, however based on the *virtual work* (but extended to dynamics), which, as well known, is equally applicable to conservative and nonconservative systems.

Here, the *Extended Hamilton Principle* (EHP) is discussed, valid for both discrete and continuous systems. It appears as a generalization of the Virtual Work Principle to the dynamics. For discrete systems, the EHP provides the *Lagrange's equations of motion*.

© The Author(s), under exclusive license to Springer Nature Switzerland AG 2023   695
A. Luongo et al., *Stability and Bifurcation of Structures*,
https://doi.org/10.1007/978-3-031-27572-2

D.2   Extended Hamilton Principle

A variational principle applicable to nonconservative systems is the *extended Hamilton Principle*. When the system is conservative, it is simply called the Hamilton Principle.

Discrete Systems

Given a discrete mechanical system of Lagrangian coordinates $\mathbf{q}$, the EHP states that its motion between two instants t_1, t_2, in which it assumes *prescribed* configurations $\mathbf{q}(t_1)$, $\mathbf{q}(t_2)$, satisfies the variational equation:

$$\delta H := \int_{t_1}^{t_2} (\delta T - \delta W_{int} + \delta W_{ext}) \, dt = 0, \qquad \forall \delta \mathbf{q} \mid \delta \mathbf{q}(t_1) = \delta \mathbf{q}(t_2) = \mathbf{0},$$

(D.1)

where $\delta \mathbf{q}$ are kinematically admissible virtual displacements. In Eq. D.1, δT is the variation of the kinetic energy; δW_{int}, δW_{ext} are the virtual work done by the internal and external forces, respectively, on the variation $\delta \mathbf{q}$ of the configuration variables. The Principle generalizes to dynamics of the Virtual Work Principle $\delta W_{int} = \delta W_{ext}$, known in the static context, including the inertia forces in the balance, according to d'Alembert Principle.[1] Equation D.1 (as discussed in Sect. D.3) leads to the Lagrange equations of motion in the coordinates $\mathbf{q}$, which are ordinary differential equations in time.

Remark D.1 It is worth noticing that while $\delta T := \delta(T)$ is the variation of the $T(\mathbf{q})$ function, and therefore it is an exact differential, δW_{int}, δW_{ext} are only small quantities (i.e., they are *not* exact differentials, and neither is δH). The case of conservative systems is an exception, in which (a) the internal forces are elastic and descend from a potential $U(\mathbf{q})$ and (b) the external forces admit a potential $V(\mathbf{q})$. Therefore, $\delta W_{int} = \delta(U(\mathbf{q}))$, $\delta W_{ext} = -\delta(V(\mathbf{q}))$, so that the Principle in Eq. D.1 can be written:

$$\delta H := \delta \int_{t_1}^{t_2} (T - \Pi) \, dt = 0, \qquad \forall \delta \mathbf{q},$$

(D.2)

with $\Pi := U + V$ the total potential energy of the system. In this form, $\delta H := \delta(H) = 0$ ensures that the Hamilton function $H(\mathbf{q})$ (also called *action*) is stationary when $\mathbf{q}(t)$ describes the trajectory actually executed by the system in the interval

---

[1] For example, for a mass m in unidirectional motion $q(t)$, the kinetic energy is $T = \frac{1}{2}m\dot{q}^2$, hence:

$$\int_{t_1}^{t_2} \delta T \, dt = \int_{t_1}^{t_2} m\dot{q}\,\delta\dot{q} \, dt = -\int_{t_1}^{t_2} m\ddot{q}\,\delta q \, dt + [m\dot{q}\,\delta q]_{t_1}^{t_2}$$

in which in integration by parts has been performed, and it has been taken into account that $\delta q = 0$ at the ends of the interval (t_1, t_2). The integral represents the work done by the inertia force $-m\ddot{q}$ in the virtual displacement δq, during the whole motion that develops in (t_1, t_2).

(t_1, t_2). Equation D.2 constitutes the *Hamilton Principle*, valid only for conservative systems (of which Eq. D.1 is, indeed, the extension).

Continuous Systems

The extended Hamilton Principle also applies to continuous systems. Given a system of deformable bodies, the configuration of which is described by the displacement field $\mathbf{u}(\mathbf{x}, t)$, dependent on the position $\mathbf{x}$ and on time t, the Principle establishes that motion of the system between two instants t_1, t_2, in which it assumes *prescribed* configurations $\mathbf{u}(\mathbf{x}, t_1)$, $\mathbf{u}(\mathbf{x}, t_2)$, satisfies the variational equation:

$$\delta H[\mathbf{u}] := \int_{t_1}^{t_2} (\delta T - \delta W_{int} + \delta W_{ext})\, dt = 0, \tag{D.3}$$

$$\forall \delta \mathbf{u} \mid \delta \mathbf{u}(\mathbf{x}, t_i) = \mathbf{0}, \ i = 1, 2, \quad \forall \mathbf{x},$$

where the symbols have been defined above. From this Principle, the equations of motion in the configuration variables $\mathbf{u}(x, t)$ follow, which are partial differential equations in space and time.

D.3   Lagrange Equations of Motion

An alternative formulation to the EHP is offered by the *Lagrange equations of motion*, valid, however, only for discrete systems. Equations can be derived directly from the EHP, operating as follows.[2]

Forces are distinguished in conservative (c) and nonconservative (nc). Regarding conservative forces (internal and external), by accounting for the Remark D.1, it is $\delta W_{int}^c - \delta W_{ext}^c = \delta(\Pi[\mathbf{q}])$. Regarding the nonconservative forces, *all treated as external to the system*, it is $\delta W_{ext}^{nc} = \sum_{i=1}^{N} F_i \delta \eta_i$, where $\delta \eta_i$ is the displacement dual of the ith force F_i. Since, from the (nonlinear) kinematics, it is $\eta_i = \eta_i(\mathbf{q})$, it follows:

$$\delta W_{ext}^{nc} = \sum_{i=1}^{N} F_i \left(\frac{\partial \eta_i}{\partial \mathbf{q}}\right)^T \delta \mathbf{q} = \mathbf{Q}^T \delta \mathbf{q}, \tag{D.4}$$

where:

$$\mathbf{Q} := \sum_{i=1}^{N} F_i \left(\frac{\partial \eta_i}{\partial \mathbf{q}}\right), \tag{D.5}$$

---

[2] There exist, however, alternative derivations, based on balance of forces [1].

are the *Lagrangian forces* (i.e., the forces expending in the Lagrangian parameters the same work done by the effective forces in the displacements of their application points).[3] The EHP, Eq. D.1, is therefore written as:

$$\delta H = \int_{t_1}^{t_2} \left(\delta L \left(\mathbf{q}, \dot{\mathbf{q}} \right) + \mathbf{Q}^T \delta \mathbf{q} \right) dt = 0, \qquad (D.6)$$

where a new quantity has been introduced, i.e.:

$$L := T \left(\mathbf{q}, \dot{\mathbf{q}} \right) - \Pi \left(\mathbf{q} \right), \qquad (D.7)$$

which is called the *Lagrangian of the system*. It is equal to the difference between the kinetic and the potential energies and therefore depends on position and velocity.

By performing the variation of the Lagrangian, integrating by parts and enforcing the conditions at the ends of the interval, one gets:

$$
\begin{aligned}
\delta H &= \int_{t_1}^{t_2} \left[\delta \dot{\mathbf{q}}^T \frac{\partial L}{\partial \dot{\mathbf{q}}} + \delta \mathbf{q}^T \left(\frac{\partial L}{\partial \mathbf{q}} + \mathbf{Q} \right) \right] dt \\
&= \int_{t_1}^{t_2} \delta \mathbf{q}^T \left[-\frac{d}{dt} \left(\frac{\partial L}{\partial \dot{\mathbf{q}}} \right) + \frac{\partial L}{\partial \mathbf{q}} + \mathbf{Q} \right] dt + \left[\delta \mathbf{q}^T \frac{\partial L}{\partial \dot{\mathbf{q}}} \right]_{t_1}^{t_2} = 0, \qquad \forall \delta \mathbf{q},
\end{aligned}
$$
$$\qquad (D.8)$$

from which the *Lagrange equations* follow:

$$\frac{d}{dt} \left(\frac{\partial L}{\partial \dot{\mathbf{q}}} \right) - \frac{\partial L}{\partial \mathbf{q}} = \mathbf{Q}. \qquad (D.9)$$

These are immediately applicable (i.e., without the need to carry out variations but only derivatives), once the Lagrangian of the system, Eq. D.7, has been written. The procedure is therefore simpler than the variational method. On the other end, this latter allows to treat discrete and continuous systems in a unified way. Examples of application of Lagrange's equations are shown in Chap. 13.

References

1. Gantmakher, F.R.: Lectures in Analytical Mechanics. Mir, Moscow (1970)
2. Lanczos, C.: The Variational Principles of Mechanics. Dover Publications, New York (1986)
3. Meirovitch, L.: Principles and Techniques of Vibrations. Prentice Hall, Upper Saddle River (1997)

---

[3] It should be noticed that in linearized kinematics, where η_i is linear in $\mathbf{q}$, the Lagrangian forces do not depend on $\mathbf{q}$. This is *not* the case of finite kinematics.

Index

Printed in the United States
by Baker & Taylor Publisher Services

Printed in the United States
by Baker & Taylor Publisher Services